AF621550

The Leopard in the Garden

The Leopard in the Garden

ANIMAL AND HUMAN LIVES IN PARIS AT THE FIRST PUBLIC ZOO OF THE MODERN ERA

Richard W. Burkhardt Jr.

THE UNIVERSITY OF CHICAGO PRESS
CHICAGO AND LONDON

The University of Chicago Press, Chicago 60637
The University of Chicago Press, Ltd., London
© 2026 by The University of Chicago
All rights reserved. No part of this book may be used or reproduced in any manner whatsoever without written permission, except in the case of brief quotations in critical articles and reviews.
For more information, contact the University of Chicago Press, 1427 E. 60th St., Chicago, IL 60637.
Published 2026
Printed in the United States of America

35 34 33 32 31 30 29 28 27 26   1 2 3 4 5

ISBN-13: 978-0-226-84397-1 (cloth)
ISBN-13: 978-0-226-84398-8 (ebook)
DOI: https://doi.org/10.7208/chicago/9780226843988.001.0001

Published with support of the Susan E. Abrams Fund.

Library of Congress Cataloging-in-Publication Data

Names: Burkhardt, Richard W. (Richard Wellington), 1944– author
Title: The leopard in the garden : animal and human lives in Paris at the first public zoo of the modern era / Richard W. Burkhardt Jr.
Description: Chicago : The University of Chicago Press, 2026. | Includes bibliographical references and index.
Identifiers: LCCN 2025029460 | ISBN 9780226843971 cloth | ISBN 9780226843988 ebook
Subjects: LCSH: Ménagerie du Jardin des Plantes (Paris, France)—History | Menageries—France—Paris—History—18th century | Menageries—France—Paris—History—19th century | Zoo animals—France—Paris
Classification: LCC QL73.F82 B87 2026
LC record available at https://lccn.loc.gov/2025029460

♾ This paper meets the requirements of ANSI/NISO Z39.48-1992 (Permanence of Paper).

Authorized Representative for EU General Product Safety Regulation (GPSR) queries: **Easy Access System Europe**—Mustamäe tee 50, 10621 Tallinn, Estonia, gpsr.requests@easproject.com
Any other queries: https://press.uchicago.edu/press/contact.html

Contents

PART 1

The Revolutionary Menagerie: Creating the Modern Zoo in Paris (1793–1815)

1

Imagining and Establishing the New Menagerie

[1.0]

Introduction

The modern metropolitan zoo traces directly back to Revolutionary Paris and the menagerie of the National Museum of Natural History. There were previous menageries; those of Louis XIV at Versailles (founded 1663) and of Franz I at Schönbrunn (founded 1752) are the most illustrious examples, but the Paris menagerie was dramatically different. Rather than having been designed at a monarch's behest, it came into being by accident, a consequence of contingent circumstances in the years 1793–94. For political and ideological reasons, it needed to distinguish itself as much as possible from its royal predecessors. It was open to the public, free, dedicated to the needs of the new nation, and totally lacking in extravagant trappings, which were not an option anyway. The Museum of Natural History was an institution of the state, and the new Republic was hard-pressed to provide even the barest minimum of funding to keep its collections and menagerie from going to ruin.

Notwithstanding the precariousness of its early years, the menagerie quickly captured the collective imagination and became the public's favorite part of the Museum. Although not its top priority, when circumstances were dire the Museum could use the menagerie's plight to call the government's attention to the Museum's needs more generally. Through good times and bad, the menagerie's fortunes reflected its institutional and political ecology. It was a conspicuous element of the world's leading natural history museum, a facility of the state, and subject to the vicissitudes of the political order, the economy, and France's place on the world stage.

By the beginning of the nineteenth century, the menagerie of the Museum of Natural History had become, like the Louvre, an emblem of the French capital's cultural and scientific preeminence. Inevitably it attracted the attention of civic leaders abroad who wanted to emulate it. The Zoological Society of London founded its own zoological garden in 1828, and other zoological gardens followed shortly (including Dublin [1831], Bristol [1835], Amsterdam [1838], and Berlin [1844]), with the word "zoo" coming into vogue later in the century. Neither London nor the later entrants were modeled exactly after Paris. Each reflected its own local circumstances and place in the wider world. But all shared the challenge of maintaining live wild animals under conditions far different from their native lands. In addition, they all embraced and exemplified the belief that it was a human imperative to master, order, superintend, and even improve the animal world for the benefit of humankind. They saw acclimatizing and domesticating new species as plausible goals.[1]

By the end of the 1800s, better funding had allowed many of the new zoos to surpass the Paris menagerie in the size of their facilities, collections, and staff. From a historical standpoint, however, the Paris menagerie remains exceptional. Its chronological priority and its importance as a role model for other zoos are enough to guarantee that status, but it is also unmatched in the wealth of archival and other resources related to its functioning and development. The extent to which it aspired to be a site for the advancement of science likewise sets it apart. More than any other zoo of the first half of the nineteenth century, the Paris menagerie aimed not only to provide zoologists with fresh materials for their anatomical and taxonomic work, but also to expand zoology's boundaries to include the study of the living animal: its faculties, its behavior, and the changes it is subject to.

The menagerie's story is also special because it was part of the world's preeminent natural history museum during the golden age of French zoology, presided over by Georges Cuvier (1769–1832), Jean-Baptiste Lamarck (1744–1829), and Étienne Geoffroy Saint-Hilaire (1772–1844). No previous work, however, has sought to integrate the menagerie's history with that of the Museum or the history of science. The present book addresses both issues while examining the menagerie's operations, development, and relations with the wider world.

Archival sources have been invaluable for this work, especially the abundant papers of the Museum of Natural History preserved at the Archives nationales in Paris in the series AJ/15. Included in this series are the microfilmed minutes of the weekly meetings of the professors of the Museum, which allow us to track the Museum's collective concerns over time.[2] Even greater treasures were buried in the eighty large cartons containing the correspondence and other documents associated with these meetings, which had largely escaped modern scholarly attention.[3] Though these papers date back more than two centuries, they still convey urgency with respect to problems to be solved and opportunities to be grasped, lives being lived, animals suffering, human friendships and rivalries in action, personal ambitions colliding with collective purposes, and more.

Among the riches in those cartons are the observations of musicians who performed for two elephants; a previously unpublished report by Georges Cuvier from 1803 on the Museum's need for more gallery space; the menagerie superintendent's weekly menagerie reports over the years; and correspondence revealing everything from administrative matters to interpersonal relations. Though the actual noises and smells of the menagerie are missing from these wonderful boxes, they are described there. In addition to these documents at the Archives nationales, I have relied on manuscript material at the Museum of Natural History itself and at the Institut de France (the site of most of Georges Cuvier's papers), along with a wealth of scientific publications, newspaper accounts, memoirs, paintings, and other sources relevant to the animals of the menagerie and the humans who engaged with them.

These materials allow us to reconstruct the early history of the Paris menagerie as an interplay of animal and human lives, situated in a special setting at a special time. We learn how the animals came into the menagerie's possession, their living conditions, the food they were fed, the illnesses they suffered, the injuries they sustained at the hands of an unruly public, and more. Reports by the menagerie superintendents; correspondence from collectors, donors, and dealers; essays by zoologists and journalists; portraits by artists; minutes from the Museum's administrative assemblies; and the insights, enthusiasms, and moralizing of guidebook writers bring different perspectives to bear on a collection that over time included lions and lemurs, bears and bison, elephants and eagles, a giraffe, an orangutan, and an electric eel. It was customary to distinguish between the "fierce" animals in their cages and the "peaceful" animals in their parks, but the archives testify that this distinction was not always reliable: some of the "fierce" carnivores were gentle and affectionate; some of the "peaceful" herbivores were very dangerous. Several of the animals gained fame and inspired poems, pamphlets, illustrations, stage productions, and, in the case of the giraffe, women's fashions, decorative porcelain, and other memorabilia. Jack the orangutan, the next great sensation after the giraffe, was an occasion for journalists' frivolity but also for weightier reflections regarding orangutans' proximity to humans.

The range of human actors involved in the menagerie's life was as remarkable as the animals' diversity. In many zoo histories, the institutions' promoters and directors are presented as the primary human actors while other critical participants, such as the keepers—the men (and sometimes women) most intimately involved in the animals' daily care—are marginalized or overlooked entirely. The keepers figure prominently in this history, however, as do the two men who served as the menagerie's superintendents for most of the years covered here. Others who shaped the menagerie's development include the Museum professor/administrators and their assistants; artists; students; families; tourists; authors of plays, guidebooks, and newspaper articles; musicians; vendors; sentries; politicians and bureaucrats; emperors and kings; princes and paupers. Some of them could always be found at the menagerie; others visited frequently; some came only occasionally. Beyond the Museum's gates but consequential for the menagerie's functioning were government ministers, voyager naturalists, foreign potentates, French consuls and other colonial agents, departmental prefects, ships' captains, animal conductors, shipping agents, diplomats, donors, and dealers. Some operated at the Museum's behest, others on their own initiative. Not always did their interests match the Museum's.

This book highlights a previously unappreciated actor in the history of science, Frédéric Cuvier (1773–1838), Georges Cuvier's younger brother. Appointed superintendent of the menagerie in 1803, the younger Cuvier sought to make the menagerie a site of scientific consequence. He took it upon himself to establish a new role: not a sedentary or office naturalist like his brother or Lamarck; not a voyager naturalist who collected specimens abroad; not a field naturalist (which few investigators of the time could claim to be); but instead a naturalist who through observations and experiments studied the lives of the animals populating the menagerie. Most naturalists assumed that the behavior of wild animals confined to cages was too unnatural to merit scientific attention, but Frédéric Cuvier sought to prove them wrong. Above all, he aspired to create a whole new science devoted to "the numerous psychological phenomena that the long series of animals presents." He looked to establish a comparative psychology of animals that could inform the psychology of humans.

This book has been a long time in the making. The scholarly practices and technologies for working through materials central to this project have evolved over the three decades I have devoted to it. The manuscript-filled cartons of the Archives nationales still require the old-fashioned practice of sifting through them, piece by piece, to find items of significance. At first I had to take notes by hand, type them on a primitive laptop, or splurge on a limited number of expensive photocopies for a special document (subject to an archive overseer's decision that the document was not too fragile to copy). Today one can photograph documents to one's heart's content. Books too can now be photographed, but many of them have already been digitized, making it possible to do research from the comfort of one's home, albeit without the pleasures of working with the original volumes in a Paris library or walking streets named for the very scientists whose lives one is studying. Digital finding aids for the French periodical press offered by Gallica and RetroNews now provide access to a range of journalistic information on the Museum and the menagerie that would have been impossible to survey effectively earlier. I have been fortunate that the time required to gain a satisfactory command of the archival materials critical for this project has corresponded with a period when so many publications have been made readily accessible online.

The Paris menagerie exists to this day. This book focuses on its formative years, 1793–1838, the period of its origin, growth, greatest prominence on the European stage, and greatest scientific ambitions.[4] The beginning date, 1793, is the year the first live animals arrived at the Museum; the end date, 1838, is when Frédéric Cuvier died. His broad scientific plans for the menagerie faded with his demise. Its territorial expansion was complete, the major animal quarters originally imagined for it had all been built, and a menagerie for reptiles had just been founded. The late 1830s also correspond to the end of French zoology's golden age. Lamarck and Georges Cuvier were gone (in 1829 and 1832, respectively) and Étienne Geoffroy Saint-Hilaire's scientifically productive years were over.

This book consists of four chapters in roughly chronological order, beginning with the menagerie's origins and early travails, followed by the challenges of acquiring animals and caring for them. The third chapter tracks the Museum's reconnection with the wider world after the Napoleonic Wars, with the freedom of the seas restored and the Museum's collections growing as commerce, diplomacy, war, and colonialism linked Europe to the four corners of the world. The fourth chapter focuses on the effort to make the menagerie a site for scientific research. Throughout the book, thematic and and narrative sections alternate with sections dealing with individual animals or species, and they are intended to complement each other.

Several of the individual animals have previously received historical attention, but most have not, and a few may seem surprising in a book about a zoo. Among the latter is a creature dead for three thousand years, a mummified sacred ibis Geoffroy Saint-Hilaire brought from Egypt that underscores the importance of dead specimens as much as live ones for the Museum's naturalists; sets the stage for the famous disagreements between Georges Cuvier, J.-B. Lamarck, and Étienne Geoffroy Saint-Hilaire on the issue of species change; and connects with Frédéric Cuvier's hope to address the species question by studying live specimens at the menagerie. Another section considers an unwanted but ineradicable animal present at the menagerie from the beginning: the brown rat. Its story testifies to the limits of Western science's ambition to order, control, and improve the animate world.

Today one of the primary rationales offered for zoos is that they can serve as sanctuaries for species in danger of extinction, but that danger was scarcely recognized two hundred years ago and was not a function the Paris menagerie imagined for itself. Nor did the Museum articulate the idea of valuing animals for their own sakes or think that keeping live animals under conditions antithetical to the development of their faculties was fundamentally cruel. To the contrary, the animals' best interests were controverted from the start. Their trajectories of subjection began when they were wrested from their natural social groupings and native lands. They were then transported great distances to a climate generally inhospitable to them and incarcerated in cramped cages or other quarters or at best kept in more commodious paddocks when the season permitted it. Only rarely, within the web of competing interests at the menagerie, were the animals' needs allowed precedence over those of the public, the Museum's scientists, or the government.

The Museum prized rare specimens, but it did so in the context of assembling, ordering, and controlling a collection that in principle was to include all of nature's productions. When it came to a new animal species, the Museum wanted to know not only its affinities with other species but also whether it could be tamed, trained, acclimatized, and domesticated. Having a menagerie in Paris was understood as a broad biopolitical project whose aim, in effect, was to civilize animals, plants, and humans. In this vision (not without providential undertones, though these usually remained unstated), plants and animals would be perfected for human purposes, the lower classes would be elevated morally through exposure to images of nature's order, and uncivilized peoples would be brought into the ranks of the civilized through contact with and tutelage from the West.

The project's accomplishments in these regards were uneven. The menagerie represented a regime of unequal power relations with humans in charge, but the animals were not wholly without agency. They could resist the roles humans tried to assign them. They could kick, bite, scratch, or attempt to escape. For diverse reasons, those deemed good candidates for domestication often proved unsuited for it. Meanwhile, the attitudes and behavior of humans in relation to the menagerie's animals ran the gamut from hostile and abusive to enraptured and reverent. Some menagerie visitors thought of animals only in terms of their utility or as sources of entertainment; others saw them as sentient beings worthy of affection, admiration, and even awe. But it is not obvious that cruelty or kindness was correlated with social class, or that public morality in the form of kindness to animals at the menagerie changed conspicuously over the decades covered by this book.

The Leopard in the Garden seeks to reconstruct the life of the menagerie when the possibilities and purposes of the institution were still being imagined, when animal and human lives intersected both within the Museum's gates and far beyond. Given the political, social, and economic turbulence of the time, it is not surprising to find that contingencies and compromises featured as prominently in the menagerie's development as did well-conceived plans. Despite so much that was unprecedented and uncertain, involving exotic and often dangerous creatures along with humans who had their own stories and interests, the menagerie became firmly established at the Museum, in the mind of the public, and in the European consciousness. How this played out for animals, humans, science, and society is the subject of this book.

Figure 1.1.1. "La panthère." Nicolas Maréchal, Simon-Charles Miger. From Étienne de Lacépède, Georges Cuvier, and Étienne Geoffroy Saint-Hilaire, *La Ménagerie du Muséum national d'histoire naturelle, ou Description et histoire des animaux qui y vivent et qui y ont vécu* (Paris: Aug. Renouar, 1801–5). R. Burkhardt collection.

[1.1]

The Leopard in the Garden

It is unlikely that Marie Antoinette on her way to the scaffold in the Place de la Révolution on 16 October 1793 caught a glimpse of Dominique Marchini's leopard. But there is evidence that it was close by. Marchini was making a living displaying his leopard and three other exotic animals (a polar bear, a civet, and a monkey) to the crowds that flocked to the sprawling city square. Wild animals, to be sure, were not the main attraction. That distinction belonged to the guillotine, the Revolution's signature instrument, looming on a platform halfway between the entrance to the Tuileries gardens and the center of the great square. Previously named Place Louis XV, the square was designed to showcase a bronze equestrian statue of the king, and in 1770 the marriage of the king's son (the future Louis XVI) to Marie Antoinette was celebrated there. The Revolution of 1792 transformed the site for glorifying the royal family into a place of regicide. Revolutionaries toppled the statue in August 1792, and five months later the guillotine was erected to behead the deposed king. Marie Antoinette's execution came nine months later.

Between Louis XVI's execution and that of Maximilien de Robespierre in July 1794, the guillotine took the lives of royalty and revolutionaries alike. More than a thousand individuals identified as enemies of the people were beheaded in the Place de la Révolution (renamed in 1795 the Place de la Concorde). Among them were the great chemist Antoine Lavoisier and such diverse agents of the Revolution as Philippe d'Orléans (Philippe d'Égalité), Jacques Hébert, and Georges-Jacques Danton.

The name of Dominique Marchini, the man who displayed his animals in the Place de la Révolution, does not appear in histories of the period; he was not an actor of any consequence on the Revolution's stage. Nor were his animals appreciably different from the small troupes of strange creatures that other entrepreneurs presented on the streets of Paris at about the same time. Yet Marchini's animals hold a special place in history. Less than three weeks after Marie Antoinette's execution, his leopard, polar bear, civet, and monkey were confiscated by the police, taken from the Place de la Révolution, and escorted through the streets of Paris to the Jardin des Plantes, the site of the new National Museum of Natural History. There they became the founding animals of the first public zoo of the modern era.

Marchini had not seen this coming. On the morning of 14 brumaire an 2 of the French Republic (4 November 1793) he had no reason to expect that two local commissioners (one from the police, the other from the civil committee of the quarter) would show up to appropriate his animals. The officers informed him that the Commune of Paris had decreed that displaying dangerous animals like lions and leopards on the streets of the city constituted a threat to public safety. The officers' instructions were to round up all such animals and deliver them to the Jardin des Plantes. The Commune promised to indemnify animal proprietors for the loss of their livelihoods.[1]

Obliged to surrender his animals, Marchini was quick to tell the commissioners how much he had paid for them and for their cages, wagons, and related paraphernalia. He also proposed that he and his assistant, a young man named Amé, be hired as the animals' caretakers, "being very familiar with them and knowing perfectly everything that could be necessary for their conservation."[2]

Marchini was not the only one whose plans were upset that day. When a member of the municipal guard appeared at the Museum's gates with Marchini, his four animals, and inevitably a train of gawkers in tow, it was the turn of the professor/administrators of the Museum of Natural History to be surprised. They hadn't heard of the Commune's decree either, and they certainly hadn't asked for any animals. Those who dealt with zoological matters were accustomed to working with dead specimens, not live ones, especially not the kind that could inflict grievous injury with the bite of their powerful jaws or a swipe of their sharp claws. At the height of the Revolution, two months after the Reign of Terror began and only five months since the Museum's founding, the animals presented the professor/administrators with unexpected problems, though perhaps with opportunities as well.

Like the Place de la Révolution, in 1793 the Museum of Natural History was a transformed space with new functions. Its predecessor was the *Jardin du Roi*, comprising the royal botanic garden and the king's "cabinet," a collection of natural history specimens. For nearly half a century, until his death in 1788, Georges-Louis Leclerc, le comte de Buffon, had directed, developed, and dominated the institution. His death, together with the onset of the Revolution in 1789, made its structural change possible, a development greatly desired by the naturalists who had served as officers at the Jardin du Roi under Buffon's dictatorial rule. In 1790 they jointly drew up a reorganization plan whereby Buffon's old position, that of *intendant*, was abolished and the former officers, as twelve wholly equal professor/administrators, were put in charge of the institution's direction. Governance would be achieved through "a fraternal and reciprocal supervision of all over each and each over all, [an] equilibrium maintained without effort."[3]

Their plan went unrealized for another three years until it was finally brought into being on 10 June 1793 by a decree of the National Convention, urged on by Joseph Lakanal, a young legislator and educator from the provinces who became the leading figure of the Committee of Public Instruction (CPI). He had been given the reorganization plan when he visited the Jardin des Plantes just the day before. With the Convention's decree, the Jardin des Plantes was reconstituted as a democratically run institution dedicated to serving the needs of the new Republic. Each of the twelve newly named professors (all of whom had served as officers of the old institution) was responsible for teaching his own area of science and developing the collections of specimens or other materials appropriate to his field.[4]

The Museum staked its claims to social legitimacy on offering courses of public instruction, maintaining and developing its natural history collections for the advancement of science, pursuing the practical projects of acclimatizing, domesticating, breeding, multiplying, and distributing useful plants and animals, and presenting ordered images of nature that helped citizens understand their place in nature and society. In effect, the Museum's purpose was to civilize specimens and citizens at the same time.[5]

But how live leopards, polar bears, monkeys, and the like might fit into this picture was by no means clear when the Museum began receiving its first live exotic animals some six months after its founding. These were certainly not the sorts of animals naturalists typically designated as worthy of naturalization and multiplication for the public good. Nor was it easy to imagine how the sight of such animals confined to cramped cages would elevate the public morale. Indeed, the basic concept of a menagerie was hard to square with Revolutionary ideals. The best-known menagerie in France was the old royal one at Versailles, conspicuous for its ostentation, wastefulness, and association with royal privilege.

Two days after Marchini's animals arrived at the Museum, the botanist René-Louiche Desfontaines, secretary of the professorial assembly, wrote to the CPI seeking guidance. He acknowledged that the Museum had a place where the creatures in question could be provisionally housed, and that it was large enough to accommodate additional animals should the CPI decide to send more. There was no doubt, he added, that "a collection of living animals could be very advantageous for public instruction and for the progress of natural history" and might also play a role in acclimating animals from foreign lands. "But," he observed, "it is up to the Committee to weigh in its wisdom if these advantages can be reconciled with the present needs of the Republic." The Museum, he explained, could not be expected to provide for the animals from its meager funds. Its most pressing need was to enlarge its galleries, and to achieve this goal it was already forced to strictly economize in all its endeavors.[6]

Desfontaines was finishing his letter when he learned that more animals had just arrived at the Museum. He identified them in a postscript: "a tiger-cat, a male polar bear, two mandrills, and three eagles." The animals (including a pair of agoutis that Desfontaines had not heard about) had been the property of animal exhibitors Henri Gueny and Bernard Louzardy, who, like Dominique Marchini, were anxious to know what their future held. The Museum decided to retain all three as the animals' caretakers. The animals were kept in cages and displayed in the open arcades on the ground level of the natural history galleries, a manifestly inadequate arrangement. The cages were cramped, and the odors emanating from them penetrated the galleries above.[7]

The Museum of Natural History and France's National Archives preserve records that shed light on the lives of Marchini's leopard and other animals that arrived at the Museum in 1793–94 and afterward. We know when they were delivered, where they were kept, what they ate, what it cost to care for them, and what happened to their remains when they died. Some information about how they behaved offers clues to their emotional states, but beyond that we can say little about the animals' perspectives. Animals can give warning cries. Some species can even practice deception (though writers around 1800 were not inclined to believe it). Animals may leave physical traces, like scat, paw prints, claw marks on trees (or menagerie walls), beehives, birds' nests, beaver dams, and such, and their bodies can be preserved in tar pits, ocean sediments, or museums; but unlike human animals, they cannot tell their stories. In the words of Henning Mankell, "a truer nomination for our species than *Homo sapiens* might be *Homo narrans*, story-telling human."[8]

One cannot imagine Dominique Marchini's leopard being interested in human events, comprehending its transfer through the city to the Jardin des Plantes, and caring or being able to communicate that experience to another big cat.[9] As for the great majority of humans who participated in the life of the menagerie, they surely reported their experiences to other humans, but their voices are largely missing from the historical record. The archives offer very little, for example, of Dominique Marchini's story. His signature in Museum ledgers shows that he could barely sign his name.[10] His fellow animal keepers were not writers either, by and large, though others sometimes recorded their needs and activities.

What we can say about Marchini's leopard, thanks to the Museum archives, is that it ate six pounds of meat per day at a cost of four livres and four sous, more than the cost of food for Marchini's other three animals combined, and more than Marchini's own daily wages of four livres. The leopard was the most valuable creature in the collection, and Marchini reported its worth as 10,000 livres (compared with 4,000 livres for his polar bear, 2,000 livres for his civet, and 200 livres for his monkey).

To assess the fairness of Marchini's and the other animal proprietors' valuations, the Museum turned to a man named Martin who ran the *combat du taureau*, a violent spectacle held outside the city limits, pitting animals (such as lions and bears or dogs) against each other in bloody fights. Martin had reason to know what a wild beast was worth, but to the former proprietors, his estimates were shockingly low. He valued Marchini's leopard at 1,500 livres, ditto for his polar bear and another belonging to Bernard Louzardy (who had valued his bear at 10,000 livres). Finding Martin's estimates too low and the proprietors' too high, the professors sent them to the CPI for its opinion.[11]

Keen to compile accurate portraits of its specimens, especially the living ones, the Museum held a competition in the spring of 1794 to select three artists as official illustrators. On 16 prairial an 2 (4 June 1794), it announced the winners.[12] The best known was Pierre-Joseph Redouté, remembered for his magnificent paintings of roses, copies of which still grace the stalls of the *bouquinistes* along the Seine. Redouté's younger brother, Henri-Joseph, was also selected. The third artist, Nicolas Maréchal, became the leopard's portraitist and the Museum's primary painter of mammals for the next decade.[13]

Maréchal's duties had a special urgency. On the day of the artists' appointment, the Museum paid 500 livres to a veterinarian to treat animals that were ailing.[14] The professors' anxiety on this score was evident a month later when they asked him to produce paintings "as soon as possible" of "the quagga, the panther, the polar bear, the serval, the green monkey, the adult mandrill, the Barbary ape, the raccoon, the agouti, the baboon, the mountain goose, and the cockatoo."[15] Maréchal understood the situation. Among the paintings and sketches he submitted in the artists' contest was one of a hartebeest that died as soon as it arrived at the Museum.[16]

While the Museum systematically sought to preserve faithful images of its live plants and animals, it made no comparable effort to study and record animal behavior. Typically, only dead animals were subjected to detailed scientific scrutiny. When Marchini's leopard died (3 November 1794) its carcass passed into the hands of Jean-Claude Mertrud (1828-1802), the Museum's professor of animal anatomy, who dissected it as part of his anatomical lessons. The anatomist's customary procedure was to preserve the organs in alcohol and have the specimen's skin and skeleton prepared for display. Maréchal made sketches of the animal's stomach and its reproductive organs.[17] We do not know what happened to the leopard's other internal organs and its skeleton, but the skin was stuffed and mounted for public display in the natural history gallery.[18]

The only record of the leopard's behavior regards how much meat it ate daily. We do have brief comments about the habits of a different leopard specimen, a young animal that came to the Museum after the first leopard died. The anatomist and zoologist Georges Cuvier (1769–1832) observed in an article of 1801 that this second leopard was happy to play with its master and its behavior was that of any young cat. Growing up in captivity had done little to soften the creature's temperament, however. Cuvier did not elaborate on this remark but mentioned a distinctive behavior that must have startled observers: "It likes to launch its urine toward those who stare at it."[19]

In Cuvier's opinion, the greatest challenge the leopard presented to natural history was distinguishing it from other species of big cats. Naturalists of the past had assigned a host of different names to these animals and provided inconsistent descriptions and illustrations, often making it difficult if not impossible to tell which creature a given author was referring to. Having a menagerie would bring new precision to the task, Cuvier explained, because it would allow the naturalists of the Museum "to describe exactly and successively what is offered to us." That, he said, is "the principal purpose of a menagerie."[20]

Identifying animals amid a plethora of names, uneven descriptions, and imprecise illustrations was a problem for zoologists in Georges Cuvier's day, and it remains so for average readers and even scientists today. The police commissioner who confiscated Marchini's animals identified them as a leopard, a civet, a monkey, and a "sea lion" (*lion marin*). The Museum's zoologists readily identified the last of these as a polar bear, just as we would today. But the animal the commissioner called a "leopard" (as I have done here), consistently appears in the Museum's records as a "panther."

This discrepancy was not accidental. Cuvier presumed that as knowledge accumulated about which big cats came from which continents, and about their appearances at different stages of development, some felines previously classified as separate species would prove to be varieties of the same one. Consequently, names like "leopard" and "ounce,"[21] Cuvier predicted, would become superfluous. The animal that had once been called a leopard, he believed, would eventually be recognized as a panther. But something of the opposite has happened. While scientists no longer recognize the panther as a species, they identify several species of leopard. What is commonly called a "black panther" is, by the reckoning of modern zoologists, a leopard with a very dark coat. But *Panthera* is used as the name for the genus.

We can regard the taxonomic problems involving Marchini's leopard as symbolic of the multiple uncertainties the Museum faced with its new menagerie. Neither with the leopard nor with the whole of the new menagerie were the professors of the new Museum sure what they had. Maintaining a collection of live exotic animals would be an unprecedented challenge, and it was by no means clear what this enterprise would mean over time: for the scientists and other workers at the Museum, for the citizens of Paris, for the French government, or for that matter for other people—and animals—throughout the world.

[1.2]

Imagining the New Menagerie

Late in 1792, nearly a year before the leopard became a resident of the Jardin des Plantes, the celebrated author Jacques-Henri Bernardin de Saint-Pierre addressed a long memoir to the National Convention lobbying for establishing a menagerie there. Not until the end, and even then only in a footnote, did he explore the political dimensions of the situation, which he introduced with a pleasantry. You could tell where a Parisian woman lived, he wrote in 1792, by the orders she gave a coachman when leaving the theater. If she commanded, "to the *hôtel*," it was the faubourg Saint-Germain. If she ordered, "to the dwelling [*logis*]," she lived in the Marais. If she said, "to the house [*à la maison*]," she was from the faubourg Saint-Denis. But the witticism did not extend to women of the faubourg Saint-Marceau, because they seldom went to the theater at all; their quarter did not have one. An attempt to establish a theater there shortly after the Revolution began failed within weeks, while five or six new theaters elsewhere in Paris flourished. Bernardin was not interested in local parlance or the habits of theatergoers. He related the anecdote to make a point about the geography of socioeconomic inequality in Paris.

Bernardin believed that poverty and radical politics went hand in hand, and as the latest *intendant* (director) of the Jardin des Plantes (appointed July 1, 1792), he was concerned. The faubourg Saint-Marceau, home to the Jardin des Plantes, was the city's poorest quarter, and its political district had recently renamed itself the "Section of the *Sans Culottes*." Representing the far left of the Revolutionary spectrum, the sans culottes were laboring poor who wore trousers instead of the fancy breeches (*culottes*) and stockings affected by the higher ranks of society, clothing that was both a political statement and a source of pride. The *sans culottes* were also known for mob action, notably in the massacres of September 1792, when hundreds of priests, aristocrats, and others confined to Paris prisons were killed as suspected counterrevolutionaries. Bernardin did not mention these events; he

Figure 1.2.1. "Le bubale" (northern hartebeest). Nicolas Maréchal, Simon-Charles Miger. One of the last animals in the old royal menagerie at Versailles, the hartebeest died almost immediately on being transported to Paris in the spring of 1794. From Lacépède et al., *La Ménagerie du Muséum national d'histoire naturelle*. Crerar Rare Books, Hanna Holborn Gray Special Collections Research Center, University of Chicago Library.

simply remarked that this neighborhood of sans culottes was "one of the most patriotic."

The poverty of the faubourg Saint-Marceau was not new. Its shabbiness left a lasting unfavorable impression on Jean-Jacques Rousseau when he first came to Paris in 1742, and five decades later the district was further run-down. The faubourg had, by Bernardin's estimation, suffered more from the Revolution than had the rest of Paris. The faubourg Saint-Germain, the Marais, and the area around the former Palais-Royal remained prosperous (even if emigration had reduced the population of Saint-Germain). The faubourgs Saint-Denis and Saint-Antoine still enjoyed their commerce and manufacturing respectively. Saint-Marceau, in contrast, had lost its monasteries and convents and many of the boardinghouses that had provided much of its pre-Revolution income. As Bernardin saw it, the quarter was "very populated and very badly off," a volatile situation. "In my opinion," he wrote, "the primary cause of rebellions in cities, even of revolutions, is when all the rich are on one side and all the poor are on the other."[1]

Few of the Convention's deputies may have read Bernardin's memoir to the end, but none would have doubted that the political situation in Paris was volatile. The Tuileries Palace had been stormed the previous August, followed by the September massacre of priests and aristocrats. With winter approaching, economic misery, military threats to the nation, and rumors of aristocratic plots against the government had the citizens of Paris on edge.

The primary theme of Bernardin's 1792 memoir was not the relation between economic inequality and political unrest but "the necessity of attaching a menagerie to the national botanic garden in Paris." He cited scientific, agricultural, diplomatic, moral, and artistic reasons for his proposal as well as socioeconomic and political ones. Beyond these broad concerns, Bernardin had a more immediate one: what to do with the animals that remained in the old royal menagerie at Versailles. Louis Charles Couturier, the general manager at the palace, raised the issue in a letter (September 1792) informing Bernardin that the menagerie was about to be destroyed and asking if the Jardin wanted any of the animals for its natural history collection—for example, the "superb rhinoceros."[2]

Bernardin hastened to Versailles, accompanied by the botanist René-Louiche Desfontaines and the Jardin's gardener-in-chief André Thouin. With tears in his eyes, the keeper of the menagerie, Charles Laimant, told them that since the Revolution began, a dromedary, five species of monkeys, and all but one of its birds had been lost to pillagers. The visitors viewed the survivors: the rhinoceros; a quagga (a native of South Africa and close relative of the zebra); a bubal (the northern hartebeest, a spiral-horned antelope from North Africa); a beautiful blue-crested pigeon from the Moluccas (the only bird remaining from the king's once extensive collection); and a lion from Senegal, living contentedly with a hunting hound as companion.

Bernardin was struck by how these animals responded to humans. When he and his colleagues arrived, the bird hurried to and fro at the back of its cage "in a perpetual agitation." The rhinoceros, bubal, and quagga, however, came forward to be petted through the bars, as did the dog and lion;

the dog wagged its tail in welcome while the lion "walked soberly the length of his bars, against which he rubbed his enormous head."[3]

"Never had I seen so much generosity in a lion," Bernardin wrote, "and so much amiableness in a dog. The latter seemed to guess that his familiarity with the king of beasts was the principal object of our curiosity. Seeking to please us in his captivity, as soon as we had addressed some words of affection to him, he threw himself gaily on the lion's mane and playfully bit his ears. The lion, lending himself to these games, lowered his head and made low roaring sounds." The dog bore a wound suggesting the lion's playful indulgence had its limits, but Bernardin was inclined to believe their relationship was amicable and their ill-tempered moments less frequent or dangerous than human quarrels. The keeper assured Bernardin and his colleagues that the lion "often invited the dog to play by lying on his back, his feet in the air, embracing [the dog] with his front legs."[4]

Bernardin, whose *Les études de la nature* (*Studies of Nature*, 1784) had won him fame and authority, declared that the friendship between the lion and the dog was "one of the most touching spectacles that nature can offer to a philosopher's speculations."[5] The lion's disposition was "a proof of what society's influence can have on the most savage character," for it was "much more cheerful than a solitary lion."[6] The Jardin's director viewed the lion's case as comparable to British prison reformer John Howard's conclusions with respect to the moral improvement of prisoners: companionship, not isolation, was essential.[7] To Bernardin, the study of animals allowed humans to understand their own nature and place in the world, a compelling reason for transferring them to the Jardin des Plantes. He had been offered the animals for their skeletons, but he wished instead to promote the study of living nature.

Bernardin was not the first to propose attaching a menagerie to a center of scientific research, but his predecessors were few. Menageries dated to nearly the beginning of recorded history, but the idea of a menagerie as an integral part of a scientific research center was much newer.[8] Francis Bacon's utopian work *The New Atlantis* (1620) made an early case for combining the two by describing an imagined community of natural philosophers dedicated to studying the natural world. As the head of Bacon's fictional "Salomon's House" explains, "*The* End *of our* Foundation *is the Knowledge of* Causes, *and Secrett Motions of things; And the Enlarging of the Bounds of* Humane Empire, *to the Effecting of all Things possible.*"[9]

Significantly, Bacon furnished his philosophers with "Parks, *and* Enclosures *of all* Sorts, *of* Beasts, *and* Birds" that were subject not only to observation but also "Dissections, *and* Trialls," aimed at determining "*what may be wrought upon the* Body *of* Man." Salomon's House members studied how to breed animals to affect their size, fecundity, color, shape, and behavior, and through hybridization they produced new animal races capable of reproducing: "*We find means to make* Commixtures *and* Copulations *of diuerse* Kindes, *which have produced many* New Kindes, *and them not* Barren, *as the generall* Opinion *is*." They also created new species through spontaneous generation: "*We make a number of* Kindes, *of* Serpents, Wormes, Flies, Fishes, *of* putrefaction; *Whereof some are advanced (in effect) to be* Perfect Creatures, *like* Beastes *or* Birds; And haue Sexes, *and doe* Propagate."[10]

Bacon's "Salomon's House" is routinely cited as inspiring the two major scientific societies founded later in the century, the Royal Society of London (1660) and the Académie royale des sciences of Paris (1666), but neither was endowed with a menagerie. Their members did not need exotic creatures for physiological experiments when dogs or chicks were readily available, and developing new breeds was more the work of rural livestock owners than urbane natural philosophers. The 1660s also saw the creation of a magnificent menagerie at Versailles for Louis XIV, which, while intended for royal pleasure, not scientific advancement, benefitted the academicians of Paris.[11] Anatomists now had the opportunity to examine the carcasses of lions, a chameleon, a camel, a bear, a gazelle, a serval, a shark, and other remarkable creatures, including an elephant given to the king of France by the king of Portugal in 1668 and famously dissected by anatomist Joseph-Guichard Duverney in 1681.[12]

A century later, academicians still focused largely on anatomy and morphology, not on the living animals' habits—what they ate, how they were caught, their medicinal properties, and such—all topics traditionally the province of natural historians.[13] Studying the behavior of animals in captivity was pointless, according to Georges-Louis Leclerc, comte de Buffon (1707–88), the most famous naturalist of the eighteenth century. *Intendant* of the Jardin du *Roi*, by royal appointment, from 1739 until his death, and author of the multivolume masterpiece *Histoire naturelle, générale et particulière* (1749–89), Buffon insisted that to understand the true nature of any given species, it should be observed in the wild, not under the conditions of captivity or domestication. Nevertheless, in 1782 he proposed adding the animals of the royal menagerie in Versailles to the Jardin du Roi as part of a plan to expand the institution's territory and natural history collections.[14]

Buffon died before his proposal could be realized, but the idea survived. In 1790, when the officers of the former Jardin du Roi began to consider its restructuring, the botanist André Thouin (1746–1824), who established the Jardin as an influential center for the naturalization, perfection, and distribution of plants, readily imagined that a menagerie could do the same for animals.[15] Thouin did not develop a detailed argument for bringing the royal menagerie to Paris, and neither did Louis Jean-Marie Daubenton (1716–1800), who had distinguished himself not only as a naturalist and anatomist but also as a sheep breeder who helped improve the French wool industry. Daubenton envisaged the Veterinary School of Alfort, not the Jardin des Plantes in Paris, as the site for animal domestication projects.[16] The case for establishing a Paris menagerie fell to Bernardin de Saint-Pierre.

When the National Convention appointed Bernardin director of the Jardins des Plantes in 1792, the naturalists there were unhappy. It was not just that they felt he lacked proper credentials as a naturalist; they did not want a director at all. They had chafed under Buffon's autocratic rule. When invited in September 1790 to present to the National Assembly a plan to reform the Jardin, their top priority was to abolish the position of *intendant* altogether. Buffon's successor, Charles Flahault, marquis de la Billarderie, knew nothing about natural history, accomplished nothing during his tenure, and had not even bothered to take up residence at the Jardin. The naturalists' reform plan called for sharing the duties of administration, offering courses of public instruction, and developing the collections. In the fall of 1790, however, the National Assembly had more urgent political and military matters to consider than reforming the Jardin des Plantes. The naturalists' hopes were rekindled late in 1791 when Flahault fled the capital to save his life, but they were dashed again in July 1792, when the National Convention named Bernardin as Flahault's replacement.

Though Bernardin's scientific credentials were slim, he was appreciated as an author who understood the moral value of attunement with nature. The fame he acquired in 1784 through his *Études de la nature* was enhanced by his best-selling novel *Paul et Virginie* (1788), about a young couple growing up in harmony with nature on a tropical island. Bernardin had witnessed nature's productions firsthand in distant climes. At twelve, he sailed to Martinique with his father, a captain in the merchant marine. After studying engineering and entering the army, the three years he spent (1768–71) at Isle de France (Mauritius, a French possession of strategic importance) formed the basis of his travel narrative *Voyage à l'Isle de France* (1773).[17] In his new position as the Jardin des Plantes' director, he decided that a menagerie would be his special contribution.

The naturalists of the Jardin des Plantes did not share the National Convention's high opinion of Bernardin, noting that his romanticism surpassed his critical judgment. Bernardin's *Etudes de la nature* maintained that nature never acts in vain and displays a special benevolence toward man. Animals harmful to man, he wrote, typically announce themselves at a distance. He remarked of fleas that their dark color makes them easier to spot against the skin's whiteness (an unselfconscious Eurocentric conceit).[18] He suggested that nature is prepared to depart from her laws for man's benefit, as for example in producing, from the mating of the ass and the mare, "the mule, which is so useful in the mountains," yet she deprives the mule of the ability to reproduce itself, thereby preserving the primitive species, which "are of a more general utility."[19] The unhelpfulness of such assertions aside, in his year at the Jardin Bernardin proved a more capable administrator than the professors probably anticipated.

Bernardin's memoir advocating the addition of a menagerie was simultaneously a paean to nature and a prescription for public education. "The study of nature," he proclaimed at the outset, "is the basis of all human knowledge." The best way to study her, he said, was to contemplate her as a unified whole. For the Jardin des Plantes and its natural history collection to properly instruct the public, all of nature's three kingdoms (animals, plants, and minerals) should be considered in terms of their dynamic interactions.[20]

Bernardin was calling for a new approach to natural history education. In his view, the soils that sustained plant life were of greater interest than individual minerals dug from deep within the earth, and living plants and animals were more instructive than their dried or stuffed remains. He called natural history *cabinets* (the rooms where natural history specimens were stored, displayed, and studied) "the tombs of the kingdoms of nature." Gardens, in happy contrast, were nature's "cradle." Without nature's "most important" kingdom, live animals, creatures "that feel, that love, that know," the Jardin des Plantes was incomplete. Their presence would help liberate zoology from comparative anatomy's dominance, for however useful knowledge of animals' physical features may be, studying their "tastes, instincts, and passions" would shed a greater light on man's own needs and nature.[21]

The Revolution reawakened the question whether there should be menageries at all, given that they smacked of "enslavement." Bernardin appreciated that a new one in Paris would have to be very different from the ostentatious and wasteful one at Versailles.[22] He knew that however charming it may be to see a lion and dog cavorting, or to pet a rhinoceros's snout, there were more important reasons to establish a menagerie in Paris at the Jardin des Plantes. He used a range of arguments to make his case, beginning with matters of international diplomacy and national dignity. When the powers of Africa and Asia made gifts of wild animals to the French nation, he wrote, France needed to be able to receive and care for them properly. To kill and make skeletons of them would insult the donors, and to refuse them on the grounds that France had no place to lodge them would be unbecoming for the nation.[23]

A menagerie would contribute significantly to the study of nature, Bernardin continued, by giving artists, poets, and writers the opportunity to base their works on living animals instead of copying the works of others. The public would be attracted, given the way people flocked to fairs to see exotic animals, and would gain a feel for the distant lands they represented. Carried away by this thought, he suggested that witnessing the lion, with its powerful claws and fearsome roar, would provide a deeper sense of its native land than seeing that land itself.[24]

Bernardin made no effort to portray the menagerie as a place where animals might enjoy a semblance of freedom. He vehemently opposed human slavery but comfortably used the word *esclavage* (slavery) in writing about animals, believing that philosophy might benefit from observing its effect on animal character. The conditions of captivity, he believed, were key. Isolating an animal would reinforce its ferocity, while treating it well and allowing it to "socialize" with others could make it happier. On a return visit to Versailles, Bernardin witnessed what he regarded as another example of the lion's socialization: "I went to see him a second time in the company of a lady who amused herself by making her fan move in front of him; he regarded it with the greatest attention and displayed all the postures of a cat who wants to play."[25]

The friendship between the lion and the dog that he'd observed on his first visit was a sign that even the most savage beasts could be tamed if treated properly, but it also led Bernardin to imagine extraordinary couplings. Had the dog been a female of the largest canine breed, he ventured, the lion and dog's "mutual affection would have redoubled, and a mating might perhaps have resulted from it." To support this fancy, Bernardin cited Pliny the Elder attributing to Aristotle the claim that "the Indians had their dogs covered by tigers and dog-tigers were born from this." Bernardin knew that dog-wolf hybrids had been produced in France, and he saw no reason why dog-lions could not be produced as well.[26]

Bernardin's optimism regarding crossbreeding did not extend to every species. The quagga at Versailles had been gelded and consequently had no reproductive future. As for the rhinoceros, Bernardin could not conceive of its being of any use, despite reports that rhinos had been domesticated and used as beasts of burden in Ethiopia. The hartebeest and the Moluccan pigeon held promise; although both were males, they could be mated with females of their species if available, or with females of domestic species, thereby producing new races distinguished by their size or swiftness.[27]

The Jardin's director believed the menagerie would provide opportunities for naturalizing animals, ameliorating their natures, and crossing them with domestic races. Only rich nations like France, he said, had the resources to move animals around the globe and acclimatize them to new settings (as had been done with plants). Exotic plants from Asia, Africa, and America had already enriched France's farms and gardens, and exotic animals could likewise find a place in France's barnyards and groves. Menageries would serve as way stations, because animals from foreign lands would need special conditions to adjust to their new surroundings.[28]

Bernardin pictured scenes of happy mating and, perhaps drawing on personal experience, believed that successful coupling depended on congenial surroundings.[29] Animals raised in the same temperatures and among the same plants as their native lands would "forget their captivity" and "give themselves up to love-making," thinking they'd never left home. Beavers surrounded by poplar trees, reindeer nibbling on lichens, delicate animals raised in greenhouses would flourish and reproduce in settings designed to seem familiar. More than encouraging animals to breed in captivity, the strategic juxtaposition of animals and plants in a shared setting would allow agriculturalists and veterinarians to experiment with new foods for the animals and new plant sources for treating animal diseases.[30]

Bernardin de Saint-Pierre presented a draft of his memoir to several colleagues before publishing. Louis Jean-Marie Daubenton, the Jardin's eldest naturalist, had long been interested in domesticating wild forms and improving domesticated animals, and he endorsed Bernardin's efforts.[31] Some botanists did not. They would have preferred, Bernardin reported, to have the animals at Versailles killed, dissected, and stuffed. He argued that flattening dead plants between sheets of paper was not studying nature, and if the difference between a living plant and a dead one was enormous, the difference between a living and a dead animal was greater still. What distinguished a wolf from a dog, Bernardin insisted, was not the structure of its teeth but rather its instincts as a living being, which were what determined the wolf's ferocious character and differentiated it from man's canine comrade. To the objection that wild animals lose their character in captivity, he curtly replied, "they lose much more by death."[32]

Ideally, according to Bernardin, exotic creatures would be studied in their native lands. Voyager naturalists, he accurately observed, were more interested in gathering dead specimens than studying live animals in their natural habitats. Plus, little was likely to be learned about the normal lives of animals from creatures that attempted to flee or cowered in fear when caught.[33]

In Bernardin's Edenic vision of the world's beginning, humans and animals lived contentedly together. Men with guns destroyed an equilibrium that survived only where primitive peoples still lived in peace with nature. In the Pacific islands, Bernardin said, penguins, seals, and sea lions showed no fear when Captain Cook walked among them. Savages and shepherds were more likely than "murderous naturalists" to understand the behavior of wild animals. According to Bernardin, man's ill treatment of his fellow man often started with mistreatment of animals; conversely, displays of kindness to animals led to kinder relations between humans. By demonstrating the benefits of treating animals humanely, a well-run menagerie could have positive consequences for human society.[34]

After insisting that a menagerie could be economically sustained and that its expenses were justified by what it would contribute to the nation's well-being, Bernardin rounded out his argument by explaining why the Jardin des Plantes would be the best location. There was ample room for expansion, thanks to the government's confiscation of neighboring properties previously owned by the church. The nearby fountain of Saint-Victor would provide fresh water for the animals. A menagerie in the Jardin des Plantes would draw visitors, benefit the local economy, and consequently diminish political volatility. Bernardin acknowledged that while the people of the faubourg Saint-Marceau might be relatively coarse, they were less corrupt than other Parisians. Prostitutes rarely frequented the quarter, he said, the air was pure, and the Seine, which entered the city at this point, was not yet polluted by urban refuse. Plus, thanks to water drawn from wells, the beer and the bread of the rue Mouffetard were the best in Paris.[35]

Having laid out his argument for establishing a menagerie in Paris at the Jardin des Plantes, Bernardin was confident that the deputies of the National Convention would find the project irresistible. Nature was indispensable to the foundation of an indissoluble republic, he insisted: "Nature alone through her benefactions will bring together people whom religions and patriotism have divided."[36]

Bernardin's confidence that his memoir would capture the minds of the Convention's deputies was not shared by the anonymous author who reviewed the memoir for the *Magasin encyclopédique*. The reviewer was not opposed to the idea of a national menagerie in Paris, but he doubted the deputies would be persuaded by Bernardin, who was not "a true naturalist."[37] He scoffed at the notion that France needed a menagerie to accept animals as diplomatic gifts from "African princes," expressing his displeasure with the diplomatic necessities that involved the country in "the ridiculous imitation" of the extravagance of foreign rulers.[38] As for Bernardin's suggestion that a menagerie in the faubourg Saint-Marceau would help the quarter's indigent class, the reviewer judged that "unworthy of being presented by a philosopher."[39] Institutions for the poor were necessary, he acknowledged, but they had nothing to do with institutions for promoting science. Calling Bernardin's memoir "a pleasant rather than a useful treatise," he predicted the Convention's leaders would dismiss the menagerie as unnecessary.[40]

The reviewer expected that a committee appointed by the Paris Society of Natural History would reject Bernardin's proposal, but he was wrong.[41] The committee consisted of Aubin-Louis Millin (editor of *Le Magasin encyclopédique*), the physician Philippe Pinel, and the chemist and geologist Alexandre Brongniart. Beginning with the obligatory rejection of the wasteful and useless menageries of kings, the committee proclaimed that "a menagerie without splendor" could be "extremely advantageous to natural history, physiology, and the economy; and the advantages that the nation would derive from it would amply compensate for the expenses it would take to establish it."[42]

For naturalists, the committee said, a menagerie would facilitate identifying and distinguishing animal species. It would also mean that voyager naturalists would no longer be the only ones to enjoy a view of living nature. A menagerie in Paris would allow the naturalists of the city to gauge "the veracity and the exactitude" of the voyager's observations.[43] The committee, like Bernardin, argued that while a menagerie would not reveal all the characteristics of an animal in its natural state, a living animal was nonetheless far more instructive than a skin or skeleton. Furthermore, a menagerie would allow naturalists to conduct experiments and study, for example, how much domestication alters an animal's character or which species are easy to tame and which are not.[44]

Physiologists also, who previously had had too few species for experimental study, would benefit from a menagerie in Paris. It would allow them to study issues like hibernation, the secretions of different animal glands, and the menstruation of females of certain monkey species. Such studies required "slow, peaceful, and continuous observations in order to make them exact and precise." They would not be left to the chance observations of voyager naturalists, who were "nearly always in a hurry, and lacking for the most part the instruments needed for experimentation."

Finally, the committee suggested, if the "enslavement" of captive animals prevented mating, perhaps artificial insemination, recently essayed with a dog by the Italian naturalist Lazzaro Spallanzani, could circumvent the problem. It could allow scientists to raise species from other lands and produce new hybrids. As in Bernardin's memoir, the report recited Daubenton's tally of the mammals, birds, and other animals that might be brought to France, acclimatized, domesticated, and integrated into the rural economy.

The report concluded by strongly urging the establishment of a menagerie in Paris, for the benefit of science and the glory of the republic.[45]

Ultimately, the Paris menagerie was not created in response to Bernardin de Saint-Pierre's memoir or the urgings of the Paris Society for Natural History. It was the result of the Paris Commune's decision, in the fall of 1793, to remove dangerous, exotic animals from the Paris streets and deliver them to the Jardin des Plantes, by then restructured as the National Museum of Natural History. The following spring, the last residents of the royal menagerie (the quagga, the hartebeest, and the lion and dog pair) were transferred from Versailles to Paris. Neither the rhinoceros nor the blue-crested pigeon that Bernardin had seen a year and a half earlier was in the convoy. The pigeon's fate is unknown, but the rhino's is well documented. It never reached Paris to have its snout petted by visitors to the Jardin des Plantes, having expired in its pool during an August 1793 heat wave. Its rapidly decomposing carcass was rushed to Paris, where the anatomists Jean-Claude Mertrud and Félix Vicq-D'Azyr (1746–94) dissected it. Artists Nicolas Maréchal and Pierre-Joseph Redouté diligently sketched the animal's internal and external organs, some of which were preserved in alcohol, while the rhino's bones were set out in the sun to bleach.

As for the hartebeest, it suffered injuries during the journey to the Jardin and died shortly after arriving. The quagga, the lion, and the dog made the trip without mishap. At the new Paris menagerie, the friendship of the lion and the dog continued to be a source of amazement and occasion for speculating about what humans might learn from studying nature. But imagining a menagerie would prove far easier than creating one that served the diverse interests of science, the public, and the new nation.

[1.3]

The Revolutionary Museum and the Menagerie (1793–1795)

The National Museum of Natural History in Paris was a product of the French Revolution, brought into being by a decree of the National Convention on 10 June 1793. Thirty years later, the Museum's first historian, Joseph Deleuze, sought to portray its founding as happening in the nick of time, with the legislator Joseph Lakanal moving quickly to forestall the plans of Jacobins bent on destroying the Jardin des Plantes. But this was only a convenient fiction in the years of the Restoration when a Bourbon king was back on the throne and the Museum was called the Muséum Royale d'Histoire Naturelle. It would have been impolitic to suggest that the Jacobins had respected nature and the Museum had thrived under their rule. But that is what historians have argued since.[1]

The study of natural history resonated with the ideology of the Revolution and the nation's need for natural resources. At the ideological level, the naturalists' cause was buoyed by the belief that contemplating nature's order would elevate public morality and help citizens recognize their place in society.[2] With respect to the nation's commercial and military needs, the naturalists (chemists included) could point to the cultivation of nonnative species and the manufacture of gunpowder as among their skills. When Lakanal made the case to the National Convention about establishing the Museum, he predicted that planting nonindigenous trees would one day allow France to avoid paying huge sums to other countries for naval timber. He likewise noted how the first coffee plants transported to Martinique came from the Jardin des Plantes, and how this had led to a branch of commerce "of the highest importance" for France and her colonies. He additionally promised, "The time will come, undoubtedly, when one will raise at the National Garden species of quadrupeds, birds, and other foreign animals that can acclimate themselves on the soil of France and thus procure her new riches."[3]

But if natural history's ideological and economic utility buoyed support for its study during the Revolution, that does not explain how the Museum

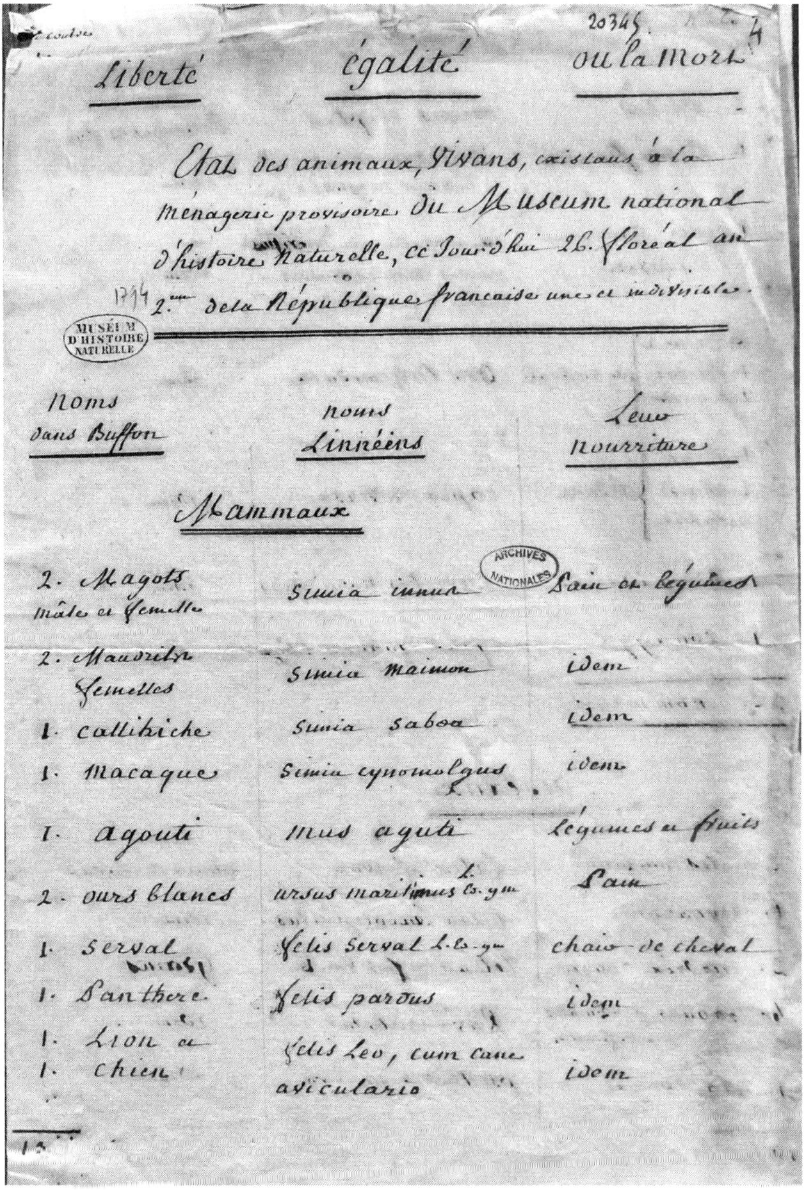

Liberté égalité ou la mort

Etat des animaux vivans, existans à la ménagerie provisoire du Muséum national d'histoire naturelle, ce jourd'hui 26. floréal an 2.me de la République française une et indivisible.

Noms dans Buffon	noms Linnéens	Leur nourriture
Mammaux		
2. Magots mâle et femelle	Simia inuus	Pain et légumes
2. Mandrills femelles	Simia maimon	idem
1. Callitriche	Simia sabaea	idem
1. Macaque	Simia cynomolgus	idem
1. agouti	Mus aguti	légumes et fruits
2. ours blancs	Ursus maritimus L. G. gm	Pain
1. Serval	Felis Serval L. G. gm	chair de cheval
1. Panthère	Felis pardus	idem
1. Lion et 1. Chien	Felis Leo, cum cane aviculario	idem

Figure 1.3.1. The Revolutionary slogan "Liberty, equality, or death" features at the top of the first page of Étienne Geoffroy Saint-Hilaire's list of the animals in the menagerie as of 26 floréal an 2 of the French Republic (15 May 1794). Archives nationales, AJ/15/844. Photograph by the author.

happened to be founded when it did. Pragmatic responses to highly contingent circumstances were also key to the Museum's creation and subsequent development. What triggered the transformation of the old Jardin du Roi into the new Museum of Natural History in June 1793 was the practical problem of accommodating a flood of new specimens arriving at the Jardin des Plantes following the confiscation of émigré property.[4]

The specimens were from Chantilly, the castle of the former prince de Condé. He had fled the country, and the government had claimed his property. The obvious destination for Condé's natural history collection was the former *Cabinet du Roi* (the king's natural history collection) at the Jardin des Plantes, but it had no room for new specimens. The galleries and storerooms of the building housing the national *cabinet* were already overflowing.[5] A solution, however, was easy to imagine, namely, developing the third floor of the building for more galleries. Lakanal presented the idea to the National Convention on 26 May 1793. On his recommendation, the deputies decreed that the minister of the interior immediately order that the third floor of the building housing the national natural history collection be made ready "to receive the natural history collection from Chantilly, together with the various other productions of nature accumulated without order in the same building."[6]

Lakanal's visit to the Jardin des Plantes on 9 June 1793 was intimately tied to the project of expanding the institution's natural history galleries to accommodate the multitude of specimens for which there was no place. Did he go on his own initiative, or had he been invited? If the latter, was it to discuss matters in addition to enlarging the galleries? Whatever the case, what followed is a matter of general historical agreement: that while he was there, he was provided with a copy of the 1790 plan for reorganizing the institution, and later that same evening, at a meeting of the CPI, he and his colleagues produced an abbreviated version of the plan that became the decree he presented to the Convention the next day. The Convention ratified the Museum's founding decree without discussion. It had more pressing problems: foreign armies on its borders and uprisings at home.[7]

It is fitting that the need to accommodate a new collection of specimens precipitated the founding of the Museum of Natural History. Specimens were the institution's lifeblood.[8] Their collection, processing, preservation, and deployment required constant attention. Some came to the Museum through naturalists working on the institution's behalf. Others came through confiscations, donations, exchanges, or purchases. They were brought under increasing physical and conceptual control through a range of interrelated practices: collecting; preserving; packing for shipment; transporting; unpacking; identifying and naming; describing; classifying;

interpreting; displaying; and sometimes distributing to other sites. This was true for plants, animals, and minerals alike, and for all of these other, specific procedures were also applicable. In the case of animals, dead specimens were subject to dissection, their organs preserved, their skeletons cleaned and mounted, their skins stuffed. Live specimens were subject to taming and training, which, together with acclimatizing and domesticating were part of the general goal of making them more amenable to human needs. All these practices served to adapt specimens to the needs of science and society and in effect progressively civilized them. The civilizing of specimens meshed with the belief that presenting citizens with images of nature properly ordered would help them recognize their own place in nature and society, thereby enhancing their morals and behavior. The projects of civilizing specimens and civilizing citizens were thus imagined going hand in hand.[9]

Not all this was articulated with respect to the specimens from Chantilly, though it was understood that their confiscation would ultimately make them accessible to the public. Meanwhile, the Chantilly specimens underscored the fact that there no place at the Jardin des Plantes to display them, as was true for many specimens already there, confined to crates where they were no doubt deteriorating.

∴

The Museum's founding decree gave the former officers of the old Jardin du Roi the two administrative reforms they most urgently wanted: the suppression of Buffon's position of director (*intendant*) and the creation of a cohort of twelve professor/administrators with equal rights and functions. The professors treasured this new, democratic arrangement and were keen to defend it as the government went from Convention to Directorate to Consulate to Empire. They displayed admirable unity in challenging government attempts to weaken their autonomy. On the other hand, their democratic organization did not prevent frictions and factions from arising over time as they pursued their scientific and personal interests within the Museum's close quarters.[10]

The new Museum's principal purpose was identified as "the public teaching of natural history," where natural history was to be understood "in its fullest extent and particularly applied to the advancement of agriculture, commerce, and the arts." The original statutes called for twelve courses of instruction and a professor for each of them. The courses thus identified were mineralogy; general chemistry; the chemical arts; botany at the Museum; botany in the countryside; agriculture and the cultivation of gardens, fruit trees, and woods; the natural history of the quadrupeds, whales, birds, reptiles, and fish; the natural history of insects, worms, and microscopic animals; human anatomy; animal anatomy; geology; and, finally, "natural iconography, or the art of drawing and painting all the productions of nature."

Each professor was required to deliver a course of instruction in his assigned area. For nearly all of them this meant approximately forty lectures over a span of about three months. The courses on human and animal anatomy needed to be taught in the winter months, when the weather and the demonstration rooms were cold and cadavers decomposed less quickly. Botany was taught in the spring and summer months when plants were up and growing. Other courses were not dependent on the season.[11] The teaching load was not onerous. Nonetheless, assembling a course of lectures could transform a professor's thinking as well as the students'.

In addition to teaching, the professors were responsible for developing the collections and arranging the galleries according to the order they adopted in their courses. To aid them in the task, their regulations called for the creation of four naturalist's assistants (*aides naturalistes*), whose duties included preparing specimens, placing them in the galleries, and supporting the courses as needed.

Beyond developing the collections and delivering courses, the professors' other primary duty was their joint administration of the Museum. In the Museum's first months of existence, the professors assembled frequently as a body, as often as every day or every other day, but by floréal an 2 (April 1794) they had arrived at a more regular schedule, with meetings on the fourth day (*quartidi*) and eighth day (*octidi*) of every ten-day week (*décade*). Five months later, they reduced the frequency of their assemblies to once every *décade* and set their length at two hours. To encourage attendance and promptness, they agreed to levy a fine of five livres on anyone who failed to arrive on time.[12]

The original professor-administrators of the new Museum all had previous connections with the Jardin des Plantes, but they varied greatly in age and experience. Daubenton, seventy-seven years old at the time of the Museum's founding, was the most advanced in both respects.[13] Buffon had named him superintendent of the Jardin du Roi's natural history collection nearly a half-century before, and Daubenton supervised it up to the Museum's creation. He was a skilled anatomist (he had done the anatomical work for Buffon's *Histoire naturelle*). He also devoted himself to improving wool production in France so that French textile manufacturers would no longer have turn to Spain for the best wool. The position he took when the Museum was founded, however, was as the professor of mineralogy.

The youngest (twenty-one years old) and least qualified of the new professors was Daubenton's former assistant, Étienne Geoffroy Saint-Hilaire

(1772–1844) swept by events into the company of the other professors and responsible for the natural history of the quadrupeds, whales, birds, reptiles, and fish.[14] The previous year, Geoffroy had rescued the crystallographer and cleric René-Just Haüy from prison just in time to escape the September 1792 massacres. Shortly afterward, Daubenton appointed Geoffroy assistant keeper of the natural history collections. The position did not put him in line for a professorship, but it fell to him in June 1793 because the naturalist intended for the post, the comte de Lacépède, had fled the capital. Some professors questioned Geoffroy's qualifications, the legality of his appointment, and the terms of his payment, but he soon had the job and the same salary as the others.[15] When the first live animals arrived at the Museum, they fell under his aegis.

The professor of zoology responsible for "the insects, worms, and microscopic animals," was Jean-Baptiste Lamarck (1744–1829). He is best known today as the major proponent of evolution before Charles Darwin, but insofar as he was already forty-nine in 1793, with a previous career in botany and limited qualifications for his new post in zoology, there was little reason to anticipate his future as a zoological theorist. His expertise in his new area of responsibility was by and large limited to the knowledge he had picked up as an avid shell collector. Had he remained a botanist, he most likely would never have developed the broad evolutionary vision for which he is now famous. Becoming a professor of zoology at the Museum of Natural History was crucial for his intellectual development. His dual responsibilities as a professor—ordering and developing the collection specimens under his charge and offering a course of public lectures on the part of the animal kingdom assigned to him—were jointly critical for transforming his thinking. As professor of the "invertebrates" (his designation), he had no compelling reason to be concerned with the menagerie when it took shape, since it was composed primarily of mammals and birds; but as a conscientious professor-administrator he took a turn supervising it (July 1801 to September 1802), and in 1809 a new acquisition at the menagerie sparked an important development in his theorizing.

The coming of the menagerie's first animals in the fall of 1793 was of more obvious interest for the professor of animal anatomy, Jean-Claude Mertrud, but that was because they would soon come under his scalpel. Known more for his anatomical skills than for his insights about animal anatomy, he conducted his most notable dissection in the summer of 1793, with Daubenton and Vicq-d'Azyr, on the rapidly decaying cadaver of the rhinoceros from Versailles. The Museum's new menagerie kept him busy dissecting its early losses and preparing skeletons and other remains for preservation and display. Mertrud's most enduring claim to fame, however, was that he allowed

Georges Cuvier to substitute for him in delivering lectures on comparative anatomy, thus providing the future star zoologist a foothold in the Museum.

André Thouin (1746–1824), the professor of plant cultivation, was disposed to be interested in the menagerie from a different perspective: his extensive experience from the botanical side of things in developing useful forms and moving them around the country and the colonies. Thouin had made the old Jardin du Roi the center of a vibrant network of plant and seed exchanges, and no professor was more active in his efforts to put nature's productions to use for the benefit of the new Republic. From the start of the Revolution, Thouin kept the needs of the Jardin des Plantes forefront in the minds of the appropriate government officials.[16] Beginning in 1794, he provided the Museum with invaluable service as leader of the Commission of Science and Arts, which followed closely behind the French armies to sequester scientific and artistic treasures from defeated territories. Among the specimens he helped confiscate for the Museum were those in the natural history collection and menagerie of the Dutch Stadtholder.

The rest of the Museum's original professors had less reason to want a menagerie, though several in administrative capacities furthered its cause. This was the case for René-Louiche Desfontaines (1750–1833), professor of botany at the Museum, and Antoine Laurent de Jussieu (1748–1836), the Museum's professor of botany in the countryside. Their own work did not involve studying living animals, but as administrators they represented the Museum's and the menagerie's needs to the government, as did Antoine-François de Fourcroy (1755–1809), professor of general chemistry. In the 1780s Fourcroy distinguished himself as a promoter, practitioner, and teacher of the revolution in chemistry launched by Antoine Lavoisier (1743–94). In the 1790s, he was an ardent supporter of the Revolution. He replaced Jean-Paul Marat at the National Convention after Marat's assassination in July 1793, and later that year he was elected president of the Jacobin Club. Fourcroy's fellow professors appreciated him as both a strong proponent of the Museum's usefulness to the nation and someone whose political connections could stand the Museum in good stead. In February 1794 the Committee of Public Safety recognized his work as essential to the nation's defense, granting 40,000 francs and other support so that making gunpowder could be taught in the Museum's chemistry lab.[17]

The Museum's professor of chemical arts, Antoine-Louis Brongniart (1742–1804) was away from Paris in the mid-1790s as chief pharmacist of the French army in Italy and had no significant role in the developments that concern us here. The same must be said of the remaining professors on the roster. Barthélemy Faujas de Saint-Fond (1741–1819), professor of geology, and Gerard Van Spaendonck (1746–1822), professor of iconography, were only tangentially interested in the menagerie. Faujas was involved in confiscating the Stadtholder's menagerie in Holland, and one of Van Spaendonck's students was Nicolas Maréchal, the Museum's main animal portraitist. Finally, the Museum's professor of human anatomy, Antoine Portal (1742–1832), was unconcerned with the menagerie and most of the Museum's other business as well. As former physician to the royal family, Portal's career proceeded unhindered through the Revolution, the Directory, the Empire, and into the Restoration, when he was physician to King Louis XVIII. At the Museum he delivered his lectures and attended meetings but otherwise left no mark. He chose not to live in the apartment the Museum provided for him but installed a brother there instead.[18]

When the professors of the new Museum convened for the first time (14 June 1793), they elected Daubenton as the assembly's president and Desfontaines and Fourcroy as secretary and vice-secretary. They then directed their attention toward ways of profiting from opportunities that the Revolution presented, such as acquiring natural history books from the libraries of suppressed monasteries. They also resolved to use the funds the Convention allotted for the renovations with the architect, Edme Verniquet.[19] In addition, they named a delegation to thank the CPI for its interest in the Museum and for the services it had rendered with respect to the National Convention.[20]

In the ensuing months, the professors finalized rules of operation; established and maintained good relations with government bodies; lobbied the government for funds, new positions, and other needs; claimed specimens, books, and paintings from nationalized properties; set priorities for construction projects; sought to create more apartments at the Museum so that all staff members could live on location; selected additional staff; responded to workers' complaints about their meager pay in the face of inflation, and much more. Not until November came the event, beyond the professors' control, that set in motion the historical developments that are the subject of this book. Two small troupes of animals, soon to be followed by others, were dropped unannounced on their doorstep.

The professors were not opposed to a menagerie. On the contrary, their regulations called for adding one when the Museum could afford it, as a way "to accelerate the progress of natural history." It was presumed that once the Museum was equipped to maintain a collection of live animals, the zoology professors would describe them, study their habits, and promote the acclimatization, multiplication, and distribution of those that were useful.[21]

The problem was that the animals arrived before the professors were ready. Money was their first concern, since feeding animals cut into funds slated for the urgently needed expansion of galleries for nonliving specimens.

The Museum sought guidance from the CPI, which responded by requesting a budget, with line items including the costs for temporary quarters for the animals; reimbursements for their former owners; feeding and caring for the creatures; and the cost of a large complex of buildings and a parcel of land next to the Museum (the "Régie") "in case the National Convention should decide to form a menagerie." It appeared that the Convention would have the final say on the menagerie, whether the Museum wanted one or not.[22]

It was easy to get a contractor's quotation on the cost of building temporary quarters and to estimate food expense, since the professors had only to tally up the cost of what the various animals were already eating.[23] Assessing the animals' value required the assistance of an expert (Martin, the "combat master"), whose estimations were far lower than the animal proprietors', so the matter was passed to the CPI. As for the Régie and the land adjoining it, the professors were interested in the property but considered it unsuitable for a menagerie. They suggested instead two *arpents* (acres) of marshland contiguous to the Jardin.[24] They estimated the cost of buying it and building animal quarters there to be about 60,000 livres. "Expenses would be greatly diminished," the professors added, if the Convention gave them materials and tools from the former menageries at Versailles and Chantilly. They also suggested there be an artists' competition to come up with a menagerie design.[25]

At the first meeting of 1794, Daubenton suggested the time was ripe for Lakanal, the Museum's champion on the CPI, to deliver a report to the National Convention about establishing a menagerie at the Museum. Daubenton hoped Lakanal would also lobby for a librarian and a third zoologist.[26] Thouin and Jussieu were delegated to confer with the CPI to enable Jussieu to report to the National Convention on "the menagerie to be established at the Museum."[27] Jussieu drafted a decree for the CPI to present to the Convention, portraying the menagerie as the missing piece of the Museum's constitution. A menagerie would provide "the means of observing living animals, of studying their habits, of multiplying and propagating useful species throughout the Republic; of crossing and improving the races of domestic animals, of presenting to the arts rare species forms to be imitated by paintbrush or chisel."[28]

Jussieu's draft outlined the necessary steps: discontinuing the menagerie at Versailles; transporting its animals to the Museum; officially establishing a menagerie at the Museum; organizing, funding, and provisioning it; and indemnifying the animal proprietors. He estimated the cost of the menagerie's construction at 50,000 livres if the materials from Versailles could be obtained. "The parks and huts for the innocent animals," he noted, "will be less costly than the solid quarters of the dangerous animals." Jussieu's draft seems not to have reached the Convention, but the project of collecting the animals and equipment from Versailles moved forward.[29] By early March the Museum was ready to receive them, or nearly so. The administration hoped Couturier could send a man who would stay a few days and provide instructions on feeding and caring for the lion.[30] In mid-March the administration charged Geoffroy with going to Versailles to expedite the animals' transport.[31] Still, it was another five weeks before the animals and their accompanying hardware were brought to Paris. The animals from Versailles—the quagga, the bubal, and the lion and dog pair—finally arrived on 7 floréal an 2 (26 April 1794).[32]

Five days later, the professors instructed their new architect, Jacques Molinos (1743–1831), to prepare signage. Entrance signs would read National Museum of Natural History. The title Galleries of Natural History was to appear on the building with the natural history specimens, and the new lecture hall would be designated Amphitheater.[33] A sign on the quarters of the live animals would read Provisional Menagerie, along with a plea to the public, *de ne rien jetter aux animaux*, that is, "don't throw anything to or at the animals."[34] As we shall see, both points needed to be made.

Besides the remnants of the royal menagerie, the Museum had to consider a larger collection of live animals from the grounds of Raincy, the estate of Philippe Égalité, the former duc d'Orléans. Égalité had been a deputy at the Convention and voted for the king's execution, but as a prince of the blood he could not escape suspicion and was sent to the guillotine on 6 November 1793.

The woods of Raincy had been populated with deer. There were also sheep, goats, a camel, a small English bull, a female water buffalo (*une vache bufflé*), swans, geese, red-legged partridges, and parrots. The state's confiscation of Raincy offered the Museum the chance to have more specimens. On the invitation of Joseph Augustin Crassous, a Jacobin representing the departments of Seine-et-Oise and Paris, the Museum sent its zoology professors, Lamarck and Geoffroy, to select the animals appropriate to natural history, the arts, or agriculture. The professors toured the estate and listed four swans, eight Perlier sheep, five red deer, six fallow deer, the water buffalo cow, the English bull, the goats, and the mountain geese. The camel had already been purchased by Merlin de Thionville, but this prominent Jacobin leader, who met Lamarck and Geoffroy on their visit, told them he would happily donate it to the Museum.[35]

The two professors reported back to their colleagues on 21 March, but the transfer to Paris of the Raincy animals, like those from Versailles, took time. They arrived some two weeks after the animals arrived from Versailles.[36] This major influx swelled the live animal collection to thirty-two mammals and twenty-six birds. At their next assembly (13 May 1794), the professors decided to send Thouin, Lamarck, and Geoffroy to the CPI "to beg it to accelerate the report that must be made at the National Convention relative to the menagerie of the Museum."[37] Two days later Geoffroy prepared a catalogue of the live animals in "the provisional menagerie of the National Museum of Natural History." He headed it with the Terror's slogan: "*liberty*," "*equality*," "*or death*."[38]

Was it the Museum's plight in the face of this influx of new animals that triggered a sudden flurry of activity by the government committees responsible for the Museum's fate? The chronology fits this interpretation. On 16 May 1794 at 6:00 p.m., a special delegation from the Committee of Public Safety (CPS) showed up at the Museum. Fourcroy had warned Thouin that the delegation was coming, and ten professors were there to greet the members and report on their work and needs.[39] Impressed, the delegation took the Museum's case back to a meeting of the whole committee, which drafted two decrees of great consequence for the institution's future.

The CPS's first decree effectively confirmed the menagerie's establishment by charging the Commission of Public Works (CPW) with arranging provisional lodgings for the Museum's live animals (in locations selected by the professors), using materials from the old menagerie at Versailles (as far as possible), and funding the animals' care and feeding.[40] The second decree was of another order entirely, stunning in its boldness and breadth. It proposed expanding the Museum's grounds to four or five times their current size.[41]

The architect Molinos was to examine the site and report to the CPS as soon as possible the status of national properties and accompanying land the Museum might annex. The territory would include the river Bièvre and extend "to the boulevard and horse market on one side and to the rue Saint-Bernard on the other."[42] The CPS asked Molinos to assess the value of the area's privately owned properties. Simultaneously, it instructed the Committee of National Domains to suspend the evaluation and sale of any national properties within that territory.

Signing this decree were Jacques-Nicolas Billaud-Varenne, Bertrand Barère, Claude-Antoine Prieur-Duvernois, Lazare Carnot, Jean-Marie Collot d'Herbois, Georges Couthon, Robert Lindet, and Maximilien Robespierre. By chance, these famous revolutionaries' support of science came only a week after the illustrious chemist Antoine Lavoisier was executed for having been a member of the hated tax-collecting agency the Ferme générale. Lavoisier's status as a scientist had not figured in the charges against him, but neither had it saved him.

The CPS's decrees set wheels turning. The professors met with Molinos the following day, asking him to draw up plans placing the aviaries and the provisional menagerie for the carnivores in the large square basin at the center of the garden.[43] Six days later, members of the CPW met with the professors and Molinos and chose a different location, in the lane of chestnut trees bordering the garden's grounds to the South, along the rue Saint-Victor. They also recommended a row of double cages to house the animals, facilitate public viewing, and ensure the project's speedy completion. The visitors and the professors jointly approved Molinos's plan.

After their visit, the CPW produced two reports corresponding to the CPS's two decrees. The first detailed the committee's findings and recommendations regarding where to locate the animals' provisional quarters.[44] The second, addressing the Muséum's expansion, expressed unbounded enthusiasm for "the construction of the most beautiful natural history garden possible." Extolling the CPS's "grand and bold views" in contrast to those of lesser souls, the CPW had no qualms about four million francs as the price tag for the properties in question. It also deemed it "easy" to reroute the Bièvre River and deliver other water to the site to make the garden "the most beautiful in the universe." The "most curious" and "marvelous" productions of the three kingdoms of nature would all be assembled there—"the rarest animals, the plants from all the climates."[45]

In waxing eloquent about the expanded, definitive menagerie, the CPW commissioners made no mention of rows of cages, either single or double. In their vision for the future, each animal lived under its natural conditions.

> Near it will be the trees it loves, the plants appropriate to it. The lion of Africa will have his lair in a rock near the shadow of date and coconut palm trees; the ferocious panther, the bloodthirsty tiger will inhabit the same areas; but the polar bear of Siberia, sheltered by some high boulder from the winds and the heat from the south, will be placed in the coolest spot; a spring of fresh water will serve to quench his thirst and cool his abode further. As for the peaceful animals, they will browse freely in the meadows, their simple huts, covered with thatch, will have the form of their native country, they will find the same trees, they will browse the same plants.

This "simple precaution," the commissioners suggested, might encourage "[the animals'] conservation and reproduction." Birds would be visited "alive and free in their vast aviaries, before seeing them dead and injected in the cabinet of natural history."

Earlier advocates of a Paris menagerie had scorned the one at Versailles, but the CPW commissioners believed it served as a reference point for improvements. The new menagerie would surpass the old by being a "living whole," with plants and animals together. The space allotted the animals would reduce space for the plants, but the loss to the public would be mitigated by "the interest and variety of the spectacle." Visitors young and old, physicians and philosophers, would come to the garden to study, love, and meditate upon nature. Nothing would be spared to produce this microcosm of the whole of nature, just as a regenerated France would be the microcosm of a new moral world.

The CPS endorsed the recommendations in the CPW's first report about where to situate the provisional menagerie. It also agreed that the design for "a definitive menagerie" should be decided through an artists' competition.[46] Meanwhile, the Museum had not given up on locating the provisional menagerie in the center of the garden.[47]

Two days after the CPW delegation's visit, the Museum professors gave representatives from the CPI a guided tour and an account of the funds that had been advanced "for the Museum, the library, the rhinoceros, the menagerie, the public courses, [and] the indemnities accorded to the workers in light of the present scarcity of provisions." The visitors promised to report favorably back to the CPS.[48]

This burst of government interest and support fueled the professors' optimism. Given the CPS's expansion plan, Daubenton had proposed to his colleagues that they double their request for funds.[49] Museum secretary Desfontaines wrote to the CPI seeking authorization to buy instructive animals. He reported that three more animals had been rescued from the streets of Paris and delivered to the Museum. Six months earlier, when the first animals arrived, his tone in describing them had been cautious. Now he wrote enthusiastically. The new mandrill was "one of the most beautiful and biggest ever seen." The raccoon was "an animal the size of a fox" that washed its food before eating it. Along with an Angora goat, the professors deemed that all three should be kept as aids to public instruction.[50] The CPI responded warmly that it would "hasten to welcome" any proposals from the Museum that served to perfect such a useful institution and contribute to national prosperity.[51]

Relations with the Revolutionary government seemed optimal, but immediately after these developments, the Committee for Public Health (CPH) intruded in Museum affairs in a profoundly disturbing way by adding a naturalist's assistant to the staff without consulting the professors.

According to the Museum's founding document of June 1793, the professors would choose the successors of colleagues who died, resigned, or otherwise vacated their posts (title 1, article IX). More detailed rules and regulations, articulated in September 1793, specified that the assembly would name "all the employees" of the Museum and fire them should they fail to perform their duties (chapter I, article VI). Chapter III, article VIII called for four naturalist's assistants, a number that could fluctuate "according to the needs of the establishment." Assistants nominated by individual professors were subject to the assembly's approval.[52]

On three occasions in 1794, however, the government threatened the Museum's autonomy by involving itself in the selection of naturalist's assistants. In February, the CPI recommended the appointment of an entomologist named Laurent. The Museum respectfully responded that the position had yet to be created by the National Convention and that if the CPI could persuade the Convention to do so, the Museum would give its request all the consideration it deserved.[53] In March the minister of the interior, Jules-François Paré, asked that Olivier Marat, brother-in-law of Jean-Paul Marat, the famous martyred "friend of the people," be named as a naturalist's assistant and that he and his wife be given an apartment.[54]

The Museum was reluctant to challenge the appointment of a Jacobin hero's brother-in-law, who was also reported to have developed a promising new means of preserving insects. Lamarck and Geoffroy were sent to examine Olivier's insect collection, and at the next assembly they delivered a favorable report.[55] A letter to the minister said the Museum was willing to fulfill his request provided he did them a favor. As Geoffroy and Lamarck put it, the Museum professors were "ready to do all that will be in our power to fill your wish with respect to Olivier Marat and to give you this proof of our devotion and our patriotism, but for that to succeed, citizen Minister, it is necessary that you support us with the National Convention."[56]

The professors reminded Paré of the budget they'd provided months back for the Museum's 1794 reorganization. Funds were urgently needed for reorganization, operating expenses (exacerbated by the skyrocketing cost of living and of labor), and finally, "to feed and care for animals that have been deposited here by order of the Commune of Paris." Hiring Olivier Marat as a naturalist's assistant was "their hearts' wish," Lamarck and Geoffroy wrote, but it was impossible without the resources the Museum needed.[57]

The minister then pressured the professors for an apartment for Marat's sister. As short of apartment space as it was short of funds, the administration postponed a decision until Daubenton and Jussieu could discuss the matter with Paré directly. But the minister's status proved more precarious than the Museum's. Before long he was ousted and arrested for his close association with Danton, who was guillotined on April 5, 1794. In its form, if not in the dramatic specifics, this general scenario would rescue the Museum on other occasions. A fortuitous change in government, typically in the person of the minister of the interior, more than once spared the Museum from complying with unwanted government demands.

The Museum dodged the first two attempts to usurp its prerogatives in choosing naturalists' assistants, but it was a different matter in June 1794 when the Committee for Public Health (CPH) officially created two positions for *naturalistes conservateurs* (naturalist conservationists) at the Museum. Their job would be the "arrangement, disposition, and conservation of the productions of nature" under the professors' surveillance. Beyond creating the pair of posts, however, the CPH filled the first of them with an individual of its own choice, a man named Jean Joseph Gautier, identified only as "a member of the revolutionary committee of the section of William Tell."[58] For the second position, the Museum was invited to present three nominees from which the CPH would make the selection. The annual salary projected for the new assistants was again 3,000 livres, roughly 200 livres more than the professors' compensation.[59]

Staffing the Museum of Natural History was a tiny item of business for the CPS amid its staggering administrative workload. The CPS directed the armies, worked with ministers of state and the National Convention, and much more. Nonetheless, shortly after announcing the imminent appointment of two conservation naturalist's assistants, the CPS revised that decree twice, even as events of profound significance to the Revolution were rapidly unfolding. On 8 June 1794 Parisians were presented with the "Festival of the Supreme Being," an extraordinary propagandist production conceived by Robespierre and organized by the artist Jacques-Louis David to celebrate "the immortality of the soul." Ironically, though the Terror was about to peak, the festival gave both the guillotine and French citizens the day off. Two days later, the National Convention enacted the notorious law of 22 prairial calling citizens to denounce anyone suspected of counterrevolutionary sentiments, greatly empowering the Revolutionary Tribunal, and ushering in what is now called the Great Terror, six weeks of "national defense" when more heads rolled than in the previous thirteen months combined.

The Museum's professors wisely judged this the wrong time to petition the CPS about its intervention in Museum affairs. They contented themselves with the committee's confirmation of its choices for two previously authorized naturalist's assistant posts (in mineralogy and zoology) and the endorsement of their request for a third such post (in botany). In the meantime, a new assemblage of live animals had been added to the existing collection: a panther, two mandrills (male and female), a female macaque, and a sulfur-crested cockatoo.[60] They were not especially important for the menagerie in themselves, but the same cannot be said of the man who escorted them: their owner, Félix Cassal, a former lion tamer who would play an outsized role in the menagerie's story.

On a larger canvas, the French army's victory over the opposing Coalition forces at the battle of Fleurus in Belgium (8 messidor an 2 [26 June 1794]) forced the Allied armies to abandon Belgium and retreat beyond the Rhine. With threats from abroad reduced, the Convention turned its attention to unrest at home, where factionalism, purges, and executions had become the order of the day. Identified correctly or not as the architect of the Terror, Robespierre came to be viewed by other Revolutionaries as too dangerous to be allowed to remain in power. The Terror came to an end on 10 thermidor an 2 (28 July 1794), when Robespierre went to the guillotine and Georges Couthon and Louis de Saint-Just from the Committee of Public Safety went with him.[61]

The Museum carried on after Thermidor, much as it had before, under the Committees of Public Instruction and Public Safety. Lamarck and Geoffroy were sent to report to the CPI on the menagerie's "pressing needs" regarding the reimbursement of former animal proprietors.[62] Jussieu was assigned to present the CPI with the Museum's needs as a whole.[63] Geoffroy, Lamarck, and Daubenton continued teaching courses that had begun back in floréal, before the most violent days of the Terror began.[64] On the professors' urging, the architect continued to design an overall plan for the Museum's expansion while making drawings for the upper galleries and the menagerie.[65] New specimens continued to arrive, despite the Museum's having no place to put them.[66]

Early in September 1794, the professors decided the monkeys would temporarily stay in a ground floor room of the building where Buffon had lived and where apartments for Lamarck and Lucas (the head of the galleries) were being prepared.[67] At the assembly's next meeting, decisions were made both about the menagerie staff and about where to put the *animaux féroces*. At Geoffroy Saint-Hilaire's recommendation, his colleagues decided to retain Cassal, Marchini, and Louzardy as keepers but let the others go. The professors approved Molinos's plan to accommodate the big cats and other fierce creatures in "the small building serving as a greenhouse on the terrace alongside the river."[68]

Molinos's reconfiguration of the greenhouse featured a row of eight cages in a long building with arched openings.[69] Two short arms extended from both ends, each with three cages. Additional cages branched off perpendicularly from the arm on the left. According to the plan, the lion would get the largest cage, flanked by the cages of the two polar bears. The large mandrill had a private cage, as did each of the two panthers, side by side. This left two cages empty in the row of eight. Birds (especially eagles) would occupy five cages of the two short arms, while the sixth would be for monkeys, with four additional cages branching off from there. The plan included a service corridor behind the main row of cages and a storeroom on one side. The Commission of Public Works approved the design on 25 September 1794, but the construction would drag on nearly another year before the animals could be installed there.

As the plans for housing the animals advanced, so did those for accommodating professors and staff. Evicting the former architect, Verniquet, would make an apartment available for the professor of animal anatomy, Mertrud. The Museum also eyed a property known as the "Maison Leger," which it identified as "an indispensable necessity for the establishment." Large enough to include several apartments for professors, the building could fulfill other needs like storing new specimens before they were placed in the galleries.[70]

Finally in mid-October, with the political heat greatly reduced and construction of the provisional menagerie and the purchase of the Maison Leger on track, the Museum revisited the one conspicuously sore point in its dealings with the pre-thermidor Committee of Public Safety: its preemptive appointment of Gautier as naturalist's assistant. Wanting him out, the professors argued that not only was he unqualified, he had never shown up for work.[71] Despite Gautier's protests, the CPS followed the Museum's recommendation.[72] Come December, the Museum was free to choose a naturalist's assistant in zoology.

Lamarck, as professor of insects, worms, and microscopic animals, was to oversee the post, and he compiled a list of possible candidates to present to the assembly. Most of those named had applied for the position; others were Lamarck's suggestions. The name at the top of Lamarck's list belonged to a man who would become one of history's most renowned zoologists. Georges Cuvier had yet to come to Paris, but his reputation as a rising star preceded him. Lamarck hailed Cuvier's talents but hedged about his suitability for the assistant's position, observing that "this citizen seems to me more of a scientist than an artist." The post required an "artist" skilled in the art of taxidermy, and Lamarck considered it his duty "to put each in his place."[73]

Proceeding down his list, Lamarck crossed out several candidates who were excellent at preparing bird specimens but lacked expertise with the kinds of animals Lamarck was responsible for. He then ruled out two more—Laurent and Olivier Marat—who did have experience with insects. Laurent was skilled at collecting but not preparing them; Marat was qualified in both, but Lamarck judged that the ideal candidate should be able to assist in all the zoological collections, not just the insects. A broadly skilled naturalist, capable of preparing mammals, birds, reptiles, fish, and insect specimens would be the best choice.

Having made his case for the candidate's desired qualifications, Lamarck announced his choice: Jean Desmoulins (1798–1875), a man nearly a decade Lamarck's senior. As a child, Desmoulins had painted and sketched animals, then learned how to preserve them in their natural form. No one had been more successful in preparing large mammals, and his skill at stuffing birds was laudable. He had prepared insects "all the perfection possible." The assembly voted unanimously in his favor.

One could imagine that Lamarck's choice of Desmoulins over the soon to be distinguished Georges Cuvier was a snub that helps explain Cuvier's later animus toward the older man, but there is no evidence that this was so. Cuvier was not yet in Paris and had not applied for the post. Furthermore, Lamarck's assessment of Cuvier as "more of a scientist than an artist" was accurate, and it was true that skill as an artist/taxidermist was an essential prerequisite for being a naturalist's assistant in zoology. Where Lamarck misjudged Cuvier was in portraying his character as "very gentle" as well as "very social." Cuvier was indeed socially adept, but as his career progressed, few would have described him as gentle. As for Desmoulins' time as a naturalist-assistant, his accomplishments would include a much-admired, life-like stuffing of the lion from Senegal.[74]

As the appointment of a new naturalist's assistant approached a satisfactory conclusion, a milestone event in the Museum's history occurred. On 11 December 1794 its mission was presented anew to the National Convention, which ratified all that was asked of it by Antoine Claire Thibaudeau, reporting jointly for the Committees of Public Instruction and of Finance. Thibaudeau reviewed all the good that had come from the Museum's founding and all that remained to be done. Bills needed paying, even as new expenses connected to the Museum's growth arose. The existing buildings could not hold the riches in natural history that had poured in from the collections of *émigrés*, *condamnés*, and sites in Belgium. To be the nation's center for endeavors in natural history and agriculture, it needed more space. That was why, Thibaudeau explained, the CPS had commissioned Molinos to draw up a grand plan for expansion of the Museum that would surpass in utility and beauty the most magnificent of ancient monuments.

Anticipating objections to the project as a burden to the nation's fragile economy, Thibaudeau called for a scaled-down version, half of what was originally planned, but still doubling the Museum's size. Two parcels of marshy land would be added. One, west of the Museum, included some lumberyards and nondescript houses but also two substantial buildings: the large grain storehouse called the "Régie" and the property known as the Maison Leger. The second parcel, to the east, held flower gardens and had the advantage of proximity to water sources, including the Bièvre River. This land, Thibaudeau said, was "indispensable" for displaying crops, trees, and flowers and developing the menagerie.

With respect to the menagerie, the legislator echoed ideas voiced by Bernardin de Saint-Pierre, that the remains of dead animals were not sufficient for their study and representation. "When one compares the lions in the majority of artists' paintings to the magnificent individual that exists at the Museum, one sees that the majority of artists, copying from each other, have not conveyed nature, and that their imitations fall far short of the model."[75]

Thibaudeau explained how the new menagerie would differ from those of the past, even the best of which were merely "prisons where the locked-up animals had the physiognomy of sadness, lost a part of their coats, and nearly always remained in positions that attested to their apathy." To instruct the public, he said, menageries "ought to be constructed so that the animals, of whatever species, enjoy all the liberty consistent with the safety of the spectators, so one can study their nature, habits and intelligence, and enjoy their natural pride in all its development."[76] These animals would not display "the stigma of slavery."

Thibaudeau wished to improve the professors' lot as well as the animals.' He asked for a third professorship in zoology and for salary increases. It was wrong, he said, that men like Daubenton, famous for their long service and accomplishments, were paid just 2,800 livres "while a crowd of inept clerks squander more than double that in idleness."[77] He called for raising their salaries to 5,000 livres annually.

Thibaudeau's address focused more on the utility of the study of natural history than on its moral value, but he invoked the latter in a rhetorical flourish near the end: "Call all men to consider the great and magnificent spectacle of the power of nature, the variety of her productions, and the harmony of her phenomena. She is the source of good laws, of the useful arts, the gentlest enjoyments, and of happiness." After remarking how the Jardin des Plantes, "the temple of nature," had survived the tempests of the Revolution while other institutions were being shut down, he concluded by urging the legislators to adopt the CPI's and the Committee of Finances' proposals to enlarge the Museum.[78]

The Convention agreed to everything Thibaudeau proposed, though some deputies, angry at the profligate spending of previous committees, were displeased with the new financial commitment.[79] Thrilled with the success, the professors at their next assembly elected to buy four hundred copies of Thibaudeau's address with the Convention's decree appended to it. They presumably intended to distribute them to government officials, landowners, contractors, and others as proof of the Museum's worth and the Convention's support.[80] The decree did not specifically mention the menagerie, but agreeing to pay for its past and upcoming expenses made it officially a part of the Museum.

On 13 January 1795, the professors turned to the task of appointing a third professor of zoology, observing the protocol of considering eligible candidates before taking a vote. With "the assembly sufficiently enlightened," they reached their foregone (and appropriate) conclusion, unanimously electing Bernard Germain Étienne de Lacépède (1756–1825) "professor of zoology for the demonstration of the oviparous quadrupeds, reptiles, and fish."[81]

At the same meeting, the Museum reviewed letters from the Committee of Public Instruction recommending how to ensure the success of the École Normale, created to prepare teachers for the new nation. Some 1,500 students from around the country had been selected and brought to Paris. Instruction would begin on 1 pluviôse an 3 (20 January 1795) in the Museum's amphitheater. It could seat only six or seven hundred students, but it was to serve as the temporary lecture hall until the desired space at the Sorbonne was ready.[82] The Museum was to be open from 9:00 a.m. until 4:00 p.m., with a knowledgeable person on hand in the library and galleries to answer the students' questions.[83]

Among the new school's distinguished professors was the Museum's director, Daubenton. Two days after the opening ceremonies, he gave his first lecture, a general introduction to the study of natural history. The next day at the professorial assembly, he hailed the École Normale's establishment as "an epoch in the annals of the Revolution." At the same meeting, Geoffroy reported on another consequence of the Revolution, the Museum's receipt of a share of the scientific apparatus and related belongings of Lavoisier.[84]

On 26 January 1795, Daubenton was back at the lectern, this time stressing the importance of presenting scientific facts in a straightforward, unadorned way. To make his point, he offered the contrasting case of a famous naturalist who had sacrificed accuracy for rhetorical effect. He did not name names, but there was no chance here for mistaking the naturalist's identity. Daubenton was clearly referring to the late comte de Buffon, former director of the Jardin du Roi and Daubenton's previous boss. To illustrate how grandiloquence may obscure truth, Daubenton cited Buffon's representation of the lion as the king of beasts and objected succinctly, "The lion is not king of the animals; there are no kings in nature."[85]

Daubenton was known as a soft-spoken lecturer, but his words electrified his audience. A footnote to the stenographic record of his lecture states that the "hands, eyes, faces, acclamations, [and] attitudes" of the students reflected their feelings. "The public spirit," the note continued, "gushed forth in all its purity and strength from the lecture hall. Never did the hatred of royalty manifest itself with more energy, and the love of the Republic with more gusto." The note's author added, "Those who do not want kings, even among ferocious beasts, will not put up with them amongst themselves."[86]

Over forty years later, Étienne Geoffroy Saint-Hilaire recalled that the students cheered Daubenton's remark for "nearly a quarter of an hour," a response that left Daubenton nonplussed. "The excellent old man was keenly troubled by it," wrote Geoffroy, and asked Geoffroy, who stood by his side as his assistant, "Why this noise, these furious cries?" The students were even more animated, Geoffroy said, because they believed that Daubenton was offering "a loyal and simple expression of his thought as a naturalist."[87]

Memory is notoriously unreliable, and Geoffroy's reconstruction of the scene in the Museum's new amphitheater is suspect on several counts. Could the student uproar have lasted as long as Geoffroy said it did? It seems an exaggeration, but the students were disposed to displaying patriotic fervor, sentimentality, and a joy in making noise for its own sake; and enthusiastic and sustained applause was a hallmark of the École Normale's early sessions.[88] More problematic is Geoffroy's claim that he was present when it happened. Geoffroy had been Daubenton's assistant, but at an earlier date. He became a professor himself in June 1793. This would not have prevented him from helping his elderly colleague on the podium two winters later, but there is no mention of it in the documents of the time or in Paul Dupuy's meticulous accounting of the École Normale's history. It seems likely that four decades after the event, when Geoffroy told his story, he conflated his earlier experiences as Daubenton's assistant with this lecture.

What is hardest to believe, however, is Geoffroy's portrayal of Daubenton as so oblivious to the revolutionary sentiments whirling about him that the students' reaction to his "no kings in nature" comment took him by surprise. What we know of Daubenton and his times indicates otherwise. As the Museum's first elected director, Daubenton had guided it through a precarious but productive first eighteen months. He had witnessed the dangers of the Revolution up close, and he knew that judicious self-presentation was vital for survival. He could play to an audience when required. Late in November 1793, when he needed to secure a certificate of civic worth from the section of sans culottes, he shrewdly consented to be identified not as a scientist but as a shepherd, a politically convenient fiction related to his efforts to improve sheep breeding in France.[89]

The Reign of Terror was over when Daubenton gave his rousing lecture, but even in the more moderate climate of the Thermidorian Reaction, enthusiasm for the Revolution ran high. The students were of varying ages and educational preparation, but their politics were relatively similar. Most were Thermidorians, happy that the Reign of Terror had ended and the Jacobins had been ousted, but full-throated in their support of the Republic and their hatred of kings.[90] Three days before Daubenton's lecture, they had been given the day off to celebrate the second anniversary of Louis XVI's execution. When Daubenton declared there were "no kings in nature," the students were primed to agree.

When the hubbub over his remark died down, Daubenton specified why the lion was not royal or noble, and neither lions nor kings fared well in his analysis. Unlike kings, the lion had no sycophants and courtesans, but its roar was not nearly as ferocious and powerful as alleged, and there was no "noble anger" or "magnanimous courage" in the way the beast captured its prey. Daubenton likewise discredited Buffon's suggestion that there was something noble about the lion's not having degenerated into diverse species. In that respect, Daubenton said, all animals would be noble, because all animals of a species share the same essential characters. Daubenton's dismissal of the lion as the king of beasts was no off-the-cuff remark. It was a premeditated attack on the ancien régime signaling the Museum's alignment with the new political order.

The Museum and its menagerie came into being buoyed by beliefs in the regenerative power of contemplating nature and the benefits to be had from subjugating nature's resources to human interests. These broad assumptions undergirded the Museum and the menagerie alike. However, contingent events featured prominently in how each came into being. The mundane problems of controlling specimens loomed large in both cases.

The Museum's founding was triggered by the need to accommodate the natural history specimens confiscated from the estate of Chantilly. The menagerie's first animals were taken from men who had been exhibiting their potentially dangerous charges on the streets of Paris in the weeks after Marie Antoinette's execution. Live-animal confiscations in the spring elicited more support from the government than the Museum's professors had dared to hope given the other demands on the new nation. Although it was not the Museum's top priority, the menagerie in its early months had a disproportionate influence on the government's support of the broader institution.

Through it all, the Museum embraced the Revolution, the Revolution embraced the Museum, and the Museum largely managed to dodge the state's challenges to its principles of self-rule. The menagerie's situation, however, was far from Thibaudeau's characterization of the Jardin des Plantes as "the temple of nature." He portrayed adding live animals to the Jardin des Plantes as necessary for displaying nature's grandeur more fully, but the animals in effect ended up not in a temple but in a prison. The plans for the quarters for the *animaux féroces*, which were still not completed in January 1795, promised no better. The best that could be said for this menagerie, with its narrow cages and its bars retrieved from the old royal menagerie at Versailles, was that when it was done it would still be "provisional."

[1.4]

The Lion, the Dog, and the Librarian

It is a curious but instructive fact that in the early years of the menagerie of the National Museum of Natural History, it was librarians, not zoologists, who paid the closest attention to the animals. The head librarian, Georges Toscan (1756–1826) observed and wrote about them at length; the assistant librarian, Jean-Claude-Michel Mordant Delaunay (1750–1816), was the menagerie's primary supervisor. Toscan provided the most detailed account of the menagerie's first lion and his dog companion, and Delaunay identified the cause of the lion's death. The Museum zoologists were concerned with the lion's transfer from Versailles and how to house him, but they did not consider the animal a rare species and saw no advantage to studying him alive. The dog interested them even less. Nor did the interactions between the pair stir the professors' curiosity. The study of live animals was not their province.

When the animals of the royal menagerie of Versailles arrived at the Museum in late April 1794, the Terror was running high, but it was also a time of cultural innovation, including the debut (29 April) of a periodical, *La Décade philosophique, littéraire et politique* (*The Philosophical, Literary, and Political Week*). The work of "a society of republicans" promoting the ideas of the Enlightenment and the cause of the new republic, the journal was published on the last day of each ten-day week of the newly instituted calendar: hence the name *Décade*.[1]

Georges Toscan, the future Museum librarian, was the *Décade*'s natural history editor, and in its inaugural issue he offered his "general views" on nature. The purpose of studying nature, he declared, was "to find more ways to love her," and his purpose in writing about nature was to encourage others to love her too. Through contemplating and embracing nature he believed humans bettered themselves.[2] He proposed that when writing about botany, his special passion, he would focus not on the achievements of the century's great botanists but instead on how botany could be linked to "morality, public education, and the love of country."[3]

Figure 1.4.1. Nicolas Maréchal's sketch "The Lion of the Museum of Natural History with His Dog." *La Décade philosophique* (30 vendémiaire an 3 (21 October 1794). Rare Book and Manuscript Library, University of Illinois at Urbana-Champaign.

Such sentiments would not have hurt Toscan when he applied later that summer to be the Museum's librarian. The idea that moral and social benefits flow from the study of nature was a theme of Revolutionary ideology that redounded to the Museum's benefit. But ideology was not enough. Specific talents, experience, and personal connections would also factor into the Museum's choice of a librarian, and Toscan referred to all of these when he submitted his name for the librarian post.

His love of nature, Toscan told the professors, derived from his youth in the French Alps. He became a pupil and friend of the region's leading botanists, Dominique Villars and Pierre Liotard, and accompanied them on excursions in the mountains of the Dauphiné. In 1780 he expanded his botanical interests abroad as assistant secretary to the French embassy in Constantinople. For two years he collected plants and seeds for the Jardin des Plantes guided by a wish list prepared for French diplomats by André Thouin, the Jardin du Roi's head gardener. Toscan's travels took him "from the shores of the Black Sea to the plains of Bythnia and to Mount Olympus."[4]

Toscan could also claim to be "one of the first" to propose a natural history library at the Jardin des Plantes. In 1793 he published a twenty-four-page memoir on the topic, in effect angling for the post of librarian before the Museum decided to create one.[5] In his letter of application, he cited his experience working with natural history books and highlighted his political suitability, observing of himself (in the third person): "Regarding his patriotism, it will suffice to say that it is not in a life devoted to meditating on nature that one nourishes servile ideas and habits."[6]

On July 26, 1794, when the professors gave Toscan the job, they referred to his "literary works," "his knowledge of foreign languages," and his memoir on the utility of establishing a library at the Jardin des Plantes. It may also have helped Toscan that the Museum's professor of geology, Barthélemy Faujas de Saint-Fond, was a friend of Toscan's mentors, Villars and Liotard.[7] The professors decided to present Toscan "to the Committee of Public Safety of the National Convention as the subject most capable of filling the place of librarian of the National Museum of Natural history." The event could scarcely appear less consequential in comparison with what was about to take place at the CPS. Robespierre was arrested the next day, and the day after that he and two other members of the committee, Georges Couthon and Louis de Saint-Just, were guillotined on the Place de la Révolution.

Botany was Toscan's first love, but the story of the lion and dog was irresistible. The first installment of his two-part article on the pair appeared in *La Décade philosophique* in October 1794, the second ten days later.[8] He began by stating that he would periodically report on the menagerie and the "tastes, inclinations, and instincts" of "the rarest and least known" of its animals. In the case of the lion, he was offering "the history of the most terrible and most powerful of all." The menagerie's lion furthermore presented a golden opportunity for philosophical reflection. By comparing it with lions in the wild, one might learn both what it had lost in captivity and what was gained "through the influence of society and its blessings."[9]

Toscan described the lion in the wild, noting that "its imposing form, its assured look, its proud bearing, and above all its terrible voice" made it easily recognizable as "the despot that nature has made reign over the deserts."[10] Though he questioned Buffon's claims that differences in climate influenced the character of a species, he nonetheless portrayed the lion's nature in much the same terms as Buffon had used nearly half a century earlier in his *Histoire naturelle*.[11] "The lion," wrote Toscan, "is never cruel; he attacks and kills only by necessity. As soon as he is satiated, he is completely at peace." Moreover, "the anger of the lion is noble and generous. Along with his very impetuous appetites, nature gave him a character susceptible of gentle and social modifications. He is irritated by mistreatment, but he remembers and appreciates kindness."[12]

Citing an account of a lioness that became as tame as a dog after a man cured her of an illness, Toscan reflected on the process of domestication, where the key, he said, was kindness. Man's heart and humanity were what had gained the familiarity and trust of "the first animals whose descendants live with us, share our labors, and provide for our needs."[13] Cheerfully anthropocentric, Toscan did not mention that domestication constituted a form of slavery, an idea advanced by Buffon and Bernardin.

Toscan did not refrain from calling the lion and dog's abode at the Museum "a prison." His focus, however, was on the lion's magnificence and the animals' friendship. Despite the lion's long captivity, "the nobility of his origin" was manifest. Toscan wrote:

> His face is always imposing, majestic; his proud, gleaming look still seems to command all who approach him; and from the depth of his prison, each sign of his will seems to be an absolute order that must not be resisted. His size is midway between that of the mid-size and large species of lions; he is six and a half feet long, and three feet, two inches tall. A thick mane covers his head and the front parts of his body, which is all nerve and muscle. The color of his coat, a fiery, fawn color on a dark background, gives still more passion to the expression of his traits and his movements, but across this terrible expression is painted the sensibility of a character cultivated by kindnesses and softened by the pleasures of friendship.[14]

Toscan omitted a comparable description of the hound as unnecessary, finding the way it played with the lion much more interesting. The animals' friendship faltered only at mealtimes, wrote Toscan, when they separated to receive their food: meat for the lion, bread for the dog. Neither interfered with the other's meal or even dared look at it. The first to finish (usually the lion) would wait for the other to finish before rejoining him. If the public threw bread into the cage, however, the dog claimed it and likewise gobbled down the lion's meat when the keeper mistakenly tossed the animals the wrong food.

Toscan believed he detected an additional "nuance" in the relationship between the two animals. "Independent on the earth, savage and proud by instinct, now solitary and captive, the lion associated with a friend; he loved the friend for the friend himself and loved him uniquely. Equally sensitive, the dog loved too, but before giving himself to the king of animals, Nature had given him to man: faithful to his instinct, it was to man that he referred the caresses that he gave the lion."

Toscan saw the dog as the lion's sole consolation. The dog, on the other hand, while enjoying the lion's friendship, still desired human affection: "With what tenderness did he not run to the man who opened the door of his prison to provide him a moment of liberty! How he bustled about him. What gaiety was in his eyes, while his poor friend, disquieted by his absence, gave off plaintive roars, paced the length of his bars, went to the back of the cage, stopped at the spot from which his companion had departed, came back, then turned around again! Would the dog have to return? The dog met his companion again with joy, but his last glance seemed to say to the keeper who had taken him away, 'it is to please you.'"[15] In the librarian's view of nature, anthropomorphism and anthropocentrism intertwined.

The dog eventually died of a gall developed from sleeping with his back pressed against a damp wall of his quarters beneath the natural history galleries. His demise left the lion morose. The staff thought to relieve the lion's despondency by providing a new dog, but the lion was infuriated at the sight of it. "He fixed the dog with a gleaming eye, rage broke out across his face, he let out a frightening roar, and with feet stretched and claws extended he made ready to spring." The staff misjudged the situation, supposing that the lion's rage was directed toward the person who held the dog's leash, preventing the lion from having a new friend. Their mistake was quickly revealed: when the dog was introduced into the lion's cage, the lion killed him on the spot.[16]

"Indeed," explained Toscan, "it was not a dog that the lion missed, it was a friend. Time eased the pain and restored the lion's health and strength, but he never fully recovered. Even now," Toscan wrote in the fall of 1794, "his sense of loss renews itself and turns bitter at the sight of a dog passing by; he rises at once, he becomes agitated, he growls dully, and becomes peaceful again only when this sad sight moves out of his vision."[17]

When Toscan first wrote about the lion and its canine companion, he knew only that a French colonial administrator had raised them together from a very young age, allowed them to roam freely in his house, fed them leftovers from his table, and petted them. Under these conditions, said Toscan, the animals developed their strong mutual affection. It was not rare, he allowed, for two animals of different species and characters to bond, but this tended to happen only among those living close to humans, and always in an atmosphere of human kindness.

Six months after publishing his account of the lion and the dog, Toscan received a letter from Jean-Gabriel Pelletan, the man who had acquired and raised them.[18] A former director of the Compagnie d'Afrique in Senegal, Pelletan, like the lion, had his life derailed by the Revolution. Suspected of being a counterrevolutionary, along with other directors of the French African and East Indies companies, he was imprisoned in St.-Lazare. Fearing he might be taken to the scaffold at any moment, he occupied himself by composing a treatise on Senegal, promoting its potential for colonial development, while arguing for the suppression of the slave trade.[19] He was released after Robespierre's execution.

Pelletan told Toscan he had acquired the lion in Senegal in the fall of 1787 when it was three or four months old. He and his household treated the cub with affection, caressing and petting it, and the animal responded in kind. The young lion recognized and answered "marvelously" to the name "Woira." Now and then he had to be prevented from doing something or be shooed away, but the animal never reacted angrily and was largely allowed to come and go as he pleased. "I noted especially," wrote Pelletan, "that [the lion] loved society; he preferred to go to the places where he found the most people. When he was chased from the offices, he went to the kitchens." The lion enjoyed sitting at Pelletan's feet in his office. When Pelletan went to dinner the animal followed him to table.[20]

The young lion was not the only animal in Pelletan's west African household, where everyone's behavior resembled that of Bernardin de Saint-Pierre's image of the world when it was new. Pelletan described the scene:

> My house was filled with animals of all kinds, and all were at liberty. Sheep, dogs, cats, monkeys, ostriches, geese, ducks, turkeys, chickens, horses, parrots—they all lived on good terms with the lion. None feared him, none was harmed by him. Their friendly relations and mutual confidence were so well established that on winter nights, which are very cool in Senegal, they slept together pell-mell, one on top of another, in the stables with the horses. The lion slept in the middle and the Negroes who served me went likewise to sleep among them, in search of warmth. It was truly a curious assemblage, this heap of animals of such different species, lying peacefully together. No accident, fight, or violence ever troubled this singular and touching harmony.[21]

The lion was seven or eight months old when Pelletan's dog gave birth to two puppies. Fascinated by the canine family, the lion looked after the puppies whenever the mother was briefly absent, taking great care to treat them gently. When one of the puppies died, the lion grew even more attached to the survivor and reportedly took nearly as much care in his upbringing as his mother did. Affectionate toward his young charge, the lion indulged the puppy's every whim. His playful bites sometimes drew blood, but the lion "never dreamt of taking revenge." As the animals matured, they "could not live without each other; they always slept together, the dog in the lion's paws, leaning against his stomach." If they ever happened to be separated, the household would soon hear the lion's roar and the hound's howl, and the animals were not content until they were reunited.[22]

Pelletan decided to give the lion away when, one day, he heard shrieks outside the house and found the lion rolling on the sand in play with a child. Though unhurt, the child was terrified. Realizing that the children and the women of the household now feared the lion despite his history of gentleness, Pelletan resolved to send him to France. He decided the dog should accompany the lion, to avoid the cruelty of separation. He arranged for the lion to be chained aboard ship, an unnecessary precaution since the lion proved as placid as ever. Allowed to move freely, it even "slept and ate with the sailors." The ship sailed to Le Havre, where the lion and dog disembarked in September 1788 and were taken to Versailles.[23]

Pelletan's letter ended at that point. Toscan's article had already provided details about the animals' 28 September 1788 arrival at Versailles, their stay in the royal menagerie, their transfer to Paris, and their experience there. Supported by Pelletan's recollections, Toscan told the tale of a ferocious creature softened by friendship and kindness. The story's happy, optimistic tenor, however, had a dark side. The lion's trajectory began in freedom and ended with incarceration and despondency. Though as gentle as the dog when they arrived at Versailles, the lion's ferocity reawakened there, "undoubtedly aggravated by his captivity."[24]

Moving to Paris made things worse. By Toscan's account, the lion disliked the horsemeat he was fed there; it made him ill. The switch to a new keeper may also have affected the lion's appetite, along with the deficiencies of his lodgings. The cramped and damp quarters beneath the natural history gallery were far inferior to the lion and dog's home in Versailles, and the urban environment was less tranquil. The crowds that visited the animals allowed them no rest. They wanted to see them in action, not sleeping or lying in their cages. They threw bread into the lion's cage and sought to aggravate him and elicit roars by yelling at him, or in the case of soldiers, by rattling their sabers on the bars of his cage.[25] Rather than just a simple story of nature's affections and the benefits of human kindness, Toscan also described a sentient creature's miseries and decline under the effects of incarceration, the physical and psychological deterioration owing to poor living quarters, public abuse, lack of wholesome food, and the loss of a companion, the lion's "sole consolation."[26]

The librarian's engagement in the lives of the menagerie animals did not go unnoticed by the Museum professors. On 4 brumaire an 3 (25 October 1794), four days after Toscan's account of the lion and dog appeared in *La Décade philosophique*, Daubenton, the Museum's most venerated professor, proposed that the artist Maréchal, who had already painted the lion's portrait, be authorized to paint one of the lion-dog pair. Daubenton, who had apparently heard about Toscan's article, wanted the "remarkable fact" of the animals' friendship preserved for history, and his colleagues agreed to the proposal.[27] They were unaware that Maréchal had already produced a sketch of the pair, the basis of the engraving that accompanied Toscan's article, and that the dog was already dead.[28] Keeping up with the goings-on of the menagerie was clearly not high among the professors' priorities.

[1.5]

The Lemur in the Laboratory

Of all the exotic animals that lived at the Museum of Natural History in the institution's early years, none was more attractive than the ring-tailed lemur, the *Maki moccoco*. With its handsome tail, foxlike face, agreeable proportions, and engaging mannerisms, it was a favorite of all who interacted with it. Few had the chance to do so, however, since the lemur was not kept in a cage for public viewing. It lived in one of the Museum's most important behind-the-scenes work spaces, the zoology laboratory. Most of the menagerie's animals passed through the laboratory when they died, to be mounted for later display in the Museum galleries. The lemur was unique in that the laboratory was its home and its freedom of movement there was relatively unconstrained, in stark contrast to the many animals forced to live out their lives in cramped cages.

The ring-tailed lemur's story connects with multiple themes in the menagerie's history, including the role of confiscations in the growth of the collection; the problems of lodging and feeding live exotic creatures; and the mismatch between the naturalists' tendency to prize rare specimens and the public's preference for visually appealing ones. In all these regards the lemur was an exception. It came to the Museum indirectly following the confiscation of émigré property, and it was spared the gaze of the public. At a time when it was impossible to find food of sufficient quality or quantity for most of the menagerie's animals, the lemur was pampered and fed delicacies. Last, while professional zoologists were disposed to value a specimen's novelty more than its beauty, the naturalists in the laboratory liked the lemur because it was a handsome and charming companion.[1]

When Geoffroy Saint-Hilaire published a long paper (1796) on the natural affinities of the different animals that had been classified as lemurs, correctly or not, the ring-tailed lemur (Maki mococo *Lemur catta*) appeared in his chart of genera and species without additional information. What excited Geoffroy were species previously unknown in Paris or at least understudied. He described two *Galéopithecus* (Colugo) specimens from

Figure 1.5.1. "Le maki mococo" (ring-tailed lemur). Nicolas Maréchal, Simon-Charles Miger. From Étienne de Lacépède, Georges Cuvier, and Étienne Geoffroy Saint-Hilaire, *La Ménagerie du Muséum national d'histoire naturelle, ou Description et histoire des animaux qui y vivent et qui y ont vécu* (Paris, Aug. Renouar, 1801–5). R. Burkhardt collection.

the Stadtholder's collection in The Hague that were incorrectly classed as lemurs. He renamed another specimen confiscated in Holland, *Loris paresseux* (slow loris), which Andreas Vosmaer had previously identified as a sloth. Further specimens from other sources helped Geoffroy distinguish the makis from the tarsiers, and based on a skull from Senegal, courtesy of the naturalist Michel Adanson, he established an entirely new species "not yet announced by any naturalist," the *galago*, or bush baby.[2]

The ring-tailed lemur, in contrast, was already well known to naturalists and the public, not only because it was relatively easy to transport from its native Madagascar, but because its beauty and gentle manners made it an attractive pet. The eighteenth-century French aristocracy and upper bourgeoisie considered it fashionable to own and coddle exotic animals such as parrots and monkeys.[3] A lemur, however, made a better pet than a monkey; as Buffon observed in his *Histoire naturelle*, because while similar in some ways, lemurs lacked monkeys' malice. Because monkeys were constantly in motion, Buffon noted, their owners often chained them to keep them from getting into too much mischief. Lemurs could be trained well enough to move freely without their owners' worrying that they might flee.[4]

The menagerie's ring-tailed lemur had been a pet well before arriving at the Museum. Its first owner was the Marquis de Nesle, who apparently acquired it young and spoiled it. By Geoffroy's account, at an early age it "developed a taste for cooked meat and spirituous liqueurs."[5] In April 1792 Nesle chose to save his aristocratic neck by fleeing the country, leaving his lemur behind. A man on the opposite end of the political spectrum, Antoine Merlin de Thionville, gave it a new home and, so far as we know, the lemur's regimen remained unchanged.

Merlin's radical credentials were substantial. He sat with the Montagnards on the far-left side of the hall of the National Convention. Throughout the Revolution, he successively advocated the death of the king, contributed to Robespierre's overthrow, helped organize the Thermidorian reaction, and won a place on the Council of Five Hundred. He was also the legislator who initiated the call to sequestrate émigré property (though presumably not with the explicit intention of getting his hands on the marquis's lemur).

Merlin's path first intersected with the Museum's in March 1794. Lamarck and Geoffroy found him hunting on the grounds of Raincy, the duc d'Orléans's former estate, when they visited to select animals for the menagerie.[6] As Geoffroy later recalled, the famous Revolutionary was furious to be interrupted midchase, but he cooperated once he saw the naturalists' authorization papers from Joseph Augustin Crassous, another ardent Jacobin, and understood their mission.[7] He eventually assisted in the transfer to Paris of the deer and other animals the naturalists selected, in addition to animals he'd purchased himself, including a live camel he'd intended to kill in order to donate its skeleton to the Museum.[8] It was not until four years later, however, that Merlin gave the Museum his lemur.

The lemur first appears in Mordant Delaunay's ledger of menagerie expenses in early January 1798, which lists one franc and four sous as the cost of wine and two kinds of cakes (*biscuits* and *échaudés*) purchased "for the maki."[9] The following week the lemur's treats were scaled down to wine and bread (cost, one franc). A special entry for the lemur appeared in Delaunay's accounts every ten days for the next fifteen months, until the animal's death, one franc (or twenty sous) each time, presumably for bread and wine. Geoffroy Saint-Hilaire, however, reported that the lemur was fed bread, carrots, and fruits (which it especially liked), and that it was also fond of eggs.[10]

The lemur's Museum lodgings were as special as its diet. The zoology laboratory was the province of the Museum's highly skilled assistants in zoology, Louis Dufresne and Jean Desmoulins.[11] There they applied their taxidermic artistry to the goal of preserving and displaying thousands of specimens, from tiny insects to large vertebrates. They brought years of experience to the task. By the mid-1790s Dufresne was in his forties, Desmoulins in his sixties. Among the tools they deployed was Jean-Baptiste Bécoeur's arsenic soap, a concoction applied to animal skins to protect them from the ravages of insects. Bécoeur died in 1777 without publicizing the soap's ingredients, but he gave the recipe to the naturalist voyager François Le Vaillant, who told it in turn to Dufresne.[12]

Despite all the paraphernalia, specimens in process, and dangers of the laboratory, Dufresne and Desmoulins chose to bring the lemur there. They evidently felt not only that this attractive, tame, and gentle creature needn't be confined to a grim, narrow cage, but also that having it as a companion would be worth whatever disturbances it might cause. As Geoffroy later reported, it was "always in motion. It examined, touched, and turned over everything within its reach."[13]

The quarters where the lemur lived and where Dufresne and Desmoulins worked in the late 1790s was an improvement over the space the men were given when they were first appointed as *aides naturalistes* in zoology (in June and December 1794 respectively). What we know of the first lab comes from a letter they wrote in July 1795 describing it as too small, too humid, too poorly lit and ventilated: "Those among you who have come to the laboratory know that upon entering, one is overwhelmed by a noxious odor." They asked that a new lab be established in the recently acquired Maison Leger. The professors agreed to allocate space on its third floor (three floors above the ground floor) and promised to try to reserve apartments for both men in the same building.[14]

Three years later, in their new laboratory, Dufresne and Desmoulins brought fresh complaints to the administration about their salaries and overall working conditions. Their pay, they explained, was insufficient to support their families and incommensurate with the arduous and dangerous nature of their work. Preparing specimens was "extremely prejudicial" to their health because it required working with twenty-five pounds of arsenic annually. They toiled from sunrise to sunset but could scarcely keep up preparing, preserving, and repairing the thirty thousand zoological specimens in the Museum's storerooms and galleries. With stored specimens deteriorating daily, the task was urgent. The skin of the quagga had been with the tanner for three years; the giraffe, two kinds of bubal, several gazelles, and other exotic species plus "a very large part of the mammals of Europe" had yet to be mounted and placed in the galleries.[15]

The stuffed specimens already in the galleries likewise required attention. Three-quarters of the birds should be replaced, they said, and the entire insect collection (some eighteen thousand specimens) needed remounting. The naturalists' assistants were additionally obliged to supply professors' requests for teaching materials. Dufresne and Desmoulins urged the Museum to hire two more workers for the laboratory. Capping their list of woes, they pointed out that the recently purchased Régie building was gravely unsuited for storerooms because it had previously served as a flour warehouse. Flour was in every crack and crevice, from which "millions" of destructive insects emerged daily to attack the animal skins.

It seems the only thing Dufresne and Desmoulins did not complain about was the lemur that had shared the bustling zoology laboratory with them for nearly six months, although it was not exactly an ideal guest. That Desmoulins and Dufresne tolerated its presence despite the pressure of their work testifies to their fondness for the animal. Geoffroy described it as "of the greatest gentleness, appreciative of caresses, friendly with everyone, a little taciturn in its old age." It would sit on visitors' knees or climb on their shoulders. At night it performed an elaborate routine, jumping back and forth in a measured way for half an hour. After its dance, it climbed directly to its bed on a shelf above the laboratory door and promptly fell asleep.[16]

Geoffroy did not write about the lemur until nearly three years after its death at age nineteen, a life span he thought proved that the species could withstand a climate considerably colder than that of its native land. Yet the lemur disliked the cold, responding to chilly temperatures by rolling itself into a ball, legs tucked up to its stomach, with its long tail covering its back. When there was a fire in the laboratory hearth, the lemur opened its arms before it and sometimes got so close it burned its face or mustaches before backing away or simply turning its head from side to side. On warm, sunny days it stretched its arms toward the sun. Geoffroy lamented that virtually nothing was known of the species' habits in the wild, only that it was said to live in trees, often in troops of thirty to forty individuals— information reported by Buffon nearly half a century earlier (1765).[17] Geoffroy did not record the exact circumstances of the lemur's death. He may not have known, because he was in Egypt at the time, a member of Napoleon's expedition. Delaunay's menagerie record book, briefly notes that the lemur was "poisoned by the preservative of the laboratory and deceased 7 floréal an 7 [26 April 1799]."[18]

The lemur's skin was subsequently stuffed, presumably by one or both of its laboratory companions, a preparation doubtless involving the same arsenical preservative that had caused the animal's death. Six weeks later, Nicolas Maréchal submitted his painting of the ring-tailed lemur for the Museum's collection of animal portraits on vellum.[19] When its mounted reincarnation was dry, the lemur was displayed in the zoology gallery, like nearly all the other animals confiscated in France or abroad in the 1790s and brought to the Paris menagerie. Unlike them, before being stuffed it had roamed freely in the zoology laboratory for fifteen months, indulging its curiosity, eating delicacies, drinking wine, warming itself by the fire, performing its bedtime dance, and sleeping through the night.[20]

Dufresne and Desmoulins shared the lemur's experience of life in the zoology laboratory. So too did the young naturalist René Maugé. Having gained experience in the lab before going on Captain Baudin's 1796-1798 expedition to the Antilles, Maugé returned to the lab in 1798 to attend to his specimens from the voyage. Among these were 700 birds, several new to science. As he prepared them for display in the galleries, the lemur would have been in the background, sometimes closer. Maugé sailed again with Baudin in 1800 as the chief zoologist on Baudin's expedition to Australia, but that voyage proved fatal to him. He died in Tasmania from dysentery in 1802.[21]

Two other voyager naturalists for the Museum bear mentioning here, for although they did not overlap with the lemur in the lab, they too experienced poisoning in their efforts as taxidermists. Pierre-Antoine Delalande (1787–1823) was sixteen when he was hired as Desmoulins' replacement (in 1803). Ten years later, the Museum managed to have him exempted from military conscription by maintaining that his work with poisons in the lab gave him severe stomach pains and bouts of vomiting. When the Napoleonic Wars ended and peace was restored, however, the Museum was happy to recommend Delalande to the government as a voyager naturalist. His death in Paris at just thirty-six years old (1823), was attributed to the diseases and hardships of his 1818–1820 expedition to the Cape of Good Hope and the African interior, not to his earlier exposure to toxic substances in the lab, though these may have affected his constitution.

The other voyager naturalist, Pierre-Médard Diard (1794–1863), was a prolific collector, initially for France (late-1810s to mid-1820s) and then Holland. In his travels in Southeast Asia, his robust constitution withstood the kinds of illnesses that killed his friend and fellow voyager, Alfred Duvaucel (1793–1825), and another talented young French naturalist, Victor Jacquemont (1801–1832). In his late sixties he was still healthy and working as a naturalist and administrator for the Dutch government in Java. He accidentally poisoned himself while attempting to develop an improved preservative for his bird and other specimens. He did not live to tell the tale.[22]

Delalande by contemporary accounts did not die from his early exposure to the poisons of the zoology lab but instead from the effects of his travels as a naturalist voyager. Diard, to the contrary, survived his travels but died in his effort to advance the taxidermic arts. Another figure in our story was a contrast with both. This was Louis Dufrense, who not only survived a notoriously ill-fated voyage of discovery but also went on to work for years in the zoological lab and to live to a ripe, old age. In 1785, at the age of 32, Dufresne sailed from Brest as a naturalist on the Lapérouse voyage of exploration, designed to go around the world. Aboard the ship, *L'Astrolabe*, he traveled to South America, around Cape Horn, and to the Sandwich Islands, Alaska, and China. After having distinguished himself by his services (which included overseeing in China, for the benefit of the crew, the sale of some 1000 sea otter furs acquired in Alaska), Dufresne was permitted by Lapérouse to leave the expedition at Macau, c. January 1787. This ultimately saved his life, for both ships of the expedition and all aboard were lost somewhere in the Pacific in spring 1788, after having been last seen in Australia that March.

The French government in 1791 dispatched the Entrecasteaux expedition to explore Australia and to search for Lapérouse, but it failed to discover the Lapérouse expedition's fate, which remained unknown for decades. Meanwhile, Dufresne was appointed the Muséum's first assistant in zoology. He went on to become the head of the zoology laboratory and Europe's leading authority on taxidermy. His wife, a skilled taxidermist in her own right, seconded him in giving lessons to young taxidermists in training. Among these was Florent Prévost, who after Dufresne's death at the age of 80, from pneumonia, succeeded him as head of the lab.[23]

The lemur's travels took it from Madagascar to Paris and ultimately the Museum's laboratory of zoology. Several humans in this story were long-range travelers as well, but whereas they had some choice in their itineraries, the lemur's trajectory was forced upon it. Still, the most salient feature of the lemur's story was the amount of freedom the animal enjoyed in the laboratory. Dufresne and Desmoulins welcomed it to their workspace and treated it as a pet despite its potential for disrupting their work. This image of the lemur and the naturalists together in the laboratory, together with that of the lemur and the naturalists similarly endangered by the poisons of the taxidermist's craft, serve here as a reminder of the manifold and sometimes ironic ways in which human and nonhuman animals are all in it together.

[1.6]

Between Glory and Ruin

From 1795 to 1800 the Museum of Natural History teetered between glory and ruin. On the glory side, stunning new specimens arrived for the collections, and the Museum enjoyed an honored place in France's end-of-century claims for cultural supremacy. At the same time, the institution struggled to keep its live specimens from dying, its dead specimens from deteriorating, and its workers from starving. Time and again the Museum implored the government for help, but stopgap assistance was all it could offer.

Although the menagerie was never the Museum's top priority, in the final years of the eighteenth century their futures became more closely entwined, though not necessarily because the Museum wanted that. Other interests were in play. Despite its glaring shortcomings, the menagerie had become a great public favorite, and in 1798 it was a matter of national pride when the menagerie received two elephants from Holland and three bears from Switzerland as trophies of war, plus a small troupe of lions and other animals from North Africa. Invested with patriotic meaning, the animals constituted a great responsibility in the straitened circumstances of those years.

Thanks to the National Convention's decrees agreeing to expand the Museum's territory, raise the professors' salaries, and create a third professorship in zoology, 1795 began auspiciously. In January the Museum's amphitheater became the lecture hall for the new École Normale; an assistant naturalist position in botany was created in February, and by mid-March the provisional quarters for fierce carnivores were reportedly almost completed.[1]

More good news came in April when the professors learned of an extraordinary windfall from the Stadtholder's natural history collection in The Hague. The information arrived in a letter to Jussieu from André Thouin, who with his colleague, Barthélemy Faujas de Saint-Fond, was part of the special commission sweeping up treasures of art and science in the wake of the French army's advances. Thouin began, "We are preparing, citizen

Figure 1.6.1. The "Festival of Liberty" parade. Abraham Girardet and Pierre-Gabriel Berthaut, *Entrée triomphale des monuments des sciences et des arts en France; fête à ce sujet: les 9 et 10 thermidor an VI (July 28–29, 1798)*. Muséum national d'histoire naturelle. IC 3607 GF.

colleague, to send you two elephants, male and female, which belonged to the Stadtholder [Dutch head of state] and which are located in his menagerie at Loo." He went on to explain that there were treasures for everyone: for Daubenton, minerals from as far away as China and Japan; for Geoffroy Saint-Hilaire, a stuffed baby elephant, many well-preserved monkeys, two giraffe hides, and more; for Lacépède, reptiles and amphibians. Lamarck could expect a wealth of shells, corals, sea urchins, and such. Mertrud's animal anatomy collection would be enhanced by an array of skeletons including a giraffe fifteen feet tall, a large orangutan, and a sizable crocodile; botanists and chemists would likewise find their holdings enriched. "Each will have his share," Thouin promised, "and everyone, we hope, will be happy."[2]

Late in June, the first boatload of specimens arrived at the far end of the Museum's gardens, ready for unloading. Propitiously, the government's news that it had finally acquired Maison Leger, a parcel of desperately needed land and buildings, arrived at the same time. The good news was tempered by the plight of Museum workers, who reminded the administration of the skyrocketing and debilitating cost of subsistence. The day workers in the gardens begged the administration for a wage increase; but, citing its own lack of funds, the administrators could only promise to present their appeal to the government.

The specimens from Holland sparked excitement in Paris. The Museum immediately mounted a partial, temporary display in the amphitheater, perhaps in keeping with a rationale for confiscating the Stadtholder's collection: that its owner had withheld these treasures from public view. In truth, Dutch scientists had already described some of them.[3]

Geoffroy Saint-Hilaire promptly utilized the new specimens in collaboration with his new friend Georges Cuvier. Twenty-two-year-old Geoffroy was a scientific novice, and at twenty-five, Cuvier's talents as a comparative anatomist were just becoming known. Specimens from the Stadtholder's collection soon featured prominently in the young scientists' publications and helped establish their professional credentials.[4] A specimen preserved in alcohol allowed them to quickly coauthor a paper on the classification of the tarsier.[5] The "large orangutan" from the Stadtholder's collection featured in a second joint paper on the classification of the apes.[6] Cuvier used two Stadtholder's elephant skulls, one from Ceylon (now Sri Lanka) and the other from the Cape of Good Hope, to argue that Asian and African elephants are distinct species.[7] Stadtholder specimens likewise figured in Jean-Baptiste Lamarck's first publication on invertebrates (1798).[8]

The acquisition of the Leger property offered possibilities for housing specimens and professors alike.[9] In the fall of 1795, the Museum acquired another major property, the large complex of buildings known as "la Régie des Fiacres" where the government had lately stored flour. The Museum charged the architect with drawing up plans to establish an orangery, galleries, sixteen apartments (including four more for professors and six for naturalists' assistants), and a barracks for the veterans who guarded the establishment.[10] It also became the new home of the zoology lab, though it was not entirely suitable because the flour remnants attracted insects.

The menagerie meanwhile limped along after a dismal performance in the second half of 1794. Of the carnivores that arrived during the menagerie's first six months, all except the lion were dead by 1795. The list of animals that came under Mertrud's scalpel reads like a *Who's Who* of early menagerie specimens. For his animal anatomy course in the winter of 1793–94, he dissected an agouti, the civet, the hartebeest (bubale), three Barbary apes, two Barbary sheep, and the serval.[11] For his 1794–95 class, Mertrud displayed the anatomies of the dromedary, the panther, two "eagles," two

Barbary apes, and "the big baboon."[12] The arrival of two live dromedaries from the estate of the former Prince de Ligne in Belgium, fetched by their keeper, Bernard Louzardy, hardly counterbalanced these losses.

The provisional menagerie for the *animaux féroces* was finally ready in September, but while the animals attracted crowds, their lodgings elicited disdain.[13] The most consequential development for the menagerie came in mid-November when it was agreed that someone other than a professor should oversee its operation. The administrative assembly concluded that the professors of zoology had too much work of their own to give the menagerie the careful daily surveillance it needed. As a result, the animal keepers were essentially "abandoned to themselves."

When Jean-Claude-Michel Mordant Delaunay (1750–1816), the assistant librarian, said he was willing to manage the menagerie, the administration gave him the job.[14] One of the first issues he faced was the embarrassing paucity of fierce animals to fill the recently completed structure intended for them. Worried that the sight of empty cages was bad for public relations, the professors charged Delaunay with finding native animals to fill the vacant cages.[15] The pair of wolves and pair of foxes he secured were better than nothing, but only just.

Meanwhile the animals, the Museum staff, and the people of Paris were hungry. The Museum raised day workers' daily wages four times from May to October, but it still wasn't enough.[16] The city's bread shortage was severe. Since August, the government had supplied the Museum, at its request, with a daily bread ration of sixty pounds, fifty for workers and ten for the bread-eating animals.[17] In mid-November the Museum's naturalist's assistants asked to be added to the bread list.[18] Their request was approved, as was a similar request from the librarians Toscan and Delaunay two months later.[19]

∴

If 1795 was a slow year for the menagerie with only two dromedaries and Delaunay's appointment, 1796 witnessed a medley of important menagerie-related events. The first was its dodging a potential disaster. It involved a plan conceived by a distinguished physician, Michel-Augustin Thouret (1748–1810), director of the École de Santé. Thouret had recently examined the smallpox pustules on a "perfectly preserved" woman's body that turned up in excavations in Paris. He proposed using some of the menagerie's monkeys and sheep for "experiments at the Museum of Natural History relative to this kind of disease."[20]

The professors recoiled, wanting nothing of the sort on the premises, "considering that the establishment is open to the public and in particular to children, for whom this disease could become fatal," according to their meeting minutes of 13 February 1796. The professors instructed their recently elected director, zoologist Bernard de Lacépède to point out these "inconveniences" and tell Thouret the Museum could not agree to his request. Had Thouret's plan been carried out and a smallpox outbreak ensued, the Museum's reputation would have been indelibly stained and the history of the urban zoo in the nineteenth century may well have been altered.

Of greater actual moment was the government's decision in early July to follow up on the Museum expansion promised late in 1794, sparked by the imminent arrival of the convoy of live animals from the Dutch Stadtholder's menagerie. Just as the government's first exuberant plans for the Museum's expansion were inspired by the arrival of new animals needing accommodation, so the Council of Five Hundred, France's legislative body, resolved in July 1796 to activate the law of 21 frimaire an 3 (11 December 1794) calling for expanding the Museum with adjacent properties exchanged for state-owned ones. The Directory's Council of Ancients approved the resolution, trumpeting how the Museum's collections represented "the zeal of explorers, the research of scientists, and the heroic courage of the triumphant armies of the Republic" and declaring that these "trophies" would be displayed "in a manner worthy of their importance for public instruction and the majesty of the French people."[21]

The legislators' visions, however, collided with the interests of local landowners, particularly the timber merchants, who refused to exchange their properties and leave the area. In September 1796 the Museum sought a resolution satisfactory to all concerned, offering to make the patriotic gesture of giving up or postponing its claims to expansion if the government would reciprocate by providing the Museum with the funds needed to complete the natural history galleries.[22]

Ironically, while Lacépède is commonly represented as a key advocate in the menagerie's early history, it was he who, during a short stint as director of the Museum, led a delegation of colleagues to the Executive Directory and delivered a speech on the Museum's willingness to drop its expansion plans in exchange for gallery construction. Citing the Museum's "total devotion to the Republic's prosperity," he explained that it did not want the government to sacrifice valuable national property for its expansion when it had more pressing needs. The Museum, he said, had its own immediate needs, above all the completion of a gallery that had become "absolutely necessary for the conservation of the rich and immense collections of the Republic." Expressing hope that in the future, when France had brought peace to all of Europe, the government would be able to provide new favors to the Museum, Lacépède closed his speech with a magnanimous nod to the importance of helping the local landowners, since the prospect of the Museum's expansion was causing their property values to decline.[23]

At the next professorial assembly, Lacépède proudly reported that he had been invited to put his oral presentation into writing. He seems to have thought the nation's leaders wished to preserve his fulsome expression of the Museum's patriotic sentiments for posterity, but they more likely wanted a signed statement of the Museum's renunciation of any immediate plans for expansion. Lacépède's colleagues were already having second thoughts about it, fearing that he had agreed to give up too much and that if the laws regarding the Museum's expansion were totally set aside, it would never get the land the menagerie needed. They decided to write to the minister of the interior saying that on further reflection, a certain amount of additional space was necessary for the menagerie's development, including several acres to accommodate the Stadtholder's ruminants and the elephants, "*if the government intends that the Museum continue to have a menagerie*" (italics added).[24]

This phrase from the meeting minutes, suggesting the tenuousness of the Paris menagerie's existence, may seem surprising, given its place in the history of modern zoos. The phrase was repeated nearly verbatim in the letter to the minister in conjunction with a damning commentary on the recently completed "provisional menagerie" for the *animaux féroces*. Signed by Daubenton (newly reinstalled as director) and Geoffroy (secretary), the letter stated, "If the intention of the government is that the Museum continue to include a menagerie of live animals," it could not be a provisional one that was "unhealthy, shabby, ridiculous, [and] unworthy of the French nation in all regards." To give the menagerie the space it needed, given the new ruminants from Holland and the two elephants still to come, the Museum needed more land, which it hoped could be acquired by swaps with other national properties.[25]

The inadequacies of the carnivores' quarters had been underscored in July 1796 by the death of the lion from Senegal. *La Décade philosophique* announced the loss with a single sentence: "The handsome lion of the Museum of Natural History of Paris, the one who lived in such good relations with a dog, his friend, has just died."[26] The *Décade* did not give a cause of death, but for Delaunay there was no doubt: it was from eating diseased meat, which was all the menagerie could get at the time. With the lion gone, the only animals remaining in the dismal provisional menagerie were a polar bear and the wolves and foxes he had acquired a few months earlier.

The rest of the menagerie fared better, boosted at the end of August by the arrival of Louzardy with his convoy of animals from Holland, minus the two elephants, which came later. Even without the great pachyderms, the haul was impressive: seven axis deer, three Angora goats, a porcupine, a cassowary, a king vulture, a crowned crane, six storks, two crested curassows (*hoccos*), and twenty-five silver pheasants.[27] Feeding them was not as difficult as providing for the carnivores, because Delaunay had contracted with a farmer for fodder and grain.

But the loss of the lion and the pathetic state of the carnivore collection were almost too much to bear for Félix Cassal, the keeper of the *animaux féroces*. Realizing there were too few fierce animals left to justify his position, he volunteered in December 1796 to direct an expedition to Tunis to replenish the collection, offering to front the expenses if the Museum would reimburse him when he returned. The administration acknowledged the menagerie's "state of penury inappropriate for an establishment that ought to honor the nation." It noted its duty to improve the menagerie by stocking it with "curious and useful animals," of which it offered a list.[28] Adding such animals, it indicated, would give the menagerie status it currently lacked, and it identified Cassal as the best person for the job.

Cassal explained that to succeed he would need to go at once, since the season when baby animals were born and could be readily captured had already begun. The Museum appealed to the minister of the interior to decide quickly. In January 1797 he agreed to fund the expedition, authorizing Cassal to make advances of 6,000 to 8,000 francs for the purchase, feeding, and transportation of a collection of African animals, expenses for which Cassal was to be reimbursed when the expedition was over. The minister additionally arranged for letters of recommendation to be written on Cassal's behalf by French counselors abroad. He also connected Cassal with the Tunisian ambassador in Paris, who was soon to return to Tunis. A glowing note in *La Décade philosophique* predicted that when Cassal returned in six months with all the animals anticipated, it would "make our menagerie one of the richest in Europe."[29]

∴

The year 1797 at the menagerie was mostly given to waiting: for animals to arrive from Holland and North Africa, for new land for the expansion, and for the architect's plans. A few animals meanwhile made their way to the menagerie, beginning with a young lioness in late February. A French corsair had found her aboard a British vessel. The corsair's owners gave her as tribute to the Directory, which passed her along to the Museum.[30] In mid-May a shipment of live birds arrived from Brittany, seventy-three individuals representing thirteen species, sent by Jean François Emmanuel Baillon, an enthusiastic amateur naturalist who had been a valued correspondent of Buffon's.[31] Summer brought the exciting addition of a pair of live rattlesnakes from North America, courtesy of Pierre-Auguste Adet, the French minister to the United States.[32]

It was just as well that the menagerie took in few new animals other than Baillon's birds in 1797, because funds and food supplies were running dangerously low. In February Delaunay reported that the menagerie had not received fodder, straw, or grain for three months because farmers were unwilling to continue supplying them without being paid.[33] Nor was the Museum able to cover the costs of its lectures. In April, Fourcroy told his colleagues he needed 250 francs to deliver his chemistry course. They replied that the government was not coming up with the funds, obliging the professors to find creative solutions, including subsidizing from their salaries "the most pressing needs of the establishment's creditors."[34] The Museum sold sixty pounds of mercury to make Fourcroy's chemistry course possible and thought of paying for gallery construction by selling precious stones and gems that were commercially valuable but "entirely superfluous for the study of natural history" or the dignity of the nation.[35] Whether or not gemstones were involved, in mid-May Geoffroy informed the minister of the interior that construction on the galleries had resumed after a period of stasis.[36]

Delaunay was fretting about how to care for the animals en route to the menagerie (Cassal's collection from North Africa and the elephants from Holland), but their arrival was not as imminent as he thought. Word came early in December that Cassal and his animals had reached Toulon, where they were undergoing quarantine and would then stay until the spring.[37] Shortly thereafter, the Museum learned that the elephant convoy had left Holland late in September but had encountered bad roads near Cambrai and was stopping there for the winter.[38]

∴

New animals made 1798 a better year for the menagerie but a more perilous one for the Museum when it came to maintaining its collections, living and nonliving. Not coincidentally, it was a year when the Museum was at pains to display its patriotism and thereby remind the government, in an era of straitened circumstances, how valuable the institution was to the Republic.

In January, in response to a citizens' campaign to create a fund to support an invasion of England, the Museum sent the legislature a "patriotic gift" of 359 francs. Many professors and staff contributed over a day's wage to the effort.[39] More dramatically, in late February the Museum with great fanfare planted a Tree of Liberty in its main courtyard. The institution's professors, officers, naturalist's assistants, artists and several invited dignitaries assembled for a ceremony to give "public testimony of its love for liberty and its fidelity to the constitution of the Republic."[40] With drums beating and music playing, the veterans and National Guard escorted the group to a grove near the riverbank where the Tree of Liberty, decorated with flowers and tricolor ribbons, was ready to be taken to the Museum's central courtyard.

The celebration included a speech by Jussieu, the latest Museum director, recounting how the Constitution and Liberty had ultimately triumphed over the "factions" that had sought to dominate the country and how the communal spirit of his colleagues had "saved us and this establishment in those times of barbarism when the sciences were proscribed, and savants were persecuted." A "fraternal banquet" followed where songs were sung, and toasts raised to "the progress of the sciences and the arts, particularly natural history, the tight and eternal union of those who cultivate them, and the glory of the government that favors them."[41]

If the festival left some residual glow, the menagerie supervisor's conscientiousness did not permit the professors to bask in it for long. Four days after the Museum planted its tree, Delaunay reminded them that Cassal's African animals would soon arrive. More than just finding a place to put them, he said, they needed the right place, food, and care "to trick these animals about their change of climate and to efface among them the memory of their lost liberty." (Liberty was not for all.) Delaunay painted a worrisome portrait of the menagerie's condition. Inadequate housing and bad food had killed animals, and the public blamed the Museum for the losses. Meanwhile, some herbivores had taken to fighting, largely because incompatible species were crammed into the same stables. There were not enough stables, and many of those were too unhealthy to house valuable animals, including those in the Régie where the Museum planned to put the elephants.[42]

A committee charged with the menagerie's future recommended improving the elephants' quarters and developing a plan for the menagerie's expansion, but the Museum did not have the leisure to focus on live specimens. Other specimens were deteriorating in storage while the new galleries awaited completion, and the contractors had begun to revolt. Some workers had finished a portion of the grand staircase, but the mason had not and insisted he would not because his payments were a year and half in arrears. Without the stairway, the public could not reach the galleries to see the collections.[43]

Jussieu wrote a long letter to the minister of the interior identifying four sets of problems demanding swift attention.[44] The first was preparations for the elephants. Second, but identified as "our principal object," was completing the galleries and rescuing deteriorating specimens. Third was the impending arrival of the animals from Africa. It was urgent, Jussieu said, to deal with the animals' placement and, more generally "with the arrangement of the menagerie, since it is wanted that we have one of them." If the menagerie was not wanted, Jussieu continued, then "the animals that [the Museum] can neither house nor care for" must be taken elsewhere lest they "soon perish before our eyes, which would cast a disgrace on the establishment, which we want to avoid." A comparable situation (the fourth pressing problem) was accommodating the live plants that Captain Baudin was bringing from the Americas, some of which were very large. The Museum would need greenhouses of considerable height to keep them.

For these reasons, Jussieu said, it was critical to get the contractors back to work. He called for special funding at the modest rate of 3,000 francs per ten-day week to pay for the construction projects. He concluded by stating, "If our request is not granted, we insist on one essential point. We will expressly request that the elephants, as well as the animals from Africa, be placed elsewhere than at the Museum. We will be obliged to make our overtures known to the public, which clandestinely accuses us of negligence and lack of foresight, and to inform it that we have unceasingly solicited the means to ensure and to hasten its pleasures."[45]

Five days later the elephants arrived in Paris, at Port des Invalides on the Seine. This stirred the minister the next day to summon Jussieu "to come confer with me today at three o'clock" about the elephants' handling.[46] Three days later (24 March 1798) the elephants' transfer to the menagerie was achieved. They were installed in their new quarters in the Régie the following morning. Crowds thronged the Museum's grounds to get a look. At their next administrative meeting, the professors decided to restrict visiting hours to 11:00 a.m. to 1:00 p.m. and 3:00 to 5:00 p.m. on the even days of the *décade*.[47] Other significant menagerie-related matters were addressed in the same session. The Museum learned through a letter from the minister of the interior of the imminent arrival of bears confiscated by French troops from the bear pits in Bern, Switzerland.[48] A letter from André Thouin sent from Toulon reported that Félix Cassal and his animal convoy had reached Avignon, where they were waiting to join a large convoy of art treasures appropriated in Italy.[49] The professors tasked the architect, Molinos, with developing a general plan for the menagerie, taking into account the marshlands and other recently added properties and giving special attention to the quarters for the elephants and the animals from Africa. Meanwhile, in a different vein but of indirect consequence for the menagerie, Geoffroy

Saint-Hilaire announced that the minister of the interior had instructed him to report immediately to Bordeaux to participate in an unnamed mission that turned out to be Napoleon's expedition to Egypt.[50]

Molinos presented his general plan for the menagerie at a special meeting thirteen days later (16 April 1798) and followed it up at the next regular assembly with his plan for a park for the elephants.[51] Delaunay thought the progress too slow. He chastised the professors in a letter, stating, "You want to conserve the signs of our armies' victories of which you are the trustees, but you will succeed in doing so only by giving the animals won from our enemies the care they require." Delaunay predicted that the bears would soon die without roomier quarters less exposed to the sun and that the elephants would do no better unless they were more protected from the public and fed more fresh greens. Pregnant deer and Angora goats about to give birth needed provisional parks, and the camels needed a new enclosure. Cassal had informed Delaunay that he was bringing two white camels, two ostriches, two gazelles, four lions, and five vultures. The time had passed for considering how to prepare for them, Delaunay said. It was time to decide.[52]

Jussieu related Delaunay's concerns and borrowed some of his language in a letter to the Executive Directory. Despite all its economies and care, he wrote, the Museum was "menaced with an imminent destruction. The rich and large collections that give [the Museum] its preeminence and its utility, these precious properties of the Republic, these glorious trophies of our victories, are close to being annihilated." The goodwill of the contractors working on the gallery had worn thin, he said, and specimens remained in places where they suffered from humidity and insect attacks. The Museum had neither the funds nor the naturalist's assistants needed to mount them.[53]

As for the menagerie, Jussieu stated, "The elephants from Loo have arrived in the Museum where they ceaselessly attract a multitude of curious onlookers. Several very rare and very precious quadrupeds and birds from Africa are also going to arrive very soon in the establishment we direct. Despite the most multiple and well-concerted efforts, these animals will soon die, like several of those that we tried in vain to keep alive in the too small and too unhealthy provisional menagerie, if one does not construct for them as soon as possible some enclosures and shelters." The professors had done everything in their power to conserve the collections, Jussieu said, but "vigilance, assiduity, and experience" were not enough. "You alone can save them," he told the members of the Executive Directory. "You alone can prevent our country, Europe, and posterity from losing the most important and most instructive of repositories."[54]

While the Museum was begging for help from the Executive Directory, it was simultaneously at loggerheads with the minister of the interior, François Sébastien Letourneux, who wanted to put all the municipal building projects in his portfolio under the authority of a single architect he identified as Citoyen Duhameau. The professors were aghast that the choice of architect was being taken out of their hands. The institution's long-term health, they insisted, should not be subject to the changes of direction that occurred when ministers of the interior succeeded one another. Furthermore, with critical work yet to be done on the galleries and the menagerie, it was no time to switch architects.[55] The Museum and Letourneux continued to disagree until the latter was suddenly replaced by Nicolas-Louis François de Neufchâteau, who had been his predecessor, proof of the Museum's observation regarding the disruptions caused by changes at the ministry. Molinos remained the Museum's architect.

Despite its dire circumstances, in summer the Museum was a key player in the public celebrations of France's recent cultural and military victories. The festivities highlighted the arrival of valuable works of art, literature, and science confiscated from the Vatican and elsewhere in Italy under the Treaty of Tolentino. André Thouin, who had once again traveled with the French armies, helping to select items to appropriate, now helped devise ways of making a strong impression on those who witnessed the treasure-laden convoy winding its way toward Paris.

Back in June 1797, witnessing the wagons from Rome passing through Livorno, Thouin had marveled at the spectacle's impact on the local populace and considered the best route through France for delivering the booty to Paris. He observed that transport by canals and rivers would be the least costly but the most time-consuming. Sailing to Le Havre and mounting the Seine by barge would be quickest, but it would be the most hazardous unless peace were made with England. Thouin shuddered to think of the Apollo Belvedere or Raphael's painting *The Transfiguration* ending up in the British Museum. The third option was the costliest, but worth considering because of "the impression it would produce in France, and especially in Paris." Thouin envisaged that

> a convoy of fifty wagons laden with the spoils [*dépouilles*] of Italy, pulled by more than seven hundred draught animals, followed by camels, beautiful Tuscan donkeys, Roman cattle with great horns, and water buffalo from the Pontine marshes would make a triumphal cortège that, crossing a large part of the Republic's territory, would give its inhabitants a lofty idea of the victories of its armies, the energy of its government, and the national strength.[56]

Ultimately the convoy traveled in France by land and river, with the expectation it would arrive in time for the festival of the Republic on July 14. When low water in the Loire slowed progress, the government rescheduled the triumphal entry into the capital to coincide with a major anniversary two weeks later—9 thermidor, the day of Robespierre's execution.[57]

The published program of the so-called Festival of Liberty shows its careful orchestration and the prominent role assigned to the Museum of Natural History.[58] The professors and other participants were to assemble on the banks of the Seine, just outside the Museum's gates, at 9:00 in the morning of 9 thermidor. Organized in three divisions (natural history; rare books, manuscripts, medals, etc.; and fine arts), they were to march in that order to the Champs de Mars right behind the cavalry and military band that led the whole parade, announced by a banner reading "HISTOIRE NATURELLE." After them came ten wagons festooned with flags and garlands and laden with specimens, flanked by students and friends of natural history of the Museum's choice. The first four wagons carried minerals, fossils, foreign seeds, and foreign plants. Next came three wagons bearing lions from North Africa acquired by Félix Cassal. The eighth wagon held one of the Bern bears and was followed by a pair of camels and a pair of dromedaries. The final two wagons carried agricultural tools from Italy and two blocks of crystal donated to the French Republic by the people of Valais.

Behind the parade's natural history division came a choir singing patriotic hymns, actors from Paris's principal theaters, and professors from the École polytechnique and the Collège de France accompanying six wagonloads of manuscripts, rare books, medals, and more. Last came twenty-nine wagons bearing the "monuments" of fine art that Napoleon had exacted in tribute from his Italian conquests. The first two bore the bronze horses of San Marco, taken from the eponymous basilica in Venice. Busts of Homer and Brutus were paraded, along with sculptural masterpieces from the Vatican, including the Laocoön and the Apollo Belvedere and paintings by Raphael, Titian, and Veronese, though these were mostly packed in crates. Award-winning artists and various administrators, professors, and students of the schools of painting, sculpture, and architecture marched in this division.

Along with solemn ceremonies and military exercises, the festival program featured the ascent of an aerostat (manned balloon) decorated with garlands and inscriptions and trailing the tricolor flag. An artists' rendering of the scene at the Champ de Mars conveyed the magnificence of the general setting and elements of the procession. Observers, including Prussian linguist and diplomat Wilhelm von Humboldt (brother of the renowned naturalist Alexander) noted the animals' prominence in the parade and how they impressed the public.[59]

The government's position was that treasures of fine art had been liberated from former owners who were unworthy of them, but this could not be argued for the lions and camels that North African rulers presented as gifts to Cassal (who had been unsuccessful at capturing animals on his own). These and other animals and plants were living proof of France's extensive power, from the isle of Trinidad to the burning deserts of Africa and the glacial regions of the North. They were intended to show how "all parts of the world have been made to contribute to enrich the most beautiful of our festivals and to make it as magnificent as the triumphal entry of [Roman consul] Aemilius Paulus into Rome [after defeating King Perseus of Macedon]."[60]

The comparison with ancient Rome was intentional, as the center of Western civilization in antiquity and, until France's victories in Italy, the place artists went to study the works of the masters. The thesis of the Festival of Liberty, however, was that the Romans had not been worthy of displaying the Greek statues taken as spoils, nor had the great Italian painters created their masterworks merely to please kings and pontiffs. Only a free people could appreciate their creative genius. An inscription accompanying the four horses of San Marco told the story: "HORSES TRANSPORTED FROM CORINTH TO ROME, AND FROM ROME TO CONSTANTINOPLE, FROM CONSTANTINOPLE TO VENICE, AND FROM VENICE TO FRANCE. HERE THEY ARE FINALLY ON FREE SOIL." Possessing these masterpieces implied that Paris had replaced Rome as the capital of Western culture and civilization.[61]

In the official ceremonies André Thouin, on behalf of his fellow commissioners, presented their lists of collected objects to the minister of the interior. Minister Neufchâteau thanked the commissioners, calling the treasures "monuments" to French military victories. Both speakers sought to disassociate the French from any charge of vandalism, though the Italians may not have seen it that way. As Thouin put it, "from the simplest soldier to the general in chief" the French had respected and shown great interest in the objects of science and art. Neufchâteau portrayed the festival as "expiating the crimes of the tyranny overturned on the ninth of thermidor" and representing "the triumph of nature, the triumph of the arts, the triumph of liberty."[62]

The Festival of Liberty surely buoyed the spirits of the professors, staff, and friends of the Museum in acknowledging their work as integral to France's claim to be the leader of world civilization. At the Festival of Arts held several days later, the verses of one song all ended with the couplet: "Rome is no longer in Rome / It is all in Paris." Another song invoked the lions from North Africa, the bears from Bern, the horses of San Marco, and the hope of a "well-bridled leopard" from England, referencing France's intention of bringing England to heel.[63]

Earlier that year, when Napoleon was named a member of the Institut de France, he spoke of the country as leading the world in the advancement of science. Hailed as a conqueror, he demurred as he accepted his membership, saying "The true conquests, the only ones that cause no regret, are those that are made over ignorance" and emphasizing that these were the honorable and useful goals for nations. "The real power of the French Republic," he proclaimed, "must consist henceforth in not permitting the existence of a new idea that does not belong to her."[64] When the Festival of Liberty took place in July, Napoleon was in Cairo on a mission he envisaged in terms of both military conquest and the acquisition of knowledge.

The Museum of Natural History was keen to enlist in France's quest for intellectual and cultural dominance. Late in 1798, when the minister of the interior asked if the Museum wished to employ an expert in modeling plants and animals in wax, it readily accepted. The intention was to have the greatest number of natural history specimens overall, displaying all the world's natural productions in their different stages of growth, and to be best in everything pertaining to natural history. "This is the purpose of the establishment," the Museum told the minister. "It is the object that public instruction and national glory require."[65]

∴

In 1799 the menagerie was not likely to bring the word "best" to the lips of an unbiased observer. In March a man named Leroi wrote to the Museum expressing distress at seeing the pitiful "boxes of stone" in which the animals were confined. Whereas exotic animals in the London menagerie had been successfully mated the previous year, he said, in Paris the animals appeared to be suffering and dying. He reported that a keeper (undoubtedly Cassal) had told him that if the lion and lioness had a small courtyard and more air, they would produce offspring. The writer also called for more parks for herbivores, so that they might not degenerate under the conditions of their captivity. The professors agreed to thank him for his letter while saying that "circumstances have not permitted [us] to do better."[66]

In April, the Museum learned from the minister of finances that a man named Talboutier wanted to exchange a property adjacent to the Museum for a state-owned property. Not wanting to appear too aggressive on the matter, the Museum cited public opinion as the reason the property should not be refused, the public having judged the elephant park to be too small. The property was ultimately acquired, but not until 1803.[67]

In October the menagerie's fodder supply was again in danger of running out unless the farmer was paid.[68] The professors panicked, and their newly elected director (Fourcroy) met with the recently appointed minister of the interior (Nicolas-Marie Quinette), to describe the situation. In response to the minister's request for a follow-up letter outlining the Museum's most pressing needs, Fourcroy painted a bleak picture. The most pressing need was "the food of the elephants and of the other animals of the menagerie," who risked "dying of starvation," Fourcroy wrote, "if you do not promptly give us the means of providing for their needs." Firewood for heating the greenhouses was also essential, lest precious plants be irreparably lost. The wood merchant, like the farmer, was owed "a rather substantial sum" from the previous year, and the Museum could not expect "a single log" without paying him something.

Fourcroy's recitation of miseries went on. For eight months, workers had been paid little or nothing. The professors had assisted them by sharing a portion of their salaries, Fourcroy said, "just as we already did to prevent the elephants from starving," even though the professors themselves were struggling to make ends meet. Wrote Fourcroy, "I cannot tell you all that we have to suffer in seeing the horrible distress torturing so many fathers of families employed in our establishment."[69] Of the Museum's other needs, Fourcroy mentioned professors' salaries and the completion of the natural history gallery. Without the latter, he said, "the magnificent national collection of birds, mammals, etc." would "perish." He requested an immediate weekly infusion of 3,000 francs, which was what Quinette and the minister before him had promised. He ended with an impassioned plea:

> Our establishment is of a kind that is unable to suffer for a longer time from the horrible penury under which it languishes; it is a question of the life of a great number of employees, of those animals that are rare and useful to the arts as well as to a great amount of knowledge; even of precious vegetables which continually multiply the riches of the republic; of the conservation of the most beautiful monument that civilized nations have elevated to nature and to the emulation of human industry. The public has not yet appreciated the penury that harms us, because of the good appearance preserved by our care of the collections, but the suffering is so great that everything is threatened to perish at once.[70]

Fourcroy waited forty-eight hours before writing to the minister again, apologizing for doing so but saying that the Museum was within days of "total ruin." "It is impossible for me, Citizen Minister, to not present it to you in all its hideous nakedness." Only four to five days' worth of fodder remained in the storerooms, he said, and his appeal to the Ministry of War for fodder had brought the response that the request would have to come from the minister of the interior.[71] The ministers worked out an arrangement whereby the military would provide fodder and grain to the Museum for twenty days.[72]

Three days later Fourcroy received a letter from the minister of the navy, who, oblivious of the Museum's plight, blithely reported that the Tunisian ambassador wanted to send the Museum a collection of live animals. Fourcroy had to explain how bad the institution's situation was: "The frightful penury in which the Museum finds itself at present, the absolute lack of food for the animals and the fear of seeing them die of starvation, absolutely prevents desiring for the moment the augmentation of this collection that [the Museum] can no longer sustain such as it is."[73]

Two weeks later, the coup d'état of 18–19 brumaire an 8 (9–10 November 1799) overthrew the Executive Directory, replacing it with the Consulate, establishing Napoleon as first consul, and essentially ending the Revolution. Despite its political importance, the event went almost unremarked at the professors' next assembly. The only hint was that Fourcroy's colleagues instructed him to petition the new minister of the interior (mathematician and physicist Pierre-Simon Laplace) for the properties "indispensable for extending the parks of the menagerie and the nurseries of the Museum."[74] Within days, on the new minister's instructions, the professors and employees of the Museum stood erect and bare-headed to jointly swear "to be faithful to the Republic, one and indivisible, founded on equality, liberty, and the representative system."[75]

∴

At the new century's very beginning, the Museum lost its patriarch, Louis-Jean Marie Daubenton. He died at age eighty-five, an hour after midnight on January 1, 1800. His colleagues gathered to mourn their revered colleague and plan an elaborate funeral for him.[76] After arranging for a deputation to offer condolences to Daubenton's widow, the professors directed the anatomists Jean-Claude Mertrud and Antoine Portal to preside over embalming Daubenton's body. It would be placed in a lead coffin, inside a wooden box, and interred on the Museum's grounds in a grand commemoration of Daubenton's life and contributions to science.[77]

Held on 14 nivôse an 8 (4 January 1800) beginning at 11.00 a.m., the ceremony, like the Liberty Tree festival, started in the library, where the professors received a host of invitees; relatives, friends, and colleagues of Daubenton, members of the Institut de France, and a range of civil and military officials. With each carrying a cypress branch, they followed the Museum's military guard from the library across the gardens to the new conservatory, which was transformed for the occasion into a temple. Tapestries from the Gobelins manufactory and the Musée Centrale des Arts adorned the interior, depicting plants, animals, sites around the globe, and moments in history. Flowering shrubs and plants surrounded the sarcophagus, on which a bust of Daubenton rose from a floral pyramid, crowned with laurels and more flowers. The attendees arranged themselves in two rows, and Lacépède mounted a platform to recite his eulogy of his colleague.[78]

After Lacépède's address, the cortège made its way with measured steps through the gardens and halfway up the hill where Daubenton's final resting place awaited. There Fourcroy delivered a second eulogy, no less eloquent than Lacépède's and nearly as long, assuring his auditors that death does not destroy men like Daubenton: "they live forever in the memory of nations." Attendees laid their cypress branches at the foot of Daubenton's sarcophagus, bringing the ceremony to a close.[79]

Daubenton's death marked the end of an era at the Museum, but the institution's circumstances remained grim on into the new century. In February Jean-Thouin, the head gardener, sent an appeal to the minister of the interior bearing his signature and those of the fourteen gardeners, lamenting "the weak cries of our languishing children, lacking bread and half naked; the tears and desperation of their poor mothers, weakened by a long and cruel misery; and finally we ourselves, most of us sexagenarians, infirm and lacking everything."[80]

The new minister of the interior was Lucien Bonaparte, the eldest of Napoléon Bonaparte's three younger brothers, and a major actor in the coup of 18 brumaire. He was receptive to many of the Museum's requests, starting with permission to bury Daubenton on the Museum grounds. The Museum in turn answered Bonaparte's routine requests for specimens for provincial schools or to light up its buildings for special events like Napoleon's victory over the Austrians at Marengo, celebrated on 21 June, and then the anniversary of the Republic on 14 July.[81]

In mid-June, however, Bonaparte made a demand the Museum could not abide. He told the professors to help themselves by charging admission to the gardens and galleries and for viewing the elephants.[82] The Museum staunchly refused, maintaining it would be beneath the nation's dignity.[83] The minister wrote back that he had grown weary of the Museum's constant demands on the public treasury, and he reminded the professors that he had recently agreed to a large sum they'd requested for the purchase of a number of impressive animals from London. Considering all he had done for them, their unwillingness to defray expenses by charging admission fees distressed him. After lecturing them, he presented his demands: that admission be charged nine days per week; that the price be set at one franc or one franc fifty centimes per entry; and that on décadi, the last day of the ten-day week, viewing the galleries and the elephants would be free. Finally, he commanded, "you propose to me your views on the reestablishment of the menagerie and the means of ending its state of degradation."[84] Still, the Museum would not budge. It sent the minister yet another refusal, and Bonaparte seems to have decided to forget the matter.[85]

If the professors were indignant at Lucien Bonaparte's suggestion that they sell admission tickets, they were traumatized by his decree at the beginning of the new calendar year, 1 vendémiaire an 9 (23 September 1800) stating that he would appoint a single administrator for each of the nation's public institutions and would deal with that individual directly. Since the Museum's founding, the professors had cherished their equal status and ability to elect their own directors, unlike the time when Buffon lorded it over them. Restoring the previous arrangement, they believed, would be disastrous. Responding to Bonaparte, they pointed out that the Museum had prospered more in the seven years since its founding than in the previous century when it labored under royally appointed directors. The proposed administrative change, they insisted, would erode fraternal relations among the professors. Each would be left to pursue his own interests over those of the institution.

Bonaparte remained firm, countering the Museum's objections and naming Jussieu as administrator. Jussieu protested strongly, then accepted provisionally, rather than see an outsider imposed on the Museum.[86] An eleventh-hour rescue came via intrigue within Napoleon Bonaparte's entourage, prompting Napoleon to remove Lucien from Paris and install him as France's ambassador to Spain.[87] The professors breathed a collective sigh of relief on 7 November 1800 when he was replaced as minister by the chemist Jean-Antoine Chaptal (1756–1832). The Museum sent a delegation and letter to Chaptal explaining how his predecessor's plans would have ruined "one of the most beautiful and useful establishments of the Republic." Chaptal received both favorably and promised the Museum's organizational structure would remain as originally determined in 1793.

Contingencies had shaped the Museum's history, up to and including this latest ministerial appointment that came on the brink of an institutional disaster. Chaptal proved to be one of the most helpful ministers the Museum ever worked with. His appointment did not solve all the Museum's problems, however, including those of the menagerie. The spring and summer of 1801 would see a race to complete the new animal lodgings before the animals destined for them died. Still and all, it appeared that the Museum had weathered an extended period of great peril.

The menagerie had not yet served the new scientific functions envisaged by its early champions. Rather than a new understanding of animal life, its contributions to science consisted mostly of cadavers, like earlier menageries. It had, however, served a crucial political function for the Museum and, by extension, for science in these difficult times. Underscoring how the visibly appalling conditions of the menagerie influenced citizens' and foreigners' perceptions of the Museum and the nation was a powerful way for the professors to call the government's attention to the impending demise of the rest of the collections if help was not forthcoming. In desperate times, the menagerie became a sort of bargaining chip for the Museum. The public would not take kindly to the loss of animals symbolizing France's heroic victories, the Museum explained, and if that were to happen, it would be obliged to put the onus on the government.

[1.7]

The Bears from Greenland and the Bears from Bern

The Paris menagerie's first two bears were polar bears confiscated from their owners by the police and escorted to the Museum of Natural History in November 1793. According to the dispossessed animal showmen, the bears were from a litter of three, acquired in Greenland by a sea captain when the cubs were small enough for him to hold them in one arm. The captain sold the cubs in Liverpool. One died and the other two ended up in Paris, one in the hands of Dominique Marchini and the other with Bernard Louzardy. In the fall of 1793, both men and bears, the bears now seven or eight years old, ended up at the Museum of Natural History.[1]

They were the only bears in the menagerie until three European brown bears arrived in 1798.[2] Brown bears were not rare; thanks to their amenability to training, they often featured in traveling animal shows. These bears excited French citizens, however, because they came from the famous bear pits of Bern, Switzerland, as trophies of war. The press hailed "the bears of Bern [that] have arrived at the menagerie of the Museum of Natural History. Few nations have had the privilege, as we have, of stocking their natural history collections via a series of victories, each more glorious than its predecessor. In peacetime a soldier of the country will not be able to take a stroll in Paris without noticing everywhere the fruits of his efforts."[3] Throughout the bears' lives, guidebooks to Paris and the Jardin des Plantes identified them as "the bears of Bern."

The bears gave no sign of being pleased with their relocation, however. When they arrived at the menagerie on the evening of 23 April 1798,[4] they refused to leave the cage they had been transported in. A scene of resistance and violence unfolded. Fire was introduced into one side of the cage to drive them out, but to no avail. Efforts to tempt them with other animals also failed. According to the naturalist Charles-Nicolas-Sigisbert Sonnini de Manoncourt, it took several hours for the bears to leave their traveling cage.[5] Then, as the *Gazette des deux ponts* later reported (Friday, 4 May), as the bears were being transferred to their new quarters, "one of them,

Figure 1.7.1. "The polar bear." Nicolas Maréchal, Simon-Charles Miger, "L'ours polaire, *Ursus maritimis*," from Étienne de Lacépède, Georges Cuvier, and Étienne Geoffroy Saint-Hilaire, *La Ménagerie du Muséum national d'histoire naturelle, ou Description et histoire des animaux qui y vivent et qui y ont vécu*, Paris, Aug. Renouar, 1801–5. Muséum national d'histoire naturelle, 419 pl. 30.

furious at being locked up again, broke the bars of its enclosure and threw itself on one of the spectators, who was said to be a deputy. He was saved only by the agility of one of the animal conductors, who succeeded in freeing him from the bear. Fortunately it seized only his coat, which it tore to shreds."[6]

The bear was subdued, but news of the incident inspired a control problem of another sort. Caroline Wuiet, royalist and *femme extraordinaire*, announced in her short-lived newspaper *Le Cercle, ou Journal des arts et des plaisirs*, "The bears of Bern have declared war on the Jacobins of Paris." She wrote, "If *some hundreds of Jacobins per week* were delivered to these savage beasts, it would be but an act of justice, allowed in expiation of the victims they delivered to the guillotine during the time of their reign. *That is my vow, and I do not doubt that it will be shared by all good citizens.*" Responding with alacrity to these sentiments, the government closed the journal and decreed that its author and printer be denounced before the criminal tribunal and "pursued as agitators of murder."[7]

Le Cercle was shut down, but Wuiet, its author and printer, thrived. She produced new journals, composed and played music, criticized art, sparkled in Madame Tallien's salon, and through the help of her friend Josephine Bonaparte, gained the police's permission to dress in public as a man.[8] The bears, on the other hand, confined to narrow cages in the "provisional menagerie," enjoyed no liberties at all. Sonnini observed that "we really don't know why" the bears were brought from Bern to their cramped quarters in Paris, and he added, "they seem to suffer greatly from their uncomfortable captivity."[9] Despite the press's language about "liberating" specimens from other countries, bringing them to "free" soil, and making them available for everyone to see, the bears were much worse off in their new quarters in Paris than they had been in their pits in Bern. In Bern they had room to move about, trees to climb, and a pool for bathing. Visitors had ample opportunity to observe them. A German tourist in Paris in the spring of 1798 underscored the difference:

> I was touched to see at the Jardin [des plantes] the bears of Bern, shut up in a rather narrow hovel, behind iron bars, next to a beautiful polar bear and other animals. I knew these good animals, having seen them in their vast enclosure in the Bern bear-pit where they often entertained the young and old inhabitants of the city with their appetite and their capers. I found them here five months later, more than a hundred leagues away, but alas!—like very confined prisoners. They turned around constantly in a narrow circle, but they were accustomed to a better regime, and they need more movement![10]

Mordant Delaunay, as overseer of the menagerie, had lately warned the Museum's professors about the inadequacy of the bears' quarters: "You want to preserve the signs of victory of our armies, of which you are the recipients, but you will only succeed in this by giving to the animals obtained from the enemies of the Republic the care they require. The bears will indisputably die if they are not quickly given a more comfortable lodging, less exposed to the intense heat of the sun, and more spacious."[11] By the spring of 1799, one of the bears from Bern had indeed expired, fighting against the bars of its cell.[12]

The other two bears lived on for some years, but not contentedly. Georges Cuvier reported that they spent most of the day lying down, and all the night sleeping, but Delaunay painted a different picture. In the summer of 1801, he told the administration that he feared that the cage of one the bears, a female, would not long endure her assaults. With claws and teeth, she had attacked one of the walls and weakened it to the point that it might soon give way.[13] The Museum authorized the architect to repair the bear's enclosure. The cage was reinforced, but it seems nothing was done to enhance the animal's well-being.

The German traveler's observation that the Bern bears "turned around constantly in a narrow circle" was an early glimpse of what scientists would later call "stereotypic behavior," which zoo animals exhibit under the stress of close confinement. Early in the new century, another German observer gave a more explicit description of one bear's movements, but he did not attribute it to the size of the animal's cage. Gotthelf Fischer (1771–1853), a young German naturalist studying at the Museum, thought the bear's movements were related to the polar bear in a neighboring cage.

Wrote Fischer:

> This proximity causes a very strange movement. Slowly and majestically, [the brown bear] glides close to the iron grille at the front, and then, when he comes to the thin wall that separates him from his relative, the white bear, [and he is standing] sometimes more, sometimes less, on his hind feet; [he] holds his nose to the cracks of the same and just as he breathes the neighbor's smell, he hurriedly jumps backwards, as if fear drives him, goes to the wall that makes up the back of his cell, and then takes the old path again, up to the dividing wall, where he always makes the same jump. He keeps moving in this circle incessantly.[14]

As of late July 1801, Delaunay had passed the mantle of menagerie superintendent to Lamarck, but in January 1802 he again took up the Bern bears' cause, warning the professors they would soon die if nothing more was done for them, for they were "thin and ill and have never shown a great vigor since they were changed from the spacious and airy pits of the city of Bern to narrow, badly placed, and unhealthy cells—the only asylum that you have been able to provide them."[15] Two months hence, a second bear died. Lamarck announced the death to his colleagues as due to "the difference of regime that it followed in Bern compared with that which one was forced to make it experience here."[16] Some months later, the third bear from Bern perished as well.[17]

The bears from Bern suffered visibly from their living conditions at the menagerie, but not as much as the polar bears, obliged to endure both cramped quarters and the heat of central France. To cool them down, the keepers splashed them with sixty to eighty buckets of water each day year around.[18] When Nicolas Maréchal produced his portrait of a polar bear, he set it against an arctic scene that was much more elaborate than the backgrounds of most of his other animal portraits. Perhaps this reflected his sympathy for a polar bear in Paris, so far from its native habitat.

In 1800, Georges Cuvier reported that the two polar bears had displayed "a singular and constant movement of the head and neck from high to low and from low to high." He ascribed this to the cramped quarters the bears occupied before coming to the Museum, stating, "It is believed that they acquired this habit because the cage where they spent their first years was too tight."[19] In 1803, describing the remaining polar bear, the playwright and guide writer Jean-Baptiste Pujoulx remarked on its "repetitive and continuous movements, which have something monotonous and tiring [about them], and which seem to be the result of the uneasiness it experiences in its narrow prison."[20] Interestingly enough, though we have descriptions of stereotypic behavior for polar bears and brown bears alike, no author at the time generalized about the phenomenon. Generalizations about the behavior of animals in captivity were a long way from being systematically developed.[21]

While the animal keepers and the Museum scientists recognized the need to keep the polar bears cool and provide them ample space, they never mentioned problems of nutrition. In the bears' first months at the menagerie, they each received eight pounds of bread and one pound of meat daily. Later their individual ration was reduced to six pounds of bread per day, a regimen under which they gained weight steadily and became very fat. Reporting in 1804 on the remaining bear, the new menagerie superintendent, Frédéric Cuvier, remarked that she would also eat meat and loved fruits, especially apples, but "contented herself with bread."

At the time of her death (18 May 1806) at about age twenty, this bear was the doyenne of the menagerie, the last of the animals that had arrived in the fall of 1793. By then she was exceptionally fat and could no longer see. The German naturalist Fischer attributed her blindness to her inability to clean her eyes by immersing them in water (the Museum still had no place for her to bathe). As a result, the bear's eyes suffered "as much from the dust as from the urine-like pungent smell, which becomes particularly strong here in hot weather." Daily dousing helped keep her clean but was not enough. Fischer could not help remarking on the superior living conditions of the polar bear at the Kaiserliche Menagerie Schönbrunn. That bear had an open-air enclosure surrounded by an iron fence and with a large cage of iron bars. Below the elevated cage was a large basin or pond where the bear could totally immerse himself and climb back up to his cage. He "likes to play with a wooden ball," Fischer wrote, "and is generally very lively. He catches particularly skillfully the fish thrown at him, even if one tries to hurl them at him on the opposite bank of the pond."[22]

Frédéric Cuvier imagined in 1804 that the bear's blindness was related to the heat, noting that animals transported from cold climates to more temperate ones often lost their sight, but diet rather than heat or the lack of the chance to swim was probably the Paris polar bear's main affliction.[23] The Museum's naturalists knew that in their native habitat polar bears ate mostly fish, and they believed this was what lent their flesh a fishy, unpalatable taste. But they did not seem to realize that the menagerie's polar bears, however fattened on bread, needed fresh fish in their diet. The Schönbrunn polar bear was fed fish, whether owing to this knowledge or for convenience. The discovery of vitamins and vitamin deficiencies was far in the future, as was any science of the nutrition of zoo animals.

The polar bears from Greenland and the brown bears from Bern exemplify the difficulties facing the Paris menagerie. Ample, airy, and clean quarters were not easy to provide, nor was healthy food in adequate quantity. Indeed, the Museum lacked a sure knowledge of what an animal's space and nutritional requirements might be. Its failings inevitably damaged the health and hastened the deaths of many of the bears and other animals that came into its possession.

In the summer of 1803, the Museum's professors gathered on the grounds destined for the menagerie's expansion to discuss improvements and additions, including the construction of a bear pit. They agreed that the bears' dwellings within the pit would be made of wood and that iron grillwork would be erected around the pit to keep the bears from getting out and prevent the public from falling in.[24] At the year's end, the Museum acquired three new bears (two young black bears from Louisiana and an adult brown bear from the Alps), plus a monkey, in exchange for a lion.[25] Within a month the big bear from the Alps had all but broken out of its cage.[26] The administration intended to move it to the pits as soon as they were ready, but that took over a year.

At last the *procès-verbal* of the professors' meeting of 21 August 1805 stated: "Two American bears have been placed in one of the pits that was destined for them. There they are agreeably situated for the enjoyment of the public and for their conservation."[27] Unfortunately, the public's enjoyment and the animals' well-being were not entirely compatible. The bears became a favorite target of the public's teasing and more serious abuse as well.[28]

[1.8]

Theory and Practice (1795–1801)

The gap between theory and practice was painfully evident in the early years of the Paris menagerie, when lofty visions of its future floated high above its awkward present realities. If the Museum of Natural History assured the public (and itself) that the menagerie's shabby state was only "provisional," this was little solace to those responsible for the animals' daily care. While the Museum's professor of birds and reptiles, Bernard de Lacépède, liked rhapsodizing about what the menagerie could become, the Museum's assistant librarian, Mordant Delaunay, was tasked with keeping the menagerie functioning in trying circumstances.

The two men could not have differed more in temperament. Lacépède radiated bonhomie and optimism, told his students that happiness was one of the rewards of studying nature, and is said to have never thought ill of anyone.[1] Delaunay was a fussbudget; he always found something to complain about. One cannot doubt that Lacépède would have made the more agreeable colleague. To reconstruct the menagerie's early history, however, Lacépède's two essays on the menagerie are no match for the dozens of letters Delaunay wrote to the Museum administration. Long buried in the archives, they provide a gritty corrective to histories that have relied too exclusively on Lacépède's published and more easily accessed essays. The productions of the two men must be examined side by side. Gustave Loisel argued more than century ago that "the hazard of circumstances" and "the energy of some men" had more to do with the menagerie's creation and development than did the writings of Buffon, Bernardin de Saint-Pierre, and Lacépède. Delaunay was prominent among the men he specifically credited. Most subsequent writers on the menagerie, however, have attended more to its theorists than its practitioners.[2]

Lacépède was born to a family of long and noble lineage (the title of count came from a maternal uncle). As a youth he received encouragement from the naturalist Buffon and the composer Christoph Willibald Gluck.[3] His early creative efforts included an opera, a treatise on "the poetics of

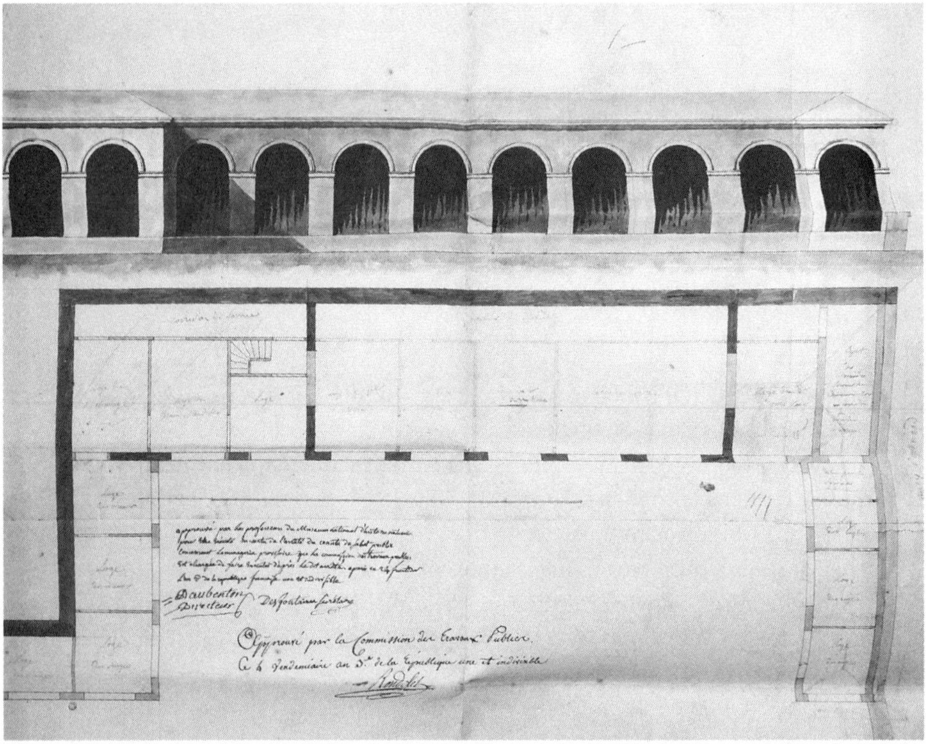

Figure 1.8.1. The architect Jacques Molinos's plan for the provisional menagerie for the carnivores and other fierce animals, approved by the professors of the Museum on 10 September 1794 and by the Commission of Public Works on 25 September 1794. Archives nationales F/13/1218/A. Photograph R. Burkhardt.

music," and a book on electricity that the Academy of Sciences abstained from approving. None of these works was a breakthrough achievement, and a career in science opened when Buffon took Lacépède under his wing, appointing him "superintendent and assistant demonstrator" of the Jardin du Roi's natural history collection. Buffon subsequently entrusted Lacépède with writing the volumes on reptiles ("oviparous quadrupeds and snakes") for his *Histoire naturelle*.

When the Revolution came, Lacépède embraced the new order. He served as a deputy to the Constitutional Assembly and then to the Legislative Assembly. When the Revolution took a more radical turn in spring 1793, he took refuge in the village of Leuville, some twenty-five kilometers south of Paris, but only after his friends insisted he leave the capital. He was sufficiently naive about the dangers facing him in Paris to seek Robespierre's permission to resume his studies there. Fortunately for Lacépède, as the story goes, Robespierre replied, "Is he in the countryside? Tell him to stay there."[4]

Lacépède was in Leuville in June 1793 when the Jardin des Plantes was transformed into the Museum of Natural History, and the chair of zoology he would otherwise have been awarded went to the twenty-one-year-old Étienne Geoffroy Saint-Hilaire. Lacépède's former colleagues felt it unsafe to call him back to Paris until long after the Reign of Terror ended. With the government's approval, the professors divided Geoffroy's chair into two professorships, one for mammals and birds, and one for reptiles and fish. The professors elected Lacépède to the latter post on 24 nivôse an 3 (13 January 1795). Three weeks later, he attended his first professorial assembly as one of them.[5]

In his writing style, Lacépède took Buffon as his master. The sweeping and inspiring views of Lacépède's first essay on menageries (1795) would have made Buffon proud.[6] Lacépède sketched the history of menageries from antiquity to the present, then elaborated on what the menagerie of the future could be. Terrible tyrants and superstitious savages had kept menageries in the past, he explained, and in eighteenth-century Europe there were still countries whose rulers, like the Caesars of ancient Rome, continued to keep menageries as a show of power. Instead of seeking to extract useful secrets from Nature, they put her in chains; instead of allowing the animals a modicum of freedom, they were locked in cramped cages that degraded and denatured them. This would change, Lacépède promised, when a free and enlightened people created a new kind of menagerie consistent with the values of a republic. He failed to mention that in the immediate present the ferocious beasts in the provisional menagerie in Paris were no better off than those in the tyrants' menageries.

Lacépède went so far as to suggest that a national menagerie should not include ferocious beasts at all, because showing such specimens to "easily seduced eyes" would conjure up images of a nature ruled by force and ferocity. Nor, he said, should an enlightened institution exhibit animals in ways that smacked of "enslavement." Most animals were "innocent and peaceful" creatures that could be put in large enclosures and given some liberty. Significantly, the public's moral education, not the animals' welfare, concerned Lacépède. In his words, "Let the images of constraint and the appearances of slavery be removed as far as possible from the eyes of a free people, and let us use all our resources for these animals that are less rare than the others and less capable of exciting the vain curiosity of the vulgar, but more appropriate for exercising the genius of the observer and satisfying the soul of the citizen."[7] He wanted the national menagerie to feature useful creatures in an instructive setting.

Lacépède maintained that the careful study of useful, innocent creatures would illuminate their habits, affections, needs, and life histories, knowledge naturalists could use to acclimatize animals, cure their ills, improve their meat, and decide which species was best for a given purpose. Seeing naturalists engaged in such studies would teach citizens to distinguish between the productive and the sensational. It would be immensely instructive for the public, he wrote, to see "the example of the lion, so proud and so superb, but so useless and so dangerous, cede its place to the Kashmir sheep, so gentle, so humble, so peaceful, and so useful."

There is no evidence that the public had the slightest interest in what Lacépède suggested. There is, however, a 1798 fictionalized guide to the menagerie for youths that has the animal keeper, Félix Cassal, lamenting the deaths of the male lion and other fierce beasts, saying, "I would give the little I possess to see this menagerie populated with all kinds of ferocious animals. When the public comes to visit, I am ashamed to have only ordinary animals to show them. I would like to be able to say: here is the tiger, here is the panther, here is the leopard, here is the jackal, here is the hyena, here is the lion."[8]

Lacépède's grand vision for the new menagerie counted on the government's endowing the Museum with an expanse of land some six hundred to eight hundred yards long and four hundred to six hundred yards wide. The whole would be divided into seven sections, each subdivided to provide space for various species. Two large canals would irrigate the area, and as far as climate permitted, the vegetation throughout the menagerie would be chosen so that each animal was surrounded by plants from its native land, offering the public a kind of "dual instruction."[9]

Lacépède detailed what would go in each section. The first would be planted with bushy trees and would accommodate hares, rabbits, squirrels, turtles, and beavers with a big pond of their own. Section two would have aviaries for various fowl: chickens, pheasants, turkeys, grouse, partridges, quail, guinea hens, agamis, hoccos, pheasants, bustards, ostriches, cassowaries, and dodos. (That no living dodo had been seen since the seventeenth century may be taken as another reflection of Lacépède's optimism.)[10] The third section would be for elephants, rhinoceroses, and perhaps even hippopotamuses. Lodged in sunlit huts and separated from each other by strong iron trellises, these "walking colossi" would bathe in tributaries of one of the large canals. Given an independence akin to that enjoyed in their native lands, they would, Lacépède surmised, readily display their natural habits.

The fourth section would house goats, gazelles, antelopes, oxen, buffalo, bison, species of deer, elk, horses, donkeys, zebras, camels, "and even the tapir." The fifth would be distinguished by its diverse species and rugged terrain of cavernous rocks, trees, and thickets where foxes, dogs, badgers, cats, ermines, and martens would live. These would be joined, for "a certain number of years," by "the lynx, the leopard, the panther, the tiger, and the lion," which would have large caves for their dens and spacious areas to move about, separated from the public by strong bars. The fifth section would also have aviaries to shelter "the tyrants of the air"—the eagle, the falcon, and the vulture—and hollows in the rocks to accommodate owls and other night birds.[11]

The sixth section, across from the carnivores and raptors, was dedicated to monkeys. With plentiful trees, it would allow them to display all the "games, capers, sprightly jumps, imitative gestures" and other habits that made them so worthy of scientific observation. Finally, the menagerie's seventh section would be a large meadow with hillocks, trees, and large aviaries overlooking ponds of varied sizes. Walkways and footbridges would permit the public to stroll comfortably through the area. The aviaries would house pigeons, turtledoves, orioles, parrots, and more; the water's edge would be the setting for herons, oystercatchers, flamingos, and other wading species or shorebirds. Some of the ponds would feature ducks, geese, swans, and such; others would be for seals, marine turtles, and shellfish. In the waterways and additional ponds would swim fish of different species that could be hybridized to produce "new races," which would be tastier, healthier, and more prolific than their predecessors, easier to introduce elsewhere, and more commercially valuable.[12]

As a final touch to the overall scene, Lacépède added insects. Beehives and other shelters would be scattered among flowering bushes and shrubs, providing homes for various useful insects representing still more sources of "industry, population, and prosperity." Everything in this menagerie, he promised, "would tend toward the progress of the sciences, the general utility, and the public good."[13] Despite his marvelous vision of the new menagerie, however, actualizing it was not Lacépède's top priority. In September 1796, nine months after his essay on menageries was published, he volunteered to the Executive Directory that the Museum would postpone its claims to an expanded territory in exchange for funds to complete the natural history galleries.[14]

∴

Unlike Lacépède, Jean-Claude-Michel Mordant Delaunay was not at leisure to imagine an ideal menagerie. As inspector of the provisional menagerie from 1795 to 1801, he confronted daily conditions that bore no resemblance to Lacépède's grand views. Far from upbeat, his reports to the administration descried the menagerie's state as unbecoming to the nation, precarious, dismal, and desperate.

Delaunay's background bore little relation to the tasks he faced as menagerie supervisor, unless some training as a lawyer was reflected in his persistence in arguing the menagerie's case. His application for the position of Museum's librarian (1794) noted that he'd planned a career in medicine and spent four years studying chemistry, pharmacy, materia medica, and botany. Delaunay's parents, however, preferred he become an attorney. Delaunay did not say whether he had ever practiced law, but he did allow that in the Revolution he had lost most of whatever wealth he had. More relevant to his application was Delaunay's long-standing interest in natural history, which had led him to perfect his Latin and Greek and to learn Italian and English.[15] When the Museum chose Georges Toscan as its librarian, Delaunay urged it to appoint him as assistant librarian, which it ultimately did.

Once installed, Delaunay sought to serve the Museum in any way possible, and he took special interest in the menagerie. In the summer of 1795, when Félix Cassal offered to sell the Museum a cockatoo and a pair of white peacocks, Delaunay consulted with Paris bird dealers to determine their value. Several months later, when he offered to oversee the menagerie's operations, the professors agreed, tasking him with daily inspections and with ensuring that the animals received wholesome food. The animal keepers were enjoined "to recognize citizen Delaunay as their boss," to take orders from him, and to direct their requests to him. The professors were charged with "forming a plan of experiments, the execution of which will be confided to Citizen Delaunay in such a way as to utilize for the public good the majority of the animals of the establishment's menagerie."[16]

Delaunay, it seems, was the only one who took these instructions to heart. The keepers continued to behave independently, and the professors never provided Delaunay with a program of experiments. They did ask him to procure native animals (wolves, badgers, boars, and the like) to occupy the menagerie's empty cages, but he was otherwise left to act on his own initiative, which he zealously did.[17] In his first months on the job, he called for the acquisition of new specimens; new or improved animal quarters; regulations to deal with an often ill-behaved public; and wage increases for the keepers.

Delaunay maintained contact with Bernard Louzardy when the animal keeper traveled to Holland to fetch the animals from the Stadtholder's menagerie, and it was Delaunay who pestered the administration about the preparations necessary for their arrival.[18] When Louzardy's convoy reached Paris, Delaunay's anxieties were not allayed, simply refocused. Getting the animals to Paris was one thing; keeping them alive and healthy was another. Public relations demanded it, as Delaunay explained in a letter (October 1796) to the professors: "The menagerie is not the Museum's most important object, but it is perhaps of the most general interest to all the classes [of society], and in this regard it merits your attention."[19]

The menagerie's problems, identified by Delaunay, included inadequate aviaries, wells filled with sludge, and a public whose misconduct placed the animals' lives at risk. People continually put dogs into the animal enclosures, and unaccompanied children sometimes found their way there as well. Some people brought their goats into the gardens to eat hay that was meant for the menagerie. Animals and public alike were sometimes endangered, as when men tormented the male fallow deer with their canes and were injured when the deer struck back with its antlers. Delaunay tried chaining the animal to a post in the middle of its pen, out of reach of malefactors, but the results were imperfect. The deer sometimes supported its misery patiently; at other times it struggled so violently that keepers feared it would strangle itself. Also, rutting season had arrived, and the chained male was unable to service the females.

Delaunay wanted a larger menagerie reconstituted on a plan that was more rational and easier to police. He imagined that hard economic times could be good for the Museum in terms of getting more land at a better price. Unlike Lacépède, who sympathized with the neighboring landowners, Delaunay supposed that the Museum could profit from their straitened circumstances.[20] Gaining confidence in his role as menagerie inspector, Delaunay addressed his concerns not only to the Museum but to the minister of the interior and the director of public instruction, compelled to do so by his "attachment to the Museum, zeal for the sciences, and love of my country." The menagerie's biggest difficulty, he explained in a letter to the minister, was that "the public that abuses everything" could not be adequately supervised when the animals were so widely dispersed. The menagerie needed to be a unified whole where the public was admitted at specific hours and could be "always watched."[21]

"Whatever injunctions one makes not to give or throw anything to the animals," Delaunay continued, "there are always people or children who have no regard for them." A female axis deer had recently choked to death on an apple that was too large and a chestnut that was too hard, both of which had been thrown to her. To enforce the rules, Delaunay allowed, the menagerie needed more than just a sentry for every animal park, it needed "a sentry for each animal." The animals were too spread out for close surveillance; the camels were at one end of the gardens, the carnivores at the other, birds were in the middle of the pond, the deer in the parks along the side, the elephants in one part of the "Régie," the new birds from the Stadtholder's collection in another. What the Museum needed, Delaunay wrote, was more room, specifically the land and buildings bordering the Seine. Without this it would be impossible to establish a menagerie worthy of the French nation.

Neither the minister nor the director of public instruction wanted to deal directly with Delaunay. Zeal was a good thing, the latter told the Museum, but only when it flowed through the proper administrative channels. Duly chastised, Delaunay henceforth directed all his reports to the Museum. Some problems were transitory, others were recurrent or continuous. In February 1797, the most immediate difficulty was provisions for the animals. For three months no new hay, straw, or grain had arrived. The minister had authorized the Museum to secure the fodder, but the farmers now wanted their money in advance.[22]

The decision to send Félix Cassal on his collecting expedition to North Africa posed the question of where to put the animals he brought back, increasing Delaunay's anxious concern.[23] In March 1798, as Cassal's return approached, Delaunay reminded the professors that they needed lodgings "to trick [the animals] with respect to their change of climate and to efface among them the memory of their lost liberty."[24] The immediate issue was that "animals originally from hot and arid countries" were not well suited to "the humid, cold, and always foggy air of the river" beside which the *animaux féroces* were currently kept. It was additionally necessary to keep the animals clean and "provide them with well-selected meat in sufficient quantity." The keepers, Delaunay allowed, were doing a good job on the first score, but securing wholesome meat was an ongoing struggle. Bad meat had killed the famous Senegalese lion.

The public's misbehavior, Delaunay told the administration, also had to be considered. Though the public tended to attribute the menagerie's "considerable losses" to a "lack of care," its abuse of the animals was an added factor.[25] Delaunay always had a list of incidents to recite. A small stag, chased by a dog, had tried to jump a trellis where it caught its hind foot; it would have died hanging there had someone not come to its aid. Stones thrown at the Brazilian stag caused a nearly fatal abscess on its back. One of the female deer had limped for six weeks from being struck by a stone. Delaunay reiterated his demand that dogs be excluded from the Jardin des Plantes, or at the very least, be put in "boxes" at the doors of the establishment, the practice employed at Paris's Bicêtre and Salpêtrière hospitals.

After calling again for a larger, more integrated, better situated menagerie, Delaunay closed with the warning: "If all the problems I have just enunciated are not remedied, and if all the needs I have described are not provided for, then the menagerie, far from being useful to the Museum, will only cause it to be seen in a less favorable light by the public, and perhaps by the government, which at the very least will be able to accuse you of not having kept it informed."[26] Two months later, Delaunay submitted another litany of the menagerie's needs, prefacing it with the prediction that the bears from Bern and the elephants from Holland, testimony to France's military victories abroad, would perish without proper care.[27]

In 1799 Molinos, the Museum's architect, drew up a plan for the "definitive" menagerie as part of a Museum master plan, incorporating elements envisaged by Thibaudeau and Lacépède: paddocks, ponds, and shelters for the "peaceful" animals, a large rocky escarpment for the lairs of the ferocious animals; streams running through the grounds; and pathways affording visitors picturesque views of animals with shelters and vegetation reminiscent of their native lands.[28] Pleased with the plan, the professors agreed to post it in the assembly room (March 1799). This seemed like progress, but Delaunay hastened to remind his colleagues that the menagerie's current condition was precarious. "There is not an animal," he wrote, "that is not badly placed, whether for itself, for the public, for the staff, or for surveillance."[29]

Only a fragment of this letter remains, but a follow-up written a week later rehearsed his demands: "I requested the very prompt repair of the ponds. . . . This point is even more pressing because the waterbirds languish on dry land in the stables. I requested that the elephant enclosure be embellished and that a pool be made there where, in bathing, they would find pleasure and health. I requested that near this enclosure new enclosures be established for the camels and dromedaries so that they could take the air there sheltered from the stones with which the public assails them."

Delaunay went on to ask for water buffalo, because the public seemed to be especially fond of them and was "always astonished not to find any at the menagerie." They too would need a pool. He requested that "the means be given to me to have the enclosures of the red deer and fallow deer repaired, which these animals can jump or break easily, which would not be done without danger for the public or at least without inconvenience for the garden." He warned that individual birds and bird species in the aviaries would decrease in number without a place for hatching chicks and allowing them to move about. He asked that a new man be hired to attend the ostrich and the animals quartered with it, given that their regular keeper (Ballet) often had to be performing tasks elsewhere. He again cautioned the professors that the public was likely to blame them for any animal losses, even if insufficient funds were the real cause. Finally, he listed abuses related to "the dogs that ruin everything in the garden."[30]

With the Museum's impoverishment, Delaunay's harangue could only worsen the general headache, but it did prompt the administration to refuse other untimely proposals, like the minister of the navy's offer in 1799 to procure animals from Tunis. The Museum could scarcely feed the animals it had. Delaunay had taken to rationing as the fodder supplies ran out because the Museum could not pay its provider. Meanwhile, the costs of feeding the carnivores were skyrocketing because the sources of cheap meat unfit for human consumption had dwindled and the Museum had to buy more expensive meat from the butcher shops.[31] Delaunay started considering which deer, goats, and cattle in the menagerie might be deemed "not useful" and sacrificed.[32]

When a man named Garnier, owner of an itinerant menagerie, offered to sell his animals to the Museum and join the menagerie staff (November 1799), he was told they couldn't afford it.[33] At year's end, when Delaunay suggested that his economies in provisioning the elephants made it possible to add two water buffalo and two Romany oxen to the menagerie, the administration refused.[34]

In April 1800 Delaunay was back on form, writing to the assembly: "Citizen Professors, I always come back to the needs of the menagerie and the means of utilizing and treating more appropriately the animals that embellish it."[35] Describing how the elephants loved to bathe, he called for building them a pool. He expressed anxiety about the security of the ostrich, the cassowary, the zebu, and the hind. Summer was approaching, he said, when the animals needed to be outdoors in the air, but because an adjacent building was being demolished, their previous summer shelters were no longer serviceable. He feared the animals would be wounded or killed during the demolition, fall prey to malefactors, or suffer from the cold. He was convinced that the inadequate state of the shelters and their poor location (not a lack of medical care) had caused numerous deaths already. "I would be able to provide you this list," Delaunay told the professors, "if I did not fear it would frighten you."[36]

Regarding the menagerie's camels and dromedaries, which were costly to maintain because they ate so much, Delaunay noted that their portraits had already been painted, and since all other necessary observations had been made, they served only to satisfy the public's "vain curiosity." The camels should earn their keep, he suggested, by being equipped with saddle bags to deliver meat and transport soil or other materials around the gardens. The only drawback, he remarked, was that each camel would need a day worker to accompany it.

The assembly refused all these proposals, citing more urgent needs, notably, paying workers' salaries, but it did not dismiss a far more expensive proposition of Delaunay's: the purchase of exotic animals on sale in London. When the government agreed to provide the funding, the administration sent Delaunay and Louzardy to acquire the collection: a pair each of tigers, leopards, panthers, and lynxes, plus a hyena. The administration also approved Delaunay's idea of selling animals that were superfluous or uninteresting; he soon found buyers for several deer, an old swan, and four Angora goats. When the remaining salable animals (several goats and a small bull) attracted no takers, Delaunay obtained the administration's permission to have them slaughtered and fed to the carnivores.[37]

∴

Lacépède meanwhile was occupied with matters largely unrelated to the menagerie, teaching Geoffroy's course on mammals and birds while Geoffroy was in Egypt and publishing his own *Natural History of the Fish* (a contribution to Buffon's great *Natural History, General and Particular*). Other than his elegant essay of 1795 on what a menagerie could be, his main contribution to the menagerie was to lend his authority to the publication of a handsome work titled *La Ménagerie du Muséum national d'histoire naturelle*.[38] Conceived by the engraver Simon-Charles Miger, the book featured Miger's engravings of Nicolas Maréchal's paintings of the menagerie animals, each accompanied by a naturalist's account of the animal's external characteristics, anatomy, habits, and country of origin. Miger would publish a four-part installment every three months. The first (November 1800) showcased the cassowary, the polar bear, the camel, and the ostrich. Georges Cuvier wrote this and most of the forthcoming text, though he remained anonymous at the time.

Lacépède wrote the official introduction, which appeared in 1801, and reserved for himself the entries on the lioness and the tiger. It was his first commentary on menageries since his 1795 essay and also his last. He repeated ideas from his original essay while developing a new theme: the correlation between a country's menageries and its level of civilization. According to Lacépède there were four types of menageries, each with a different motivation and corresponding to a different stage in the development of civilization. The earliest type was "the product of *need*." It originated when primitive tribal chiefs confined the animals they used for hunting and warfare in large enclosures. Then came menageries motivated by pride, when powerful rulers wished to signal their might by showing how they subjugated ferocious beasts. Later, when military victories won a population the independence, prosperity, and leisure time necessary to cultivate science, a new type of menagerie became possible, motivated by scientific curiosity.

These menageries featured rare species as well as those found in the previous types of menageries.

The fourth, most advanced and least common, type of menagerie, applied scientific knowledge to the needs of society. By Lacépède's account, civilization's continued progress led to new needs for food, clothing, agriculture, and transportation, which required the perfecting of useful species of animals. Only "a very advanced civilization" could produce such menageries, and only lately, inspired by the writings of naturalists such as Buffon, Linnaeus, and Daubenton, had several nations begun to establish them. This, in brief, was how menageries evolved over time: "by the need of the hunter-warrior, the pride of the dominator, the curiosity of the scientist, and the commitment of the citizen."[39]

Lacépède credited himself with a small role in the history of the new Paris menagerie, having been among those who helped keep the flame of science alight, even while forced to shelter from the tempest of the Revolution. "Foreseeing the calm they would enjoy today," these scientists had "redoubled their efforts to apply human knowledge, freed from all shackles, to the public good." His contribution had been to describe the ways menageries contributed to the public good and to propose "some ideas on the subject," which his colleagues at the Museum were eager to adopt. "They perfected them, in applying them to the establishment confided to their administration. They developed them in a menagerie plan worthy of a great people, through which they proposed to replace what they had been forced to construct in haste. . . . They resolved no longer to allow the existence of these narrow enclosures in which animals have been condemned for so long to all the suffering of captivity." He and his colleagues sought instead to erect a monument "on the ruins of these prisons."

The "proportions, the beauty, and the decency" of this new menagerie "would announce the greatness of the nation, the dignity of the history of nature, and the importance of the results desired." Patriotism, science, and philosophy would embrace the new menagerie which would render obsolete the first and second types. Rejecting "the arsenals of force or the inventions of pride," the new menagerie would display "the product of knowledge and the work of a humane policy, carried to the highest degree of perfection that the present state of civilization can bring into existence."

As in his earlier essay, Lacépède emphasized the importance of menageries not only for agriculture and commerce but for the public's moral elevation. This could not happen if visitors were confronted with "multiple images of constraint, boredom, anxiety, chains, and torments," sights that wounded "the eyes of the citizen, the naturalist, and the feeling man" and "that no utility could justify." In the new menagerie, images of independence and liberty would abound. The overall layout would resemble "a varied and cheerful countryside," where different species would "enjoy all the liberty it is possible to allow them" (without endangering a public that might sometimes act imprudently). Surrounded, insofar as possible, by plants from their native lands, the animals would feel "neither their exile nor their loss of independence." They would "represent to the eye of the observer the faithful picture of the productions of nature" from "the most remarkable lands of the globe."[40]

Aside from such rhapsodizing, Lacépède participated in several menagerie-related activities in 1801. In January his colleagues asked him to examine the places intended for the nine carnivores recently purchased in London and to arrange for signs with their scientific and common names.[41] In July, also at the administration's request, he joined Lamarck and Jussieu in developing a new set of regulations for the menagerie's organization.[42] Later that month, he joined the delegation to the minister of the interior to discuss the architect's plan for a rocky escarpment (*le rocher*) with lairs for the *animaux féroces*.[43]

∴

In the meantime, Delaunay soldiered on. In February 1801 he carefully reviewed Molinos's plans for the menagerie, but he was not wholly satisfied and worried that the construction process endangered the lives of some specimens.[44] In April he complained to the administration that the small pens where the cassowary, the ostrich, and other animals had enjoyed fresh air and exercise the previous summer were demolished, according to Molinos's plans and despite his objections. The alternative pens Molinos had promised had yet to materialize, Delaunay said, but hot weather was coming, and confining precious animals in their winter quarters (the camels' stable) was unhealthy.[45]

The professors consulted Molinos about relocating these animals, but if any action was taken, it was not in time. The cassowary and the pair of lynxes were dead by June.[46] Fearing further losses, the administration beseeched the minister of the interior for more funds. New quarters were desperately needed, the professors said, for the existing carnivores and for new animals arriving from Africa. By the professors' account, the carnivores' quarters were stifling in summer and were so close to each other that infection spread between animals from breathing the same air. To display the new lion cubs to the public, the Museum had to put them in a portable cage outside the regular quarters.[47]

Curiously, amid all Delaunay's complaints to the administration, he said little about the keepers, aside from supporting their requests for better wages and for jobs for family members or testifying to how diligently they cared for the animals. From other quarters, however, reports of reprehensible behavior had been trickling in. At the same meeting where the professors learned of the deaths of the cassowary and the lynxes, a letter from Minister Chaptal passed along an allegation that Cassal had withheld the newborn lion cubs from public view, showing them only to people willing to pay for the privilege. The professors sent a committee of Jussieu, Lamarck, and Lacépède to investigate. Their report was so damning that the administration decided a professor must take responsibility for the menagerie.

The most obvious person for the job was the professor of mammals and birds, Geoffroy Saint-Hilaire, but he was still in Egypt. The next best choice for menagerie superintendent would have been Lacépède, based on his publications, but he'd been spending little time at the Museum. He did not live there, and as a member of the Senate he was increasingly involved in government affairs. Instead, Lamarck, professor of invertebrate zoology, was asked to oversee wayward keepers and manage the menagerie's daily functioning, taking over in July 1801. Although he maintained an interest in the menagerie for at least another year, Delaunay was back to being the Museum's assistant librarian, a position he held until his death in 1816.[48] The Museum at that point chose not to replace him, deciding that the money allotted to his position was better spent on new specimens or on repairs.

As for Lacépède, one of his last acts on behalf of the menagerie was leading a delegation to win the minister of the interior's support for the plan to lodge the *animaux féroces* in a rocky cliff, an idea described in his 1795 essay. Chaptal's initial response was positive, and he promised to provide the funds, but on second thought he decided the structure would be too expensive to build and maintain, unhealthy for the animals, difficult for the keepers, and hopeless for the viewing public. The problem of housing the fierce carnivores was back to square one, and the poor beasts were left in their cramped, fetid cages in the old converted greenhouse by the Seine.

There were two ways the gap between Lacépède's early theorizing and the daily realities of the menagerie had nonetheless narrowed. For one, the herbivores' quarters had improved, with pleasing new parks and shelters. For another, Lacépède had walked back the idea of replacing lions with sheep to benefit the public's moral development. In his 1801 text "the lioness" in Miger's *Les Animaux de la ménagerie*, he rejoiced at how many lions and lionesses graced the menagerie. Lions were back in Lacépède's favor, as they'd always been in the public's. Nor did it seem to bother him that power and pride, motives for menageries of a more primitive sort, were fully present in Paris, where citizens especially liked the animals acquired as trophies of war.

Flexibility helped see Lacépède through France's successive political upheavals. He became a senator after Napoleon's coup d'état of 1799, and Napoleon named him grand chancellor of the Legion of Honor four years later. At this point Lacépède handed his teaching responsibilities over to his assistant, André Marie Constant Duméril (1774-1860), and turned down an invitation to be the Museum's next director. His contributions to science were basically over as his attention turned to the needs of the state. Under the Empire, he was raised to the exalted positions of count of the empire, minister of state, and president of the Senate (twice). Similar posts awaited him in the Restoration. Lacépède's fanciful visions of the new menagerie's possibilities were far removed from the business of its maintenance and development, but his rhetorical and administrative talents kept him in good standing with one regime after another.

[1.9]

The Concert for the Elephants

On the twenty-ninth of May 1798, a fourteen-piece orchestra performed at the Museum of Natural History for an unusual audience, two elephants taken as trophies of war and recently arrived at the menagerie. The concert was not a lark by the musicians but an experiment undertaken in earnest to see how the elephants would react to music played and sung in different keys, at different cadences, and on different instruments. The idea made sense, since elephants were reputedly the most intelligent of all animals (other than humans) and (like other higher animals) credited with having the same sorts of emotions as humans, albeit lacking the faculty of reason.

Considering the wondrous events unfolding in Paris in spring 1798, a concert for elephants was not out of place. Parisians were still celebrating the peace with Austria that Napoleon secured through the Treaty of Campo Formio (17 October 1797); the newspapers were reporting that the last of the art treasures from Venice had reached Toulon and would soon be coming to Paris; and Napoleon, whose successes in Italy lent him authority as both a conqueror and a peacemaker, was about to set off on an expedition of unknown destination and purpose.[1] A French fleet and a multitude of soldiers had assembled in Toulon, but why? Was it the long-rumored invasion of England? Or were they headed to Ireland, Scotland, Lisbon, or the Black Sea (with the intention of making Poland a republic)? Or off to Egypt to establish a colony as a first step toward challenging the British in India?[2]

Readers may not be familiar with the revolutionary song "Ça ira," which featured in the concert for the elephants, but most will have heard the parable about three blind men who each touched a different part of an elephant and came to very different conclusions. While the identity of the elephants in the Paris menagerie was not in question, they meant different things to different people. Their long-time keeper, the Museum staff who had fetched them from Holland, and the government and military officials involved in their transport were just a few of those whose lives were interwoven with these massive and marvelous sentient beings. In addition to the musicians

Figure 1.9.1. "The elephants represented in the instant of the first caresses they made when music was played to them." J. P. L. L. Houël, *Histoire naturelle des deux élephans, mâle et femelle, du Museum de Paris, venus de Hollande en France en l'an VI* (1803). Muséum national d'histoire naturelle. Bibliothèque centrale. Fonds général. 1094.

who performed for them, writers recounted the elephants' history and actions, artists reproduced their likenesses, Museum naturalists eyed them as scientific specimens, citizens came to gawk at them, food vendors and pamphlet writers profited from the crowds, and vaudeville performers contrived a lighthearted musical comedy loosely based on the concert.

The elephants generated excitement partly as trophies of war, but more significantly for how breathtakingly novel they were to almost everyone who visited them. No elephant had been exhibited in Paris for a quarter-century, and sixteen years had passed since the royal menagerie at Versailles possessed one. Since Hannibal marched his war elephants across ancient Gaul (218 BCE), the number of times elephants had been seen in France could be counted on two hands.[3]

Nearly a thousand years before the concert for the elephants, the French experience with pachyderm gift giving began. Charlemagne received one from Harun al-Rashid, the caliph of Baghdad. In 1254, Egypt's Mamluk rulers sent Louis IX an elephant, which he passed on to his brother, Henry III of

England. In 1591, when Philip II of Spain sent an elephant to the new French king, Henry IV, he regifted it to Queen Elizabeth. Elephants in France include the one of unknown origin that the polymath Nicolas-Claude Fabri de Peiresc reported seeing in Toulon in 1630 as it was touring the South of France. The king of Portugal sent Louis XIV an elephant from the Congo in 1668, about the same time an Indian elephant was making the Paris rounds with a troupe of entertainers. The Indian elephant's fate is unknown, but the Congolese one died in 1681, having spent thirteen years at Versailles. That was the last France saw of elephants for nearly a century. When Buffon published his account of elephants in his *Histoire naturelle* (1754) he was forty-seven years old, and he had yet to see one of the fabled animals alive.

Buffon finally got his chance in 1771 when a male elephant was exhibited in Paris at the Saint-Germain fair. In 1773, he saw his first female at the same venue, describing her as gentler than the male, "as in all species in Nature."[4] That year another elephant, sent from India, arrived at Versailles, where she lived for nine years until she broke out of her cage, fell into a pond, and expired.[5] Despite this brief flurry of elephants in the 1770s, by 1795 there were none to be found in France and perhaps no more than three in all of Europe: one in London and two in a menagerie outside Amsterdam belonging to the Dutch Stadtholder, Willem V, who was forced into exile in England as the French armies approached.[6]

The news that the Stadtholder's elephants were destined for the Museum in Paris arrived by letter (April 1795) from André Thouin in The Hague. He described their character as it had been represented to him: gentle rather than mean (*méchant*), but also mischievous (*espiègle*) and sly (*malin*). If tormented, they bore grudges and sought revenge. "In the season of love" they were not easily managed, and precautions were needed in approaching them, but they had never injured anyone. Thouin hoped the elephants' caretaker would accompany them to Paris and remain long enough to teach the keepers at the menagerie how to handle them. He forewarned his colleagues that the elephants would need accommodations of considerable size and offered explicit suggestions.[7]

While waiting for the elephants to arrive, a Parisian could consult ancient and modern sources for information on the animal. Pliny the Elder was the most famous of the former, Buffon the clear favorite among the latter. Pliny's *Naturalis historia* (first century AD) identified the elephant as the largest of the land animals and the closest to humans in intelligence. "It understands the language of its country," Pliny reported, "it obeys commands, and it remembers all the duties it has been taught." In addition, it was "sensible alike of the pleasures of love and glory, and, to a degree that is rare among men even, possesses notions of honesty, prudence, and equity;

it has a religious respect also for the stars, and a veneration for the sun and the moon."[8]

Georges Cuvier disparaged Pliny as "a compiler," but for a full fifteen hundred years Pliny's *Natural History* constituted the single most important collection of facts about the world, ranging from astronomy and physics to geography, agriculture, commerce, the natural history of plants, animals, and minerals, and the histories of different peoples.[9] Pliny provided information about when elephants were first seen in Italy, how they were made to fight other animals and humans in the Colosseum, their ability to walk a tightrope, and the different methods for capturing them. Pliny echoed the notion of other ancient writers, that elephants were so modest they would copulate only in private, away from potential onlookers.[10] Buffon expressed amusement at Pliny's credulity about some of the elephant's qualities, but he doubted neither its intelligence nor its modesty.

By Buffon's eighteenth-century reckoning, there were four animals (dog, monkey, beaver, and elephant) whose attributes approximated those of humans. The dog's chief traits were sociability and capacity for companionship and loyalty, which had led it to associate with humans in the first place and had developed further over time. The monkey's appearance and dexterity led people to credit it with human characteristics (though the monkey, unlike the dog, tended to be disagreeable). The beaver's faculties seemed inferior to those of the dog and the monkey, but it far surpassed them as a social animal in planning and executing communal projects. But the elephant, said Buffon, was superior to them all, possessing their best traits and other qualities besides. Beyond "the intelligence of the beaver, the dexterity of the monkey, and the feelings of the dog," it had prodigious size and strength and mental capabilities such as "courage, prudence, *sang-froid*, exact obedience."[11]

Buffon stressed that animal intelligence and human intelligence were not identical, embracing the Cartesian position that animal intelligence is a function of material organization and never involves the immaterial mind or soul that humans alone possess. Yet if he denied the elephant reason, he credited it with intelligence and character. The creature's even temperedness and judiciousness impressed him: "It maintains its moderation, even in its most ardent passions; it is more constant than impetuous in love; in anger it does not mistake its friends; it attacks only those who have offended it; it remembers good deeds for as long as injuries; having no taste for meat and nourishing itself only on vegetables, it is not born the enemy of the other animals; finally, it is loved by all, since all respect and have no reason to fear it."[12]

When Buffon characterized the elephant as "a miracle of intelligence and a monster of matter" he did not refer simply to the animal's size. He found the proportions of the elephant's body parts all wrong: the eyes too small, the legs too massive, and the ears and especially the nose too large. As for the elephant's reproductive organs, the male's penis was disproportionately small ("no larger and scarcely longer than that of the horse"), while the female's vulva was located higher ventrally ("nearly in the middle of the belly") than in other quadrupeds, making it unlikely that the male could successfully mount the female from behind like other quadrupeds.

Elephants, Buffon surmised, mated face-to-face.[13] This helped explain why they did so only when at liberty, away from prying eyes: "The female must not only consent, but it is further necessary that she provokes the male by an indecent posture that she apparently takes only when she believes herself without witness." Buffon thus credited Pliny's assertions about the elephant's modesty and offered physical and psychological explanations in support. Modesty, as Buffon further explained, was a general trait of the female sex (along with "gentleness, moderation, and temperance").[14]

Buffon further credited the elephant with a sense of virtue absent in other domesticated species. Humans, he observed, had succeeded in capturing elephants, taming them, and making them perform a range of services, but this was not the same as domesticating them. An animal was not fully domesticated until it was bred in captivity. The elephant was not one of "these born slaves that we propagate, mutilate, or multiply for our use." Though an individual elephant might be made to work, the species remained independent, "constantly [refusing] to grow at the profit of the tyrant." The tamed elephant retained a sense of disgust regarding its servitude, so that even when feeling "the most vivid attractions of love, it never produces nor copulates in the state of domesticity." Such thwarted passions could have consequences; "indignation over never being able to couple without witness" infuriated the elephant, making it "more dangerous than any other untamed animal."[15]

Buffon accepted the testimony of a certain Marcel Bles who had lived in Ceylon, witnessed elephants mating, and reported that they do so like other quadrupeds, but never in captivity.[16] Based on this eyewitness account, Buffon corrected his earlier claims about how elephants copulate, but long after his death, when the menagerie in Paris received its first elephants, the questions of their mating position and breeding in captivity, remained unsettled.

The elephants that drew musicians to the Museum were captured very young in Ceylon (Sri Lanka) in 1785 and transported to Holland under the care of a captain of the Dutch East India Company. They debarked in Amsterdam in July 1786 and were delivered to Het Kleine Loo ("The Little Grove"), the menagerie of the Dutch Stadtholder Willem V at Voorburg, near The Hague. Their next stop, a few months later, was Het Loo ("The Big Grove") at Apeldoorn, some eighty miles from the capital, where they joined a menagerie consisting mostly of birds and herbivores. At the Stadtholder's court they were treated well; the male was named "Hans" and the female "Parkie." Two of the leading Dutch naturalists of the period, Arnout Vosmaer and Petrus Camper, had the opportunity to observe them.[17]

In 1788 the elephants' first keeper in Holland died and was replaced by a James Thompson.[18] When French troops swept into the Low Countries early in 1795, the Dutch contemplated killing the elephants, worried about the cost of feeding them. Thompson pleaded for their lives, promising to care for them if he could keep their remaining provisions. After the Dutch agreed, Thompson had to persuade first British and then French troops not to harm the animals.[19]

Arranging their transportation was more difficult than anticipated.[20] In June 1795 the representative of the French Republic in Holland, Charles-Jean-Marie Alquier, provided directions for ensuring the safety of the Stadtholder's menagerie and bringing it to Paris, but funds were scarce, and the Committee of Public Instruction (CPI) had to be convinced that the animals were worth the 10,000 florins for transport. Joachim Lebreton, head of Museums and Lycées for the CPI, argued that the elephants would appeal to the public and scientists alike. Having a male and a female, he said, would provide an unprecedented opportunity to learn about the animals' social relations, and such knowledge might lead to the successful naturalization of the elephant in France, rendering it of use in public works and agriculture. To strengthen his case, he raised the specter of France's hated rival, Britain, "that nation so jealous of carrying off from its neighbors everything that could bring them some glory." The British, he claimed, had already offered to buy the Dutch elephants. As it was, he warned, the Dutch might decide to keep them for themselves.[21]

Shortly after Lebreton made the elephants' case to the CPI, the National Convention was dissolved. The Directory became France's new governing body, a new minister of finances was installed, and months passed before funding for the elephants' move was assured. But aside from earmarking funds for the project, there were psychological and logistical obstacles to overcome, the former related to the elephants and their keeper. The elephants clearly needed to travel with someone familiar, namely Thompson. Thouin suggested from the beginning that Thompson was needed both for getting the elephants to Paris and for establishing them in their new home. Thompson, however, did not want to come to France.

Securing Thompson's participation took time. On behalf of the reluctant elephant caretaker Delaunay told the Museum in Fall 1796, "a good welcome and fair conditions are the only means of attaching to us this citizen whom past circumstances have terrified but whose fright has to be dissipated."[22] The following summer, as the final plans for moving the elephants coalesced, François-Joseph-Michel Noël (1756–1841), minister plenipotentiary of the French Republic at The Hague, described Thompson as "a weak head on an invalid and decrepit body," yet he insisted the man was essential: "he is absolutely necessary for the transport of the elephants, and the Museum will have to make some sacrifices to retain him."[23] In March 1798, after the elephants finally arrived safely in Paris, the minister of the interior, François Sébastien Letourneux, told the Museum: "With respect to citizen Thompson it is my intention to retain him for the care of the elephants. . . . He can be assured that nothing will be spared to improve his lot."[24] Letourneux set Thompson's salary at 2,400 francs, a figure well exceeding the 1,500-franc annual salaries of the other animal keepers, Cassal and Louzardy. The Museum arranged to have Thompson's family escorted to Paris.

Among the logistical hurdles the elephants' transport presented, the need for exceptionally strong cages and wagons was clear from the start, along with horses to pull the wagons, good roads, food for the animals and their attendants, and funds as needed. Bernard Louzardy was sent to The Hague in autumn 1795 to collect the elephants and other animals from the Stadtholder's menagerie, but funding for the project failed to materialize. In spring 1796 the Museum sent Louzardy back to Holland accompanied by Claude Lassaigne, a skilled carpenter and mechanic who had packed confiscated natural history specimens for transport to the Museum on previous occasions, including the Chantilly collection (1793) and the specimens Thouin and Faujas collected in Belgium and Holland (1795).[25]

Elephants presented a greater challenge. Their size and strength proved too much for the cages and wagons Lassaigne built for them, thwarting the plans for a June 1796 departure. By one account, the male elephant triumphantly destroyed his cage.[26] By another, the wagon wheels could not withstand his weight.[27] Louzardy left for France with a convoy of other animals from the Stadtholder's menagerie, leaving Lassaigne to find a sturdier solution for moving the elephants. Louzardy returned in November, but the new convoy covered little over a mile before the wagon carrying the male irreparably collapsed. Lassaigne did his best to care for the elephants in this time of crisis, but he later felt that Hans associated him with the trauma of the period.[28]

In the months that followed, the provisional commander of the northern army, General Jean François Aimé Dejean, lent his support to the elephants' transport. In 1797, after consulting with other army officers and engineers, he concluded that the kind of wagon used for hauling heavy artillery was what was needed. He had some built, then called for experiments to determine if a wagon loaded with a weight greater than an elephant's could be pulled along a sandy road. The wagon passed the test, and Dejean wrote to assure the minister of the interior that all was ready: "the roads, the means of introducing the animals into their cages, the means of embarking and disembarking these heavy masses," and the special boats needed to convey the elephants at specific stages of the journey. He promised to write to the minister of war to request the horses needed for pulling the wagons from Cambrai to the village of la Fère. He also identified the worst sections of road in that stretch.[29]

On 24 September 1797 the elephants were loaded, each into its cage on its own robust wagon. The convoy set off the next day (or the day after) with one hundred horses. Fourteen pulled Hans's wagon; sixteen pulled Parkie's. At Deventer, the elephants were moved in their cages from the wagons to boats. Traveling by waterways and then again by land, they arrived in Cambrai, France, in mid-November. They were only fifty miles north of Paris, but the road ahead was impassable until spring. They settled for the winter in the Cambrai military barracks. The elephants remained imprisoned in their cages on their adjacent wagons.

Letters from Laissaigne to Delaunay and the Museum refer to the elephants' behavior in these highly constrained conditions as well as to how devoted Lassaigne was to the animals and the mission. On 1 December 1797, he reported that the elephants had room in their cages to turn around and lie down to sleep. The female did so every night, he said, but the male only once a week. His more concerning news was that the female had come into heat three to four days earlier and had been rampaging in her cage throughout the night. Wrote Lassaigne, "She is always trying to put her trunk to the ear of the male to tell him of her love and to incite him to disruption, but the male does not make much of it." Lassaigne reassured Delaunay that the elephants' health was at least as good as it had been back at Het Loo, and he pledged to do everything he could to conserve it. He also expressed his greatest fear: that the elephants would expire and be left to the anatomists and their scalpels.[30]

Three weeks later, Lassaigne wrote that while things were going well "for the moment," the elephants had begun stamping in their cages, perhaps "a little for pleasure but also out of boredom." He and Thompson worried that the longer the elephants remained prisoners, the more likely it was that they would become unhappy and angry and harm themselves in an accident.[31] Thompson's anxieties had reduced him to tears, but even while suffering a stomach illness, he spared no effort to ensure the elephants' health and safety. At night, when the animals grew restless, Thompson lit a candle and talked to them to calm them down.[32]

On 26 December 1797, Lassaigne wrote that the situation had worsened. The female had taken to bashing her head furiously against the wooden bars of her enclosure. The bars could withstand the assault, Lassaigne said, but the elephant was likely to injure herself seriously. The male was less agitated, but he too seemed a candidate for a serious accident.[33] Lassaigne did not mention it, but Hans had broken a tusk before they left Holland, and Parkie had broken one in Gand.[34]

Lassaigne did what he could to expedite the convoy's departure. He inspected the roadbeds out of Cambrai to identify the many large holes that needed filling for the wagons to make their way. He wrote to engineers and other officials about the road work and about having good horses ready for the task. In February he reported to Delaunay that the holes were being filled with stones from the demolition of the Cambrai cathedral.[35]

On 14 March 1798 the convoy ventured cautiously from the barracks and beyond the city gates, led by a battalion chief and engineer named Boucher.[36] The next day it traveled sixteen kilometers and spent the night at le Catelet. When they reached St. Quentin (16 March) Lassaigne wrote to Delaunay that they were finally on their way. All was well, he said, except that the male had been angry and refused to eat and both animals were "a little tired" from the severe jolting of their wagons. Boucher had intended to travel all the way to Paris by road, as it was cheaper than traveling by waterway, but the first bumpy days changed his mind.[37] At Noyon they embarked on the Oise River and then the Seine, reaching Paris's Port des Invalides on 20 March 1798.[38]

It took four days to disembark the elephants and haul them to the Jardin des Plantes, where they arrived at 8:00 p.m. on 24 March.[39] The process undoubtedly attracted crowds, but no account exists of the disembarkation and trip to the Museum. There are, however, two eyewitness reports of the elephants' behavior on the morning of 25 March 1798, when they were introduced to their new quarters and together for the first time since leaving Het Loo eight months earlier.[40]

Georges Toscan, the Museum librarian who had written extensively on the menagerie's lion and dog, provided the most widely circulated account. He described the quarters created for the elephants in the Régie as "a spacious hall, well-ventilated and bright, with a stove for heating it in the winter." The hall was divided into two living sections with a sliding door between. The male elephant was released from his cage and allowed to enter first. He lifted a bar that kept the sliding door in place, opened it, and entered the second space, where he began eating. The female was then brought into the first section. She and the male did not immediately catch sight of one another, but when they did, they "ran to each other and set about making such piercing and loud cries of joy that the whole hall shook." With their trunks they also made a sound like "a violent wind."

Wrote Toscan, "The joy of the female was especially keen; she expressed it especially with a precipitous batting of her ears, which she made move like a bird makes its wings move at a high speed. She passed her trunk over the body of the male with the greatest tenderness and voluptuousness. She brought it especially to his ear, where she held it for a long time. Often too, after moving it over the whole body of the male, she brought it back amorously to her own mouth. For his part, the male did the same thing over the body of the female, but his joy was more concentrated, and he seemed to express it by the tears that flowed in abundance from his eyes."[41]

In the *Chronique universelle*, a writer named Lacoste reported that there was "nothing more touching" than seeing the pair reunited: "Their eyes shining with sentiment, the clacking of their ears, their interlaced trunks, announced their intense satisfaction and all their happiness to see each other." After that, he said, "Nature had her way. The female announced it first by cries at intervals, very strong and yelping, but not at all frightening, because they only expressed love and desire. She caressed the male with her trunk, who feeling somewhat too vividly the impression of a rough and calloused skin displayed impatience with it by some kicks, but gently directed." Lacoste, aware of the view that elephants reproduce only in liberty, wanted to see what would happen next, but Thompson interrupted by offering them food. The female disdained it, but it captured the male's attention and diverted him from their interaction.[42]

Geoffroy Saint-Hilaire was present, and Lacoste discussed with him Buffon's mistakes with respect to the size of the male's penis and the female's position in copulation. "It is certain," Lacoste said, that she maintains the same posture "as the majority of other mammals." He also dismissed Buffon's notion that elephants would not mate in captivity because the species shunned enslavement. That could not apply to this pair, Lacoste observed, because they had been captive since birth and had known no preferable state. He supposed they would yield to the impulses of nature the same as a human couple raised in isolation without "any idea of modesty and moral obligation."

Lacoste suggested a romantic setting might aid nature's course. He proposed removing the divider separating the two cages, upholstering the enclosure with moss, surrounding the whole cage with leafy branches, providing both the elephants but the male especially with abundant and tasty food, and "even exciting him with the eau-de-vie that he loves, or other warming and prolific liqueurs." With an eye to how long the pair had been separated and to the coming of spring, Lacoste predicted that a productive coupling would soon take place and that the Museum naturalists would be able to learn about the elephant's gestation period and whether baby elephants nurse with their mouths or their trunks.

Once installed, the elephants drew large crowds.[43] Shortly after their arrival, members of the public irritated them to the point that the administration feared their injury and decided to limit public access to the even days of each *décade*, from 11:00 a.m. to 1:00 p.m. and from 3:00 p.m. to 5 p.m.[44] Crowds attracted the usual profiteers; merchants jockeyed for vending space outside the Museum's gates. The Museum's records include a mid-April plea from "Citizen Reneaudu," a fruit seller based outside the portals of the St. Victor Abbey for the past two years whose sales, she wrote, had grievously declined since the elephants' arrival owing to the new competition. She begged the Museum to allow her to sell fruits, lemonade, and herbal tea (*tisane*) in the courtyard adjoining the elephants' lodging, but the administration refused, saying it wished to limit the number of vendors on the grounds.[45]

The minister of the interior was also paying attention to the crowds. Asked by the Executive Directory in mid-April if it was true that visitors were feeding the elephants, Letourneux wrote to the Museum, "I am convinced you have made the most explicit prohibitions in this regard, and the rumors that have reached the Directory are without foundation. However, since allowing everyone to feed these animals would cause the gravest harm, I invite you to prevent this by taking very severe measures to make sure that, in accordance with the custom wisely practiced in all menageries, the elephants receive no other food than what the people to whom their care has been confided should give them." He suggested that the Museum post a notice to this effect.[46]

∴

The concert for the elephants took place on 10 prairial an 6 (29 May 1798). The musicians hailed from three nodes of the French music scene: the Conservatory of Music, the Théâtre Feydeau (known for comic opera), and the Théâtre de la République et des Arts. The instruments were two violins, a bass, two flutes, a piccolo, a bassoon, a clarinet, and a bugle. Among the fourteen performers were François Devienne (a flutist/composer known as "the French Mozart") and the singer/composer Arnold-Michel Adrien, who together had composed a score for Robespierre's Festival of the Supreme Being four years before.[47]

Explanations regarding the concert's motivations differ. One is that the professors, fearing the elephants were melancholy, decided music would raise their spirits.[48] Another suggestion is that the concert was held to determine how to induce the elephants to mate under domestication, and to observe their mating position.[49] The evidence supports neither. It appears instead that the musicians had the idea of testing how these marvelous creatures, known to be highly intelligent and endowed with emotion, would respond to different kinds of music.[50]

As the concert proceeded, the musicians recorded on two sides of a single page what happened. The unpublished handwritten document begins by identifying the musicians' names and the compositions the concert began with. Two columns hold the data the musicians considered relevant. On the left are the titles and key signatures of the pieces, the names of the performers and their instruments, and occasional bits of other information. On the right are the musicians' observations on the elephants' reaction to each piece. The report was delivered to the Museum and reviewed at a professorial assembly two and a half weeks after the event.[51]

Georges Toscan's report of the concert, published two months after the event took place, did not mention the existence of the musicians' report, but strong similarities between his account and theirs indicate that he wrote with their document in hand. However, rather than having the musicians' interest in the elephants' response to different instruments, cadences, and keys, Toscan, the self-styled "friend of nature," took the concert as proof that noninvasive, painless (and even pleasure-giving) experiments could be conducted on living creatures. He also concluded that music could help bring about the monumental achievement of getting the elephants to breed in captivity.[52]

The elephants were housed on the ground floor of the Régie, the large structure the Museum acquired in 1795. The concert was held on *décadi*, the last day of the ten-day week, when public viewing of the elephants began at 11:00 a.m. The concert likely took place earlier in the morning before the public could complicate the scene. According to Toscan, the musicians set up in a gallery above the elephants' quarters and out of the animals' sight. Thompson kept his charges busy, giving them treats to eat to distract them from anything unusual happening in their vicinity. A trapdoor in the roof of the elephants' cage was then quietly opened and the musicians, stationed around the door, began to play.

> They commenced with a trio of light airs in B major for two violins and a bass. When the music started, the elephants stopped eating and hurried to the spot from which the sound issued. Everything about the strange situation, the glimpses of the instruments and the musicians through the trapdoor above, the music, the silence of the spectators, the stillness of their keeper, became for them a matter of "curiosity, astonishment, and unease." They lifted their trunks toward the opening; they reared up on their hind legs. Thompson's reassurances initially failed to calm them: "they looked at the assistants and seemed to examine if some trap was being laid for them." Eventually they lost their fear, and the chief influence upon them seemed to be the nature of the music.[53]

The first of the airs was relatively gentle in character, but for the second piece the violinists and bassist played an energetic dance tune in B minor from the first act of Christoph Willibald Gluck's *Iphigénie en Tauride*. According to the musicians, the effect of this piece was "infinitely more appreciable" than the previous one, and the elephants appeared "greatly agitated."[54] Toscan suggested that they seemed to move in rhythm with "the savage and strongly pronounced character" of the piece; they "bit the bars of their cages, clasped them with their trunks, pressed the weight of their bodies against them, as if there were not enough room for their frolics, and they would have liked to push back the boundaries." Asked whether the animals' piercing cries and whistles were a sign of joy or anger, Thompson responded in his broken French, "them, not angry."[55]

The great beasts calmed down during a bassoon solo in C minor of a little ballad, "O ma tendre musette." The musicians noted only that the elephants paid great attention to the piece, but Toscan stated that once again they moved in time with the music: "Their motions were slow, measured, and partook of the gentleness of the song," but the two responded differently. "While Hans limited himself to his normal prudence and circumspection, a passionate, caressing Marguerite [Lassaigne's name for Parkie] stroked him with her long and flexible trunk, passing it over his back, over his neck, then bringing it back to herself, pressing her own mammaries with the tip of her trunk, and as if the tip was thereby imprinted with a more urgent and tender feeling, she immediately brought it back to her mouth, then put it in the ear of Hans, who did not hear, or perhaps was still unaware of this language."[56]

The musicians then played "Ça ira" (in the key of D). Like that other famous revolutionary anthem, the "Marseillaise," "Ça ira" calls for vigorous and violent action—in this case hanging aristocrats from lampposts—and was ubiquitous in Paris at the time. A few years earlier, in his symphony celebrating the French victory at Jemmapes (1794), François Devienne tracked each stage of the battle, culminating the piece with "Ça ira."[57] No one suggested that the elephants understood the words to "Ça ira," but no one doubted they were affected by its energetic performance and by the piercing sound of the piccolo. The elephants emitted what appeared to be happy cries, some deep, some shrill, together with whistling noises. Their every movement seemed impelled by the rhythm. Parkie again flirted with Hans, caressing him, running away from him, coming back, and occasionally kicking him to get his attention.

The next piece, a choral duet in E flat from Antonio Sacchini's opera *Dardanus*, calmed the elephants. The musicians' manuscript indicates that as it was sung, accompanied by the entire orchestra, "the male appeared to reflect profoundly."

Next came a reprise of "Ça ira," this time in the key of F, but played less vigorously than before. The musicians concluded that animals experience music both according to its general character and to the key it is played in; the elephants were "infinitely less agitated by this air in F than the first time, when it was in D." The flute duet that followed produced no excitement, but the subsequent solo vocal of Henri IV's "Charmante Gabriele" in C major so moved the elephants "that the whole gallery noticed that the female shed tears."

The entire orchestra then played the overture to Jean-Jacques Rousseau's opera *Le Devin du Village* in D major, which "agitated" the elephants, the musicians noted, "but less than 'Ça ira.'" Then came second renderings of "O ma tendre musette," (this time in A minor) and "Charmante Gabriele" (once again in C major). Singing both songs slowly and tenderly, with light accompaniment, the vocalist Adrien elicited "approximately the same" response that "Charmante Gabriele" had produced previously.

After these tranquil pieces, the orchestra played "Ça ira" for a third time, again in the key of D, but with Adrien, another musician (Guichard), and many orchestra members singing. The results were extraordinary. Said Toscan, the female seemed unable to contain herself. She trotted and jumped to the rhythm, and she added her trumpeting to the chorus of instruments. With ears flapping rapidly, she approached the male and ran her trunk sensuously over his entire body. As before, she kicked him, but this time, "often in her delirium, she fell to earth on her rump, waving her front feet in the air and pressing her back against the bars of the cage. She was heard in this posture to emit cries of desire, but immediately afterward, as if she were ashamed of an action with so many witnesses, she got up and began trotting around again."[58]

The musicians decided to call intermission. When they returned, they set up in full view of the elephants and played four more pieces beginning with a selection from a Haydn symphony. The musicians identified this as a "large piece of Haydn in C major, very sparkling and executed by the whole orchestra." In response, they noted, "the female [was] generally much more agitated than the male." Toscan ignored the female's behavior while noting that the resounding bursts of music from this rousingly performed piece did not attract the male's attention in the least: "He showed neither curiosity nor surprise."

The second postintermission offering had greater effect: a combination of two pieces for clarinet, the overture to the comic opera *Nina,* in F, and "O ma tendre musette" for the third time, now in D minor. The musicians noted "the tender character of these two pieces where the amorous sound of the clarinet brought the male to the most remarkable attention and produced on him the most vivid impression." Toscan elaborated that as soon as he heard the clarinet, "he sought the sound that caressed him and stopped when he was facing the instrument, extending his trunk toward it. Attentive, immobile, he listened. . . . The fires of love insinuated themselves into his veins. Betrayed by the external signs of this, and as if astonished himself by this new sensation, he took several steps backward, and when the symptom diminished, or when it had entirely passed, he approached the music again, listened, and found himself in the same state—passing fires which burned for an instant only to disappear and did not even serve to guide him toward his companion."[59]

The orchestra than played "Ça ira" for a fourth time, to lesser effect than before, when the orchestra was out of sight. The musicians suggested that "perhaps having the orchestra under their eyes, the sight of the objects that produced these effects left them calmer than when they were struck by the sounds without perceiving the cause of them. Perhaps too they had begun to be tired."[60] Toscan agreed that the elephants had wearied, otherwise they would surely have noticed the blaring hunting horn that ended the concert. "This instrument, which they had not heard before, would probably have made more impression on them if it had been put to use earlier."[61]

Testimony that the event aroused the elephants romantically was provided independently by the Museum's professor of iconography, Gerard Van Spaendonck, an eyewitness. A week before the Museum had the musicians' report in hand, the painter reported to *La Chronique universelle* the elephants' joy on hearing the music and how "the hilarity of these animals reached its peak and manifested itself by a kind of delirium, when the 'Ça ira' was played. The female was conspicuous by the continued movement of her ears and trunk and by that kind of cry that Buffon calls the cry of love."[62] This was the first published mention of the concert.

The musicians offered no general conclusion after the performance, instead asking that they be allowed to continue the experiment with another concert. The professors expressed their appreciation of the musicians' efforts and invited them to conduct another test when the new elephant park was ready and the male had recovered from a slight ailment.[63] Toscan for his part was thinking that the idea that elephants refuse to breed in captivity might soon be proved wrong. He concluded his essay with the encouraging news that several nights after the concert, Thompson heard a noise in the elephants' enclosure and quietly went to check. He found the female lying on her back with the male standing above her, his legs apart and making a motion. "As soon as they were discovered," Toscan recounted, "the male, very adroitly disengaged his legs, made a half turn to the side, making it easy for the female to get up." As Toscan saw it, "This observation, joined to the induction that I have derived from the spontaneous movements of the female during the experience of the concert, can no longer leave in doubt and controversy the copulating position of these animals."[64]

Thompson reported this was the first time he had surprised the elephants in this position. Toscan ascribed this development to "the agitation and excitement in which the music had put them," maintaining that under the proper conditions, in their new park, enjoying more freedom, the elephants might be induced to mate. In addition to music, the elephants' mood would be influenced, Toscan imagined, by ample and well-chosen food, by "the joy of seeing each other after a more or less extended separation," and by "the season of spring, which invites all beings to lovemaking." He did not mention Lacoste, who had said much the same thing. Toscan offered more refinements to set the romantic stage: "I would especially like to have the experiment made during a beautiful night illuminated by the soft rays of the moon; that the most absolute silence and solitude would be able to surround them; that they should see neither the musicians nor the *cornac* [mahout] himself; that no speaking voice struck their ears; and that only the melody of the songs and instruments made themselves heard."[65]

The musicians and Toscan were not the only ones to record their observations of the elephants' response to music. The young German naturalist Gotthelf Fischer, a friend of Alexander von Humboldt's and a student of Georges Cuvier's, provided an additional account in his two-volume description (in German) of the Paris Museum (1802–3).[66] Noting Toscan's account of the first concert, Fischer described "another time" when fewer musicians performed and the orchestra was arranged differently, at an angle and farther from the elephants. This new setup, Fischer presumed, allowed the elephants to focus less on the origins of the sounds and more on the pursuit of their "lusty sensations." He identified no song titles or keys and only one instrument, a flute. His perspective was that of a zoologist watching the elephants become sexually aroused by the music. Whereas Toscan had written of how "the fires of love insinuated themselves in [the male's] veins," Fischer was more explicit.

When the orchestra began playing, according to Fischer, a gentle harmony filled the room; the elephants stopped eating and displayed "attention and astonishment. They stood motionless as if they distrusted the sounds. Their trunks stretched out to the air, which was blowing the sweet tremor to their ears, to ask, so to speak, what was the cause of this strange emotion."[67]

The elephants settled down for a moment, but as the music continued, they emitted unusual, harsh noises from their trunks. In Fischer's blow-by-blow account: "The movements of their trunks became more varied. The female is now touching all the parts of the male's body; her ears flap like the wings of the male pigeon when it approaches its female. The sensations of the male are not absent either. His penis stretches straight to the ground. This is not hidden from the female. Her sensations now also reveal their seat more precisely." Fischer described the changes in the female's vulva, noting how it lengthened, protruded toward the rear between the female's hind feet, "and swelled more and more with beating and twitching movements."[68]

After debating whether the elephants' emotions were greater on this occasion than another (it is not clear if he meant the first concert), Fischer reported that in this case the female, aroused by the music, "probed the ears, the back, and even the belly of the male with her trunk." The male, "not in the mood to reciprocate her caresses," stepped away, sniffing and blowing powerful blasts of air through his trunk strong enough to spread lime from the wall far and wide. Then the male returned, the two animals "lay close to each other, and now the male showed the female the same attention. Now he touches the female's trunk, and both quickly embrace each other like snakes trying to mate."[69]

The flutist now played a single note in crescendo, while the spectators held their breath. By Fischer's account, the male experienced an erection and "the female caught fire. Her trunk probed softly, but quickly, all parts of his male body; even the penis that reached like a fifth foot to the earth. Now the female could no longer comprehend it. She showed herself to the male from all sides, even sat down, either to invite her companion more, or really to dull a little on the cool ground the unfamiliar stimulus brought out in her reproductive parts." Fischer's description of the effect of the music on the elephants ends at this point, other than to say that it gave the Parisian naturalists the hope of witnessing "a performance which, even in the wild, elephants conceal from their own kind."[70]

The artist Jean-Pierre Houël mentioned what appears to have been yet another concert, on 26 prairial an 6 (14 June 1798), sixteen days after the first one. In this case musicians from the School of Trumpets on the rue du faubourg Saint-Denis came to test the elephants' responses to wind instruments played individually and together. According to Houël's very brief account (published in 1803), the musicians produced a variety of sounds, including screeches painful to the human ear, which failed to move the elephants. Only when the musicians offered "harmonious and tender modulations" did the elephants respond, surrendering to the gentle inclinations these sounds awakened, "expressed by lavish caresses given mutually and without mystery."[71]

Houël, who wrote about and illustrated the menagerie's elephants more than anyone else, had much to say about their sexuality and, more generally, about them as living creatures. With the Museum's permission, he devoted two months to studying and drawing the animals, even spending the night in the menagerie so he could observe them sleeping, "lying on their sides with their legs outstretched and their heads resting on their ears."[72] He published a brief notice about them in 1798, followed by his 1803 volume, which featured illustrations, observations, and Lassaigne and Thompson's accounts of the elephants' history. He chose to portray these great beasts in nature rather than behind bars, because, he said, any "friend of liberty" would prefer to see them living free.[73]

Among Houël's illustrations was a speculative one of the elephants mating face to face, with the female on her back and the male on top, as Toscan (and Buffon earlier) had proposed.[74] Houël suspected, however, that they also mated like horses, cattle, and other quadrupeds, and he had seen Parkie present herself to Hans in both positions.[75] He concluded that Hans was simply too young and inexperienced to know how to respond to Parkie's advances, and that modesty had nothing to do with it. Parkie had offered herself to Hans in a most immodest way, on her back with her legs extended, in the presence of Houël and more than a dozen additional witnesses.

Humans might promote modesty for reasons of public morality and maintaining social bonds, Houël wrote, but there was no reason to imagine the elephant would be more scrupulous than other creatures like "the vigorous bull that, under the eyes of the innocent shepherdess, mounts the cow she is leading; the donkey that in the middle of towns shows to everyone the astonishing signs of the gratification of his passions; the dog that in the streets disputes twenty rivals for the object of his love. . . ." His list went on, concluding with the admonition: "Let us banish these absurd suppositions about modesty in the acts of animals that, while endowed with intelligence, have not received the blessing of reason."[76]

Although he upheld the distinction between intelligence and reason,

Houël comfortably credited Hans with substantial mental faculties. When Hans extended his trunk through the bars of his cage toward him, Houël reported, "I interpreted this gesture as a request. I looked around to see if I would have the means to satisfy it. A box of carrots caught my eye. I took it and chose on purpose to give him first the smallest one. He took notice of it, and did not disdain it, but he wanted me to appreciate that such a small object did not at all offer a meal sufficient for his appetite; that it was not worth the effort of the considerable movement he would have had to make to carry it to his mouth, and that he had higher hopes." Houël detailed how Hans held this carrot with part of his trunk, did the same with a second carrot, and carried these to his mouth only when he had a third. With larger carrots, wrote Houël, Hans could be satisfied with two or even one at a time.[77]

Houël also described the elephants' moods, reporting how they responded with an almost drunken joy to being led outside in the garden on warm, sunny days. He interpreted these playful movements as an index of "this love of liberty, innate in all beings, and which the habit of captivity knows not how to extinguish."[78] He told how Hans became furious when Thomson gently chastised him for responding too roughly to Parkie's amorous attentions, and how Parkie liked to amuse herself by annoying the people who lived on the other side of her wall.[79]

A work of great care, the volume Houël published in 1803 came well after lesser elephant-inspired publications. One author had his pamphlet ready for sale at seventy five-centimes three weeks after the elephants were installed in the Régie.[80] L. F. Jauffret, who already had a children's guidebook/storybook about the Jardin des Plantes forthcoming, added an elephant chapter in time for the work's publication in June 1798.[81] A retired lawyer named E. J. B. Vignier wrote, "Moral virtues of the two elephants, male and female, newly arrived at the national menagerie of the Jardin des Plantes," followed by "Epistle to the Elephants," featuring a flowery poem and, like his previous booklet, elephant lore lifted from Buffon and others.[82]

The elephants' park was ready by the end of August 1798, connected to their stables by a passageway so they could move freely back and forth.[83] At last they could get some healthy exercise, the press observed. The *Moniteur* noted that they enjoyed spraying themselves with water from a trough. *La Décade philosophique* remarked that planting the park with grass and trees would enhance the elephants' "feeling of liberty" and give their instincts more play.[84] Not every day in their park was happy. In June 1799 the press reported that someone had let a dog loose there, causing Hans to slip and injure himself.[85]

Absent from press coverage were the Museum's dire warnings to the government in 1798 and 1799 about the difficulties of finding food for the elephants and the professors' having to pay for it out of their own pockets. In July 1799, the most frequent elephant-related items in the press were reviews of a "bluette" titled "The Concert for the Elephants" performed by a troupe of five actor/singers at the Théâtre du Vaudeville. The *Almanach des spectacles de Paris* praised the piece's delicate poetic couplets, gentle barbs, comical scenes, happy dénouement, and eulogy to Buffon at the end. The *Mercure de France* was equally enthusiastic, noting that if it cited all the lyrics that won applause, it would have to publish the whole piece. The *Journal de Paris* called the opening night a complete success, despite the simplistic story line. Others agreed.[86]

∵

The elephants were the first of several charismatic menagerie animals to be invoked for theatrical fun and commentary. The professors, however, took the elephants very seriously as symbols of the Museum and its ability to care for its collections in a manner that favorably impressed the public and government. But they did not endeavor, as scientists, to study the animals' behavior, test their intellectual abilities, or offer suggestions on how to get them to mate in captivity. Geoffroy Saint-Hilaire was the naturalist most closely connected to the animals in terms of his professorship, but for nearly four years he was unable to study them. They had scarcely arrived when he announced (3 April 1798) his departure on a secret government mission. When the elephants heard the strains of "O ma tendre musette," Geoffroy was sailing from Toulon as part of Napoleon's scientific expedition to Egypt.

Nor had Geoffroy returned when Lamarck, as acting supervisor of the menagerie, reported to the professors early in November 1801 that the male elephant had experienced a "rather serious indisposition." A caustic fluid flowing abundantly from the openings beside his eyes had enraged him and taken away his appetite.[87] Though he seemed to recover, the "indisposition" returned two months later, and on the morning of 17 nivôse an 10 (7 January 1802), Hans breathed his last.

The Museum prepared an official report for the first consul, the minister of the interior, and the press.[88] It stated that days before the elephant's death, he suffered from the flow of fluid from his temples to his cheeks and had also released quantities of semen by making a certain motion with his thighs, a habit he had developed over the previous two years, despite the Museum's efforts to make him stop.[89] He ate no breakfast the day before he died, though he ate a little hay and some potatoes in the late afternoon. At sundown, he "broke one of the iron bars of his cage and used it to strike violently the gates of his enclosure." When given water, he drank some and used part of the rest "to bathe the openings around his temples, where he seemed to have more irritation and pain than ordinarily." The end was rapid: "During the night he made much noise; around four o'clock in the morning he let out some sharp cries and toward six o'clock he fell down dead."[90]

The Museum stressed that the elephant's death had not resulted from inadequate attention from the institution. His cage seemed warm enough for comfort, he had been receiving the same quantity and quality of food as usual, and there had been no slackening of care. The Museum was undertaking anatomical investigations in hopes of shedding light on the cause of death and, more generally, to provide the first detailed anatomy of the Indian elephant. It had "confided to Citizen Cuvier the anatomical examination of this animal and put at his disposition all the means appropriate for aiding his research; accomplished painters will represent all the parts that are still little known, and this work will be communicated to the Institute."

The Museum told the minister of the interior, "This event has caused us much pain. But we attest to you that it was caused by no lack of attention either on our part or on that of the persons and employees of the menagerie."[91] The male's companion, according to the official report, had also suffered: "She seemed strongly affected by the death of her male. She tried to lift him up again from the place where he had fallen. She shed abundant tears and emitted cries different from those heard from her up to this moment." The Museum went on to say that she had been separated from the male, and she had begun eating again, "but she often looks into the enclosure that her male inhabited." A professor proposed that a camel be put there so that it might soothe her suffering, and his colleagues agreed.[92]

No one was better prepared to perform a scientific examination of the elephant's remains than Georges Cuvier, who six years earlier had assumed the duties of Jean-Claude Mertrud, professor of animal anatomy, and would replace him later in the year. Not only was Cuvier a skilled comparative anatomist, he was also an expert on elephants, a species that had figured prominently in establishing his reputation. Two elephant skulls from the Stadtholder's natural history collection, one from Ceylon (Sri Lanka) and the other from the Cape of Good Hope, had allowed him to make the case that Asian and African elephants are sufficiently different in their skull shape and the structure of their molars to be considered distinct species.

He delivered his argument (January 1796) in a memoir on living and fossil elephants as a newly elected (and youngest) member of the new Institut de France. His first paper for the Institute made such an impression that his colleagues invited him to present it again in April on the prestigious occasion of the Institute's first public session, attended by a host of dignitaries.

In addition to arguing that the Asian and Indian elephants were different species, Cuvier identified the mammoth of Siberia and a fossil elephant found in Ohio as distinct species as well. Comparing other fossil and modern forms, he proposed that a whole world of prehistoric creatures had once inhabited the Earth but had since vanished completely. Petrus Camper and Johann Friedrich Blumenbach had already identified the African and Indian elephants as separate species based on differences in their teeth, but Cuvier went further, comparing the remains of modern and ancient forms to gain insights into the Earth's distant past.

Another significant feature of Cuvier's lecture for the history of science is that it represented the first time he clearly rejected the idea that one species could be transformed into another. After explaining how the differences between Asian and African elephant skulls and teeth proved they were separate species, he pronounced: "To say [that climate] could change all the proportions of the bony frame and the deep texture of the teeth would be to maintain that all the quadrupeds could derive from a single species; that the differences they display are only successive degenerations. In a word, this would reduce natural history to nothing since its object would consist only of variable forms and fleeting types."[93] It was bold to suggest that the validity of his practice as a scientist depended on the immutability of species, and ultimately he would be proved wrong on both counts—both that species are immutable and that natural history would be adrift if the idea of species change came to be generally accepted—but that proof was not clearly established in his lifetime.

While Cuvier's work in comparative anatomy and vertebrate paleontology depended largely on skulls, teeth, and bones, he paid a certain amount of attention to animals' lives, their other distinctive features, and their "histories" in the essays he wrote for the handsome grand in folio illustrated work, *The Menagerie of the National Museum of Natural History, or Description and History of the Animals That Live or Have Lived There*, which was published in installments. Cuvier's erudite, elegant, and judicious piece on the Indian elephant appeared in 1801 in the second installment.[94] He discussed the elephant's anatomy at length and provided an overview of its natural history based on the best writings from antiquity to the present, including information on how elephants are captured. Though he focused largely on dead specimens, he also noted the food and water consumption of the menagerie's live elephants and gave a concise report of their mutual affection:

> These two individuals have the most tender attachment to one another. When one of the two experiences some fright, the other comes running

> to its side. It is especially when they are struck by some object that is new to them that their caresses redouble in intensity. Then they run from one side to the other, they cry out, they caress each other with their trunks. The male gives the signs of an ardor that is ordinarily very strange to him. These motions were never more marked than upon the elephants' arrival in Paris when they found each other again after a long separation. There was even a moment of hope that they would arrive at a real union, but this hope has been disappointed up to now.[95]

Cuvier did not read too much into the effects of music on the elephants, saying "they felt exactly the same things when they were given a concert as when a bathing pool was constructed for them." (Alas, no account remains of their response to the pool!) The sounds of the instruments apparently affected the elephants, but no more than the effect such sounds made on dogs. The elephants' responses "were limited to some roars, some cries, and some rhythmical jumps." He nonetheless acknowledged that their "romantic attempts that day were very strong and frequently repeated, albeit without success."[96]

Cuvier disabused his readers of the idea that elephants adopt a special posture when mating. He cited a paper in the *Philosophical Transactions* of the Royal Society of London saying that the coupling of elephants was (in Cuvier's words) "entirely like that of the horse and lasts approximately the same time."[97] With Buffon's disparaging comments about the male elephant's organ surely in mind, Cuvier went on to observe that "the penis becomes lengthened so much that it nearly drags the ground and has six to eight inches in diameter. Those who have pretended that it was not proportioned to the magnitude of [the elephant's] body, have undoubtedly never seen it in this state."[98]

What Cuvier did not mention, and perhaps did not know, was that as of 1800 Frederick II's menagerie at Schönbrunn had a pair of elephants, purchased from the animal exhibitor Antonio Alpi who had bought them in 1799 from Gilbert Pidcock in London.[99] Not only did this pair match what those in Paris had done, in five years they proceeded to do what Hans and Parkie did not. They mated in human view, destroying any vestige of the notion of elephant modesty and demonstrating that elephants copulate in the same manner as other quadrupeds.[100]

Cuvier's essay on the Indian elephant appeared the year before Hans's death. His demise gave the anatomist the chance to thoroughly examine an elephant's remains, though Cuvier was not the first to do so. Joseph-Guichard Duverney famously dissected the elephant that died at Versailles in 1681. The story has it that the king, upon arriving on the scene, failed to see the anatomist and asked where the man was. Duverney emerged from inside the carcass, scalpel in hand, and thanked His Royal Majesty for his interest.[101]

Cuvier did not climb inside the elephant's belly, but he oversaw the anatomization, and he concluded that the cause of death was a form of pneumonia. His naturalist's assistant, Emmanuel Rousseau, performed the messier aspects of the job and received a two hundred franc bonus (more than a month's salary) for "the dangers he ran at the time of the dissection of the male elephant." The insupportable stench of the decaying body was part of the affair.[102] Cuvier had the artist Maréchal make careful drawings of the animal's parts, and over eighty illustrations were produced.[103]

Dissecting the elephant took about forty days, but preparation of the remains for long-term display took longer still. The detached skin, weighing a prodigious 568 pounds, had to be sent to a tanner for preservation. The bones were cleaned in maceration barrels where chemicals ate away any flesh, a process not without "inconveniences." The Museum responded to complaints about the stench by transporting the materials "to a place where they cannot bother anyone."[104] Louis Dufresne, the Museum's immensely capable naturalist's assistant and taxidermist, oversaw the mounting of the prepared skin in a lifelike form, working closely with Claude Lassaigne, who built the wooden frame needed to support the elephant's skin.[105]

In October 1803 Geofroy Saint-Hilaire announced to his colleagues that the elephant's skin had been placed on the wooden manikin and the result was perfectly lifelike.[106] Six weeks later, Geoffroy reported that the elephant was "completely prepared" and asked permission to place it in the Museum galleries. The professorial assembly invited him "to take the swiftest measures to let the public enjoy the sight of this great animal."[107] Geoffroy asked permission to purchase a pair of tusks to replace the elephant's broken ones. He also asked that a new workshop be built for Lassaigne because his current one would have to be demolished to get the mounted elephant out.[108]

As arrangements for the specimen's move to the natural history galleries were finalized in January 1804, Lassaigne sent every member of the Museum administration an engraving of a picture of the wooden manikin he had made, noting that its artifice was "totally invisible" (thereby enhancing the lifelike illusion). The assembly ordered that a copy be deposited in the collection of *vélins* alongside the picture of the male elephant. In addition, a copy of the engraving was to be framed and placed beneath the mounted animal when it was installed in the galleries, an event that took place a full two years after Hans's death.[109]

The *Journal des débats* described the achievement: "The skin of the elephant has been placed on a body of wood resembling perfectly that of this enormous animal; all the proportions are well observed, as the carpenter charged with this work spent ten months achieving it. This model elephant now only lacks eyes to appear alive. It is on its feet and seems ready to walk. It can be seen at the end of the quadruped room, between several rhinoceroses and a giraffe."[110] Here, then, was Hans the elephant, lifelike in his "afterlife," identified not as an individual but as a representative of the species, and sporting a new pair of tusks. Soon he would have eyes of porcelain as well.[111]

As soon as Hans died, the Museum and the government made replacing him a priority. The new elephant, which was finally acquired in September 1803 after several false starts, lived only until July 1804, when it too died and was dissected.[112] Cuvier's anatomical studies of both elephants appeared together in a long memoir (1806) comparing elephant species of the present and past.[113] Though he identified prosaic causes for the deaths of the menagerie's two elephants, he thought such causes were insufficient to explain the total annihilation of species that had once walked the earth. They had perished, he said, either by a "revolution" that killed them all "or by a change of climate that prevented them from propagating." Whatever the cause, he insisted, "it must have been sudden."[114]

Among the varied lenses through which the first elephants of the Paris menagerie were seen, Georges Cuvier's was grounded in comparative anatomy. He viewed the elephants as representatives of their species rather than as individuals. For his purposes, their bones mattered more than their behavior. In contrast, Jean-Pierre Houël, who documented the elephants in drawings of scrupulous scientific accuracy, thought of them as individual beings with personalities, and in writing he referred to them by name. Georges Toscan, for his part, perceived the elephant pair as sentient, emotional creatures, moved by music and by love for each other. He recognized that they enjoyed what liberty they were permitted, but rather than picturing their greater enjoyment in the wild, he wanted to see them domesticated. Though he did not put it that way, he understood humane treatment as the key to human dominion over the animate world.

Houël and Toscan's accounts of the elephants are familiar to historians, and Georges Cuvier is a figure of great renown. Much less is known about the carpenter/mechanic Claude Lassaigne, who had the longest relationship with the elephants of anyone connected to the Museum, even James Thompson, who was their keeper from 1788 to 1803.[115] Lassaigne's connection to the elephants lasted twenty-one years, beginning in 1796, when he joined the mission to transport them from Holland to Paris, until 1817, when Parkie died.

Though he initially served as a carpenter and wheelwright building the wagons for the elephants' transport, while moving the elephants to Paris Lassaigne ended up caring for the great beasts, securing food for them, befriending them, observing their emotions, and worrying about their preservation. In the depths of winter in Cambrai, he displayed not only his dedication to the mission and the public good but also his love of the elephants. He even sought to give Parkie a new French name, Marguerite.[116] After his return, the Museum rewarded his efforts with promotion to a salaried post.[117] In Paris, Lassaigne continued to help care for Hans and "Marguerite" and serve as an interpreter for Thompson. When Hans died, Lassaigne designed and built the manikin that made it possible to display the elephant in a lifelike posture.

When the time came to move Parkie/Marguerite from the Régie to the newly completed rotunda (1814), she refused to be led on foot, so the Museum called on Lassaigne to build a wagon for her, as he had once before. When she died (1817) he built her the same kind of wooden frame he had made for Hans. Having served the elephants during their lives and benefitted from them in his career, he helped ensure their posterity. In the natural history gallery, the realistic rendering of Hans and the framed engraving of the manikin spoke to a taxidermic achievement that represented one of the many intersections of animal and human lives in the history of the menagerie. Lassaigne had touched the elephants, and they had touched him in manifold ways.

[1.10]

The Keepers

The first animal keepers at the Paris menagerie were an independent, enterprising lot. Having made their livelihoods by exhibiting exotic animals at regional fairs and on city streets, they were savvy about both animal and human nature. They knew how to care for their creatures and how to extract money from people who came to see them and hear stories about them. When the Museum hired these men, from fall 1793 to spring 1794, they had to adapt to being employees. Their habit of soliciting tips from visitors, for instance, went against the Museum's rule that all exhibits were free and open to the public.

When the first former animal proprietors and their animals showed up at the Museum's gates, the professors decided to have them continue to care for their own animals. In this way the Museum gained expertise, the keepers earned their living, and the continuity in caretakers helped ease the animals' transition. The Museum paid the keepers according to the work and resources their animals required. Dominique Marchini and Bernard Louzardy each received four livres per day, Marchini for his leopard, polar bear, civet, and monkey, Louzardy for his polar bear, two mandrills, and a vulture. Henri Gueny was paid three livres per day to care for his serval, two eagles, and two agoutis. Marchini's assistant, a young man named Amé, received two livres daily.[1] This practice of differential remuneration continued as more animals were confiscated and more former proprietors were added to the payroll.[2]

Initially each keeper was responsible for acquiring food for his animals, but that proved inefficient and difficult to monitor. The post of "keeper of the animals' provisions" was created in May 1794 to address the problem. The most striking thing we know about Pierre Guichenot, the man chosen for the job, is that he and his wife had twenty-six children. He was obviously a man who knew something about provisioning.[3]

Félix Cassal, a rugged former lion tamer, joined the staff in June 1794.[4] Early in his career he had worked in Orléans for the animal fights (*combat du*

Figure 1.10.1. "Animal keeper tending camels operating the pump for the Museum's well." Andrew, Best, Leloir after Karl Girardet. From François Boitard, *Le Jardin de Plantes* (1842). Author's copy.

taureau). Lately he had been in Tours, exhibiting his own mini-menagerie: a panther, a pair of mandrills, a female monkey, and a yellow-crested cockatoo. When the Revolutionary Council closed his show and ordered him to deliver his animals to the Museum in Paris, Cassal did as instructed, evidently sensing an opportunity. The Museum hired him at four livres per day.

Several other animal owners became keepers at the menagerie, but most did not stay long. One, named Cochon (pig), was on the books for only a month. Marchini's helper Amé left after eleven weeks to seek his fortune in North America. A keeper named Dallair stayed for nearly three months but was thanked and dismissed in August 1794 (along with Henri Gueny) when Geoffroy Saint-Hilaire proposed that Marchini, Louzardy, and Cassal be designated the Museum's official animal keepers. Marchini and Cassal were elevated to the status of salaried employees, with Marchini named "first keeper" and Cassal "second keeper." Louzardy remained on the books as a day worker. All three were paid four livres per day.[5]

Marchini resigned in June 1795 for unspecified reasons but possibly because his leopard, civet, and monkey had died and only the polar bear remained of his 1793 troupe.[6] Other likely factors were the escalating cost of subsistence in Paris and the difficulty of dealing with the volatile Félix Cassal. After Marchini's departure, Louzardy became in effect the keeper of the "peaceful animals" while Cassal was charged with the care of the *animaux féroces*."[7]

Cassal and Louzardy were barely literate, but they left an indelible mark on the menagerie's early history. They found others to record their words when needed, and their actions and ideas drew comments from other writers. More keepers were added to the menagerie's staff in the last years of the century, but Louzardy, Cassal, and Guichenot were the mainstays of the menagerie throughout its first decade.

∴

The Museum's early records of Bernard Louzardy concern his wages and the value of his animals. The first personal information appears in a letter written on his behalf in April 1795 informing the administration that as a father of four, Louzardy was finding it "absolutely impossible to live with the modest wage of four livres that you are giving him." While pleading for financial assistance (but only, he said, for as long as the cost of living remained so high), the keeper sought to convey his broader usefulness and resourcefulness. He offered to travel to Belgium on foot and bring back a pair of dromedaries confiscated from the former prince de Ligne's château at Beloeil. In addition, he recommended two animals available for purchase: "a superb wolf from the Pyrenees" and "a genuine Angora he-goat," but he cautioned that the price demanded for a *hocco* (crested currasow) was exorbitant.[8]

The Museum and government approved Louzardy's travel to Belgium to collect the dromedaries. On his return, he asked for a small apartment at the Museum and reimbursement for extra expenses incurred on the expedition.[9] The Museum agreed to both.[10] Soon he was sent on a larger mission to Holland to fetch the animals from the Stadtholder's menagerie, elephants included. He did not report frequently. As he noted on a later trip, "I won't write to you from all the towns I reach because I don't know how to write."[11] When he did send letters, even with someone else writing for him, his messages could be short on pronouns and wholly lacking in punctuation, as exemplified in the following passage, translated verbatim from a letter he sent from The Hague in October 1795:

> I am today Thursday at the Hague I presented myself before the representatives of the people and told me that had not received any letter from the Committee of Public Instruction nor from the Jardin des Plantes that the representative of the people told me that he had written and that he is surprised himself to have no response and he told me that I had to wait while the order came and I have no money he neither and told me that I await his colleague who must come from Bruxelles and he will see if he has more power than he and then one will decide if I can lead the animals. . . .[12]

However patchy, the letter conveyed the government ineptitude that resulted in the failure of Louzardy's first trip to Holland; money for the mission was not released.

In April 1796, when funds finally materialized, he set off to Holland again, this time with Claude Lassaigne. Louzardy returned to Paris without the elephants (or Lassaigne) but with a ten-wagon convoy of other animals. These he had moved primarily by night to prevent them from suffering unduly from the summer heat. He chose land rather than water travel to spare them from being tossed about too much in their cages. Ten days from Paris, writing from Valenciennes, he reported to menagerie supervisor Delaunay that the convoy would arrive soon, that the animals were in good health, and that he had done everything in his power to fulfill his mission in a laudable way. But not everything had gone smoothly: "You have written me that I could collect money in the villages I passed through not at all I have been refused at Valenciennes." Fortunately Louzardy had a cousin who lent him money and traveled ahead of the convoy to make advance arrangements.[13]

Louzardy returned to the Museum on 31 August 1796 with seven axis deer, three Angora goats, one porcupine, and thirty-six birds of six different species.[14] In October the Museum considered Lassaigne's request to have Louzardy sent back to Holland to assist the departure of the elephants.[15] That departure was delayed, and it is unlikely Louzardy went.[16] He was not part of the trip that finally brought the elephants to Paris. Instead he stayed in Paris, where he substituted for Cassal as keeper of the *animaux féroces* while Cassal was on his mission to Africa.[17]

However marginal his writing ability, Louzardy's ear was tuned to news of exotic animals that might be had for the menagerie. It was he who learned of the dromedaries in Belgium in 1794, and it was he again, evidently, who called Delaunay's attention in 1800 to the collection of big cats that London's most illustrious animal dealer, Gilbert Pidcock, was offering for sale. In August 1800 Louzardy was sent to London with Delaunay to buy the animals in question. Hoping to defray some of the expense, the administration allowed the men to display the animals for profit on their way back. Their first attempt was during a several-day stop in Brussels. Writing to Fourcroy, the Museum's director, Delaunay admitted having attempted "to excite the curiosity of the Bruxellois" by putting up placards that had "a bit of charlatanism" about them. He feared being seen at the door, should people later recognize him in Paris, worrying that it might be bad for the Museum's reputation. But Louzardy was in his element, in Delaunay's words, a "true man of the fair," putting on a show "in a manner as amusing as it is bizarre."[18] What Delaunay would not have dared came naturally for Louzardy. It was undoubtedly Louzardy who thought of hiring a drummer to announce the animals' arrival in town, a crier to draw an audience, a cashier to take money at the door, and, inside the hall, someone to make sure those who had paid for the cheaper seats did not sneak into more expensive ones.

A few months after the trip, Louzardy offered to correct what Delaunay had considered its main shortcoming, the failure to acquire kangaroos. Louzardy told the Museum that a London agent had purchased for him "two kangaroos of the large species, male and female," and that he was sure they would serve "the embellishment of the menagerie and the progress of natural history." The Museum rejected the idea, finding the 160 guinea asking price too high.[19] When another purchasing expedition to London was set up in 1802, it developed through Louzardy's contact with Pidcock, and Louzardy this time joined the taxidermist Dufresne and the carpenter Lassaigne for the journey.

Louzardy was put in charge of the elephant, Parkie, when Thompson retired in the summer of 1803. Then, when Cassal was fired at the end of the year, Louzardy was awarded the post of head keeper of the *animaux féroces*. Always alert to improve the menagerie, Louzardy and his assistant (Magloire Rousseau) purchased an American black bear on their own initiative in 1809. The Museum acquired it from the two keepers.

Louzardy remained as head keeper of the carnivores until his death in 1813. The Museum's papers mention his wife (who died in 1808) and their six children. The Museum employed his three sons at one time or another in various capacities. At least one helped his father in caring for the carnivores and had his hand mauled by the tiger when he made the mistake of trying to pet it while it was eating. Another was employed in the zoology lab but was dismissed when he was suspected of theft. All three were conscripted into the army.

Though the administration recurrently complained about the keepers' independent behavior (and Louzardy was surely not without such transgressions), no specific complaints about him from the administration or the public are to be found the Museum's papers, which is not the case for many of the other keepers, particularly Louzardy's counterpart Félix Cassal.

∴

We know somewhat more about Cassal than about Louzardy, including his appearance, thanks to a portrait of him with his lions made in 1800. We have letters he wrote (or, more precisely, had written for him) to the Museum administration. Various authors discussed his talents and techniques as an animal tamer and cited his observations on animal behavior. He was mentioned in guides to the Museum and in the memoirs of one of Josephine Bonaparte's friends. He was the subject of complaints from individuals outraged by one thing or another he had done.

An instructive introduction to Cassal's interests and character comes from a book for young readers, Louis-François Jauffret's 1798 *Voyage au Jardin des Plantes* (*Trip to the Botanic Garden*). Offering an overview of the wonders of the Jardin des Plantes as seen through the eyes of a fictional boy named Gustave, the guide presents a conversation with Cassal, who explains that there are two types of animals: ferocious and peaceful. "The menagerie is a prison," he says, "where only those individuals who abuse their liberty are locked up." In contrast, the "innocent and peaceful" creatures can be allowed more freedom, separated from the public by simple fences. When Gustave remarks that his job seems dangerous, since the prison is filled with creatures that might revolt at any time, Cassal replies that he wishes there were more and is ashamed to have only ordinary animals to show the public. He tells Gustave that what he most desires is to be sent to Africa to collect lions and other fierce creatures for the menagerie. Confident of his success, he says, "I have spent my whole life with the most ferocious animals. I know how to tame them, and besides, it is less difficult to tame tigers than it is to tame certain men."[20]

Though Jauffret invented this dialogue, it is in keeping with Cassal's character and intentions. He thrived on his work with wild animals, reveling in the way his entering a dangerous animal's cage inspired spectators with fear and awe, and he recognized that his services would not be needed if the menagerie had no more dangerous animals. These motives (plus an expectation of financial gain) led him to propose in 1796 that the Museum send him to North Africa on a collecting trip.

The expedition did not go as well as planned. When the minister of the interior agreed to support Cassal and the Tunisian ambassador proposed that Cassal travel with him to Tunis, the prospects seemed excellent; but to Cassal's dismay, the Tunisian ambassador delayed the departure from Paris by nearly two months and then humiliated Cassal by treating him as a domestic servant.[21] The wounding of his pride was but a minor mishap, however, compared with the situation Cassal faced in Tunis. Plague ravaged the city, multitudes were dying, and neither the French consulate nor any inn would give him lodging. He was left to find a small apartment and told it would be another two to three months before the pestilence abated.

Writing to the Museum, Cassal expressed his exasperation: "If the captain of [the ship that brought me] had wanted to take me back on board, I would have returned with him, but since I had already disembarked, he would not allow it."[22] The French consul in Tunis wrote to the Museum and to the minister of foreign relations confirming that Cassal had arrived at a bad time: "Tunis and its environs are afflicted by the plague, which makes communications very difficult and dangerous. Algeria is experiencing the same calamity with more devastation." It was already early May, and it seemed that Cassal might not be able to go anywhere before mid-August.[23]

Cassal tried to keep the Museum informed of his travails in a series of letters, but some of them were rendered useless when the ink was washed off by the vinegar used to disinfect them.[24] Impatient for action, Cassal set off for La Calle in mid-July and was well received by France's principal agent there, a man named Peron, who made him quarantine for eighteen days. Peron meanwhile wrote to the leaders of neighboring countries, asking them to assist Cassal's mission. Cassal wrote to the Museum with cautious optimism: "The country abounds in porcupines and hyenas. We will do everything possible to get some. I will go to Bône and from there to Callo, where I will find monkeys. At Constantine, where I hope to have the protection of the bey of Constantine, to whom the agent promised to write in my favor, I hope to get some lion or tiger." But Cassal continued to fret about lost time, fearing that if he did not get a lion or a tiger in Constantine, he would not get one anywhere: "The season is already well advanced, and the Arabs do not dare confront lion cubs that already follow their mother. If I had not waited for the ambassador of Tunis, it is certain that I would already have in my possession a large number of the animals you want for the menagerie."[25]

Had he arrived sooner in Tunis, Cassal later told Delaunay, he would have had the chance to acquire animals from the bey of Tunis, but as it was, the bey had already sent them to Constantinople. When Cassal reached Constantine, he found he was again too late. Two weeks earlier, a local Arab chieftain (or perhaps the bey of Constantine himself: Cassal's account is unclear on this score) had sent two panthers and a leopard to Mascara. Cassal managed, nonetheless, to purchase a young lioness. Subsequently, with Peron's help, he received a remarkable pair of three-year-old lions, male and female, from the bey of Constantine as gifts to the French Republic. Later, as a gift to France the prince of Sala-Bey presented Cassal with a small but extremely savage lioness from the Sahara Desert.[26]

When Cassal returned to Tunis, the Tunisian government presented him with more animals for the menagerie: a pair of ostriches, a pair of white camels, and a gazelle. Awaiting him were four vultures he had acquired earlier. He then purchased three monkeys and yet another lioness before sailing for France in November 1797. His collection consisted of seventeen animals for the Museum plus three horses and a gazelle he had acquired on commission for other parties.

After the twenty-day crossing from plague-ridden North Africa, Cassal was forced to undergo a forty-day quarantine in Toulon, where he chafed at the delay and begged Delaunay to help him leave for Paris as soon as possible.[27] The administration wanted him to remain in Toulon until spring to avoid the dangers of winter travel, but as he told Delaunay, "I know the precautions one must take to guarantee these kinds of animals, and I hope the Museum will leave it to my experience and to the procedures I would use." He promised to deliver the animals to the Museum "safe and sound."

Cassal's appeal was refused, and his stay in Toulon was prolonged for another forty days after quarantine. Told to detour to join the convoy laden with the art treasures from Italy, he protested, saying that time made no difference to artworks, but animals were another matter.[28] Ultimately he was not made to change his route but was obliged to wait in Avignon for the convoy to catch up with him. André Thouin saw him there and relayed the government's and the Museum's admonition "not to expose his animals to the regard of the curious." Thouin was unsure if Cassal would follow these instructions but was confident he would safeguard the animals in his charge.[29] After the convoy from Italy reached Avignon, Cassal and his animals traveled with it by river to Paris, arriving on 27 messidor an 6 (15 July 1798), eighteen months and twenty-two days after he had left.[30]

Thirteen days later, five of Cassal's animals (the lion pair, the Saharan lioness, and two camels) featured in the "triumphal entry of the objects of science and art collected in Italy" that launched the "Festival of Liberty."[31] Cassal was doubtless in the parade, walking proudly beside a lion-bearing wagon. His expedition to Africa required an extraordinary combination of self-confidence, courage, skill, entrepreneurship, and knowledge. It was also unprecedented. After 1815, the Museum had agents abroad keeping an eye out for live specimens as well as dead ones, but it had no one quite like Cassal collecting animals in their native lands and bringing them home himself. The era of big-game hunters filling commissions for European zoos did not begin until the 1830s at the very earliest, when new zoos needing animals began coming into existence.

However substantial his contribution to the menagerie, Cassal admitted that his mission had not had "all the success I would have desired." He had wanted to capture animals in the wild with the help of local hunters, not just receive gifts from African rulers or purchase specimens from animal dealers. Too many factors, in addition to the plague, had conspired against him. He offered three "excuses" to Delaunay. The first was timing: arriving in North Africa as late as he did, when the young animals he had hoped to capture had grown too large to be readily trapped, "an evident danger which the Moors did not want to risk despite my promise and my example." Second, long-standing internecine warfare among local tribes had driven the animals to more distant, inaccessible mountains. Third, the Arabs had set fire to the woods in the mountains around Calle, making it impossible to venture there.[32] Cassal insisted that he had pursued his goals with the utmost zeal and that had he not endured flaming deserts and braved "the plague and the brigands that infest this country," he would have returned empty-handed.[33]

Excuses and disappointments aside, the collection Cassal brought back from Africa was of genuine interest. The three-year-old male lion, in Cassal's opinion, was "superb," and the lioness of the same age looked to be a good candidate for cohabitation with the male if the menagerie could arrange the animals' cages appropriately. The smaller lioness, he said, was, "the most ferocious I've seen in my life." A female lion cub completed Cassal's collection of big cats. The Museum considered this more lions than necessary, but it called on the minister of the interior to reimburse Cassal for all his expenses: the purchase and transport of animals, cages, and other equipment; the cost of feeding and lodging the animals; and the five francs per day allotted for Cassal to feed and lodge himself. Cassal's receipts were jumbled, but Thouin and Jussieu combed through them and concluded that after subtracting the 1,100 francs he had received in advance, he was owed 7,397 francs, 95 sous. The two professors proposed selling or trading one large and one small lioness, given that the menagerie already had a beautiful lioness, it did not have space for all the new arrivals, and big cats were expensive to feed.[34]

When Cassal's reimbursement proved slow in coming, he confessed to the Museum that his own resources had been insufficient to front the trip's expenses, and he had had to resort to borrowing money from "friends." Now that he was back in Paris, they were threatening him. They were not satisfied with the interest he was paying, and they wanted the capital back immediately. Cassal begged the Museum to intervene with the government so he could be reimbursed and satisfy his creditors as soon as possible.[35]

Cassal was likely the only animal keeper in this kind of financial fix, but the others all knew general impoverishment. The highest earners as of 1800 were Thompson, at 2,400 francs a year, and Cassal and Louzardy, who each had an annual salary of 1,500 francs. Others had to manage with much less, as the case of Ballet demonstrates.

Ballet, after first working in the gardens, moved to the menagerie in the spring of 1796 to fill in when Louzardy went to Holland. He continued there after he was injured by a stag and after Louzardy returned, but with his old wages as a gardener.[36] The following year he asked for a raise, saying that a keeper's job was more dangerous than a gardener's; it required more labor and diligence, was harder on his clothing, and left him less time for rest. He wanted to be paid like Cassal and Louzardy, or at least to get a raise sufficient to support his wife and children. He also wanted an apartment, to be "close to my work" and thereby to be on hand "to prevent or to take care of accidents," or "to receive the groups that present themselves or that are often sent by the professors when I have gone to lunch or dinner."[37]

Ballet received an apartment, but as of January 1798 he was still earning a gardener's wage and still filling in for Louzardy (who for the previous ten months had been minding the carnivores while Cassal was in Africa). The former gardener asked to be added officially to the menagerie staff, noting that "everyone who cares for the animals has, in addition to his wages, the dung that the animals leave, and I do not have this sort of profit."[38] That request was granted, but Ballet's argument that he should be remunerated at the same level as Louzardy and Cassal was unsuccessful. His work was as arduous as theirs, he said, and it also required "more vigilance and moving about, since the animals confided to me are dispersed throughout the parks, stables, and aviaries."[39] Supporting a renewed plea by Ballet late in 1800, Delaunay observed that when the man's widespread duties were too much for him, his wife had to share the burden.[40] Finally, in September 1801, the Museum put Ballet's name in the account books as a salaried worker rather than as a day laborer, but still at a much lower rate (1,000 francs annually) than Louzardy and Cassal.[41]

Under these conditions, it is not surprising that the keepers did what they could to supplement their wages. This was already the case in 1795, when Delaunay was appointed to oversee the menagerie and the keepers, who had been left to their own devices, nor was he able to entirely suppress their entrepreneurship. Their sidelines were undoubtedly behind the Museum's decision in January 1797 to post a notice every six months forbidding staff to request or receive tips.[42] The wives of keepers and gardeners were likewise resourceful, as indicated by the administration's move in October 1797 to prevent them from taking in laundry for profit and hanging it out in their gardens.[43]

In July 1801, realizing that the situation was once again out of hand, the administration appointed the zoologists Lamarck and Lacépède and the botanist Jussieu to investigate. They identified a series of infractions, beginning with the keepers' feeding their own animals food meant for the menagerie animals. In addition to soliciting tips or bribes from the public, the keepers grumbled audibly when such perks were not forthcoming. They had taken the liberty of making changes to the animals' quarters without prior approval and were mating animals without giving the Museum's naturalists the chance to observe the process or the results. The problem, the report concluded, was that "each [regards] himself as virtually the proprietor of the animals in his charge." The solution, the Museum decided, was to relieve Delaunay of the job of policing the keepers and turn the responsibility over to a professor.[44]

In short, the attitudes and behavior of the keepers had not changed much since the menagerie's earliest days. The Museum cherished industry in its workers, so long as it was properly directed toward the institution and not their self-interests. The professors had basically had no choice in hiring the original keepers, needing them to care for the animals they had owned. Likewise, there had been no question when it came to hiring Thompson to get the elephants to Paris and oversee their establishment in their new surroundings. But even later, when the Museum had a freer hand in appointing the keepers, it found it difficult to keep them in line, given that the work was arduous and dangerous (as Ballet's case makes clear) and the pay meager. Under such conditions it was not easy to establish a cadre of keepers whose enthusiasm and attention never flagged and who were not tempted, when no one was looking, to break the rules.

[1.11]

Lions, Cubs, and Their Keeper

Daybreak at the Jardin des Plantes in 1800 was announced by the roaring of Marc, the menagerie's magnificent male lion. Four lionesses including his mate, Constantine, joined him in a chorus that lasted nearly ten minutes and could be heard across the river. According to the zoologist Lacépède, they also roared after they were fed, but other than that, they did so only when the weather was about to change or when their keeper was away from them.[1]

Implicit in Lacépède's observations on the lions' roaring was an acknowledgment of their keeper's dominant role in their lives. The male lion may have led the dawn chorus, but it was Félix Cassal who had charge of the lions' care and feeding. Cassal was also responsible for the lions' roaring when he was absent; he had made Marc and Constantine so psychologically dependent on him that they were wont to express their unease when he wasn't there. Unmentioned by Lacépède was an additional scenario in which Cassal prompted lions to roar, in this case angrily. Here Cassal's targets were the other two lions he had brought back from North Africa. One was the lioness from the Sahara that had impressed him from the beginning as the most ferocious he had ever seen. The other was a female cub whose ferocity he cultivated through mistreatment. To impress the public with his talents as a lion tamer, and for his own pleasure, he provoked the animals, elicited their roars, then dominated them with his shouts and the crack of his whip.

Cassal loomed large in the lives of these lions, who were deprived of the social contacts and groups they would normally have experienced in the wild. He aimed to be the focus of their attention and sentiments. He owed much to them, for they justified his position as keeper, and furthermore they gave him the opportunity to develop and sustain the multiple interrelated images of lion tamer, lion lover, and lion expert. In dominating the fiercer of them, he played king of the beasts. With the gentler Marc and Constantine, he played head of the family. The first role involved force, the

Figure 1.11.1. “The lioness nursing her cubs.” Maréchal, Simon-Charles Miger, “La Lionne, Felis leo fœmina,” from Étienne de Lacépède, Georges Cuvier, and Étienne Geoffroy Saint-Hilaire, *La Ménagerie du Muséum national d’histoire naturelle, ou Description et histoire des animaux qui y vivent et qui y ont vécu*, Paris, Aug. Renouar, 1801–5. Muséum national d’histoire naturelle, 419 pl. 21.

second emphasized love, but each demonstrated mastery and control. His campaign to head a lion family was gloriously consummated when Marc and Constantine reproduced. In the eyes of the Museum’s scientists and of the public alike, the birth of lion cubs at the menagerie was a sensation, and it would not have happened without Cassal’s devotion to the project.

Contemporaries wrote much about the interrelations of Cassal, Marc, Constantine, and the cubs. They said little about Cassal’s relations with the other lionesses from North Africa and nothing about the fifth adult lion in the menagerie in 1800. This was the young lioness found aboard a captured British ship in 1797 and given to the menagerie while Cassal was away on his collecting expedition. He likely never felt the close connection with her that he had with the others.

Cassal gave the names Marc and Constantine to the young male and female pair that the bey of Constantine presented him in 1797 as a gift to

France. At the time, they were about three years old. Captured together in a trap in 1795, they were presumed to be from the same litter.[2] In Paris, they came to be known for their love for their keeper. Constantine was gentle and trusting with everyone, but her affection for Cassal was exceptional. His attention elicited from her "the frisky joy of the dog and the delicate and velvety movements of the cat." Marc's demeanor was more solemn, and he was more aggressive with people except for Cassal, with whom he was docile.[3]

The year after the lions were brought to the menagerie, Georges Toscan, the nature-loving librarian, related an anecdote about Marc's affection for Cassal, "his keeper, or rather, his friend." Toscan said nothing about how that affection had been established, only about the way it manifested itself when Cassal was ill for several days and had to have a substitute. The lion was upset:

> Sad, lonely, he stayed crouched at the back of his cage, not wishing to receive the care of the stranger, whose presence was hateful to him. From the back of his cage, he threatened him with muffled roars. Even the company of his female displeased him, or at least he no longer paid attention to her. The restlessness, the interior suffering of this animal, led one to believe that he was truly ill; but no one dared approach him.

Cassal's return was a happy occasion for all:

> Finally, the day when Félix was well again, he visited the lion. Wanting to surprise him, he glided very quietly along the side of his enclosure and put just his head at the edge of the bars. The lion caught sight of him and leapt to him in one and the same movement. He stood up against Félix, pressed his feet against him, licked his hands and his face, and roared with pleasure. The female, joyous, came running too, but the male pushed her back; he was angry, fearing she would rob him of the favors of which he was jealous. They were about to fight, but Félix came into the enclosure to satisfy them both, caressing them one after the other and receiving their caresses in turn.

Toscan continued:

> One sees [Félix] frequently in the middle of this formidable couple, whose power he knows how to constrain. He talks now with the male, now with the female; he strokes them; he kisses them on the mouth. If he wants to separate them and make each retire to its cell, he has only to say a word. If he desires that they lie on their backs, to show onlookers their paws

> armed with terrible claws and their jaws bristling with menacing teeth, at the slightest sign they lie down, offer their paws calmly, one after the other, open their jaws, and, as a reward, get the favor of licking his hand.[4]

Cassal had hoped the lions might mate. He sought to facilitate this by darkening their cage in the daytime, having noticed, by one account, that "these two beautiful animals only gave themselves over to intimate caresses in silence, obscurity, and solitude."[5] Intimate caresses were not the order of the day, however, when Constantine came into heat. An initial calm was followed by a fierce struggle lasting an entire week. The lions' quarters, built of strong, thick planks and masonry, needed reinforcement to contain the courtship. Constantine surrendered only by force. Meanwhile, "the male's fury was carried to the highest degree, and it surpassed all the limits of the imagination when one showed oneself to him; it was impossible then not to tremble with fear at the violence of his fury and the force of his roars." It was one of those "great spectacles of nature" that left the viewer speechless.[6]

Constantine became pregnant, but she miscarried two months later. One explanation was that she injured herself in her habitual bounding about her cage; another was that a visitor had irritated her, and she wounded herself in her anger.[7] Though the cubs were stillborn, she sought to care for them. It was only with great difficulty that they were taken from her, to be preserved in alcohol for the Museum's collection.

The cubs' deaths notwithstanding, the fact of a lioness giving birth in captivity and in France evoked outsized praise in the press. One of the Museum's assistant naturalists portrayed the event as "new proof" of how man could "conquer the laws and change the influences of climate, habit, and often organization." The case, he said, gave hope that animals from Africa or "the East" could be tamed and acclimatized to become "of the greatest utility" to the people of Europe.[8]

Some two to three weeks after Constantine's first litter, she again went into heat, and again she became pregnant. Cassal chained her this time so she could not leap about her enclosure. Three and a half months later, on the night of 18–19 brumaire an 9 (9–10 November 1800), with Cassal serving as midwife, she gave birth to three healthy male cubs.[9] It was propitious timing, the first anniversary of the coup d'état that overthrew the Directory and made Napoleon first consul. The Museum sent a long account to the press for publication, touting how scientists now had a clear sense of the length of lions' gestation.[10] Cassal named the first of the cubs Marengo after Napoleon's victory on that Italian battlefield just five months earlier. He named the other two cubs Jemmapes and Fleurus for earlier French military triumphs in the Low Countries in 1792 and 1794. Maréchal, the

Museum's animal painter, memorialized the births with a wonderful painting of the mother and her offspring, achieved after a campaign to earn her trust.[11] Jean-Baptiste Huet executed several additional illustrations: notably, a splendid portrait of the whole lion family; a finely detailed drawing of the mother and three cubs, executed in black chalk with white highlights; and an ink-on-paper engraving of the three cubs, their parents, and Cassal. As the historian Katie Hornstein has observed, the family portrait (fig. 2.4.1), portrays Cassal as the head of the family. Holding one cub, he is clearly the authority figure in the painting. Marc looks off into space, uninvolved.[12]

Crowds came to see the new cubs. So did Napoleon and his wife, Josephine, who brought friends, among them the duchesse d'Abrantès, whose memoirs include an account of the visit.[13] According to the duchess, Cassal entered the cage and took the cubs from their mother while they were nursing. Constantine appeared untroubled by Cassal's action. "She seemed to love him and proved it by turning on him her very soft yet very attentive eyes." When Cassal let Josephine hold one of the cubs (outside the cage) she growled, and Cassal assured both the lioness and the ladies that no one was in danger (since the lioness was behind strong bars and not at full strength). Constantine rolled over and seemed indifferent, but when Cassal handed one of the cubs to the duchess, she stood and gave a tremendous *meow* that shook the whole place, at which point Cassal took the cub back and held it himself.

The duchess found the animal keeper amusing. "This Cassal," she related, "was an extraordinary man in his way. He recounted such astonishing things that he claimed to have seen in his travels that it was impossible to believe them. However, since he had in effect traveled a great deal and consequently gathered many observations, even the most ordinary, regarding the countries he had traversed, he was always interesting. Not everything he said was a lie, if not everything was the truth, and one found it amusing and entertaining."[14]

By the duchess's account, Cassal told the ladies how Napoleon had visited Constantine and caressed her and had been kept informed of the cubs' births and of what the lioness ate and drank. Cassal said that a general who had accompanied Napoleon had given Cassal "a beautiful gold piece" so that the lioness could "drink to the health of the Republic." Was the story true? In a later volume of her memoires, d'Abrantès wrote of Cassal again, saying, "there was never a bigger liar than Félix."[15] By her testimony, however, Napoleon's familiarity with Cassal was real. Before the ladies left, Napoleon rode up on horseback. Cassal gave the first consul a report on Constantine's condition, assuring him "that the lioness had drunk to his health and that she was in marvelous shape." The duchess recounted, "Napoleon caressed the lioness and chatted with Félix about all his animals with such great ease, a grasp of everything regarding their hygiene and their conduct, that in truth one would have believed that he had made a particular study of this part of knowledge." The author then witnessed Cassal try to push the credulity of his auditors too far:

> Félix, feeling so well encouraged, began to tell one of his most beautiful stories, but as he reached what he called the most astonishing part, Napoleon said to him, while striking him on the head: "Félix, you lie, my boy, there are no crocodiles in the place that you say, and there never were any. But it is all right, continue your story." That was easier said than done. Poor Félix was so disconcerted by the reproach of the First Consul that he lost the thread of his adventure. "Very well! This can be for another day," said Napoleon generously. "Only you will remember that crocodiles do not eat those who bathe in the Bosporus."[16]

The duchess described how tenderly the lioness cared for her cubs, licking them continually with her large tongue, making comforting sounds, and responding to their cries of discomfort. When bothered by the sun or wind, she picked the cubs up in her jaws to move them, "just like a cat picks up her babies." What the duchess did not remark, but the artist Maréchal did, was that visitors made Constantine anxious. Maréchal's portrait of the lioness and her cubs shows Constantine staring watchfully at someone peeking into her cage. Her postpartum anxiety was also noted by the young man who would later become the menagerie's superintendent. Frédéric Cuvier interpreted Constantine's taking her cubs in her jaws and moving them around as motivated by a sense of danger, not discomfort from sun or wind. "Impelled to remove her babies from all danger and to hide them in some deep, secluded place," he recalled, "she picked them up in her jaws and carried them around in this way for entire hours, turning about in her cage and manifesting the greatest restlessness."[17]

Constantine nursed the three cubs for two and a half months, then took the initiative of feeding them meat. When they were four months old, her attentions began to wane. Once again, she came into heat and mated. The ensuing pregnancy resulted in the birth of two more live cubs, both females.

Although it was Cassal who dreamed of and oversaw the mating of the lions, it is not entirely clear whether Cassal or the administration was responsible in the spring of 1801 for the idea of castrating the most unruly of the cubs to see how this would affect its behavior. We are inclined to credit it to Cassal, however, judging from the professors' complaint about the keepers just a few months later that they "attempted mating and other experiments made without order and without informed witnesses."[18] The purpose of this experiment was to see if castration would make the cub more manageable, but the reverse happened.[19] The longer-term effects of the operation could not be studied because none of the cubs lived to maturity. Around the age of one year, they all succumbed to difficulties associated with teething.

When Lacépède wrote "The Lioness" (1801) for Miger's large illustrated work on the menagerie's animals, he proudly supposed that the number of lions in the collection at that point was the greatest ever possessed by any menagerie devoted to the advancement of science. There were ten all together, five adults and five cubs. With this many specimens, he said, it had been possible to discover new facts about the species. One was that the lion had a gestation period of 108 days. Another involved how the cubs' coloration changed as they matured.[20] Much of what he had to say, however, echoed the writings of predecessors, especially Buffon. Ignoring Daubenton's "no kings in nature" formulation, Lacépède followed Buffon's lead in portraying the lion as the king of beasts, indeed as a "species king," with an "empire." Like Buffon before him, he described how man had challenged the lion's domain and forced the species to retreat to inhospitable wastelands, but he also attributed the shrinking of the lion's range to climate change.

In addition to reiterating the familiar story of the lion and dog from Versailles, Lacépède cited the case of a lioness currently living at the menagerie with a dog as a friend, offering it as an example of how companionship helped captive animals endure their loss of independence as "prisoners."[21] He did not say so, but this was the lioness Cassal brought from North Africa as a cub and mistreated to make it fiercer. He said nothing about Cassal's harsh treatment of the animal. What he offered instead was the claim that when an animal was not exposed to suffering and fear it tended to lose its ferocity. He acknowledged, on the other hand, that no matter how kind its treatment, the species' original disposition lay just beneath the surface: "Prudence must never allow us to forget that with a very strong animal with very intense appetites, ardent affections, violent movements, and terrible weapons, a sudden and unexpected impression can bring it back at once to the natural character of its species." The keepers had learned to be cautious, Lacépède noted, when the lions were mating and likewise when the lioness was protecting her cubs.

Georges Cuvier's essay "The Lion" appeared eleven months after Lacépède's. Cuvier dispensed with grand statements about the species' character, instead providing a judicious overview of accounts of lions from antiquity to the present, including "some facts we owe to the keeper Félix Cassal." Legend had it that the lion feared the rooster's crowing and the squeals of pigs, but Cassal reported that "[the lion] has been seen to devour two or three [roosters] in a few minutes." He also allowed that the boar was the lion's main quarry in North Africa and said he had seen a lion carry one away "like a wolf carries away a sheep." The duchesse d'Abrantès might have suspected Cassal was embellishing, but Cuvier reported the claim without comment.

Cuvier wrote further that lionesses in North Africa were known to bear their young in marshy places, where it was easy to prey on animals that came to drink, and that males helped provide for the young, which seemed to suggest that lions are monogamous. This conclusion, Cuvier noted, was supported by the example of the Museum's male lion, "who has refused all the females offered to him, with the exception of the one to which he is attached."[22]

What the behavior of animals in captivity revealed about their behavior in the wild was an ongoing question. However, while writers offered the public descriptions of the lives of animals in their native haunts, what the public wanted to see on visiting the menagerie was activity on location. A sleeping animal was scarcely better than a stuffed one in the galleries, but the sight of Constantine caring for her cubs was to be cherished, as was that of Cassal commanding his lions to do his bidding. Both, furthermore, assured onlookers of *Homo sapiens*'s dominance over the rest of the creation.

In the summer of 1802, when Georges Cuvier wrote his essay "The Lion," he was prepared to credit Cassal as an authority on the ways of lions. The administration considered Cassal an essential member of the menagerie staff. His tenure at the Museum, however, would not last indefinitely. Visitors' complaints about him began to mount. Among the accusations, surprisingly, was that he mistreated the lion cubs. The Museum never believed that charge, but when Cassal plotted to have a fellow staff member killed, the administration concluded he had to go.

[1.12]

Buying from the Enemy

The themes of conquest, liberation, and civilization were enmeshed in France's justification for confiscating the art and natural history treasures of other countries.[1] In Holland in 1795, Alquier, the representative of the French government, explained the fate of the Stadtholder's natural history collection by saying, "the French have conquered less for themselves than for the human race, the happiness and liberty of which are attached to the propagation of useful knowledge."[2] Likewise in 1798, when the bronze horses from the cathedral of San Marco in Venice were displayed in Paris at the Festival of Liberty, they were said to have been liberated so a free people could enjoy them. Similarly, Napoleon's Egyptian expedition was hailed as an "association of sciences and weapons, this beautiful idea" of a hero "who had undertaken the conquest of a country that had become barbarous, only to bring it to the level of civilization."[3] In this way the conquerors reassured themselves that conquering, liberating, and civilizing were part of the same process.

But not all the major additions to the Paris menagerie in its first seven or eight years fit this mold, particularly the specimens France purchased in 1800 from Britain, her sworn enemy. A pair each of tigers, leopards, panthers, and lynxes plus a single hyena arrived from London in November 1800. Adding to the luster of the haul was the claim that the animals had originally belonged to the late Tipu Sultan, the monarch of Mysore, a longtime friend of France and foe of Britain. The impressiveness of the specimens seems to have mitigated the awkward fact that the French had to pay Britain for the collection.

Although repeated in numerous histories of the menagerie, the Tipu Sultan aspect of the story turns out to have been false, but as an example of animals' bearing weight as political markers, the case deserves attention.[4] Also worth scrutiny is the establishment that sold the animals to the Museum, Gilbert Pidcock's "Royal Menagerie" at Exeter Exchange (or "Exeter 'Change") in the Strand in London. Pidcock bought the menagerie in 1793

Figure 1.12.1. 1795 broadside advertising Gilbert Pidcock's display of live animals at Exeter-'Change, the Strand, London. © Victoria and Albert Museum, London.

from its founder, Thomas Clark, and quickly established himself as London's most prominent buyer, exhibitor, and seller of exotic animals.[5]

Pidcock's operation stood in marked contrast to the national menagerie in Paris. He relied on sales of admission tickets, but the Paris menagerie was free and open to the public. Whereas the Museum sought to avoid anything that smacked of the commercialism and vulgarity of fairs and street shows, Pidcock embraced it, even while advertising his exhibitions as suitable for royalty, gentry, families, and children. His featured attractions in the 1790s included a two-headed heifer and "the Irish Dwarf" (Peter Davies, a man who at thirty-six was three and a half feet tall). Pidcock also had numerous exotic species, including a rhinoceros he toured with in fall 1792, cheerfully describing it as "the REAL UNICORN" and as "that most wonderful of all animals."[6]

In fall 1793, as the new owner of the menagerie at Exeter Exchange, Pidcock trumpeted his exciting latest acquisitions. "Just arrived, in the last fleet from India, and to be seen alive, in the Great Room, over Exeter 'Change, a most stupendous LIVING MALE ELEPHANT," plus "two beautiful KANGUROOS [*sic*], male and female, brought over in the *Atlantic*, the vessel in which Governor Phillips arrived from Botany Bay."[7] In January 1794 he described the elephant as "so gentle that Ladies and Children may stroak [*sic*] him with the greatest safety." By this time he had also acquired "the Royal Bison, from the river Missisippi [*sic*], near the Gulph [*sic*] of Florida; the only one of the kind seen in this kingdom for upwards of 40 years. . . . his head is larger in proportion than any other beast." The animals were displayed in three rooms. The price of admission for each was one shilling, but a client could see all three for two shillings sixpence and enjoy a special bonus: "The Guillotine will be shewn gratis to those who view the three Exhibitions."[8] Previously Pidcock had advertised the item as the "French Beheading Machine."[9]

In addition to touting new additions to the menagerie, Pidcock delighted in mentioning visits from royalty. In March 1795 he reported that the Duchess of York and members of the Dutch royal family had visited the Grand Menagerie together. *The Times* reported that they "expressed their great satisfaction" at seeing Pidcock's animals. The newspaper did not mention that the family had come to England because of France's conquest of Holland or that the animals that had graced the Stadtholder's menagerie would soon be sent to Paris.[10]

Pidcock was obviously not shy about his menagerie. Where Thomas Clark before him had boasted of having the finest collection of exotic birds and beasts in Europe and P. T. Barnum would later promote his circus as "the greatest show on earth," at the turn of the new century Pidcock touted his "Grand Menagerie" as "superior to every other place of the same kind in the universe."[11] His collection of living animals in 1800 was more than a match for the Royal Menagerie at the Tower of London and arguably superior, in number and variety of animals, to the Kaiserliche Menagerie of Schönbrunn on the outskirts of Vienna—and to the menagerie of the Museum of Natural History in Paris.[12]

Pidcock's creatures reflected Britain's imperial reach: tigers and elephants from India, kangaroos from Australia, a zebra from South Africa, and more.[13] When an animal in his collection died, it was not necessarily lost to the public's gaze; Pidcock had his splendid cassowary stuffed so the public could continue to pay for the privilege of viewing it.[14] A sixty-six-foot skeleton of a sperm whale was another of his attractions.

All of this suggests that Pidcock would have jumped at the chance to own an animal that once belonged to Tipu Sultan. But so would the French, who had been enamored of Tipu since 1788, when the sultan sent three ambassadors to France to elicit Louis XVI's assistance in his struggles with the British. The French king did not fulfill Tipu's request, but his diplomatic envoy generated a wave of "Indomania" in French society. Enthusiasm for Tipu and his kingdom ran high, fueled by French economic interests, daily press reports about the ambassadors' activities, and a slew of paintings, porcelain, and other artistic productions inspired by the ambassadors' exoticism.[15]

France's rivalry with Britain went back centuries. In the second half of the eighteenth century alone, they were opponents in the Seven Years' War and the American Revolutionary War, and there was no end in sight for the hostilities that arose in 1793 related to French Revolutionary expansionism. When Napoleon launched his expedition to Egypt in 1798, the British interpreted it as the first step in a plan to drive the British East India Company out of India. When they learned that the French were entering into an alliance with Tipu, a long-standing thorn in the side of the British East India Company, they decided to deal with Tipu once and for all. In 1799 they laid siege to Mysore's capital, Seringapatam. In the battle where the British colonial army was commanded by Arthur Wellesley (soon to be the Duke of Wellington), Tipu was killed, and his troops were forced to surrender.[16]

The British seized a variety of remarkable war trophies from Tipu's palace, many of them bearing the sultan's emblem, the tiger. Among the most extraordinary was an elaborately painted, nearly life-size, automaton of a tiger engaged in devouring a European. Inside the wooden tiger was a pipe organ. With the turn of a crank, *The Times* of London reported, the organ produced "the cries of a person in distress, intermixed with the roar of the Tyger." *The Times* continued: "The machinery is so contrived, that while the organ is playing, the hand of the European is often lifted up to express his helpless and deplorable condition."[17]

Interpretation of the symbolism of this marvelous machine was a matter of perspective. At Tipu's court it would have signified fierce and successful resistance to the incursions of the West. Exhibited in London by the East India Company, it could be read as an index of Tipu's brutality that justified Britain's actions and increasing control over India.

Pidcock probably would have loved to have the spectacular "man-tyger organ," a device even more appealing than the "French Beheading Machine," but he acquired neither this nor apparently any animals from Tipu's menagerie. In a notice of 1799 he proudly announced four Bengal tigers but made no claim that they had once belonged to Tipu.[18] A booklet (1800) promoting Pidcock's collection said only they had "arrived at Exeter 'Change at different times within the last four years." The publication boasted of the range of sensational animals in the collection, all of them potentially for sale. Aside from the four tigers, these included two elephants (a male and a female), two lions, one lioness, six kangaroos, a panther, an "ounce" (described as a lynxlike cat), a leopard, and three hyenas. The pamphlet additionally noted that Pidcock's two-headed heifer had recently died just as he was about to sell it, a financial loss he greatly lamented.[19]

It was evidently through Louzardy that word reached Delaunay of the animals Pidcock had to offer. Delaunay told the administration this was a chance "to add some *éclat* to the Museum." Naturalists as well as the public, he predicted, would want to see the new animals, and Pidcock's prices seemed reasonable. The Museum could have a beautiful "Royal Tiger" for 150 guineas. Pidcock was also offering "kangaroos at forty guineas apiece, and other animals almost as rare that have come to him from the menagerie of the Cape of Good Hope, Batavia, [and] the islands of the South Sea." Even in peacetime, said Delaunay, one would expect to pay much more, and the chance to get them at such a price was "a unique opportunity" of which the Museum would be wise to take advantage.[20]

The Museum initially tabled the request, but one of the professors suggested sending it to the minister of the interior, Lucien Bonaparte, who surprisingly agreed to fund a purchase, though not without grumbling when the Museum was less cooperative with his requests of it: "To pull the menagerie from the state of nothingness of which you complain, I have done all that you have asked me."[21]

In August 1800 the Museum sent Delaunay and Louzardy to London to buy the animals from Pidcock. Delaunay characteristically groused about numerous things, including the cost of living, the food, and the weather, but he met the famous British scientist Joseph Banks (1743–1820), who gave him several books for the Museum, and he was also well received by William T. Aiton, the head gardener at Kew, who prepared a large case of plants to be shipped to Paris. Pidcock, however, was off touring the summer fairs with his animals and would not return until after the St. Bartholomew's Day fair on 4 September. Meanwhile word reached Delaunay that four of Pidcock's animals (a lion, two tigers, and a cougar) had died suddenly on the same day. Pidcock maintained they had been poisoned and announced a fifty-guinea reward to anyone who could identify the culprit(s).[22]

Delaunay had wanted the cougar but retained high hopes of returning with an exciting collection: a pair of tigers, a leopard, a panther, a pair of lynxes, a pair (or more) of kangaroos, and a hyena. These would cost the Museum 13,000 francs, not counting travel expenses. The kangaroos had especially caught Delaunay's fancy. "It is upon them," he wrote the Museum, "that I stake our fortune. Imagine a clown who bends his head and turns his arms forward, and advances only by making great leaps on the tips of his feet."[23]

Delaunay's efforts to buy kangaroos did not pan out. The King's Garden at Richmond had twenty-one, Delaunay learned, but there was no way to acquire them. Pidcock refused to part with a wonderful kangaroo family: a male and female plus three babies in the female's pouch. The only specimen he was willing to sell seemed too feeble to survive the trip to Paris. When Delaunay and Louzardy left London, they had eight big cats and a hyena but not a single kangaroo.

On the way home they stopped to exhibit the animals to defray their travel costs, giving Louzardy the chance to employ talents the Museum did not wish to see displayed on its own grounds. The myth that the animals came from Tipu's menagerie apparently originated in Brussels. Writing to the Museum director (Fourcroy) about preparations for the show, Delaunay remarked, "it is good to have it announced in the papers that this menagerie already has the reputation of having belonged to Tipoo-Saïb and to have it placed on posters on the street corners."[24] It was undoubtedly Louzardy's idea to promote the animals this way, but Delaunay was complicit. The story reached the French capital weeks before the animals did. *Le Journal de Paris* picked up a notice in the Brussels press saying the animals were "a part of the menagerie of Tipoo-Saib, purchased in England."[25]

Delaunay, Louzardy, the big cats, and the hyena arrived in Paris on 19 brumaire an 9 (10 November 1800), the same day the lion cubs were born.[26] The professors were focused on resisting Lucien Bonaparte's plans to restructure Museum governance, and the menagerie's new animals were not even mentioned at the professors' next meeting. But the animals' arrival obliged others to spring into action. Cassal had to care for them. Maréchal had to begin painting their pictures. E. J. B. Vignier, who had written about the elephants, dashed off a new pamphlet titled "Abbreviated description of the animals from the menagerie of Tippoo Saïb, newly purchased in London to enrich the menagerie of the Museum of Natural History, followed by an account of the progeny of the lioness, which one can regard as a phenomenon of natural history."[27]

The confident ring of the booklet's title and the continuing celebrity of Tipu in France made a persuasive combination. The claim about the animals' provenance was repeated in the *Magasin encyclopédique* and showed up later in guides to the Museum.[28] It is difficult to imagine Félix Cassal not embellishing and perpetrating it. But Vignier apparently discovered that it was untrue. In an updated version of his pamphlet, he named the new specimens but never mentioned Tipu. The British, he wrote, may have sold their tigers to France at a high price, but it was the tigers' French keeper(Cassal) who tamed them, and soon the first consul (Napoleon) would subdue the British, forcing them to hand France the olive branch of peace—for free.[29]

The French still keenly felt their lack of kangaroos, while Pidcock reveled in his own possession. He crowed that his Grand Menagerie was "so incessantly visited by the respectable Families in Town," and even more so "since the appearance of the young Kangaroo, which familiarly makes its appearance almost every minute from a pouch in the belly of the Female, whose curious formation, as well as that of the Male, can only be thoroughly conceived by inspection." The kangaroos performed "very entertaining tricks" and wrestled with their Keeper."[30]

The Museum of Natural History finally acquired its first live kangaroos in 1802, thanks to a second trip to London, but the kangaroos were not the primary goal of the trip. Its purpose was to replace Hans, the dead male elephant. The minister of the interior, Chaptal, agreed that having elephants was prestigious for the menagerie and the nation and furthermore that a male was needed "for conservation of the female." Knowing that Pidcock had a male elephant for sale for 750 guineas, the Museum asked Louis-Guillaume Otto, France's minister plenipotentiary to the British king, to arrange to buy the elephant at the best price possible, once Chaptal authorized the purchase.[31]

According to the 1800 description of Pidcock's collection, he had acquired the elephant in 1793 at the jaw-dropping price of "one thousand guineas." The same pamphlet vaunted the "things almost incredible" that the elephant had learned since coming to England:

> At the command of his keeper, he will take up the smallest piece of money, a watch, &c. and lodge it in the pocket of any lady or gentleman in company, although they are the greatest strangers to him. He will likewise carefully take it out again and return it to his keeper. He will take up a tankard of any kind of liquor, but particularly ale, with his trunk, blow it into his mouth, and in return for favours received, make a bow to the audience, which shews a strong sense of his gratitude. He will take a piece of money from any person in company, and present it for bread, fruit, &c. to a boy who attends for the purpose, and return his benefactor a printed bill, with his description. He has been observed, while his keeper was cleaning his den, to take up a broom with his proboscis, and sweep after him, as if willing to lend all the assistance in his power. When his keeper leaves him, he follows him to the door, shuts it, and seems highly pleased with what he has done.[32]

To advance the elephant transaction, ambassador Otto sent one of his secretaries to Exeter Exchange on the pretext of looking at Pidcock's collection, but Pidcock recognized why he was there. Otto reported back to Paris what the secretary learned: "This elephant is eleven years old, eight English feet tall, and very healthy. He is fed hay and potatoes, of which he is very fond. His ordinary drink is water, but he is extremely content with beer. He is extraordinarily intelligent and very docile. Mr. Peacock [*sic*] speaks of £1,000 sterling (approximately 24,000 francs) as the only price to which he would agree. He swears that is what he paid to have it." Otto considered that exorbitant.[33]

Pidcock soon wrote to Louzardy, to whom he had previously quoted a price of 750 guineas, explaining that he now wanted 1,000 guineas because the elephant had grown considerably since his earlier offer. He told Louzardy he had no intention of parting with his kangaroos, but he was interested in one of the lionesses in Paris (if he could not get a male lion), and he would be willing to count that in partial exchange for the elephant. In a follow-up letter, Pidcock told Louzardy he would reduce the price to 750 guineas and pay 100 guineas for the lioness if Louzardy brought her to England, "where I hope to have the pleasure of seeing you again." He claimed he had received better offers and would be making a sacrifice, but he wanted to honor his original offer of 750 guineas. The sum, said Pidcock, was to be paid "as soon as [the elephant's] exit from the doors of Exeter-'Change."[34]

If the Museum had any doubts about Pidcock's trustworthiness, it did not register them. By the end of June the administration and the minister were ready to finalize arrangements. The Museum put Louis Dufresne in charge of the mission, with Louzardy and Lassaigne as assistants, and instructed him to take advantage of any other opportunities that might arise in London with respect to live animals, plants, or other natural history specimens.[35] For the time being, thanks to the Treaty of Amiens of 27 March 1802, France and England were no longer at war, and it seemed an auspicious time to travel.

Letters to Paris from Pidcock as well as Dufresne, plus Dufresne's official report, detail the highs and lows of the whole affair. Dufresne and his companions were foreigners abroad, unfamiliar with the English language, seeking to do the Museum's bidding from afar, too distant by mail to know whether it would approve the initiatives they were obliged to make in the face of various contingencies.

Louzardy, Lassaigne, and the lioness left Paris by wagon for Calais on 17 July 1802, a seven-day trip. Dufresne arrived there by stagecoach in three days. The crossing to Dover on July 25 was rough, but it caused Dufresne less anxiety than his experience with the customs officers in Dover. He risked the lioness's confiscation if he estimated her value too low, but too high an estimate would make the customs duty more than he could afford. Unable to communicate in English, he nearly lost the animal, but with the aid of a French commissioner, he ended up having to pay only eight guineas of duty in Dover, with the final reckoning to be made in London.[36]

From Dover, Dufresne set out for the British capital, again by stagecoach with his assistants traveling by wagon. The day after his arrival, he visited Pidcock's "Grand Menagerie" at Exeter Exchange, where an awkward farce played out. Instead of announcing himself as representing the Paris Museum, Dufresne came incognito with an interpreter. After paying to enter, he asked to see Pidcock. Pretending no special interest in elephants, Dufresne asked Pidcock about other animals in the collection, but Pidcock's mind was fixed on the elephant for the distressing reason that it had died three days earlier. Pidcock claimed it had been poisoned. His loss was immense, he told his visitor, for he planned to sell the animal to the French government for 750 guineas. Dufresne, who had not yet revealed his identity, asked to see Pidcock's kangaroos. He was permitted to do so, but only after paying admission to another room. Ultimately he admitted he was the Museum's representative, come to London for the elephant.

It was an inauspicious beginning: a Frenchman not speaking English, yet trying to conceal his purpose, meeting an animal dealer whose elephant had just died, purportedly in nefarious circumstances. Nor did the situation improve when Dufresne revealed himself and lamented Pidcock's misfortune. Pidcock still wanted the lion for 100 guineas. Dufresne refused, maintaining that the elephant's death had voided the contract. He was suspicious of Pidcock's account of the elephant's demise and its causes. To confirm that the animal was dead, he went to see the body, which Pidcock had sold to a surgeon.

Dufresne then had to decide what he might salvage from the trip. He presumed he could sell the lioness or trade her for other animals. Pidcock said he wanted the lioness so that no one else in England could have her. Though his elephant was gone, he had plenty of other live animals for sale that the Paris menagerie lacked. Dufresne mailed the list to the Museum, with his commentary. Heading the list was a nilgai deer that Dufresne deemed "very healthy" and "much larger" than the one in the Museum's gallery. Next came the creatures that attracted Dufresne the most: the "very beautiful male and female kangaroo, the female carrying a baby in her pouch the size of a hare." Included among the more than two dozen other specimens were "a superb mandrill like the one that lived at the menagerie five years ago" and "the male bison (it will be possible to procure a female)." Attentive to the condition of the animals, Dufresne noted of "Ursus, the wolverine, the size and shape of the brown badger, with black extremities," that "this animal is very sad and almost always sleeps." Of the two ocelots he wrote, "I believe they are slightly ill." Dufresne told the Museum that other London dealers also had interesting animals to sell, but he was in an awkward position, needing authorization to purchase them. He was clear about what he wanted most: "I hope in one way or another not to leave here without a pair of kangaroos. These singular animals will amply compensate the public for a second elephant and will be more useful to science."[37]

Dufresne still needed to deal with the Customs Office, which again proved troublesome. The ensuing scene involved the customs officers' asking Pidcock to estimate the lioness's value, Pidcock slipping away in the maze of Customs House offices with a paper he wanted to use as evidence that the lioness was already his, and a French businessman (brought in by Dufresne for assistance) chasing Pidcock down and sorting things out. In the end Dufresne had to pay only nine guineas more than the eight he had already paid at Dover, and he still had the lioness.[38]

Dufresne's problems were compounded by how long it took for his letters to reach Paris and for him to get a reply. A whole month elapsed from the time he arrived in England (July 25) until he received his first letter from the Museum (August 24). He wrote four letters to the Museum in the meantime. In the second through fourth he described Pidcock's kangaroo family, worried that he might have to settle for the smaller (and cheaper) kangaroos at Bath, and detailed some minor purchases of shells, birds, and other specimens that he hoped the Museum would welcome. He closed his fourth letter by stating, "I have the honor of assuring you that that I await your instructions with much impatience. One is not happy in a beautiful city without money, especially when one knows neither the language nor the streets."[39]

When Dufresne's first letter (of July 30), was read at the Museum on August 9, the professors registered "the greatest desire of possessing at least two kangaroos" and instructed Geoffroy to indicate the other specimens that "by their rarity, forms, and habits would most attract the attention and serve the progress of science." The director was then to seek Chaptal's permission to purchase the specimens chosen, either with the funds originally destined for the elephant or with new funds.[40] But Chaptal was still focused on a male elephant. He had been told of one for sale in Beaucaire in the South of France and wanted to reserve his funds for it. He authorized selling the lioness for 100 guineas and using the funds to purchase a pair of kangaroos and any other animals the Museum lacked, but he said no additional money would be forthcoming.[41]

Learning this news, Dufresne chose to economize and buy the kangaroos at Bath from a man named Maille, who had previously worked for Pidcock. Pidcock was furious when he learned of the transaction and wrote to the Museum to complain about how "hurt" he was. Louzardy, said Pidcock, had given his word that he would sell the lioness only to him but had "collusively" sold it to one of Pidcock's former "servants," receiving only "two *small* kangaroos" (Pidcock's emphasis), in contrast to the two large ones and their baby that Pidcock had offered for 150 guineas (plus the lioness). Protesting "the ill treatment I have experienced from one of your confidential servants" and lamenting "much misrepresentation," Pidcock suggested that the Museum "take steps" regarding Louzardy.[42]

Dufresne's account of the affair makes it clear that Pidcock had been the one responsible for misrepresentations. Dufresne had been advised to bring a lawsuit against Pidcock for planning to sell an elephant that was not in good health, contrary to what he'd said, and had been ill for months. "More than ten persons," Dufresne reported, had offered to give depositions against Pidcock, since the elephant had "had a wound on its thigh for a long time, and when it was lying down it could be made to stand only with the aid of a pulley that is still attached to the ceiling of its quarters."[43] Pidcock for his part claimed that the elephant was "in perfect health" on the morning of Friday, 23 July, "performing his admired tricks as usual," whereas in the evening "to my great surprise and no less grief" the animal was "in a very bad state" and died "in great agony" at 3:00 a.m. the next day.[44]

Having learned that Chaptal would provide no additional funds for the expedition, the Museum instructed Dufresne to gather whatever information he could on natural history objects that might be acquired through future exchanges. Dufresne was then to return to Paris, making sure his expenses for animals and transportation did not exceed the price of the lioness plus the 3,600 francs allotted for travel expenses. He was given the option of exhibiting the animals on the return trip to defray his travel costs.[45]

Dufresne arranged for the transportation of the kangaroos plus several plants from Kew. While the cages were being constructed, he consulted Pidcock and Pidcock's animal keeper about four rare mammals at Exeter Exchange, not to buy live but to receive as specimens after they died. He identified them for the Museum as:

1. The hunting tiger of Tipoo Saib.[46]
2. The black tiger. This animal is very vicious [*méchant*] and smaller than the ordinary panther; its fur is black, and its spots stand out from the background by an even deeper tint.
3. The spotted hyena.
4. A bear from Hudson Bay. Although it is an adult, it is only roughly half the ordinary size. Its fur is fawn colored and the end of its nose a beautiful red.[47]

On 13 September 1802, Dufresne, Lassaigne, and Louzardy sailed from Dover to Calais with their kangaroos and plants, plus four rattlesnakes and a pair of flying squirrels Louzardy acquired on his own initiative. They tried exhibiting the animals on the way home, but with disappointing results. Three days of shows in Amiens netted only 120 livres. Dufresne considered heading straight for Paris, but with expedition expenses in mind he decided to stop in Rouen, hoping to find an audience there.[48]

Beginning on 30 September they exhibited the animals for eight days. Writing to the Museum three or four days into the stopover, Dufresne described his efforts to drum up business.

> I have neglected nothing to pique the public's curiosity. I rented the most beautiful hall in the city on the most frequented street. This hall costs 12 livres per day. I had 150 posters and 3,000 fliers printed, which have been distributed at the Comedy and among the people. Finally, I have neglected nothing to try to defray the expenses of my voyage. I oversee the receipts as much as possible, with the precaution of not vulgarly appearing as the director of this spectacle, a point which you had the goodness to recommend to me.

Notwithstanding these efforts, the average daily receipts in Rouen were only about fifty livres. Delaunay on his trip back from London in 1800 had lamented not having any kangaroos to show, but Dufresne found that two kangaroos (plus Louzardy's rattlesnakes and flying squirrels) were not enough to stimulate the public's imagination. "Kangaroos," he explained, "interest only informed people, of whom there are not a great enough number."[49]

Dufresne arrived in Paris by coach on 8 October 1802, with Louzardy, Lassaigne, and the small collection of animals following two days later. The four-month expedition had been frustrating for Dufresne, but he had done his best in difficult circumstances. On Monday, 11 October 1802, Dufresne wrote a short letter to the Museum announcing the kangaroos' arrival and explaining that, hoping the government would allow the Museum to use some of the funds designated for the elephant to improve its holdings, he had purchased some Australian specimens the Museum lacked. "These objects, which existed only in England and which, as the result of my acquisitions, exist there, for the most part, no longer, will add notably to the growth of your collections." Acknowledging he had spent more than the Museum wished, he offered to display his acquisitions immediately and state their price so the professors could judge what he had accomplished.[50]

Their judgment proved harsher than he had anticipated. The professors told their director to "express to him the administration's little satisfaction" with what he had accomplished.[51] It was an unworthy reaction on their part. No one at the Museum had been better qualified than Dufresne for the trip to London. He was a skilled zoologist who knew the Museum's holdings intimately, and his devotion to the Museum was total. He had a substantial natural history collection of his own, and he was well experienced in acquiring specimens. It was not his fault that he hadn't come back to Paris with an elephant. The Museum was redirecting toward him, it seems, the minister's displeasure with it.

Dismayed, Dufresne prepared a thirty-seven-page report documenting his activities. He included the texts of his original charge, the letters he had written and received, and other information. He concluded with a *cri de cœur*:

> The administration's interest presided over all my actions. Amid obstacles, difficulties, and perplexities of all sorts, the love of science alone was able to sustain and encourage my zeal. By what unhappy misunderstanding have I merited the assembly's disapproval? If in quitting the soil of England a painful feeling remained with me, I must admit that its only source was regret not to have had the means to bring back for the augmentation of our natural history collections a great number of objects from New Holland [Australia] which have never appeared in France and of which the English take glory in being the sole possessors.[52]

The story of the Museum envoys dispatched to London to buy specimens for the menagerie illustrates how possessing exotic animals could be a form of international competition. The Museum and the minister of the interior were keen to replace the menagerie's male elephant, seeing a pair of elephants as a mark of distinction for a European nation. Though the ventures to London brought the Museum new specimens, they were also a reminder that Britain's rule of the seas gave her naturalists a great advantage in acquiring new and rare forms. Dufresne was proud to think that his acquisitions had bolstered the collections in Paris to the detriment of collections in Britain, but he regretted not being able to bring home more. It pained him to think that insufficient support from the French government might render French natural history collections inferior to those of rival countries. And it distressed him deeply to think his efforts had gone unappreciated by the naturalists at the Museum.

Significantly, while the difficulties Dufresne experienced as the result of communication delays between France and England were substantial, they were nothing compared with those faced by the voyager naturalists who traveled farther afield during the Restoration, after the freedom of the seas was restored. Some sailed to the New World, others around the Cape of Good Hope to Madagascar, India, the Dutch East Indies, or Australia. Out of touch with Paris for months and sometimes years on end, they were uncertain about the status of shipments they had sent or funding that might be coming their way. They were beleaguered by illnesses and by multiple mishaps and, like Dufresne, left to improvise with the limited resources they had.

[1.13]

The Tiger

Wicked, treacherous, bloodthirsty, cruel, the world's most dangerous mammal; that was how Buffon portrayed the tiger in 1761. Lacépède said much the same in 1803 in his essay on the tiger for the volume he, Cuvier, and Geoffroy wrote on the animals of the menagerie.[1] But if Lacépède thought Buffon had already painted the tiger's essential character in "everlasting colors," he believed there was still information to impart about the tiger in captivity, because since November 1800 the menagerie had been home to the pair of tigers acquired in London.

Buffon had structured his discussion of the tiger by comparing it with the lion, much to the tiger's disadvantage. "In the class of carnivorous animals," he wrote, "the lion is the first and the tiger the second; and as the first, even in a bad genus, is always the largest and often the best, the second is ordinarily the most wicked." Whereas the lion joined "nobility, mercy, magnanimity" to "pride, courage, strength," the tiger's chief traits were a fundamental ferocity and gratuitous cruelty. There was a lesson to be learned here, Buffon maintained: that when status is determined by force, "the first, who is capable of everything, is less of a tyrant than the [second], who, unable to enjoy plenary power, avenges himself by abusing the power he possesses." Whereas the lion hunted only when hungry, the tiger, even when sated, seemed "always thirsty for blood." Not satisfied with devouring one victim, it would rip another apart with the same rage. It devastated the regions it inhabited, fearing "neither the sight nor the weapons of man." Its fury was only temporarily interrupted when preparing its next ambush.[2]

One could read the characters of the lion and tiger, Buffon believed, in their physical appearance. The lion had "a noble air" with the length of its legs "in proportion to the length of its body." Its impressive mane, "assured look," and "grave gait" all announced the animal's "proud and majestic intrepidness." In contrast, the tiger was "too long of body, too short on its legs." The tiger's "bare head," "haggard eyes," and blood-red tongue (which always seemed to be hanging out of its mouth) were likewise disturbing.

Figure 1.13.1. "The tiger." Nicolas Maréchal, Simon-Charles Miger, "Le tigre, *Felis tigris*," from Étienne de Lacépède, Georges Cuvier, and Étienne Geoffroy Saint-Hilaire, *La Ménagerie du Muséum national d'histoire naturelle, ou Description et histoire des animaux qui y vivent et qui y ont vécu*, Paris, Aug. Renouar, 1801–5. Muséum national d'histoire naturelle, 419 pl. 38.

Buffon interpreted them as signs of "low wickedness and insatiable cruelty." All the tiger's instincts, he maintained, were "a constant rage" and "blind fury," leaving it "incapable of knowing or distinguishing anything" and often leading it "to devour its own infants and tear apart their mother when she wants to protect them."[3] Buffon repeated his view of the tiger's blood lust: "Often [the tiger] abandons the animals it has just killed to slaughter more of them; it seems it seeks to taste their blood, it savors it, becomes drunk on it."[4]

Lacépède agreed that the tiger's nature was to rage against the world. A fiery physiology forced it to seek constant refreshment in its victims' blood. Its frightful roar left its prey terror-stricken; its gnashing teeth promised them sudden death. As it ripped open their flanks it trembled with rage, ceasing this wanton destruction only when no more prey was within reach. Neither its mate nor its offspring were exempt from attack. In Africa, the Indies, and China the animal was a "scourge," infesting the forests and laying treacherous ambushes beside the riverbanks.[5] Allowing that it feared "neither sword nor fire," Lacépède deemed it "the most dangerous of mammals."[6]

But Lacépède had to admit that the menagerie's tiger and tigress were not so fierce. The essence of a species, he explained, could be realized only by fullgrown, adult animals living freely in the wild, and on both counts the local pair fell short. They were still young, and they had been raised in captivity. Nonetheless, by observing them at the menagerie, much could be learned regarding what was alterable in their nature and what was fixed. For example, by watching how they responded to the shouts and whip of their keeper one could learn how much they were affected by fear.

Félix Cassal demanded obedience from the tigers in his charge. He also wanted them to mate successfully, as his lions had. There were grounds for hope, since the male tiger had reputedly fathered a cub in London with a different female. In Paris, Cassal began by placing the male and female in adjoining cages separated by a single rank of bars, allowing the animals to see and become familiar with each other. After eight days, the male showed signs of agitation, including running his paws rapidly across the bars restraining him. Lacépède reported what ensued:

> Félix removed the barrier all at once. In an instant, the male dashed upon the tigress, abandoning himself to the fire that had ignited him; but instead of sharing his ardor, she fought him with such force and tenaciousness that Félix was obliged to pull them apart. He succeeded only with very great difficulty. He ran a great danger. He even feared for the life of the male, who, uniquely occupied with the object of his desires, suffered, without defending himself, the female's cruel bites.[7]

In other situations, Lacépède wrote, the female tended to be less ferocious and fearful than the male, but both hated their captivity. This was not always the case with the lions: "Sometimes one sees the lion and the lioness forget their chains, one after the other, to abandon themselves to a feeling of well-being, to give themselves up to a playful gaiety, and to roll over and bound about." The tigers, in contrast, remained nearly immobile, either standing or lying down. They appeared, as Lacépède interpreted it, "isolated, sad, with an air of thinking ceaselessly of carnage or their escape," in a kind of "sinister reverie." This "state of anxiety and dark silence" was broken only by "their ferocious joy" on seeing their food arrive every evening around four or five o'clock. "Then," wrote Lacépède, "they become furious.

All their cruelty reawakens, they throw themselves on their food as if on living prey; and never ceasing to fear that one might snatch this food away from their bloodthirsty voracity, they seek to drive every enemy away with a frightening roar." Each tiger ate four to five kilograms of raw meat daily and drank three liters of water, driven by "the force of their inner heat."[8]

The tigers roared while eating, and they roared again when they were finished. They also roared after defecating. The male roared whenever someone approached him. As Lacépède described it, "he utters a sudden cry, and not remembering that the bars stop the efforts of his rage, he prepares to rip apart the person who disturbs him." The male also roared when he caused himself to ejaculate, a "depraved habit" that Lacépède explained as a function of the beast's solitude and imprisonment. In this case, Lacépède believed, the male's "frightful roar" was an expression simultaneously of "the torment that oppresses him and the desire that consumes him." Like the leopard, the male tiger sometimes sprayed urine at people who drew too near to his cage. At other times, "plunged into a sort of mournful sadness," he ignored everyone in his vicinity.[9]

By the summer of 1803, when Lacépède signed his article on the tiger, the menagerie's tigers had been in Paris for more than two and a half years, but Cassal had made little progress in taming them. He could make them lie down or stand up, but they did so reluctantly. Wrote Lacépède, "They yield only while growling, they obey only slowly; and to impose this involuntary submission on them, he has to raise his voice, shout forcefully, and menace them with a fearful whip." Lacépède's unsurprising conclusion was that bondage had bent the tigers' character but not broken it. The moral he derived from this was that "among animals, as with humans, only fear makes tyrants bend."[10]

Lacépède may have been selective in the facts he used to make his point. Already in 1802, Simon-Charles Miger's small guide to the menagerie reported that Cassal had tamed the tigers to the point that they let him pet them.[11] Beyond that, the new generation of zoologists at the Museum was not disposed to indulge in such political moralizing, and Lacépède's channeling of Buffon was passé. When Georges Cuvier's younger brother, Frédéric, wrote about the tiger in 1820, he insisted it had no special instinct for cruelty and no greater blood thirst than the lion or the jaguar. Its ferocity arose when humans encroached on its territory, but conversely, if tigers were taken very young and treated with kindness, they became tame "just like lions and all other species of cats." They sought and gave caresses "with the same eagerness, the same abandon as our most domestic cats." Nature had endowed the tiger, like "all the other intelligent beings," with "faculties that develop in proportion to its needs," wrote Frédéric Cuvier, and whether they remained weak or became exaggerated depended on "what the conservation of the individual requires."[12]

Whereas Lacépède sought a moral lesson in the tiger's character, Frédéric Cuvier was interested in establishing a science of animal behavior that studied the ways species could be developed and improved by training and the ways they could not.

[1.14]

The New Menagerie Takes Shape

In the summer of 1803, the *Magasin encyclopédique* announced the "astonishing metamorphosis" of several acres of land that had been added to the Jardin des Plantes. Lumberyards and market gardens were replaced with a charming "rural valley." Surrounded by grassy hills, enclosed by wooden trellises, and interlaced with walking paths, it extended from the Museum's amphitheater northeast toward the Seine. Visitors to the new area were treated to the sight of both exotic and familiar animals in picturesque parks, each enclosed by chestnut latticework constructed "in a very artful way, in the Swiss style," and featuring an attractive hut, unique in form and color. The strollers who frequented this new setting called it la vallée suisse (the Swiss valley).[1]

The writer was charmed as much by the site's design features as by the animals he encountered there. A handsome bridge made of hefty tree trunks spanned the center of the valley. Similarly striking were the new shelters for the goats. Tall, covered with thatch and reeds, and embellished with narrow staircases and turrets, they allowed the animals to move freely and display their agility to perfection: "One sees she-goats hanging on these stairways surrounded by their little ones; one sees he-goats climb up smooth inclines where our most intrepid roofers would not be able to hold on." Other enclosures held deer from Europe and "the Ganges." In the site's center, shaded by weeping willows and other trees, was a pond with swans. At the far end of the "valley," a mock ruin with "a half tumbled down tower" served as a refuge for deer.

What the reporter failed to mention was that the new accommodations for the "peaceful animals" made the old quarters for the fierce animals look more unsuitable than ever. While the herbivores grazed in their new parks, the carnivores remained confined to the cramped quarters of the converted greenhouse by the Seine. J.-B. Pujoulx confronted the problem in his 1803 guide to the Museum. Instead of beginning his menagerie tour with a stroll through the Swiss Valley, he started at the carnivores' dismal

Figure 1.14.1. The "Swiss Valley" at the Jardin des Plantes. Nicolas Huet, François Aubertin, *La vallée suisse, jardin anglais, Vue du Jardin des Plantes*, [s.d.]. Muséum national d'histoire naturelle. IC 2693.

home, exclaiming, "What? Is it there that the most powerful animals live? Is it there that the tiger, which reigns as a tyrant in the desert, and the lion, equally terrifying but more generous, receive their callers?" He hastened to assure his readers that no one had imagined this building would be anything more than provisional. "It would be far better to have no menagerie," he wrote, than to have one that reminded people of the hastily constructed quarters seen at animal fairs.[2]

The Museum had said much the same thing the year before when it presented a program for the menagerie's development to the Ministry of the Interior's Council for Public Buildings (CPB). It called for giving the animals enough space "to allow them to enjoy as much liberty as possible without endangering the public, so one could have an accurate idea of their forms, agility, and habits." Not doing so, the Museum noted, was out of the question, given that "there is so little difference . . . between stuffed animals and those imprisoned in cramped cells that it is not worth the costs of maintaining the latter."[3]

Pujoulx told his readers that to imagine the new menagerie, one had only to turn to "the paddocks and the picturesque cottages where the camels, the kangaroos, and other animals . . . have been placed." Pleasing to the human eye and agreeable to the animals, these new structures were part of the master plan for the menagerie conceived by Molinos, the Museum's architect, and all the menagerie's animals were eventually to be lodged according to this plan.[4] No one anticipated at the turn of the century that two decades would pass before the menagerie's *animaux féroces* had a new home, or that that home would look nothing like what Molinos had planned for it in the first (or second) place.

The menagerie circa 1800 was conspicuously flawed with respect to the animals' wide dispersion across the Museum grounds and the inadequacy of their quarters. Gotthelf Fischer identified the problems in his description of the Museum (1802–3).[5] The carnivores were confined to narrow cages in the "provisional menagerie" close to the Seine; some herbivores were situated in airy parks, but others remained in cramped quarters near the Régie. Birds were scattered even more widely, waterbirds in the big basin in the middle of the garden; birds of prey in an aviary next to the carnivores; guinea fowl, pheasants, and the like in narrow cages by the Museum's entrance on the rue de la Seine; and ostriches, cassowaries, and larger birds at the Régie.

Fischer was pleased to describe Molinos's master plan to expand and unify the menagerie and present the public with enchanting and edifying scenes of nature. Visitors would enter the menagerie through a gate near the amphitheater. Once inside, they would encounter several small, enclosed groves for sheep, rams, and goats and then similar enclosures for deer. The elephants would continue to have a courtyard adjoining their quarters in the Régie, but their park would be expanded. From there, strollers would follow a walkway through a small vale and be treated to "a very picturesque view" of a pyramidal aviary, partially hidden by foliage. Beneath a bridge spanning the vale, they would find the park for the camels and dromedaries, enhanced by trees and plants from the animals' native land and a shelter designed to evoke the same geographic area. Next was the park for the water buffalo, two additional groves for the gazelles and the ostriches, and, last, enclosures for cattle.[6]

Molinos planned a wilder setting for the bears, lions, and other carnivores, consisting of caves in a *rocher* (a large rocky escarpment). The animals' dens would be carved into the cliff, with the lions on the south, the bears on the north, and the wolves and foxes on the east and west. To make the lives of the animals and their keepers easier, the quarters for the lions and bears would include two chambers extending through the cliff to the opposite side, allowing the keepers to clean one chamber while animals were in the other. This design would also help the animals adjust better

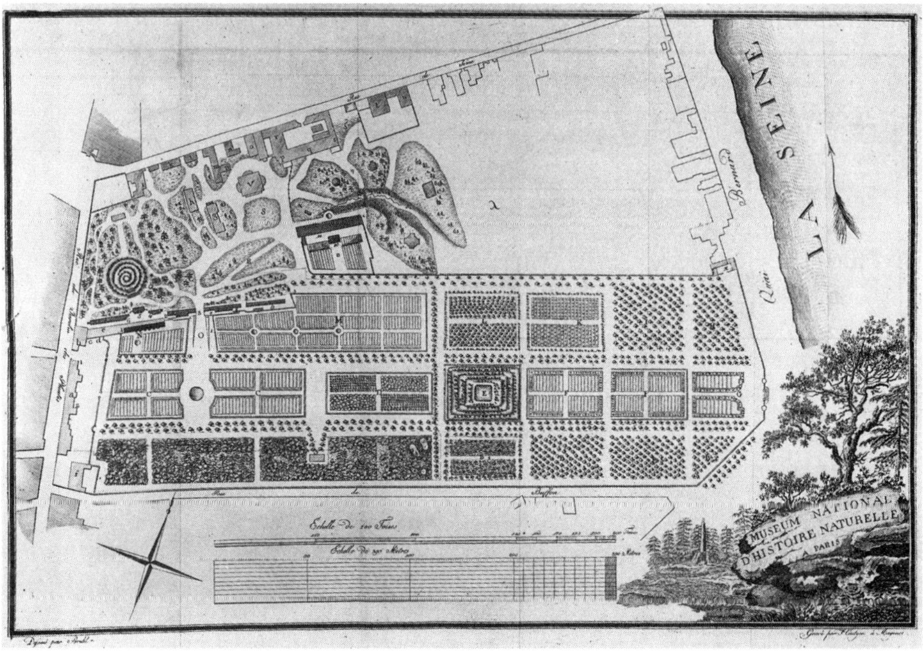

Figure 1.14.2. Map of the Museum of Natural History c. 1802 showing gardens, buildings, animal parks and shelters, and territory yet to be acquired by the Museum. From Gotthelf Fischer, *Das Nationalmuseum der Naturgeschichte zu Paris* (1802–3). Museum national d'histoire naturelle. Bibliothèque centrale. Fonds général. 7665—1, 2.

to seasonal changes. In summer the lions could move to the north side for shade and a cooler temperature; in winter the bears could relocate to the south side for sun and warmth.[7]

Fischer was impressed by the design elements Molinos introduced and with the way he had managed "to multiply the spaces" in a relatively small area. Visitors would be charmed by the contrasts between "the most fearsome" animals and "the most beautiful" ones and between wild and more rural settings. Beyond the picturesque was the educational dimension of displaying fauna and flora from the same lands.[8] Monkeys, raptors, and other creatures would have living spaces (some heated) along the walls to the west, north, and south. A fountain on an evergreen-covered hill would fill a large reservoir from which water would cascade to a pool in front of the amphitheater, then to another pool at the botanical school and elsewhere on the grounds as needed, with any surplus flowing into the Seine.

Fischer enthused that "the execution of this beautiful plan" was "in full swing," and he summoned the image of "the beautiful moment when the picture is fully finished," when one could pause and reflect on all the wonderful, communal human projects achieved through the work of many busy hands guided by reason.[9] But that beautiful moment never arrived; the plan was significantly revised before Fischer's book was even published, and there was never a point when one might say the menagerie was definitively complete.

In October 1801, when Chaptal, the minister of the interior, scrapped the idea of the craggy cliff with homes for the lions, bears, and other fierce creatures, he did the Museum another large favor. He had already spared it the administrative reorganization that his predecessor, Lucien Bonaparte, had demanded. Weeks later, wishing to help the Museum and provide winter jobs for laborers from the public workhouses, Chaptal invited the Museum to state its construction priorities.[10] It identified developing land for the menagerie as its most pressing need and requested the assistance of two hundred laborers working nine days out of ten for the next 150 days. The property owners who had been warned to vacate their land were finally ordered to move, and terrain long earmarked for the menagerie began to be developed. The creation of the "Swiss Valley" was under way.

Molinos's rocky crag for the *animaux féroces*, however, never got past the design stage. He produced a clay model, then revised his design to reduce the projected building cost from 700,000 francs to 100,000 francs. Chaptal liked the idea when the professors first pitched it, but he soon changed his mind.[11] The design, he told them, had "several inconveniences that I had not at first noticed and that have led me to renounce the project." He worried that the caves would be too damp for the animals and that the public would be unable to see them should they retreat too deep into their lairs. Creating the *rocher* would be expensive, given the quantity of boulders it demanded. The vegetation covering them would gradually grow roots between the rocks and ruin the binding cement, adding to the annual maintenance costs.

Chaptal requested a more basic building, one where the quarters for the animals would be as healthy as possible. He instructed the Museum to inform the architect of his critique and urged the professors to "invite several artists" to offer new ideas. Did it embarrass the professors that Chaptal identified serious defects in the original plan that they had missed? The minutes from their meetings are silent on this score, but Molinos, apparently not wishing to be upstaged by another architect, had produced a new plan by the middle of November. The earliest reference to his design speaks of "circular cages."[12] This was the project that evolved into the structure called the "rotunda."

Early in December 1801, Molinos told the Museum that Chaptal had approved his new plan and was prepared to provide 80,000 francs to have some of the animal cages completed in the next five months.[13] But Molinos was slow to produce detailed drawings, leaving the professors anxious about the cages' dimensions. The Council for Public Buildings told Chaptal that it was "not possible to form an idea, even close," of Molinos's project given the superficial sketch it had received. The CPB felt it needed from the Museum, independent of Molinos's estimates and construction plans, "the program of this new menagerie's needs." Ten days later the Museum had its "Program for a national menagerie" ready for the CPB's consideration.[14]

The program was basically a statement of guiding principles. With a broad brush, it portrayed the kinds of spaces and shelters needed for the diverse animals that the menagerie wanted to house: herbivores, carnivores, monkeys, terrestrial and aquatic birds, insects and fish. A more detailed plan, it said, would need to address the animals' health, the mechanics of caring for them, and the public's enjoyment. The animal quarters would be designed "to offer safety to the animals so they are not disturbed by the public, [or] poisoned or stolen by the malicious." The menagerie's elements would be coordinated to present "a picturesque whole." Strolling its winding pathways, visitors would take pleasure in seeing the animals, the little huts, fishponds, greenery, and plantings, and would never fear danger from the animals. The biggest challenge, the Museum acknowledged, would be the building for the *animaux féroces*, but the Museum expressed pleasure in Molinos's new design, "a rotunda that has all the characteristics of an ancient arena."[15]

According to Molinos's plan, the animals' cages would be arranged not in the interior circles but around the outer circumference. Each cage would have a small, outdoor park "where during good weather the animal would be able to breathe fresh air and give the public an idea of its agility." An interior gallery would allow visitors to view the animals in bad weather when they were in their cages. Helpfully, the cages that were finished first could be occupied before the whole construction was complete.[16] It is often stated that the rotunda was designed to resemble the "Grand Cross" of the Legion of Honor, but nothing in the Museum's records supports this claim; in fact, Molinos's design preceded that of the Legion's medal.[17]

The desired site for the rotunda overlapped with a property belonging to a man named Talboutier, who was willing to exchange it for one belonging to the state. The Museum told Chaptal that Bonaparte had examined the area and said he would "gladly lend himself to all that can make the Museum more worthy of the French nation, of which it is one of the principal monuments."[18] After Chaptal and the minister of finance endorsed the acquisition and Talboutier agreed to one of the properties offered him, the exchange awaited the Tribunat's approval.

The case was presented by Mathurin Sédillez, who described the

Museum of Natural History as "the most interesting of all our monuments." Whereas other sights in Paris "often left one's eyes tired and one's heart empty," he said, everyone could find "pleasures suitable to his tastes" in the Museum, which offered "samples of all the beings of nature." Here was "nature herself abridged," a "terrestrial paradise," a "divine Elysium," a "delicious Eden" where one could believe oneself "present at the creation." Expanding the Museum's territory was a matter not only of "communal and public utility," but also of "national glory" involved in "perfecting a monument destined for the instruction and pleasure of all French, [and] worthy, more than any other, of the admiration and envy of foreigners."[19] Sédillez did not, however, overstate the case with respect to the animals themselves. Each could be seen as the proprietor of its own domain, he said, receiving as it were "all the advantages of civilization," but these advantages came "in exchange for its liberty" and were but a "feeble indemnity, calculated on the pleasures of the master, rather than on the interests of the slave."

The acquisition of Talboutier's property in August 1803 allowed Molinos to begin laying the rotunda's foundations. It also inspired the Museum's zoologists to draw up their own plan for making the menagerie larger and more harmonious.[20] Ignoring key features of Molinos's design, notably the bear pits and the rotunda, the professors of zoology drew up a broad proposal calling for the creation of ten animal "villages," five for mammals and five for birds, each with its own kinds of animals and shelters.[21] The assembly instructed the zoologists to consult Molinos about their plan and the work it would require, but nothing came of it. In retrospect, the exercise demonstrates how uncertain the Museum's vision of the menagerie's ultimate layout remained at the time.

The ceremonial setting of the rotunda's first stone took place on 9 February 1804, but work on the building proceeded fitfully at best.[22] In April 1805 the structure was far from completion, and Geoffroy told his colleagues that the 20,000 francs remaining of the menagerie's construction funds for the year would be insufficient to finish it. He proposed spending 13,000 francs on projects that could be completed that year: the bear pits, an aviary, new monkey cages, and other small improvements.[23] His colleagues agreed, despite director Fourcroy's observation that the remaining 7,000 francs would scarcely advance the rotunda's progress.[24] As it was, finishing the bear pits was an excellent idea. The facility, designed to accommodate three pairs of bears in three pits, opened in the summer of 1805 with two American black bears as the first residents. The site suited the public's enjoyment and the bears' health, and it became one of the menagerie's main attractions.[25]

From 1805 to 1808, work on the rotunda essentially ceased. Then, in June 1808 the administration decided to repurpose the structure. Parkie, the female elephant, needed new housing. When a recently acquired building was judged unsuited for the purpose, the Museum cast its eyes on the rotunda and instructed Molinos to revise his plans so the building could house herbivores instead of carnivores.[26] The structure was subsequently designated for "the peaceful animals."[27]

A new purpose and name did not ensure that work on the building would resume. It took a confluence of contingent events to put things in motion. From one direction, Napoleon's military success in Austria ironically illuminated the menagerie's shortcomings. Concurrently, the government's offer in May 1809 to purchase a powerful and expensive water pump for the Museum led the administration to think that the money could be better spent elsewhere. The assembly appointed Thouin and Desfontaines to identify the institution's greatest needs.[28]

Previous French military victories had brought animals to the menagerie as war trophies, with the elephants from Holland and the bears from Bern the most prominent examples. In May 1809 the menagerie seemed poised to benefit again from the advance of French troops. Vienna was occupied, and Napoleon had taken up residence at the Schonbrunn Palace, whose menagerie was said to have aurochs (wild oxen). Acquiring them for the Paris menagerie was an attractive possibility. Aurochs went unmentioned in Thouin and Desfontaines's report on the Museum's needs, though they may have had them in mind in observing that the herbivores' parks were ill equipped to receive the animals needed "to complete this useful and beautiful part of the Museum."

The two professors' primary concern was that the numerous French citizens who were visiting Schönbrunn would see that its menagerie and greenhouses were far superior to their counterparts at the Museum in Paris. One could only look "with pity" on the latter, they said—not the sentiment the French government wanted to evoke. If the government was willing to provide special funds for an expensive water pump, it was their duty to invite it to direct its largesse to ensuring that the Museum was a match for any institution in the world devoted to the progress of the sciences and national glory. What the Museum needed was the completion of the rotunda. Other pressing needs included new quarters for the *animaux féroces*, expanded animal parks, and the construction of a greenhouse that could accommodate large plants from the Tropics and Torrid zones. The professors framed finishing the rotunda and attending to the other items as matters of national honor.[29]

Two weeks after the report by Thouin and Desfontaines, Geoffroy presented his thoughts on how to make the rotunda suitable for the elephant and the other animals. His colleagues followed up with additional suggestions in a report to the architect.[30] In August the Museum wrote to the minister of the interior to urge that work on the rotunda resume and that the structure house herbivores instead of carnivores. It would be cheaper to use the rotunda for herbivores, the Museum said, because having carnivores there would require more iron bars and guardrails and a long, expensive sewer running to the Seine. Plus, the current location of the elephant's quarters and the herbivores' winter stables was unsatisfactory. These were in the Régie directly under Cuvier's comparative anatomy gallery, and the animals' emanations were rotting the gallery's floorboards. Moreover, with the animals in this space, the comparative anatomy gallery, "the most precious in the world," could not be expanded.[31]

In January 1810, when the Museum again pressed the government to complete the rotunda, it was dealing with a new minister of interior, Jean-Pierre Bachasson, comte de Montalivet (the fourth minister since Chaptal was replaced in August 1804). Montalivet tasked Molinos with drawing up the plans. The architect envisaged the structure as being ready for occupancy by the coming winter, but when the CPB reviewed the plans and investigated the site in July, it found the project severely wanting. It reported to Montalivet that the construction resembled "a veritable ruin," and that the rotunda's design could not serve its intended purpose. It recommended demolishing the unfinished structure and having Molinos produce a design better suited for herbivores.[32]

Writing to Montalivet, the Museum countered the CPB's observations and objections, while curiously noting that the rotunda's design with its "bizarre forms and appearance like a fortress" had "never been to the administration's taste." It also recalled (incorrectly) that while the government was pressing for "the erection of a building for the animals," the Museum had only two designs to choose from—the cavern-filled *rocher* and the rotunda—and it chose the latter as "the less defective." The rocky escarpment design had in fact been rejected by Chaptal, and the rotunda was designed as an alternative. Now (August 1810) the administration told Montalivet they felt it was best to continue with the rotunda, which was already under way, but it was prepared to leave the decision up to his "superior knowledge."[33]

Montalivet instructed the professors, Molinos, and representatives of the CPB to meet on the site to pursue the matter. Inspecting the unfinished structure, they concluded that its foundation was sound but some features needed revision. The CPB representatives, who also visited the existing stables for the elephant, camels, and other herbivores in the Régie, confirmed that the animals' exhalations threatened the comparative anatomy gallery above. Face was saved all around. It was collectively decided to proceed with the rotunda, and Montalivet promptly authorized Molinos to recommence work.[34]

The rotunda was not ready to house animals that winter.[35] Eighteen months later (May 1812) the building was still not finished.[36] Another year went by, and the professors were again exhorting Molinos to complete the work.[37] It was not until March 1814 that they charged him with adding the finishing touches that would make the rotunda "entirely habitable" and allow the elephant's transfer there by mid-May.[38] At that point the Museum received another reminder that animals do not always follow the scripts humans write for them. The rotunda was ready for the elephant, but Parkie was unwilling to traverse the unfamiliar territory between her quarters and the new domicile.

The staff supposed that if they did not feed her on the night of 30 June, she would be easier to entice with food the next morning, but on 1 July 1814, when the doors of her quarters were opened and she was gently encouraged to venture forth, she became anxious and then frightened. As reported in the minutes of the professors' next meeting, she displayed "an extreme distrust and finally a firm resolve to remain in her old abode."[39] The report did not say so, but an elephant with firm intentions was not to be trifled with. Six weeks earlier, Parkie had nearly killed a man allowed into her stable by a keeper, against all rules, in exchange for a tip. Menagerie superintendent Frédéric Cuvier proposed transporting Parkie to her new quarters by a wagon that would have to be built, and as winter approached he recommended waiting until spring to attempt the move.[40] Consequently Parkie did not enter the rotunda until the end of May 1815.[41]

Aside from the bears that had the relatively good fortune to be moved to the bear pits, the fierce animals remained in their cramped cages by the Seine. After the administration's decision in 1808 to house the herbivores in the rotunda, it was almost a year and a half before new housing was planned for the carnivores. In November 1809 Geoffroy Saint-Hilaire presented a detailed description to his colleagues of the features an appropriate new building would comprise.[42] In August 1810 Montalivet reported to the Museum that 400,000 francs had been granted by the emperor Napoleon "for the construction of a beautiful menagerie at the Museum of Natural History."[43] Molinos and the professors produced a new plan, but when Napoleon saw it he was not impressed. As Montalivet told a delegation of professors in late January 1811, "He does not find it magnificent enough."[44]

The menagerie for the *animaux féroces* was not achieved under Napoleon. The preparations for war with Russia in 1811, the disastrous invasion of Russia in 1812, and the country's dire financial circumstances in 1813 precluded any thought that the new menagerie, magnificent or otherwise, would be realized anytime soon.[45] The question was not revisited until 1817, two years after the Bourbons were restored to the throne, and the *animaux féroces* in the old, provisional menagerie did not get a new home until 1821.

As for the rotunda, its occupation in 1815 marked a milestone in the menagerie's early development. It was not the rocky escarpment that Lacépède, the architect, or the administration had first envisaged. Nor had it ended up serving the purpose for which it was originally intended (housing the *animaux féroces*). Its design was purportedly not to the administration's taste, only the least defective of its available options. That the rotunda stands today as the menagerie's most iconic structure may be ironic, but it is historically fitting. The missteps, meanderings, contingencies, and compromises in its story were characteristic of the first decades of the menagerie's existence.

2

Life in Motion

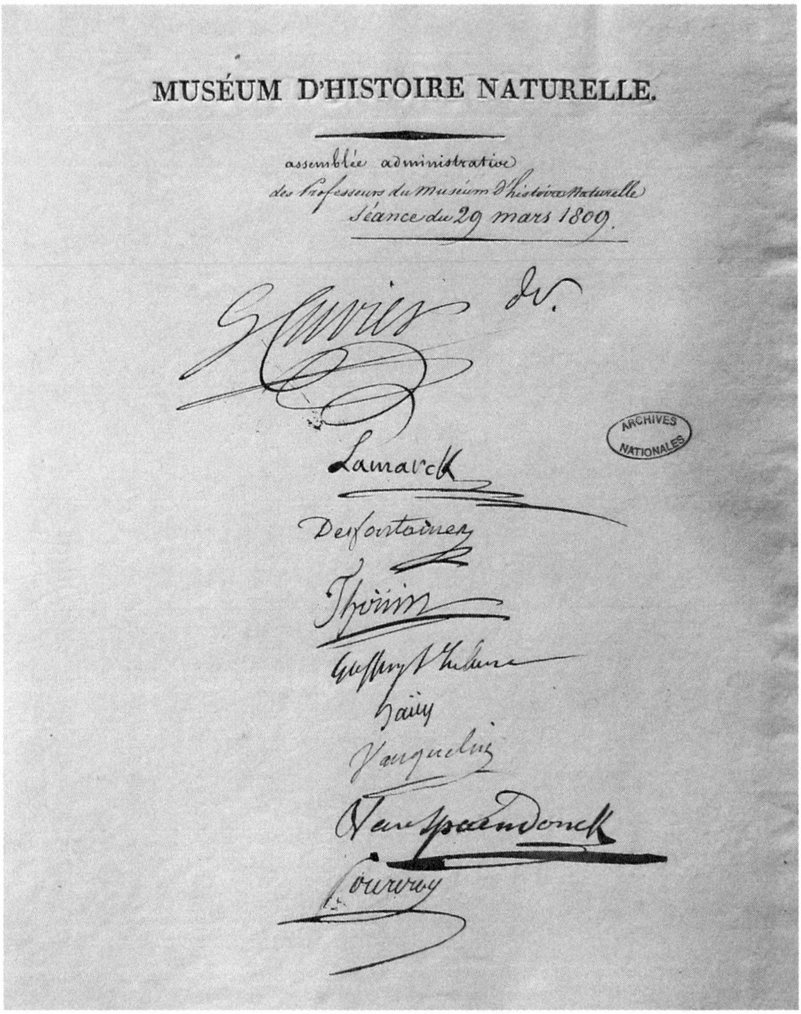
MUSÉUM D'HISTOIRE NATURELLE.

assemblée administrative
des Professeurs du Muséum d'histoire Naturelle
Séance du 29 mars 1809.

G. Cuvier
Lamarck
Desfontaines
Thouin
Geoffroy St Hilaire
Haüy
Vauquelin
Van Spaendonck
Fourcroy

ARCHIVES NATIONALES

Figure 2.1.1. Attendance sheet for the professors' weekly administrative meeting of 29 March 1809, signed by Georges Cuvier, Jean-Baptiste Lamarck, André Louiche Desfontaines, André Thouin, Étienne Geoffroy Saint-Hilaire, René Just Haüy, Louis Nicolas Vauquelin, Gerard van Spaendonck, and Antoine-François Fourcroy. Archives nationales. AJ/15/601. Photograph R. Burkhardt.

[2.1]

Naturalists' Practices and the Menagerie in the Golden Age of French Zoology

The Treaty of Amiens, signed in March 1802, established peace between France and Britain and brought joy to citizens on both sides of the Channel. Now British tourists could return to Paris to see the city's renowned old sights and new ones as well, including the Louvre's recently appropriated Italian paintings and the live animals at the Museum of Natural History's menagerie. Naturalists came too, attracted by the Museum's collections and by its faculty, three of whom (Georges Cuvier, Jean-Baptiste Lamarck, and Étienne Geoffroy Saint-Hilaire) were ushering in a golden age in French zoology.

Cuvier's focus on animals' internal anatomy had brought about a reformation of zoological classification. Lamarck, building on Cuvier's main idea, was identifying new classes of invertebrates based on their internal organization. Meanwhile the two were laying the foundations for vertebrate and invertebrate paleontology. In addition, Cuvier, though not yet a professor, had built the Museum's comparative anatomy collection into one of the institution's most extraordinary assets. In 1802 Cuvier and Lamarck were rejoined by Étienne Geoffroy Saint-Hilaire after his nearly four years away on Napoleon's expedition to Egypt. He returned with a host of new specimens, adding to what the Museum could already boast was the world's best natural history collection. With its new but once again bursting galleries and the presence of Cuvier, Lamarck, and Geoffroy and their talented assistants, the Museum stood at the vanguard of zoological study.

The fall 1802 inauguration of *Annales*, the Museum's first journal, signaled the institution's vitality.[1] Appearing monthly, it featured the best and latest work in natural history. In its first volume, Lamarck began publishing his studies of fossil invertebrates from around Paris; Cuvier brought anatomical insights to the classification of mollusks; and Geoffroy presented papers on fish he had studied in Egypt: a new genus from the Nile, a flat fish from the Red Sea, and a comparative study of the electricity-producing organs of three species (one of them an eel).[2] The same volume included the Museum's official report on Geoffroy's Egyptian collections and, among many other items, a notice about the recent purchase from England of two kangaroos, "perhaps the most singular animals the national menagerie has yet possessed."

Aside from the kangaroos, the *Annales'* first volume gave little ink to living animals. The menagerie's first role in French science was as a source of cadavers for anatomical study and for gallery specimens. The menagerie's residents were front and center, however, in Simon-Charles Miger's elegant publication pairing Nicolas Maréchal's attractive animal portraits with text by the Museum's zoologists. Appearing in installments from 1800 to 1805, Miger's work ultimately included thirty-seven essays, thirty-eight masterful hand-colored engravings of Maréchal's paintings, and an introduction where Lacépède maintained that menageries reflected a civilization's advancement. Cuvier wrote thirty-two of the animal essays, Geoffroy three, and Lacépède two. Two of Geoffroy's pieces, on the Egyptian goose and the mongoose, reflected the expedition he had just returned from.[3]

While all three leaders of French zoology's golden age had important interactions with the menagerie, they did not view it as a major site for research. Their preferred scientific habitats were rooms where they could assemble and compare animal remains, aided by scientific publications and illustrations. In addition to this focus on dead specimens, another feature of French zoology's golden age looms large in retrospect, though it was little appreciated at the time. This was Lamarck's elaboration of the first comprehensive theory of biological evolution. I will describe that theory in due course, but for now let me note how two events occurring on the same day underscore the way the life of the menagerie, Lamarck's evolutionary theorizing, and French zoology's primary concerns proceeded largely independent of each other.

On 16 May 1802 (27 floréal an 10), Lamarck made two important announcements, one shortly after midday in the opening lecture of his invertebrate zoology course, the other early that evening in a report to his fellow professors at the Museum's weekly administrative assembly. In the lecture, he outlined his bold new view of the evolution of all the different forms of life on Earth, beginning with the simplest and proceeding gradually to the most complex, and he told his students there was no issue in natural history more deserving of their attention.[4] At the assembly of professors, he announced that the female elephant's recent digestive difficulties had been happily resolved. The former event stands as a landmark in the history of biology. The latter was just a minor episode in the history of the menagerie. Ironically, we know how Lamarck's colleagues reacted to his news about

naturalist, one that takes his observations from the subject itself, and not from the writings of others."[11] White acknowledged the utility of naturalists who produced systems of classification, without which "the field of Nature would be a pathless wilderness," but he insisted that classification should be a means, not an end, of studying nature.[12] "Faunists," he said, were "too apt to acquiesce in bare descriptions, and a few synonyms," which could "be done at home in a man's study." What interested White was "the investigation of the life and conversation of animals, [which] is a concern of much more trouble and difficulty, and is not to be attained but by the active and inquisitive, and by those that reside much in the country."[13]

The best French example of someone paying attention to the behavior of wild animals was Charles-Georges Leroy (1723–1789), the keeper of the king's game at Versailles and Marly. Calling himself "a dedicated hunter," he sought to illuminate the intelligence of common animals (rabbits, foxes, deer, wolves, and such). Writing that he had "taken his course of philosophy in the woods," he urged that the naturalist, after learning about animal anatomy, "should give up his scalpel, abandon his cabinet, and plunge into the woods to study the ways of these beings."[14]

An example of a voyager naturalist who spent time observing the lives of birds rather than just collecting them was François Levaillant (1753–1824). In recounting his efforts as a naturalist in South Africa from 1780 to 1784, he often complained that he suffered pains and deprivations to advance ornithology, while cabinet naturalists, who claimed to teach the secrets of nature, remained comfortably at home and consequently ignored the wealth of "true facts" that "simple folk" who lived close to nature could have told them.[15] His experience in nature, Levaillant insisted, helped him discern relations between forms that Buffon had missed. "Let us study the ways of life of animals," he wrote. "Let us follow them in their private lives and in their habits. It is then that the affinities they have with one another will not escape us, and we will know the true places that nature has assigned them."[16] But Levaillant was another naturalist whose career encompassed more than one regime of practice. He spent five years making field observations and collecting specimens in Africa, but then he spent twenty-five years back in Europe organizing his ornithological cabinet.[17]

Emmanuel Baillon (1742–1801) was yet another naturalist of the era who displayed the sensitivities of a field naturalist but was adept in other contexts as well. Residing in the Breton countryside, he spent hours, no matter how uncomfortable the conditions, studying animals in their habitats. He cultivated a close relationship with the Museum, but his behavioral observations were not the feature of Baillon's efforts that the Museum valued most. His primary importance to the institution was as a supplier of specimens, especially waterbirds, not only for the galleries but also for the menagerie.

A final type to add to the taxonomy of naturalists circa 1802 was yet to be identified, though it was near to being so. This was the naturalist who chose to study the lives and behavior of wild animals in captivity, as in the menagerie. Many contemporary zoologists viewed this as a quixotic project, believing that the unnatural conditions menagerie animals lived under rendered studying them useless. Frédéric Cuvier, Georges Cuvier's younger brother, would seek to prove otherwise. Yet he did not limit his research to menagerie studies. One of his accomplishments was strictly cabinet based: he developed a system of mammalian classification at the genus level, based on teeth. With such distinctions and qualifications in mind, we return to the three leaders of French zoology, Georges Cuvier, Jean-Baptiste Lamarck, and Étienne Geoffroy Saint-Hilaire.

In his capacity as chair of invertebrate zoology, Lamarck had no professional reason for involvement with the menagerie. In the Museum's early days, as one of the Museum's two zoologists, he was often charged with menagerie-related tasks, usually in conjunction with Geoffroy; for example, in 1794 when they were sent to Raincy to choose live animals to bring to the Museum.[18] He was relieved of most of these tasks when Delaunay became menagerie supervisor in 1795, until July 1801, when the administration again decided to appoint a professor to the job. Geoffroy was in Egypt, and Lamarck agreed to undertake "the surveillance of the animals and their keepers" and report regularly to the assembly on "everything concerning the menagerie."[19]

Whatever Lamarck's colleagues may have thought of his theorizing, they were happy for him to assume numerous administrative duties, from reporting on specific questions to serving on the Museum's "bureau" as its secretary or treasurer.[20] As menagerie inspector, he began considering actions to ensure that the carnivores would survive the coming winter.[21] He was the administration's contact with the keepers, whose flouting of Museum rules had caused the menagerie's surveillance to be reassigned to a professor. For example, he had to deal with a complaint about Félix Cassal, while also transmitting Cassal's helpful recommendation that the menagerie be closed early enough in the evenings for the animals to be fed without the public's interference.[22]

Lamarck concerned himself with the design of the new menagerie for the *animaux féroces*, including the appropriate dimensions for their cages, the requisite construction materials, and the means of ensuring that naturalists, artists, and the public alike had a good view of the animals throughout the year.[23] He reported on animal illnesses, including those of both the male and the female elephant.[24] He likewise announced the death of the last cub from Constantine's three-cub litter of 1800, a victim of teething problems.[25] In addition he proposed means of thwarting thieves, repairing pavement, improving storage for fodder and meat, culling overabundant herbivores, and providing provisional housing for the latest litter of lion cubs, which were growing stronger and beginning to misbehave.[26]

When Lamarck was appointed menagerie supervisor, it was understood the job was temporary and would be assumed by other professors. After a year, he pleaded that the "poor state of his health" prevented him from continuing and asked to be relieved. Home from Egypt, Geoffroy replaced Lamarck on the first day of the new Revolutionary year: 1 vendémiaire an 11 (23 September 1802).[27]

Lamarck is not remembered for his efforts as menagerie overseer. He is known instead for his work as an invertebrate zoologist, paleontologist, and evolutionary theorist. But he had been with the menagerie for a year when he published his first major evolutionary treatise, *Researches on the Organization of Living Bodies* (*Recherches sur l'organisation des corps vivans*), a book he'd rushed into print to forestall any misrepresentation of the ideas he had lately announced to his students.[28] It's worth asking whether the animals of the menagerie were in any way an inspiration for Lamarck's evolutionary thinking. The answer is almost certainly no for the earliest expression of his ideas on species change but yes for a last-minute addition to his later *Zoological Philosophy* (*Philosophie zoologique*, 1809).

Lamarck delivered his first statements about species change in the introductory lecture for his course at the Museum in 1800 and expanded his theory in the comparable lecture for 1802. In both, he boldly told his students he could "prove" that habits give rise to structures rather than vice versa, and that gradually, over time, habits, ways of life and environmental conditions had been responsible for producing the animal forms and faculties of the present.[29] To support this assertion he cited the correlation between birds' habits and the structure of their feet or legs. By his account, the webbed feet of swimming birds, the curved feet of perching birds, and the long legs of wading birds were all consequences of habits maintained over great stretches of time. The wading bird, he explained, does not like to swim but must draw near the water to find its prey. Obliged to walk in the mud "but wishing to behave in such a way that its body does not plunge into the water," it develops the habit of extending its legs. "For the generation of these birds that continue to live in this manner, the result will be that individuals will find themselves elevated, as on stilts, on long naked legs."[30]

What made Lamarck think about wading birds? One possibility is that he observed the ones Emmanuel Baillon had sent the Museum the previous year.[31] Another is that he contemplated the specimens recently installed in the ornithology collection on the second floor of the natural history building. But he may also have consulted the many mid- to late eighteenth-century ornithology and natural history texts where the correlation between birds' feet and their habits was already discussed. The Abbé Noël-Antoine Pluche's immensely popular *Spectacle de la nature* (1741), a celebration of the Creator's wisdom, explained, for example, how the heron's "bizarre" appearance, its long neck, long, featherless legs and thighs, and long, sharp-notched beak, corresponded with the way the bird wades and catches its prey.[32] Aside from texts like Pluche's lauding divine design, others looked to the structure of birds' feet as a primary character for avian classification.[33] There is nothing to suggest that Lamarck derived his ideas from firsthand observations at the pond for waterbirds or the galleries where stuffed birds were displayed. Standard discourses about animals' habits and structures and their relation to living conditions were all he needed as a starting point.[34]

Lamarck famously cited the giraffe as an example of evolutionary change in 1802, years before the Museum in Paris received its first live specimen (1827), at which point he had been blind for nearly a decade. He referred to the giraffe to illustrate how changes in habits produce changes in structures as an apparent afterthought in the index to his *Researches on the Organization of Living Bodies*. Under the heading "animal habits," he wrote: "To the examples I have cited I would add that of the form of the giraffe (*camelopardalis*), a herbivorous animal that, living in arid lands without herbage, is obliged to browse the leaves of the trees and to strive continually to reach them."[35]

We can only speculate about why Lamarck added the giraffe to the bird examples he had presented earlier, but he would have been familiar with the giraffe specimen installed in August 1800 in the hall of mammals on the upper level of the Museum's natural history gallery. Towering in the center of the hall, surrounded by a balustrade, it was the most striking sight for all who entered. Voyager naturalist Levaillant had killed the animal in South Africa and preserved its skin. Naturalist's assistant Desmoulins mounted the specimen with the help of the taxidermist Antoine Delalande and the carpenter Lassaigne. The press hailed their achievement of "preparing the remains of animals to the point of giving them the appearance of life and movement."[36]

Lamarck left no hints about what prompted his use of the giraffe as an example of species change, whether the specimen in the hall of mammals, the skeleton from the Stadtholder collection, or an illustration or discussion in a book. He never saw a live giraffe, and his bird examples were evidently stock cases from texts. But there was a captivating case to come involving observation of an animal in the menagerie. In 1809 the Museum acquired a live seal that led Lamarck, the naturalist philosopher, to ruminate not only on the influence of habit on form but also on the path nature had taken in the development of the vertebrates.

∴

The menagerie directly contributed to Georges Cuvier's success as a cabinet naturalist by providing him with specimens for anatomical observation. The deaths of the menagerie's first male elephants in 1802 and 1804 were a boon to him. In addition, his essays on the menagerie's animals, paired with Miger's engravings of Maréchal's illustrations, gave him the opportunity to display his writing skills as well as his mastery of the zoological literature. In one respect, however, Cuvier's stance toward the menagerie bordered on the adversarial. He saw galleries as the primary site for zoologists and comparative anatomists to display the breadth of their collections and the power of their approaches to comprehending the order of nature. The Museum's funds were limited. More money spent on the menagerie meant less was available for the galleries. And then there was the problem of the vapor rising from the stables of the elephant and other herbivores, situated directly below Cuvier's comparative gallery. The moisture was rotting the beams supporting the gallery floor.

Compared with Lamarck and Geoffroy, Georges Cuvier was the most resolute of the Museum's zoologists in extolling the practices of the cabinet naturalist. In 1798, when he and Geoffroy were invited to join Napoleon's expedition, Geoffroy jumped at the chance, but Cuvier decided to stay put. Historians of science have often cited his shrewd assessment of which choice would better serve his career. As Cuvier recalled years later, "My calculation was soon made. I was at the center of the sciences in the midst of the finest collection, and I was sure to do the best work there, more coherent, more systematic, and containing more important discoveries than in the most fruitful voyage. . . . I have always congratulated myself on this decision."[37]

Two fascinating reports on galleries that Cuvier wrote in 1803 exude the self-assurance he had gained from his appointment as a professor at the Museum (October 1802) and his election as a perpetual secretary of the Institut de France (February 1803).[38] In the report appearing in the Museum's *Annales*, Cuvier described his success in developing the comparative anatomy collection and its gallery.[39] In the other, never published, he described on behalf of a Museum committee the institution's desperate need for more gallery space.[40] Previously unremarked by historians, this invaluable document goes beyond the rationale for expanding the galleries to expound on the Museum's broader intellectual purpose, the cabinet naturalists' need to compare and manipulate specimens, and the politics of building special collections at an institution where many professors had a specific collection and gallery to promote.

In the *Annales* report Cuvier recounted how he rescued the comparative anatomy specimens from dilapidation and disarray and transformed them into the world's greatest collection. The process began in 1795 when he took over most of aging anatomist Jean-Claude Mertrud's duties and rented Mertrud's apartment, which abutted the recently acquired Régie building. Cuvier had a door cut between the apartment and the Régie, gaining access to two huge rooms that he partitioned into several smaller ones. He gathered specimens the Museum already owned but had been keeping elsewhere: animal skeletons scattered high and low beneath eaves and in cellars; organs in alcohol that were not needed in the zoology collection. New specimens arrived through donations, purchases, or deaths at the menagerie. In January 1797, after the naturalist Emmanuel Baillon praised him for so admirably developing the new space and the collection, Cuvier half playfully "scolded" Baillon for his effusiveness, writing: "Believe seriously that no one knows how to love the collection [*cabinet*] more sincerely than I do; it seems to me that it belongs to me. I have the pleasure of adorning it, as a lover dresses his mistress. My wishes would be fulfilled if I had the certitude of seeing it one day complete."[41]

In his 1803 paper Cuvier said of his expansion into the Régie: "One henceforth had all the space necessary and nothing stood in the way of the collection's undergoing the most rapid growth."[42] Years later, preparing notes for some future biographer, Cuvier's account of this achievement had more bite: "It was in pursuing these enterprises, with the help of some Professors, and against the hindrance of others, that I succeeded in making my collection so important that no one dared oppose its further enlargement."[43]

Cuvier did not elaborate on the politics of collection building in his 1803 paper, but he addressed it candidly, even cheerfully, in July of that year when he reported for a five-person committee on the Museum's need for more gallery space.[44] He characterized the Museum's administration as a system that happily combined "an absolute division of the scientific work" with "a communal sharing of economic decisions." "The zeal of each professor," he said, "animated by the ardor of personal glory, makes him ceaselessly plead the cause of the objects he is charged with studying. The disinterested wisdom of the whole assembly checks this zeal only when it begins to harm the other parts."[45] Cuvier admitted that the comparative anatomy collection's location in a building separate from the other collections was fortunate, as it obviated the need to compete for space and enabled the collection "[to arrive] in a very few years at the point that all the world knows."[46] He in effect excused any aggressiveness on his part by insisting that individual enterprise was good for the whole. He saw the quest for glory as an entirely appropriate stimulus for the naturalist intent on expanding the boundaries of knowledge.

Cuvier's triumphs in assembling the comparative anatomy collection were more than a function of his ambition and talents or of having adequate space. He benefitted from his very capable assistants, starting with Pierre Rousseau, twelve years Cuvier's senior, who dissected animals for the collection.[47] Cuvier went on to recruit and pay for additional helpers from his hometown of Montbéliard, including his younger brother, Frédéric, followed by Georges Louis Duvernoy and Charles Léopold Laurillard.[48]

Despite the assistance of these and others enlisted as correspondents, Cuvier liked to portray himself as working alone in his office, surrounded by specimens, contemplating and handling them, allowing them to guide him to unimagined insights. His discourse on the galleries situated the Museum as an ideal site for the exercise of "*le génie*" (meaning "one's genius" here, but potentially referring to a person as well). "The great utility of a cabinet consists especially in the numerous means that it ceaselessly offers to *le génie* to arrive at new results." The Museum should represent "nature herself brought near and presented in her entirety to all minds."

Cuvier acknowledged that seemingly unremarkable specimens could "change the face of the globe or humanity's destiny," but he saved his best rhetoric for the considerations he felt most worthy of the human mind. One needed to go beyond a "knowledge of particular species and their properties," he said, to consider "their mutual comparison, their reciprocal relations, the place each occupies in the great system of being, the phenomena of life and their development, the role they play in the vast machine of the universe and in the maintenance of this order of nature, always in movement, yet always immutable." These were subjects for "profound meditations" and "sublime conceptions," and all were accessible "in a collection like ours."[49]

How would this function in practice? The savant's workspace, Cuvier indicated, should be disposed in such a way that "one could without pain, at every instant, manipulate, bring together, [and] compare the reassembled objects, study them under all their aspects, [so] that their order could be varied according to all the combinations, that each new idea could be verified and followed at the moment it arose." He continued, "it is only by living amid the objects of research, pursuing them by all the senses, nourished so to speak by their perpetual contemplation, that *le génie* prepares itself for the journey. The illumination of a happy chance will reveal the route; it is thus necessary to multiply as much as possible the combinations where chance can arise and from which illumination can spring forth."[50]

All this related to the need for more gallery space. Unlike philosophers, whose meditations were not tethered to their material situation, Cuvier said (of himself and his colleagues), "we cannot do without the eye or the hand." How was one to gather and prepare specimens when there was no place to put them? How was one "to care for and conserve objects when one is not able to manipulate them?" And how was one "to study them when the eye cannot seize them from all sides?" Finally, how was one "to arrange them in a reasonable order when one is forced to keep them all in a narrow locale?"[51]

Having set the stage for the basic facts of the committee's report, Cuvier described in detail the galleries' history since 1793 and the need for further expansion. Since the Museum's founding, the space allotted to the display and storage of specimens had nearly tripled, thanks to the construction of the third floor of the natural history galleries, the purchase of the Régie, and the acquisition of a storage site for specimens not on display. These new spaces also made it possible to accommodate more students in the public courses.[52]

The story of the gallery of mammals was particularly striking. Before the natural history building's expansion, only about fifty mammal specimens were on display, more than half of them poorly mounted. Most were in a side gallery on the first floor, others in the room that had since become the library, and their total area was just thirty-three square meters. The new gallery on the building's second floor gave the mammals 138 square meters of floor space, and the number of specimens on display swelled to over five hundred, all in good shape. But now the space was packed so tightly that using the gallery was "very inconvenient," and the specimens could not be displayed in a rational order. Some huge animals were there, mounted with a "formerly unknown art," but there was no place to put fifteen additional ones, including the camel, the dromedary, the beautiful elephant that had died the previous year, and a recently mounted species of antelope, "big as a horse."[53]

Cuvier told a similar story regarding the ornithology gallery. The display space had grown from 45 to 162 square meters. There had been 1,034 specimens in the collection in 1793, but only eighty-eight had been kept because the early techniques of conserving them had proved more destructive than preservative. Despite this loss, by 1803 the collection grew to 2,359 specimens, thanks to additions from the Stadtholder and Levaillant collections, Baillon's birds from Brittany, and birds collected by others abroad. Here again more display space was needed because the collection still lacked a great many species and it was important to have multiple specimens to show differences relating to sex, age, and season. Ideally, eggs and nests would be displayed as well.[54]

Having offered comparable information on the other collections, Cuvier concluded that the Museum needed an additional 600 square meters to meet the needs of science, provide the conditions necessary to preserve the specimens and maintain the Museum's reputation. The committee report in effect called for an increase in display space twice as great as the gains made in the previous decade. Not all Cuvier's colleagues were enthusiastic about his formidable mix of talents and ambition. The minutes of the meeting where he presented the report simply state that they listened with great interest and then forwarded it to the architect. No mention was made of the professors' response to his thoughts on the Museum's greater goals, the life of the cabinet naturalist, or the competition involved in collection building within the institutions.[55]

In discussing the Museum's galleries, Cuvier practically ignored the menagerie, mentioning it only as a source of specimens. As a comparative anatomist and cabinet naturalist, he valued the galleries above all and did not view the menagerie as a site for sustained research. The competition with naturalists who did most of their work outdoors, however, was not easily dismissed. The same volume of *Annales* that carried Cuvier's report included letters from the most famous voyager naturalist of the day, Alexander von Humboldt (1769–1859), who had been traveling in South America for nearly five years.[56] His accounts of his adventures and scientific accomplishments were inspiring. If Cuvier embodied the cabinet naturalist's quest for glory, Humboldt had a claim to fame that was even more readily recognizable.

Cuvier confronted the question of the relative advantages of the voyager naturalist's approach versus that of the "sedentary" cabinet naturalist in a long review of Humboldt's *Tableaux de la nature* (*Pictures of Nature*, 1808).[57] "Traversing with more or less rapidity a multitude of diverse regions," Cuvier wrote, "[the voyager naturalist] is successfully struck by a great number of interesting objects and beings. He observes them in the very locations nature placed them, in true relations with their surroundings, and in all the fullness of their life and activity." Yet, while witnessing living beings in their natural context was a special advantage, such opportunities came only in brief intervals that the voyager was "not master of prolonging at his will." He was thus "deprived of the means of comparing each being with those that resemble it and of rigorously describing its characters, deprived even of books and the possibility of consulting those who have seen the same things before him." Courage and energy notwithstanding, if he was not gifted with an extraordinary memory "and the high intelligence [*génie élevé*] that quickly grasps the relations of the most distant of things," all that would remain of the voyager's observations were "isolated and fleeting traces."

By Cuvier's admission, the "sedentary naturalist," lacked firsthand knowledge of distant lands. He could not have as vivid a sense of nature's grand scenes as the eyewitness. Likewise, he missed the details of the habits of living things "that would have struck him if he had been in these places." These "inconveniences," however, were mitigated by certain advantages. "[Though] he does not see nature in action, he can pass all her products in review before him. He compares them among each other as often as necessary to arrive at sure conclusions. He chooses the questions to consider. He limits their extent at will. He gives all the time to examining them that he wants. He gathers from everywhere the relevant facts."

As Cuvier represented it, the cabinet naturalist's efforts were as courageous as the voyager naturalist's (or more) when considered intellectually: "The voyager follows but a narrow route," he wrote; "it is truly only in the cabinet that one can traverse the universe in all directions, but that demands another kind of courage, one belonging to an unlimited devotion to the truth and an unwillingness to abandon a subject except when, by way of reflection, observation, and erudition, one has illuminated it with all the rays that the present state of knowledge can provide."

Cuvier stated that he was not criticizing Humboldt, a rare individual who embodied the merits and qualities of both the voyager and the sedentary naturalist. Nevertheless, highlighting the cabinet naturalist's advantages over the voyager naturalist was a curious way to begin a review of Humboldt's *Tableaux de la nature*. Contemporaries might have interpreted it as a sign of jealousy. Perhaps Cuvier recognized this and so left the review unpublished.

Humboldt's *Tableaux de la nature* did not dwell on the relative advantages of the practices of the voyager versus the cabinet naturalist. He did say that to appreciate the character of the various large plant groupings typifying different landscapes around the world, one should be an artist who saw them in real life, as opposed to in greenhouses or in books on botany.[58] Time and again he referred to the grand scenes of nature that touched one's innermost being, noting that painters should capture them. Describing the thrilling sight of South American natives driving horses into a pond to agitate large electric eels, he wrote, "I would have wanted a skilled painter to have been able to capture the moment when the scene was the most animated."[59] What Cuvier wanted most from painters of nature, in contrast, was the accurate representation of individual specimens.

Cuvier's scientific authority was based on a thorough knowledge of specimens and books alike. He organized his library at the comparative anatomy gallery for maximum efficiency, with ornithology books in one room, ichthyology in another, osteology in a third, and so forth. Admiring this arrangement on a visit to Paris in 1829, English geologist Charles Lyell told his sister that it displayed "that extraordinary power of methodizing which is the grand secret of the prodigious feats which [Cuvier] performs annually without appearing to give himself the least trouble."[60] After Cuvier's death, G. L. Duvernoy also remarked on this efficiency: "No other author has written as many original works and took so little time in doing it."[61]

Cuvier displayed this seemingly effortless efficiency in the first years of the century in writing about the animals of the menagerie. For Miger's elegant enterprise Cuvier wrote essays on the camel, polar bear, ostrich, cassowary, Indian elephant (male), red-fronted gazelle (*la gazelle Corinne*), dromedary, panther, hyena, brown bear, white-nosed monkey, civet, quagga, zebu, hartebeest (bubale), green monkey, agouti, lion, serval, mandrill, Indian elephant (female), porpoise, female axis deer, capuchin monkey, one-horned (Indian) rhinoceros, wild goat, American black bear, llama, genet, zebra, and small zebu without horns.[62] Of these, all but two species, the rhinoceros and llama, were animals the menagerie had possessed. Cuvier's rhinoceros essay drew on anatomical illustrations made on the specimen that died at Versailles and was dissected at the Museum. His llama essay referred to a pair of animals living at Malmaison.

Cuvier's essays for Miger's project relied more on texts and dead specimens than observations of live animals at the menagerie. Live animals could not be held in one's hand, scrutinized from every angle, and precisely measured. Nonetheless, he was able to provide masterful descriptions of each species, overviews of the zoological literature (including ancient sources), and judgments on classification. He wrote economically, not padding his texts with anecdotes. Occasionally he mentioned information gleaned from his colleague Geoffroy or the keeper Cassal.

Some menagerie animals died so soon after arrival that there was little to say about their behavior. Of a white-nosed monkey (*le blanc-nez*) Cuvier wrote that it "lived only a few days at the menagerie, and it was not possible to observe exactly its habits, though it seemed gentle and affectionate." The ostriches lived long enough for him to report on their copulation, and he described the cassowary's cry and noted that the menagerie's specimen was sometimes mean (*méchant*) and irritated by people who were poorly dressed or wearing red. He reported that a she-ass in heat had been brought to the quagga, which treated her well and covered her several times, "without the need of painting her [with stripes]," which was what was done to a she-ass in England before a zebra would mate with her.[63]

He noted that the dromedaries provided the energy for the Museum's water pump and that the "male would try to crush against a wall anyone to whom it had taken a disliking." His essay on the Indian elephant recited a history of the menagerie's elephant pair, including their amorous behavior when reunited in Paris, and their responses to the concert and to a bath that was built for them. He described the genets' copulation, noting that the male killed the baby on the day it was born. His essay on the mandrill elaborated on "the love of monkeys for women," a point previously made in his essay on the green monkey. The menagerie's mandrill, Cuvier said, grew frenzied at the sight of young women:

> It was clear that he chose those on which he wanted to exercise his imagination, and he did not fail to give preference to the youngest. He picked them out in the crowd; he called them by voice and gesture, and there was no doubt that if he had been free, he would have resorted to violence. These well-established facts, confirmed by thousands of intelligent witnesses, render well worthy of belief all that voyagers report of the dangers posed to Negresses by the large monkeys inhabiting their countries.[64]

Passages with such behavioral detail were rare in Cuvier's contributions to *La Ménagerie*. Typically he said little more about an animal's life in the menagerie than what it ate and excreted and whether it was gentle or not. Occasionally, as with the civet, he remarked how little was known about the species in the wild:

> One has no details at all of the life of civets, their reproduction, their number of offspring, the time of their birth [their gestation period], the term of their growth and their life, nor of the resources nature provides them to feed and defend themselves; all one knows is that if they are not tamed when they are young they display a savage character, and even a kind of ferocity.[65]

Cuvier did not aspire to illuminating behavioral matters in his work. The essay he wrote to accompany Maréchal's illustration of the female Indian elephant was emblematic of his infinitely greater interest in anatomy.

Feeling he had said enough about the Indian elephant as a species and the history of the menagerie's pair, he focused on the anatomical features of his dissection of the male.[66] Readers hoping to learn more about Parkie's reaction to the death of Hans would have been sorely disappointed.

∴

Étienne Geoffroy Saint-Hilaire signed up in 1798 for Napoleon's secret mission to Egypt, motivated partly by the desire to enhance his credentials. He was only twenty-one when appointed to his Museum professorship, and his critics complained not only of his youth and inexperience but also that he had never been a voyager naturalist. He expressed his self-consciousness about this in a letter to Cuvier from Cairo in August 1799, a year after his arrival in Egypt. Preparing to travel up the Nile, he wrote, "I will undertake a very painful trip in order to merit returning to your midst in a manner worthy of you. Oh, this time if someone reproaches me for not being a naturalist because I have not traveled, they will be very wrong."[67] Yet Geoffroy was already proud of what and he had collected or discovered. Writing to Cuvier he jested: "Prepare yourself to make to me the greatest sacrifice. I demand nothing less than the throne of anatomy. If you hesitate, I will reply: Have you found in a single fish the organization of the mammals and the cuttlefish? Have you explained the action of the very admirable organs of the tetrodons [puffer fish]? No, you will say, I have made other discoveries deserving of the first rank; . . . if so, grant me at least second place."[68]

In his more than three years as a scientist in Egypt, Geoffroy dramatically expanded his credentials and horizons. He collected fish from the Nile and from the Red Sea and the Mediterranean; took mummified animal specimens from the tombs of the pharaohs; studied the behavior of live creatures and the anatomy of dead ones; and discovered species previously unknown to zoologists. He observed and verified animal habits, noting, for example, that what Herodotus and Aristotle had said about plovers' venturing unharmed into the mouths of crocodiles to collect insects was entirely true.[69] He was inspired to think as a zoologist, anatomist, archaeologist, geologist, and biogeographer. He also learned to appreciate local knowledge. The fishermen who supplied his specimens impressed him both by their ability to catch fish (four fishermen accompanied him from Cairo to Suez to fish in the Red Sea), and by their implicit zoological acumen.[70] "The fishermen in Egypt," he wrote to Cuvier in April 1800, "distinguish perfectly all the species and employ a nearly Linnaean nomenclature in the names they have given them. These names always capture their most apparent and distinctive character, or one of their most remarkable habits."[71]

The locals' understanding of species differences was more than skin deep. Initially Geoffroy refused to accept their distinction between two species of the same shape, color, and number of dorsal fin spines, but when dissected, the fish proved to have markedly different internal organs.[72]

At the same time, Geoffroy felt keenly the disadvantages of being far from home. Without books, he was often unable to ascertain whether or not a fish he had found was new to science, and he bemoaned not hearing from his colleagues back home.[73] When after writing many letters to Cuvier he finally received from Cuvier what Geoffroy described as "two lines," Geoffroy welcomed the contact but did not fail to notice the unevenness of the exchange.[74] A few months later he admitted to Cuvier that he felt "some resentment" toward him because "you have never written me, and my father has been able to get his letters to me."[75] Almost a year and a half later, as his scientific mission in Egypt neared its end and yet another ship from France arrived with no news for him, Geoffroy wrote Cuvier, "Is it because I inundated you with my letters that I deserved your never writing to me?"[76] He complained again about missing books, and he summed up their respective situations: "While I vegetate and sleep here, you are advancing these sciences."[77]

Added to Geoffroy's anxieties about his limited scientific resources and Cuvier's failure to respond in kind with news or psychological support were his fears about the fate of his collections. In the fall of 1800, he warned his colleagues that insects were destroying his bird skins and that other specimens were endangered by the evaporation of the alcohol they were preserved in. He begged the Museum to have the government recall him from Egypt. He had seen everything in natural history there was to see, he said, and his health was grievously diminished. At year's end, he reported almost constant bouts of dysentery plus a malady of the eyes that left him blind for twenty-nine days.[78]

Geoffroy rebounded from his illnesses. In the winter of 1800–1801 he delivered a wide-ranging series of papers at the Institut de France in Cairo, concluding with an analysis of crocodile anatomy delivered at the Institute's closing session (21 March 1801).[79] He and his fellow academicians then left plague-ridden Cairo for Alexandria, where they would embark for France. Ironically, it was while he was in Alexandria waiting to sail, and while the bombs of the British siege of the city were bursting around him, that Geoffroy had the good fortune to study two live electric fish of very different species that he had long sought: the *malaptérure* from the Nile and the *torpille* from the Mediterranean Sea. Experimenting feverishly on how the fish produced their shocks, Geoffroy was led to the boldest of speculations. He believed he was on the verge of understanding the fundamental forces governing the universe.[80] In a brief letter to Cuvier, Geoffroy said that as soon as he reached quarantine in France he would send him a long work on "physics, chemistry, and physiology." He explained, "The discovery of the nervous fluid and the vital principle has led to me to a very vast theory. I hope to return to France worthy of you and my illustrious colleagues."[81]

In his last days in Egypt, while attending to his research and final preparations for the voyage to Paris, Geoffroy had to defend his papers and specimens from confiscation by the British. With the surrender of the French forces in Alexandria, France's last foothold in Egypt, the British general demanded that Napoleon's retinue of scientists relinquish their specimens and papers. Geoffroy insisted that he and his colleagues would destroy them first, leaving the British to share the infamy of those who had burned the Library of Alexandria some two millennia before. The British relented.[82]

When Geoffroy arrived in Marseille in mid-November 1801, he entered a month-long quarantine, yet another hardship for the voyager naturalist. He wrote to the Museum that he and his companions had forty to fifty crates of specimens to ship to Paris, with half of the specimens in alcohol that needed replenishing.[83] He likewise wrote to Cuvier asking for news, including whether his apartment had been kept for him (it had not).[84] In a subsequent letter to Cuvier, he outlined his grand new theory of how the operations of the subtle fluid (caloric) could explain the phenomena of galvanism, electricity, magnetism, the nervous fluid, and germination, plus the formation, nutrition, and generation of the organs. He was unaware of how unreceptive Cuvier was to such speculating. In a postscript, Geoffroy said he'd heard that Cuvier's younger brother was living with him and seemed interested in the sciences. "Give him, I beg you, my compliments, and indicate the keen desire I have to make his acquaintance."[85] Geoffroy's subsequent relations with Georges Cuvier's brother, Frédéric, would be of the utmost consequence for the history of the menagerie.

Geoffroy returned to Paris in early February (1802) and attended his first professorial assembly in nearly four years. The meetings' minutes report his choosing a new apartment recently acquired in exchange for a state-owned property near the Museum's amphitheater, putting him near the Cuvier brothers.[86] The prospect of being so close to his friend Georges Cuvier surely heartened Geoffroy, who in Egypt had reminisced about their early work together in "that beautiful period of my life, when we occupied ourselves together with amassing new facts and lived without noticing the vexations of society and the trifling passions of men."[87] That idyllic time was never to be recreated.

Back at the Museum, Geoffroy saw to his specimens from Egypt, delivered and published papers on his findings, and handled other business relating to the zoology collection. His engagement with the menagerie was limited primarily to giving it a tame male mongoose he had brought back with him. He was in no hurry to replace Lamarck as the menagerie's overseer. The first item he presented to his colleagues was a papyrus manuscript he took from an ancient tomb in a necropolis in Thebes (West Bank, Luxor).[88] More exciting for zoology were the embalmed animals he turned over to Cuvier. Some of Geoffroy's specimens were soon displayed in the zoology and comparative anatomy galleries, but seven months elapsed before the administration, at Geoffroy's urging, appointed a committee to produce an official report on the fruits of Geoffroy's efforts. Its members were Lacépède, Lamarck, and Cuvier (who at that point was still an adjunct professor).[89]

The report was completed in a week. It reads as if Lacépède were its author and presenter, for it begins, "You have charged us, the citizens Lamarck, Cuvier, and me, with examining the collections brought from Egypt, given to the Museum by our colleague citizen Geoffroy, and reporting on them to you; it is this joint report that I will present to you." In fact, Lacépède neither wrote it nor presented it. Cuvier wrote it (the manuscript's handwriting and the ideas in it are distinctively his), and A.-L. Brongniart, the Museum's secretary, read it in Lacépède's absence. It was signed by Cuvier, Lamarck, and Lacépède, but it seems that their signatures were all that Lamarck and Lacépède contributed to it. The assembly accepted the report and agreed both to send it to the minister of the interior and to publish it in the Museum's *Annales*. In the volume's table of contents, it appears under Lacépède's name.[90]

The report sounded themes well practiced in the previous wars of the Revolution. Geoffroy's work had been "protected by our victorious weapons and by generals friendly to the sciences and arts." He had liberated specimens from the tombs of the pharaohs for the good of science and civilization. The benighted practices of the ancient Egyptians had turned out well for modern science: "These bizarre people, in embalming with so much care the brute beings that they made the objects of their stupid adoration, have left in their sacred grottos nearly complete zoology collections."[91]

Geoffroy's specimens (wrote Cuvier) offered an unprecedented opportunity to answer the question naturalists had long wanted to know: "If species change form over the course of time." Thanks to the "great number of species" in Geoffroy's collection and the "several thousands of years" they represented, naturalists were better able to answer the question than ever before. From ancient tombs Geoffroy had taken human and animal mummies. He had the embalmed remains of cattle, crocodiles, mongooses, monkeys, and ibises. "One cannot control the flights of one's imagination,"

Cuvier enthused, "when one sees, still preserved, the smallest bones, the smallest hairs, and perfectly recognizable, such an animal that two or three thousand years ago in Thebes and Memphis had its priests and altars, but without going astray amidst all the ideas that this connection gives rise to, let us limit ourselves to explaining to you that the result of this part of citizen Geoffroy's collection is that these animals are perfectly similar to those of today."[92] So much for respecting the voyager naturalist's right to be the first to present the implications of his collections!

The report went on to review Geoffroy's findings regarding modern animals. He had identified a range of new mammal species (hare, fox, hedgehog, four species of rats, and eleven species of bats), plus several new species of reptiles, but he had been most successful in collecting fish from the Nile, the Red Sea, and the Mediterranean. The report noted that by preserving a great number of specimens in alcohol, Geoffroy made it possible to study specimens' internal organs and skeletons as well as their external characters, a triple collection, in effect. Several skeletons had already been installed in the comparative anatomy gallery, among them one of an ancient ibis.

One imagines Geoffroy was pleased with this praise, but the report was sketchier than he had had in mind. He had wanted the report to provide not just a general sense of the nature of his collections but more specific information about the condition and total numbers of specimens he had brought back. Did he mind also that some of his specimens were already escaping his control and being enlisted in a debate he was not yet ready to join? With Geoffroy remaining on the sidelines, they became a flashpoint of the conflict between Lamarck and Cuvier on the topic of species mutability.

All of this might seem to have been well removed from the business of the menagerie, but in diverse respects it was only a step away. Less than two weeks after the report on his collections was presented, Geoffroy resumed oversight over the menagerie. In the months that followed, he conceived the plan of making Georges Cuvier's brother Frédéric the menagerie supervisor (under Geoffroy's direction). Lamarck would later find inspiration at the menagerie for a further development of his evolutionary theory, and Frédéric Cuvier would hope to develop the menagerie into a site for confirming that the variability of species was limited. Cuvier also sought to create a new role for a naturalist, neither voyager nor cabinet naturalist but rather a naturalist who studied live animals in a menagerie.

[2.2]

The Sacred Ibis

The ancient Egyptians raised ibises near temples in honor of the god Thoth. Fed and otherwise cared for until they died (or were sacrificed), the birds were embalmed and placed in royal tombs or animal cemeteries.[1] Three thousand years later, Geoffroy collected their mummies from tombs in Saqqara (ancient Memphis) and Thebes, packed them for shipment to Paris, and resisted British efforts to confiscate them as he was about to leave Alexandria. In Marseille, the ibises and other specimens were quarantined two weeks longer than Geoffroy was because their cotton packing was a suspected source of contagion.[2] Once in Paris, they had to be unwrapped carefully owing to their extreme fragility. Next came the delicate and difficult job of reconstructing them for display and determining their place in the galleries. The reconstructed remains and individual bones were compared with those of other forms to assess the species' place in nature's order. Comparison with modern forms also elevated the ancient specimens to a position of great importance in the latest discussions of the history of life on Earth.

Georges Cuvier had brought the topic to the fore, arguing persuasively that various large mammals found only in fossil form represented species that had disappeared from the earth. He insisted that the differences between prehistoric and living vertebrate forms could not be explained solely by the former's changing into the latter as the result of climatic influences. He proposed instead that a global catastrophe had wiped out the entire vertebrate fauna of a previous age.

When Cuvier first presented these views in 1796 at the first public session of the Institut de France, he claimed that "several learned conchologists [shellfish specialists] maintain that none of the shellfish now living in the sea are to be found among the numerous petrifactions of which the continents are full."[3] When his paper was published in 1799, however, he left out that claim, likely because he'd learned that Lamarck, arguably France's leading conchologist, considered some species of fossil shells to be perfectly

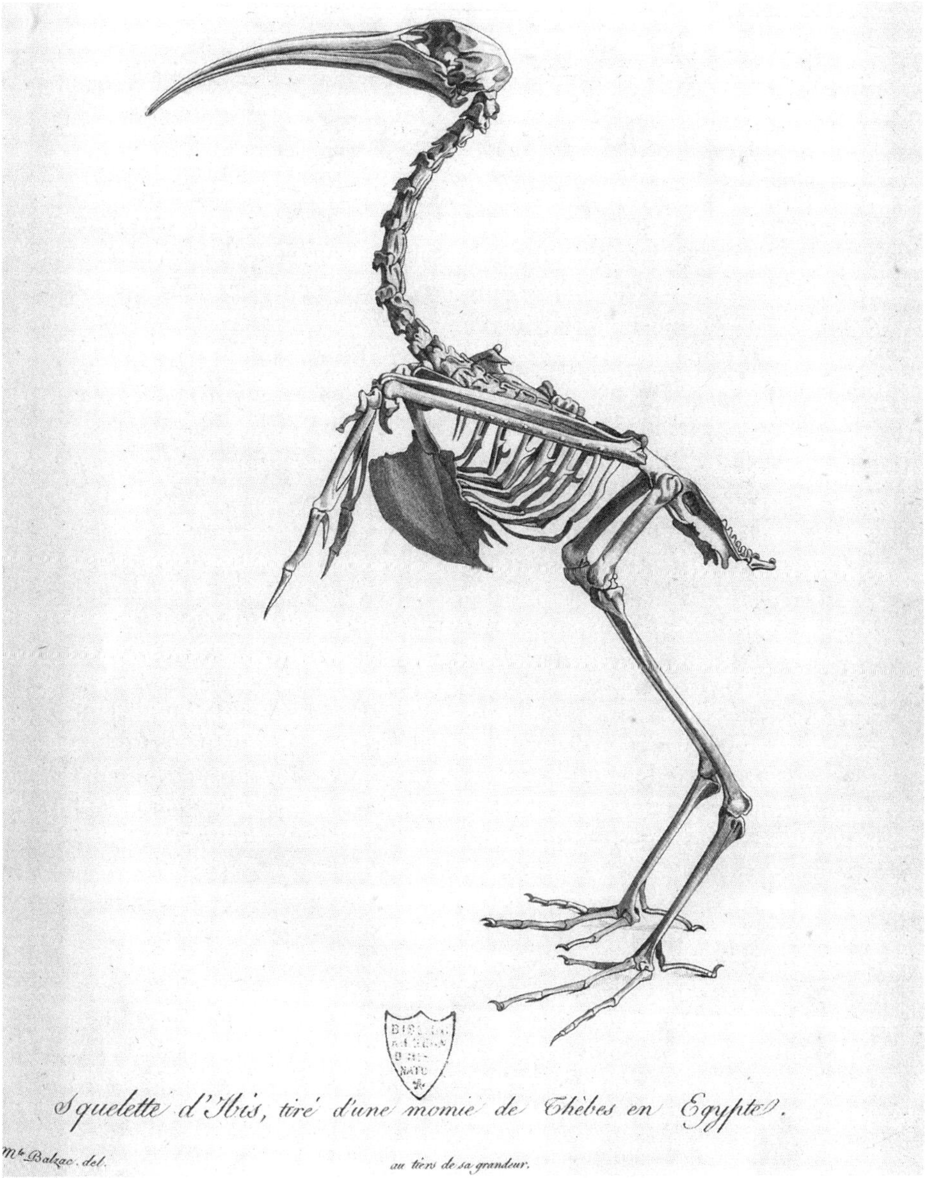

Figure 2.2.1. "Skeleton of the ibis, taken from a mummy from Thebes in Egypt." Georges Cuvier, "Mémoire sur l'ibis des anciens Égyptiens," *Annales du Muséum d'histoire naturelle* (1804). Muséum national d'histoire naturelle. Bibliothèque centrale. Pr 260.

analogous to modern forms. In retrospect, Cuvier and Lamarck's competing views illustrate the difficulty of working with the evidence and preconceptions of their day. Cuvier's view of species extinction was one of sudden, wholesale destruction.[4] He denied the idea of species transformation. Lamarck resisted the idea of extinction, rejected as unscientific Cuvier's idea of a global catastrophe, and chose to explain the differences between fossil and modern forms as the result of gradual change over time. The only kinds of animals Lamarck could imagine possibly becoming extinct were a few large mammals that reproduced slowly and had few offspring and could become the victims of human actions.[5]

With Cuvier's denial of species change in his memoir of 1799 and Lamarck's explicit endorsement of the idea in 1800, the stage was set for a confrontation, triggered in September 1802 when Cuvier wrote the report announcing that Geoffroy's specimens had a direct bearing on the question of species change. By stressing that the forms from ancient Egypt were "perfectly similar to those of today," Cuvier threw down a gauntlet. Lamarck did not register an objection to the committee report when it was read; he cosigned it with Cuvier and Lacépède. But he was not about to let the similarity between modern specimens and ancient mummies undercut his idea of species mutability. The committee report on Geoffroy's specimens was published in the third installment (December 1802) of volume 1 of the Museum's *Annales*. The next installment (January 1803) had Lamarck's rebuttal, which appeared in his introduction to a long series of papers on fossil shells.[6] Observing that three thousand years was nothing compared with the vast expanse of time nature had at her disposal, Lamarck took an acerbic swipe at anyone who suggested otherwise:

> I fancy I hear those small insects that live for only a year, that inhabit some corner of a building, that one would suppose busy consulting tradition among themselves to pronounce on the age of the edifice in which they happen to be. Going back in their meager history to the twenty-fifth generation, they would decide unanimously that the building sheltering them is eternal, or at least that it has always existed, for they have always seen it the same, and never heard it said that it had a beginning. Great magnitudes, in space and time, are relative.[7]

This was not an obscure place to register his complaint. The *Annales* were the premier locus for the newest, best work on natural history. A few months later, he repeated this passage verbatim in introducing students to his invertebrate zoology course at the Museum for 1803. His introductory lecture was "On the question: What is the species among living bodies?"[8]

Likening Cuvier to a short-lived insect was hardly politic, but Cuvier did not react publicly, whatever he might have said in private about Lamarck's theorizing. He stuck to his position that silence was the proper response to speculative "systems" and that the naturalist's duty was simply to report well-constituted "facts." He scrupulously followed this prescription in his 1804 memoir, "On the Ibis of the Ancient Egyptians." Published in the Museum's *Annales,* it matched the sacred ibis of ancient Egypt with a species of modern times; but unlike his committee report on Geoffroy's specimens, it offered no hint of the implications for the history of life on Earth.[9]

Cuvier's sacred ibis memoir was a tour de force by a master cabinet naturalist. It displayed a compelling command of texts and specimens, ancient and modern. Scrutinizing the writings of classical and modern observers, untangling knots of nomenclature, comparing multiple modern specimens with mummified birds from the crypts of the pharaohs, Cuvier built his case. He identified a living curlew from Africa, rather than any of the birds that modern ornithologists had identified as ibises, as the same species as the sacred ibis of the Egyptians.

Cuvier offered special thanks to Geoffroy, "my learned friend," for "sacrificing" some of his specimens to benefit Cuvier's research and the comparative anatomy gallery.[10] In fact, Geoffroy had ample mummified ibises to share. The Egyptians had stacked them up in their sepulchres, and Geoffroy had come away with a great many. There were other instances where a voyager's collections contained only one or two representatives of a sought-after species and their use became a source of contention, but this was not one of them.

That said, some of the mummified birds were in better condition than others. Geoffroy had collected his specimens from two locations. Those from Saqqara proved extremely fragile, disintegrating on the slightest contact, rendering it impossible for Cuvier and his assistant, Pierre Rousseau, to assemble an ibis skeleton from them. Fortunately, the birds mummified in Thebes were in better shape. With "patience, skill, and ingenious and delicate procedures," Rousseau stripped away the remaining tissue, reassembled the bones, reattached them with "a very thin copper wire," and succeeded in reconstructing an entire skeleton. Examination of the left humerus revealed that the bird had sustained a wing fracture that had subsequently healed, leading Cuvier to conclude it had been domesticated in the temples, because a wild bird, unable to secure its prey or escape from predators, would likely have died before healing. Cuvier triumphantly reported that the ibis had been placed in the comparative anatomy gallery, where it stood out as a gem of the collection.[11]

Just as French army victories in the Low Countries had enabled the sequestration of the Stadtholder's natural history collection, so had Napoleon's campaign in Egypt made Geoffroy's collection of artifacts possible. Moreover, the African curlew that Cuvier identified as the ancient ibis's modern equivalent was a specimen from the Stadtholder's collection. He found it in the Museum's ornithology gallery among the birds put "in such good order" by Lacépède. The specimen's native land was unknown, but the naturalist Desmoulins testified to having seen two other such birds that he knew were from Senegal.[12]

Eight years later, in the introduction to his trail-blazing *Research on Fossil Bones* (1812) (*Recherches sur les ossemens fossiles*), Cuvier used the equivalence of the ancient ibis and a modern African bird to counter the idea of species change, presenting the reasons mammals of the present should not be viewed as descendants of fossil forms of the past. Over the course of human history, he explained, the influence of different climates and the efforts of breeders had produced nothing more than superficial differences in animal species, and there was no reason to expect more change over time. Without naming Lamarck, he wrote, "I know that some naturalists count greatly on the thousands of centuries they accumulate with the stroke of a pen, but in similar matters we can hardly judge what a long time would produce except by multiplying it by what a lesser amount of time produces."[13]

Cuvier refrained from broaching these subjects in his 1804 ibis memoir. Instead, he devoted himself to placing specimens side by side, scrupulously measuring their beaks and bones, comparing their proportions, researching what authors from Herodotus to the present had said about ibises and similar birds, examining copies of ancient illustrations depicting ibises in Egyptian ceremonies, and more. He sorted out the different species of birds previously represented under the name ibis, showing that the ancient ibis of the Egyptians was a different species altogether (Cuvier named it *Numenius ibis*). He concluded by admonishing naturalists to be more critical in identifying, naming, and characterizing species, and his memoir was a prime example of how to do it. He was practicing what he preached, which included, at least for the time being, holding his tongue on Lamarck's speculations about organic change.

[2.3]

Guiding the Public

At the beginning of the nineteenth century, guidebooks to Paris described the Jardin des Plantes as the city's healthiest and most agreeable site for a leisurely, contemplative stroll. The gardens and conservatories with their remarkable plants, the menagerie with its gentle herbivores and ferocious carnivores, the galleries with their fascinating specimens, the picturesque views from diverse vantage points, and the relative quiet combined to make a fine setting for walking, gazing, and conversing.[1] Visitors could enjoy all sorts of refreshments. A café sold food, wine, and various liquors. Porters at each of the Museum's three gates offered food and beverages and, along with the café proprietor, were responsible for keeping the exuberance of drinkers and musicians in check. A *laiterie* on the high knoll (*grande butte*) overlooking the gardens provided milk and pastries. In addition, a few wives and widows of Museum workers were allowed to sell baked goods, and three fruit vendors had stalls in the courtyard between the galleries. Here too were a toy seller and a shop selling physical and optical instruments and finally, near the *orangerie*, a bookstore.[2]

Yet visits to the site could still be less than idyllic. The cries of unauthorized vendors hawking honey cakes ("*plaisir*") often pierced the air. According to an internal report of 1800, these vendors were a great nuisance: "nothing is more irksome for citizens who enjoy the pleasures of strolling and conversation than to be followed, interrupted, and importuned by the bawling of these vendors."[3] Menagerie supervisor Mordant Delaunay additionally complained of the self-appointed guides who attached themselves to visitors, "ruthlessly" exploiting those naive enough to trust them. This sometimes led to quarrels, and there was one "very indecent scene" when the *conducteurs* brawled among themselves. Delaunay urged banning them from the grounds.[4]

Delaunay also suggested posting signs directing visitors to the animals, but the administration failed to follow up, leaving visitors to find their own way, which was no straightforward matter.[5] The *animaux féroces* and a few

Figure 2.3.1. Hercules in the form of a baboon, waltzing with the goddess Omphale. From E. J. B. Vignier, *Phénomène d'histoire naturelle. Récit de la deuxième gestation de Constantine* . . . (an IX). Muséum national d'histoire naturelle. Bibliothèque centrale. Fonds général. 80 Res 30.

raptors were lodged in the provisional menagerie by the Seine. Most of the other birds were at the large pond or in the aviary at the foot of the labyrinth. The herbivores were in their newly constructed parks in summer, but in winter they and the ostrich and cassowary were kept with the elephants in the Régie.

The Museum was scarcely more helpful in providing information on specimens, which, whether in the menagerie, gardens, or galleries, were typically identified solely by a simple *etiquette* (a label, tag, or plate) bearing a Latin name. Many visitors also wanted to know the specimen's common name, country of origin, and uses. For animals they might also want to know their habits and history. But the Museum resisted Lucien Bonaparte's proposal that the Museum produce a visitor's guide to provide in exchange for an admission fee. It objected that charging admission would be unseemly for a great national museum; that a simple guide to the collections would be too superficial to be worth the effort; and that a serious, *catalogue raisonné* would be too costly to produce.[6] The Museum did not even produce a simple map of the grounds.

From 1798 to 1803, at least seven authors moved to fill the void: Louis-François Jauffret, a social reformer and writer of children's books (1798); Edme Jean Baptiste Vignier, an elderly former lawyer (five guides between 1798 and 1801); Georges Toscan, the Museum's head librarian (1800); Simon-Charles Miger, a skilled engraver (1801, 1802); Gotthelf Fischer von Waldheim, a German naturalist (1802–3); an anonymous writer (1803); and J. B. Pujoulx, a playwright (1803). Inevitably, some were better than others in helping the public navigate the grounds and learn about the specimens. All subscribed to the view that the study of nature and the moral elevation of citizens went hand in hand, but their other motives and backgrounds were diverse.

Jauffret was a young father interested in instilling in children a respect for nature.[7] His *Voyage au Jardin des Plantes* (1798) was a fictional account of a series of visits to the Museum by a boy named Gustave. Small enough to be carried on an outing, the book could just as easily be enjoyed at home. In the section devoted to the menagerie, Gustave and his older companion (Jauffret) meet Félix Cassal, and both adults help Gustave appreciate the menagerie's moral economy. Cassal explains the difference between the "innocent" and "peaceful" animals, which can be allowed to live in parks, and the "ferocious" ones that need to be restrained behind bars. Jauffret ascribes Cassal's success in taming animals to his love for them: "It is by loving them that he makes them love him."[8] Impressed by the diversity of the menagerie's creatures, Gustave learns that the Creator is responsible for their marvelous structures and instincts. Informed of the stork's reputation for devotion to its young and to enfeebled elder storks, Gustave vows "never in my life to kill a stork or even cause it the slightest grief."[9]

Ironically, while Jauffret preached the rewards of virtuous behavior, he was somewhat less than virtuous in composing his book. Some twenty pages of it he appropriated nearly word for word from Georges Toscan's earlier essay on the lion and dog. His stratagem was to introduce Toscan in the book and have him recount to Gustave the tale of the lion and his canine companion.

Toscan would later say that the appropriation of his writings by Jauffret and others was what induced him to collect a set of his essays and republish them in 1800 under the title *The Friend of Nature*.[10] He appended to this book a forty-one- page "catalogue" of all the animals in the menagerie. This can be considered the first full guide to the menagerie. Toscan grouped the animals according to the five parts of the establishment where they were to be found, listing their scientific and common names accompanied by a sampling of information, some gleaned from the writings of naturalists and travelers, some drawn from his own observations. Anecdotes about individual animals, such as the two elephants and the lion and lioness, offered readers an impression of the animals' personalities. Such information enabled visitors to look beyond the animal's immediate situation in a cramped cage to imagine its life in the wild. It also gave Toscan the chance to acknowledge the menagerie's shortcomings. He longed for a time when the scientific study of animals would involve situating them "in sites more appropriate to their nature" where, with greater liberty, they would "be able to give flight to their instincts."[11] His misgivings related primarily to the fierce animals in their quarters by the Seine. He saw the deer, in contrast, as enjoying considerable freedom, even more than they experienced in the wild, where they were subject to "the sound of the bugle," "the cry of dogs," and "the cruel joy of the hunter."[12]

E. J. B. Vignier's menagerie guides had a different flavor, reflecting his impoverished circumstances. The Revolution had ended his legal career and dramatically reduced his income.[13] Thinking he could earn money writing about the menagerie, he wrote a three-hundred-page manuscript about the animals, only to set it aside after judging it would be too expensive for the modest budgets of the "Republicans" who were his intended readers.[14] He produced instead a series of cheap booklets with long titles such as the first: *Moral Virtues of the Two Elephants, Male and Female, Newly Arrived at the National Menagerie of the Jardin des Plantes, preceded by a treatise on the nature of these in every way precious and singular animals, taken from the celebrated Buffon, and from the best authors who have written on natural history*.[15] His *Epistle to the Elephants of the National Menagerie* featured a four-page ode of rhyming couplets of his own composition, information and anecdotes about elephants (mostly from Buffon), and a smattering of details about the menagerie's pair.[16] This was followed by another booklet on elephants that lacked the ode but added a one-page list of the common French names of a dozen species of mammals and twenty species of birds in the menagerie. He titled the list "the fashionable stroll" ("La Promenade à la mode"), but he did not specify where the animals were to be found.

New and exciting arrivals at the menagerie in the late fall of 1800 prompted Vignier to publish his *Abbreviated Description of the Quadruped Animals from the Menagerie of Tipoo-Saiab, Newly Purchased in London to Enrich the Menagerie of the Museum of Natural History, followed by an account of the progeny of the lioness, which one can regard as a phenomenon of natural history*.[17] In the "corrected and augmented" "new edition" of this booklet that soon followed, Tipu was no longer mentioned (Vignier having apparently learned that the animals in question had nothing to do with the famous sultan), but an *Account of the Second Gestation of Constantine, One of the Lionesses of the Menagerie of the Jardin des Plantes* remained, along with information about the various ferocious creatures under the care of Félix Cassal. To further engage readers he offered bits of information about the natives of the animals' homelands. Of the panther he wrote, "The voyagers allow that its flesh is not bad to eat. The Indians and the Negroes find it good, but it is true that they find that of the dog still better and treat themselves with it as if it were a delicious dish."[18] Of the baboon he allowed, "There are several peculiarities of this animal that we have omitted so as not to offend the decency of young persons into whose hands our work might fall." He nonetheless related that "[these animals] are very ardent in love, and take the most virile liberties, when in private and free, with regard to sex; for travelers report (according to Buffon), that the white [baboons] are strong enough to violate women when they find them alone, and often abuse them to the point of death." Concerns about potential younger readers aside, he included a rather salacious illustration of Hercules as a baboon, waltzing with the goddess Omphale.[19] He wanted his booklets to sell.

The professors ignored Vignier's offerings until the summer of 1801, when he complained that Félix Cassal had ridiculed and attacked him and his writings in public and driven him from the menagerie.[20] Vignier wanted the administration to intervene on his behalf, but the professors chose not to take sides, noting that Vignier and Cassal were in competition, both selling menagerie guidebooks that were poorly written, riddled with errors, and "harmful to the progress of science." The Museum had kept "too long a silence," it said, with respect to those "diverse brochures sold inside the garden and in the vicinity of the menagerie under such titles as "Description of the Galleries," "Catalogue of the Jardin des Plantes," or "Description of the Menagerie." From the Museum's perspective, "the majority of these pretended catalogs" had merely provided "inexact and erroneous copies of the *etiquettes* of the galleries, gardens, parks and menagerie."

When the administration decided to enforce its existing regulations forbidding the sale of unauthorized items inside the institution's gates, Vignier felt unjustly treated.[21] Several hundred copies of his booklet had been printed, he told the Museum, and his financial loss was unbearable. Instead of profiting from his labors, he had to resort to giving away copies.[22]

Meanwhile, Cassal was selling a guide produced by the engraver Samuel-Charles Miger, whose career had also suffered during the Revolution. Having lost his status as a member in the Royal Academy of Painting and Sculpture when it and other royal institutions were suppressed, Miger hoped that his large folio project on the animals of the menagerie would reestablish his reputation and his income. After the first installments began to appear, he conceived the idea of printing a small, inexpensive guide to the menagerie as a means of advertising the larger work. He needed to attract subscribers, given that the high quality of the paper, the time involved in executing the engravings, and the printing of consecutive installments made the enterprise very costly. Akin to a *catalogue raisonnée* of an art exhibition, the guide would provide information on each animal displayed: its name, age, provenance, how it arrived in Paris, its character, and its habits. The Museum told Miger he was free to publish it but turned down his request for its official endorsement.[23]

Miger's ninety-two-page guide was published anonymously under the title *Notice of the Animals Presently Living at the National Museum of Natural History, and a Description of Their Way of Life and the Habits of Each of Them; With the Portrait of the Lioness Feeding Her Cubs*.[24] It cost fifteen sous and could be purchased at the Museum from "Félix" (Cassal) as well as from the printer (Patris) and the bookseller (Gilbert). Cassal's ridiculing of Vignier's guides apparently began once he had the chance to make a few sous selling Miger's offering.[25]

Introduced by an elegant frontispiece (Miger's engraving of Maréchal's compelling portrait of the lioness nursing her cubs), Miger's guide was finer than anything Vignier had produced. It was also more helpful for finding individual animals and describing their characteristics. He borrowed freely from numerous sources, but the only work he cited specifically was his own folio publication as he explained that each installment would have four new plates with corresponding texts. Installments would appear every three months and cost eight francs.[26] In a new edition of his guide in 1802, Miger included "some details on the death of the male elephant" and kept readers abreast of the new installments of the folio work.[27] That the title page still indicated the guide could be purchased at the Museum "*Chez Félix*" suggests that the Museum had decided in this case to overlook its recent prohibition of selling unauthorized guidebooks on the premises.[28]

Gotthelf Fischer's *The National Museum of Natural History in Paris* (two volumes, 1802–3) was a guide to the Museum, but not one an average visitor would carry on location.[29] It was written in German for naturalists and others wanting to learn in detail about the world's greatest museum of natural history. It no doubt helped Fischer himself in his next career move, from a post at the University of Mainz to the professorship of natural history and directorship of the Museum of Natural History at the University of Moscow.

Much more useful for Francophone visitors were two guides to the Museum and menagerie that appeared in 1803. The first, published anonymously, was titled *The Guide to the Jardin des Plantes, or Description of All the Most Curious Things That the Galleries, Greenhouses, and So Forth Contain*. Some one hundred pages long, it identified its audience as the general public, particularly inquisitive youths, citizens seeking entertainment as they perused the gardens, and foreigners who, "at a glance," would become familiar with France's riches in natural history.[30] What distinguished this guidebook from contemporary offerings was not so much the information it imparted about specimens and sights as its enthusiastic portrayal of the emotions that visitors to the Museum would experience. The sentiments of patriotism, fear, joy, admiration, gratitude, and awe would all be evoked. Here one would find the "unique and curious objects" that were "the fruit of French conquests in Holland, Germany, and Italy." Here was nature "at every instant beautiful and surprising." Contrasting images beckoned: "The proud lioness raises her cubs near the timid lamb that bounds lightly after its mother; the ferocious tiger drinks blood, and the nimble deer browses on the blades of grass, laden with morning dew." The casual visitor would be captivated by many sights: not only the plant, animal, and mineral specimens but also "a rich library, a magnificent amphitheater, well-maintained greenhouses, delightful walks, and picturesque views." Committed nature lovers would be even more satisfied on seeing the Museum's scientists surrounded by their students. One could only admire, the guide said, the wise government that spared nothing to enrich this national museum, whose work was to enable man "to know and appreciate himself by comparing himself with all that surrounds him."[31]

The guide began by escorting the visitor through the gardens, entering by the gate of rue Saint-Victor, passing along a walkway lined with linden trees, ascending a ramp to the School of Botany, and reaching a point where the visitor could enjoy a "most picturesque view" that included the magnificent cedar of Lebanon planted years earlier by Guy de la Brosse, a tree so large that three men together could not reach around its trunk. From this spot one could look down on the professors' abodes, each with its own little garden, and each displaying the individual proprietor's favorite subject, rare plants in one case, unusual birds in another. The tour continued, climbing past the tomb of Daubenton to the Jardin's highest point. Then, after an overview of the whole, it was time to descend for a detailed visit, stopping first at the amphitheater with the new orangerie across from it, and between them the huts and shrubs where Barbary sheep and other animals lived.[32]

Throughout the tour, the guide enthused about the attractiveness of the new buildings, the marvelous nature of the animals (the "very rare" kangaroos; the enormous crow "said to be more than 100 years old"; the mongoose, which did the great service of eating crocodile eggs and attacking snakes); the quality of the collections; the talents of the staff; and more. Arrival at the elephant stable provided the occasion to retell the story of the elephant concert and the later death and anatomization of the male elephant.

From there the tour proceeded to the gallery of comparative anatomy, where one could see the immense skeleton of the dead elephant and "the no less astonishing" skeleton of the giraffe brought to France from Holland. Viewing additional skeletons of ox, deer, lion, dog, bird, and fish, one would pause in wonder at the "wise foresight of Nature" they manifested. "What energies, what forces, what machines to aid and make precise the least movements of the animal from the largest to the smallest!" Readers were assured that, thanks to the science of comparative anatomy, taught by Citizen Cuvier, the organs of sensation, motion, circulation, respiration, and nutrition were all becoming better understood. As for the skeletons, it was "to the talent of Citizen Rousseau," the guide explained, "that we owe these perfectly mounted and preserved bony masterworks."[33]

Moving on from the comparative anatomy gallery, the visitor would be treated to other remarkable sights, like a female ostrich walking gravely in its park and, elsewhere, camels, including one from Egypt that General Bonaparte was said to have used as a mount. "At the extremity of the garden, on the banks of the Seine," were the *animaux féroces*, "confided to the care of Citizen Félix Cassal, whose familiarity with them is truly frightening. At his voice, the leopard cringes and the king of the animals, the lion, seems to forget his voracious instinct."[34] The guide told anew of the lion and dog, the tigers from the menagerie of Tipu Sultan, the polar bear that suffered in the summer heat, the notorious male and female wolves that ate their young, the wicked baboon, and more.[35]

The guide also evoked the emotional journey accompanying a stroller's walk: "Upon leaving the lair of the ferocious beasts, one skirts the banks of the Seine, where there is an entry to the garden, then follows a long alley of linden trees. . . . With what pleasure one then encounters other animals of a gentle and peaceful nature. One's heart, tightened by fear and fright by the appearance of those one has just left, can occupy itself with less terrible objects. One can approach these latter animals with confidence. They come running full of joy at the least invitation. If they are busy caring for their little ones, your presence does not in the least scare them; they let you play with their young, innocent ones, and passing their heads through the grillwork that separates them from the public, seem to solicit caresses."[36] The *Guide to the Jardin des Plantes* thus instructed readers not only about the institution's treasures but also how to respond to them.

The author of the second guide published in 1803 was Jean-Baptiste Pujoulx (1762–1821), who had come to Paris as a young man intent on becoming a writer and had fulfilled that ambition. He had begun with Molière as his model, writing comic operas poking fun at the follies, foibles, and fashions of his day.[37] His aims went beyond entertainment, however. Opposing what he regarded as the increasing decadence of contemporary theater, he maintained that the playwright's job was "to turn the power of the imagination to the benefit of morality."[38] The moral improvement of the populace was the goal he brought to his first book, *Paris at the End of the Eighteenth Century, or A Historical and Moral Sketch of the Monuments and Ruins of This Capital, of the State of the Sciences, Arts, and Industry in This Epoch, as Well as the Morals and Ridiculous Behavior of Its Inhabitants* (1801).[39] In this work he proposed replacing the capital's street names with names more suited to public instruction. Regarding the existing names as outdated, ridiculous, and in some cases almost obscene, Pujoulx called for renaming all the streets, cul-de-sacs, squares, and quarters of Paris after the cities, towns, and villages of all of France in such a way as to conserve their general positions and relative placement with respect to one another, turning Paris into sort of miniature map of the country.[40] In the same year, Pujoulx produced a book for adolescents titled *The Second Age: Entertaining Knowledge about the Natural History of Animals*, a product of his belief that natural history constituted "the most useful, the most agreeable, and the most varied" kind of scientific knowledge youth could acquire.[41]

By this time Pujoulx had become, with his wife, a compulsive visitor to the Jardin des Plantes. Over time, he developed a series of favorite itineraries there and recommended them to friends, inspired by the questions people asked him when they found him taking notes on specimens. Encouraged by a publisher, Pujoulx set about writing *Strolls at the Jardin des*

Plantes, at the Menagerie, and in the Galleries of the Museum of Natural History.[42] Designed to be "agreeable and instructive for the general public of both sexes and all ages, who wish to gain a little something from their visits," it consisted of two pocket-sized volumes describing eight detailed itineraries that would each take a thoughtful walker a few hours to complete and could be modified to suit the visitor's interests.[43] The guide's helpful index allowed one to look up information on any of the menagerie's animals, regardless of whether one encountered it in its summer or winter quarters or in some different location.[44]

Pujoulx was a talented writer who anticipated visitors' questions and empathized with their likely sentiments, for example, their disappointment in seeing the pitiful "provisional" menagerie for the *animaux féroces*. He offered pertinent information to pique a sightseer's interest and imagination. He advised anyone who wished to witness the voraciousness of the two African vultures to come at five in the afternoon to watch them tear apart the meat thrown to them. He also explained that however disgusting the birds' preference for rotting flesh, vultures performed a service to people in hot climates by ridding them of "dead animals that would corrupt the air and spread afar the germ of contagious maladies." He also explained that the African vulture was not the largest of such birds; in Peru the condor attained a wingspan of eighteen feet and could prey on sheep and cattle.[45]

Pujoulx undoubtedly took this exaggeration of the Andean condor's size from Buffon, but his description of the polar bear's stereotyped pacing, which he attributed to the conditions of its confinement, was based on his own observations.[46] His attention to how the reader, book in hand, would respond to the menagerie's animals was evident in his discussion of the lions. He gave no physical description, because the animal would be right before the reader's eyes and because previous writers (notably Buffon, Lacépède, and Bernardin de Saint-Pierre) had already explained its character so well. The animal the reader would see in the cage was "still the lion," Pujoulx prompted, even if "degenerated by enslavement, oppressed in its narrow prison, [and] weakened by rest." He invited readers to imagine themselves in the lion's own land, North Africa, "alone, lost, [and] deprived of all help" at just the moment when the lion made the air resonate with his terrible roar. Such roaring, Pujoulx wrote, caused most animals, including humans, to be seized by a kind of torpor, for they recognized it as "the signal of death."[47]

Having given the reader a frisson, Pujoulx provided the reassurance (also to be found in Buffon and others) that man was the one creature that could vanquish the lion. "Feeble in appearance, without natural weapons," but possessing a high intelligence to which he owed everything, man could conquer the lion by kindness or by force. Indeed, the same animal that could terrorize and kill could be dominated "by a Negro of modest stature, of ordinary strength."[48]

Comforted by this reminder of humans' ascendancy over the rest of the animal kingdom (and, implicitly, of the superiority of some human races over others), the reader could enjoy Pujoulx's recitation of the stories of the lion and dog, the birth of the lion cubs, and more. They would also find the erroneous claim about the Museum's acquisition of tigers and other beasts that had once belonged to Tipu Sultan. As an author who wanted his writings to be a stimulus to moral improvement, however, Pujoulx preferred not to elaborate on how the baboon, in certain seasons, seemed "to make a game of displaying his lewdness."[49] Unlike Vignier, he did not trot out stories of monkeys and apes ravishing African women. As for the bonnet macaque that lived in a dark cage next to the baboon, Pujoulx borrowed the claim that this species had been observed on riverbanks fishing for crabs with its tail. It would place its tail between a crab's claws, and when the crab pinched, it adroitly pulled its catch to shore, "thus executing the movement that the fisherman makes with his line when he feels that the fish has taken the hook."[50] Buffon had told the story before him, but Pujoulx added the fisherman analogy.

Pujoulx sought to be judicious with respect to what others had written about various animals. He balked at the claim that swans could live more than three hundred years, but he thought a life span of a century was plausible. As for the idea of the "swan song," he considered it much exaggerated, and he identified eating swan meat as simply a matter of vanity. He allowed it was not as good as goose but was better than peacock.[51] With respect to the latter, he was underwhelmed by the peacock's beauty, considering it counterbalanced by the bird's irritating cries and the damage it caused. He denied the superstition that the peacock's cry predicted rain or, worse still, the death of a neighbor.[52]

In addition to such evaluations, Pujoulx sought to be up to date on the menagerie's news, in some cases, even before it was news. For example, he described the talents of a male elephant, newly on display in Paris, that was being proposed as a replacement for Hans, and he expressed the hope that if the government did buy this elephant for the national collection, the Museum would study its intelligence and evaluate how much confidence naturalists could have in travelers' accounts of the animal's abilities.[53] Pujoulx also signaled that a pair of previously unknown creatures from Australia, brought back by the Baudin expedition, would soon be available for viewing. What Geoffroy had named *phascolomes*, Pujoulx said, looked rather like marmots but had pouches like kangaroos. We know them as wombats.[54]

In concluding his tour of the menagerie Pujoulx explained why he had paid no attention to the sheep and goats with extra or bizarrely shaped horns. Such specimens were of no consequence, he said, if these curious features were simply due to chance. People could see stranger things, he said, simply by visiting the zoology or comparative anatomy galleries. He noted further that such anomalies occurred in humans as well as animals. Whole families had hands with six fingers, a trait that persisted over multiple generations. Likewise, some infants were born with two heads, but such monstrosities served only for the "amusement of the lazy and the admiration of the ignorant." Pujoulx viewed these aberrations as the simple result of nature's laws operating under diverse circumstances. He urged viewing "these fleeting variations with eyes free of prejudice," but he did not spell out the kinds of undesirable conclusions that prejudiced eyes might derive from them. An author dedicated to the moral improvement of the public through an appreciation of the order of nature was certainly not going to introduce readers to ideas that questioned the standard view of how the Creator had brought into being all the diverse forms of life. He concluded his stroll through the menagerie with the simple observation that the Creator had given man the means to change the forms and properties of the different forms of life according to his needs.[55] He was evidently referring to the "perfecting" of plants and animals through domestication and breeding. He did not allude to any more than that.

The guides to the menagerie and the Museum written from 1798 to 1803 reflect one of the major themes of this book, that individuals came to their engagement with the institution promoting different if sometimes overlapping interests. When Frédéric Cuvier as the new superintendent of the menagerie came in 1804 to write a new guide to the menagerie, his offering would be appreciably different from what had come before.

While far from perfect, the early guides performed a valuable service that the Museum failed to offer. Accompanying visitors on explorations of the grounds, they transformed casual strolls into expeditions, enchancing the visitors' appreciation of natural history while providing education and entertainment alike. Their portrayals of what a visitor could experience on an outing to the Museum cannot be taken as a perfect mirror of what visitors did derive from these occasions, but direct testimonials of this sort are rare and fragmentary at best, leaving guides like Pujoulx's and the less solid but enthusiastic anonymous guide of 1803 as perhaps our best instruction on the subject.

The guides identify opportunities for their carriers to be educated, entertained, awed, and thrilled, but also to be disappointed. As Pujoulx presented it, one's first impression on seeing the quarters for the *animaux féroces* was that they were a disgrace. He softened the blow by saying these lodgings were only provisional, and he likewise pivoted as quickly as he could after noting that the menagerie's lions were suffering from their degeneration in captivity, their cramped conditions, and their lack of exercise. His eloquent testimony to the lion's magnificence in the wild, however, may not have distracted the visitor from feeling some unease with respect to the sorry state of the animals in front of him. But the guidebooks did not advise the public on where to find rocks to throw at the animals or how to tease the bears by dangling food on strings and pulling it away when the bears tried to reach it. People figured that out on their own.

The guidebooks additionally allow us to reiterate the menagerie's integral relation to the Museum. Most of the guides dealt with the menagerie as one site within the Museum as a whole. It was clear enough from them that the menagerie was intimately related to the zoology and comparative anatomy galleries as a source of fresh specimens. Animals that died at the menagerie commonly had "afterlives" as reconstructed specimens in the galleries. As it was, a visitor might well find in the galleries an animal that Pujoulx's book had described as in the menagerie. Thanks to the book's index, the reader could readily track down where Pujoulx had described it or the species it represented.

In other ways too, the distinction between the menagerie and other parts of the Museum was relatively fluid. In the anonymously written *Guide to the Jardin des Plantes* of 1803, the visit to the comparative anatomy gallery occurs while touring the live animals. The dead specimens serve to illuminate the live ones and vice versa. This seems to have been more a consequence of physical proximity and opportunity than of didactic intent, but it serves as a reminder that visitors to the menagerie often viewed a variety of collections during their trips to the Museum.[56] The Paris menagerie was not a stand-alone operation, and it was certainly less so than most of the famous nineteenth-century metropolitan zoos that came after it.

[2.4]

Firing Félix Cassal

Not all visitors to Félix Cassal's domain enjoyed the experience. In the spring of 1801, a certain comte de Bontemps wrote to the minister of the interior to say that Cassal was hiding the menagerie's lion cubs from people who were unwilling to pay to see them.[1] Months later guide author Vignier protested to the Museum about Cassal's ridiculing him and effectively banishing him from the site. The next year, an army battalion leader named Boutin complained about Cassal's shameless solicitation of bribes and his brutal treatment of animals.

The Museum administration and Minister Chaptal shrugged off the first two complaints but took the third more seriously. Boutin reported that he had arrived on the grounds early one morning, before the menagerie's opening hours, and observed "several citizens admitted by a certain door," though they appeared to have "no other title than their purses, which they began to untie." When he stepped forward to be admitted with them, Cassal brusquely turned him away. Bystanders informed the officer that this happened daily, often with greater rudeness, and even to people with entry passes signed by a professor. Cassal was civil, Boutin learned, only to those who purchased expensive permits from Cassal's wife.[2]

Had this been his only grievance, wrote Boutin, he might have kept it to himself; but Cassal's brutality toward the animals was too great to overlook: "His way of taming them is to beat them inhumanely, so much, that I'm told he killed his little lions, which he says died in teething." It was alleged that the zoology lab assistant charged with the cubs' taxidermy found them to be thoroughly bruised. Boutin urged the administration to reprimand the keeper.[3]

On this score, the Museum's records report only the administration's decision to close the door to the walkway next to the animal cages. No one would be allowed there when the menagerie was open; only professors and their guests would have access when it was closed; and fresh signs announcing Museum rules to visitors would be posted. It was left to Desfontaines as

Figure 2.4.1. Félix Cassal, the lions Marc and Constantine and the three lion cubs. Jean-Baptiste Huet, Jean-Baptiste Chatelain, *Marc et Constantine, Lions amenés d'Afrique et apprivoisés par Félix Cassal leur gardien, avec leurs petits nés au Muséum d'histoire naturelle, le 18 brumaire, an 9*, [n.d.]. Muséum national d'histoire naturelle. IC 4718 GF.

Museum director to admonish "those who had strayed from their responsibilities."[4]

Dissatisfied with the Museum's response, Boutin wrote to the minister, Chaptal, testifying that he had seen Cassal beat the animals again. He repeated his accusations about Cassal's extortion of visitors, violence toward his animals, and responsibility for the lion cubs' death. It was Cassal's brutality, Boutin insisted, rather than alleged teething problems, that had killed them. The professors, Boutin wrote, were letting the minister down, perhaps because they feared that if they angered Cassal, he might poison the other animals in his charge.[5]

Disposed to believe the battalion leader's story, Chaptal wrote to the Museum, "Other persons, worthy of faith, have denounced this keeper verbally to me. They have witnessed the mistreatment he has made the animals suffer and the bad food he gives them. The losses that the menagerie has recently experienced support these diverse witnesses." Accordingly, Chaptal recommended that the professors "watch more closely than ever over a man against whom so many complaints are accumulating. If a single one of these alleged facts is verified, you should propose his replacement to me at once."[6]

Cassal did not know how to put pen to paper, but he knew how to find someone who did. The result was a polished letter denying all the charges against him, with Cassal's crabbed signature affixed at the end. Moderate in tone, reasoned in its arguments, and thoroughly respectful toward the Museum's professor/administrators, the letter cast Cassal in the best possible light. Why would he want to harm the creatures that were the basis of his employment at the Museum? Cassal suspected that the individual who had lodged the complaint against him was a certain soldier "in a green coat" who had caused a scene one night at closing time. By Cassal's account, when he informed him that the menagerie was closing, the soldier, who was drunk, responded with insults and threats. Cassal admitted that in the exchange that followed he might have used some expressions that were "a little rough," but these were only in response to the "threats and uncouth words" that had been showered upon him. Cassal invited the professors to visualize the awkwardness of his situation, placed "between my obedience to your orders and the demands of a sometimes unreasonable public." He explained, "The public never likes to see the animals put away. Often I receive insults for not responding to the solicitations that are made to me."[7]

Cassal named two Museum employees who could vouch for him: Grancourt, the head of the company of veterans who patrolled the Museum and saw Cassal several times a day, and Pierre Guichenot, the menagerie's provisioner. Guichenot had observed the altercation with the soldier, Cassal said, and his word should count: "His veracity is recognized, and I demand his testimony."

The professors believed the cubs had died from teething problems, not mistreatment, and were impressed by the character and credibility of Cassal's letter. They nonetheless instructed Desfontaines, as they had several weeks earlier, to remind Cassal of all that his role as animal keeper entailed, and they finally got around to composing a notice to post near the *animaux féroces*: "In consideration of the public, and to prevent any accident, the assembly has decreed that the keeper of the menagerie will not allow any person inside the enclosure surrounding the animal cages, unless some circumstance or the good of science demands it."[8]

What Desfontaines said to Cassal is not known, but he wrote to Chaptal to rebut the suggestion that the Museum might have been lax in its oversight of the lion cubs and their keeper. Boutin's story was groundless, said Desfontaines. The cubs' death, he explained, had been shown by autopsy to result from afflictions of their gums and jaws at the time of teething. Even in their native countries, he allowed, "approximately seven-eighths" of young lions did not survive this critical moment in their development. The Museum's director likewise rejected Boutin's charge that the Museum had not reprimanded Cassal for fear he might poison his charges. If the Museum had ever suspected Cassal of such a thing, Desfontaines wrote, he would have been fired instantly. The director further noted that the meat fed to the cubs was carefully selected by an individual whose job was to ensure its quality, and that one of the Museum's professors (Lamarck) visited the menagerie every day in an oversight capacity. As for Cassal, Desfontaines succinctly summed up the administration's feelings: "He has coarse manners which correspond to his occupation. We have often reproached him, and we have strongly done so following your letter. But he is intelligent, attentive, full of courage, he likes his animals very much, he comes to love them, and we believe that despite his defects (perhaps inseparable from the position he performs), he would be very difficult to replace."[9]

Desfontaines acknowledged that the Museum had been losing animals, but that was because of the climate and the lack of a new building for the *animaux féroces*. In the current structure, the cages were too small, the animals were unable to exercise, and the air they breathed was unhealthy. The Museum's director could not deny that Cassal had taken bribes from people who wanted to get closer to the animals and that this had led to "murmurs on the part of the public, and even sometimes quarrels." Nonetheless, witnesses had confirmed that in the squabble that led to the soldier's denunciation of Cassal, the keeper had not been in the wrong. As for Boutin's claim that the Museum was declining day by day, Desfontaines deemed this unworthy of reply. "You know," he assured Chaptal, "how we respond with zeal and alacrity to your benevolence."

Cassal weathered the storm, but what took place the following year was more than the Museum could forgive. His undoing began with an accusation from Guillaume Guichenot, one of the many children of the menagerie's provisioner, Pierre Guichenot, whom Cassal had praised for his truthfulness. When Pierre Guichenot died in February 1803, the Museum appointed his son to replace him. Guillaume Guichenot's qualifications included having served in the corps of veterans that guarded the Museum, having helped to supply the menagerie during his father's final illness, and according to his mother, having the same admirable and honest character as his father.[10] In his new role, he impressed the administration by enthusiastically scouring the city for the best meat at the lowest price.

Guichenot *fils* was charged with acquiring the animals' food and also with monitoring what happened to it after he turned it over to the carnivores' keeper. One day a few hours after delivering to Cassal half a sheep "of rather good quality," Guichenot found that the two better parts of the meat had gone missing, though feeding time had not yet arrived. There was also the issue of a slab of beef that grew smaller after delivery but before feeding. More damningly, over an eleven-day period a total of 689 pounds of meat disappeared unaccountably. The provisioner concluded that Cassal was selling meat destined for the carnivores to local restaurants. Guichenot denounced Cassal to Geoffroy, and the zoologist proceeded, as best he could, to sort out the facts.[11]

Cassal had an answer for each of Guichenot's charges. The half sheep, he told Geoffroy, had been eaten by the menagerie's carnivores despite what Guichenot had said about their normal feeding time. As for the slab of beef, the butcher had cut off a chunk for dog food, and Cassal claimed to have a receipt to prove it. Finally, the 689 pounds of missing meat could be explained in two ways: first, the menagerie's carnivores ate more than Guichenot realized, and second, meat that is not immediately consumed loses weight as it dries out.

Guichenot did not believe Cassal's excuses. He told Geoffroy that next to Cassal's house was a cabaret where the food was exceptionally cheap, and the proprietors were Cassal's good friends. Guichenot suspected that the cabaret's meat was coming from the menagerie. The provisioner was sure that in the winter of 1794 Cassal regularly sold meat to a restaurant on the rue Jardin des Plantes. Guichenot produced a former waiter of that establishment named Maillard, who testified to Geoffroy that Cassal had arrived "nearly every other day with big hunks of meat under his clothing." Geoffroy knew, however, that Maillard and Cassal had been enemies dating back to when Maillard provided bread to the menagerie and Cassal's complaints about its quality caused the Museum to choose a new supplier.[12]

Geoffroy reported all this at the professors' meeting of 30 November 1803. He concluded that Guichenot had done his duty in calling out Félix, but since the allegations had not been proved, he recommended that Cassal be retained. He promised to keep a close eye on the keeper and said he believed that Cassal, having been alerted to the precariousness of his situation, would see the light and perform his duties faithfully. But that was wishful thinking, and Geoffroy had reason to know better. He did not mention to his colleagues that Cassal had already given him a statement threatening to quit if he were not relieved of Guichenot's surveillance. Geoffroy had not shared the document because he believed its abusive tone would hurt Cassal's cause. The professors, lacking that information, agreed to keep Cassal in his job, but with a "redoubling of surveillance of his conduct."[13]

When he learned that the administration insisted he be watched, Cassal was furious. He told Geoffroy he could no longer tolerate Guichenot's presence and was about to quit. Geoffroy sought to dissuade him, but Cassal submitted his letter of resignation to the Museum's director, Fourcroy. Almost immediately, however, he had second thoughts and asked for his letter back. Fourcroy obliged, after consulting Geoffroy.[14]

What happened some two weeks hence was reported by Geoffroy at the professors' assembly of 21 December 1803:

> Last Wednesday, at half past four in the afternoon, Guichenot went as usual to be present at the feeding of the carnivores. Two soldiers from the First Battalion of the Parisian guard, including the drum major, approached him, and after being assured of his name proposed they go have a drink together. Guichenot refused on the pretext he was busy. The drum major insisted on the grounds that they were soldiers. Guichenot replied that that was no reason to abandon his duties. The drum major, unable to lead him outside, heaped insults upon him there in the menagerie and proposed to him that they fight. Guichenot remarked that he did not know him, that he had never had occasion to insult him. He hesitated as to whether he could arrest this man who provoked him without cause but reflecting on the reproach of cowardice that had already been made to him and to what he owed himself as a soldier, he accepted the challenge.[15]

Geoffroy told his colleagues this was a true account of events, as confirmed by three bystanders (two carpenters and "the wife of Louis the gardener"), all of whom were impressed by Guichenot's behavior and praised "the moderation of his responses." Geoffroy continued:

> Guichenot called for Dumontel, a soldier in the company of commander Grandcour, to serve him as a second, then went to drink with his aggressors and his second, prior to the action, in Jouani's cabaret, located across from the veterans' barracks. There he insisted on knowing from the drum

major why he had provoked him. The only response he obtained was: "You will know it on the field of battle."

The fight took place. With the same blow, the aggressor was wounded in the arm and the thigh. Guichenot then addressed himself to his aggressor's second, announcing that he would call him to combat if he did not say why he had been insulted by the drum major. The man swore that he and his comrade had spent part of the day drinking with Félix Cassal, who told them about all his complaints against Guichenot and stirred them up by his accounts. The man indicated further that the drum major was a relative of Félix's wife, that he had promised her he would get rid of Félix's enemy, and that he had been given three livres as a tip. The drum major also said that he was a relative of Félix, that he had worked with him at the *combat du taureau* [animal fights] in Orléans, and that he had the scar from a dog bite to show for it.[16]

Extraordinary as it may seem, "the combatants then went to drink a second time at the same cabaret." There were thus ample witnesses to the whole affair. Dumontel (Guichenot's second) and the cabaret owner (Madame Jouani) had testified to all that Geoffroy reported. Madame Jouani added that the drum major came back later looking for a cane he had forgotten and "continued to say that he had promised Félix to kill Guichenot for him and was resolved to achieve this after his recovery." Geoffroy told his colleagues there appeared to be no doubt in the matter:

> The whole of these facts you will judge, no doubt, as I do, my dear colleagues, present Félix Cassal as the instigator of the duel which just took place. I made this observation to him, but he limited his justification to denying formally to me that he had a relative in the Paris guard. He swore he had no connection with the soldiers of this corps and protested that nothing I was told was accurate. I contradicted him with my own observation. I had seen him at eleven in the morning speaking to a soldier of the first regiment of the Paris guard, with an air of mystery and secrecy, of which I had already become somewhat suspicious.[17]

It didn't matter whether or not Cassal was present for Guichenot's attack, Geoffroy told his colleagues, because he was nonetheless aware of the situation, "like Agamemnon who covered his eyes with his cloak when his daughter was sacrificed." Said Geoffroy, "Félix allowed the fight to take place" to satisfy "his hate and spirit of vengeance." Sending an assassin to attack Guichenot, Geoffroy said, was too atrocious an act to be excused, and given the other offenses the lion keeper was already accused of, the professors had no choice but to resort to the most severe punishment at their disposal, namely, "destitution." Geoffroy admitted that however painful it was to call for this, it was his duty. He requested, though, that the assembly not vote on the matter for another week, in case information favorable to Cassal should come to light.[18]

Cassal continued to deny the charges, but no extenuating facts emerged. When the assembly held its next regular meeting, Geoffroy reread his report, and the matter was discussed at length. Then, by secret ballot, ten professors voted for Félix's dismissal, and one voted to adjourn the question. Félix Cassal's connection with the Museum of Natural History was thereby officially severed.[19]

On 4 January 1804 Geoffroy was able to report that Cassal's departure had taken place without mishap and that Cassal's nephew, who had helped at the menagerie, had also resigned. Wishing to give some thanks to the former keeper for going quietly, Geoffroy provided Cassal and his nephew with a certificate bearing what Geoffroy described as "measured" words on "the degree of satisfaction I had with their services." Fourcroy added his signature to the document.[20] Now unemployed, Cassal envisioned returning to North Africa to assemble a menagerie to bring back to France. Geoffroy urged his colleagues to ask the government to provide Cassal the visas needed for a trip to the Barbary Coast. The professors agreed, and Geoffroy drafted a letter to Talleyrand, the minister of foreign relations, asking for his benevolence toward Cassal, the "former keeper of the *animaux féroces* in the menagerie of the Museum of Natural History."[21]

After his firing, Cassal continued to plead his case, but the Museum's secretary, André Thouin, on behalf of his colleagues, told Cassal: "The administration can only keep the employees that have its confidence; you have lost it."[22] Bernard Louzardy was appointed head keeper of the *animaux féroces* in Cassal's place. But if the Museum was happy to be rid of Cassal, the lion, Marc, responded differently. He began to languish, and before the year was out he was dead.

Had Cassal succeeded in making another trip to North Africa, it would surely have appeared in the Museum's correspondence or other records, yet no evidence has turned up. The records do show that in 1808, when Geoffroy was about to travel to Portugal to appropriate specimens, Cassal asked to accompany him. The Museum replied that it did not think his presence necessary for the expedition's success.[23] When Bernard Louzardy died five years later, Cassal applied to fill the vacancy, but the Museum refused to consider his request.[24]

[2.5]

The New Superintendent

When Étienne Geoffroy Saint-Hilaire returned to the Museum in February 1802 after nearly four years in Egypt, he found the supervision of the menagerie in the hands of his colleague Jean-Baptiste Lamarck and was in no hurry to take back the responsibility.[1] But once he resumed the position (September 1802), he quickly advocated restructuring the menagerie's operations.[2] No text of his proposal survives, but its central feature was the creation of a new post, a chief of menagerie operations. Geoffroy mentioned the idea to Georges Cuvier's younger brother, Frédéric, suggesting he take the job.[3] Frédéric's qualifications were not obvious. What Geoffroy found most appealing about Frédéric was that his brother, Georges, had been Geoffroy's best friend and closest colleague before he left for Egypt. Geoffroy was eager to resume relations with Georges, and cultivating a friendship with Frédéric seemed a good way to do it.[4] Georges was thirty-three, Étienne thirty, and Frédéric twenty-nine.

Frédéric arrived in Paris in the spring of 1798, summoned by Georges from Mulhouse, where he worked as a watchmaker, to assist with his brother's Museum-related work.[5] Frédéric was the gentler, unassuming brother, having grown up in Georges's shadow in their hometown of Montbéliard, part of the Duchy of Württemberg before it was annexed by France in 1793. In the Cuvier family, Georges received the larger share of his parents' emotional investment.[6] His intellectual talents and passion for study attracted the attention of the Duke of Württemberg, who provided him a scholarship to his Caroline Academy (Karlschule) in Stuttgart. Frédéric did not display the same brilliance, ambition, and sense of direction, and there was to be no Karlschule for him. His formal education ended at age fifteen when his father, aware of his son's interest in things more mechanical than academic, apprenticed him to a watchmaker.

In these circumstances, Frédéric might have felt some resentment toward his brother, but sibling rivalry seems not to have entered his mind;

Figure 2.5.1. Frédéric Cuvier (1772–1838) pictured roughly two decades after his appointment in 1803 as superintendent of the menagerie. Anonymous artist (n.d.). Muséum national d'histoire naturelle. Po 1324.

he wanted only to appear worthy in his brother's eyes. Yet Georges's eyes were focused elsewhere. The historian Dorinda Outram notes that in his account of his childhood, Georges neglected to mention having a younger brother at all.[7]

In Paris, Frédéric was Georges's multipurpose assistant, running errands, writing extracts of scientific books and articles, handling aspects of Georges's business when he was away, and helping prepare the catalog of Georges's burgeoning comparative anatomy collection. As young men of Montbéliard, part of an annexed territory, the brothers were fortunately not subject to military conscription, which would have interfered with their work. In addition to assisting his brother, Frédéric studied physics and chemistry. He assisted with the physicist Jean-Baptiste Biot's studies of the role of oxidation in generating electric current in the voltaic pile.[8] He also took on editing the multivolume dictionary of natural sciences (*Dictionnaire des sciences naturelles*) that his brother had initiated.

The uncertainties Frédéric felt regarding his future are evident in letters he wrote to Georges in 1801 and 1802 when one or the other was away from Paris. The first, sent in October 1801, spoke of his work on the dictionary (all the entries under the letter A) and of important personal news as well. He could not see himself editing much longer, Frédéric said, because he would soon have to leave Paris. He needed to be more serious about his future, because back in Montbéliard there was a young woman he planned to marry, "the daughter of Madame Macler."[9]

Georges would have known the name well. The Cuviers were Protestants, and Madame Macler's late husband, Jules Frédéric Macler, had been the local Protestant minister. Frédéric confided that the young woman shared his affection and had been waiting for him for the past five years, turning down marriage proposals from others who could offer brighter prospects than the "miserable lot in life" Frédéric pictured for himself. He hadn't told Georges sooner, he confessed, fearing his disapproval. "I have often reproached myself for this lack of confidence in your regard; but a hundred times I have been on the point of opening my heart to you, and my timidity has always held me back." For the moment, Frédéric felt bolder and unwilling to mislead Georges about how much time he could give to the dictionary. He wanted Georges to know how things stood and how happy he would be if he greeted this confidence kindly.[10]

Georges's response is absent from the record, but Frédéric's side of the correspondence indicates that he supported his younger brother's marital intentions. The following spring he aided Frédéric's cause by writing a flattering letter to Madame Macler, which evidently pleased her. Georges was the native son who became a famous scientist in Paris, and a letter from him was considered a great honor in Montbéliard. When Frédéric visited home in May 1802, he wrote to Georges in Paris that mother and daughter had greeted him warmly and everyone in the village came to see Frédéric's copy of Georges's recent portrait. Some of the Cuvier cousins set about collecting birds to send to Georges, and Frédéric went fossil hunting on Georges's behalf with another Montbéliard native, Louis George Duvernoy. Later that year Duvernoy went to Paris to collaborate with Frédéric and Georges on the development of Georges's comparative anatomy collection.[11]

Frédéric devoted more than half of his letter to recounting his efforts to help Georges scientifically before telling him more about Christine Charlotte Macler, his fiancée. She was "cherished by everyone for her kindness of heart, her sense of humor, and the beautiful gaiety she always maintains." These qualities were sufficiently endearing, but she also had "a rather nice face, without being really pretty." In addition she knew how to manage a household without outside help, she made her own clothes and embroidered her own handkerchiefs (all virtues, Frédéric apparently believed, worthy of his brother's approval).

The problem, as Frédéric explained, was that while he and Christine were happy to contemplate their life together, they no longer happily imagined it in Montbéliard. Five years in Paris had changed Frédéric's mind about village life. "When I knew only my own birthplace," he explained, "I believed its ways to be very pure compared to those of the big cities." Now he recognized the attraction of the latter, where one could choose the people one wanted to associate with, whereas in small villages "one is obliged to live with everyone." Christine too had come to feel the circumscribed nature of life in Montbéliard. The atmosphere of her household had changed since her brother had married and taken charge, and she was no longer comfortable living there.[12]

Georges invited Frédéric, who already lived in George's apartment at the Museum, to bring his wife there once they were married and to stay for as long as it took them to arrange more permanent accommodations. Frédéric did not want to be a burden, but he was eager to return to Paris to try to establish a scientific career. Earlier, he told Georges, he had thought of a career in industry (manufacturing sulfuric acid) but lacked the resources to start an enterprise. Now, with Georges's encouragement, he wanted "to devote myself entirely to the study of the natural sciences," particularly physics and chemistry, sciences for which "I always feel an irresistible inclination." He admitted that he did not yet know enough about these fields and was "frightened," but he had studied them "for more than two years" and made them "the object of all my cares and thoughts." His heart's desire, he said, was to be a teacher of science: "If I could only merit a place in public instruction! That's what I am thinking since I no longer dream of staying in Montbéliard. What do you think? Tell me what I must do and hope. In any case it will always be to you that I owe the happiness that I experience today in my work, and I will always have for you the deepest gratitude."[13]

Although relieved to see a clearer path before him, Frédéric still had concerns, including worrisome chest pains. He expressed anxiety about whether his wife-to-be would please Georges and whether Georges would treat her kindly. In the fall of 1802, when Frédéric was back in Paris, he wrote to Georges (then in the South of France helping set up a nationwide lycée system) remarking his fiancée's "extreme gentleness, excellent heart, and gaiety . . . she is not worldly, but could become so. She is far from lacking in intelligence." Frédéric asked his brother's "indulgence, and especially when you see her, one of those encouraging smiles that you employ with so much success when you wish; but with which you are sometimes too sparing. For I, who am so often in need of them because of my timidity, have seen them only rarely. If you had known how much it pleased me, how much my heart opened in confidence when you gave me this indication of friendship, you would have bestowed this favor more often, and I would not have merited the reproach of being reserved."[14]

Nine days later, Frédéric wrote to Georges about Geoffroy's idea of putting him in charge of the menagerie, but that became another source of uncertainty for him in the weeks that followed, when Geoffroy no longer mentioned it.[15] Early in December Frédéric fretted in a letter to his brother, "[Geoffroy] has spoken no more to me about charging me with [the menagerie] as he proposed recently," despite his efforts "several times" to bring Geoffroy back to the subject.[16]

On December 30, 1802, Frédéric married Christine Macler in Montbéliard in a Protestant ceremony, followed by the civil ceremony at the mayor's office.[17] The newlyweds went to Paris at the beginning of January and took up residence in rooms in Georges Cuvier's apartment at the Museum. Georges, still in the South of France, was not expected back until the end of March or early April.

In February 1803 Geoffroy presented his plan for the menagerie staff to his colleagues; establishing the post of "*aide-naturaliste et sous-chef*" was the primary innovation. As Geoffroy explained, despite his enthusiasm for the task, he was too much occupied with teaching, advancing science, and administering the galleries to give the menagerie the direct and daily attention it needed. Continued oversight of the keepers was essential to prevent them from behaving in a "semi-independent" fashion, leading to "very serious inconveniences" with respect to the animals and the visiting public. The menagerie needed a "special head" (*un chef spécial*), he said, who was devoted to overseeing the keepers "at every moment of the day." The *chef* would also be charged with keeping track of the changes at the menagerie, gathering observations about the animals, pursuing experiments requested by the professors, and managing the menagerie's stores. Three keepers would work under him (for the carnivores, herbivores, and birds) as would a commissioner of provisions. Geoffroy had specific articles to this effect that he was prepared to have his colleagues ratify. They responded by asking him to develop his ideas more fully.[18]

Toward the end of March, without specifying that Geoffroy had given him renewed reason for hope about the job, Frédéric wrote to Georges enthusiastic about the menagerie's recent good news: "We received in these last days some of the animals destined for the menagerie: the civet, two genets, a superb lion, and eighteen large-tailed sheep with some she-goats and he-goats from Judah." There was also bad news; "I fear that we will lose a kangaroo," wrote Frédéric, "it had violent convulsions yesterday. It would really be too bad if it died before you arrived. Its anatomy would be nearly lost."[19]

Frédéric also recounted his young wife's reactions to Paris:

> At first, she was stunned by it: everything she saw differed so much from what she had seen before that she did not know whether she should find it agreeable or not. But after several days, truly beautiful things soon seemed so to her. It was thus with women's clothes, except for what was too forward in some of them and capable of alarming the modesty of a woman from the provinces. The public gardens and the French theater have charmed her equally, but the others have not yet been able to find favor, except for the Grand Opera because of its decorations. At the Feydeau and the Vaudeville [theaters], she is especially shocked by the mixture of song and ordinary speech.[20]

Frédéric reported that his wife felt somewhat isolated, for she was unused to mixing with people she had not known for a long time. Madame De Candolle impressed her as the nicest of the ladies she'd met, but Christine saw her rarely, and only on social occasions of a more ceremonial than friendly sort. Frédéric supposed she would eventually learn to enjoy the company of people who gathered for reasons other than mutual affection, and that she would grow fond of Parisian life. "She is young," he wrote, "and a woman always yields rather easily to the taste of the world in the midst of which she is living."[21]

We do not know whether Georges Cuvier approved of the woman who greeted him on his return to Paris, since the two brothers could now talk instead of writing letters. Likewise, we do not know if Georges praised the improvements the newlyweds had made to rooms in the apartment, or how soon he was told, or perceived, that his brother's wife was in the early stages of pregnancy.

Scientists leave traces of themselves in many forms: scientific papers, books, notes, and illustrations; by constructing instruments, collections, or displays or establishing practices that others adopt; in the ideas and attitudes they convey to their students, and so forth. These rarely offer much information about their personal lives. Letters are more illuminating but are unnecessary when people can meet and speak. Within settings like the Museum of Natural History in Paris or the town of Montbéliard, news of personal joys and tragedies was typically communicated face to face. Information regarding births, marriages, and deaths was documented in official registers, but personal reactions were not. Autobiographies can be revealing in this regard, but Frédéric Cuvier left none. The official records report that Christine Charlotte Marie Cuvier died in childbirth on 9 October 1803, after bringing a son into the world. Frédéric named him Frédéric Georges Constant Fortuné Marie Cuvier. Although a great deal is known about Frédéric Cuvier's career, of his personal life, including how he managed to raise his son, we know very little.

The turning point in Frédéric Cuvier's scientific life occurred late in December 1803, some two and a half months after his son's birth and his wife's death. The prelude occurred at the professorial assembly of 14 December, when Geoffroy proposed creating the position of menagerie superintendent (*garde de la menagerie*) at the Museum, but without mentioning Frédéric's name. Geoffroy's sixteen-article list of regulations defining the post encapsulated his view of the menagerie's potential and problems and the control he wanted to continue to exercise. Most of the articles were of great consequence for the menagerie's subsequent history.[22]

The first five articles specified the *garde de la ménagerie*'s place in the Museum hierarchy with respect to his immediate superiors and to the keepers he was meant to supervise:

> Article 1. The officer employed with respect to the oversight of the animal keepers . . . will hold the title of *garde de la Ménagerie.*
>
> Article 2. He will be placed under the command of the professor in charge of teaching the natural history of mammals and birds and will communicate with the administration through this professor as intermediary.
>
> Article 3. However, when animals exist at the menagerie corresponding to the provinces of other professors of zoology, he will take his orders and instructions regarding these animals from those professors.
>
> Article 4. He will have under his command all the animal keepers, the day workers assisting them, and the commissioner who distributes the provisions.
>
> Article 5. The oversight of these employees will have as its object the good health of the animals and the maintenance of good order when the public is admitted and near them.

Of these articles, the second became a flashpoint for Geoffroy's anger years later when he felt Frédéric Cuvier was disobeying his orders and communicating directly with the administration without his approval. The third article was never observed because there were virtually no animals in the menagerie besides mammals and birds (other than an occasional reptile) and only Geoffroy to issue any explicit instructions regarding them.

The next ten articles addressed the superintendent's duties with respect to caring for the animals, conducting experiments, keeping records, and reporting to the professors:

> Article 6. He will recommend the greatest cleanliness as one of the most effective means of preventing the majority of illnesses.
>
> Article 7. He will be present as often as possible at the feeding of the animals to make personally sure no abuse slips into the consumption of provisions and to judge, according to the animals' greater or lesser appetites, the state of their health or illness.
>
> Article 8. In case of [an animal's] illness, he will observe its treatment.
>
> Article 9. He will train and use in the work [of the Museum] those animals that are capable of it.
>
> Article 10. He will neglect no opportunity to attempt the acclimatization and naturalization of foreign species that would be of some benefit for the rural or domestic economy.
>
> Article 11. He will be charged with conducting experiments with respect to which the professors of zoology will deliver a program.
>
> Article 12. The study of the habits of animals will be part of his province. He will record with care the times which begin and end rutting, gestation, laying, molting, and generally all the changes of state that happen to animals each time of the year.
>
> Article 13. He will have for this purpose a register which will contain first the names of all the animals of the menagerie, and where will be transcribed daily all the facts of their activity that he observes. These notes on each animal will be preceded by an account making known its origin and as much as possible about it, the different places it had been transported, and the different uses to which it had been put. A summary of this register will be presented each month to the administration.
>
> Article 14. The *garde de la ménagerie* will be charged with managing all the material of the establishment. He will be responsible to the professor

> who is his director and to the administration for all the provisions deposited in the storehouses of the Museum. He will account for its consumption by receipts from the keepers who employ it. This record keeping will require him to maintain a special account book where he will mention the incoming and outgoing of all the foods that are deposited in the storehouses and consumed by the animals.
>
> Article 15. All the changes at the menagerie as much with respect to staff as to materials will, by means of a written report, be placed each morning before the professor who is primary chief of the section.

The care and feeding of the animals, efforts to acclimatize and naturalize foreign species, and the study of animal habits will be discussed later. There was only one instance of putting animals to work at the menagerie: camels were used to draw water from a well. Surprisingly, article 11 was not enacted, as the professors of zoology never submitted a program of experiments for the *garde* to conduct. Frédéric did his best at first to adhere to the three articles on record keeping, but several of the prescribed practices quickly fell by the wayside, such as preparing daily reports for Geoffroy. Over time, a weekly menagerie report seemed to suit all parties best.

The final article read as follows: "Article 16. The zoo being open to the public every day, the *garde* is obliged to be present there likewise. He will not be able to leave the Museum without the permission of his chief."[23]

There was more to article 16 than meets the eye—indeed, more than Geoffroy's fellow professors appreciated when he proposed it. Not until twenty-four years later, when his relations with Frédéric Cuvier had deteriorated to the point of acrimony, did Geoffroy reveal that he had, with Frédéric's approval, added the rule to increase Frédéric's chances of securing the post, since he already lived on the Museum's grounds.[24]

At their meeting of 14 December 1803, the professors discussed Geoffroy's proposal in general and scheduled a more in-depth discussion for the coming week. That next meeting was momentous. Amid the usual announcements and items of business, Geoffroy took center stage three times: to read again his list of regulations for the menagerie (which his colleagues approved); to recommend the firing of Félix Cassal (a decision that was postponed until the next week); and finally to recommend someone for the post of menagerie superintendent.

Geoffroy had prepared his remarks on the new appointment carefully. He began by saying that his recommendation would be guided by the regulations his colleagues had just approved. These called for someone who was at once an animal doctor, an able naturalist, and a capable administrator; but none of the applicants, he said, was adequately qualified.[25]

Working through his list, Geoffroy quickly dismissed one applicant whose natural history credentials consisted only of "an extreme passion for raising pigeons" and a talent in drawing and sculpting. A young veterinarian with no knowledge of natural history who had not been heard from for months was also readily rejected. A more serious candidate was Professor Dupuis of the Veterinary School at Alfort. He appeared "to combine all the qualities you ought to want for the place of superintendent of the menagerie: knowledge of animal medicine, the taste for comparative anatomy that will have led him to the study of natural history, a character that the animal keepers would find imposing, etc." Acknowledging that André Thouin thought highly of Dupuis, Geoffroy said he would have happily supported his candidacy, but Dupuis had told him he intended to hold the post temporarily, then turn it over to one of his students. Furthermore, Dupuis seemed to think the post only required regularly checking the animals' health. Having discovered that Dupuis was unable "to devote all his time to our establishment," Geoffroy had decided to thank him and look elsewhere. Dupuis introduced Geoffroy to a student candidate whom Geoffroy found "so young and so new to the ways of the capital that I was forced to continue my searches."

These applicants convinced Geoffroy that rather than a veterinarian seeking to establish his credentials, the Museum really needed an administrator to actively oversee the menagerie's operations. A "physiologist" (naturalist) would be just as good as a veterinarian, since he would have some familiarity with the animals' digestive systems and over time learn how to care for sick animals.

Having made a case for choosing a naturalist, Geoffroy turned to three who had applied for the superintendent's post. One knew how to prepare specimens and had experience as an administrator, but Geoffroy judged these qualifications insufficient. Another had some ability in determining species and had worked as Geoffroy's secretary, but what Geoffroy knew of the man's character made him unsuited for the position. That left the candidate Geoffroy had been saving for last, Frédéric Cuvier.

Noting Frédéric Cuvier was so modest that one might have heard only of his scientific work in chemistry, Geoffroy pronounced, "I know him to be very well informed in comparative anatomy, in natural history, in general medicine. It is he who writes the greatest number of zoology articles for the new *Dictionnaire d'histoire naturelle*. He has practice in bookkeeping. He knows how to manage people while treating them gently. I am happy to do him this justice." Geoffroy continued: "If you did not know him yourselves, citizen colleagues, I would have to speak to you about the details of his morality, but there is none among you who does not know very well that he has a character worthy of the spirit that animates you. He is perfectly acquainted with the duties he would have to fulfill if you were to accord him your votes."

To demonstrate that Frédéric had considered the matter, Geoffroy deposited on the assembly room's desk two memoirs Frédéric had composed, one on the uses of a menagerie and the other on the functions of a menagerie supervisor.[26] Geoffroy concluded: "I am persuaded that Monsieur Cuvier has the qualities appropriate for the job. It would be agreeable to me to work with a person of such pleasant intercourse. I have the honor of presenting him to you for the place of superintendent of the menagerie."

Geoffroy crafted the recommendation shrewdly. He did not need to say that Frédéric was the brother of Georges Cuvier or to appeal to his colleagues' sympathies by mentioning that Frédéric was a widower with an infant son. As for the question of whether Frédéric, with his gentle personality, would be able to keep the keepers in line, the professors understood that the imminent dismissal of Félix Cassal would make the job of menagerie superintendent much easier. Like the decision on Cassal, the vote on Frédéric was postponed until the next assembly, 28 December 1803. At that meeting, Cassal was fired and Frédéric Cuvier was elected unanimously as menagerie superintendent at a salary of 2,000 francs per year. He began work the next day.[27]

Frédéric took up his new position energetically. To systemize the distribution of food to the different parts of the menagerie (and eliminate keepers' abuses), he developed tables indicating the kinds and quantities of food the menagerie's animals ate. He called attention to animal quarters that needed repair or reconstruction, identified animals needing relocation, and reported on animals that were ill. He became acquainted with the keepers and learned of frictions between them even with Cassal removed from the scene.[28] His experiences in his first six months also impressed on him how much he still needed to learn, a fact he lamented in August 1804 in a new menagerie guide that he published anonymously:

> Who can one consult [about] constructing the most suitable shelters for wild and often dangerous animals mostly coming from climates different from our own? What nourishment do most of them require? How must they be treated to keep them in good health and to manage them?[29] To what diseases are they subject? What remedies can one administer to them? What would be required to do that?

There was practically no way, he said, of answering these questions, or a hundred like them, because "nothing has been written, almost nothing has been seen, and everything remains to be done."[30]

Over the next thirty-five years, Frédéric Cuvier formulated tentative answers to these questions as he supervised the development of the modern era's first public zoo. His achievements would include a more comprehensive and broadly informed assessment of mammalian behavior than ever attempted previously. His successes and frustrations would be intimately related to the peculiarities of the niche where he found himself. The material conditions of the zoo, the kinds of research that could be conducted on captive animals, the zoo's functions amid the purposes of the Museum as a whole, interpersonal relations, and the dependent status of the superintendent in the institutional hierarchy would all have a bearing on what he was able to accomplish and what he was not.

[2.6]

Kindness in Context

The young Frédéric Cuvier had likely never seen a live lion until he arrived in Paris in spring 1798 to assist his brother with the Museum's comparative anatomy collection. At the time, there was only one representative of the species at the menagerie, the young female taken from an English vessel by a French corsair the previous year. Then, in the summer of 1798, Félix Cassal returned from his collecting trip to North Africa with four new lions. In the months that followed, Cuvier had the opportunity not only to observe the lions in their cages but also to witness Cassal's treatment of them. What he saw left a lasting negative impression on him. He first wrote about Cassal's relations with the menagerie's animals in 1804 in his menagerie guide, but not until a decade and a half later did he describe how Cassal had achieved mastery over the lion, Marc. As Cuvier recalled with distaste, "The lion ended up being tame and very docile for his master, but this submission had been obtained only by violence." Cuvier elaborated:

> In its youth, [the lion] was inclined to ferocity, and good treatment softened it without bringing it under subjection. However, its keeper wanted it to obey his voice, and with kindness not succeeding, he had recourse to punishment. His first efforts were painful; he was several times on the point of renouncing his project; but he finally succeeded in breaking the character of the lion, in making it know fear through pain, and in rendering himself entirely the master. Henceforth this animal learned to obey certain commands and could be offered in show by its master, which the public was pleased to compensate, but this lion carried all its life the traces of the blows it had received, the scars of the wounds that the blows of the whip had made.[1]

At the end of June 1799 Toscan published his anecdote about the lion's love for his master. He never mentioned the brutal training regimen through which Cassal had achieved the lion's obedience and affection. That

Figure 2.6.1. The new ménagerie for the animaux féroces, finally completed in 1821. From Augustus Pugin and Charles Heath, *Paris and Its Environs, Displayed in a Series of Two Hundred Picturesque Views* (Jennings and Chaplin, 1831). R. Burkhardt collection.

apparently occurred during the lion's first nine months in the menagerie. And Frédéric Cuvier never wrote about the lion's affection for Cassal, only Cassal's success in rendering the lion docile. Cuvier mentioned that success in his 1804 menagerie guide, but only in passing, not to promote the menagerie as a site for learning how to manage animals. Rather, he emphasized the menagerie's importance for zoology. He explained that by comparing diverse animal forms side by side, in all phases of their development, naturalists would be able to distinguish different species with a precision that was otherwise not possible. He additionally suggested that menageries could help with domesticating and acclimating animals. Unlike Bernardin de Saint-Pierre, Toscan, and Lacépède before him, however, he made no claims for the menagerie as a site for elevating public morality. He did not suggest that seeing the benefits of treating animals kindly would lead people to be kinder to animals and to each other. Nor did he suggest that being kind to animals was a good in itself, or even an issue. The questions that needed resolving, he said, were about feeding and lodging the animals, keeping them healthy, and managing them; but concerning the last of these, he did not say what the questions were.

Not until he published a new edition of his guide in 1809 did Cuvier spell out the animal management question, "Should one use force or kindness?"[7]

Already in 1804, however, in writing about specific animals in the collection, he made his conclusion clear: Kind treatment makes animals gentler, while cruel treatment makes them more savage. His conclusion emerges in his accounts of the menagerie's wolves, the new male elephant, and three of its lions. Of the story he would later tell about the lion Marc, however, he made no mention.

CASSAL'S LIONS REVISITED

Though the Museum in 1802 rejected the rumors about Cassal's injuring the lion cubs, and Cuvier never contradicted them, he acknowledged in his 1804 menagerie guide that treating animals harshly was a hallmark of Cassal's regime as carnivore keeper. About one of Cassal's lionesses Cuvier wrote, "If she seems savage it comes from the bad treatment she received." He explained, "Her first keepers [made] a kind of glory for themselves in having under their direction animals that appeared to be very ferocious and intractable." They mistreated the animals to enhance "the defiant and independent character" that was already natural to them.[3] Regarding "the lioness from the Sahara," Cuvier reported, "when Citizen Cassal was in charge, she seemed to have the most intractable nature: at the mere sight of her master, she bounded about her cell and menacingly gnashed her teeth in the most frightful manner." After Cassal's departure, however, she showed "much less ferocity" toward her new keepers. They found that it helped to speak to her gently: "Her ferocity was not at all natural to her," wrote Cuvier, and she "would perhaps be very friendly today, if she had experienced gentler treatment [earlier]."[4]

The hyena was another menagerie animal whose aggressiveness seemed to have been exaggerated by Cassal's ill treatment. Georges Cuvier said of this individual in 1802, "it seems to hate its keeper more than any other person; it is enough for it to notice him to become furious."[5] Six months after Cassal's firing, Frédéric Cuvier could offer the animal as "a very strong proof of the effects of good treatment on the character of fierce animals." It had begun to show its new caretakers an unexpected trust, even taking bread from a keeper's hand.[6]

The story of gentle treatment making ferocious animals less fierce was complicated, however, by the examples of individuals that displayed allegiance to earlier keepers and failed to warm to new keepers who treated them well. This was the case for two male lions newly arrived at the menagerie in 1804 as gifts to Napoleon from the dey of Algeria. Whenever their new guardians approached them, they became furious. Both, however, had been fond of the Spanish conductor who brought them from North Africa to Paris "and obeyed his voice without resistance." When he came back to visit them, they "testified with lively caresses all their joy in being together again with their old master."[7]

The lion Marc also preferred his keeper (Cassal) to all others, though Cuvier did not explicitly identify the animal's sentiments toward his keeper as affection. What Cuvier said was that Cassal had succeeded in taming the male and making it "extraordinarily" docile toward him.[8] After Cassal's dismissal from the Museum Marc became ill, but Cuvier did not correlate the events. In a notebook devoted to the menagerie's animals, on a page headed, "the lion named Marc," Cuvier described the animal as "languishing."[9] Likewise in the daily menagerie reports he prepared for Geoffroy Saint-Hilaire, he identified the lion as one of several animals whose health was of concern. For a while the lion seemed to have regained its appetite, but Cuvier's report of 1 February read, "The big lion has refused to eat and he has vomited again." He interpreted the lion's illness as indicative of the problems of keeping exotic animals healthy in a hostile climate. "The bad state of the carnivores" had no other causes, he said, than "the season and the way they are lodged." He warned, "one will never be able to preserve animals from the south if one does not shelter them from the inclemency of our climate.[10]

Later that summer, in his guide to the menagerie, Cuvier noted that the lion was not yet comfortable with his new guardians, "despite the care they take for him, although he receives food from them and sees them at each instant of the day, [they] have not yet been able to inspire his confidence.[11] A month and a half later, the procès-verbaux of the professors' assembly (12 September 1804) announced, "The handsome lion of the menagerie named Marc died last Sunday between six and seven o'clock in the morning."[12] Dufresne's zoological lab report explained that the lion was "dead as the result of a long illness." A heat wave forced the taxidermists to work quickly. They preserved Marc's skin and the other parts needed for mounting him. The job was "nearly finished" in three days.[13] Fifteen years after Marc's death, Frédéric Cuvier called him "the most beautiful individual of its species that our menagerie ever possessed."[14]

In 1799 Georges Toscan had recounted how Cassal's absence left the male lion suffering as if "truly suffering from illness." Cuvier mentioned that illness but did not ascribe it to Cassal's being away.[15] Nor did he connect Marc's decline in 1804 to Cassal's dismissal. Knowing what he did of Cassal's brutal character and his cruel treatment of the lion, Cuvier could not have been comfortable thinking Marc was grieving the lion keeper's absence.

Cuvier's thoughts after Marc's death took a different direction. His menagerie notebook mentions that tubercles were found in the lion's lungs, liver, and colon, and "the testicles especially were in a state of inflammation that had nearly destroyed them." Although the lion often copulated with the female, "he had been separated from her for some time." Cuvier wondered if this had contributed to the animal's death and if it was dangerous to develop needs in animals without providing the conditions to satisfy them.[16] He appreciated that the conditions of captivity could influence an animal's mental as well as physical well-being. He was sympathetic with Constantine after the birth of her three cubs, recognizing the "discomfort she was in, the narrow limits of her cage, [and how] the people ceaselessly moving around her cage" made her seek instinctively to remove her cubs from danger.[17]

The anonymous 1803 guidebook to the Jardin des Plantes that described Cassal's "familiarity" with the fierce carnivores as "truly frightening" went on to say, "At his voice, the leopard cringes and the king of the animals, the lion, seems to forget his voracious instinct."[18] Subsequent editions of this guide bore the same passage about the *animaux féroces*' obeying their keeper's voice and simply inserted the names of the new keepers in place of Cassal's. But none of Cassal's successors ever affected a lion tamer's persona. Frédéric Cuvier would not have permitted it.[19]

THE WOLVES

The menagerie's wolves gave the new superintendent another opportunity to comment on how methods of treatment could modify an animal's nature. The menagerie had wolves from early 1796 when Delaunay acquired a pair to occupy empty cages. Few Frenchmen of the time would have had any doubts about the wolf's character. Buffon portrayed it as a coarse and cowardly creature prone to killing sprees (1749), and the rash of wolf attacks that bloodied the French countryside in the 1760s reinforced his view, whether one believed "the beast" that ravaged the Gévaudan region was a wolf or something even more monstrous.[20]

The government answered the lupine onslaughts of the 1760s with a campaign to exterminate wolves throughout the kingdom.[21] The results were impressive but not definitive, and by the 1790s the wolf population had rebounded. The government's elimination of the Office of the Wolf Hunt (*louveterie*) in 1787 as a cost-cutting measure no doubt contributed to the comeback, as did the Revolution, which saw the emigration of aristocratic hunters, the disarmament of country dwellers, and other wartime disruptions. In 1797 the government responded to the "frightening" multiplying of wolves in the countryside by establishing bounties for dead ones: 150 livres for a wolf that had attacked adults or children, fifty livres for a pregnant she-wolf, and twenty *livres* for a young wolf that had reached the size of a fox. That year, 5,351 bounties were paid.[22]

The wolf's notoriety would have made it at least a minor attraction when Delaunay acquired the menagerie's first pair in 1796. By the summer of 1798, when L. F. Jauffret published his Jardin des Plantes guide, another specimen had been added to the collection. The wolves were small, but a reader could still be gripped by Jauffret's assertion, lifted directly from Buffon, that the wolf "loves human flesh, and perhaps, if it was the strongest, it would eat no other."[23]

The behavior of the menagerie's first wolves did nothing to rehabilitate the species' dark reputation. When the female gave birth early in June 1799 to six wolf pups, the menagerie took away three and left the others for the parents to raise.[24] A few weeks later, the parents ate them. Félix Cassal interpreted it as punishment for the pups' misbehavior. As Georges Toscan recounted: "One of these wolf pups occasionally went out through the bars of the cell to play with the dogs of the menagerie and then returned. The keeper maintains that this company displeased the father and mother of the wolf pups. Whatever the case, one fine morning they fell upon the three little ones and devoured them. All that remained of this wolf dinner were some skin and bones."[25] It seems Toscan was not entirely persuaded by Cassal's explanation, and his reference to wolf "fathers and mothers" shows he was not fully conversant with the case, since there was only a single adult male wolf in the menagerie at the time.[26] Miger's 1801 guide to the menagerie did not speculate about the parents' motives, saying only that "the two [wolves] of the menagerie have had several litters, and their first babies lived; but for two years they have taken the habit of killing their babies and eating them; even after they have nourished them for two months."[27]

Did the behavior of the menagerie's wolves become notorious beyond the pages of the guidebooks? Apparently not enough for Jacques-Louis David to send one of his students to see the menagerie's wolves when the student needed to incorporate a wolf in a painting. One of the benefits claimed for the menagerie was enabling artists to gain a better grasp of a species' character, but David indicated that a stuffed wolf would in this case be sufficient, and he asked the Museum to provide one.[28]

The story of the wolves' eating their pups was repeated in two guidebooks of 1803, the anonymous *Guide au Jardin des plantes* (which repeated Miger's account) and Pujoulx's *Promenades au Jardin des Plantes*, which embellished the Cassal/Toscan version and cited the wolves' actions as proof that the species should be exterminated. Pujoulx acknowledged that wolves could be tamed to a degree, but he denied the training would last. A single incident, he said, could cause a tame wolf to revert to its natural ferocity.[29]

Frédéric Cuvier introduced a new perspective about wolves in his guidebook. Of the male and female pair, he stated, "Each year they produce babies that they kill and ordinarily eat. It would be difficult to find the cause of an act of such ferocity, but it is a fact that one scarcely finds examples of this except among animals diminished by captivity. It is also a fact that these examples are not rare and are found even in animals that are not led, by their nature, to fierce actions, such as rabbits, guinea pigs, etc."[30]

Cuvier acknowledged that the two wolves were "very mean" (*très-méchants*), but he attributed this to their mistreatment. In contrast, the menagerie's third wolf was nearly as friendly as a dog (a consequence of being well treated). Cuvier admitted, however, that friendliness did not prevent the she-wolf from trying to escape from her cage and threatening those who sought to recapture her.[31] Writing about her years later, he did not recall these bids for freedom, reporting instead that she had lived familiarly with the dogs of the menagerie and that "had she been corrected of her taste for fowl, one would have been able to let her enjoy, without inconvenience, the greatest liberty."[32] It was a rosy recollection of the wolf in her captivity.

In contrast to Pujoulx's call for the extinction of the entire wolf species, Cuvier offered a nuanced view of how heredity, individual differences, and kind treatment contributed to an animal's character. In 1821, with seventeen years of experience as menagerie superintendent, Cuvier told the stories of two more wolves, a male that displayed extreme affection for its owner and a female that was friendly to everyone. He did not dispute that wolves displayed "fear, defiance, and hatred" when living in places dangerous to them, but he believed that in the wild, without enemies and with plentiful prey, they were bolder in character, stronger in body, and finer in form, and that when they were well treated in captivity they could become tame and affectionate.[33] Though he characterized the wolf as "condemned by its nature to have cruel appetites," he recognized that it was "endowed with the intelligence to adapt its behavior to all the different circumstances necessary for satisfying these appetites," and that when circumstances permitted, the wolf could develop "gentleness, confidence, and affection."

In this regard Cuvier took the wolf as emblematic of carnivores in general. "There is no species of carnivorous animal that cannot be tamed by appropriate treatment, that is to say, that is not susceptible of becoming loving and friendly to a certain point toward those who feed and care for it." It was nonetheless important to appreciate that "these sentiments do not manifest themselves in all [species] to the same degree, and one finds differences between individuals."[34] Already in 1804, Cuvier recognized that wolves' eating their pups was not a fault of the species but aberrant behavior in animals suffering in captivity.

JOUBERT'S ELEPHANT

In his first weeks as superintendent, Cuvier was confronted with multiple cases of animals' suffering from the conditions of their confinement. It was the cold, it seemed, that led the female tiger and the civet to develop pneumonia and die. A paca died trying to escape its cage. A jaguar had grown too large for its cage and was mutilating itself.[35] The male elephant had taken to pressing his tusks sidewise against the vertical beams and bars of his stable, producing such a misalignment that the tusks' tips nearly touched, making it impossible to bring his trunk between them to his mouth.[36] Cuvier had the jaguar moved to a larger cage, to good effect, and saw to straightening the elephant's tusks, but he failed to solve the emerging problem of the elephant's keeper mistreating his charge.[37]

The administration appreciated that elephants can be particular about their handlers. They put a high premium on getting Thompson to accompany the menagerie's original elephant pair from Holland to France and remain their keeper in Paris. Six months after the January 1802 death of the male elephant, Hans, Thompson asked permission to retire, but he agreed to stay until the female, Parkie, had become familiar with someone else. When Thompson finally retired in July 1803, the Museum put Bernard Louzardy in his place.

At that point the prospects of acquiring a new male elephant were finally looking up. Earlier possibilities in London, Basel, and Toulouse had come to naught (owing to death, gender, and asking price, respectively), but in May Geoffroy went to Rouen to see a young male Asian elephant that was touring France. Based on his positive report, the Museum urged the minister of the interior to fund the purchase.[38] Encouraged by the possibility of the sale, the owners brought the elephant to Paris in June and put it on display. As the press reported, the elephant's mahout had trained it to perform tricks at his command. The elephant would lie down, get up, or start walking as instructed. It would uncork and empty a bottle of wine, lower its head to help its mahout climb on its back, carry him in its tusks, give him his hat, or collect money.[39] Encouraging the government to buy the elephant, the press cheerfully observed this would end the female elephant's "widowhood."[40]

After considerable negotiating, the male was acquired in September for 16,000 francs, less than its owners hoped, but more than the cost of the entire collection of big cats acquired from Gilbert Pidcock in 1800. Younger and smaller than Parkie, the male was frightened when first brought near her. She in turn was cautious, but when she discovered her new companion was a male that intended her no harm, she signaled her joy with gestures and cries, and was the more expressive of the two when they later exchanged caresses.[41] The new male came in the care of Joseph Joubert, son of its former owner. Joseph had been the elephant's mahout, and the Museum hired him to continue in that capacity. The young man and the elephant seemed a good combination, but that was not to last.

By January 1804, when Frédéric Cuvier set to work as superintendent, a fraught relationship had developed between Joubert *fils* and the keeper Antoine, who had been assigned a role in the elephants' care. Recognizing their antipathy, one of Cuvier's first acts as their new boss was to recommend to Geoffroy that the two keepers' responsibilities be separated. Antoine would collect the elephants' food, remove their dung, and oversee their park. Joubert would handle the elephants, oversee them in their stables, and show them to the public. Cuvier suggested a raise for Joubert (from 2 francs per day to 2.5 francs) in recognition of his good conduct to date and the risks he ran in working so closely with the elephants. Cuvier hoped the raise would help attach Joubert to the institution. Geoffroy endorsed the proposal.[42]

After less than three weeks on the job, Cuvier turned to the male elephant's tusks. Not only had pressing his left tusk against the bars of his enclosure brought the elephant's tusks too close together, but the elephant had also begun bringing his trunk to his mouth over his displaced tusk, moving the tusk even farther from its normal position. Cuvier reasoned that if such forces caused the displacement, then similar forces applied in reverse might bring the tusks back into position.[43] An iron bar with a length-adjusting screw in the middle was placed between the tusks. When the administration learned of this, it had doubts and called for the bar's removal. Fortunately the brief experiment in elephant orthodontia had pushed the tusks far enough apart for the elephant to pass his trunk between them again. It was a small improvement for the elephant, but it did not change his fundamental situation as a captive.[44] Nor did it spare him from the suffering visited on him by his keeper, Joubert.

The same day Cuvier announced the successful tusk realignment to Geoffroy, he reported that Antoine and Joubert were quarreling again, He acknowledged that all the menagerie's employees hated Antoine because of his behavior toward them, but Antoine, wrote Cuvier, was at least conscientious in his duties. The same could not be said of Joubert, who had grown so negligent as to require reprimands. Cuvier predicted Joubert would sooner or later quit his post.[45] As it was, the female elephant's stable was not clean. Her urine was not draining from the paving stones, and she was damp in the morning from lying in the wet straw at night. Extra straw was being used, but the paving stones needed replacement. Cuvier had the stable floor repaired in a matter of days, but he was unsuccessful in reforming Joubert's behavior.

By May, Joubert's relations with the male elephant had deteriorated dangerously. The previously docile and gentle animal had become unmanageable and subject to fits of rage. Cuvier ascribed its state to physiological changes associated with its transition to adulthood, but Joubert's mistreatment, which was egregious enough to make the papers, was also to blame. *Le Journal de Paris* reported (3 May 1804): "A few days ago the mahout hit the male elephant's trunk too hard with his iron-tipped baton, unintentionally injuring him. The animal is angry and holds a grudge against him. One does not know how to succeed in calming his anger."[46]

Years later, Cuvier elaborated on the story. Joubert's bad habits commenced as soon as he was no longer under his father's eye. He took to drinking and striking the elephant. The animal in response became "dejected and taciturn," and though it performed its customary tricks, it no longer did so playfully, as if they were games or amusement. Joubert was oblivious to the effects of his violence and humiliated to have lost authority over the animal. Cuvier insisted that good treatment alone could return the animal to its gentler state, but his advice went unheeded. Wrote Cuvier:

> [Joubert's] irritation was increasing, and one day when he found himself less susceptible to reason than customary, he struck his animal with such brutality that the latter, provoked beyond endurance, gave such a cry of fury that its master, frightened by this cry that he heard for the first time, hastened to flee, and it was good for him that he did; for the elephant no longer even permitted him to approach it. At the mere sight of him, the elephant became furious, and all the means tried thereafter to bring back better feelings in the animal were unsuccessful. Hate had replaced love, intractability replaced obedience, and while this animal lived, these two feelings dominated it.[47]

At the time, Cuvier recorded of the elelphant in his menagerie notebook: "He all at once stops obeying his master, who beat him, and becomes very mean [méchant]."[48] The situation was bad enough for Cuvier to mention it in his menagerie guidebook, written when the elephant was still alive. He noted that the keeper's efforts to control it through force and mistreatment had worked when the animal was younger, but only awakened its rancor as an adult. The elephant was continually seeking revenge and would "infallibly kill its master, if it could catch him."[49]

Accompanied by various warning signs, the story hurtled to its unhappy conclusion, but it was the elephant, not Joubert, that died. The professors at their meeting of 27 June 1804 learned that the elephant had broken its left tusk in a rage.[50] In a menagerie report two weeks later, Cuvier noted that all the staff were performing well except for Joubert, who was "out of control" (*déréglée*).[51] Cuvier's notebook of observations on the menagerie's animals records the day-by-day sequence of events:

> He breaks the end of his other tusk in trying to open the door separating him from the female.
> He falls ill, stops eating and drinking as usual, and his excrement is harder.
> He refuses all nourishment, he drinks, and he bathes. One dares not approach him for fear of his malice.
> Same state as the preceding day. He lies down in his park. He returns.
> He is found dead in the morning [of 30 messidor an 12 (19 July 1804)].

The autopsy took place the next day, and the considerable inflammation of his pleura and intestines were found to be the cause of his death.[52] The press reported these dates accurately, but the earliest announcements mistakenly said the female had died. *Le Journal de Paris* got the gender right but could not resist noting, "It is said the male's death, was caused by an excessive love for his female, who was much too strong for him."[53]

Joseph Joubert was dismissed, his services having become superfluous.[54] No one blamed Frédéric Cuvier for the course of events, but it must have been sobering that his first major attempt at personnel management had ended so disastrously. Joubert was not accused of causing the elephant's death, but his heinous treatment had cost the Museum a prize specimen for which the government had paid dearly.[55] It was a fiasco no one wished to repeat, and there was no rush to find a new male elephant.

KINDNESS IN CONTEXT

The Joubert affair presented an instructive sampling of the challenges Cuvier would face overseeing the interrelations of animals and people. He continued to regard kind treatment as the best approach to winning animals' trust and securing their cooperation. His first six months on the job offered proof that incremental changes could lessen the animals' suffering. The menagerie was nonetheless a place where animals were constrained and controlled, where violence was never far from the surface, where suffering and illness were generally the animals' lot, where affection and brutality often existed side by side, and where the interests of the public, the keepers, and naturalists were commonly privileged over those of the animals.

Cuvier promoted kindness, but he did not question the regime of animal subjugation that the menagerie represented.[56] That this was unkind to animals was barely thinkable in his day. Societies for the prevention of cruelty to animals were decades in the future, and when they were founded, their first concerns were the brutal treatment of domestic animals and such entertainments as staged animal fights.[57] Paris had both. One of the attractions just outside the city walls was the *combat du taureau*. Not of the Spanish kind, these "bull fights" typically pitted dogs against each other or against a bear, wolf, panther, lion, bull, or donkey. Posters for the show pictured its purported highlights, including a bull tossing half a dozen disemboweled dogs into the air, with their blood flying.[58] Held on holidays, Sundays, and Mondays, the show concluded with fireworks and a bulldog lifted high in the air by a cord clenched in its jaws. One guidebook to the sights of Paris (1807) allowed that this "atrocious amusement" was frequented only by the lower classes, but a wise person could go there "to study the habits and character of the multitude [that was] always curious to see blood and scenes that strongly disturb the soul." Commented the author: "It is the tragedy of the faubourgs."[59]

A stroll at the menagerie at the Jardin des Plantes was understood to be a more refined kind of entertainment. The views of nature it offered were intended to calm the passions, soothe the soul, and elevate public morality. Illustrations of the Museum typically pictured bourgeois adults and families already behaving properly. However, while differences and the distance between the menagerie of the Jardin des Plantes and the "combat du taureau" were substantial, they were not infinite. At the latter, butcher boys brought their dogs to fight bears and wolves, but at the former, the sentries had to be on constant lookout to keep dog walkers from sending their dogs after the menagerie's animals. In addition, Cuvier on more than one occasion sent or sold to the combat master a menagerie animal that had become superfluous or too difficult to handle.[60]

When Frédéric Cuvier sought to improve the lives of the animals of the menagerie by moderating their physical miseries and promoting kind care for them, he was part of a long tradition that was less concerned with the animals' welfare per se than with how promoting their welfare would further human interests. If the animals were trained to be gentle and cooperative, that would make the keepers' work easier and the animals more amenable to scientific study. A main reason for studying them was to learn how to improve animal races for the benefit of mankind via training, breeding, and domestication.

The subjugation of the menagerie's animals to human interests was apparent not only in the superintendent's overall view of human-animal relations but also in the way the plans for the menagerie of *animaux féroces* evolved. In the early visions of the "new" menagerie, the animals enjoyed settings resembling their native habitats, their sense of captivity was minimal, and they behaved as they would in nature. The first plan called for lodging the carnivores in a rocky escarpment. After this design was judged impractical and its cost exorbitant, Molinos conceived of the rotunda, but it was never put to its original purpose, ending up instead as a structure for the elephant and other herbivores. In the fall of 1809 Geoffroy Saint-Hilaire offered his thoughts on a new structure to house lions, panthers, tigers, hyenas, and other such beasts from torrid zones (the bears, having their pits to live in, were no longer part of the equation). Nothing about his design reflected the earlier dream of giving the animals enough freedom to forget they were captives. Rather, he explicitly called for the opposite.

Geoffroy proposed a long, rectilinear structure with the animals in double cages arranged in a row. The "principal" cages would have southern exposure and individual courtyards in front. Behind each principal cage, inside the building, would be a secondary cage, protected from the cold winds of the north. The two-cage arrangement gave the animals quarters for day and night, and likewise for summer and winter. The design benefitted the keepers by making the quarters easier to clean. A narrow service corridor ran the length of the building between the outer and inner cells. The keepers could move the animals from the front to the back and vice versa via a system of sliding doors. Behind the interior cells, a corridor ten feet wide permitted public viewing of the animals in the winter. The design also called for two cellars, one for the heating system, the other for the sewer for carrying away the animals' urine and excrement and the wastewater from cleaning their cages.[61]

This arrangement was intended to be good for the animals' health, but Geoffroy made it clear that there was more to it: "It is not enough in an institution like the Museum of Natural History for the animals themselves to be well off, they must also be good for the public." In his design, the animals could never escape the public's gaze. As for the size of their quarters, Geoffroy was not generous. He proposed that the principal cells and their adjoining courtyards each be ten feet square. The secondary cells would be ten feet wide and seven and a half feet deep. Regarding how suitable this space would be, he tersely observed: "One must fear making the cells too big. The animals destined to live in captivity must feel at every moment the harshness of their condition, so that they submit better to the bridle."[62]

Geoffroy's plan underwent certain modifications as different architectural designs were successively presented during the Empire and Restoration, but its basic character prevailed. The result was a long, straight, neoclassical edifice with animals lined up one after another like specimens in a gallery. Ready at last for occupancy in 1821, the new pavilion for the *animaux féroces* was the antithesis of the naturalistic picture that Lacépède had painted a quarter of a century earlier, with the animals scarcely recognizing they were captives at all. With its twenty-one barred double cages arranged in a line, there was no disguising that the carnivores were prisoners, and there was no reason to think they were being tricked into feeling otherwise. In Geoffroy's formulation, it was better that they knew it, so they would submit more readily to their masters' wishes.

As menagerie superintendent, Frédéric Cuvier was in closer contact with the menagerie's animals and their sufferings than Geoffroy was, and he consistently preached that in the training of animals, kindness was the first approach to try. That kindness must be understood in context, however. In Cuvier's calls for kindness, his empathy for animals in the constrained circumstances of the menagerie remained secondary to the project of animal management. Kindness was a means of getting animals to cooperate with human plans for them.

[2.7]

The Empress, the Kangaroos, and the Black Swans

Elation and anxiety stirred the professors of the Museum as they gathered in special session on 31 March 1804 in response to exciting news. The *Géographe*, the second of the two ships from Captain Nicolas Baudin's "Expedition of Discovery" to Australia, had returned to France and was riding at anchor in the port of Lorient, heavily laden with specimens. This was cause for rejoicing but also concern, as the collection was not yet under the Museum's control. Word from Lorient was that the plant specimens (preserved and living) had suffered grievously in transit. The professors had yet to hear about the zoological side of things.

The Baudin expedition was an extraordinary undertaking. Proclaimed a voyage of geographic and scientific investigation, it comprised a twenty-two-person team of naturalists, astronomers, cartographers, gardeners, and artists. Great fanfare attended the 17 October 1800 departure from Le Havre of the *Naturaliste* and the *Géographe*, but by the time the latter returned, nine months after her sister ship, the sense of glory had long since ceased. From a political standpoint the enterprise was a bust, having failed to establish clear territorial claims for France on Australia's western coast. Furthermore, illness, desertion, and death took a heavy toll on the naturalists, crew, and officers. Only four members of the scientific team made it home on the *Géographe*. Captain Baudin succumbed to tuberculosis at Isle de France (Mauritius) on the return trip, but his loss was not widely lamented. He was accused of mismanagement, and Napoleon allegedly remarked that he "did well to die; if he had returned I would have hanged him."[1]

The Museum mourned the loss of former staff members, the expedition's head gardener, Anselme Riedlé, and its head zoologist, René Maugé, but it did not share the government's dim view of the expedition's accomplishments.[2] When cataloged and counted, the collection amounted to the greatest zoological haul the Museum had ever received.[3] The expedition's surviving gardener, the young apprentice Antoine Guichenot (another son of the menagerie's first commissioner of provisions), reported that many

Figure 2.7.1. "Cygnes noirs du détroit du Bass." Leon de Wailly. Collection des Vélins du Museum national d'histoire naturelle. Portfolio 84, folio 12.

live plants had perished and that the dried specimens had been ravaged by insects and rodents. He feared further losses between Lorient and Paris.[4]

The professors were acutely aware that specimens had been stolen from the *Naturaliste* while the ship was docked at Le Havre. To prevent a similar dispersal of the *Géographe*'s treasures, they sprang into action, with the botanist Antoine Laurent de Jussieu (1748–1836) and Geoffroy Saint-Hilaire hastening to the ministers of the navy and the interior to stress the importance of safeguarding the new specimens.[5] A professor was required in Lorient to see to the Museum's interests, a "commissioner" who would ascertain the nature and condition of the collections, handle the necessary formalities, and find the best means of ensuring that "all these objects reach the Museum in the best state and condition possible." To ease the way, the professors wrote to the maritime prefect at Lorient, Admiral Antoine Thévenard, urging him "to employ his good offices for the integral conservation and prompt transfer" of the expedition's collections.[6]

Geoffroy volunteered to go. He was the youngest of the professors, an experienced and hardy traveler, and willing to make the grueling trip of at least six uncomfortable days each way to relieve his colleagues of the trouble. He set off on 2 April 1804, while François Péron (1775–1810), the only zoologist to survive the expedition, was on his way to Paris. With these two concurrent journeys, the control of the voyager naturalists' hard-won specimens was about to pass into the hands of the Museum's cabinet naturalists.[7]

Péron harbored a special anxiety regarding the specimens he had assembled. An important dimension of his industry had been the pains he took to identify and record where he collected each specimen. He entered this precious biogeographical information in notebooks keyed not only to his fifty-four crates of specimens but also to the order in which the items were packed inside. He realized that if the order was disturbed during unpacking, he would be unable to identify the provenance of individual specimens, which would consequently lose much of their value, and his biogeographical notes would be worthless. There is evidence, alas, that what Péron feared did happen, and at least some of the specimens became dissociated from his information about where they came from.[8]

Péron's efforts as the expedition's de facto zoologist included overseeing the care and transport of an impressive collection of live animals. Those that made it to France included kangaroos, parrots, and cassowaries from Australia; lemurs, a panther, two axis deer, a dove, a cassowary, and two tortoises picked up at Isle de France; monkeys, lionesses, more panthers, a gnu, a zebra, various birds, and a large complement of tortoises brought aboard at the Cape of Good Hope. On the *Géographe*'s first morning in Lorient harbor, Péron compiled a list of seventy-nine live animals, some of which were evidently on their last legs. Later in the day, when the animals were disembarked, there were six fewer among the living. Two kinds of lemurs, two kinds of tortoises, a Cape hyrax, and a giant kangaroo had expired.[9]

When Geoffroy arrived in Lorient, he was generally enthusiastic about what he found. Admiral Thévenard had already arranged for wagons to take the live animals to Paris, and some of them appeared to be quite unusual, perhaps more than Péron had realized. What Péron had called genets, Geoffroy identified as small Cape mongooses ("charming and very tame"). The *chien-tigre* ("dog-tiger") was a spotted hyena, a species Geoffroy had been trying to buy for the menagerie. As for the "black tiger," Geoffroy believed it "could well be a black variety of the panther." The monkeys seemed to represent "either a new species or a previous undetermined variety." Of the two kangaroos, which were of different species, Geoffroy liked the second because of the mix of gray and white in its fur. He believed it belonged to the new species that Paris naturalists had identified from a dead specimen brought back by the *Naturaliste*. "Our colleague Cuvier," predicted Geoffroy, "will know how to recognize it alive."[10]

Geoffroy's enthusiasm was mitigated by something he learned only after reaching Lorient. His colleagues heard it first, but not until he had left

Paris. The minister of the interior, Chaptal, informed them that "some [animals] are destined for Madame Bonaparte. They will be delivered to her upon their arrival."[11] The minister of the navy and colonies subsequently informed the Museum that an agent of Madame Bonaparte would travel to Lorient on her behalf and that "Citizen Geoffroy ought to be forewarned that he must not extend his claims to the whole of the collections, but only to the part that is destined for the Museum of Natural History."[12] The plants and seeds were likewise to be divided equally between the Museum and Madame Bonaparte. Thus the demise, deterioration, or theft of specimens from the *Géographe* was not all the professors had to worry about. They also had to deal with the wishes of the first consul's wife. They may well have seen the problem coming, having previously acquiesced to the demands of this powerful figure whose desires they were helpless to deny.

Born in Martinique in 1763, Josephine left home for France at sixteen for an arranged marriage with a rakish aristocrat, Alexandre de Beauharnais. She bore him a son and a daughter, and the couple separated in 1785. Beauharnais was executed during the Terror, and as the vicomtesse de Beauharnais, Josephine was imprisoned and nearly suffered the same fate. Released when the Terror ended, she became the mistress of Paul Barras, who had assisted in Robespierre's ousting, then served in the Executive Directory (1795-99). Finding Josephine too expensive to maintain, Barras introduced her to the young military hero Napoleon Bonaparte, who fell in love and married her in 1796.[13]

Josephine's interactions with the Museum commenced around the time of her 1799 purchase of Malmaison, a property some eight miles from central Paris that she secured with a down payment while Napoleon was in Egypt. On his return, he obligingly covered the balance of 200,000 francs. Over the next decade, she lavished attention on the property, with Napoleon footing the bills. Architects helped transform the old farmhouse into a small neoclassical château, while landscape gardeners reworked the grounds along English lines, embellishing the land with flower beds, ponds, statues, a temple, and a grotto, making Malmaison a botanical mecca for visitors from all over Europe.

In the spring of 1800, Josephine offered to provide the Museum's menagerie with much-needed fodder in exchange for exotic plants for her orangery and greenhouses.[14] Further transfers and exchanges followed. In 1801 she presented the Museum with a pair of swans.[15] In the spring of 1802 she donated a rare parrot; in fall of the same year she invited the professors to choose plants from her greenhouses that their collection lacked, trusting her generosity would be reciprocated with "several plants that would be agreeable to her."[16]

What began as a symbiotic relationship became more competitive in June 1803 when the *Naturaliste* returned from Australia. Anticipating that a conflict might arise over the plants and seeds on board, the professors sought to make a friendly gesture. Acknowledging "the good intentions that Madame Bonaparte displays toward the Establishment entrusted to our care," they decided to deputize representatives to visit her collections, inform her of the Museum holdings that she might find new and interesting, and propose exchanges.[17] But Josephine was not content to hear about duplicates the Museum was willing to offer. She wanted a direct share of the plants and seeds from Australia. Furthermore, she wanted the most elegant of the expedition's live animals, a pair of black swans.

Native to Australia, black swans had never been seen in Europe, and the Museum was keen to have them. The *Naturaliste*'s commander, Captain Hamelin, had reported to the maritime prefect in Le Havre that the swans were destined for "the national collection," that is, for the Museum, but Madame Bonaparte's wishes trumped the Museum's, and the minister of the interior, Chaptal, had to tell the professors that she would have them.[18] Shortly thereafter, he instructed the Museum to deliver to Charles Brisseau de Mirbel, Josephine's botanist, whatever plants she wanted for the gardens at Malmaison, except those essential to the Museum's work.[19]

In December 1803 the Museum received from the bey of Tunis a shipment of live animals consisting of two lions, a lioness, a dog that served as the lioness's companion, a gazelle, a doe, and three ostriches. All were designated as gifts for Madame Bonaparte, but she did not care to have dangerous big cats or even the gazelle or doe at Malmaison and claimed only the ostriches.[20]

All this was a prelude to the March 1804 arrival of the *Géographe*, with many of the animals on board intended as offerings to Madame Bonaparte. When the *Géographe* stopped at Isle de France, General Charles Decaen earmarked expressly for Joséphine the "black tiger," a panther, a cassowary from the Moluccas, and two crowned pigeons from Java.[21] Likewise, at the Cape of Good Hope, Jan Willem Janssens, the governor general of the French colony, reserved the gnu, the zebra, two more panthers, and several smaller mammals and birds for her pleasure. To further complicate matters for the Museum, Chaptal acknowledged that Captain Baudin had received special instructions to reserve a share of the collection for Madame Bonaparte.

Which of the live animals did she want? Her agent gave Geoffroy a provisional answer. First, the kangaroos and the gnu, the creature that in Geoffroy's eyes was "the most beautiful of the expedition, the most inconceivable relative to its form." She also claimed the forty or so tortoises. Although she had not planned to take the zebra, Geoffroy suspected she might change her mind if she discovered it was tame enough to ride.[22] Powerless to dispute these claims, Geoffroy could only oversee the distribution of the animals designated for either Malmaison or the Museum. The nine-wagon convoy set off on 11 April 1804, eighteen days after the *Géographe* arrived. Back in Paris, Frédéric Cuvier saw to the preparations for housing the new animals and made ready to go to Versailles to meet the convoy and bring back whatever creatures Madame Bonaparte did not want.[23]

When Josephine learned that the zebra had been intended for her, she decided to take it, along with the gnu. She also wanted the cassowary, Javan pigeons, and lemurs sent by General Decaen. The Museum could have the rest of the animals from Janssens and Decaen. With respect to "the national collection" (the specimens collected by the expedition), she would take the kangaroos, the water hens, the secretary bird, and the few surviving tortoises, noting she would relinquish some to the Museum if "the interests of science" required it.[24]

Josephine's claim to the kangaroos put the Museum in an awkward position. Leaving Australia, Captain Baudin had as many as eighteen healthy kangaroos aboard, but they fared badly on the long voyage. When the *Géographe* arrived home, only three kangaroos were alive, and two of these, the "giant kangaroo" (gray kangaroo) pair, were unwell. The female died before she was disembarked; the male lasted long enough to be loaded on a wagon, but five days out of Lorient, near Rennes, he expired. The cadaver was rushed to Paris by stagecoach so the Museum anatomists could dissect it before it decomposed.[25] This left only a single kangaroo from the expedition, a new species from King Island that Péron named *Kanguroo kingiensis*.

The botanist Mirbel conveyed Josephine's disappointment to the Museum: "She believed when she asked for the kangaroos some time ago, that they were of the same species. She flattered herself that they would arrive safe and sound at her estate and that, thanks to the profound peace animals enjoy in this vast domain, they would be able to multiply there."[26] Instead, she had received one lone individual. But Josephine's hopes were soon restored. Not long after taking the gnu and the zebra, she announced that the Museum could have them after all. Recognizing the need to make a noble gesture in return, the Museum gave her the pair of kangaroos it had purchased in England a year and a half earlier.[27]

Fourcroy, then serving as Museum director (and counselor of state), wrote an obsequious letter to Josephine stating that the kangaroos were on their way and that in establishing them at Malmaison, she would be doing natural history—and the kangaroos—a great favor: "Everything leads me to believe, Madame, that the new abode will render them infinitely more interesting for natural history than [did] the one that they inhabited up to the present. For this, all they need is a larger space and a little tranquility."[28] Frédéric Cuvier escorted the kangaroos to Josephine's estate and collected the gnu and zebra for the return trip to the menagerie.

Instead of ending up with new live kangaroos from the Baudin expedition, the Museum lost the two Dufresne had worked so diligently in London to acquire. As with the black swans, the Museum's professors knew better than to complain, and instead they lavished praise on the wife of the first consul. Georges Cuvier hailed her gift of the zebra and gnu as reflecting her willingness to sacrifice personal enjoyment for the public good. She knew her gifts would further the naturalists' studies, he wrote, and it was consequently "a duty for the persons attached to this establishment to seize every occasion to express to this princess their respectful recognition for the constant and much sought-after benevolence with which she honors it."[29]

In 1804 Napoleon crowned himself emperor, thereby making Josephine empress. At Malmaison, construction began on a magnificent greenhouse, fifty meters long. Completed in 1805, it made a stunning setting for a paradise of tropical plants. There and in the other greenhouses and gardens, many new plants were introduced to France. The painter Pierre-Joseph Redouté found Josephine a generous patroness and was paid handsomely for his renderings of flowers, including the roses that won him enduring fame. Mirbel, appointed director of the park and gardens with a salary of 6,000 francs per year (1,000 francs more than a Museum professor) might also have thrived there indefinitely, but his flagrant overspending attracted Napoleon's attention and eventually got him fired.

As the empress took her pick of the Baudin expedition's animals, she told Mirbel to inform the Museum that she appreciated that France was unrivaled in its possession of skilled naturalists. She intended for the nation to maintain its status, she said, aided by "the immense quantity of rare objects" the naturalists would be studying.[30] What the Museum received from her after 1804 in terms of specimens for the menagerie or the zoological galleries were the items that did not especially interest her. They consisted of a pelican, three ostriches, and three cassowaries (February 1805); two eagles (March 1805); a dead peccary (November 1805); the recently deceased kangaroo from King Island (February 1806); a dead llama (February 1806); two sheep, two male Angora goats, two female chamois, three Louisiana red deer, and a female axis deer (May 1806); a dead white blackbird (July 1807); two dead lemurs (August 1807); all of her stuffed animals (October-November 1808); a lion from Tangiers (February 1809); a seal (October 1809); a marten (November 1809); a dead male axis deer (January 1812);

a dead kangaroo and its dead baby (February 1812); and a dead monkey (December 1812).[31]

The Museum continued to express its gratitude and to please the empress when it could. At her request, it had its artist, Léon de Wailly, make her a copy of his painting of the foal that resulted from the zebra's mating with a donkey. Likewise, when Josephine asked for specific seeds, plants, shrubs, or trees, the Museum graciously provided them. In a letter of November 1808, she asked for a female Carolina duck, a request that excited the professors because Josephine also expressed her hope that in the coming year she would be able to offer the Museum two black swans.[32] The duck was quickly provided, but no swans were forthcoming. The pair from Australia reproduced twice at Malmaison, but the Museum did not profit from either occasion.

Napoleon divorced Josephine in 1809 and remarried, hoping his new wife would bear him an heir. Josephine received a substantial yearly stipend and was allowed to keep Malmaison. When Napoleon was defeated in April 1814 and exiled to the island of Elba, many allied leaders made their way to Paris. Some visited the Museum and the menagerie. Some also visited Josephine. One of these was the Russian czar Alexander I. It is said that in escorting Alexander through her gardens, Josephine caught a cold that developed into fatal pneumonia. She died May 29, 1814.

The Museum's records do not say if the professors mourned the empress's death, but they did indicate their eagerness to know what would become of Malmaison's black swans and kangaroos. Three days after Josephine's demise, at their regular assembly, the professors agreed to inform Aimé Bonpland, Malmaison's head botanist, that they wanted "to take care of the animals of Malmaison." The minutes of the following week's assembly noted: "Measures will be taken to try to obtain from the proprietors of Malmaison the kangaroos and the black swans that are deposited there and that had belonged to the Museum."[33] But the professors failed to extract swans or kangaroos from the estate's new owner, Josephine's son, Prince Eugène de Beauharnais. The closest they came to having black swans from Malmaison was when the Museum served as a temporary way station in the transfer of a pair to the Bagatelle Garden for the duchesse de Berry, the wife of King Louis XVIII.

Writing of the Museum's kangaroo pair fifteen years after their transfer to Malmaison, Frédéric Cuvier offered only a terse and inaccurate observation: "The individuals that our menagerie possessed had been brought to Europe by the expedition to Australian lands of Captain Baudin. They remained in our possession for just a few days and were sent to Malmaison, where they reproduced; unfortunately, no observations were made of them."[34]

It is curious that Cuvier misremembered a central part of the story: that these were the kangaroos Dufresne purchased in London and that Captain Baudin had nothing to do with them. Did he recall that in May 1804 he had drafted a letter for Fourcroy to send to Malmaison professing that the grounds there offered opportunities to observe animals in ways that were impossible at the menagerie?[35] Fourcroy reprised the theme in writing fawningly to Josephine that at Malmaison the kangaroos would be "infinitely more interesting for natural history" than they had been in their Paris quarters. But the prediction was not fulfilled. No naturalist went to Malmaison to study the kangaroos. Frédéric Cuvier himself was in no position to do so, even had he wanted to. His job was to be at the menagerie overseeing the animal keepers and their charges. But it was not just a matter of not having the time to undertake a long-term, on-site study of the behavior and lives of kangaroos living in semifreedom at Malmaison. Field studies were simply not the kind of work professional zoologists did in the first half of the nineteenth century. In any case, the kangaroos at Malmaison, like the black swans, were not under the Museum's control. The Museum was an institution of the state, dependent on the government's largesse, and ultimately subject to its orders.

[2.8]

Collecting for the Menagerie Before 1815

When Félix Cassal proposed in 1797 to the Museum professors that he travel to North Africa to collect animals for the menagerie, he imagined capturing them personally. He pictured himself in the wild, aided by local hunters, tracking animals in their native haunts. He wanted very young ones, knowing they are easier to catch and tame than adults, but circumstances prevented him from hunting in the right season. Ultimately, all the animals he brought back to France were acquired as gifts or purchases.

Had Cassal been directly involved in capturing a live animal, it would have been an exceptional event for an agent of the Museum in the institution's first fifty years. The Museum generally entered the acquisition process at a much later stage, far removed geographically, chronologically, and emotionally from the violence involved in wresting an animal from its natural habitat. It was not a gunshot, the snapping of a trap, the baying of dogs, or an animal's cry of pain that announced to the Museum's naturalists that a new creature was on its way to the menagerie. The news almost always came by letter.

The letter writer was typically the minister of the interior, relaying information from a prefect in the provinces, a French consul overseas, or the minister of the navy. Sometimes the writer was a ship's captain returned to port, a shipping agent announcing a specimen recently arrived from overseas, or a citizen wishing to pass along an exotic pet that had become too troublesome to maintain. Occasionally the letter was from an animal dealer with something to sell. Sometimes the news came from a Museum professor or other member, such as André Thouin when he reported on the animals he and Faujas de Saint-Fond were confiscating from the Stadtholder's menagerie in Holland (1795); or Félix Cassal when he reported on his expedition to North Africa (1797–98); or Mordant Delaunay (1800) and Louis Dufresne (1802) when they recounted their experiences purchasing animals in London. The letter writer almost never knew whose hands these animals had previously passed through or the details of their capture.

Figure 2.8.1. "The ruff 'in love.'" *Abraham Jacobsz Hulk after Jacques de Sève.* Buffon, *Histoire naturelle des oiseaux*, new ed. (Amsterdam: J. Schneider, 1775). R. Burkhardt collection.

There was one major exception in the menagerie's early years: Jean François Emmanuel Baillon (1742–1801), an amateur naturalist from Brittany whom the Museum honored as "First Correspondent." From the mid-1790s until his death in 1801, Baillon supplied the institution with shipments of dead specimens for the natural history and comparative anatomy galleries and with live specimens, primarily birds, for the menagerie. He contributed more live animals than any other source, the confiscation of the Stadtholder's menagerie, and the Baudin expedition combined. More than a supplier of specimens, Baillon was a distinctive type of naturalist for his era. Besides being an assiduous collector and a skilled taxidermist, he spent days watching animals in the wild to gather information about their lives and habits. He also kept animals in his garden and succeeded in raising some wild bird species from eggs.

Later in the nineteenth century when animals became a traded commodity and hunting them in exotic climes a profession, descriptions of their capture were commonplace, but that was not the case in the early 1800s.[1] The zoologists in Paris understood but did not dwell on the fact that acquiring a wild animal typically required a brutal act, separating a young animal from its mother. When Georges Cuvier wrote about the menagerie's two polar bears, for example, he noted only that they were purchased in Greenland as very young cubs. Elsewhere in his essay he stated that mother polar bears are known to defend their offspring ferociously. An attentive reader might put two and two together.[2]

That the Museum's zoologists realized capture involved violence is evidenced by Frédéric Cuvier's forgetting what he once knew about the menagerie's first wild sheep (*mouflons*), a mother and her young son that arrived in November 1808.[3] Writing ten years later, he incorrectly recalled them as a pair of orphans, explaining, "The two of them were taken while fleeing with their mother, who was killed."[4] His memory seems to have been reshaped by his recognition of how young wild animals were typically captured.

The primary scenes of capture recorded in the early 1800s concern kangaroo hunting during Baudin's expedition to Australia. The captain's journal shows that as the time to leave Australia approached, he considered how many live kangaroos he needed onboard to ensure that some survived the journey to France. His 19 January 1803 entry reads: "During the day, with the help of the dogs, the hunters caught twelve giant kangaroos of various sizes. Seven were taken alive and put in pens aboard the ship to be kept. Among these we hope to carry back to our country are three females that have offspring and may prosper. I shall try, before leaving, to obtain a full twenty live ones, so that we shall have better hope of keeping some throughout the voyage."[5]

Three days later Baudin noted more captures: "There were two males and one female, [all] caught unharmed by our dogs. One tried to escape by throwing himself into the water, but fell into our hands after putting up magnificent resistance." On 25 January he wrote, "I sent some men to hunt kangaroos, but we had no luck, for only one was taken alive. It was brought back to the ship with three others that had been throttled by the dogs."[6] On 28 January he recorded having fifteen healthy kangaroos on board. Two had died, "having been too badly wounded by the dogs to recover." Another specimen was added the following day and three more the day after, though one was "severely wounded."[7] On 13 February, regarding the loss of another kangaroo, Baudin remarked, "its death resulted from the wounds the dogs had given it, quite apart from its having been hit by small shot."[8]

Baudin's notes refer primarily to methods and the number of captured animals. François Péron, author of the final account of the expedition, remarked on the behavior of individual kangaroos seeking to escape the hunters. When the expedition arrived on the islands where the banded kangaroo lived, each of the adult females was seen to be carrying a "rather large" infant in her pouch, he wrote. When pursued, the mothers sought to save their offspring "with a courage that was truly admirable."

> Wounded, they fled, carrying their baby in their pouch, and never abandoned it except when, too overwhelmed by fatigue, too exhausted by the loss of blood, they could carry it no longer. Then they stopped, and squatting on their hind feet, helped it with their front feet to leave the maternal sack, and sought in a way to show it the hiding places where it could hope to save itself. They themselves then continued their flight with as much speed as their strength permitted.[9]

If the mother escaped the hunters, Peron said, she would return to the bush where she had left her baby, call it to it, and caress it, "as if to dissipate its fears," then seek a safer hiding place. Péron further observed:

> The same proofs of intelligence and affection reproduced themselves in an even more touching manner on the part of the mothers when they felt themselves mortally wounded. They directed all their cares to the health of their infant; far from trying to save themselves, they halted under the shots of the hunter, and their last efforts were given to the conservation of their babies.

It was the kind of "generous devotion," Péron remarked, "of which the lives of animals offer so many examples, and that we are often reduced to envying!"[10]

∴

Jean François Emmanuel Baillon, the bird man from Abbeville, kept the menagerie stocked with shorebirds and waterbirds, providing more than two hundred live specimens in the late eighteenth and early nineteenth centuries. There are no descriptions of capture for Baillon's birds, unless one counts his report on collecting shelduck eggs that he brought home for domestic ducks to incubate. Most of his birds were caught by hired hunters using the various methods at their disposal, especially netting.[11]

Though largely overlooked in the history of science, Baillon was greatly appreciated by the Museum professors, and before them by Buffon and his colleagues at the Jardin des Plantes. A dedicated field naturalist, he watched birds, sometimes from specially designed blinds, acquiring extensive knowledge of their arrival and departure in his region, their feeding and mating habits, their means of defense against predators, and the seasonal changes in their plumage and coloration. He was also skilled at raising wild birds.[12] Such practices distinguished him not only from the Museum's cabinet naturalists but from most voyager naturalists, who were more devoted to collecting animals than to studying them as living beings. Baillon was adept at both, and if he happily obliged the Museum's naturalists with dead specimens (fish, for example, for Georges Cuvier), on his own initiative he saw to supplying the menagerie with live birds.

Before the Revolution, Baillon lived at Montreuil-sur-Mer, working as crown prosecutor and bailiff while cultivating his fascination with natural history. By 1778 he was sending bird specimens to Buffon, France's best-known naturalist, and Buffon was asking for information on the habits of the species he sent.[13] Two years later, Baillon's name appeared in Buffon's magisterial *Histoire naturelle des oiseaux*, in the article on the bittern. "No observer," wrote Buffon, "has given us better information than M. Baillon on the habits of this bird."[14] Buffon reproduced in his text Baillon's detailed description of the bittern's life. Later in the same volume Buffon included Baillon's account of the mating habits of the ruff:

> I have often followed these birds in our marshes, where they arrive in the month of April with the redshanks, but in fewer numbers. Their first concern is to pair, or rather to contend over the females. With little cries the latter enflame the ardor of the *combatants*.[15] Often the struggle is long, and sometimes bloody. The vanquished retires, but the cry of the first female he hears makes him forget his defeat, ready to enter the fray again, if some antagonist presents himself. This small war recurs every day, morning and night, until the departure of these birds, which takes place in the course of May.[16]

Baillon observed birds he kept in his garden after they were captured in nets and pinioned (sandpipers, plovers, lapwings, oystercatchers, various species of gulls and ducks). He collected the shelduck's eggs from the rabbit burrows where they normally laid them and used domestic ducks to hatch the eggs and rear the young.[17] Some birds did not take to captivity. He could not keep terns for long, he told Buffon, because of their persistent crying "even during the night." He found them a sad sight, noting they "lose nearly all their gaiety; built to frolic in the air, they are hampered on the ground, their short feet encumber them in all they encounter."[18] Baillon reported that the avocet evaded all his ruses to take it alive.[19] At Buffon's request, Baillon sent him live birds to observe, including two shelducks, a brant goose, a black-backed gull, and a herring gull; Buffon kept the gulls in a garden for nearly fifteen months.[20]

In 1781 as Buffon worked on the ninth and last volume of his *Oiseaux*, he urged Baillon to supply additional information on the birds he intended to describe (including swans, geese, ducks, geese, guillemots, and petrels), "for which I will be infinitely obliged to you." In the volume, Buffon acknowledged that "[Baillon] has taught us more facts about the waterbirds than all the naturalists who have written about them."[21] Buffon did not, however, privilege the study of behavior over the possession of additional specimens. At one point, Lacépède instructed Baillon on Buffon's behalf to "please suspend your research relative to living birds . . . and only send us the animals we ask of you."[22] Baillon also wanted specimens. For birds that displayed sexual, seasonal, and age differences in their plumage and structures, he wanted the complete series. At the same time, he sought to combine his knowledge of animal behavior and taxidermic talents by mounting his specimens to display their characteristic "attitudes."

When the Revolution came, Baillon lost his government posts and relocated to Abbeville, some fifty kilometers from Montreuil-sur-Mer. His finances suffered, but the move presented opportunities; he was now near the Baie de la Somme, one of northern Europe's largest tidal estuaries, where waterbirds and shorebirds abounded. Baillon continued observing, collecting, taming, and rearing live birds and mounting dead ones, while developing a regional supply network of hunters and correspondents.

In fall 1794, Baillon asked the professors of the new Museum of Natural History to commission him as "correspondent" of the institution, a title he had enjoyed previously at the Jardin des Plantes, thanks to Buffon.[23] He noted how he'd assisted Buffon and asked them to help, as Buffon had, in getting him gunpowder. He was willing to pay hunters a good price for birds, he said, and if he could give them powder, they would do whatever he asked no matter the weather. Storms were good for collecting, he explained, because they made migratory birds seek refuge on a leeward shore. That was how he acquired a snowy owl, "one of the most beautiful birds in nature." To furnish his cabinet with fine specimens, he said, he spared no expense.[24]

Baillon's vision of natural history studies extended well beyond collection building. On 31 December 1794, he wrote to the Museum about the various species of birds that the cold had driven down from the north. The joy he experienced watching animals in the wild shone through his account. He spent hours lying on his stomach behind piles of ice by the sea, watching a "charming" seal gambol on the beach. He saw two flocks of white-fronted geese in flight and a pair of eider ducks. He had no complaints whatever about tarrying in the cold some twelve leagues from home. He planned to head back to the sea the next day, he said, because good times for watching were rare, and "you have to profit from them."[25]

Baillon outlined his methods and his hopes of cooperating with the Museum. By offering bounties to local hunters, he said, he had the unexpected good fortune of acquiring a white-fronted goose, an adult female in fine plumage. He pledged to give it to the Museum if it lacked an exemplar. He had meanwhile promised the hunters fifty francs for the seal he'd seen on the shore, dead or alive, because existing illustrations of seals were imperfect. He recommended that scientists investigate the causes of the winter migrations he was witnessing, and he remarked more generally, "I believe everything relating to natural history ought to be communicated to the committee of the French Museum. I beg you to inform me if you are of this opinion, which I hold strongly. In this case I will not send anything to any scientist without communicating it to you." He wrote that he and a fellow citizen of Abbeville were planning to establish a natural history collection, botanical garden, and library for the community. He imagined the public good of having such centers throughout the country reporting their discoveries to the Museum in Paris, "the common center of the sciences."[26]

Baillon initially set out to refresh the Museum's collection of native waterbirds, indigenous and migrant species alike.[27] He had asked the professors to name all the birds they wanted him to find. He began by sending preserved birds for the zoology gallery. By early 1796 he was planning to send live birds to the menagerie. He had nine brant geese, seven common pochards, and twenty shelducks and sheldrakes tamed and ready to dispatch, and he anticipated being able to send tame wigeons, pintails, gadwalls, teal, and other species.[28] "I love birds" he told the Museum in the spring, "I never stop studying this part of natural history. I know a little about them."[29] His collection continued to grow, and by winter he was eager to transmit it to the Museum.[30] When neither the Museum nor the government was able to help with the transport costs, Baillon footed the bill, and seventy-three live birds representing thirteen species made their way to the menagerie.[31]

During the winter of 1798–99 Baillon sent the Museum 161 dead and 56 live specimens "of the most interesting species," as he told the minister of the interior. He had tamed them all himself, he said, adding proudly, "nearly all the birds that adorn the pond [of the Jardin des Plantes] are from me." He announced that he was writing a memoir based on his experience of "how to tame and multiply these beautiful species for our poultry yards."[32] Delaunay calculated that Baillon had sent the menagerie some 170 birds representing thirty or so species.[33]

Baillon did not ask the Museum to reimburse his expenses, but he did request its patronage to help maintain his hunting privileges, get gunpowder, and secure a position in the new forestry administration.[34] Eventually he sought the Museum's support to obtain a government pension. What he wanted most, however, was for his twenty-year-old son Louis to be exempted from military conscription. The dreaded news of the draft arrived late in September 1798, when Baillon and Louis returned to Abbeville after eight days watching birds by the sea. Baillon was beside himself as he petitioned the Museum. The Revolution, he said, had led to his wife's death, and now it would take his only child, a blow he could not withstand. He wondered if his son could be posted to a military division stationed in Paris where his drawing skills could be used, or if he might pay some of his pension to keep his son in the city. He begged the Museum to think of other ways of "saving my child from the dangers of the war." He closed his letter by saying that his tears prevented him from writing further.[35]

The order for Baillon's son to report for service arrived on the morning of 1 November 1798. Baillon wrote to the Museum, "I would have preferred to receive a shot from a pistol." The gendarmes, he said, were taking away the draftees and shot them if they offered the least resistance. "My son will leave," Baillon reported, "but I will not live long after his departure." The Republic has taken "my wife, my brother, and the greater part of my fortune." Only his son had been left to console him. "What should I do? Is it possible to survive so many misfortunes? At the beginning of the Revolution, I was offered a fine position in the Museum of a powerful foreigner. I loved my country too much to leave her. I am well repaid for it!" He concluded: "I am sending you by the public coach the drawing my son just finished today. There will be no more of them. He is going to his death."[36]

The Museum came to the rescue by putting Louis Baillon to work in the zoology laboratory. It also wrote to the Executive Directory asking that the young man's military service be attached to a Paris-based cavalry unit so he could continue to help with the Museum's collections when his military duties did not demand otherwise. The secretary general of the Directory and the minister of war agreed.[37] Baillon and his son came to the administrative assembly of 14 November 1798 to thank the professors. The Museum's director hailed Baillon for all he had done for the Museum's ornithological collection, and Baillon presented the assembly with a species of duck the Museum lacked that was very rarely seen in France. He offered to provide all the waterfowl of northern France needed for the collections of the Museum and the Écoles centrales. The administration accepted the proposal "with gratitude."[38]

Months later, when the minister of war refused to grant Louis Baillon additional leave to work, the Museum wrote to insist that his unique talents were urgently needed in the zoology laboratory and to ask that he be permitted to offer a substitute, as the law allowed, so that someone else performed his military service in his place.[39] Whether or not the claim about his unique talents was precisely true, Louis remained in the Museum's employ.

While continuing to send the Museum specimens, in the fall of 1800 Baillon presented a highly original suggestion. His plan, as the Museum characterized it, was to annex to the Museum "a pond by the seashore where a great number of aquatic birds are found and the waters are filled with many fish of different species." Situated near the village of Aule some fifteen miles from Abbeville, the pond covered approximately forty-five acres (counting all its tributaries), was connected by a channel to the sea, harbored green spaces, and was a resting place for an immense number of waterbirds and shorebirds. at migration time it attracted as many as two hundred hunters. From them, Baillon told the Museum, he purchased at a relatively high price "nearly all the specimens that I send you each year."[40] The pond, called "l'étang du hable d'aule," belonged to the government. The government had no use for it, but Baillon believed it could be of great utility to the Museum. He promoted the project primarily on the grounds that it would furnish a steady supply of fresh specimens for the galleries and menagerie, but he also had his studies of bird life in mind, saying that annexing the pond to the Museum would accelerate discoveries through experiments and regular observations. He proposed overseeing the property by controlling access, supervising staff, and ensuring that its income covered its expenses.[41]

Charged with assessing Baillon's idea for their colleagues, Lacépède and Lamarck voiced confidence in Baillon and recommended asking the minister of the interior to incorporate the pond as part of the Museum. The professors agreed, and they instructed their director to pursue the matter with the minister, but nothing ever came of it. Neither the Museum nor the minister had cause to view it as a top priority. The Museum could expect Baillon to send them fresh specimens regardless, and studying live animals in their natural habitats had no real place on the Museum's agenda.[42]

Baillon wrote to the Museum in October 1801 expressing pride in the collecting network he had established, whereby hunters and others informed him of things of interest and he gave instructions about the specimens he wanted. He expected a productive winter, "especially if I am seconded by the cold and the snows." Bad weather, he reminded the professors, benefitted the collector, who could capture birds more easily when they hunkered down in a storm.[43] Announcing that he was shipping several dead specimens that day, Baillon noted four dippers (a species he knew the Museum had long lacked) and, most important, "a species of phalarope that I believe has escaped the investigations of naturalists." It was much smaller, he said, than the one previously described, and its feet and beak were different. Baillon remarked on its behavior and capture: "This phalarope swam in the middle of the mouth of the Somme. It dived continually, making a thousand twirls while emitting an absolutely unknown cry. It disappeared continually, coming back up a little farther away in the twinkling of an eye. After a pursuit of several hours, [the hunters] surprised it at the moment when it resurfaced. A piece of lead shot went through its neck. They brought it to me this morning."[44]

The hunters, Baillon noted, appreciated his recognition of their "perseverance in pursuing a bird of so little interest for the common man." He hoped Lacépède would examine the bird and describe it at the Institute. The successor of Pluche and Buffon, he said, ought to be the first "to describe a bird that no naturalist has seen up to this day."[45]

Baillon hoped to obtain a seal for the Museum. The season favored it, he said, because the herring had arrived, and as always, seals, dolphins, and porpoises pursued them. However, fishermen dared not go to sea for fear of being captured by the British. "In peacetime," Baillon vowed, "nothing will escape me, of whatever you should desire." But peace did not come until the following spring, and Baillon did not live to benefit from it. He died 24 October 1801, three weeks after the letter just cited, taken by "lethargic fever."[46] Grieving the loss of their ardent correspondent, the professors instructed the director to convey their condolences to Baillon's son, Louis. They hoped Louis would continue to assist in the zoology lab, in the position previously held by René Maugé, who had perished during the Baudin expedition.

Louis Baillon (who had wished to join the Baudin expedition himself but chose not to leave his father alone) returned to Abbeville to act in his father's stead as the Museum's correspondent.[47] On 21 November 1801 he sent a ray and a flounder by stagecoach, knowing their skeletons would fill gaps in the comparative anatomy collection.[48] Over the next half century, he filled the Museum with specimens for its galleries and menagerie, while assembling his own renowned ornithological collection. The bird "Baillon's Crake" is named after Louis Baillon.[49]

The Baillons' service to the Museum was unparalleled. No one enriched the galleries and menagerie more. Knowing them so well, the professors could rely on their understanding of the Museum's needs and desires and on the quality of their specimens. Not all collectors or specimen suppliers were as accomplished. The case of a "young tiger" and a prefect at Brest offers a good example. It began in December 1807 when the minister of the interior wrote to inform the Museum that a French corsair operating out of Saint-Malo had captured a Portuguese ship carrying a "young tiger." The animal was delivered to the maritime prefect at Brest, who offered it to the Museum. Responding immediately through the minister, the professors accepted the prefect's offer and told the minister that 500 francs would cover the animal's food and transport to Paris. The acquisition would be particularly useful, they said, "at a moment when it is impossible for us to communicate with foreign countries." If the prefect knew the animal's place of birth its value would be greater, but without its provenance the specimen "would be of lesser interest."[50]

Eight weeks later, the minister reported that the animal in question, now described as a "young tiger," had died. Undaunted, the Museum requested its skin and skeleton, indicating that the prefect would be rendering a service to science by sending them,[51] but he sent only an account of the creature's death and physical particulars. Identifying the animal as a six-month-old female "panther," he attributed its death to its inability to withstand the cold. Upon its demise, he proudly said, it became a subject of local instruction with its dissection at the town amphitheater providing a lesson in comparative anatomy. Following a description of the animal's internal organs, the prefect remarked that "the persons charged with stuffing the panther, attracted by the whiteness of its flesh and provoked by a curious appetite, ate some of this flesh; they report that its taste was agreeable."[52]

The prefect praised the taxidermic procedure. With the animal's head and feet left attached, the skin was stuffed, presented attractively "in the posture of a cat lying on its stomach, with its front paws stretched out forward and its back legs flexed." The quality of the work convinced the prefect that the specimen belonged in the museum at Brest. Paris probably had many such specimens already, he reasoned, while the local museum's meager collection deserved improvement. Regarding the Museum's request for the panther's skeleton, it arrived only after the bones had been discarded. Once again, a specimen eluded the Museum's control.

∴

Two of these cases of naturalists' attitudes toward animal life remind us that perspectives change over time. Péron's compassion for the mother kangaroos hunted by the Baudin expedition, for instance, suggests a naturalist sympathetic to the living animal, but his engagement with a baby banded kangaroo taken aboard the *Géographe* reveals the limits of his sympathies. One of several captured taken aboard ship, only this kangaroo took to captivity and survived, eating bread "with pleasure" and responding to sugar water "with delight." But when it died of an unspecified accident while the *Géographe* was docked at Timor, Péron did not lament its loss as his last specimen of the species, precisely because on its own it wasn't useful. He needed both a male and a female to naturalize the banded kangaroo in Europe. He had already found that the species had the best-tasting meat of any kangaroo species the expedition had encountered, and he was content to conclude from his experience that the species responded well to human care and promised to be "a precious acquisition for our barnyards."[53]

Today we don't think of naturalists as discovering new species for the dining table or of zoos as way stations naturalizing animals as food, but in 1800 the first was common, and the menagerie was projected to help with the second. When Jussieu surveyed the products of the Baudin expedition, he identified the top prospects: "several species of kangaroos, of which the fur is good and the meat excellent"; the wombat ("equally good to eat"); the cassowary (with flesh reminiscent of turkey and suckling pig); and the black swan, decorative on water, but on land a source of feather down and tender meat.[54]

Emmanuel Baillon was attentive to living animals and joyfully observed them, but his field naturalist side was not as developed as that of British field naturalist Edmund Selous exactly one hundred years later, who admitted he had once belonged to "the great, poor army of killers." "Now that I have watched birds closely," Selous confessed, "the killing of them seems to me as something monstrous and horrible; and for every one that I have shot, or even shot at and missed, I hate myself with an increasing hatred."[55]

Baillon was enchanted by both the novelty and the vitality of his new phalarope. Its twirling, diving, and unknown cry were part of its very being. However, as with the "charming" seal he had seen gamboling at the beach some years before, he had no compunction about his hunters' pursuing and killing it. He looked forward to having the specimen described in Paris. Baillon was an exceptional observer of bird life, but even for him the study of the living bird took a backseat to the imperative of having all the birds of northern France on display in the galleries of the Museum of Natural History.

[2.9]

The Gnu

For its first fifty years, the menagerie of the Museum of Natural History had only one gnu (wildebeest). A female from the Cape of Good Hope, she was embarked aboard the Baudin expedition vessel *Géographe* in January 1804 by Jan Willem Janssens, the governor general of the Batavian possessions at the Cape, as a gift for Josephine Bonaparte. The gnu excited the Museum zoologists because it represented a species that had never been seen alive in Europe. It was also a taxonomic challenge, difficult to place zoologically. While resembling various species, it wasn't quite like any of them. What the keepers noticed most about the gnu was its temperament. It proved dangerous to those who cared for it, yet another example of the inappropriateness of characterizing herbivores as "peaceful animals."

Geoffroy, who declared the gnu "the most beautiful animal of the expedition, the most inconceivable in its form," told his colleagues that "the illustration published by Gordon does not fully express its anomaly with respect to the family of antelopes in which it has been placed." His first impression of the young animal's disposition was that it was "gentle and caressing."[1] He was disappointed when Madame Bonaparte claimed the gnu for her estate, but she decided almost immediately to send it to the Museum. Two days later, the Museum's animal painter Léon de Wailly completed a portrait of the "Antilope gnu" for the Museum's vellum collection.[2]

Naturalists had placed the gnu in the antelope family, but the Museum's zoologists questioned the classification. Frédéric Cuvier thought the gnu might be closer to the ox family, albeit not decisively so. In his menagerie guide he wrote, "it seems, above all, to serve as an intermediary between these two families." As the "link" in the "chain of beings" uniting the two families "in a single series," Cuvier said, the gnu occupied a "remarkable place" in the effort to understand nature's general laws.[3]

Travelers in Africa reported that the gnu was neither particularly docile nor susceptible to taming, and Cuvier was of the same opinion. He saw in its oxlike form the suggestion of an intractable character, far less gentle than

Figure 2.9.1. "The female gnu." C. de Lasteyrie after Jean-Charles Werner. É. Geoffroy Saint-Hilaire and F. Cuvier, *Histoire naturelle des mammifères*. General Research Division, New York Public Library.

a gazelle's. "The [menagerie's gnu]," he observed, "without being mean, is rather brusque in its movements and easy to irritate, so that it is necessary to take precautions in approaching it." The caretakers had to worry especially about its horns. The gnu would willingly lick the hand that approached its muzzle, but Cuvier saw "nothing affectionate in this apparent caress." Nor did he find anything distinctive about how the gnu capered when let loose in its park, attributing this to its youth rather than to a characteristic playfulness.[4]

Cuvier's critical assessment of the gnu's disposition, formed in the months immediately after Geoffroy's first happy impression of it, was borne out a few years later when the gnu sank one of its horns into the thigh of the keeper, Ballet. The wound was as severe as the one that a stag had dealt the unfortunate man some years earlier. Ballet never completely recovered from the gnu's attack, which left him with a weak leg. He struggled on as an animal keeper until 1808, when he was granted a post as gatekeeper. In recommending Ballet for the less physically demanding position, Cuvier observed that caring for herbivores was "much more dangerous than caring for the *animaux féroces*" because the herbivores were free in their pens, not caged, and their keepers tended to be weak or old. Cuvier had in mind both Ballet's interests and the need for a stronger, younger herbivore keeper.[5]

While Cuvier and the keepers were focused on the gnu's temperament, visitors were struck most by its bizarre appearance. King Friedrich I of Württemberg, touring the menagerie late in December 1809, was so taken with the animal that he asked for its picture. The Museum obliged by instructing Nicholas Huet to produce a copy of de Wailly's portrait.[6] The gnu continued to astonish visitors to the menagerie until its death in 1811.[7]

Writing of the gnu nine years later, Cuvier summarized his observations on its management: "Good treatment affects it little; it is by force alone that one succeeds in getting it to submit, otherwise it is impossible to act on its intelligence."[8] He used the creature as an example of how zoologists classified animals. The gnu appeared to be a type unto itself, he told readers, in the way it combined traits of the goat, ox, and horse without being any one of them. To reach a more certain classification, one had to go beyond the animal's external features and compare its internal organs with those of other animals. This was the practice by which his brother Georges had reformed zoological classification over the past quarter-century, and it was the foundation of Georges's magisterial *The Animal Kingdom Distributed according to Its Organization* (1817). Describing the gnu, the elder Cuvier acknowledged that "at first glance" it appeared to be "a monster composed of the parts of different animals," but he confidently put it with the antelopes and continued to call it l'*Antilope Gnu*.[9] That was where it was placed physically after it died and was stuffed: among the antelopes on the third floor of the natural history gallery.[10]

The grouping of specimens in the galleries corresponded as far as possible with what the professors considered their natural affinities. However, while the consideration of internal organs was crucial for forming the different animal classes and the orders within these classes, when it came to distinguishing among genera and the species within them, "less important," external features increasingly came into play. In the antelope genus, As Georges Cuvier explained in *The Animal Kingdom*, naturalists had been obliged to distinguish the different species primarily by the form of their horns. Also of use in identifying species was knowing where they came from. That is why the Museum's zoologists were suspicious when they heard from the minister of the navy and colonies in 1825 that a gnu had been acquired on their behalf not in southern Africa, but in India, by the Comte Du Puy, governor of the French footholds there.

The minister's letter included a report by the assistant navy commissioner at Pondicherry, a man named Bédier. He related that an Indian missionary had acquired the animal in question virtually at birth, raised and tamed it for two and a half years, and was offering it for sale. Comparing the animal with the Dutch naturalist Jean-Nicolas-Sébastien Allamand's classic description of a gnu from the Cape of Good Hope, Bédier acknowledged the differences but presumed they were "modifications" reflecting differences in climate.[11] He had learned about capturing gnus and keeping them healthy, he wrote, noting that they were said to be born always in pairs, one male, one female, and it was best to capture them when newly born and still lethargic (presumably like their mother) after their first meal of mother's milk.

The professors annotated the minister's letter containing this report with the terse observation: "This gnu is a nilgai [Asia's largest antelope]." The Museum's letter of response was less curt, explaining to the minister that based on the animal's description and provenance, it was not a gnu. Zoological matters aside, the Museum was concerned with the expense of the animal's purchase and transport. The professors said they were flattered by Du Puy's interest in their institution and would gladly accept the animal, but only if it were an outright gift. They could not shoulder the cost of specimens purchased without prior consent. Unanticipated expenses, wrote the Museum administration, could force it to forgo critical projects or overspend its budget.[12] The professors had been clashing with the government on this issue for nearly a decade. The minister reassured the Museum that Du Puy had undoubtedly intended the animal as a gift.

A year elapsed before news arrived that the nilgai had died on embarkation from Pondicherry, but its skull, bones and skin had been entrusted to the health officer of the royal corvette *La Pomone*, which had just docked at Toulon.[13] The minister asked if the Museum wanted the remains. Otherwise they would be deposited in the natural history collection of the port of Toulon.[14] The professors replied that while a living nilgai would have been preferable, they were very interested in the animal's remains. They promised to contribute an equally interesting specimen to the natural history collection in Toulon and urged the minister to send the nilgai's skin and bones to the Museum via stagecoach.[15] The skin was expertly stuffed in Louis Dufresne's zoology laboratory, recreating the nilgai for display in the natural history gallery among the antelopes, near the gnu.[16]

[2.10]

Caring for the Animals of the Menagerie

Frédéric Cuvier's 1804 lament about having nowhere to turn for guidance on the care of the menagerie's animals was partly rhetorical. The menagerie had already had a decade of experience dealing with the challenges of sheltering and feeding animals and keeping them healthy. Some of the challenges, furthermore, were much the same as caring for barnyard animals. When it came to horses, cows, and sheep, French veterinarians knew (even without the germ theory of disease, which was still in the future) that animals suffered when crowded in sheds and stables where the air was stagnant and "infected" by the animals' breathing. Likewise understood was the importance of keeping the animals clean and giving them water that was uncontaminated and food that had not spoiled.[1] But recognizing basic challenges was not the same as solving them. Securing enough "wholesome meat and good fodder," as the professors instructed Mordant Delaunay in 1795, was easier said than done in times of punishing scarcity.[2] Likewise, designing living quarters for the animals required balancing the animals' needs with those of their keepers and the public and having the funds, space, material, and workers for the construction. Years would pass before the herbivores and carnivores occupied new structures. And even with new structures, the menagerie's record with respect to preserving hot-climate animals from lung diseases would be poor. Then too, getting a wild animal to submit to the kinds of medical procedures humans endured (bloodletting, for example) was a risky business for the attending physician/veterinarian and the animal alike.

Regulating the public's contact with the animals was another feature of the animals' care the menagerie supervisor had to attend to. Frédéric Cuvier as well as Mordant Delaunay before him knew very well that animals suffered at the hands of the public, but preventing this was difficult, and even setting the hours of public admission was complicated. When it came to the diverse needs of the animals, the public, the keepers, and others, the animals' best interests often fell by the wayside.

Figure 2.10.1. The menagerie's slaughterhouse. In the 1790s and first years of the new century, a critical problem for the menagerie was providing its carnivores with wholesome meat. This later picture of the menagerie's slaughterhouse depicts a time of plenty, very different from the early years when the menagerie depended on meat from old or diseased horses or tainted meat confiscated by the police in the markets. Andrew, Best, Leloir after Karl Girardet. From Boitard, *Le Jardin des Plantes* (1842).

∴

The kinds and amounts of food the menagerie needed at any given time depended on the animals in the collection. The menagerie's population grew and diversified in the volatile years from 1793 to 1800, making provisioning increasingly costly and demanding. Carnivores dominated the menagerie's first months. The leopard, serval, civet, white-tailed sea eagle, and two golden eagles needed meat. The other animals (two polar bears, two agoutis, and three monkeys) were fed bread, vegetables, and fruit.[3] When the animals from Versailles and Raincy arrived in spring 1794, the menagerie suddenly had nineteen herbivores needing hay and other fodder.[4] From then onward, fodder was the largest item on the menagerie's provisioning budget.[5] The bird population of twenty-three in May 1794 (partridges, peacocks, parrots, swans, and geese), was likewise rising. In September 1797, when the next census was taken, there were well over one hundred birds requiring seeds and grains for their meals.[6]

In 1795, new herbivores joined the more than thirty mammals in the menagerie, among them two camels Louzardy fetched from Belgium and the various deer and goats he escorted from the Stadtholder's collection in Holland in 1796. More deer and goats arrived from the veterinary school at Alfort, and a deer was born, perhaps the menagerie's first birth. The bird collection quadrupled, thanks primarily to specimens from Holland and the waterfowl and shorebirds Baillon sent from Brittany. The years had not, however, been kind to the carnivores. Not one from 1794 was alive in 1797. The serval, leopard, and dog died in 1794. The menagerie's prize specimen, the lion, died in July 1796, and by September 1797 one of the polar bears, the last agouti, and all the monkeys were dead.[7]

The diminished collection was dramatically refreshed in 1798 with the addition of two elephants from Holland (March), three bears from Bern (Apri), and Cassal's sizable troupe from North Africa (four lions, two camels, one gazelle, two ostriches, and four vultures, July). The elephants' appetite for bread created a provisioning challenge at a time when the price of bread and other staples was skyrocketing. The new lions required meat, as did the two tigers, two lynxes, and the leopard, panther, and hyena purchased in London in 1800. The problem of feeding all these animals was at that point acute.

In the late 1790s, the sale of meat in Paris was officially regulated. Vendors were not allowed to sell the flesh of animals that had died natural deaths. Nor, because of worries about depleting the nation's livestock, could they sell the meat of cows, ewes, and sows suitable for reproduction. The prohibition of greatest importance for the menagerie, however, forbade selling the public meat that was putrefying and therefore unfit for human consumption. To reduce abuses, police commissioners kept an eye on the meat in markets, butcher shops, and charcuteries, and they confiscated any that had gone bad. They worked out a deal with the Museum whereby they made some of this meat available to it. Guichenot, the menagerie's commissioner of provisions, was on call to retrieve it.[8]

For a time, the carnivores' primary fare had been horsemeat. Horsemeat could not be sold for human consumption, but people could buy it for their dogs, and the Museum fed it to its carnivores.[9] Guichenot acquired horsemeat from a knacker (*équarrisseur*) who slaughtered and dismembered horses that either were too old to work or were diseased. This arrangement became untenable early in 1796, when the Museum could no longer afford the *équarrisseur*'s services but still needed meat for its carnivores. Delaunay tried unsuccessfully to get the meat of animals that had died or been killed at the veterinary school at Alfort. He also asked the minister of the interior to reserve for the Museum, if possible, the bodies of military horses that died at the depots of Paris, St. Denys, and Versailles.[10] The consequence of being unable to acquire satisfactory meat came that summer with the lion's death. It was "unquestionably" caused, as Delaunay later explained, by bad horsemeat. The poor beast's body was found to be "verminous," its skin riddled with "farcinous and purulent pustules."[11]

How the menagerie was provisioned in its first four years is documented only imperfectly. On the other hand, beginning on the first day of the sixth year of the French Republic (1 vendémiaire an 6 [22 September 1797]), and for the next five years, the details are as complete as one could dare hope, because Delaunay recorded the menagerie's expenditures down to the last *sous* and *denier* from one ten-day week to the next.[12] In September 1797, seized meat was the only kind the menagerie could get. Guichenot collected it an average of three or four times every ten days. Six months passed before the menagerie obtained two horses from the knacker. Two months later it purchased meat from a butcher, eight pounds of beef for the lioness.

In July 1798, when Cassal brought four lions back from North Africa, the Museum began buying butcher's meat regularly, over two hundred pounds per month. However, that was only one-sixth of what the lions needed, given they each ate about ten pounds a day. Significantly, the rationale for buying butcher's beef was not that it was better than seized meat, but that there was not enough seized meat to be had.[13] On one occasion the police furnished the Museum with an abandoned cow, but the menagerie was rarely so lucky.[14] Delaunay told the administration in 1799 that he hoped he would not have to resort to feeding "lions and other precious animals with horsemeat," because it had "certainly killed the last lion."[15] Nonetheless, he continued to buy it. Some was for the wolves, but some may have gone to the lions.[16]

When Delaunay and Louzardy returned from London late in 1800 with seven new carnivores, the menagerie's meat requirements spiked. Beef purchases increased 70 percent, coming to represent nearly 80 percent of the menagerie's annual meat expenditures but still constituting only a fraction of the meat consumed.[17] Indeed, purchased beef accounted for less than 16 percent of the total. All the rest came from spoiled meat seized in the markets and a smaller amount of horsemeat from the knacker.[18] Imagine, then, the crisis when the police confiscated no meat in the markets for twenty-one days in November 1801, or at least reported none to the Museum. The professors urged the prefect of police to command that all meat seized in the markets be designated for the menagerie, but the supply of confiscated meat continued to be precarious.[19] Years later it was all but off the menu, but in 1801 the menagerie depended on seized beef, and Pierre Guichenot and his son, Louis, traveled all over Paris to get it.[20]

∴

The herbivores generally fared better than the carnivores, but they too had hard times. They ate bundles of hay and bushels of oats, bran, barley, lucerne, and wheat, and straw was purchased for their comfort and for keeping their quarters clean.[21] In February 1797 Delaunay warned the administration and the government that the stores of fodder, grain, and straw were running out. The military, he explained, did not have supplies to spare, and the farmers were unwilling to provide any more if they could not count on prompt payment.[22] Delaunay saved the day by establishing a relationship with a farmer named Degennes who lived twenty-five kilometers east of Paris. For two and a half years he supplied fodder and was patient when the Museum fell behind in paying him. But after ten months and a bill that had grown to 4,000 francs, Degennes had had enough.[23] He said he would send the menagerie a wagonload each of hay and straw, but that was it until he was paid. Delaunay began reducing the herbivores' rations, and he identified twenty-nine herbivores the menagerie could do without, but this did not produce funds soon enough for the farmer.[24] In October 1799 Degennes insisted that the Museum would have nothing from him until it settled its account.[25]

Patchwork measures saw the menagerie through the crisis. The ministers of the interior and war collaborated to provide fodder and grain for the short term, and the Museum was permitted to purchase fodder from the company that supplied it to the army.[26] Degennes was soon delivering fodder again, and Josephine Bonaparte gave the Museum some as well (later receiving a selection of plants in return). By April 1800 Delaunay had sold a dozen superfluous herbivores, enabling him to pay part of the debt to Degennes.[27] When excess herbivores proved of no interest to the butchers, they were fed to the carnivores.[28] Even with the herbivore population considerably reduced, however, Delaunay's account book shows that by 1801 the annual cost of fodder and straw was greater than all other provisioning expenses combined.

∵

The elephants' arrival in early spring 1798 brought a sharp rise in menagerie food costs. Previously the main food purchases other than fodder and meat were carrots, apples, and greens, but the pachyderms added new items to the shopping list: potatoes, turnips, cabbage, and bread. Judging by Delaunay's account book, their appetite for bread surprised him. He bought 180 pounds in their first ten-day week at the menagerie. The following *décade* he bought 300 pounds, with weekly purchases swelling to 360, 390, 420, and finally 540 pounds.[29] Ninety pounds of this went to the three bears from Bern, which arrived in late April and each ate three pounds of bread daily.

We now know it's unhealthy for elephants (or bears) to eat so much bread, but they enjoyed it, unlike turnips. All mention of turnips disappears from Delaunay's ledgers a month after the elephants' arrival.[30] Greens, on the other hand, were a significant addition to their diet, though Georges Cuvier failed to mention them in the fall of 1801 when he listed their daily food as "one hundred pounds of hay, eighteen pounds of bread, some bunches of carrots and some amounts of potatoes, without counting what the curious constantly give them."[31] But Cuvier's estimate of the elephants' daily bread consumption was low. Delaunay at this point was feeding them twenty-five pounds per day, and a few years later, when Frédéric Cuvier was menagerie superintendent, it went up to thirty-six.[32]

Whatever the exact figures, the elephants were big eaters, and in the late 1790s to early 1800s they severely strained the Museum's limited resources. Bread was expensive, and its wildly fluctuating price compounded the Museum's budgetary difficulties.[33] In the summer of 1798, the Museum paid less for beef for Cassal's lions than it paid for bread for the two elephants.

∵

In autumn 1797, seized, tainted meat was the basic fare for the menagerie's carnivores, but veal was purchased twice from a butcher for the king vulture. It is tempting to take this as indicative of the quality of the carnivores' food—too spoiled for a vulture!—but the notation in Delaunay's register, "veal for the vulture," is better understood as evidence that the bird was sick. Loss of appetite was often the first sign that an animal was ill. Getting an animal to eat by providing better or different food was recognized as a way of restoring it to health. The relation of the animals' food to their health could thus go either way. Food could make an animal ill, or it could make it well. Then too, food was not always the deciding factor. When the vulture died in January 1798, Delaunay ascribed its demise to the cold and damp of its cage and its proximity to the river.[34]

Wherever Delaunay's account book indicates a special food purchased for an animal, the animal was likely unwell, precious, or both. Milk was given to "the serpents" (twice in 1797); to the young lioness (several times between fall 1797 and spring 1798); and to the lion cubs (in March 1802). The cubs were also fed mutton. Other special purchases in 1797 included "salad" for the pheasants, carrots for the cassowary, rice and millet for the painted buntings, and hay for the opossums.

The lemur was treated to wine with its special meals of cakes or bread. The elephants had wine but presumably for medicinal purposes (two bottles in April 1798), as did the camel (half a liter per day for six days in May 1802 and again, with honey, in June). The little fallow deer received both wine and eight days of *eau-de-vie* (in September 1802). A gazelle was given honey (mixed with sorrel and dandelion, in January 1800), as was a bear (at least twice in May 1800). The elephant also had honey (more than 100 pounds) in the spring of 1802 to treat her digestive problems, and fish oil was used to grease her skin. Special ointments were likewise purchased for the lion cubs and camel. When the menagerie experienced a rash of deaths at the turn of the century, Delaunay insisted that it was not because "the help of medicine" was lacking. Nor was it due to any lapse of attention from the keepers. The problem, he felt, was the unsuitability of the animals' quarters.[35]

Article 7 of the rules for the *garde de la ménagerie* instructed the supervisor to attend the animals' feeding regularly to judge, "according to the animals' greater or lesser appetites, the state of their health or sickness." The regulations also called for daily reports to the professor of birds and mammals on menagerie developments. Extracts from Frédéric Cuvier's early reports show how appetite was taken as an index of an animal's health. Only after the animal's death was its illness determined by autopsy.

> ***pluviôse an 12*** **(26 January 1804)**
> . . . The tiger female continues not to eat; she seems to breathe today with difficulty. The civet has eaten, its health seems completely restored, and it is the same with the male lion.

6 pluviôse an 12 (27 January 1804)
. . . The female tiger has eaten the entire heart of a cow; she seems a little better. The civet continues to be well; it eats and walks around. All the other animals are well.

7 pluviôse an 12 (28 January 1804)
There is nothing new regarding the lion, the civet, and the female tiger; the first two are well, the last does not seem to have grown worse. The phascolome is dead, and only refused to eat a few hours previously. One did not in the least suspect that its illness had become so serious.

9 pluviôse an 12 (30 January 1804)
The tiger female died in the night of the eight to ninth. Her autopsy showed it was pneumonia. Her lung was full of pustules, and a considerable deposit of blood was seen there. The civet eats from time to time but is always listless.

10 pluviôse an 12 (31 January 1804)
The civet seems always extremely feeble and its illness shows by no other symptom.

11 pluviôse an 12 (1 February 1804)
The civet is dead. It has been sent to the naturalists' laboratory. The big lion has refused to eat, and it has vomited again.

At this point, Cuvier attributed the "bad state" of the carnivores to quarters that did not protect them sufficiently from "our climate."[36]

While it was obvious that animals should not be fed rotten or unsuitable food, the idea that nutritional deficiencies caused sickness was largely unrecognized (notwithstanding that lemons and fresh vegetables were understood to prevent scurvy in sailors long at sea). In his menagerie guide, Cuvier remarked that polar bears in the wild live on fish and other marine animals, he described the menagerie's polar bear as "content" to eat six pounds of bread a day, though it also ate meat and showed a keen taste for fruit (especially apples). The menagerie's brown bear was occasionally given fruits (for which it showed "an extreme avidity") as well as root vegetables and meat, but its usual diet was ten pounds of brown bread daily. The two black bears would eat meat and fruits as well as bread, but bread was "nearly" all they were given. Cuvier noted their health problems. Though still young, the black bears had "almost entirely lost their sight." He explained this as "undoubtedly the result of their captivity." He did not remark any correlation between their vision and their diet.[37]

In an 1805 report, Cuvier recorded several symptoms of apparently the same ailment afflicting five bears at the time of molting. The animals lost almost all their fur, the control of their limbs, their eyesight, and finally their appetite. One died from exhaustion. Two were unable to stand upright and to see, but they afterward recovered both their strength and their eyesight. Cuvier attributed their cure to "a more substantial nourishment than the bread they normally have and especially to keeping them in a dry place where they have always fresh straw to lie down in."[38] It was perhaps also at this point that carrots were introduced into the bears' daily regimen.[39] Cuvier's report was deposited at the professorial assembly but not discussed.[40] The opportunity to consider the ramifications of nutritional deficiencies went unrealized.

Just as no systematic thought was given to nutritional deficiencies, none was given to the animals' psychological needs, even though animals were commonly recognized as sentient beings capable of emotion. Observers noted signs of their suffering under the conditions of their confinement: boredom, depression, rage, stereotypic pacing, and self-mutilation.[41] The idea of tending to the animals' psychological well-being, however, never rose to the level of being identified as part of the menagerie supervisor's job.

∴

Brooms were almost always at the top of Delaunay's weekly list of menagerie expenditures.[42] Nothing was considered more important for the animals' health than cleanliness, and the broom was the principal tool for the job. In the late 1790s, the keepers went through forty brooms every ten days. Cleanliness was prominent in the 1803 rules outlining the responsibilities of menagerie superintendent. Immediately after the enumeration of his administrative obligations, he was instructed to "recommend the greatest cleanliness as one of the most effective means of preventing the majority of illnesses." Whereas Delaunay's reports to the administration had emphasized the keepers' zeal in cleaning the cages, Frédéric Cuvier, after several years on the job, held a different view. In a report of 1810 he wrote, "Men, such as those whom we employ, only do well what they can do easily. The regularity of the duties confined to them always depends much more on its facility than on their intelligence or zeal."

Ideally, Cuvier continued, the keepers' caretaking chores would be simplified to the point that they became, "so to speak, mechanical." As he saw it, it was "extremely rare" to find men who were so "rigorous and meticulous" in their duties as to attend to details and to set right unsanitary conditions unless grossly conspicuous. "An animal keeper," he wrote, "will sweep the animals' quarters as well as his broom will allow and he will never do more; if the cells are not constructed in a way that the broom alone can clean them, they will very rarely be clean; the filth will accumulate in the corners, where it will be difficult to clear away, the walls will become impregnated with unhealthy odors that will vitiate the air, and in a short time the animals will breathe nothing more than a pestilential atmosphere."[43]

Cuvier was not campaigning for better brooms or better workers. He made these remarks in discussing the requirements for a new building to house the *animaux féroces*. A fundamental problem with the menagerie, not corrected in the latest plan, was that the animals were lodged in single, not double, chambers.[44] With single cages, an animal was not easily removed at cleaning time, whereas with (adjacent/interconnected) dual cages, it could occupy one while the other was being cleaned.

Moving an animal from its cage was a demanding task. The keeper had to fetch from as much as several hundred paces away a heavy portable cage Cuvier called a *trape*. The keeper had to haul the *trape* back to the animal's cage, slide the apparatus into place along specially constructed tracks, and open a door so the animal could enter the portable enclosure. The keepers avoided this dangerous, time-consuming process if they could. The frequent alternative was to wash the animal's cage while occupied. After tossing in some bucketsful of water (half of which might hit the animal), the keeper used an iron scraper to collect the excrement. "Cleaning done this way," Cuvier insisted, was "very dangerous for the animal" and "very imperfect." It was unhealthy for the animal to get soaked and for noxious filth to remain in the cell's corners, where the grooves for the *trape* were the greatest challenge to a broom's efficacy. Cuvier called for the new structure to have adjoining cages for each animal. Double cages would have multiple advantages, including being easier to clean.

∴

The germ theory of disease was not established until the later nineteenth century, but air and water quality, like cleanliness and decent food, had been recognized as critical for human and animal health since antiquity. The famous Hippocratic treatise on "airs, waters, and places" dates to approximately 400 BC. When the menagerie's supervisors worried about the air, they had in mind the effects of cold and damp on animals from hot climates, and of air that had become "vitiated" by being breathed or "corrupted" by bad odors.

The idea of vitiated air gained new meaning in the later eighteenth century through Antoine Lavoisier's demonstration that the greater part of the atmosphere consists of two substances: "vital air" (oxygen) and *gaz azote* (nitrogen). Vital air, he estimated, made up about 25 percent of atmospheric air and was essential for animal and human respiration and for combustion. *Gaz azote* made up 75 percent of atmospheric air but could support neither respiration nor combustion and was not altered by either. A guinea pig or sparrow placed under a glass jar containing only *gaz azote* would die instantly; likewise, a burning candle would go out. In addition to vital air and *gaz azote* Lavoisier identified a third airlike substance, "carbonic acid gas" (carbon dioxide). This was the "fixed" or "vitiated air" produced when the healthy part of atmospheric air was sullied by respiration or combustion. Fixed air, like *gaz azote*, would not support combustion or life.

Lavoisier experimented with animals, observing their reactions while breathing under bell jars in the laboratory. He went on to study what happened to the air people breathed in confined, crowded spaces, believing this would have implications for how to design hospice rooms or how to think about refreshing the air in poorhouses, theaters, prisons, and mines. He was additionally concerned by the "miasmas of an infinity of kinds" that filled the air. In a memoir of 1777, he wrote: "One is frightened when one thinks that in a large assembly the air that each individual respires has passed and passed again a great number of times, either in whole or in part, through the lungs of everyone present, and that it has had to load itself with more or less putrid exhalations." He wondered what illnesses could be caught in this way, and what precautions could be taken "to neutralize or destroy the dangerous influence of these emanations."[45]

Lavoisier was condemned to death and guillotined on May 8, 1794, twelve days before the animals from Versailles arrived at the Museum, the event that all but assured the menagerie would become permanent. In the years that followed, no hardy chemist ventured into a lion's or tiger's cage to take air samples as Lavoisier had done in public places like the l'Hôpital général and the Comédie française, but the Museum was appropriately apprehensive about the animals' breathing vitiated air. In 1801 Jussieu, as Museum director, told Chaptal, the minister of the interior, that the menagerie's fierce animals were so crowded in their narrow cages they could "harm each other in infecting the air they breathe."[46] Unfortunately, bringing fresh air into the cages was dangerous for animals unaccustomed to the cold. In January 1806, Geoffroy reported that keeping the monkeys warm had protected them from the lung diseases they were susceptible to, but keeping the cold out of their cramped quarters left them breathing "a foul air."[47] When Frédéric Cuvier studied the 1810 plans for the new menagerie, one fault he identified was that the animals would be exposed "to air currents that would infallibly kill them."[48]

Bad smells as well as breathing, it was believed, could vitiate the air, and the Jardin des Plantes had plenty of them. Powerful odors came from the Museum's animals and from the slaughterhouse, zoology laboratory, maceration vats for cleaning bones, and cesspools. Other noxious stenches came from beyond the Museum's gates, especially from the uncovered sewer of the nearby hospice of Salpêtrière (Lavoisier's *l'hôpital générale*) and the Bièvre River, also known as "the river of the Gobelins." Citoyen Leger, who owned property near the Museum, wrote in 1798 to ask for help with respect to the vile stink all around him. He characterized the smell from the Salpêtrière sewer as a "putrid odor" and "the air of death." From the Bièvre came the foul, chemical/industrial reek of the tanning, leather making, and other businesses on the river's banks.

As for the malodorous menagerie, Leger believed that better access to fresh water was the only solution. He warned that the animals of the menagerie would die if nothing were done to spare them from the noxious air they were obliged to breathe.[49] He called for a connection with the Seine that would allow "pure, always fresh water" to flow to the Museum. He was not just thinking of the public good. He calculated he would get enough water to operate twenty mills and grind eight hundred sacks of flour a day. He also called for covering the Salpêtrière sewer and rerouting it directly into the Seine. He additionally urged that measures be taken to make the river Bièvre healthier.

The Museum agreed to support Leger's suggestions, but nothing came of them, and two years later he wrote again. Walking in the gardens, he said, one had to cover one's nose with a handkerchief because of the "infected air" coming from the Salpêtrière sewer. "It is the sewer from the hospice," he insisted, "that gives rise to this putrefaction that will kill all your animals and that can tarnish and spoil the most precious objects." Breathing this "air of death," he maintained, had killed residents of the area "every year." At his own business, he said, "we are all sick from this smell." He could not "keep meat for four hours without its becoming spoiled," and the air turned silver "black as ink."[50]

The professors thought the tanning and other industries of the Bièvre were responsible for the worst of the smells and tabled Leger's request to reroute the Salpêtrière sewer into the Seine.[51] By spring 1804, however, the sewer's stench had worsened, and the Museum urged the government to do something about it, saying the fetid and injurious exhalations were weighing heavily on the area's inhabitants and causing visitors to avoid that part of the gardens.[52] The Museum had its own sewer that carried the menagerie's wastes to the Seine. It too was uncovered, and when blocked it produced a "fetid odor" harmful to visitors.[53]

Not all the animal dung was discharged into the river. Mordant Delaunay protested in 1813 about the "morbid and nauseating emanations" rising from the "mountains" of dung and leaf piles next to his apartment. He could not use his own garden without suffocating, he said, and almost never was able to open his windows. Begging the administration to stop depositing excrement there, the ever-precise Delaunay identified a hierarchy of offending smells: "the dung of the elephant, the dung still worse of the camels, [and] the dung even worse from the menagerie of the *animaux féroces*, which is always mixed with the remains of their dinner, bits of rotten meat."[54]

Alongside the problem of noxious smells was that of providing the animals with clean drinking water and decent troughs. Some animals needed water for bathing. Urging the administration to provide a pool for the elephants, Delaunay explained how much they enjoyed having access to water. They sprayed it all over their bodies, then stamped in it to create a mire they wallowed in.[55] The Museum had various wells, the most important of which was operated by camels powering a pump, but some wells, as Delaunay reported in 1796, produced no more than sludge unfit for drinking, washing the cages, or cooling off the bears. Cassal had to fetch water from the river for his big cats, making his work "extremely arduous and perhaps imperfect at certain times."[56]

The Seine was the menagerie's primary water source. The Museum was fortunate to be situated on the east side of Paris, before the river was sullied by the full flood of urban pollutants. But it still behooved the Museum to beware of locally produced contaminants like the Salpêtrière sewer, the Bièvre River, and its own wastes. The new menagerie plan called for the carnivore quarters to be near the Seine to facilitate getting water for cleaning the cages and to allow the building's sewers to the Seine to be shorter.[57] Surely the planners also thought of having the clean water source upstream from the sites where the filth of the menagerie, the hospital, and the river Bièvre met the Seine.

∴

When the Museum early in 1802 composed a general "Program for a national menagerie," it specified that the animals' housing should be designed to keep them from being "disturbed by the public, poisoned or stolen by the malicious." This was not hypothetical. The animals were frequently disturbed, and poisoning and theft were "not without example."[58] As menagerie superintendent, Delaunay complained about the harming of the animals. A sentry for every animal enclosure wasn't enough, he had said, one was needed for every animal. Frédéric Cuvier likewise feared the public; in 1804 when four previously healthy deer died and four more fell seriously ill, he was sure they were deliberately poisoned.[59] Three years later, he reported the death of an otter struck by a stone.[60]

An issue bearing on the animals' well-being where the interests of the keepers and the public were intertwined concerned the menagerie's opening and closing hours. The Museum's position evolved over the years. Sometimes hours were set primarily with the animals' needs in mind; at other times they were adjusted to accommodate the public's interests or the keepers' duties. The most remarkable change of hours came with the purpose of halting the keepers' flagrant practice of taking bribes to admit people when the menagerie was officially closed.

Early in the Museum's history, the public's access to the natural history galleries was limited to specific hours and days of the week. Students and others with special permission cards could visit the collections on additional days. The first regulations do not mention specific hours for the menagerie, but such hours were instituted when the elephants arrived and the crowds that flocked to see them were too unruly for the animals' well-being. The public's viewing was restricted to the even-numbered days of each ten-day week from 11:00 a.m. to 1:00 p.m. and 3:00 to 5:00 p.m.

The keepers' malfeasance undermined that schedule, obliging the administration to scrap it. Visitors had protested that the keeper of the elephants (Thompson) and the keeper of the carnivores (Cassal) were preventing them from seeing the animals during opening hours but letting people who paid see them when the menagerie was closed. To thwart the keepers' after-hours activities, the administration decided in 1801 to keep the menagerie open daily from 10 a.m. to sunset in summer and eleven a.m. to sunset in winter. The keepers were instructed to clean the animal cages early in the morning before the public arrived.[61]

The new regulations were well intentioned but flawed. The closing time needed to be earlier, Cassal explained to Lamarck, because the big cats defecated after eating, and the keepers needed sufficient daylight to clean up after them. Furthermore, it was dangerous to feed the animals while the public was still there because "there [were] always some people" who enjoyed teasing the animals while they were eating. Soldiers were the worst offenders: "The dragoons with their sabers excite [the animals'] anger to amuse themselves, and from this rage it has often resulted that the animal eats and swallows precipitously and runs the risk of choking on the bones, which can end up in the throat in a bad position."[62]

Lamarck recommended that the menagerie close at least thirty minutes before sunset. He also noted a problem that occurred in the morning. Artists were allowed to paint or draw the animals before the public was admitted. This meant the artists and keepers got in each other's way during the morning cleaning. Lamarck had no solution. The artists and keepers needed to work it out between themselves.

These regulations proved largely unsatisfactory even when amended to usher the public out before sunset. Staying open all week was trying for the keepers and exhausting for the animals, afforded no tranquility for scientific studies, and was burdensome for those policing the menagerie's grounds. In summer 1804 Geoffroy proposed opening only on Tuesdays, Thursdays, and Sundays, from 11:00 a.m. to 6:00 p.m. Students, artists, and others with special admissions cards could visit during the same hours on the days the menagerie was closed.[63] The administration approved this, but the next year Geoffroy recommended opening from 2:00 to 7:00 p.m. on the same days, thinking the public would find this more convenient.[64]

This schedule remained in place, officially, until the spring of 1816, when Frédéric Cuvier called for the larger part of the menagerie, devoted to the herbivores, to be open throughout the week. He acknowledged that this didn't help the animals, since it would interfere with their rest, but they weren't getting much anyway, he said, because of the "scandalous" way the keepers solicited bribes to let people in when the menagerie was meant to be closed. The administration followed the superintendent's advice, setting the new daily hours for viewing the animals at 10:00 a.m. to 6:00 p.m. in summer and 11:00 a.m. to 3:00 p.m. in winter.[65]

The adjusted hours are a prime example of a less than satisfactory compromise made in the face of competing pressures. The animals, constantly under the public eye, had less time to rest. In addition, the hours left artists and students little time to conduct their studies unhindered, and the same had to be true for the menagerie superintendent himself. The extended hours and days open to the public presented an obstacle to Cuvier's goal of making the menagerie a plausible site for observing and experimenting on animal behavior.

What was true about setting the hours of public admission to the menagerie applied to caring for the animals as well. It was a balancing act, and the results were typically less than ideal. The professors knew that animals from warm climates needed shielding from cold drafts but also needed fresh (not "vitiated") air. Provisioning in times of severe food shortages and limited funds required compromises and constant adjustments. Tasked with caring for hungry beasts, the professors explored their options, purchasing quantities of quality meat from butchers to supplement the not quite sufficient amount of not so good meat confiscated by the police. They identified one winning combination for the Museum's budget, feeding surplus herbivores to the carnivores. It was not a win for those herbivores, but it provided fresh meat for the carnivores, the butchering could be done on site, and it meant there were fewer herbivore mouths to feed.

These and many other matters were decided, for better or worse, by the menagerie's successive supervisors: Geoffroy, Delaunay, Lamarck, Geoffroy again, and finally Frédéric Cuvier, who served by far the longest in this capacity. For thirty-four years he worked to keep the animals healthy, their provisions ample, the facilities in good shape, the public satisfied and well-behaved, and the keepers efficient and rule abiding. He offered suggestions about facilities and about building the collection. He arranged for the transport of specimens from afar. He kept accounts of what the menagerie had on hand in the way of food as well as specimens.[66] Ultimately, however, he wanted to do more than manage the menagerie. Frédéric Cuvier wanted to be a zoologist in his own right. That too would require a balancing act.

[2.11]

Martin the Bear

The most illustrious bear in the menagerie's history was a European brown bear named Martin. Born in the Pyrenees in the spring of 1808, he was captured as a cub, brought to Paris, and purchased by the Museum in 1808 or 1809.[1] An 1811 guide to the menagerie lists him as inhabiting the first of its three bear pits. The second pit housed two black bears, one named Tautin, and the third pit held a black bear named Jacques.[2] Martin was the one who gained special fame. His intelligence, dexterity, and antics delighted visitors. One writer (1814) rehearsed a song describing him as "a courtesan escaped from the woods," like "the flatterers of the kings" but less costly to maintain. "For a chestnut, for an apple . . . he pays reverence, kneels, bows, balances, changes form and demeanor, and turns his back on you if he no longer hopes for anything."[3] Yet Martin's celebrity had a darker side. In 1813 he killed a man; in 1820 he killed another. The first had unwisely ventured into Martin's pit. The second threw himself there deliberately.

The first death occurred in the early hours of 15 December 1813, according to the minutes from the professors' weekly meeting that day:

> A veteran of the Museum guard, Dutch by origin, named Jean Loyaler, . . . having yesterday been posted as sentinel beside the pit of the bear called Martin, noticed there a large, white button, which was still there the next day, and which he took to be a five-franc piece. He came three times, after his sentry duty, to inform the menagerie's keepers. Understanding his jargon only with difficulty, and suspecting his account to be untrue, they ignored it. Apparently this unfortunate man imagined going to find the coin himself. It seems that having sentinel duty this morning at 3:00 a.m. near the administration, he left his post, found a ladder which was tied to a tree in the square across from the menagerie of *animaux féroces* and with it descended into the bear pit. When at 5 o'clock the corporal came to relieve him, he was greatly astonished not to find him. He

Figure 2.11.1. "Martin the Bear of the Jardin des Plantes." M. E***, *Promenades instructives et amusantes d'un père avec ses enfans dans Paris et ses environs* (1817). Bibliothèque nationale de France.

> looked in the vicinity, but unable to foresee what had happened, he occupied himself with putting a new man in his place and warning the commander. It was only at daybreak that the man named Loyaler was found dead and torn apart in the pit of the bear called Martin.[4]

The administration hastened to provide the minister of the interior, the commissioner of police, and other authorities with an account of the tragedy. The Museum also sent the account to the editors of the *Journal de l'Empire*, the *Journal de Paris*, and the *Gazette de France*, stating that it was not engaging them to publish it but rather warning them against crediting other accounts or rumors that might reach them. The *Journal de l'Empire* and *Journal de Paris* nonetheless published the account the next day.[5]

The commissioner of police summoned a surgeon to inspect the body and prepare an official report, which was horrifying. It indicated "there was almost no part of the body that had not been mutilated." The man's left leg had received the bear's particular attention: "All the soft parts of the left leg were destroyed, baring the two bones, as well as the right plantar region, where the muscles and tendons were shredded."[6] The body was transported to the barracks for burial.

To avoid other such disasters, the administration decided that a keeper would sleep in Bernard Louzardy's old room near the bear pits, where an alarm bell would be mounted. When a poisoned ball was found in Martin's enclosure several days later, menagerie superintendent Frédéric Cuvier concluded it was an act of revenge for Loyaler's death and ordered that "the animal be closed up day and night to protect him from the actions of evildoers who want to poison him." Having already warned the administration that the bear pits were too easily accessed, Cuvier urged them "to take measures so that a new misfortune does not take place."[7]

Brown bears were not rare in Europe. They lived in the wild in the Alps and the far reaches of Poland and Lithuania. Animal dealers and menagerie owners were eager to feature them in crowd-pleasing animal shows, since they could be trained to dance and perform tricks.[8] The first representatives of the species to appear in the Paris menagerie were the three Bern bears that arrived in 1798 as war trophies. Their narrow cages prevented them from moving about and performing as they had in their spacious Bern pits. The embarrassing contrast between those accommodations and the cramped Paris cages probably helped persuade the Museum to construct bear pits of its own, as did the 1803 arrival of three more bears (one brown, two black).[9]

The brown bear was captured in Poland as a cub and toured by an animal trainer, who, after inserting rings in the bear's lips and attaching a rope to them, succeeded in teaching the bear to obey his commands.[10] However cruel this treatment, the bear remained fond of its owner, greeting him with a special cry whenever he came to visit. The bear's new Paris quarters distressed it greatly, as Frédéric Cuvier reported:

> In the first days after his entry into the menagerie, he seemed sad and did not allow anyone to approach him. He is beginning to familiarize himself with those whom he sees every day. However, his state of servitude seems insupportable to him, at least to judge by his continual devastation of his cage. When he is not eating, he is ordinarily occupied tearing at the walls that surround him; we have already been obliged several times to reconstruct part of it, and to prevent any accident, we have covered with iron bars the weakest walls, through which he would have been able to escape.[11]

Fear that this large, ferocious bear might escape was another motivation for the Museum to finish building its bear pits.

Once occupied (August 1805), the bear pits became one of the menagerie's greatest attractions. There were three in a row, one leading into the next, each some twenty meters long, eight meters wide, and five to six meters deep, with limestone walls and floors.[12] Each pit had a den where the bears could shelter or be locked behind iron bars when the pits needed servicing. Grillwork on three sides of the pit and a thick parapet on the fourth prevented people from tumbling in. The public flocked to the site for an unobstructed view of the bears padding about, standing on their hind legs to beg for food, and climbing the tree trunks installed to encourage exercise.

An illustration of Martin in his tree served as half of the frontispiece of a guidebook (1817) titled *Instructive and Amusing Walks of a Father with His Children in Paris and Its Vicinity*.[13] A different picture of Martin, likewise in his tree, was the frontispiece for an 1820 guide to the "remarkable" things to be found at the menagerie and galleries of the Jardin des Plantes.[14] Climbing trees was not hazard-free for the bears. Museum records from December 1815 indicate that Martin "fell from his tree and injured himself rather dangerously."[15]

The bear recovered and continued to be a subject of both entertainment and cautionary instruction. In the 1817 guidebook, the son remarks how ordinary Martin appears, how grateful for the pieces of bread and cake he catches so adroitly, how he climbs his tree and walks on his hind feet like a human. In response, his father relates two incidents: one involving a child who escaped from his nurse, fell into Martin's pit, and "was devoured in an instant"; the other about the guard who descended into the pit to fetch a coin and died a horrible and painful death. Hearing these tales (only the second was true), the boy and his sister conclude that Martin is eager to eat them too, and they leave the spot posthaste.[16]

Such were the two sides of Martin, who as Frédéric Cuvier observed, acquired his celebrity "as much for his agility and his intelligence as for his cruelty."[17] Songs were written about him. Plays were written about or associated with him. At the Théâtre des Variétés (February 1820) the author of a clever farce titled "The Bear and the Pacha" credited the piece to "Martin, employee of the Jardin des Plantes."[18] Later that year, reporting on an elephant that was about to arrive at the Museum from Bordeaux, the *Journal des théâtres* joked that "all the animals of the menagerie will be at their windows to see him pass by, and it is Martin the Bear who is charged with greeting him."[19] The pleasantry was ill timed. Two days later (September 8) Martin again featured in a gruesome death.

In broad daylight, with children, nannies, and others on hand, an old man entered Martin's pit. Whether or not he died instantly from the fall of some fifteen feet to a limestone floor, Martin pounced on his body and mauled it. Once again the Museum sought to control the narrative by sending a notice to the press, but this time competing versions abounded. There had been witnesses, at least ten newspapers covered the tragedy, and variations on the story began circulating the next day.[20] The Museum's account, published by the *Journal de Paris*, *Gazette nationale*, and *Le Constitutionnel*, stated that at 10:30 in the morning of 8 September 1820, an old man, carrying no identity papers, leaped into the bear pit, and the bear killed him on the spot. A sentry had been on hand to ensure public safety, but the old man, choosing a moment when the sentry's back was turned, swung his legs over the parapet and dropped into the pit, thereby providing "every reason to think that the act was the result of a voluntary choice, and not the result of an accident."[21]

The *Journal des débats* the same day gave a more vivid account. "A man of about sixty, rather well put together, was looking, as were so many others, at Martin the bear, who at that moment was in his den, not responding to the invitation of nannies and children who were throwing him cakes. Suddenly, and to everyone's surprise, the man sat upon the parapet and launched himself into the pit. The noise woke the bear, which pounced upon the man, bit him in the neck, and then devoured him."[22]

The next day (10 September), *Le Constitutionnel* revisited its first account with added details. Martin was in his den (at 10:30 a.m.) when the old man (about sixty-five years old) put his legs over the parapet, slid down the pit's sloping wall, landed on his knees and flattened out on his stomach. Martin darted from his den, seized the man by the neck, and killed him. The man did not cry out. After that, "the ferocious animal made two tours around the pit while holding his prey as he had seized it and seemed to present it to the terrified spectators who uttered frightful cries." The bear carried the man into his den but soon returned, dragging the cadaver by one leg. Stones were thrown at the animal; dogs were released but dared not advance. Some bystanders urged the sentinel to shoot the bear, but his gun was not loaded and "besides," as *Le Constitutionnel* put it, "the man was dead." Finally, the bear was locked in its den (driven there by a courageous dog, according to *Le Drapeau blanc*), and the corpse was recovered and taken to the morgue.[23]

Of the ten press reports, six maintained that the man jumped; four said that he fell. Of the latter, the *Gazette de France* gave no explanation for his fall, but the other three each cited a reason: he was throwing cake to Martin (*Journal de Commerce*); his handkerchief slipped from his hand, and he reached to grab it (*Le Drapeau blanc*); and a seizure or some other illness gripped him (*Journal des débats*, in its second account of the event).[24] The papers also differed on whether the man died immediately from his fall, was killed instantly by Martin, or suffered longer.[25] While most accounts cited the time as between 10:30 a.m. and noon, one stated confidently that it happened at 2:00 p.m. Accounts also differed on what Martin did with the man's body, the response of spectators and authorities, and how many dogs brought the situation under control. The papers reporting the man's death as suicide were faced with the question of what "excess of despair" (*Le Courrier français)* could lead someone "to destroy himself in such a frightful manner" (*Journal des débats*). Neighborhood inquiries yielded information that the man was a *rentier* (i.e., had some income-producing property) but was nearly broke.[26]

Eventually the Museum learned the man's full name (Isidore Joseph Bernard), his age (eighty), and address (129 rue Saint-Antoine). According to his neighbors, Monsieur Bernard was despondent, having lost his wife six weeks earlier. He had no living children, and financial difficulties had reduced his income to 1,500 francs annually. These sorrows, the neighbors were convinced, had unsettled the man's mind and made him suicidal. The Museum assured the minister of the interior that no negligence was involved, and no one had been at fault: "there are some accidents that full human prudence cannot foresee."[27] The Museum explained that a sentry was posted at the bear pits from six in the morning until nightfall but had turned away to warn people not to lean over the wall when Monsieur Bernard made his move. The Museum acknowledged that opinions differed as to whether the death was an accident or a suicide, but that the police, based on all the evidence, had concluded that the man had thrown himself into the bear's pit voluntarily.

One cannot know what Monsieur Bernard expected to happen when he entered the bear pit. Perhaps when he left his apartment on that September morning, dressed respectably, with his cane in his hand and a small vial of medicine and fifteen francs in his pockets (*Le Courrier français*), he was not planning to end his life. When his walk from rue Saint-Antoine brought him to the Jardin des Plantes, where children and their nannies were tossing bread and cakes to the bears, he may still not have made up his mind. *La Quotidienne* reported that when he first approached the parapet of Martin's pit, the sentry shooed him away, but when the sentry's back was turned, the old man mounted the parapet, swung his feet over, and launched himself into the abyss, brushing the steeply inclined wall as he fell, landing hard on his stomach and face. One can only hope he was unconscious, if not dead, when Martin reached him.

Though the Museum was not accused of wrongdoing, there were

multiple calls to prevent such a thing from recurring. The press, the minister of the interior, and the prefect of police had their say, demanding that iron grilles be installed on the parapets of the bear pits.[28] The Museum assured them the necessary measures were under way. Martin was confined behind bars in his den; he would be transported to a cage in the new pavilion for fierce animals once that building was finished. The other bears would also be taken there until an iron guardrail was installed on the parapet.[29]

Some profited from the tragedy by selling brochures like the four-page pamphlet titled *The Exact Details of the Unfortunate Event Occurring at the Jardin des Plantes in the Bear Pit*. Beyond what the newspapers had said, this offered the new tidbit that after the man's body was taken to the morgue, "an immense crowd filed by to see this unfortunate victim of a new form of suicide." The pamphlet treated the Museum with deference, allowing that the "wise and distinguished professors" of "this magnificent establishment" had spared no precaution in making it possible for the public to visit the Museum daily. The death resulted from "the insanity of this unfortunate person who wanted to put an end to his days." The text discussed what bears are like in the wild, how they are hunted and trained, and how their meat tastes. The brochure concluded by cautioning visitors to the Jardin des Plantes to "follow exactly the instructions that are posted there, especially concerning the wild beasts, not to irritate them, nor to approach too near to them, in order not to become victim of a moment of imprudence."[30]

A similarly inspired two-page, quarto-sized broadsheet presented a fable reminiscent of Aesop or La Fontaine titled "Deliberation of the animals. Judgment against the bear, Martin, rendered by the animals of the menagerie of the Jardin des Plantes, under the presidency of the lion. His interrogation and his replies." In this tale, the lion president, "indignant because of the conduct of the infamous bear Martin," calls for a tribunal of animals to pass judgment. The lion does so only after securing his keeper's permission, a sign from the author that animals had best do what their human masters want. The problem, the lion predicts, is that Martin's behavior will make life hard for the rest of the menagerie's inhabitants, who had been treated well because they were well behaved. The lion laments that because of Martin's heinous act, "the happiness is eclipsed for us; and through the mistake of a malicious animal, we are all going to suffer." Martin proves unrepentant, and it does not help his cause that his attorney is the hyena. At last the monkey reads the mocking sentence: "We have condemned and do condemn the so-called Martin to perpetual captivity, not wanting to deprive the curious of the spectacle of his kindnesses."[31]

Aside from such literary productions, reactions to Martin's behavior include the attempt to poison him; the precautionary measures demanded by the minister and the police; and the measures taken by the Museum. Additionally, six weeks after Monsieur Bernard's death, a certain "Marquis" (for whom we have no further identification) proposed to the Museum that Martin be returned to liberty in his bear pit under the care of a Sr. de Vielk, the sound of whose voice would be sufficient to encourage him to climb his tree and once again amuse the public. The Museum rejected the offer, noting its plan to enhance the bear pits' security.[32]

Martin was locked in his den for nearly six months, until his transfer to the new pavilion for fierce animals, a move announced in late March 1821. As the *Journal de commerce* cryptically reported, the relocation "was not made without some difficulty. Martin had become fond of his domicile, and only left it with pain."[33] We are left to imagine a vicious scene where frightened menagerie staff deployed large mastiffs, used by the keepers when entering the bear pits, to force Martin's compliance.

For four years the bear pits were home to wild boars. When the animals they were built for were reintroduced (1824), Martin was no longer among them, having died several months before.[34] Frédéric Cuvier reported Martin's death in one of his weekly menagerie reports: "On 4 July the bear (Martin) from the Pyrenees died from a violent inflammation of the throat."[35]

The press did not remark his death, but Martin lived on in the popular imagination. Indeed, his posthumous reputation began to be cultivated even before he died. In late February 1824, the Franconi brothers offered a "burlesque melodrama" with live bears as part of the entertainment at their "Cirque olympique." Their show, titled "The Bears," drew an appreciative audience. When the author appeared for his curtain call, he told them they had witnessed "a posthumous work of Martin, who in his lifetime had been one of the best pampered (*les mieux léchés*) and most amiable bears of the faubourg Saint-Marceau."[36]

A year later, Martin was portrayed as a more sinister character in a somewhat obscure two-act parody called "Martin the Bear, as Jocko," which satirized a successful play titled "Jocko, or The Monkey of Brazil." Its author cast Martin as a hybrid to evoke the dangers miscegenation posed to civilization, with the former French colony of Saint-Domingue as the backdrop. The press seems not to have reviewed the show, nor to the best of our knowledge did anyone other than its author suggest that the brown bear resulted from a cross between the polar bear and the black bear.[37]

Martin was resurrected twenty-three years later in a satire directed at the Museum professors. The sketch, titled "The Bear and Human Justice," has Museum faculty sentencing a bear named Martin (identified as the *fourteenth* in a long line of bears at the menagerie with that name) to death, guilty of the crime of being *de trop* (too much). A politician and physician,

(Jacques) Maissiat, offers a spirited defense, arguing for the preservation of bears and their native forests, but the professors are unconvinced, their judgment biased by their desire to get their hands on Martin's skin, flesh, and bones.[38]

∵

Frédéric Cuvier's comment about Martin the bear's being as famous for his agility and intelligence as for his cruelty came in the context of Cuvier's explaining that American black bears were not as easily trained as European brown bears (like Martin) and therefore not as interesting to the public. He did not comment on the two species' relative ferocity, but the menagerie had experience with violent behavior from both. In 1816 Cuvier sold a black bear because it had grown so aggressive that its keepers could scarcely go near it.[39]

Among other bears in the menagerie's collection was a very large "gray bear coming from the Rocky Mountains," most likely a grizzly, acquired in 1828 and referred to in the Museum's records as "the terrible bear."[40] With this beast in the menagerie, together with Martin's infamous deeds, it is not surprising that on October 31, 1831, when one of the bears in the pits died, the press suggested that a fellow bear had killed it. Frédéric Cuvier discredited the story, reporting to the administration that the bear had been ailing for some time, and it was "[certainly] not killed by its comrade, as the papers have said."[41]

In judging Martin's acts "cruel," the public was attributing to a bear the ability to make reasoned choices rather than simply doing what came naturally to it. Human visitors to the menagerie, in contrast, chose deliberate cruelty, and the bear pits were a prime site for manifesting it.

Throwing bread and cakes to the bears was a favorite pastime of visitors to the menagerie. Often this involved dangling food on a string, just out of the bear's reach, and snatching it away when the bear tried to grab it. Such teasing was considered basically harmless, but malicious variations arose. In 1830 Cuvier requested that a sentry be posted at the bear pits to safeguard the cubs, which risked injury by visitors who amused themselves at the animals' expense.[42] He did not identify the specific abuses at the time, but in a warning to the professors five years later, he spelled one out:

> The bad treatment of the bears by the public will end up compromising the lives of these animals. Some persons have amused themselves by passing a lasso around the neck of a young bear, by attracting it with bread. Fortunately, the cord broke when they wanted to choke the animal, but

> the skin of the neck of the poor beast was seriously cut, and it was only on exposing themselves to the danger of being injured themselves that the staff of the menagerie succeeded in disencumbering the animal from the rest of the broken cord.

While asserting that further abuses should be severely condemned, the superintendent ruefully added, "the menagerie has no means of preventing them." The administration offered no solution except to send a letter to the commander of the guard calling for more active surveillance at the bear pits.[43]

In November 1837 the Museum purchased a polar bear from a traveling menagerie. Frédéric Cuvier installed it in one of the bear pits but feared it would suffer the public's mistreatment. He urged the administration to forbid the practice of dangling bread on the end of a line, noting a malevolent turn the practice had taken: "One suspends the bread on fishhooks, with curved iron pieces, to enjoy the animal's difficulty when it is caught. The bears have already been wounded often by this cruel game, which could compromise the life of the polar bear."[44] Despite the ferocity that Martin and other of the menagerie's bears sometimes exhibited, they had as much, or more, to fear from the public as the other way around. While some visitors fed or encouraged them to perform, others enjoyed torturing them.

Human dominance over wild animals was confidently flaunted at the bear pits, but there was always a potential for other outcomes. In the case of the octogenarian named Bernard and the bear named Martin, a man delivered himself to a bear and the bear proceeded to have a bear's way with the man's body.

[2.12]

The Seal

As correspondent of the Museum of Natural History, Louis Baillon was as enterprising as his late father Emmanuel, delivering a steady supply of specimens from the Breton coast. He mostly sent birds, but in February 1808 he loaded the cadaver of "a very beautiful seal" on the stagecoach to Paris, proudly reporting that it weighed eighty pounds and he had "never seen such a big one."[1] Baillon paid twenty-four francs for the dead animal and worried the Museum might think it too expensive, but he explained that he was keen to outbid a fur merchant prepared to pay twenty francs, and thought that some of the expense might be recouped. The oil that could be rendered from the seal's flesh was worth fifteen to eighteen francs, and the skin was worth four more.

Baillon would have preferred to send the Museum a live seal, and he must have communicated this to his network of hunters, judging from his success in 1809 in procuring three live young ones in three and a half months, all much more expensive than the dead seal. The first, which arrived in late June, cost 150 francs.[2] The second arrived several weeks later, and both were still alive when the third arrived in mid-October. But this seal survived just a week, and by mid-November the other two were dead as well.[3]

However brief their time at the menagerie, the seals of 1809 had a greater impact on the history of science than any other animal that lived there. When the first seal arrived, not only did it seem to afford Lamarck new evidence for his theory of how species change, it triggered his articulation of a significantly new version of his understanding of the path nature had followed in bringing into existence all the different forms of life. Specifically, it brought him to *branching* evolution. Frédéric Cuvier set off in a different direction. He promoted the menagerie's new seals as central exhibits for the claim that animal intelligence depends more on the brain than on sensory acuity. He hoped that studying the seals' mental faculties would help

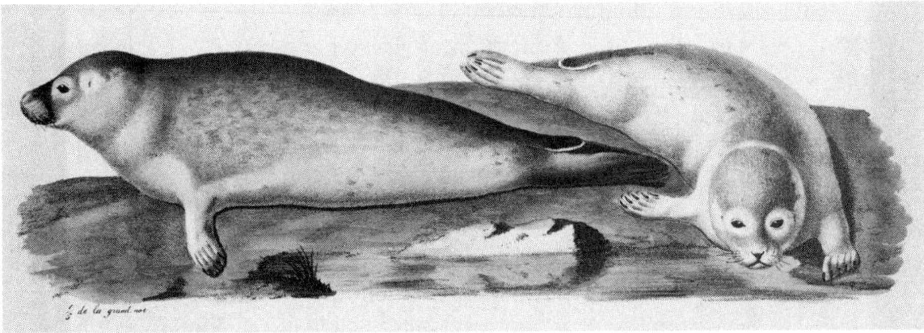

Figure 2.12.1. "The common seal." C. de Lasteyrie after Jean-Charles Werner. From É. Geoffroy Saint-Hilaire and F. Cuvier, *Histoire naturelle des mammifères*. General Research Division, the New York Public Library.

him create the foundation of a whole new science: comparative animal psychology.

∴

Geoffroy Saint-Hilaire announced the arrival of the menagerie's first seal at the professors' weekly assembly of 28 June 1809. Lamarck was present, and at that time he was putting the finishing touches on what became his most famous work, *Philosophie zoologique*. It may have been Geoffroy's announcement of the seal's arrival that prompted Lamarck to pay it a visit. On seeing it, he took it to be an important new example of how changes in habit lead to changes in structure. He swiftly composed a set of "additions" that he inserted at the end of his *Philosophie zoologique* just as it was going to press.[4] He presented a copy to his colleagues at the administrative assembly of 16 August 1809.[5] His observations on the menagerie's seal featured prominently in the fifteen-page section of additions at the back of volume 2.

Lamarck was especially interested in how the seal used its feet and limbs, especially its extremely agile hind feet/flippers, which were very short and clearly separate from its tiny tail. When swimming, the seal joined its hind limbs together and spread the digits and membranes of its feet/flippers to form a vertical paddle that, when moved from side to side, propelled the animal through the water. Critical for Lamarck's account was his claim that seals use their hind feet for other purposes; "they can even seize objects," he reported, "like true hands." The seal used these posterior "hands" as well as its mouth, said Lamarck, to seize prey. It used its anterior hands not to seize prey but instead to tear apart prey it held in its mouth as well as for moving in water.

Lamarck found it highly constructive to compare the seal's habits and structures with those of the walrus, or at least what he imagined the walrus's to be, for neither the menagerie nor the zoological gallery had any walrus specimens. He maintained (incorrectly) that the walrus's hind limbs, hind feet, and tail are joined to form a caudal fin. He further claimed (also incorrectly) that walruses feed on plants.[6] By Lamarck's account, if seals had used their hind limbs and feet exclusively for swimming, they would have become fused into a single fin like that of the walrus. This had not happened, he explained, because seals use their hind feet for grasping prey as well as for swimming. Differences in habit had led to the differences in structure exhibited by the two forms. The seal offered "new proof of the power of habit over the form and state of organs."[7]

Lamarck's contrast of the seal and the walrus was flawed. Walruses bring their hind flippers together to serve as a caudal fin when swimming, but the flippers are not physically fused with the tail in a single structure. As for his claim that the seal uses its hind feet to seize prey, the menagerie's specimen may perhaps have done so, but observers of the common harbor seal have not identified this behavior as characteristic. Frédéric Cuvier reported that the seals used their hind flippers only for swimming. That flippers can also help regulate body heat, like an elephant's ears, was not appreciated in his day.[8]

In his last-minute additions to the *Philosophie zoologique* Lamarck did more than present the seal as a prime example of the influence of habit on form, he awarded it a special place in mammalian evolutionary history. As he portrayed that history, one branch of the reptiles gave rise to "those aquatic mammals we call amphibians," while another branch gave rise to the birds. Then, from adopting different habits, the "amphibian mammals" split into three types: the cetaceans, the ungulates (distinguished by hooves), and the unguiculates (distinguished by nails, claws, or talons). According to Lamarck, the cetaceans were the amphibian mammals that "contracted the habit of never leaving the water, only coming to breathe at the surface." The amphibian mammals that went ashore divided according to what they ate:

> Some among them, becoming accustomed to browse on plants, such as walruses and sea cows, led little by little to the formation of the ungulate mammals, such as the pachyderms, the ruminants, etc. The others, such as the seals, contracting the habit of nourishing themselves from fish and marine animals, led to the existence of the unguiculate mammals, by the means of races which, in diversifying themselves, became entirely terrestrial.

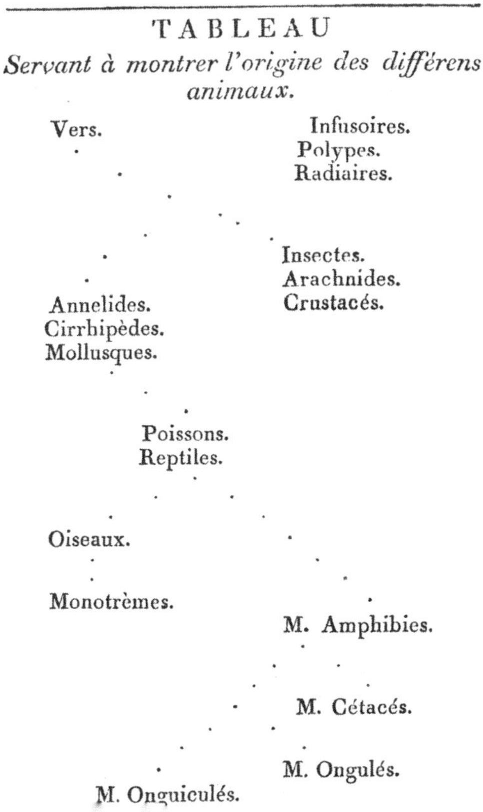

Figure 2.12.2. Lamarck's 1809 diagram briefly outlining his view of how the different major forms of life developed over time.

He acknowledged that these ideas might "seem only to be simple conjectures," but he was confident they would come to be recognized as having "a probability of the highest degree" if one paid close attention to his arguments and to "the habits and environments of the animals in question."[9]

Lamarck offered a simple sketch of the evolutionary relationships he had in mind.[10] However simplistic and problematic, it was the first evolutionary tree of the animal kingdom ever drawn. It also represented an important advance over his previous statements about the path nature had followed in bringing all the different forms of life into existence, insofar as it featured *branching.*

Before this moment, Lamarck had portrayed nature's basic path as a steady, straight-line increase in the anatomical complexity of the various animal classes. At the same time, he acknowledged that species (and the other taxonomic levels below the classes) could not be arranged in a single scale of being, attributing their "deviations" from the general order of progression to the influence of diverse environments. As he identified new invertebrate classes, he continued to arrange them in a single series. This was the case in volume 1 of his *Philosophie zoologique*, where he presented a list of ten invertebrate and four vertebrate classes ranked in order of complexity.

The diagram in the additions at the end of volume 2 was a very different matter. No longer a simple ranked list, it has zigzags and branches. The reptiles give rise to the birds and mammals separately. The cetaceans split off from the "amphibian mammals," which split again into two terrestrial forms, the hooved mammals and the nailed mammals. Also included in the diagram is the radical suggestion that animal life originated via two distinct series, beginning on the one hand with the infusorians and on the other with the intestinal worms. (The simplest forms in each of these branches, he maintained, arose through spontaneous generation.) The diagram additionally shows the intestinal worms giving rise to two branches, one leading to the insects, arachnids, and crustaceans, the other leading to the annelids, the cirripeds, and the mollusks.

Stephen Jay Gould traces the origins of Lamarck's branching view of evolution to Georges Cuvier's identification of a fundamental difference in the internal organization of parasitic worms versus annelid worms, announced by Cuvier in a memoir read at the Institut de France on 11 nivôse an 10 (1 January 1802), which Lamarck attended.[11] Historians have long appreciated that the structural differences between parasitic and annelid worms were what ultimately led Lamarck to conclude in 1809, in the *Philosophie zoologique*'s "additions," that these creatures represented separate branches in the animal scale. In elaborating on this, Gould makes the additional assumption that this conclusion preceded Lamarck's recognition of the other branching scenarios presented in the additions. As Gould puts it, "Once the mental floodgates open, the tide of reform must sweep to other locales." Lamarck, in Gould's interpretation, next claimed that reptiles branched into birds on the one hand and mammals on the other. "Finally, and still in the new spirit," Gould writes, "[Lamarck] even posited a three-fold branching in the transition to terrestrial mammals."[12]

The evidence, however, indicates that "the tide of reform" in Lamarck's thinking swept in the opposite direction, and that it was the serendipitous appearance of a seal at the menagerie that opened "the mental floodgates" for the French zoologist. When Lamarck wrote the main text of the *Philosophie zoologique*, he was still ordering the animal classes in a single series of decreasing complexity. Though he had been mulling the distinction between intestinal and annelid worms for six and a half years and put them in different classes, he was yet to put them in different series. There is no reason to think that further mulling was sufficient on its own to lead him, early in the summer of 1809, to his insight about branching.

All Gould offers on this score is that "honesty eventually trumped hope" (i.e., Lamarck finally admitted that his idea of a single series had to be abandoned).[13] Lamarck's own testimony makes it clear, however, that a contingent event, the arrival of the seal, was instrumental in developing his thinking. His introduction to the "additions" states unambiguously what inspired his writing: "In the last days of June 1809, the menagerie of the Museum of Natural History, having received a seal . . . , I had the occasion to observe the movements and the habits of this animal."[14] These observations led to Lamarck's reflections on the differences between seals and walruses, and from there to branching, with the "amphibious mammals" giving rise to whales and two divisions of terrestrial animals, those with hooves and those with nails. It bears further note that while Lamarck portrayed the infusorians and intestinal worms as different branches in the animal scale, that does not constitute branching in the sense of one lineage splitting into two. In Lamarck's diagram, branching is portrayed in the splitting of the intestinal worms (into annelids and insects), the reptiles splitting (into birds and amphibious mammals), and the amphibious mammals (as just described).

∴

Like Lamarck, Frédéric Cuvier published a book in August 1809, a new edition of his menagerie guide, but unlike Lamarck's it had no last-minute updates. Seals are not in it, no doubt because he composed it before they arrived.[15] It did include a strong pitch about the menagerie's utility for scientific research. Countering the notion that menageries were "useless for a knowledge of the most important phenomena of the animal kingdom: the habits, the industry, the character, in a word, the intellectual faculties in general," Cuvier touted the menagerie as a place where one could experiment on animals and in effect "compel them to offer us observations." This, he wrote, would allow the naturalist "to understand, explain, and connect the observations that animals in liberty present to us."[16]

Experimentation played but a small and imperfect role in Cuvier's study of the three seals that lived at the menagerie between late June and mid-November 1809. Two years later, anatomical and other considerations were at the forefront of the memoir on seals that he published in the Museum's *Annales*.[17] The piece treated the seal's physical characteristics, the functioning of its sense organs, and its mental faculties.

Displaying his zoological skills, Cuvier began by explaining why, despite the three seals' differences in coloration, he regarded them as members of the same species, the common seal, *Phoca vitulina*. Next he focused on the acuity of the seal's senses, concluding that although its vision, hearing, smell, taste, and touch were relatively weak, the seal derived from its sensations "results infinitely superior to those obtained by animals seemingly better organized." Cuvier concluded that sensory and intellectual acuity do not necessarily go hand in hand, and that the brain, not the senses, is the principal source of ideas.[18]

Despite his remarks about the menagerie as a fine site for conducting experiments, the experimenting Cuvier reported was minimal and unpersuasive. He attempted to see if seals could learn by association the names he chose to call them: "I had the habit, each time I gave a fish to one of the seals, to call it by a name; but when it did not see me, this name did not recall to it the presence of its food." All he thought he could legitimately deduce was that the seal's hearing was relatively weak, a point he had already confirmed based on its auditory anatomy. Cuvier was confident his results would have differed had he been able to continue the experiment longer, given that others (Pliny) had taught a seal to answer to its name.[19]

Among the difficulties of caring for seals was catering to their finicky appetites. Cuvier reported, "I have never been able to make the animals I observed eat anything other than the species of fish with which it was first fed. One wanted only herrings and the other only sole. The first even preferred salted herrings to other fresh fish, and the second starved in the absence of sole because of the seasonal storms that suspended fishing temporarily." The force of habit was also displayed in the place seals chose to eat their food. One always took its fish to the bottom of its pond; the other "only wanted to eat on land."[20]

Cuvier likewise stressed the importance of habit in his discussion of the seals' mental faculties or "intellectual functions." Addressing the question why seals in some regions did not fear humans, while those elsewhere fled when they approached, he explained this as a case of acquired habit. Seals that had been hunted by humans learned to fear them, a response passed down by heredity that seals in areas untouched by humans had not acquired. Cuvier felt this proved "in an incontestable manner that intellectual faculties developed accidentally are transmitted over time by generation and become hereditary."[21] Biologists today associate the idea of inheriting acquired characteristics with Lamarck and his theory of evolution, but Frédéric Cuvier firmly advocated it, albeit without embracing the idea of species change.[22]

Cuvier attributed various features of his seals' behavior to the fact that in nature seals live in groups and are highly social. Animals that "need to live in society," he explained, are much more amenable to training than those living in relative solitude. It did not surprise him that one of the menagerie's seals lost its fear of being petted in just a few days. This same seal lived contentedly in the company of two small dogs and did not object when they jumped on it, barked at it, or even bit it. Its response was to cuff the dogs with its feet (an action Cuvier interpreted as encouragement of the dogs' behavior). If the dogs ran away, the seal chased them down a stony, muddy path, notwithstanding its difficulty in doing so. In cold weather the animals "slept very close together to keep warm."

Cuvier concluded his study by acknowledging its incompleteness. His observations had been limited, he said, by his seals' extreme youth, the difficulty in feeding them, and "the impossibility of making them live long enough for sustained experiments." What he presented were "some preliminary notes" to serve as the basis for new research later.[23] Interestingly, Frédéric Cuvier said nothing in his memoir about Lamarck's observations on the menagerie's first seal, perhaps because they hadn't come to his attention, or perhaps because he knew his brother Georges viewed theories such as Lamarck's unworthy of public consideration.

Cuvier soon had the chance to observe a different species of seal, a monk seal (*Phoca monachus*), which he believed had never been described.[24] He saw the seal, a female, on display in Paris in 1813. Captured in March 1811 on the Dalmatian coast of the Adriatic and taken on tour, she was astonishingly still alive and healthy, despite severely constrained living conditions. She occupied a tank no more than a foot longer and two feet wider than her body, which measured seven to eight feet from the tip of her muzzle to the end of her back flippers. Every morning her owner fed her several pounds of river fish and gave her ten inches of clean water in which to lie for the next nine or ten hours. At day's end, the water was removed so she could be dry for the night.

Though evidently discomfited by the conditions of the seal's "enslavement," Cuvier took the occasion to focus on seals' intelligence, which he believed surpassed even that of dogs. Taming seals was no problem, he said; "they recognize those who feed them, attach to them, understand them, obey them, and seem in the end to forget entirely their former independence and make a second nature for themselves in their association with man." He saw this as proof that animals living in a social state in nature are most susceptible to domestication.[25]

The seal in question displayed great affection for her keeper, responding to his voice, seeking his caresses, turning side to side and opening her mouth at his request, "without ever showing fear or impatience." The animal did not even object to being touched by strangers.[26] Despite her friendliness, Cuvier dismissed this seal as a candidate for advancing his study of mental faculties. She slept through the night and had to be stimulated to awaken in the morning. Moreover, Cuvier admitted, her constrained state did not leave much room for her to exercise her will or intelligence.

Cuvier wrote about seals again in two essays he contributed to the illustrated, *Natural History of the Mammals* that began appearing in installments under his name and Geoffroy's in 1819.[27] The first (1819) was essentially a repeat of his seal paper of 1811. The second (1824), written primarily to reflect advances he and his brother had made in seal classification, included his thoughts after attending the performance of a trained seal. It was the kind of entertainment, he said, that animal exhibitors used "to excite the curiosity of the public" by the routines a trained animal would enact on command:

> It was told to get up on its hind feet, to take a baton between its front feet and stand like a sentinel, to lie down on its right side or its left side, to roll over, to present one or another paw, to bring its muzzle up to the face of its master or otherwise kiss him, and it conformed at once to the wish of he who spoke, who had trained it, and to which it seemed singularly attached.[28]

But however impressive the seal's performance, Cuvier explained, it was merely the result of simple associations developed through a long course of training, whereby a particular movement of the hand, or the sound of a voice, or something else entirely could serve as the sign to which the animal reacted. "[The sign] can even be so chosen or so calculated," he continued, "that it is perceptible only to the animal, and the spectator is not struck by it." At that point, the spectator was likely to go astray, Cuvier warned, and to admire the animal's intellect based on appearances rather than facts.

Alongside his research concerning animal behavior, Cuvier undertook taxonomic studies. Aimed primarily at the level of the genus rather than the species, he maintained that the most natural way of forming and arranging mammalian genera was to study the mammals' teeth. Focusing on tooth structure produced general groupings that could be subdivided successively by considering other characters relating to the sense organs, locomotion, and reproduction. Approaching animal organization this way, following rules whereby some organ systems were treated as less important than others, allowed one to obtain "a more natural and methodical classification of mammals than those accepted up to this day."[29] Cuvier published on the topic from 1807 to 1812, then returned to it in the 1820s, finally publishing in 1825 a whole book on teeth as diagnostic characters in classifying mammals.[30]

The year before his book appeared, Cuvier used the same approach in a long memoir on the classification of seals.[31] He proposed that instead of placing seals in a single genus with multiple species, they be reclassified as a distinct order with multiple genera and multiple species. Despite his confidence that tooth structure was the proper basis for pursuing a natural classification of seals, as he neared the end of his paper he made a pitch for studying living animals. The anatomical considerations he had relied on, he said, ultimately had to be coordinated with other kinds of information, notably about living seals. The various kinds of seals needed to be compared according not only to their physical organization but also to their dispositions, penchants, and faculties. These behaviors, he said, could likewise be used to sort them into genera and species, because differences in organs were always accompanied by differences in the animals' needs and their means of satisfying them. He offered hyenas as an example. Hyenas had long been classified with dogs, but their habits distinguished them from dogs just as much as do anatomical features like their digits or dentition.[32]

The problem for the naturalist, Cuvier explained, was that observing an animal's habits was "infinitely more difficult" than observing a dead animal's parts.[33] Animals in the wild were disposed to flee rather than submit to another being's will. Nevertheless, zoology needed, he said, to go beyond the study of dead animals to observe the whole, living animal and the functioning of its parts. In this way "animated nature will be offered to our eyes as she is in reality, not only rich in an innumerable variety of parts, in a mechanism that confuses our mind, but also powerful with life and intelligence."[34] Cuvier wished that the principles of studying animal behavior were better established.

If these were curious sentiments to find near the end of a long classificatory work depending almost exclusively on the consideration of seal teeth and other body parts, Cuvier, equally curiously, had nothing specific to suggest. He was not calling for studying seals in the wild, a project too impractical to contemplate, but neither did he mention the possibility of studying live animals in a menagerie, perhaps because the menagerie had not had a seal since 1809.

The menagerie did not acquire another seal until 1830. Purchased for 100 francs, it arrived at the menagerie on 29 August and died on 25 September.[35] Five years passed before two new seals, captured near Hamburg and much larger than their predecessors, were added to the collection. Their owner brought them to Paris with the wildly unrealistic hope that the Museum would pay 5,000 francs for them. Frédéric Cuvier was away for the summer in the service of the Université de France, and his stand-in at the menagerie,

Geoffroy Saint-Hilaire's son, Isidore, told the administration that the seals' owner was prepared to drop his price to 1,000 francs, or half of that, or possibly even less.[36]

Whatever the final price, the seals were added to the menagerie on 1 August 1835. Ten days later Cuvier, back in Paris, reported that they were unwell: "The seals, which are not eating and which the public overwhelms with stones, have been transported to the pond in the pheasantry, where they can be treated with more care."[37] Whatever care they received was insufficient; one died on 21 August, the other a week later.[38]

The following summer the menagerie acquired yet another seal, purchased on the recommendation of Geoffroy Saint-Hilaire. It arrived on 22 June 1836 and became the responsibility of Isidore, again standing in for Cuvier as menagerie superintendent. Isidore informed the administration: "The seal has been placed provisionally in the pheasantry, where the pond contains an amount of water large enough for it to swim easily and small enough to be renewed frequently. [The water] has already been changed twice this week, and will be tonight for a third time, and always without harm or trouble for the animal, because the change takes place during the night when the seal has the habit of sleeping in its box."

The problem with this arrangement, Isidore acknowledged, was that while it suited the seal's needs, the waterbirds suffered, since their laying, incubation, and raising of young were disrupted by all the seals' visitors. The best solution, he believed, would be to create a new pond for the seal in one of the existing parks, but sufficiently removed from the public, because otherwise the visitors would "not fail to kill this seal like all the others, by blows from stones thrown with the purpose of making it move." The administration named a committee to study the question; but if any action was taken, it did not come soon enough. The seal died on August 14.[39]

However few and short-lived, seals loomed large in the history of the menagerie. They were of greater theoretical consequence for the Museum's zoologists than any other animal. The first to arrive in 1809 inspired Lamarck to articulate a critical addition to his understanding of how life forms developed over time not simply by evolving into new forms, but by branching into different types. For the first time he pictured lineages bifurcating and sketched his first tree of life. The seals of 1809 served a different but important function in Frédéric Cuvier's theorizing. He used them to argue that animal intelligence depends more on the brain than on the senses.

But seals also serve as prime examples of how hard it was to keep animals alive in captivity in Paris. This was partly due to the difficulty of providing clean water and fresh fish to their liking, but also owing to the public's cruelty, despite the creatures' attractiveness and peaceful demeanor. An 1837 guide to the Museum recalled the seal of 1836 as a "very gentle" creature that "knew well its keeper, to whom it even gave kisses."[40] Far from their natural habitats, deprived of their normal nourishment, the seals of the menagerie were hastened to their deaths by visitors' demands for entertainment.

[2.13]

Prussians at the Gates

As a state institution, the Museum of Natural History illuminated its facades whenever the minister of the interior ordered it. Lucien Bonaparte did so in June 1800 to honor "the memorable victory won by the French commanded by the First Consul over the Austrians at Marengo in Italy."[1] The buildings were lit again two weeks later to celebrate Napoleon's return to Paris.[2] More displays followed over the years: to celebrate Napoleon's birthday, his 1810 marriage, the birth of his son the next year, his military victories, and more. In his last full year as emperor (1813), there were at least five such occasions: his birthday (15 August); a joint commemoration of his coronation and his victory at the battle of Austerlitz (celebrated 2 December); and celebrations on 19 May, 13 June, and 19 September for victories at the battles of Lützen, Bautzen, and Dresden. As part of these festivities, the *Te Deum* was sung at Notre Dame in the morning, and government buildings and monuments, including the Museum of Natural History, were lit up at night.

By 1813, however, public enthusiasm for such displays was hard to muster. The costs of war and the British blockade of French commerce had brought severe economic hardship to the country. Crop failures added to the misery. The French were war weary even before their armies were decimated in the disastrous Russian campaign of 1812. The victories at Lützen, Bautzen, and Dresden only delayed the inevitable. In October 1813 the combined forces of Russia, Prussia, and Austria dealt the French troops a crushing blow at Leipzig, a loss the French army could not recover from.

The effects of war were palpable at the Museum: budgetary constraints and uncertainties, conscription of staff members, a greatly reduced student body, and few if any specimens arriving from overseas. In 1810 the government trimmed the Museum's budget from 300,000 francs to 275,000, an amount that had to cover salaries of professors, staff, and day workers plus the escalating costs of maintaining the gardens, galleries, and menagerie. The crimped budget was stretched thinner by unexpected events like a violent hailstorm in 1811 that damaged roofs and broke some two thousand

Figure 2.13.1. Poster written in French on the left and German on the right with Prussian Commander von Pfuel's 9 July 1815 order to Prussian soldiers not to touch any objects or to approach or irritate the animals of the menagerie. Archives nationales. AN AJ/15/611 (12 July 1815). Photo R. Burkhardt.

panes of glass in the greenhouses, galleries, amphitheater, and other buildings.[3]

When the Museum overran its 1811 budget by 12,626 francs, the administration reported to the minister of the interior that menagerie expenses constituted most of it: 3,000 francs from increased food costs and an additional 7,327 francs for repairs, including reinforcing a barrier to contain an increasingly uncooperative elephant.[4] Meat purchases were a significant expense because less was being seized in the markets, which meant more had to come from butchers. Once again the Museum fed superfluous herbivores to the carnivores (in this case five sheep, four rams, eight goats, and five water buffalo) to help cut costs.[5]

Losing staff to conscription was another feature of the Museum's wartime experience. In 1811, twenty-three-year-old Emmanuel Rousseau, who had assisted his father, Pierre Rousseau, in Georges Cuvier's comparative anatomy lab, was conscripted as a military surgeon. He later was a prisoner in Russia.[6] In 1813 a call for 100,000 new soldiers jeopardized two key assistants in the zoology lab, nineteen-year-old Florent Prévost and twenty-six-year-old Pierre-Antoine Delalande. The Museum succeeded in retaining both, arguing that their skills were essential to the institution and that anyway they were unfit for military service. Prévost was too short, and Delalande was subject to stomach pains and vomiting because of the poisons he'd used to preserve specimens. But the Museum did not try to prevent the conscription of Bernard Louzardy's son, Marie Joseph, who had substituted for his father at the menagerie at times his father was ill. The young man's services, the Museum acknowledged, were not "absolutely necessary."[7]

War had its way with course attendance at the Museum. Only 800 students registered in 1813, whereas some 2,400 could be expected in peacetime, a third of them from abroad.[8] With travel impeded, so too was the acquisition of specimens from overseas. The last appreciable batch arrived in 1809 (a mother and baby mouflon from Corsica; a lioness from Algeria; a lion, lioness, and hyena from Morocco; and a lion and lioness from Senegal). An additional lioness arrived from Tunis in 1810, but nothing more came from foreign shores until months after Paris fell to the allies and freedom of the seas was restored. Meanwhile, new animals came from the provinces (especially from Baillon fils in Brittany); from births at the menagerie; and from gifts, purchases, and exchanges. The most significant influx from outside France came from the royal menagerie in Amsterdam, where Napoleon in July 1810 confiscated all the animals the Museum wanted (a lion, a tiger, a panther, a zebra, six porcupines, two raccoons, a white hare, and eleven monkeys), leaving only an ordinary wolf and an American black bear. The convoy from Amsterdam that reached Paris on 3 November 1810 brought twenty-three mammals and thirty-three birds.[9]

Paris fell on 31 March 1814. Czar Alexander I of Russia rode into the capital on horseback preceded by a guard of Cossacks. Karl Philipp, prince of Schwarzenberg of Austria rode on Alexander's left and King Friedrich Wilhelm III of Prussia rode on his right. By the time they reached the Tuileries, crowds were cheering enthusiastically. Alexander promised to preserve the city, indicating that the peace of Europe was his greatest desire and only Napoleon was his enemy. The conquerors' alacrity in visiting the Museum is testimony to its prominence on the Parisian cultural horizon. Just two days after the allies' triumphant arrival, Wilhelm III, his sons, and a nephew came to view the collections. Four days later Czar Alexander toured the menagerie, expressed his satisfaction, and said he would return to see more of the establishment.[10]

Although flattered by this royal attention, the professors worried about the future of their collections, which included specimens confiscated from other defeated countries. On the eve of Paris's occupation, the Museum wrote to the commander of the allied forces emphasizing that the collection was an integral whole, of interest to "all the scientists of Europe," and asking him to provide "all the security and safeguards necessary for its conservation."[11] The Museum also feared it might become a site for quartering allied troops. Recognizing that personal intervention might be critical in this regard, Georges Cuvier appealed to Alexander von Humboldt.[12] The famous naturalist immediately went to see General Alexander Wilhelm von der Goltz, the Allied commander of the city, who obliged him and the Museum by sparing the institution such an intrusion.[13]

The Museum was quick to seek the good graces of France's new governors. Two weeks after the fall of Paris, the professors wrote to Talleyrand, the president of the new provisional government, thanking that body for everything it had done to save the capital and the institutions situated there. They expressed their "particular joy" in anticipating the coming peace that was "so necessary for the sciences."[14] But life at the Museum was not peaceful. The institution had dodged the threat of foreign troops camped on its grounds, but foreign visitors came in droves to see the collections. The mix of royalty, military commanders, soldiers, and other sightseers caused extra work for staff.[15]

On 20 April 1814, the emperor of Austria spent three hours at the Museum. A week later, Czar Alexander was back. The director and professors met him ceremoniously at the main stairway of the natural history galleries. He received a tour of the galleries and was presented with a complete twenty-volume set of the Museum's *Annales*. Afterward, he expressed his satisfaction on seeing collections "so rich, so well ordered, and consecrated so liberally to the instruction of the whole of Europe."[16]

Meanwhile, the occupation posed a fodder problem for the menagerie. To ensure the subsistence of occupying troops, farmers outside Paris were ordered not to transport supplies from their granaries without permission. It took over a month before the farmer who supplied the menagerie's fodder could resume deliveries.[17]

Approximately a month after the allied troops entered Paris, Louis XVIII, the new king of France, made his entry to the city. Judging from contemporary accounts, the public greeted him less enthusiastically than they had greeted Czar Alexander, but a *Te Deum* was chanted at Notre Dame in the morning, and state sponsored edifices, including the Museum, were lit up at night. Keen to secure the monarch's good graces, the professors obtained an audience through the new minister of the interior on May 17 at the Tuileries.[18] Thouin, as Museum director, expressed his and his colleagues' happiness at the king's return. Underscoring the institution's royal origins, Thouin invoked the memory of Buffon, humbly allowing that "we have not inherited Buffon's talents, but his spirit animates us. We are imbued with the same ardor for the discovery of useful truths, with the same devotion for the king and for the state."[19]

The king thanked the professors for their sentiments, and his niece, the Duchess of Angoulême, permitted them to render her homage. After Thouin expressed the Museum's profound respect and devotion, he invited her to visit the institution. She accepted with pleasure and reminded him that he had given her a guided tour twenty-two years before.[20] Although her cordial response surely pleased the professors, it also made them anxious about the impression the Museum might make on Her Royal Highness were she to visit immediately. The animal keepers' misbehavior was their main concern. Antoine, the elephant keeper, had lately taken a bribe from a French army officer who wanted to be let into the elephant's park. The officer suffered a broken jaw and shoulder when the elephant jammed him against the bars of the enclosure.

The administration decided to post signs warning visitors that tipping the staff was forbidden and, more specifically, that animal keepers must not allow anyone inside the barriers separating the public from the cages or the animal parks. Meanwhile the keepers were letting their chickens run free in the animal parks and hanging their linens out to dry on the fences. The Museum forbade both practices.[21] The chickens and the washing were swiftly made to vanish, judging by the "satisfaction" the duchess expressed regarding her visit to the menagerie and other parts of the Museum several days later. In the same week, a small delegation of professors visited the new minister of the interior and returned with his assurances that the Museum would be "conserved in its integrity."[22]

By June's end the allied troops had departed, and life at the Museum began settling down, with the professors teaching their usual summer courses. Director Thouin, as was his custom, met his class at 6:00 a.m. on Tuesdays, Thursdays, and Saturdays. Furthermore, the rotunda for the "peaceful animals" was at last declared ready for occupancy. But what captured the professors' attention most in the summer and fall of 1814 was the peacetime restoration of open seas, enabling access to specimens from distant lands.

The initiatives to reconnect with the colonies and other foreign lands began in earnest in August 1814. When the new intendant general of French possessions in India asked how to make the French presence in India useful to science, the professors decided to prepare a standard set of instructions for voyagers.[23] Plans were also under way to reoccupy Senegal, where the Museum considered sending a naturalist.[24] In addition, the Museum heard from Félix Louis L'Herminier, director of the botanic garden at Guadeloupe; from Lieutenant General Douzelot in Martinique; and from the naturalist Joseph Martin, who requested hunting gear to collect specimens in Cayenne. A voyager naturalist named Patoire, back from the East Indies, Isle de France, and the Cape of Good Hope, sent the Museum 222 diverse specimens and requested some in return. The minister of the navy and colonies found a man on his way to Île Bourbon willing to undertake research on the Museum's behalf, and the English governor of the Cape of Good Hope visited the Museum and promised to send live animals.[25]

There was special cause for excitement in January 1815 when Captain Pierre-Bernard Milius returned from Guadaloupe and Martinique. He had sailed from France to the Antilles in August 1814 as an agent of the state to officially take Guadaloupe and Martinique back from Britain, in accordance with the Treaty of Paris (30 May 1814). From his earlier experience with the Baudin expedition he knew the great importance that natural history specimens held for the Museum and the nation. Full of pride he announced his return: "*Le Lys*, under my command, arrives from the Antilles [with] its hold filled with the richest, most beautiful collection of plants ever introduced to France or Europe; that which I brought back on *Le Géographe* cannot be compared with it."[26]

But Milius's return to France was eclipsed by Napoleon's. Exiled to the island of Elba by the Paris Treaty of 1814, Napoleon regained French soil on 1 March 1815. Accompanied by an ever-swelling troop of soldiers, he marched from Cannes to Paris, arriving on 20 March, just hours after Louis XVIII fled. So began "the Hundred Days," during which Napoleon sought to rally popular support while the Allies gathered forces with the intention of defeating him definitively.

After two and a half weeks back in Paris, and notwithstanding his other concerns, the emperor visited the Museum of Natural History for two hours and expressed "the intention of having new greenhouses built." The new minister of the interior, Lazare Carnot, then invited the professors to discuss the Museum's current state and its construction needs.[27] Jussieu prepared the response. He identified new quarters for the *animaux féroces* as the top priority, but the Museum also needed more greenhouses, further additions to the galleries, and a larger acquisitions budget for the library so that it, like the library of the University of Göttingen, could be "always up to date with all the discoveries."[28] Geoffroy prepared a separate document on the menagerie's history and future.[29]

The busy spring of 1815 saw Parkie, the female elephant, transferred from her old quarters to the new rotunda. As superintendent, Frédéric Cuvier relied on a strategy sketched out the previous summer to overcome her reluctance to leave her old quarters.[30] Early in April, the new cage and wagon that Lassaigne had built were placed near the door of Parkie's stable so she would grow accustomed to seeing them.[31] A month later, with both its ends open, the cage was positioned so she had to walk through it as she passed from her shelter to her park and back.[32] Two weeks later still, and once again the week after that, she allowed herself to be closed up in the cage without showing signs of distress.[33] Finally, on May 26, 1815, she was delivered "in very good health" to her new abode in the rotunda. According to the architect Molinos, she seemed pleased with the arrangement. The workers from the garden received a bonus of a franc and a half each for their help.[34]

The elephant's relocation raised new issues. For one, the rotunda was too big to heat properly in winter. A smaller enclosure within the building would have to be prepared. The trickier question was what to do with the park attached to her previous quarters, because it led to a territorial dispute between Frédéric Cuvier and André Laugier, the professor of chemistry. It was not of the greatest moment for Museum business in general, but to Cuvier, the scientific future of the menagerie, and by implication his own scientific future, hung in the balance.

At the professors' meeting of 24 May 1815, anticipating the elephant's relocation, Professor Laugier asked to use part of the elephant's old park for a private garden. It was not an unusual request (other professors had gardens), and the assembly granted Laugier's wish.[35] Appalled by the decision, Cuvier penned a five-page letter to the professors, beginning with the reminder, "When you created your menagerie, you felt it would be of great utility to zoology through the experiments to which animals could be subjected." As things stood, he said, zoologists' knowledge of animals remained "very imperfect." There were some "notions" about animal habits, but "nearly nothing" had been done to study their intellectual faculties. Aware that these questions might not be of primary interest to his colleagues, he focused on one that was: the species concept. Zoologists had acquired very little information about the extent to which animals could be modified, either physically or in their mental faculties, but only by establishing the limits of such modifications could one bring clarity and precision to species classification.[36]

How could one establish these limits when nature only rarely showed what they were? No single animal could represent the unadulterated type of its species because no animal developed independent of its modifying circumstances. One could learn the functional limits of an animal's bodily organs and "organs of intelligence" only by experimenting. The elephant's former park had been destined for such experiments, Cuvier said, and he was claiming it. There was no comparable place in the menagerie, in that it had southern exposure, was large enough and far enough from any habitation, had a pool, and could be joined to the former stables for the camels. Cuvier was convinced that if this opportunity was lost, the menagerie would never have a place devoted to the kinds of experiments the science of animals required.

Cuvier apologized for making a request that ran contrary to Laugier's, noting diplomatically that the professor had undoubtedly not realized that the elephant's park had already been designated for scientific rather than personal purposes. Sharing the park would not work, Cuvier said, unless Laugier's portion was made entirely separate from the experimental site. Experimentation required removing all extraneous influences, making experiments on animals "very delicate."

Rarely did Frédéric Cuvier make remarks of a personal sort in his communications to the administration, but now he did. Uncharitable comments had evidently been circulating with respect to his plans for the park. Wrote Cuvier, "It is said that if this locale was accorded [to me], far from using it for its proper purpose, I would amuse myself by making it a garden. I confess I would scarcely have expected that my conduct in this establishment since having had the honor of being admitted to it, would attract such an accusation." For seven years, he said, his living quarters had consisted of a bedroom in his brother's apartment. For the past year, his office had been a room so humid and unhealthy that it had made one of the animal keepers ill and had then been abandoned. Yet Cuvier had not complained. "Is this then," Cuvier asked, "the conduct of a man anxious to make his lodgings comfortable, pleasant? No Messieurs, I happily do not have time to amuse myself caring for a garden. I have occupations that better suit my tastes. My time is divided between that of the university and the study of my animals, and this study leads me, I assure you, much more to change gardens into menageries than menageries into gardens."[37]

In conclusion, Cuvier wondered whether the professorial assembly doubted his abilities: "Perhaps you fear that I lack the talent necessary for making observations. I have nothing to oppose this doubt. If it exists, it proves that the memoirs I have had the honor to submit have not received your approval. But in this case, gentlemen, it remains for you to consider the thing in itself; it can be useful independent of my person."[38]

Cuvier's letter is preserved with the papers related to the 31 May 1815 assembly, but the meeting's minutes do not indicate that it was read there. Two weeks later, the assembly officially invited him to suggest what to do with the former elephant park.[39] Cuvier's seven-page proposal was presented at the professorial assembly of 21 June 1815 and referred to Geoffroy Saint-Hilaire. Then events of a different order intervened promptly and dramatically. The assembly adjourned at 7:30 p.m. Later that night, Napoleon arrived in the capital, as did the news that he had suffered a decisive defeat at the Battle of Waterloo in Belgium three days earlier. Napoleon abdicated on 22 June. Within two weeks, 150,000 allied troops occupied Paris.

The Museum was greatly unsettled. Staff members stood guard.[40] Their inability to speak German complicated efforts to prevent Prussian soldiers from damaging the property or putting themselves in danger by exciting the animals. A group of soldiers tried to force their way into the elephant's quarters. Had they succeeded, the professors supposed, she would have crushed them. In the face of such difficulties, they again called on Alexander von Humboldt, who went straight to the commander of the Prussian troops, General von Pfuel, and secured an order that was posted in French and German on the gates of the establishment.[41] Prussian soldiers were commanded not to touch any objects at the Museum, so as not to damage them, and not to approach or irritate the animals, which could be very dangerous.[42] This order did not stop two thousand Prussian soldiers from showing up at the Museum gates two days later demanding to bivouac on the grounds. Yet another intervention by Humboldt saved the day.[43]

As happened during the 1814 occupation, the menagerie experienced shortages of fodder and straw, and the Museum had to appeal to the appropriate allied commanders to allow the farmers' wagons to reach Paris with a military escort.[44] The Museum's greatest anxiety was that it might have to surrender natural history specimens taken from other countries over the previous two decades of war. The Allies, having had to fight Napoleon a second time, were feeling less magnanimous toward France than they had been the year before. Art masterpieces may have been the most conspicuous items the previous owners wanted back, but natural history specimens were also at issue.

The earliest recovery initiatives merely called for exchanges between foreign museums and Paris. In mid-July, after a three-hour tour of the menagerie and greenhouses, the Austrian emperor ordered one of his botanists to come from Vienna with a catalog of the specimens in the imperial collection so that the Museum's botanists could choose from duplicates there and identify duplicates in the French collection to offer to Austria. A naturalist representing Prussia then asked specifically for the repatriation of certain specimens that the French had confiscated in Cologne in 1794. In addition, the Museum granted the request of the Russian crown princes, Nicholas and Michel, for seeds and bulbs for their mother, the empress (though there were no issues of repatriation in this case)

The professors took these relatively minor requests in stride. What sent them into shock (and two emergency administrative sessions) was a letter from the Dutch naturalist Sebald Justinus Brugmans, acting as special commissioner for William of Orange, king of the Low Countries, demanding the return of the Stadtholder collection, the treasure trove of specimens France had confiscated in 1796. Brugmans, who had been a disgruntled observer of the confiscation, indicated that while his instructions were to proceed "in a friendly manner," his goal was the return of the entire Stadtholder collection.[45]

The Museum professors pledged cooperation, but the Dutch demand was beset with difficulties. Except for a few specimens, they claimed, it was no longer possible to identify which came from the Stadtholder collection. Moreover, it had sent many Stadtholder specimens to museums in other countries or the provinces. How could one replace these with specimens of similar worth without locating and evaluating them? The Museum proposed instead assembling for the Dutch a major collection that would be at least as important as the Stadtholder's. It would comprise specimens collected by the Baudin expedition and other French voyagers in parts of the world where the Dutch lacked the ready access they had in their own colonies (where most of the Stadtholder specimens had originated). This, the Museum said, would be best for both countries and for science in general.[46] Brugmans replied that, while he appreciated the attractions of the Museum's plan, he had been specifically charged with recovering for Holland "the principal and most distinctive objects" remaining from the original Stadtholder collection. This, he said, had to remain the baseline of his negotiations.[47]

Georges Cuvier communicated the situation to Alexander von Humboldt, who passed the Museum's proposal to his brother Wilhelm, in Paris as the Prussian minister charged with hammering out the peace treaty that brought the long era of Napoleonic Wars to an end. Wilhelm passed the Museum's proposal to the Dutch prime minister, Hans Christoph Ernst von Gagern, who was unreceptive until Wilhelm pointed out that he and his brother Alexander approved of it. Alexander told Cuvier that Wilhelm underscored the collegiality the Museum's professors showed the naturalists of other countries, and Von Gagern softened. Alexander told Cuvier he would meet with von Gagern at once and arrange for Cuvier to see him as well, noting that he and his brother were confident Brugmans would cooperate.[48]

It was a great boon to the Museum to have the von Humboldt brothers as allies. Alexander was the most famous naturalist *cum* scientific statesman in all Europe, and Wilhelm was even more famous as an international diplomat and founder of the University of Berlin. With its collections left basically intact, the professors thanked Alexander profusely for his help. His self-deprecating reply suggested that the happy conclusion was solely due to the Museum's generous past behavior and its consequent good standing in the European scientific community. He recalled the moderation the Museum had urged some years earlier with respect to the treatment of Prussian collections and summed up by saying, "The integrity of the Museum of Natural History is of a truly European interest, not only because of the riches accumulated at the Jardin des Plantes but especially because of the noble liberality with which, Messieurs, you communicate with the scientists of all nations."[49]

The Museum's records contain details of the arrangements made with representatives of Russia, Prussia, Austria, the Low Countries, and the Italian states.[50] By the end of October, all the returns, substitutions, and exchanges had been negotiated. The Dutch demands for the return of the Stadtholder collection, although initially the most worrisome, ended up proceeding so amicably that, as the historian Pierre-Yves Lacour has observed, the word "restitution," which appeared in an early draft of the agreement, was later replaced with "compensation."

The Museum reported to the latest minister of the interior, the Count of Vaublanc, how well it had managed the situation. It had provided the Dutch with a collection that was larger, better identified and organized than the Stadtholder collection had been, and therefore more useful for the instruction of science students. Moreover, every specimen sent to Holland was at least a duplicate of the collection in Paris. "Among the numerous objects we have handed over there is nothing unique, so that the collection of His Majesty has lost nothing essential," the Museum crowed.[51] Confident that the minister would be pleased with this outcome, the Museum took the liberty of asking for funds. Its cabinets had been thinned out a little, it said, but they were being refilled, and "if your Excellency could dispose some sums in our favor, we could procure quite quickly as many objects as we have given up."[52]

When Vaublanc responded, he did not mention additional funding, but expressed his "complete satisfaction" with the Museum's careful maneuvering to retain "all the principal pieces that came from the Stadtholder's cabinet" while giving up only duplicates and triplicates.[53] Among the "principal pieces" retained were the giraffe skeleton, "Wurmb's pongo" (to be discussed later), and Parkie the elephant, recently moved to the rotunda. Vaublanc said he could not praise the Museum enough for the way it had "procured such great advantages" in its dealings with the representatives of the Low Countries, the Vatican, Prussia, Austria, and Russia.[54]

In the same week, the Museum received what remained of the plants Captain Milius had announced with such fanfare when he returned to France in January from his mission to Guadeloupe and Martinique. Unfortunately, the five hundred specimens that survived the transatlantic crossing had fared poorly in France. Their months in Brest took a heavy toll, as did their subsequent transport by sea to Rouen and from there to Paris via the Seine. Only seventy-five plants arrived at the Museum, alive but "very worn out" from their latest journey.[55] The collection was too diminished to make the dramatic impact on the Museum's collections that Milius hoped for. Yet its arrival, coinciding with the conclusion of the Museum's negotiations over contested specimens and the 20 November 1815 signing of the second Paris Peace Treaty, signaled a new chapter in the Museum's history. Napoleon's defeat had not been disastrous for the institution, and the restoration of freedom of the seas promised a new era for the Museum and its menagerie.

PART 2

Colonial Ambitions, Star Attractions, and the Quest for a Science of Living Animals: The Menagerie in Its Prime (1815–1838)

3

The Menagerie in the Wider World

Figure 3.1.1. *Le callitriche* (green monkey). Simon-Charles Miger after Nicolas Maréchal. From Lacépède et al., *La Ménagerie du Muséum national d'histoire naturelle.* R. Burkhardt collection.

[3.1]

Captain Milius's Green Monkey

The Museum of Natural History welcomed the restoration of the freedom of the seas as decreed by the Treaty of Paris of 1814. France could now reconnect with what remained of her former colonies, and the Museum naturalists again had access to specimens from distant climes. Fittingly, one of the first animals to arrive at the menagerie thanks to the Restoration came from France's first postwar colonial mission, reclaiming the islands of Guadeloupe and Martinique from the British. Captain Pierre Bernard Milius was mission commander. The creature he brought home and eventually gave to the Museum was a green monkey.

As a species, the green monkey's connection with colonialism was centuries old. A native of west central Africa, it became an incidental traveler in the dire business of the slave trade. In the ports of Gambia and Senegal, as ships were loaded with captives destined for plantations in the New World, green monkeys were sometimes brought along as pets for plantation owners or ship captains. Some were liberated or escaped at ports of call or areas where they served as pets and ended up thriving in their new settings.[1] That was how in the fall of 1814 the commander of a French ship bound for the Antilles could purchase a green monkey in Cape Verde. That was also how they became an invasive species, their populations burgeoning in lands where they had no natural predators and could outcompete native species for local resources.

Milius had been a friend of the Museum ever since his role in the Baudin expedition. More than a decade later, as commander of the *Lys*, he found new ways to help the Museum's naturalists. Not limiting his efforts in the islands to diplomatic matters, he also attended to natural history. Most important, in Guadeloupe he took on board a stunning plant collection assembled by Félix Louis L'Herminier, the pharmacist/naturalist who directed the island's colonial botanic garden. When he first returned to France and was anchored off the port of Lorient, he wrote to the Museum boasting about the collection. He mentioned that he had collected live animals in the islands but that most had perished in the storms the *Lys* encountered as she neared the French coast. He did not at the time mention his green monkey.[2]

Milius's plant cargo illustrates how networks of commerce, colonialism, and science moved plants around the globe. Based on Milius's catalog of specimens from Guadeloupe, André Thouin reported to the Paris Royal Society of Agriculture that the collection contained more than 160 species, nearly half of which were not represented in the Museum's greenhouses. But not all were native to Guadeloupe. Some originated in Peru, Japan, Africa, or Madagascar. Some were of great economic importance, like the breadfruit tree, the species Captain William Bligh brought from Tahiti to Jamaica in 1793 in the hope that it might serve as inexpensive food for enslaved laborers on the island's cane plantations.

Ironically, the breadfruit tree specimen in Milius's collection destined for the Museum had arrived in Guadeloupe not from Tahiti or Jamaica but from Paris. The Museum had received a specimen from Jamaica in the 1790s and sent a root of it to Guadeloupe, with the expectation that cultivating the breadfruit tree would be a boon to the colony. As fate would have it, the Museum's own breadfruit tree later died. The arrival in Paris of a breadfruit tree from Guadeloupe, grown from a cutting of the original Paris specimen, would have been a happy return had it happened.[3] The breadfruit and other plants aboard the *Lys* were transferred from Lorient to the marine botanical garden at Brest. This seemed the best place to care for them until spring permitted their safe transport to Paris, but the Brest winter proved too harsh. By March the breadfruit had expired, along with many other plants in the collection.[4]

The remains of Milius's once-marvelous plant collection reached the Museum in November 1815 at the same time as a letter from Milius saying he wanted to give the institution "a very pretty monkey from Cape Verde." He continued, "I would experience a real grief separating myself from this animal if I were not convinced that it will be well treated and that you will recognize in this sacrifice, Messieurs, a new proof of my concern to do everything that can be agreeable to you." Of the animal he wrote, "My monkey is eighteen months old. He has been with me thirteen of them. He has a very gentle nature. He recognizes his master and shows him affection. He is called *Monsieur le Mounar* [Milius's emphasis], and I honor him often with the title of my friend, to which he responds with much kindness in his manner. He has constantly been fed milk, fruits, and bread."[5]

Milius wrote that the monkey was home with him in Paris and that the Museum was welcome to collect the animal there any morning between eight and nine o'clock. He did not say why he wanted to give it up, but one possibility is that it was reaching sexual maturity, when monkeys often begin giving their owners grief. Perhaps the famous commander's donation was less of a sacrifice than he admitted.

The naturalists of Paris were already familiar with the green monkey as a species. Six decades earlier, the French voyager Michel Adanson had encountered the animal in abundance in Senegal in the forests along the Niger River. The monkeys' misfortune was that their travels through the treetops broke branches, and the noise alerted Adanson to their presence. Once he spotted them, he slaughtered them without compunction:

> I killed at first, one, two, and even three of them without the others' seeming frightened. However, when the majority felt themselves wounded, they began to seek shelter; some of them hiding behind large branches, others descending to the ground, and still others (the greater number of them), launching themselves from the tip of one tree to the top of another tree. Nothing was more entertaining, when several of them jumped together on the same branch, than to see it bend and let fall the last of them while the first reached the ground and the others remained suspended in air. During this small maneuver, I continued always to fire above, and I killed up to twenty-three of them in less than an hour and in the space of twenty *toises* [120 French feet] without any of them uttering a single cry, although they would several times gather in companies, frowning, gnashing their teeth, and looking as if they wanted to attack me.[6]

The menagerie's first green monkey arrived in March 1794 after being confiscated from an animal exhibitor along with two other monkeys. Maréchal immediately painted its portrait.[7] Miger then produced an engraving and published it in *La Ménagerie du Muséum d'histoire naturelle* with an essay by Georges Cuvier. The zoologist acknowledged that naturalists of his day knew no more about the green monkey's behavior in the wild than what Adanson had written decades earlier. As for its behavior in captivity, Cuvier had found nothing to recommend it. It remained aggressive for the two years it lived, often biting its keepers. Male visitors irritated it, but as Cuvier delicately put it, "women cause it a furor of another kind, which it displays in the coarsest manner."[8]

The menagerie's next experience with a green monkey was no happier. The animal had an illustrious donor, General Honoré Vial, famous for his victory in Egypt at the Battle of the Pyramids, but its tenure at the menagerie was short, and its demise was swift and brutal. In May 1804 it managed to open its door and escape. By Frédéric Cuvier's account, the monkey "grievously wounded several people who sought to recapture it, and one could only manage to catch it by mistreatment." Its death was the product of "the violence that had to be used to seize it."[9]

The relatively gentle creature that Frédéric Cuvier described in an 1819 article on the green monkey as a species was probably the specimen Captain Milius gave to the Museum in 1815. It liked to be scratched by persons it knew; it only rarely sought to cause harm, and when it felt content "it made a particular, rather soft growl," sounding like "grou," with the *r* prolonged.[10] Always interested in breeding species that had not yet reproduced in captivity, the menagerie's superintendent hoped the green monkey might be a new success story. The problem, as he put it in a menagerie report of 1824, was that it was very rare to have a male and a female of this species at the same time.[11]

For four years in the 1830s, the menagerie had both a male and a female green monkey, but they failed to reproduce, and both died late in 1837. The Museum acquired six more in 1838. They were available owing to several factors: the number of ships arriving from parts of the world where green monkeys thrived; the dealers cultivating the market for exotic animals in Paris; the attractiveness of young green monkeys as pets; and perhaps also the difficulty of keeping these animals as pets once they became adults. An animal of "malice and intelligence" was how Frédéric Cuvier characterized the adult green monkey in captivity. For someone seeking a pet, this was not a promising combination.[12]

The green monkey was an instructive bit player in the histories of Western colonialism and the modern zoo. As incidental passengers on slave ships that were hellholes of misery and disease, they ended up in parts of the world they had never been before. Unlike the human cargo on the ships, however, they often found freedom and thrived in it. Such was not the lot of those that ended up in the menagerie. Wrested from their native or adoptive lands, confined to cramped and poorly ventilated quarters, and inadequately shielded from the cold of temperate northern Europe, the monkeys failed to reproduce and quickly succumbed to disease.

[3.2]

Colonial Ambitions and the School for Voyager Naturalists

Late in July 1802, during a brief period of peace after all the blood spilled in his efforts to quell the Haitian revolution, General Charles Leclerc, Napoleon's brother-in-law, wrote from Saint-Domingue to the minister of the interior that he was sending the menagerie a five-month-old "American tiger" (jaguar) from Caracas, noting, "I don't remember having ever seen one at the Museum of Natural History." Leclerc's interest in the menagerie was further shown by his report that he had tried (unsuccessfully) to send a live caiman and had sent an officer to South America in search of llamas and vicuñas. He asked to be informed of any minerals, plants, or animals from Saint-Domingue or the neighboring islands that the Museum lacked so that he might supply them. He also proposed establishing two botanical gardens on Saint-Domingue, called for naturalists to staff them, and outlined the benefits of having mineralogists study the island's mines and chemists make healing drugs from local substances.[1]

The Museum selected naturalists for Saint-Domingue as the general wished, but they never embarked. Leclerc's confident call for them was one of the last gasps of the French in Saint-Domingue. When the populace realized the French planned to restore slavery in the colony, the insurrection was rekindled, and Leclerc's increasingly brutal attempts to extinguish it failed. He died of yellow fever in November 1802 and was replaced by his second-in-command, General Jean-Baptiste Donatien de Vimeur, comte de Rochambeau. The atrocities Rochambeau committed in trying to subdue the Haitian revolutionaries are as much a part of his biography as his help to George Washington in the American Revolution twenty-two years earlier, defeating the British at Yorktown.

In March 1803, when the French campaign was all but lost, Rochambeau embarked for France from Le Cap "two young tigers [jaguars] from the Province of Caracas" that had been acquired at Leclerc's request.[2] At least one survived the trip and was installed in the Paris menagerie. Frédéric Cuvier's 1804 menagerie guide credited it to Rochambeau but could not offer

Figure 3.2.1. Victor Jacquemont (1801–32), one of several young voyager naturalists for the Muséum who died young in a distant land. Chronicle/Alamy Stock Photo

it as an example of France's colonial reach.[3] As of 1 January 1804, Saint-Domingue had freed itself from France and reconstituted itself as the new and wholly independent Republic of Haiti.

A little more than a decade later, the end of the Napoleonic Wars and the return of the maritime peace presented France with the urgent task of reconnecting with her remaining colonies. Gone were Saint-Domingue and the revenues generated by its sugar, coffee, and indigo plantations. Additionally lost, under the Paris Peace Treaty of 1814, were Isle de France (Mauritius), the Seychelles, and small islands in the West Indies. France was permitted to keep La Réunion, Martinique, Guadeloupe, and French Guiana. She also kept her commercial outposts (*comptoirs*) in Senegal (Saint Louis and Gorée) and India (Pondicherry, Chandernagor, Mahé, Yanaon, and Karikal), plus the tiny North American islands of Saint-Pierre and Miquelon. None of these colonial sites could fill the void left by Saint-Domingue, particularly with the slave trade abolished, but the government hoped that naturalists could help make France's remaining colonies more habitable, productive, and profitable than before.[4]

As France renewed her colonial initiatives, diplomatic and scientific interests proceeded in tandem. When the French ambassador to the United States sailed for America (September 1, 1815), a fellow passenger was Jacques Milbert, an artist whose accomplishments would soon include the successful embarking to Paris of such enormous mammals as the bison, moose, and elk.[5] Likewise, when the frigate *L'Hermione* left port for Rio de Janeiro on April 1, 1816, her passengers included the French ambassador, the duc de Luxembourg, the botanist Auguste de Saint-Hilaire, and the zoologist Pierre-Antoine Delalande (who was no longer as ill as the Museum had claimed three years previously to save him from conscription).[6] In May, when the king's corvette *La Licorne* sailed from France, she carried not only the administrators charged with retaking possession of French outposts in India but also the botanist Jean-Baptiste Leschenault de la Tour, the "naturalist of the king" and director of the royal colonial garden at Pondicherry.[7]

Not all efforts to connect with distant destinations proceeded smoothly. In June 1816 the ship *Méduse*, back from lending support to the reacquisition of Guadeloupe and Martinique, sailed from France at the head of a four-ship convoy dedicated to Senegal's colonial development. The ships carried administrators and military men but also civilians planning to farm the peninsula of Cap-Vert. Among the latter were a naturalist and two "cultivators," one a European plant specialist, the other a colonial plant specialist. Their presence signals the understanding that scientific knowledge would be integral to the colony's success.[8] Today the *Méduse* is remembered instead for the disastrous shipwreck immortalized by Théodore Géricault's monumental, gruesome painting of the last survivors on a wooden raft left to drift on the ocean.

Two developments of the late 1810s initiated by the government with the advantages of the colonies in mind were of great benefit to the Museum. The first was the production of a set of general instructions for voyager naturalists and colonial officials. The second was a new item in the Museum's annual budget for a "school" for voyager naturalists and support for their efforts abroad.[9] The menagerie stood to benefit from both, even if the Museum's interest in live animals remained secondary to its desire for more dead specimens for its galleries.

∴

In December 1817 the minister of the navy and colonies, the comte de Molé, came up with the idea of a generalized set of collecting instructions for voyagers and colonial officials. He had previously given colonial administrators oral instructions and a printed circular about collecting specimens and had passed along to them specific requests from the Museum. Now he wanted a detailed set of collecting instructions for his administrators and agents abroad, confident they would be useful for his ministry, the Museum, and colonial administrators alike.[10]

The professors responded enthusiastically, flattering the minister that he alone could establish relations between the colonies and that the Museum would enrich its collections and advance the study of natural history. The Museum aspired to having "as complete a series as possible" of all the productions of nature. Such a collection would attract naturalists from all of Europe while serving French agriculture, commerce, and the arts. The international admiration of the Museum would contribute to the glory of France.[11]

It took the professors six months to prepare and publish the instructions. They appeared in July 1818 as a forty-seven-page booklet titled "Instructions on the research that could be made in the colonies, on the objects that could be collected there, and on the manner of preserving and transporting them."[12] The Museum also published the instructions in its *Mémoires*.[13] On receiving one hundred copies of the booklet, Molé asked for fifty more to give to ships' commanders as well as colonial officials. The Museum sent him another hundred.[14]

André Thouin edited the instructions, assembling them from his colleagues' contributions. The final document described how to collect and preserve specimens and prepare them for shipment to France. It specified the kinds of information the Museum needed with the specimens: where they were found, the various uses to which the natives put them, animal habits, and more. To ensure that the specimens were well preserved, the document offered details on how to make and use Becoeur soap, the arsenic-laced preservative used to protect animal and bird skins from insect attack. "The use of this preservative requires great care," the instructions warned, "it must only be put under the [specimen's] skin and not on the outside, because in touching it or in disturbing the skins when mounting them one could experience bad effects."[15]

Readers were reminded of natural history's usefulness: "It must not be forgotten that several foreign plants which are widely spread today were first cultivated at the Jardin du Roi. The whole world knows the coffee plants that populate the islands of America all come from one shoot of coffee raised in our greenhouses; and recently it is from our greenhouses that the breadfruit tree has been sent to Cayenne."[16] As for animals, the document's zoology section named the vicuña, llama, kangaroo, and cassowary as foreign creatures that might one day "render services to the domestic economy or to agriculture."[17]

Shining through the instructions was the Museum's appetite for new specimens. The document's title page gave the impression that the instructions focused on collecting in and for the colonies, but the text showed that the Museum was thinking of the whole world. The zoology section attended to Senegal, India, the Antilles, and Cayenne, where France had a colonial presence, but it also identified the Cape of Good Hope, Madagascar, Indonesia, the Moluccas, the Orinoco River basin, and New Holland (Australia) as especially promising places for finding and collecting new animals.

The menagerie featured prominently in the instructions. Zoology, readers were told, encompassed not just animal forms and anatomy but also the study of animals' habits, development, and instincts (studies the menagerie made possible). Thanks to the menagerie, Parisian naturalists no longer had to depend on travelers' accounts for information regarding animals from abroad but could observe them firsthand. The menagerie enabled naturalists to "follow animals through all the degrees of their development and compare their manner of being during their life with their organization, which anatomy makes known after their death; acquire reliable knowledge of the very important phenomena of copulation, gestation, [and] birth; distinguish between the variations related to age and those produced by climate, food, [and] the crossing of races; and determine with certitude the difference that really exists between species."[18] This was the rationale for the menagerie that the Museum had been offering for the past twenty years, distilled for collectors abroad.

While acknowledging the importance of identifying animals that might be useful to the economy or to agriculture, the instructions encouragingly stated that virtually any live foreign animal would be welcome: "Considered in relation to science, there are few animals foreign to Europe that it would not be useful for us to study. Except for the Asian elephant, the royal tiger, and the African lion, the history of all the others is more or less incomplete. Even that of the lion has become well known only since the lioness of the menagerie had cubs; and it is also since two elephants died at the Museum's menagerie that an exact knowledge has been acquired of this great quadruped."[19]

These words were surely Frédéric Cuvier's. The theme of using the menagerie "to determine with certitude the difference that really exists between species" was distinctively his. The optimism with respect to collecting specimens reflected simultaneously his general disposition and his lack of experience as a voyager naturalist. He went on to portray collecting wild animals as quite simple: "One will easily procure animals by asking the natives of the country, who know their location, and who in their travels have often encountered them. They will catch them in traps and bring them back alive. Nor will it be difficult for them to capture in their youngest days some mammals whose hiding places they know, and birds whose nests they have seen."[20]

He went on to underscore the importance of collecting young, tamable animals:

> The younger the animals, the easier it will be to accustom them to living in cages. They will require at first some particular care. It is always necessary to feed them for several weeks before embarking them, and one cannot take too much trouble with respect to taming them. An animal not frightened by the sight of those who care for it is always in better health and withstands much better the stresses of a sea voyage than one that remains wild, and there is nearly no animal that one cannot make gentle by way of good treatment.[21]

Cuvier did not picture this as overly daunting. Though the collector/sender would need to give careful instructions to whoever was chosen to care for the creatures in transit, the basic requirements were uncomplicated: one had to keep the animals clean and not overfeed them. As for selecting a caretaker, it would not be a problem: "One will always find someone on the vessel who will care for them, either for a small recompense or as a matter of amusement."[22] Cuvier noted that part of the job would be protecting the animals from passengers who might tease or torment them, but his general message was that by following the instructions, collecting live animals from distant lands for the Paris menagerie would not be difficult.

∴

The idea of a school for voyager naturalists emerged from exchanges between the Museum and the minister of the interior (Joseph Vicomte Lainé) in the fall of 1818. As with the instructions for voyagers, concerns about the colonies inspired the government's thinking, and the Museum turned the colony-oriented questions to its advantage. Late in October, Lainé forwarded a letter from Molé asking the Museum to send a botanist to French Guiana to oversee the cultivation of useful trees. Lainé followed up a week later, repeating the botanist question, inquiring about the prospects of acclimatizing the *quinquina* (quinine) tree in Guiana and Senegal, and inviting the professors to think more broadly about collecting specimens for acclimatization in France or the colonies.[23]

André Thouin and his botanist colleague René Louiche Desfontaines were the perfect duo for crafting the Museum's response. Desfontaines had more experience as a voyager botanist than any of his colleagues, having spent two years collecting plants and animals in Tunisia and Algeria in the 1780s, and Thouin had been a central figure in plant exchanges for the colonies' benefit for nearly half a century. The two commissioners did not limit their discussion to the good of the colonies. They reformulated the minister's request as "the project of sending naturalists and agriculturists to the two hemispheres to procure the most interesting productions of the least known countries" and hailed it as "well worthy of the grandeur of the French government." They detailed what was needed for it to succeed.[24]

The first requirement was identifying energetic and healthy young men who were knowledgeable about natural history and collecting, preparing, preserving, and transporting specimens. State or commercial ships would be their most straightforward means of transport, but it was important that the collectors' ventures not be restricted by the schedules of government or commerce, because these could lead to shipping plants and animals to Europe when the weather was hazardous for the specimens' health. It would be better to establish foreign bases where, over extended periods, large collections could be assembled and housed. Ideally, these bases would operate long enough to allow for multiple shipments of specimens to France, thereby insuring against the losses in transport that happened all too frequently.

The Museum's botanists went on to identify the places most likely to yield plants and animals that could be acclimatized in France. The most promising in the Old World were the shores of the Caspian and Black Seas, Asia Minor, Armenia, Persia, the Caucasus Mountains, and the plateaus of the Tatars and Tibet. These were the original sources of France's present rural economy, the professors said, and additional useful productions were likely to be found there. In the New World (North America, South America below the Tropics, Australia, and New Zealand), naturalists knew primarily about specimens collected near the coasts, but additional useful productions of nature would likely be found in the interiors of these countries.

Desfontaines and Thouin did not belabor this last point, but it was significant, for it signaled a difference between most past collecting projects and those they were now proposing. Maritime expeditions had greatly augmented the Museum's collections. Now the professors were thinking about internal, continental projects rather than coastal ones. These would require a new kind of existence for the naturalists, with special benefits but also special hazards. The naturalists traveling to the interior could observe and collect a greater variety of specimens. They could better control their own schedules and settle in productive spots, instead of being forced to return to a ship that needed to sail, as was typically the case with maritime expeditions. On the other hand, naturalists in the field would not have the safety, medical help, or scientific and other resources that a ship could provide.[25]

Turning to the minister's request for a botanist to be stationed in Guiana, the professors identified Antoine Poiteau as the ideal candidate. Regarding the minister's questions about naturalizing the *quinquina* tree in French Guiana and Senegal, Desfontaines and Thouin expressed optimism that the project could be accomplished, while acknowledging difficulties to overcome or sidestep, the greatest of which was "the jealousy of the Spanish. They no more allow foreigners to take away their *kina* than to approach their goldmines."[26] Minister of the Interior Lainé responded graciously to the professors' report, thanking them for their insights and information while stressing that finding the cinchona tree remained his most urgent concern. He asked for specific instructions to communicate to administrators in Guiana and Senegal to help them identify the tree in their areas.[27]

Two months later (January 1819), there was a new minister of the interior, Élie, comte Decazes, former minister of the police, who was prepared to carry the support of naturalist voyagers further, undoubtedly encouraged by his longtime general secretary, botanist Charles-François Brisseau de Mirbel, who had close ties with the Museum, having begun his career as a naturalist's assistant to Desfontaines. A member of the Académie des Science, Mirbel was the first to inform the Museum that the minister wanted to establish a school for young voyager naturalists.[28]

Decazes proposed creating "a school of young naturalists destined for voyages in diverse parts of the world" and planned to allot the project 20,000 francs per year.[29] The Museum swiftly formed a committee to draw up rules for the school. Exploiting an ambiguity with respect to what the minister intended to fund, the committee chose to presume that the 20,000 francs would be used both for training young voyagers and for funding them abroad. It decided the best use of the funds would be to reserve 6,000 francs for training in Paris and assign the balance to support collectors in the field. The committee sent its proposed rules and budget to the minister.[30]

The sum of 20,000 francs was a substantial windfall, an 8 percent increase in the Museum's annual budget. The government had previously supported numerous voyages, but what the minister was proposing (as the Museum defined it) was novel: an ongoing commitment to support Museum-initiated expeditions and the training of voyagers-to-be. As described to the assembly by Geoffroy Saint-Hilaire, who wrote the committee report, this was "an innovation for our social relations that cannot fail to be welcomed with feelings of very deep gratitude," since it tended "toward nothing less than augmenting, and in a more methodical manner, the sum of material without which natural history would not know how to make great progress."[31] Nothing warmed the professors' hearts more than the prospect of having a continuing supply of new specimens.

The report addressed the students' training. They would have to attend the appropriate courses at the Museum, becoming knowledgeable about the various areas of natural history; they would learn how to extract minerals, to prepare dead animal and plant specimens, and to transport live animals and plants; and they would gain experience in the laboratories and galleries, develop drawing skills, and become competent in appropriate foreign languages. (The committee proposed teaching English at the Museum because of its utility to voyagers.) Student candidates would have to submit applications and be interviewed by a committee of professors.[32]

In March 1819, when the Museum received Decazes's approval of the proposed regulations, it moved quickly to announce the creation of the school for voyager naturalists and to invite applications.[33] On 2 April 1819, a formidable selection committee composed of Georges Cuvier (age forty-nine), André Thouin (age seventy-two), and René Just Haüy (age seventy-six) began interviewing applicants, most of whom were in their early twenties. Over three weeks, the professors considered twenty-nine aspiring voyagers. Georges Cuvier's notes from the interviews suggest that he and his colleagues were lucky to find their three best prospects in the first two evenings of interviews. In that same period, they also talked to some of the worst.[34]

Among the latter was a twenty-three-year-old law student who considered himself very talented but had never taken a course in natural history. Cuvier noted: "Natural history pleases him because one talks there of flowers, which he likes very much." The credentials of a twenty-one-year-old notary's clerk were no better. Observed Cuvier, "[He] wants very much to travel and has no other qualification." A twenty-three-year-old applicant from the municipal toll office had taken a few courses in natural history and anatomy, but his knowledge of zoology and botany was limited to generalities. Cuvier concluded, "[He] has no mineralogy at all, knows no foreign languages, doesn't know how to prepare [specimens], doesn't know how to draw."

A later interviewee, thirty-four-year-old Marie Antoine Rabaroust, seemed of interest as a former army surgeon and a survivor of the shipwreck of the *Méduse*. He had not been consigned to the ill-fated raft but instead put aboard a small boat that reached shore, whence he made a grueling trek across the desert to Saint-Louis. The professors appreciated his toughness and experience as a voyager and surgeon, but as they finally summed things up, Rabaroust seemed to think "that a strong liking for stuffing birds, some knowledge of drawing, and the declaration of great zeal would earn him the goodwill of his judges." That was not enough.

The committee eventually accepted six students as the first class in the new school. They also found what they wanted more urgently: three naturalists ready to travel without further training. The most impressive was twenty-three-year-old physician Armand Étienne Havet, who had studied all the branches of natural history and responded well to the committee's questions regarding zoology, botany, and making observations as a voyager. He had experience preparing collections, knew how to draw, and spoke English, Spanish, and Italian. The professors rated him "the best of all." Also chosen was Félix François Godefroy, aged twenty-one, who had a degree in pharmacy, had taught a public course on plant physiology, and had presented memoirs at the Academy of Sciences and the Société philomatique. The committee's third selection was thirty-two-year-old botanist Auguste Plée, who expressed his hope that the industry he displayed abroad would persuade the Museum to aid in his goal of developing a new flora of France.

The common denominator among the six students was hearty youth, but they appealed to the committee for different reasons. With Marcel Cousinery it was not his knowledge of natural history but his fluency in Turkish and modern Greek. Charles Sauvigny had a lively mind, a good background in natural history, drawing skills, and spoke Italian. With a year's experience in the laboratories, he would be well prepared to travel on the Museum's behalf. Pierre Nicolas Gerdy was already capable in comparative anatomy but did not know any foreign languages or how to sketch. The other three (Jean-Baptiste Letellier, Isidore Bourdon, and Victor Fontanier) showed promise but needed further instruction. All six would receive stipends to allow them to live in Paris and study at the Museum. No mention was made of spending time in the menagerie to learn to deal with live animals. Although training was originally envisaged as lasting a year, an additional six months would be offered to students who displayed aptitude but needed more time.

The professors meanwhile decided on destinations for the naturalists ready to travel: Havet to Madagascar, Godefroy to the Philippines, and Plée to the shores of the Black Sea. Georges Cuvier explained to the minister of the interior the importance of these regions. Madagascar was "of all the parts of the world the one where one can hope to collect the most new things." The Philippines were attractive because "like all the isles of the archipelago of the Indies," they were "rich in singular plants and animals," some of which would surely be "useful to our colonies." Likewise, on the shores of the Black Sea, the Museum believed a naturalist would be able to "collect many things useful for our agriculture."[35]

Baron Portal, minister of the navy and colonies, had other ideas. He urged Decazes to order that two of the voyager naturalists be dispatched to French Guiana. To the Museum, Portal pompously declared, "few explorations of this kind can be more interesting for science in general than those that would take place in this country; and few countries are as susceptible of offering to the work of naturalists so vast and rich a field." He wanted the two voyagers to depart with the new commandant of Cayenne the very next day. The Museum viewed this as impossibly soon, and it held its ground with respect to sending Havet to Madagascar and Godefroy to the Philippines. It agreed that Plée could go to the Caribbean instead of the Black Sea.[36]

Tragically, Havet and Godefroy spent more time in transit than at their destinations. Each accompanied by a younger brother, they sailed from Rochefort on 27 January 1820 aboard the government ship *Panthère*, arriving at Île Bourbon (Réunion) after stops in the Canary Islands and at the Cape of Good Hope. The governor of the island, Baron Milius, appointed Havet his special envoy and charged him with taking gifts to Ramada, Madagascar's most powerful ruler. Havet wrote to the Museum explaining that this assignment would enable him to travel 180 leagues inland from the Malagasy port, Tamatave (Toamasina), to the capital of Ramada's kingdom (Antananarivo) in the highlands. He estimated it would be five months before he could send the Museum specimens, but he assured the professors they would be "new objects," adding, "may the notes I have the honor of offering you have some interest!"[37]

Havet disembarked at Tamatave on 8 June. The next week he met with the local ruler and prepared for his inland journey. On 16 June, he set out with his teenage brother, Nicole, plus an interpreter and forty-five native bearers. They traveled modest distances each day and spent the nights in villages. Havet collected plants and noted what he learned about their uses and about the customs and manners of the people. His brother sketched people, fauna, flora, and sites. Two weeks later Armand Havet died of a violent fever that left his brother gravely ill. Two months passed before Nicole was well enough to visit his brother's grave in Tamatave.[38] Armand Havet had written to the Museum en route to Madagascar that as a physician he'd heard about the contagious diseases said to rage in Isle de France, Île Bourbon, and Madagascar, and that he hoped to help care for the unfortunate islanders. The Museum warned in reply, "take good care that your zeal is not fatal to you."[39] By the time this advice was penned, however, the young man was already dead.

Havet's loss was lamented by Baron Milius at Île Bourbon, by the Museum professors, and by the French press. Milius's report to Portal portrayed

Havet as "the victim of his zeal," and the reports in the *Moniteur* and the *Constitutionnel* suggested the same.[40] Years earlier, Antoine de Jussieu accounted for the deaths of the botanists Riedlé and Sautier and the zoologists Maugé and Levillain of the Baudin expedition in similar terms.[41] The students in Paris in 1821 preparing to be voyager naturalists might have questioned the diagnosis, but they could not have escaped the sobering message that scientific fervor alone was no guarantee of success or survival for a naturalist far from home.

Godefroy's career in the Philippines was nearly as brief as Havet's in Madagascar, and its end was more gruesome. Godefroy and his brother disembarked at Manila on 28 August 1820, having collected specimens at ports of call along the way. They concentrated their efforts in the vicinity of Laguna de Bay, the great lake southeast of Manila. Disaster arrived early in October. A fierce storm with high winds uprooted countless trees; twenty-four hours of rain flooded the rivers and washed fish and shellfish ashore, and a cholera outbreak claimed an untold number of lives. Believing foreigners caused the epidemic by poisoning the waters, angry natives on a rampage (9–10 October 1820) massacred eighty-five Chinese, an American naval officer, and fourteen Europeans, among them the young naturalist Godefroy.[42] News of the tragedy did not reach France for nearly seven months, arriving in Bordeaux on 2 April 1821 on a ship from Manila.[43]

The last of the Museum's three new voyager naturalists of 1819, August Plée, had a longer career. He died in Martinique in August 1825 at age thirty-nine, while making final arrangement to return to France. Despite nearly five and a half years in the Caribbean, however, he had not proved especially effective in providing the Museum with the kinds or the quantity of specimens it expected.[44]

Of the six young men selected for training as voyagers, only one, Charles Sauvigny, had any appreciable success, and he never finished the appointed courses.[45] When the minister of the navy and colonies needed an agricultural botanist in Cayenne, the Museum named Sauvigny and recommended his fellow student, Jean-Baptiste Letellier, for a post in French Guiana. When Portal proposed in April 1820 that the funding of the two naturalists be shared between his ministry and the ministry of the interior, and that the latter use money from its 20,000-franc annual allotment to the Museum, the administration protested. The professors told the minister of the interior (now Joseph Jérome, comte Siméon) that they had not asked for Sauvigny and Letellier to be sent to Guiana, had identified them only on request, and should not have to subsidize missions the Museum had not initiated. Moreover, the Museum's funds were wholly committed to the three naturalists already abroad (Havet, Godefroy, and Plée) and to the students in the voyager naturalist school.[46] While the professors were pleased to have the government support naturalists abroad, they evidently did not intend to sacrifice the interests of natural history to French commerce and colonialism.

Neither Sauvigny nor Letellier ended up in Guiana. In May 1820 Portal told Siméon that an informed and fervent agricultural botanist was needed to serve on a commission to explore Africa.[47] Sauvigny, who preferred Africa to Guiana, was chosen, and in August he wrote enthusiastically to the Museum's entomologist, Pierre André Latreille, about the trips he planned to take (by land, sea, and river) and his intention of supplying Desfontaines, Thouin, and Georges Cuvier with a variety of specimens they had asked of him. He had already collected, he wrote, some insects, crustaceans, reptiles, and shorebirds, plus a pair of live gazelles and a monkey. He was sure he could provide the Museum with young ostriches, ostrich eggs in different stages of development, adult and juvenile hippopotamuses, and gazelles with horns curving forward. He would even try to get a manatee, though he expected that might be more difficult. He burned with impatience to begin, he said.[48] Yet illness forced him to delay his explorations for seven months.

Although nominally an agricultural botanist, Sauvigny was eager to collect zoological specimens, including animals for the menagerie. In October 1821 the Museum wrote to tell Portal that Sauvigny had assembled a "precious collection" of live animals and to ask him to transport them on a ship of state when an occasion arose.[49] Sauvigny managed to embark on the *Panthère* a large gray ostrich, two marabou storks, two eagles, a crowned crane, two yellow-billed storks, five spur-winged geese, a "species of fox," a "large species of cat," and a gazelle. Most of the birds and the gazelle died en route, but the crowned crane and two mammals reached Paris. Frédéric Cuvier identified the "fox" as a jackal and the "cat" as a caracal.[50]

The minister of the navy and colonies (now Aimé, comte de Clermont-Tonnerre) gave Sauvigny a new assignment that kept him in Africa several more months. Returning home, as a former student of the school of voyager naturalists, he hoped that the Museum would help sponsor his participation in a government expedition to Indochina. Instead, the administration told Sauvigny and Tonnerre that its voyager funds were all but totally committed to the naturalists Pierre-Médard Diard, Alfred Duvaucel, and Jean-Baptiste Leschenault (all of whom were in India), and Charles d'Orbigny (who lived in La Rochelle and collected specimens along the Atlantic coast).

While the Museum benefitted greatly from the government's yearly 20,000 francs, the sum was not enough to support all the voyagers the administration would have liked to commission, or to provide the naturalists already abroad with the needed assistance. The money had to be sensibly distributed to the voyagers who stood the best chance of supplying the kinds of specimens the Museum wanted most. When Decazes established the voyagers' school and its funding in 1819, Leschenault (director of the colonial garden at Pondicherry), and Diard and Duvaucel (who had ventured to India on their own initiative) were already abroad and not on the Museum's payroll. Delalande (on a government expedition at the Cape of Good Hope) and Charles Lesueur and Jacques Milbert (collecting in North America supported by the French ambassador to New York) had likewise traveled without help from the Museum. Before long, all would receive at least some assistance from the Museum's voyager fund.

In the summer of 1820, after Sauvigny's departure for Africa, the Museum decided to examine the five students remaining in the school to see if they merited continued funding. Fontanier and Letellier performed well and had their tuition extended, but the Museum dismissed Cousinery because he had missed classes and shown no intention of becoming a naturalist. The professors postponed judgment on Bourdon and Gerdy because each had a legitimate reason for not being examined just then. Bourdon later recounted that Georges Cuvier had recruited him for the school but, after the deaths of Havet and Godefroy, exempted him and Gerdy from having to continue as voyagers. They both went on to distinguished careers in medicine.[51]

By this point the school for voyager naturalists was nearly nonexistent. The Museum used its 20,000-franc budget almost exclusively for collectors in the field and never again advertised for students. It did, however, fund two more young naturalists as they prepared to be voyagers for the Museum. Alcide d'Orbigny (son of Charles d'Orbigny, the Museum's correspondent at La Rochelle) received 100 francs monthly for five months, beginning in November 1825, before his departure on what was meant to be a three-year mission to Peru and Chile.[52] Beginning in November 1827, Victor Jacquemont received the same compensation as "student voyager of the Museum" so he could attend the zoology courses and learn how to prepare animal specimens.[53] In the years when d'Orbigny and Jacquemont were abroad, their combined allocation used up most of the voyagers' fund. In 1830, for example, they received 8,000 francs each.[54]

Jacquemont became another martyr to science, dying of cholera in December 1832 in Bombay. D'Orbigny (1802–57) survived his nearly eight years of travel in South America and returned to France in March 1834, an exception to the general rule. In 1834, when reporting to the Academy of Sciences on the importance of d'Orbigny's botanical collections, Alfred Brongniart claimed that in the eighteen years since the restoration of the open seas, seven of the eight naturalists who had taken long expeditions for the Museum had died because of their travels. Godefroy, Havet, Plée, Duvaucel, and Jacquemont had "perished far from their country," and Delalande and Leschenault had succumbed to voyage-related illnesses a few years after returning to France. Only d'Orbigny, said Brongniart, lived to work on his collections at home and to announce his results to the scientific community.[55]

Ready as it was to help the government with colonial projects, the Museum's primary focus continued to be acquiring more specimens for its own collections. If this was displayed in a conspicuous way in the Museum's disagreement with the minister of the navy about where to send its first voyager naturalists, it showed up in a much smaller way in an exchange between the Museum and the voyager Jacquemont. Alert to the government's interest in identifying foreign species or varieties that could be naturalized in France or the colonies for commercial purposes, Jacquemont reported back to the Museum in 1831 from the Kashmir valley saying he'd learned of four wild species of goats or sheep whose undercoat was used for making shawls. With the promised help of the king of nearby "Little Tibet," he hoped to get two pairs of each to send home. The Museum's response expressed less interest in the goats than in a yak, or at least a yak skeleton or skull, for zoological study.[56]

The voyager naturalist most successful contributing to France's colonial interests and to the Museum's scientific interests was Leschenault. He played a preeminent role in the late 1810s and early 1820s in the intercolonial transfer of plants. At the same time, he sent the Museum diverse shipments of specimens, including live animals for the menagerie, the most impressive being an elephant. If there was one other French naturalist voyager who could compare with Leschenault as a colonial administrator and contributor to Europe's natural history collections, it was Pierre-Médard Diard (1794–1863). The hitch was that his main contributions in both regards came in the service of Holland, not France.

Diard experienced acutely the frustrations of being too far away and too out of touch with the institution and government he wished to serve. In 1825, unable to sustain his efforts as a naturalist in the East Indies collecting for France, Diard began working for the Dutch. The governor general of the Dutch East Indies put him in charge of agricultural development in Java. As a colonial administrator, Diard worked to make the colony more profitable. He promoted the development of plantations, which involved importing plants from Bengal and Ceylon.[57] It is not surprising to find him engaged in projects that benefitted the colonizers more than the colonized. He was instrumental in bringing the Javanese agricultural system under the more direct control of the government, an effort that had already begun and led to the economic and social dislocations that fueled the Java War (1825–30). In the late 1820s he was entrusted with suppressing uprisings in the coal mines of Borneo and surveying the tin mines of Bangka. His activities as a colonial administrator deserve closer study.[58]

As a naturalist, Diard introduced the silkworm to Java, improved the cultivation of indigo plants, experimented on rice, coffee, and cinnamon, and paid special attention to the sugar industry. With the approval of the governor of Java, he transported to the English colony of Mauritius and the French colony of Bourbon a variety of sugar cane resistant to an insect that was devastating cane production on those islands. In the meantime, he sent natural history specimens to Europe, primarily to the natural history museum in Leiden; to Paris he sent duplicates. Diard survived for decades in Southeast Asia and lived to age sixty-eight. He died in Batavia in 1863 by accidentally poisoning himself while attempting to perfect a new arsenic-based process for preserving bird and animal skins. After his death, the number of bird and mammal specimens in the Paris Museum's galleries attributed to Diard totaled 955.[59]

In conclusion, the French government had a clear interest in enlisting the Museum in its efforts to rebuild France's colonies in the postwar era, and the Museum professors sought to use the government's interests to their advantage. The tracks of colonialism and commerce facilitated the Museum naturalists' travels to distant lands in search of specimens. I have found no instance, however, of the menagerie being treated as a triumphant illustration of France's colonial reach in the Restoration or the 1830s. The collection was viewed with national pride, but the animals were not described as emblematic of France's colonial successes the way the elephants of the Dutch Stadtholder or the bears of Bern in the 1790s were hailed as symbolizing French military victories.

This omission may have been because France's colonial strength in the Restoration period did not match her military strength during the Revolutionary and Napoleonic Wars. With Saint-Domingue lost and the colonial powers of the British and Dutch eclipsing her own, France was perhaps less disposed to look to the menagerie as an affirmation of standing vis-à-vis other European nations. But the menagerie animals may not have been associated with French colonial endeavors simply because the most popular ones in this period were not from French colonies. The star attractions and biggest crowd draws were Martin the bear, the great herbivores sent from North America by Jacques Milbert, a giraffe from the viceroy of Egypt, and an orangutan named Jack, all from parts of the world outside French control.

[3.3]

The Egyptian Mongoose and the Brown Rat

The menagerie's first mongoose came from Egypt courtesy of Étienne Geoffroy Saint-Hilaire. In December 1801, halfway through quarantine in Marseille after his Egyptian odyssey, Geoffroy wrote to the director of the Museum expressing joy that soon he would soon be seeing his colleagues again. He spoke of the specimens he would bring, including "a pretty animal, the ichneumon, so famous for its combats with the crocodile." He described it as "tame to an extraordinary degree; its gaiety is completely remarkable."[1] Six weeks later, Geoffroy and his mongoose/ichneumon arrived in Paris. Nicolas Maréchal quickly executed a painting of the animal, portraying it in an Egyptian setting doing what its reputation suggested it should: slaying a crocodile after destroying a crocodile egg. Maréchal must not have discussed this depiction with Geoffroy, however. Geoffroy's mongoose essay, paired with Maréchal's illustration in Miger's *La Ménagerie*, debunked the myth of mongoose as crocodile slayer. It did say, however, that the mongoose liked eggs, those of crocodiles included.[2]

Geoffroy began his mongoose essay by distinguishing the three species known at the time. He addressed the Egyptian mongoose's habits but admitted he'd had little chance to observe the animal in the wild. Though common in Egypt, it was wary and furtive. It crept along the canals and furrows of farmland, advancing with great hesitation only after its keen sense of smell signaled the absence of danger. Nonetheless, this naturally timid creature could be tamed to the point of being "gentle and caressing," and it would follow its master like a dog. Most important, it could rid a house of mice and rats in short order because it was tireless in pursuing whatever prey it detected, and it killed more than it needed to eat. The name ichneumon, Geoffroy explained, signified a creature ceaselessly occupied with finding quarry, while "Pharaoh's rat" was what Europeans living in Cairo called it.

Geoffroy identified rats, snakes, birds, and eggs as the mongoose's pri-

Figure 3.3.1. "The Egyptian mongoose." Nicolas Maréchal, Simon-Charles Miger, "L'ichneumon, Viverra ichneumon," from Étienne de Lacépède, Georges Cuvier, and Étienne Geoffroy Saint-Hilaire, *La Ménagerie du Muséum national d'histoire naturelle, ou Description et histoire des animaux qui y vivent et qui y ont vécu*, Paris, Aug. Renouar, 1801–5. Muséum national d'histoire naturelle, 419 pl. 18.

mary food. When the annual Nile flood caused it to seek higher ground near human habitations, it fed on hens and pigeons, but the Egyptians seemed content to let foxes and jackals keep its numbers in check. A large lizard also helped control the mongoose population, just as the mongoose checked the crocodile population by eating its eggs. Geoffroy gave no details as to how he acquired, cared for, or transported his mongoose. He did note that it liked to sit on the shoes of its visitors at the menagerie, a behavior he attributed to the animal's penchant for cooling itself by bringing a small pocket above its anus into contact with colder objects.

Just weeks after Geoffroy's essay appeared, the menagerie received a second mongoose, a female, from Joseph Bonaparte, Napoleon's older brother.[3] Her entry to the male's cage ignited a fierce combat that she won. She exploited her victory by menacing the male whenever he tried to leave the dark corner where he had taken refuge. At mealtimes, she allowed him to eat only after she had her fill. This situation changed dramatically two months later with the onset of the mating season. The male's persistence made the female defensive. For four days and nights, he gave her no rest. "I have never seen an animal more ardent in love," wrote Geoffroy, without saying whether the male succeeded in his efforts. "Despite all this violence," Geoffroy continued, "he preserved his gentle character with his keepers and the curious who visited him." Even when pulled away from the female, he did not complain or try to bite.[4]

Frédéric Cuvier included the male mongoose in his menagerie guide (1804), writing, "This individual is extremely gentle; he has lived a long time in complete liberty, without ever trying to escape. He wages a very active war on rats and mice, and in this regard the ichneumon could replace cats. He is fed on raw meat and is in very good health."[5] Cuvier made no mention of the female, which suggests she was no longer there. Presumably she was dead. The menagerie had added, however, a pair of mongooses of a different species brought back by the Baudin expedition from the Cape of Good Hope.[6] Mongooses do not reappear in the Museum's records until the early years of the Restoration, when France reconnected with her former colonies and was confronted with the disruptions of nature's equilibrium that colonialism had helped create.

In December 1818 comte de Molé, the outgoing minister of the navy and colonies, forwarded the Museum a report on the desperate situation in Guadeloupe, where coffee growers' revenues had been halved owing to a plague of rats. "They devour the sugar cane as soon as its first buds appear. Every known means of destruction has been employed in vain against this animal. Continual hunting, traps, all kinds of poisons, bounties accorded to the Negroes and carefully paid, nothing has been able to produce an appreciable diminution, and they still take half the harvest." The report suggested that the best way to destroy the rats would be to cultivate the fallow areas where they were thriving, but this would take eight to ten thousand new workers.[7]

Molé's successor, Baron Portal, returned to the issue in September 1819 asking if the mongoose, which flourished on the Island of Cape Verde, could be usefully introduced into France's colonies, especially Martinique, to destroy the venomous snakes and rats that were a scourge to the colonists. He asked whether the "natural habits and instinct" of the mongoose were suited to this task, and he inquired whether the species' multiplication in the colonies might cause "inconveniences." He also asked if there were other species that could be introduced in Martinique to do the job.[8]

Responding to Portal's question, the Museum allowed that one disadvantage "would be the loss of some fowls and birds of which [the animal] is equally greedy," but this "would probably be of little consequence in comparison to the good that would come of it." If the number of mongooses became so large as to be a nuisance, the Museum judged, it would still be preferable to having too many snakes, because reducing the mongoose population would be easier than reducing the number of snakes.[9]

Portal transmitted the recommendation to the colonial administration in Senegal and reminded the Museum of his question about other species that might control snakes and rats.[10] Geoffroy responded by identifying and evaluating diverse creatures that might be up to the task. Of the different species of mongoose, he believed, the Egyptian mongoose or ichneumon was the best. Other mongoose species had the same indefatigable instinct to pursue prey, but of those he knew, only the ichneumon, which was the size of a large cat, was big and strong enough to possess the confidence to defeat "a legion of rats." Geoffroy recommended asking Bernardino Drovetti, the newly elected French consul general in Egypt, for help procuring the Egyptian species if the mongoose species at Cape Verde proved too small. He described Drovetti as "full of zeal for everything that is connected with his country" and having very great power because of his intimate connection with the Pasha of Egypt."[11]

As alternatives to the mongoose, Geoffroy recommended the secretary bird of Africa, four birds species from South America (including the jabiru, two caracaras, and a bird Buffon mistook for an ibis), and a large South American lizard that lived in freshwater and seemed a good prospect for rat-infested areas near waterways. Museum records do not say whether Portal asked Drovetti for Egyptian ichneumons, but in 1821, the year after receiving Geoffroy's report, the minister informed the Museum that he had just ordered the French consul at Rio de Janeiro to send to Martinique the alternative species that Geoffroy suggested.[12]

Coincidentally, in 1819 Frédéric Cuvier wrote about a new mongoose in the menagerie that greatly liked eating birds. What he called a *mangouste* was from the Malay Peninsula. Donated by a "Madame N***," it was a male, tame and gentle, that loved to be caressed. However, it was ferocious toward its prey. Wrote Cuvier:

> The birds especially are what it seems to love most: when some are put in its cage, which is very large, and where they can easily fly, it darts, and in an instant, with movements so fast that the eye cannot follow, it seizes the birds, breaks their heads, and when it has assured itself they will no longer escape, eats them with great voracity. When satiated, it lies down in the most obscure corner of its nook.[13]

The Museum had mongooses of various species over the years, including one from Java sent by the voyager naturalist Pierre-Médard Diard and another from Egypt sent by Drovetti. The Museum traded the latter in 1826 for a *vansire* (marsh mongoose), for which Frédéric Cuvier created a new genus, *Atilax*.[14] Diard's specimen lived roughly eight years at the menagerie;

Figure 3.3.2. "Le surmulot" (the brown rat). C. de Lasteyrie after Jean-Charles Werner. É. Geoffroy Saint-Hilaire and F. Cuvier, *Histoire naturelle des mammifères*. General Research Division, the New York Public Library.

the *vansire* lived over eleven. In September 1837, when a sea captain worried it might be too cold to send a mongoose to Paris from Toulon, Cuvier observed, "It is not a delicate animal, and we have reason to think it will suffer little from the cold. A half pound of meat per day should suffice for its nourishment. It will also eat bread."[15]

Considering the mongoose's subsequent history as an invasive species (i.e., one introduced to an environment without natural predators that multiplies unchecked), the phrase "not a delicate animal" was a major understatement. The Indian mongoose, not the larger Egyptian one, was the most devastating. In 1871 W. Bancroft Espeut had nine Indian mongooses brought from Calcutta to control the rats on his Jamaican sugarcane plantation. Ecstatic about his project's success, he wrote after a decade, "I question much if such enormous benefit has ever resulted from the introduction and acclimatization of any one animal, as that which has attended the Mungoos in Jamaica and the West Indies."[16] Espeut's enthusiasm inspired cane growers elsewhere, Hawaii included, to import the animal.[17]

A commission appointed by the Jamaican government eight years later concluded that the destruction of domestic animals and local fauna by the mongoose was far greater than the benefit of its controlling rats. The mongoose has since wreaked disaster wherever it was introduced, leading to the decline and sometimes extinction of native birds, mammals, reptiles, and amphibians. Today the Indian mongoose is identified as "one of the world's 100 worst invasive species."[18]

The Museum professors who advised the government on introducing the mongoose to French colonies to control unwanted animals were virtually helpless when it came to protecting their own grounds from invasion by foreign species. The brown rat (the *surmulot*, also known as the "wharf rat" and "Norway rat") was remarkably adaptable to new environments. When the Paris sewer system was constructed in the early 1800s, the *surmulot* gained another name, the "sewer rat" (*rat des égouts*). By then it had inserted itself into the menagerie's ecology and was impossible to eradicate without risking injury to other animals on site.

Frédéric Cuvier appreciated the irony of the situation. Purposely acclimatizing foreign animals was seldom successful, as they often died "after a more or less prolonged struggle," but the brown rat easily established itself, unbidden, in new lands. Native to Asia, it traveled the world as a stowaway on ships. By the middle of the eighteenth century it had reached Paris, where it thrived not because of the climate but because humans unwittingly provided it shelter and food. Cuvier did not credit the rat's success to intelligence. Instincts and a "natural defiance" permitted its survival, he said, in the face of "the small number of means we employ to harm it without bringing harm to ourselves." The only hope, he believed, was to limit its numbers to mitigate the damage it caused.[19]

Three consecutive menagerie superintendents (Delaunay, Lamarck, and Frédéric Cuvier) complained to the administration about rat-related damage. Delaunay asked permission to saltpeter a room next to Bernard Louzardy's guardroom because rats had dug "such a great number of holes" as to make walking there dangerous (1798).[20] Lamarck worried about similar holes under the paving stones of the carnivore cages, and while calling for repairs he asked that Guichenot be authorized to buy rat poison (1801). The assembly agreed to both requests and instructed Guichenot to purchase several pounds of a product unsubtly called "death to the rats" (*la mort aux rats*).[21] Frédéric Cuvier in his early days on the job reported that the pheasant house was devastated by rats and the birds' lives were in danger (1804).[22] Five years later, still struggling with rat problems, Cuvier asked Geoffroy to recommend that a ratcatcher be hired annually. As Cuvier explained, the workyards near the menagerie harbored a huge rat population, and those that found their way to the menagerie dug holes with such industry that daily repairs could scarcely keep up with them. They attacked the menagerie's meat supplies and menaced the birds. "In a single night," Cuvier reported, "one has seen rats eat over sixty pounds of meat, and in the last three years they have killed a great number of ducks, including, among others, two broods of Carolina teals."[23] The menagerie staff tried to poison the rats, but the ratcatchers' results were better: "Whether they have a particular way of distributing their poison, or whether what they use is more effective than what they sell, often, at a single time, they have killed up to

60 or 80 rats." For twenty or twenty-five francs per month, one could hire a man to exterminate the rats and protect the menagerie from new invasions, which was what other institutions were doing. The assembly agreed to hire a ratcatcher, at least provisionally, for a month or two.[24]

Despite constant attempts to control them without harming other animals or people on the premises, the menagerie continued to have rats. Achieving a balance was complicated by the range of animals involved, human and otherwise. In 1837 Isidore Geoffroy Saint-Hilaire, standing in for Cuvier as acting superintendent, expressed dismay on learning that poisonous balls (*boulettes*) had been distributed throughout the menagerie. He worried that the dogs, which went out once a day, could become accidental victims. Children were clearly at risk as well. There was another, less obvious, scenario that Isidore felt compelled to mention. Malicious persons, he feared, might find the *boulettes* and use them against specific animals, notably the bears or even the lions. "We have often had people with bad intentions make such *boulettes* and bring them here," he explained. Someone had attempted, unsuccessfully, to poison the hyena last year. "If such persons found *boulettes* already made," Isidore observed, "one cannot doubt the use they would make of them."[25]

The Museum's experiences with mongooses and rats are a reminder that human attempts at ecological engineering do not always turn out as planned. However confident the administrators of the colonies and the menagerie may have been, their resourcefulness was not a match for that of the animals themselves.

[3.4]

The New Collectors: Jacques Milbert and J.-J. Dussumier

Once maritime peace was restored, two individuals excelled in providing the Museum with living and preserved specimens from abroad. Neither Jacques-Gérard Milbert (1766–1840) nor Jean-Jacques Dussumier (1792–1833) was a trained naturalist, but both became highly knowledgeable and productive collectors. Milbert traveled to North America and made New York City his base. Dussumier's business ventures operated between the Mascarene Islands (Mauritius and Île Bourbon) and ports of call in India, Southeast Asia, and China. The two supplied the Museum's naturalists with treasures from all three kingdoms of nature: plants, animals, and minerals. The preserved fish specimens were invaluable to Georges Cuvier, who sought to identify and classify all the fish in the world. Live specimens ranged in Milbert's case from small seedling trees to such massive mammals as bison, elk, and moose. From across the world, Dussumier sent exotic creatures including a tiger, a hyena, a long-lipped bear, a lion-tailed macaque, and half a dozen species of deer and antelope. The Museum welcomed these animals, since they enhanced the menagerie's appeal and could provide zoological insights. A few were even prospects for acclimatizing. But the arrival of these magnificent new live animals at the menagerie was more a function of the collectors' initiatives than what the Museum consistently urged upon them. What the Museum craved was the constant inflow of new dead specimens for the galleries.

Jacques Milbert was a landscape painter who loved animals and travel. His appointment (1795) as professor of design at the School of Mines of Paris took him to the Pyrenees and the Alps, but he yearned to go farther.[1] Circumstances prevented him from joining Napoleon's expedition to Egypt in 1798, but in the fall of 1800 he joined the Baudin expedition to Australia.[2] That adventure began badly owing to the conditions of life at sea. In March 1801, when the expedition reached Isle de France (Mauritius), Milbert's legs were so swollen he could barely stand, owing, he believed, to the meager and unhealthy food he had aboard ship. When Baudin sailed on, Milbert

Figure 3.4.1. "The zibeth." The first of many live mammals sent to the menagerie by the voyager naturalist J. J. Dussumier. C. de Lasteyrie after Jean-Charles Werner. From É. Geoffroy Saint-Hilaire and F. Cuvier, *Histoire naturelle des mammifères*. General Research Division, the New York Public Library.

remained behind in the Saint-Louis hospital.[3] He spent two years recovering his health, exploring his surroundings, and drawing and painting landscapes.[4] On his return voyage to France, he purchased a mongoose at the Cape of Good Hope and kept the animal as a pampered pet until its death some fourteen years later. His success with the mongoose augured well for his later efforts with other wild animals.[5]

Back in France, Milbert supervised the illustrations for the government's report on the Baudin expedition. Then he published a two-volume illustrated account of his personal experiences at Isle de France, the Cape of Good Hope, and Tenerife.[6] That voyage, and his reading of Humboldt's accounts of expeditions in the Americas, reawakened his wanderlust. In 1815, shortly before his forty-ninth birthday, Milbert left family and friends and sailed for the United States. He hoped to serve France as an explorer but lacked a commission, so he settled in New York City, supporting himself by painting portraits and giving art lessons.[7] Then, late in 1817, the new French ambassador to the United States, Jean-Guillaume, baron Hyde de Neuville, hired both Milbert and Charles Lesueur (another member of the Baudin expedition) to collect specimens for the Museum in Paris.[8] Milbert sent his first shipment on 10 February 1818: four crates of fish, shells, reptiles, birds, and mammals. Nearly half the specimens filled gaps in the Museum's collection, and some were previously unknown to science.[9]

Georges Cuvier, as Museum director, wrote to thank Milbert and promised to inform the president of the Council of Ministers, the Duke of Richelieu, of his efforts. He specified what he expected from Milbert as a voyager naturalist. Responding to a question about geology, Cuvier recommended that Milbert collect fossils and note the veins and beds and depth where he found them, though warning that reconstructing the strata "requires sedentary work to be well done and can scarcely be the object of a rapid voyage."[10] Cuvier repeatedly returned to this theme, that voyagers should devote their time to collecting specimens and leave their interpretation to the cabinet naturalists back home.

Cuvier told Milbert that if he went to the Great Lakes he should "try to procure the skins and heads of the large animals of the interior, such as the buffalo or bison, the large deer or wapiti, etc." Small mammals and freshwater fish, reptiles, and shells were also wanted. Indeed, wrote Cuvier, "we possess so few productions from the interior that you ought not neglect any of them." Concluding that "our pecuniary means are a little limited," Cuvier expressed confidence the Museum would reimburse Milbert's expenses "within reasonable limits," and he promised also to urge "M. de Richelieu's" support.[11] Soon after receiving the famous zoologist's letter, Milbert decided he could do better than sending the skins and heads of North America's largest animals. He would send the animals alive.

The idea of procuring live buffalo (bison) first came up (June 1818) when ambassador Hyde de Neuville informed the Museum that hunters were offering him three to send to France. Hyde estimated the cost of acquisition and transport at 3,000 to 4,000 francs, lamenting, "it is often at great distances and in the least inhabited places that nature holds in reserve those of her riches and phenomena most worthy of attention. These excursions are very expensive in countries of such a vast extent."[12] Since his office could not fund the project, Hyde suggested the Museum ask Richelieu to foot the bill.

Milbert reported in October that he was embarking ten crates of specimens plus a live eagle and a live opossum from "the frontier of the State of Pennsylvania." He assured the Museum he would obtain the large animals it desired thanks to the network of correspondents he had established.[13] In November he wrote again saying he had just sent a case of acorns and would soon be sending trees, an elk, and several deer.[14]

When the Museum wrote to Richelieu in December it tallied and praised Milbert's recent shipments. The collection sent in June consisted of fourteen mammals, 128 birds, eleven reptiles, fifty-fish, forty-two mollusks, two crabs, and seeds of a variety of early maizefrom Massachusetts. The crates sent in October contained "six quadrupeds, fifty-eight species of birds, six reptiles, fourteen fish, a great number of freshwater shells, seven species of crustaceans, and 181 species of insects." It was testimony to Milbert's commitment, the administration averred, that everything (including species new to science) had arrived in excellent condition. As for the three buffalo Hyde de Neuville proposed, the Museum said it would be very pleased to receive them, but the funding would have to come from Richelieu because its own funds for the year had been exhausted by the transportation costs of certain animals it needed and others that it did not, notably, the lions given by the king of Morocco.[15]

A letter to Milbert sent the same day lauded his efforts and informed him his shipments had arrived "in very good shape," but it gave less space to praise than to requests for more specimens. Drafted by Cuvier, it urged Milbert in collecting birds to focus on their differences in age, sex, and season, features that would help museum-based naturalists overcome uncertainties in distinguishing species. "It is the observer who is on location," Cuvier acknowledged, "who must enlighten us on these doubts." Yet the Museum did not want redundant specimens (like the birds of Milbert's latest shipment, which were mostly species he had sent before). Rare birds were preferable, or juvenile birds or females accompanied by "a note that will indicate to us the species with which we must associate it."[16]

Cuvier also asked for more shells, small mammals (such as "the mole of Virginia"), and reptiles, specifically "a large salamander described by Dr. [Benjamin] Barton," preserved in alcohol, plus specimens of fish that had been described by "Dr. [Samuel] Mitchell." Cuvier closed by asking Milbert to thank Mitchell for his assistance and to ask him which European productions he might want in exchange.[17]

Milbert paid careful attention to the Museum's requests for specimens and its instructions regarding specimen preparation and transport. He was keen to acquire a live bison, and his early inquiries brought him one in March 1819, but it was in poor health when it arrived in New York and died only days later. Instead of loading the animal on *La Ville de Rouen* bound for Le Havre as he'd intended, Milbert could only send its skeleton. Along with dead specimens he embarked two live owls and an agouti, describing the latter as "very tame and caressing" and "my only society," commendable for its "amiability."[18] Having only alluded to his lack of friends, Milbert was more insistent regarding his lack of funds. He told the Museum he much preferred being a voyager naturalist to teaching painting, but to do so he needed financial assistance.

In spring 1819 Milbert was planning an expedition to the Allegheny Mountains and beyond, as far as Lake Michigan and "the land of the Illinois," where he expected to find animals "not yet troubled by man's presence" and rivers filled with "precious productions." He would visit Lake Erie and Lake Saint Claire, head down to Kentucky, Tennessee, and Carolina, then return up the Atlantic coast to Chesapeake Bay and the Delaware River.[19] Despite the Museum's requests to the minister of the interior on his behalf, funds for the proposed journey never materialized.[20]

Milbert continued sending specimens, embarking in April 1819 on the American ship *Nimrod* various preserved items and three live animals: two young "Louisiana deer" and a crocodile. Eight days into the crossing, a fierce storm tossed the animals about in their cages violently enough to kill them. The deer's bones were so badly broken that the captain chose not to keep them, but the crocodile's remains eventually reached the Museum.[21]

In May 1819 Milbert successfully shipped two live raccoons to the menagerie. Then in July, he began preparing for his greatest triumph to date, embarking a healthy bison brought to him by a keeper who led it on a long cord attached to an iron ring in its nose. Milbert sought to further tame the bison by using food as a reward, tempting it with hay and chopped potatoes and adjusting how much he gave it according to whether its behavior toward him was gentle or aggressive. Eventually the bison allowed Milbert to hold it by the horns and comb its mane:

> By means of steady and prolonged efforts, I brought the bison to the level of docility to which I had undertaken to reduce it. Certain thereafter of its obedience, I was not afraid to lead it morning and night to the banks of the Hudson River, and to have it plunge into the clear waters. The only precaution taken in this case was to hold it by a long rope, because it would have been easily able to cross the river and escape in the neighboring forests. . . . Sometimes I took it walking in the city to accustom it to not being frightened by seeing a great number of people, and the docility of this fine animal was no small subject of interest for the inhabitants.[22]

Milbert's account of his experience with the bison all but erased the man who was its keeper, even though he was the one who had brought it to New York from the North American interior and saw to its safe transport all the way to Paris. Milbert neither named the man nor described how the bison behaved in his company. Nor did he say whether the man was present on the excursions to the Hudson (as he probably was). Milbert did report that the keeper watched while Milbert trained the bison with food.

Milbert was surely pleased when he received the Museum's letter that told him "The buffalo has arrived in Le Havre in good shape. It will be directed to Paris where we will receive it."[23] This news came as a note at the very end of a letter primarily intended to inform Milbert that many fish in his two previous shipments had arrived badly damaged. In a barrel of fish specimens preserved in a briny liquid, many specimens had apparently been ruined by contact with the sharp, bony armor of a sturgeon. The fish in the second shipment were only slightly better. The letter instructed Milbert to wrap the fish individually to keep them from rubbing against each other and to counteract the tendency of the liquid to dissolve fish cartilage.

The letter's author, Georges Cuvier, encouraged Milbert by crediting him for his choice of specimens and assuring him that by following the recommended precautions, his future specimens would be more perfectly preserved and therefore of greater importance for science. Cuvier listed the species and categories of animals he wanted. The letter's emphasis on preserved specimens, contrasted with the quick note at the end about the live buffalo just arrived in Le Havre, highlights the comparative anatomist's priorities.[24]

In mid-November 1819 Milbert dispatched the most ambitious collection of live specimens he had yet attempted. Along with three crates of preserved specimens, he embarked on *La Jeune Louise* some small trees (American oaks that he hoped could be naturalized in France); a young cougar; two raccoons (male and female); two Louisiana deer (male and female); a black bear; an eagle; a hawk; some tortoises; and some small fish in barrels. The fish represented a considerable effort on his part. He painstakingly acclimatized fish caught on the New York coast, accustoming these saltwater creatures to freshwater over the course of several weeks before entrusting them to a captain who designed special barrel holders to prevent their suffering from the ship's pitching and rolling.[25] In the letter accompanying the shipment, Milbert said he hoped the raccoons could be naturalized in France because their flesh was edible and their fur highly prized, especially for making hats. He reported on the raccoons' diet; noted that the cougar was gentle enough to go about on a leash; and admitted he bought the bear because its price was too good to pass up. He recommended that the live fish be put in a pond on arrival so they could be subsequently introduced into France's rivers and streams. All the animals were in good condition, Milbert said, and the ship's surgeon had specific instructions on their care and, if necessary, their dissection.[26]

When *La Jeune Louise* reached Le Havre late in December 1819, the shipment remained in good condition, but a yellow-fever quarantine prevented the immediate disembarkation of the specimens on board, and a spell of bitter cold killed the fish in their barrels.[27] Frédéric Cuvier sent his most trusted keeper, Magloire Rousseau, to bring the surviving animals to Paris.[28] When they reached the menagerie, Frédéric Cuvier wrote to Milbert reporting that the bison from the previous shipment had adapted to its new home and was as tractable as one might expect a bull to be, while the "pretty deer," raccoons, bear, cougar, and birds of the latest shipment had all made the journey to Paris in excellent shape. He did not mention the tortoises or the fish. The deer was much appreciated, Cuvier explained, because it represented a species the Museum had known only from illustrations. The raccoons and the black bear were not new to the Museum or menagerie, but the cougar was possibly a member of a new species, different from the South American cougar. One would be able to tell by watching how its color changed as it aged.[29]

Frédéric Cuvier complimented Milbert's efforts at length, but like his brother Georges, he also pressed Milbert for more specimens. The Museum's naturalists, he said, did not know the deer of North America well. They had only vague notions of the wapiti or "moos deer" [*sic*] and the caribou. Live specimens, he explained, "would be of the highest importance for us, because several of their characteristics wear away when they are skinned." The menagerie superintendent also wanted American sheep, the gray bear, and the small rodents and omnivores living beyond the Allegheny Mountains. He wanted Milbert to learn if there were badgers in America and, if so, if they were the same as the animal called the "carcajou." He asked for North American foxes and the small species of the cat family that change color through the seasons, creating "much confusion" among naturalists.

Milbert had requested Georges Cuvier's book on mammals, "something on insects," and a good, two-shot gun from the Versailles factory. The Americans would appreciate this last item, he suspected, because "*tous sont armés*."[30] Frédéric informed Milbert that the Museum was sending him his brother's *Règne animal*, multiple copies of Geoffroy Saint-Hilaire's *Philosophie anatomique* and related publications, and the gun. He advised Milbert to share the publications with Charles Lesueur and distribute any extras as he saw fit. The Museum appreciated that it served their interests to keep both naturalists abreast of the latest developments in nature studies.[31] It likewise served their interests that Milbert have a good gun for hunting.

In the courtyard of his New York home, Milbert had aviaries and huts built to accommodate the birds and large mammals sent to him. He had smaller lodgings for small quadrupeds (especially opossums) and special arrangements for quail, turtles, and tortoises. He wanted the animals constantly under his observation so he could learn their habits and prepare them for domesticity or, as he sometimes put it, for "enslavement" in the menagerie.[32] Deer and elk proved particularly amenable. Of the first male elk he later wrote: "[He] stayed with me for a long time and became so familiar that standing on his back feet he rested his front feet on my shoulders and licked the top of my head. Before sending him to Europe, I concentrated on learning his tastes and habits. He preferred potatoes and newly cut grass above all other food, and liked only the clearest water."[33]

Milbert embarked the elk in March 1820. The Museum alerted the commissioner of the marine at Le Havre and the Museum's agents, Eyriès frères, to its arrival and told the latter to send the elk to Paris by wagon, noting that it was "precious" and one could not take too much care in ensuring its conservation. Late in May the Museum reported its safe arrival to Milbert, adding that it had written a second time to the ministers of foreign affairs and the interior requesting funding for his proposed trip to the Alleghenies.[34]

After his success with the male elk, Milbert sought a female, hoping elk might one day inhabit France's royal forests. In August 1820 he found another male at such a bargain rate (thirty dollars) that he felt he had to buy it and send it to Paris.[35] He found a female elk in late October and chose to keep her through the winter.[36] He named her "Betzy" and provided her with a companion, a female Louisiana deer. Both became so tame they would enter his house. The Louisiana deer, he wrote, "more alert than her companion, climbed up to my room and positioned herself near the fireplace to warm herself."[37] Milbert waited until spring to send them off.[38] His efforts to keep flying squirrels ended less happily; they were easily tamed (they even slept in his bed), but they eventually fell "under the paw" of his cat.[39]

Milbert kept his promise to secure a female bison and a moose for the menagerie. The bison came from the banks of the Mississippi; the moose came from Maine. Two years old and weighing twelve hundred pounds, the moose was not yet fully grown. Although it was tame when he received it, Milbert wished to train the moose further, taking it for morning walks to give it exercise and study its gait.[40] In August 1821 he embarked the moose, the female bison, and a collection of preserved specimens on the English ship *Betsy* bound for Le Havre. When they arrived in mid-September, Eyriès frères consulted with the Museum and judged that steamboat was the most comfortable means of transporting the animals to Paris.[41]

Milbert worked assiduously to supply the Museum with the information and specimens it requested. When Geoffroy Saint-Hilaire wanted details about opossums, Milbert reported everything he witnessed personally and sent the Museum preserved and live specimens.[42] Yet Milbert's dreams extended beyond the Museum's collections. He wanted to see North American species thrive on French soil and become integral to the landscape and economy. In January 1823 he wrote to Geoffroy saying he believed elk and bison could multiply in France. He imagined the same for the wild goose (an excellent bird "for the table"); turtles that the Americans and British made into soup; and the Louisiana deer, so renowned for its meat and the uses of its skin that even in deep forests it was not safe from local hunters.

He had high hopes for quail, sending specimens to the Museum as early as January 1820 and continuing until the government's support for him ended in 1823.[43]

In the summer of 1822, intent on developing his collections, Milbert trekked into the mountains of Vermont and Massachusetts and contracted yellow fever. He made his way to Boston, where the French consul, Joseph de Valnais, took him in and arranged his treatment by John Colins Warren of Massachusetts General Hospital and Harvard Medical School.[44] For six weeks Milbert was unable to leave his bed. His convalescence was long and painful.[45] When the Museum received his account of his illness (early 1823), it wrote a single sentence congratulating him on his recovery before laying out "the objects to which we wish you to particularly direct your attention." Frédéric Cuvier wanted small species of rodents (and especially their teeth). Georges Cuvier wanted more fish.[46]

Milbert followed up on everything asked of him, correcting the misconceptions the Paris professors had gained from the writings of naturalists who hadn't seen things firsthand.[47] Just as the Museum's letters revealed what excited the professors most, so did Milbert's letters convey his priorities: the quail he wanted to send. They needed a protected place aboard ship, he explained, and he was waiting for a captain who could provide it. He specified their need for a suitable habitat conducive to their multiplication, noting the male's "extremely gentle and melancholy" call in mating season. Quail were easily tamed, he reported, though "a miserable cat killed the pretty male that I had for some time in my room."[48]

In March Milbert sent eight more crates of specimens.[49] In April he announced he intended to profit from the new season to procure "entirely new objects" for the Museum.[50] In May he sent quail and voiced hopes of procuring a male and female caribou and a female moose. He was eager to serve the Museum, he said, if it could secure him the needed resources.[51] Long before his letter reached Paris, however, the minister of the interior refused funding for Milbert altogether. After getting the bad news, Milbert prepared to return to France, sailing (1 October 1823) on the *Paris*, bound for Le Havre. The voyage ended disastrously when a gale drove the ship onto the rocky Normandy coast. Milbert survived, but it was an inauspicious homecoming.[52]

At Milbert's request, the Museum prepared for the minister of the interior a report of Milbert's achievements in his years as a voyager naturalist in France's employ.[53] In all, he sent fifty-eight shipments of specimens from North America to France, enriching the Museum's mineralogical collection with two hundred kinds of minerals and seven hundred specimens of rocks (many of considerable size). For the botanical collection he sent crates full of seeds of numerous species in addition to six hundred live trees. His enhancement of the Museum's zoological collections was his greatest triumph: specimens of approximately two hundred mammals, two thousand birds, six hundred reptiles, and twelve hundred fish. He collected a great many freshwater shells, plus about four hundred species of crustaceans, arachnids, and insects. Many of the species he provided were either unknown to the Museum or absent from its collections.

The most dramatic of Milbert's contributions were the live animals he sent to the menagerie, with some species represented by both sexes: twenty-six reptiles; forty-nine mammals, including bison, elk, moose, several species of deer, the American cougar, and many opossums (owing to Geoffroy's special interest); seventy birds, including eagles, other birds of prey, ducks, geese, and twenty-two quail. Hailing Milbert's animals as "the principal ornament of the menagerie," the Museum was all the richer for Milbert's efforts, but he was all the poorer.[54] Government funding cobbled together over the span of seven years, 38,200 francs total, had been insufficient to cover his expenses.

Milbert wrote to the minister of the interior (April 1824) that his wages had amounted to roughly fifteen francs per day, scarcely more than those of a skilled French worker. He estimated he had spent an additional ten to fifteen francs daily, placing his personal loss at over 25,000 francs. He asked that these expenses be reimbursed in a way that would compensate for his sacrifices as a collector abroad and allow him to maintain himself and his family modestly. He had given up the school of painting he established in New York, and he had refused a job at the state military school. He had endured fevers and countless privations in his travels in North America, self-funded with the sale of art objects he had brought from France and loans he'd taken to sustain the live animals in his care. Returning home after a brush with death at sea, he was exhausted and still feeling the aftereffects of yellow fever, which would leave him "suffering for the rest of my days."[55]

The minister wanted the Museum to pay him, but the administration respectfully explained that only he could compensate Milbert and noted "how discouraging and inappropriate it would be" for a man who fulfilled his mission "with zeal, in part at his own expense, and to the detriment of his health, to return home and be obliged to spend the rest of his life in a state bordering on penury."[56] Eight months passed before the minister authorized Milbert's reimbursement at far less than the requested sum, leaving the Museum to supply the balance. It gave him 5,500 francs from the voyagers' fund and 1,500 francs from the budget for the galleries, leaving him still well short of the 25,000 francs he had spent. At their assembly of March 29, 1825, the professors nonetheless decided that their support for Milbert was concluded.

Several months later, Milbert proposed to travel on the Museum's behalf to North America, where a great canal was being dredged and quantities of valuable fossils were bound to surface. He promised to acquire any indigenous animals the Museum wanted for the menagerie, plus the remains of anything that could not arrive alive. The professors decided that the Museum lacked the funds to support Milbert's new venture, perhaps feeling that his previous efforts had made North American fauna and flora sufficiently well known or that Milbert, nearing sixty, would fail in his mission, as had far younger voyager naturalists.

Milbert continued to lobby the government for the full amount he was owed. He asked for a decent pension and to be named to the Legion of Honor. In a moment of bitter angst, he implored the minister of the interior to read the Museum's report of his accomplishments. If the minister could then honestly say he didn't deserve full recompense, Milbert said, he would "go hide in the wastelands of America his half-bald and whitened head, his humiliation, and his disgust."[57]

Milbert was made a chevalier of the Legion of Honor on 12 March 1831; his portrait by Rouillard was exhibited that year at the prestigious salon at the Louvre.[58] In 1834 he sought government support to travel to North America again, to collect seeds of trees to naturalize in France. The Museum provided a recommendation, while noting that his project was not directly in its interests and the Ministry of Commerce was the appropriate funding agency.[59] Milbert died in 1840, age seventy-two. The Museum urged the government to award his wife an annual pension, as he had wished. Milbert had asked for 2,500 francs. She instead received 1,000.[60]

A maritime merchant, ship owner, and sea captain with a passion for natural history, Jean-Jacques Dussumier's (1792–1883) circumstances were substantially different from Milbert's. Between 1816 and 1838 he made eleven long-term commercial voyages, spending more time at sea and abroad than in France while becoming one of the Museum's most consequential collectors.[61] Dussumier was twenty-six in 1818 when he introduced himself to the Museum by offering it specimens collected on his first commercial voyage, to Manila. The Museum selected eleven birds, a monkey, and some shells, giving Dussumier in return "other birds recently come from America" (undoubtedly from Milbert).[62]

Natural history specimens moved along the same pathways of commerce, colonialism, and war as did other products of nature like spices, sugarcane, coffee, cotton, and opium. Dussumier's second voyage (1818–20) aimed at furthering commercial relations with India, China, and the Philippines. In Bengal, he purchased cotton and opium to sell in Canton; in Manila, he loaded cargo for his return; and in his leisure time, Dussumier collected specimens.[63] He returned from the voyage with "a great number of interesting birds" from China and the Philippines.[64] Among the live specimens was a mammal from the Philippines that Frédéric Cuvier identified as a zibeth (*Viverra zibetha*). Having it allowed Cuvier to determine that the zibeth and the civet were distinct species, contrary to Buffon's opinion but consistent with that of Linnaeus and Frédéric's brother Georges.[65]

Dussumier's gifts offered an occasion for Frédéric to remark on the Philippines' potential for new species and new zoological ideas: "Similar to all the great isolated lands, to Madagascar, Australia, and New Guinea, one would call [the Philippines] the result of a new creation; the majority of their productions display new characters and announce unknown principles of existence; thus the naturalists who explore them will be certain to make numerous and important discoveries in all the branches of zoology." Frédéric was right that island fauna would be of great consequence; he had no inkling that the idea of evolution would have anything to do with it. Ironically, while naming the Philippines as a source for previously unknown species, Frédéric did not appreciate that his "zibeth" specimen constituted one. A dozen years later, the British zoologist John Edward Gray established Cuvier's zibeth as a new species, *Viverra tangalunga*, distinguishable from *Viverra zibetha* by the markings on its tail.[66]

Dussumier's generosity was not selfless. Having presented the Museum his zibeth and his latest set of preserved birds, he asked to be recommended to the minister of the navy and colonies so he could appeal to him to be named to the Legion of Honor. He acknowledged that the "feeble services" he had thus far provided the Museum might not warrant it, but he felt such a distinction might impress foreigners and encourage their cooperation in his collecting. The Museum obliged, sending its recommendation to the minister of the navy, who forwarded it to the minister of the interior.[67] Nothing came of it at the time, but Dussumier's enthusiasm for natural history was not dampened, and by 1825 the government had named him honorary consul to China.[68]

On his third voyage, Dussumier brought from India birds and fish for Georges Cuvier and varieties of live mulberry trees for André Thouin to propagate.[69] His fourth trip produced more birds and fish, especially from the Seychelles, many entirely new to European naturalists. Business kept him close to India's Malabar Coast on his fifth journey, but he managed to acquire mammals, reptiles, and what Georges Cuvier described as "one of the most beautiful and richest collections of fish yet received."[70] Dussumier had kept his fishermen busy. The 550 specimens represented two hundred different species, whose scientific value was enhanced by the notes Dussumier made on the fishes' colors when freshly caught. He also closely observed and collected specimens of six or more species of dolphin.[71]

Also arriving in Bordeaux with Dussumier from this fifth trip were some live animals to offer for the menagerie: a sheep from Moka; two males and a female jackal; a mongoose pair; a Chinese bonnet monkey pair; and thirteen tortoises. Hearing from him of his total haul, the Museum wrote back immediately that it awaited "with the greatest impatience the new riches that the Museum's collections will soon owe to your enlightened love for science, your indefatigable zeal, and your inexhaustible generosity." It noted that the fish were exciting "a particular interest."[72]

Dussumier did not linger in Bordeaux, departing in December 1827 on a ship he had christened the *Buffon*. He now bore the title of "correspondent," which the Museum had bestowed on him.[73] The voyage, his sixth, was also his longest (thirty-three months). He sailed first to the Seychelles, then Bombay, then on to Bengal and the Ganges Delta, stopping en route at Pondicherry. From the Ganges Delta he sailed back to Île Bourbon and Île de France for a rest. Bad weather thwarted his collecting early in the journey, but he did his best, keeping his crew fishing at open sea, catching various specimens, including dolphins, and finding many interesting smaller specimens inside the stomachs of larger ones.

Dussumier then sailed to Bengal a second time, collecting new fish, animals, and live large mammals, including a four-horned antelope. Returning to Île Bourbon, he entrusted his animals to a domestic servant's care. When business called, he sailed yet again to Bengal, where his associates had more live animals waiting for him. Then in Bombay, where he stopped to refit for the return voyage to Europe, he purchased a "long lipped" or "juggler" bear (sloth bear). From there he proceeded to Île Bourbon, adding the sloth bear to a collection that included a lion-tailed macaque, a striped hyena, a royal tiger, two lion cubs, two Timor deer, a female muntjac deer, six axis deer (males and females), the Indian antelope, a four-horned antelope, and a female gray gazelle. A young Indian sailor under Dussumier's chief assistant's direction cared for the animals so well that none were lost in transit, notwithstanding the navigational perils of the Cape of Good Hope.[74]

The Museum welcomed Dussumier's animals on 21 October 1830 and immediately signaled his contributions to the minister of the interior, urging Dussumier's induction into the Legion of Honor. Later that year, "in consideration of the services he has rendered the natural sciences and of the gifts he has made to the Museum in Paris," Dussumier received the coveted medal.[75]

In addition to the live animals he collected on his 1827–30 voyage, Dussumier returned with a stunning number of preserved specimens. They included 1,500 fish, which by Georges Cuvier's count represented 481 species, of which at least a hundred were new to science and many others could now be better known thanks to Dussumier's field observations and excellent preservation work. Among the other specimens were mammals (including thirteen species of dolphins), birds, reptiles, mollusks, and crustaceans. In reporting to the Academy of Sciences about the results of Dussumier's latest voyage, Cuvier's made the case for government support of his future efforts. It proved to be the zoologist's last gesture on behalf of this valued collector; Cuvier did not live to see Dussumier's next contributions to the Museum.

Merchant shippers operating in Bourbon and India in the early 1830s were beset with commercial difficulties, and Dussumier seems to have been no exception. His seventh voyage left no traces at the Museum. At the New Year 1833 he asked the Museum for financial support for a new commercial and collecting voyage that would take him to Île Bourbon, the Malabar Coast, through the Sunda Strait between Sumatra and Java, then on to China and the Philippines. He planned to devote time and resources to collecting, but he wanted assistance purchasing specimens and paying the exorbitant anchorage fees—35,000 francs for commercial ships of his size in Canton. He asked the Museum for 12,000 francs per annum for two years.[76]

The Museum appealed to François Guizot, minister of public instruction, its new intermediary, reminding him of Dussumier's contributions to its galleries and menagerie, but he was unmoved. Dussumier, he told the Museum, was "not, properly speaking, a voyager naturalist" but was instead a "rich ship owner." As Guizot put it, "he makes his long sea voyages for commercial interests, and since he loves and knows natural history, he collects, *en passant*, so to speak, all that seems curious and worthy of being collected." The amount he requested nearly matched what Victor Jacquemont was getting (15,000 francs) for performing "a special mission for the government as a voyager naturalist." Guizot promised to write to the minister of commerce and public works, who had previously funded Jacquemont, to see if he had something for Dussumier, but he doubted it.[77] What neither Guizot nor anyone else in France knew at the time was that the illustrious young voyager naturalist Jacquemont had died of cholera on 7 December 1832 in Bombay. News of the thirty-one-year-old's death did not reach Paris for months.[78]

Dussumier set off on his eighth voyage uncertain whether the government would provide support, which ultimately it did not.[79] The Museum, however, decided to assist him with 4,000 francs from its voyagers' budget for 1833 and soon extended the assistance to 8,000 francs for 1834 and 1835.[80] The fruits of Dussumier's eighth voyage were fish, reptiles, and four live mammals: three Abyssinian sheep and a kevel, or gray gazelle. It was not the largest of his tributes to the Museum, but still it was significant. The sheep seemed good candidates for acclimatization in France. Beneath their rough, woolly coats was a fine undercoat that held promise for the manufacture of fine cloth.[81]

Dussumier's next voyage was brief and less wide-ranging than usual. He spent less than a year on the Malabar Coast but returned (in May 1835) with as many live animals as he had collected on his 1827–30 voyage. A female hemione (the wild horse of India, or onager) was his greatest prize. In addition, he had a male and two female hog deer; a male muntjac (a small deer); and the female of a new species of gazelle, a form he previously thought was a female muntjac. He also acquired a male long-lipped bear to go with the female he had provided five years earlier; a female wild boar; a male Abyssinian sheep, and three lion-tailed macaques.[82] After the animals arrived at the menagerie (7 June 1835), Isidore Geoffroy Saint-Hilaire, filling in for Frédéric Cuvier in the superintendent's absence, confirmed that the wild boar, hemione, hog deer, muntjac male, and the "new species of antelope similar to the Grimm and the Guévri," were firsts for the menagerie. The male Abyssinian sheep was also important because the previous one had been castrated and was useless for reproductive purposes.[83]

The Museum issued a press release calling Dussumier's latest contributions to the menagerie "one of the most beautiful collections of live animals that this establishment had yet received from India." The hemione was "one of the menagerie's most precious scientific acquisitions since the arrival of the giraffe."[84] The piece listed the various categories and numbers of preserved specimens, noted Dussumier's appointment to the Legion of Honor, and mentioned the Museum's help in defraying the considerable expense of maintaining so many live animals aboard ship.[85]

Dussumier spent scarcely a month in France before departing on his tenth voyage (1835–36). Before leaving, he requested that the Museum's artist, J. C. Werner, make sketches of the male muntjac and female gazelle to send to India to help Dussumier's contacts there identify and procure mates for these animals. His voyage was hampered by bad weather, but he managed to bring back twenty-two live animals, his largest collection yet. Only one, a duiker antelope, died en route from Bordeaux to Paris; the others arrived safely at the menagerie (27 September 1836). Frédéric Cuvier tallied them as four four-horned antelopes (one male and three females); two muntjacs (one male and one female); two hog deer (one male and one female); one male wild boar; two mongooses; one lioness; two chacma baboons from the Cape; two Chinese bonnet macaques; and three sheep.[86]

Dussumier asked another favor from the Museum. He intended to stay in France for three months and was worried that the Bureau of the Marine might call for a levy of sailors and take some of his crew. His men had been with him for years, were used to helping him prepare specimens, and were "indispensable to the success of my voyages from the scientific standpoint." He hoped the Museum could persuade the minister of the navy to tell the marine commissioner at Bordeaux to leave Dussumier's nine-man crew intact.[87] Meanwhile, Dussumier was building a larger ship. He wrote to the widow of Georges Cuvier for permission to name it after her husband, who had so appreciated his efforts to advance the scientific study of nature.[88]

It is uncertain exactly when Dussumier left on his eleventh voyage or the name of his ship at the time, but he returned in July 1838 on the *Georges-Cuvier*.[89] A journalist hailed his delivery of live animals, live plants, and hundreds of preserved birds, fish, insects, and mollusks. The writer admired a pair of hemiones for their elegant form, attractive color and potential as "a new conquest for our agriculture." There was "a very beautiful species of deer" (five individuals, males and females); "a bearded monkey" (a lion-tailed macaque); two squirrels of a little-known species; "a magnificent tiger" (actually a female long-tailed panther), and "a wild cat, a boar, and several other animals." Among the plants were representatives of the four palm species of the Malabar Coast, "twelve young coconut palms, as fresh as if they left their country of birth yesterday."[90]

Dussumier apparently took at least one more trip, from which in August 1839 he presented the Museum with a nilgai, a pair of Chinese bonnet monkeys, and a black sheep. The administration replied with thanks, noting that the nilgai not only was the menagerie's first, but that it was also "one of its most precious acquisitions" and "an individual of remarkable beauty."[91]

Unlike Milbert, Dussumier did not end his traveling career impoverished. Comfortably retired in Bordeaux, he helped the Museum acquire several South American animals that arrived there, including a puma and two llamas.[92] In 1855 he wrote about the hemione for the *Bulletin* of the Imperial Zoological Society of Acclimatization, of which he was an honorary member (and in 1857, co-vice president). In 1883, more than four decades after concluding his adventures on the high seas, he died in Bordeaux at age ninety.[93]

∴

Milbert and Dussumier were the menagerie's greatest suppliers in the post-Napoleonic era. In this respect they acted more on their own initiative than on promptings from the Museum, which applauded their efforts but still felt the voyagers' time and resources were best spent collecting and preserving specimens. This attitude was voiced in the third edition of the Museum's *Instructions for Voyagers* (1829) where the section about collecting live animals was amended with the caveat that, given the difficulties of transporting live animals, "there is an easier harvest, and of which the results are more far-reaching: the remains of dead animals."[94] The amendment might have also noted, "and less expensive." The Museum's limited funds were always a concern, and in the years of the Restoration they became a matter of contention with the government in settling the transportation costs for animals arriving as diplomatic gifts.

Economies aside, the Paris zoologists far preferred the remains of dead animals, as Georges Cuvier stressed time and again. Reviewing in 1825 the results of the Duperrey Expedition to the Pacific, he wrote that it was "a great error" for voyagers to do "anything other than collect the means for [further] study." They should devote themselves to preparing specimens and attaching information such as where they were collected, local names and uses, natural colors and forms. They should not waste time worrying about nomenclature or distinguishing among species. That was the business of the cabinet naturalists at the Museum, where the goal was to possess, name, classify, and display all of nature's productions.[95]

In 1830, reviewing for the Academy of Sciences J.-J. Dussumier's latest haul of specimens, Cuvier emphasized just how much natural history depended on the efforts of voyager naturalists, though he still reserved for the cabinet naturalist the role of making thorough sense of the voyager's specimens. He wrote, "Without the dedication of voyagers an essential part of the work of naturalists would soon stop, for lack of sustenance." Describing natural history as "this vast enterprise of registering all natural beings, each one in its own rank, each with a name by which to designate it with surety to learned men of the whole world," Cuvier allowed that it "could not have even been conceived without the help of courageous men willing to sacrifice their health and even their lives to collect the productions of climates far from the places where one could make a deep and comparative examination of them."[96]

For of the history of the menagerie in Paris, what was so remarkable about Milbert and Dussumier was that they went out of their way to collect live specimens alongside the dead ones the Museum wanted most, though this entailed the extra costs that so burdened Milbert and that saw Dussumier giving space aboard ship to live animals instead of a more lucrative cargo. They did this because they saw themselves as serving France as well as the Museum. For Milbert, the landscape painter and nature lover, his "ardent desire" was to have American trees, birds, and mammals embellishing the French countryside and American fish swimming in France's rivers and lakes.[97] Dussumier found fewer specimens in his travels to acclimate in France, but he had the highest hopes for the hemione, which he was confident could be domesticated.[98]

The public appreciated Milbert's and Dussumier's contributions to the menagerie. Visitors enjoyed the spectacle of the live animals, and the professors understood that the Museum's reputation benefitted from that pleasure. Today Milbert and Dussumier are remembered less for their contributions to the Museum than for the species named after them. "Milbert's tortoiseshell" is a common North American butterfly well known to young collectors.[99] "Dussumier's surgeonfish" (*Acanthurus dussumieri*), one of his many animal namesakes, is an Indo-Pacific fish remarkable for its vibrant colors.

A problem with Dussumier's surgeonfish from the cabinet naturalist's standpoint is that its beautiful colors while alive are largely lost after the fish dies and is preserved in alcohol. When Achille Valenciennes named the fish after Dussumier, he had to rely on what Dussumier had reported about its vivid orange and gray or violet stripes and its golden yellow eye band, since they were no longer there to see.[100] The cabinet naturalist constantly faced the question of how much was lost in a preserved specimen. But one might also ask what was lost when a live animal was plucked from its homeland, made captive in a menagerie, and deprived of its normal social congeners and the stimuli of its natural habitat. It was a question for viewers of the menagerie's animals as well as for Frédéric Cuvier as he sought to make the animals the subject of scientific study.

[3.5]

The Bison

Among the specimens Georges Cuvier wanted most from North America was the bison, the "buffalo of the Anglo-Americans." He wanted a skeleton, or at very least a skull, to compare the bison with the aurochs, the huge ancient ox that once lived in temperate Europe but was now found only "in the great marshy forests of Lithuania, the Krapaks [the Carpathian Mountains], and the Caucasus." Having the two forms side by side would allow him to evaluate their affinities as part of his study of the history of life on Earth.[1] Cuvier had an aurochs skeleton, thanks to Napoleon's triumph over the Austrians in 1809. When it appeared that the French might choose from Vienna's natural history treasures, Cuvier wrote to Count Daru, French intendant general of the Austrian occupation, naming the aurochs as "the most precious object for us to acquire."[2] Cuvier believed there was a live aurochs in the menagerie at Schönbrunn, but his information was out of date. The animal had recently died, just after Daru's arrival in Vienna. The best Daru could do was ship the animal's skeleton and skin to Paris. Bones were Cuvier's element, so he was content.[3] When he compared the aurochs bones with those of bison and domestic cattle, he concluded they were sufficiently distinct to regard the animals as three different species, not just a single species, as Buffon had proposed.[4]

Whereas a military triumph brought Cuvier his first aurochs skeleton, Napoleon's ultimate defeat and the subsequent maritime peace enabled the arrival of the Museum's first bison. It came after Cuvier had published his *Règne animal* (1817), where he stated that the bison "lives in all the temperate parts of North America." France's new ambassador to the United States, Hyde de Neuville, disabused the Museum of that notion, reporting that the species was not to be found anywhere near the east coast. He called the animals by their American name, "buffalo."[5]

Jacques Milbert offered little information about the provenance of the two bison/buffalo he sent to the Museum other than that the second (a

Figure 3.5.1. "The bison." C. de Lasteyrie after Jean-Charles Werner. É. Geoffroy Saint-Hilaire and F. Cuvier, *Histoire naturelle des mammifères*. General Research Division, the New York Public Library.

young female) was from the banks of the Mississippi. Of the first bison he embarked for Paris, a male, all that is known of its early history is that it reached Milbert in New York under the care of a keeper who led it with a rope attached to an iron ring in its nose.

Milbert gave the male bison ample time to recuperate from its journey to New York before sending it to Paris. While he had it, he sought to tame it using a program of rewards. He then embarked it with its keeper on *La Bonne Henriette*, commanded by Captain Jappie, bound for Le Havre. Jappie charged 1,000 francs for the operation: 200 for the bison's water and 300 and 500 respectively for the passage of the keeper and the animal.[6]

Neither in his letters to the Museum nor in his account of his time with the bison in New York did Milbert ever mention the keeper's name. Nor does the man's name appear in the procès-verbaux of the professors' meetings. This was not unusual; the meetings typically recorded the names of animal donors but not the conductors who performed the menial task of caring for the animals on the way to their destination. Nonetheless, the name can be excavated from the archives. It was Peter Johnson. He was most often identified, however, not by his name but by his race. Johnson was a free American black man whom the Museum and its shipping agents in Le Havre for the most part referred to simply as "le nègre."

In the eighteenth century, philosophers and naturalists generally maintained that racial differences (anatomical features, skin color, or levels of civilization) could be explained by the different environments people inhabited.[7] This did not prevent naturalists (Linnaeus and Buffon included) from stereotyping different races and implicitly if not explicitly ranking them, but it allowed for rising in the ranks under improved circumstances. By the early nineteenth century, a harsher perspective was emerging, as exemplified in Cuvier's *Le Règne animal.* Cuvier endorsed the idea that there was but a single human species, as interracial fertility demonstrated, but insisted there were intrinsic differences "which seem to arrest the progress of certain races, even in the middle of the most favorable circumstances."[8]

Cuvier distinguished three main and "eminently distinct" races: the white or Caucasian, the yellow or Mongolian, and "la nègre" or Ethiopian. With supreme confidence (and apparent self-satisfaction) Cuvier wrote, "The Caucasian, to which we belong, is distinguished by the beauty of the head's oval shape, and it is the one that has given rise to the most civilized peoples, to those who have the most generally dominated the others. It varies by complexion and by the color of the hair." The Mongolian race was characterized by "its high cheekbones, its flat face, its narrow and slanted eyes, its straight and black hair, its thin beard, and its olive-colored complexion." He acknowledged that this race had given rise to empires in China and Japan and sometimes pushed its conquests further, but he asserted that its level of civilization "has always remained stationary."[9]

As for "*la race nègre*," Cuvier judged it to be unattractive in its physical features and unaccomplished in terms of civilization: "Its complexion is black, its hair wooly, its skull compressed, its nose flattened; its jutting muzzle and its large lips manifestly bring it near the monkeys. The tribes that compose it have always remained barbarous."[10] Cuvier's 1817 evaluation of the racial characteristics of the "Hottentot Venus," the Khosian woman Saartje Baartman, was similarly severe. Countering a hypothesis that the ancestors of the ancient Egyptians were black, he insisted that the real question was not about skin color but about skull and brain shapes. The skulls of Egyptian mummies proved, according to Cuvier, that the ancient Egyptians "belonged to the same race of men as us" and were "no exception to that cruel law that seems to have condemned to eternal inferiority the races with depressed and compressed skulls," including the race of the Hottentot Venus.[11]

When Peter Johnson arrived in Le Havre in 1819 and was referred to as *le nègre*, the categorization was freighted with prejudice. The French consul general in New York, Charles, le comte d'Espinville, was nonetheless fully confident in Johnson's abilities, providing him with the necessary papers and passport. He wrote to the commissioner of the marine in Le Havre with suggestions about how to get Johnson and the buffalo to Paris. Since Johnson spoke no French and the buffalo was difficult to lead, he recommended he be given a traveling companion (someone from the Museum or hired in Le Havre). The buffalo, d'Espinville advised, should be allowed to rest for a few days before setting off for Paris. Furthermore, "the conductors will have to make it travel almost only at night, and in short stages, so as not to tire it and attract the curious, who would frighten it." D'Espinville told the commissioner that the animal would be unique in France and that the government ("the Minister") and Georges Cuvier had specially requested its acquisition.[12]

Chabannon, the marine commissioner, sent d'Espinville's letter to Eyriès frères, the Museum's agents in Le Havre, who took charge of the transportation. Alexandre Eyriès's letter to the Museum (11 August 1819) gave the exciting news: "We announce to you with pleasure that Captain Jappie, commander of the ship *La Bonne Henriette*, arrived this morning from New York, is bringing to you a *live* buffalo that we are going to disembark." Eyriès emphasized the importance of the animal's keeper, "*le nègre*." The *procès-verbal* of the professors' assembly summarized the basics: "a live buffalo and a Negro, its keeper" had arrived in Le Havre from New York.[13]

D'Espinville's letter to the marine commissioner variously referred to Johnson as "un nègre conducteur," "the man named Peter Johnson," "le nègre Johnson," and "Johnson." In contrast, the early letters of Eyriès frères to the Museum almost always called Johnson simply "le nègre."[14] Likewise, in its first reply to Eyriès, the Museum called Johnson "le nègre" or, in conjunction with the buffalo, "son nègre" (its Negro).[15]

However, if Alexandre Eyriès's early news about the buffalo only rarely mentioned Johnson's name, he stressed from the start that the man was indispensable for the buffalo's care and transport to Paris. "The Negro," Eyriès wrote, "is very knowledgeable about making the animal obey." Reporting that the buffalo was "very handsome" and in "very good shape," Eyriès attributed its health to its keeper. "It was ill during the crossing," Eyriès wrote, "but the Negro took care of it, bled it &c." He went on to say, "We have lodged this man with a landlady, and he can come to care for the beast when he wants. If you cannot send someone to accompany the Negro, we will be able to give him a man who speaks English and French." In a postscript Eyriès noted that at Johnson's request the agency had provided him with clothing, to be charged eventually against his overall remuneration for the trip.[16]

Eyriès became anxious after four days passed with no reply from the Museum. He wrote again, this time saying that the buffalo was now rested from its overseas journey and could begin its trip to Paris as soon as a companion for the conductor was found. Meanwhile, the Museum administrators had written Eyriès suggesting he might send the buffalo by boat up the Seine, as he had recently done with a bear. The professors wondered if "the Negro" was really needed for the trip, but they left the decision to Eyriès.[17]

Eyriès pursued the idea of sending the buffalo by riverboat, but no captain was willing to board it. They said they lacked deck space for the kind of hut the buffalo had enjoyed on its ocean voyage. Eyriès anyway thought that the buffalo's restlessness and the way it had "tried to play some pranks on us" could have been dangerous on a boat. Plus, the animal was suffering from the heat, and Johnson indicated it needed to walk. Eyriès concluded that the buffalo would travel by land and firmly disabused the Museum of the idea that Johnson might not be necessary for the trip: "He alone," Eyriès explained, knew how to master the buffalo in its "whims" (*lubies*).[18]

Eyriès hired a man named Nozière from a local barrow company to travel with Johnson, serve as interpreter, purchase the buffalo's hay and straw, rent places for the animal to rest (uninterrupted by the curious) en route, and keep receipts of all expenses. He was to pay no more than two francs per day for Johnson's food (the sum cited in Johnson's contract with d'Espinville in New York). Beyond that, Nozière's contract stated: "He will see to it that Johnson does not drink strong alcohol and become drunk. He will oversee the Negro's conduct. But he will not minister to anything in the feeding and care of the animal, leaving the Negro the absolute master in this regard."[19]

Nozière was directed to escort his charges to Paris by the shortest route, to not exhibit the animal along the way, indeed, to travel by night so as not to attract attention. These instructions made sense. Locals may well have been disconcerted by the sight of a black man leading a dark, shaggy, brooding, oxlike creature across the countryside. How far and how long they traveled on any given night would be up to Johnson and the weather. When they reached Paris, they were to go directly to the Jardin des Plantes by way of the Austerlitz Bridge.

Johnson, Nozière, and the bison left Le Havre on the evening of 16 August 1819. They reached Paris nine nights later, waiting until dark to enter the city to avoid notice and spare the animal the traffic and frightening street noises of daytime.[20] Neither man cared to linger in Paris, spending two nights before their two-day return trip by stagecoach to Le Havre.

The Museum was supposed to settle accounts with Nozière and Johnson, but Nozière worried Johnson might rashly spend the whole 190 francs due him if he were paid all at once. It was decided to give Johnson ninety francs in Paris with the balance to be paid by Eyriès frères in Le Havre.[21] In a memorandum regarding the men's payment, Nozière, referred to as "the conductor of the Negro to Paris," was "S. [Sieur] Nozière." Johnson was not "S. Johnson" but rather "Le Ne Johnson." Johnson and Nozière were to get thirty- and twenty-franc bonuses respectively, not because Johnson's services were more valued, but because his month at sea was factored into the amount.[22]

When Eyriès frères reviewed the Museum's memorandum, they added two francs to Johnson's account because the Museum's reckoning of his travel time was a day short. When "the poor devil" discovered that the captain who had promised to take him back to the United States had already sailed, they requested the American consul in Le Havre to give Johnson a free berth on a different ship.[23] The agents' solicitude seems to have reflected their attention to detail and their appreciation of Johnson's competence. Their final letter to the Museum suggests a modicum of respect; no longer referred to as "le nègre," Johnson was "le nègre Johnson" or simply "Johnson."[24]

The Museum wrote to thank those who had helped acquire and transport the bison: Milbert and d'Espinville in New York and the Marine Commissioner and Eyriès frères in Le Havre. To d'Espinville it testified that the bison had reached Paris safely in good condition and that its "very gentle" character demonstrated how well Johnson had followed the consul's instructions. Informing d'Espinville that the animal "is known by naturalists under the name bison" (rather than buffalo), the Museum cheerfully added that it had "never been seen before in Europe." Georges Cuvier, who composed the letter, urged d'Espinville to follow up this historic first with another: "We hope that you would now want to take the trouble to procure a female for this male, whose robust health promises the conservation and propagation of this species" (in France).[25]

The Museum's early impression of the bison as "very gentle" was at odds with Eyriès's observations about Johnson's skill in handling its "pranks" and "whims." It was also belied by its behavior at the menagerie, where Frédéric Cuvier soon concluded that the bison was dangerous and malicious. Three months after its arrival, Cuvier wrote an essay on the animal for his and Geoffroy's *Natural History of Mammals*. He noted that the bison had been captured young and then "tamed and brought to Europe conducted by a Negro who had always had care of it, and whom it obeyed, albeit with the aid of a ring in its nose, like those used to master water buffaloes and bears." Cuvier continued:

> It was accustomed to being washed, combed, and rubbed down; but it put up with these attentions only from those who fed it and whom it saw each day; any other person who approached would have irritated it, and it would have attacked him, particularly when his back was turned. Its anger was not at all blind; chastised vigorously when it manifested a desire to butt with its head, it had learned that it would arrive at its goal more surely by attacking from behind than from the front; as long as you looked at it fixedly it held itself apart, especially if you looked at it with a cane in your hand; but as soon as you turned away it advanced on you.[26]

Fortunately for the bison's keepers, Cuvier explained, the animal was not very bright. It hated having a rope put through the iron ring in its nose, but it made no connection between the sight of a keeper with a rope in his hand and the discomfort it felt when the rope was slipped through the nose ring.

Visitors to the menagerie, Cuvier wrote, were struck by the bison's appearance, which was both familiar and unsettling. It resembled the ox, but it was distinctive in "its stocky form, the thick mane of its head, its small but bright eyes, the height of its withers, [and] its dark color," all of which made it look "ferocious and bad-tempered."

In spring 1823 the bison grievously injured its keeper, Warmet. The physician who cared for Warmet visited him once or twice a day for weeks, seventy-one times in all. Warmet recovered his health, but his enthusiasm for his job was gone. A year later, the Museum decided his work was so deficient he had to be dismissed.[27]

Cuvier thought it would be worth attempting to cross the bison with a domestic cow, a project Milbert had strongly endorsed. The Museum wanted first to see if the bison could be naturalized in France, and for that it needed a female bison. Milbert reported that procuring one would not be easy, since bison typically lived "well beyond the Wabash [River]." In his next letter, he noted it was only to be found in "Kentuckey" [*sic*] beyond the Ohio River. Although unsure of the bison's geographical range, Milbert was confident that multiplying it in France would be an invaluable exercise. The bison was infinitely stronger than the ox in plowing (it was said to be used as a beast of burden in northern Louisiana), and female hybrids of the bison and the cow were alleged to produce twenty bottles of milk (of a slightly yellowish hue) daily.[28]

In 1821 Milbert managed to send the Museum a female bison and a young moose. They arrived in Le Havre in mid-September and then traveled to Paris by steamboat.[29] The very young, small female bison was given her own pen and stable, but she failed to thrive in her first year. She lacked appetite, scarcely grew, molted poorly, and the menagerie feared her loss. When placed in a different stable with a domestic cow, however, she underwent a sudden change. As Frédéric Cuvier recounted, "her appetite returned, her hair became shiny, she grew considerably."[30] Cuvier began plans for her mating.

On 9 March 1824, he reported: "The female bison, apparently fully grown, I have placed in the rotunda, so that from the park where she is now, she can get accustomed to the male, and he with her, in order to reunite them for their mating, as soon as that can be done without fear of accident."[31] Three weeks later Cuvier wrote: "The bison is beginning to put up with his female, that he menaced for several days and even butted. Hopefully next week they will be completely reunited."[32] The reunion did not take place until June, and the keepers still had to see that the male did not butt the female, but the animals seemed to grow more familiar. In October it appeared that the female was pregnant. When the male became increasingly aggressive toward her, the keepers returned her to the stable with the cow.[33] On 12 March 1825 she gave birth to a small, healthy male calf, to which she gave "the greatest care."[34]

The female bison did not survive the year, dying of a combination of a stomach swollen from eating too many greens and lungs filled with hydatid cysts,[35] but the successful reproduction of the bison led Cuvier to the experiment of hybridizing the male bison with a domestic cow. In mid-August 1825 he reported, "The cow has been brought together with the bison to attempt their mating. These animals have seemed from the first to familiarize themselves with each other, and everything gives reason to hope that their mating will be next."[36] The next week Cuvier related: "Since the bison has been with the cow he has become so furious that instead of obeying, he attacks with fury everyone who approaches him, so that new measures will perhaps be necessary to avoid accidents, and I will propose them if the need becomes pressing."[37]

In January 1826 Cuvier announced that the cow was pregnant and would soon give birth to a new hybrid.[38] Unfortunately, in June he was obliged to report that "the cow that was covered and impregnated by the bison was unable to bring her baby into the world. After eight days of suffering, she had to be delivered by force. The baby died in the process and the mother survived only four days. The baby completely resembled a young bison. It has been stuffed."[39]

This left the menagerie with two bison: a father and a son. Both made Cuvier anxious. When just over a year and a half old, the son began testing his strength against the fencing of its enclosure and was soon breaking parts of it daily.[40] Wrote Cuvier, "this morning (December 27, 1826), without the presence of Guichenot, [the young bison] would have escaped, and some great misfortune would have surely arrived." Cuvier said he could no longer be held responsible for the situation unless stronger barriers were built to restrain the bison.[41] The following June, after it had broken through the door of its cage, Cuvier informed the professors, "it is impossible to dream any longer of making it inhabit the rotunda; not only does it break every obstacle one puts in front of it, it goes after the life of the keepers. An exterior shed is now indispensable for repressing this animal, which is not bothered by the winter cold."[42]

The two male bison lived on at the menagerie for another half decade. The father died in July 1835. The younger one became ill with a chest inflammation in October of the following year. Cuvier called in Alexis Dupuis, formerly a veterinary school professor (and candidate in 1803 for menagerie superintendent), who decided bloodletting was necessary, but whose attempt, with an assistant, was "without success." Cuvier did not describe the scene to the professors but noted that their colleague, Henri de Blainville, was present and could supply details.[43] It is unknown whether anyone recalled that Peter Johnson had successfully bled the menagerie's first bison on a ship at sea. The Museum was in no hurry to procure a replacement for the younger bison, and the menagerie had none until a pair of young females was added in 1845.[44]

With the bison, as with many other animals brought to the menagerie, the Museum's success was mixed. Getting the animal to Paris had been a triumph, involving a team of actors operating more or less in concert: the professor at the Museum who most wanted the animal (Georges Cuvier); two French ambassadors eager to help the Museum (Hyde de Neuville and D'Espinville); an enterprising naturalist on site in North America (Milbert); the unnamed correspondents and hunters who secured Milbert a viable specimen; the highly capable keeper who brought the bison to Milbert and cared for it from New York to Paris (Johnson); a cooperative ship captain (Jappie); the marine commissioner in Le Havre (Chabannon); the Museum's agents (Eyriès frères); and Johnson's interpreter (Nozière).

Getting the bison to reproduce in captivity was also a success. The male and female produced a perfectly healthy male calf that before long became, if anything, too vigorous. In contrast, the experiment of mating the bison with the domestic cow failed. Visions of quantities of rich, yellowish milk produced by bison/cow hybrids evaporated as Cuvier witnessed the agony of the cow trying to deliver a calf too large for her frame. Georges Cuvier announced in *Le Règne animal* (1829) that the American bison "reproduces with our cows," but the Museum did not pursue it.[45] Meanwhile, the menagerie's experience with the bison's temperament, epitomized by Warmet's grave injury, had to deflate hopes of employing the bison as a draft animal in French farmlands.

In terms of Frédéric Cuvier's project of developing a science of animal behavior, he wondered how the bison could tell when a keeper was able to chastise it but not when he was about to put rope through the ring in its nose. Was the small difference between these two circumstances, he asked himself, a measure of the animal's intelligence? He pursued the question no further. An ill-tempered bison was not a good candidate for experimental studies.

The bison can serve, nonetheless, as an example of Cuvier's assumptions about how the lives of animals in captivity compared with those in the wild. In a throwaway observation in his account of the bison for his supplement to Buffon's *Histoire naturelle*, Cuvier noted that the menagerie's male and female mated only at night. He described this as "a circumstance that is commonly observed among wild animals and seems to be caused by a special instinct of conservation in an act where all senses and forces are concentrated on one object."[46] Perhaps the menagerie's bison mated exclusively at night, but today on the plains of North America, well beyond the Wabash, bison mate in broad daylight.

[3.6]

Donors, Dealers, Showmen, and the Growth of the Menagerie

In the years of the Restoration, naturalists with government support like Milbert and Dussumier greatly enriched the menagerie. But more specimens came as donations or purchases, and the collection grew from trades and new births as well. The donors ranged from foreign potentates, aristocrats, and government bureaucrats to various travelers abroad and pet keepers at home. Those who sold live exotic animals to the Museum ranged from professional dealers with large inventories to small-time animal showmen, to the sea captains and other travelers who fancied they could make a profit acquiring an animal abroad and selling it once they returned to France.

DONORS

The animal dealers' desire for financial gain was clear, as was that of the travelers abroad who hoped that the costs of acquiring an animal and bringing it home would be more than offset by what the Museum would be willing to pay for it. The donors of animals were a heterogeneous lot, and their offerings as well as their motives likewise displayed considerable diversity. In none of these regards, however, was the diversity unlimited. Certain general patterns can be discerned both in the kinds of animal gifts the menagerie received over time and in their donors. These patterns reflected the interplay of donor initiatives on the one hand and, on the other, the amount of say the Museum had in accepting a gift or declining it. The Museum was obliged to accept diplomatic gifts whether it wanted the specimens or not. With other specimens that might or might not have been of value to it, the Museum was disposed to favor those from individuals whose good graces were most likely to be important. Meanwhile, the Museum's quest for specimens new to science was always in play.

The handful of animal donations the menagerie received in its earliest years exemplify the range of information we have regarding donors and their gifts generally. The very first such gift was the camel given to the

Figure 3.6.1. "The cheetah from Senegal." C. de Lasteyrie after Jean-Charles Werner. From É. Geoffroy Saint-Hilaire and F. Cuvier, *Histoire naturelle des mammifères*. General Research Division, the New York Public Library.

Museum in 1794 by the Jacobin legislator Merlin de Thionville. Biographical information on Merlin is readily accessible. Insofar as we can tell about his motivation, he appears to have been glad for the opportunity to make a conspicuous donation to an important institution of the new republic. We are less able to judge what led him to donate his pet lemur to the Museum in 1798, and the same can be said for three other animal gifts that arrived the same year: a raccoon from the French physicist Charles-Augustin de Coulomb; a stork from a "Citoyenne Trouillat la Roche"; and a macaw from a "Citoyen Chabert."[1] More is known, on the other hand, of a donation in 1801 from a professor of natural history at the École Centrale of l'Oise. Professor Debrun sent a hawk he'd had with him for three years, having nursed it back to health after it was wounded by a hunter. He had studied its habits, paid close attention to its vocalizations and movements, and observed how it responded to visitors. Over time this fierce wild bird (to which Debrun gave the gentle name "Coco") allowed itself to be handled like a pigeon.[2] Debrun's fondness for the bird is evident in his communications with the Museum. Not so clear is why he chose to donate it, but his self-identification to the professors as "one of your students" suggests a desire to affirm a relationship that was important to him. Reporting his observations on the hawk's behavior together with giving the animal to the national collection were ways for him to contribute to the cause of science.

Professor Debrun's donation is one of the better-documented gifts the menagerie received. The opposite extreme was a convoy of twenty-three animals that arrived at the menagerie late in March 1803. The Museum knew that the director of customs in Marseille had put the convoy en route but that he was not the source of the animals. The young lion in the collection was alleged to have come from Tunis, but its donor was unnamed. Most of the convoy's other animals (primarily sheep and goats) were identified as Egyptian, but again no donor was identified. Two genets were correctly attributed to a French diplomat serving in Egypt and Tunis, Jean-Baptiste Adanson (nephew of the famous French naturalist Michel Adanson), but he had not specifically intended them for the Museum. He left them with the customs director, who chose to send them on to Paris with the other animals.[3] In contrast, the provenance of another collection of North African animals also arriving in 1803 is clear. The two lions, a lioness (and dog companion), three ostriches, a gazelle, and a female deer were all gifts from the bey of Tunis.[4]

Though foreign rulers and French consuls were among the menagerie's early donors, French army generals outnumbered them. In addition to Generals Leclerc and Rochambeau, who respectively initiated and achieved sending the Museum South American jaguars, at least six more generals were involved early on in contributions to the menagerie. In a single year (1804), the Museum received two big cats (a "black tiger" and a panther) from Île de France, sent by General Charles Decaen as gifts for Josephine Bonaparte; a green monkey from General Honoré Vial; a pair of Siam pigs from General Louis-François Foucher de Careil; a male chamois from General Jacques-François Menou; a golden eagle from General Géraud Duroc; and a golden eagle and another eagle from General François Joseph Lefebvre.[5] That these military leaders took the trouble to send animals reflects the visibility the menagerie had achieved and the generals' recognition that it was befitting for them to assist the national collection.

Some nudging from the Museum helped develop that recognition, as least in Menou's case. He is known in military history for taking command of the French forces in Egypt after General Jean-Baptiste Kléber's assassination, but his interest to the Museum in 1804 was that he had been appointed French administrator for Piedmont, and with the Italian Alps so close to him, he might be able to procure a *bouquetin* (ibex) for the collections in Paris. The Museum wrote him to say that the *bouquetin* lived near Courmayeur in the Valle d'Aosta. It asked him for a skin, a skeleton, or best of all a live specimen. Menou promised he would send out "all the hunters" to find it. Meanwhile, he sent the Museum the male chamois and the eagle.[6]

Whether the Museum accepted a proposed gift could depend as much on the status of the donor as on the Museum's interest in the animal. Their power, elegance, and relative ease of transport made eagles attractive gifts from a would-be donors' perspective, but it meant the Museum was offered more eagles than it needed. It frequently turned down the offers on the grounds that the menagerie already had enough or that the Museum could not afford the cost of their transport and food, but these excuses were invoked selectively. Between 1800 and 1820, eagles offered by generals, princes, princesses, or the empress were accepted, but those offered by average citizens and mid-ranking bureaucrats were politely rejected.[7]

The generals' prominence as donors faded when the Napoleonic Wars ended and freedom of the seas was restored. Foreign potentates, French consuls abroad, and sea captains became the menagerie's primary donors. The rejuvenation of French maritime commerce additionally meant that exotic animals would once again be arriving in French ports, revitalizing the animal trade. Captain Milius's green monkey gift in November 1815 was the first new addition to the menagerie from overseas.

In 1816 the menagerie received an agouti from Brazil and the first major diplomatic gift of the Restoration era, a lion and two lionesses from the emperor of Morocco. A beaver from Newfoundland was delivered by a sea captain in 1817, followed by the less welcome news that the emperor of Morocco was sending two more lions, which the Museum neither needed nor wanted.[8] In 1818 the captains of the government ships *Bellonne* (back from a thirteen-month voyage to the Antilles) and *Cybèle* (back from eighteen months in the Far East) gave the menagerie respectively a monkey with a prehensile tail and a pair of sheep. From London, the zoologist William Leach sent a Newfoundland dog.[9]

In the years that followed, the most illustrious gifts came from foreign rulers, but the flow of other donations was crucial for maintaining and developing the collection. In 1819 the menagerie received a young chamois, a leopard, two bateleur eagles from Senegal, and an eagle donated by the duc d'Orléans; 1820 witnessed the arrival of a leopard and a new species of wild boar, and on it went in successive years. Some animals came from colonies (Senegal and Cayenne), but others arrived from Argentina, Spain, Russia, the Cape of Good Hope, Mexico, and more. There was a purported grizzly bear cub from North America (courtesy of General Lafayette in 1826), and a pair of ostriches from South America (couresy of the duc de Leuchtenberg).[10]

From this stream of new animal donations for the menagerie, I have selected three that illustrate different motives for animal donations: a chamois from the Italian Alps given in 1819 by M. Magol, the director of the telegraph in Lyon; a cheetah from Senegal given by the Baron and Baroness Le Coupé on their return from the colony (1822); and two young lions from North Africa given by the famous French admiral Henri de Rigny (1827).

The story of Magol and his chamois is unique in the Museum's records for its information about the donor's motivation for making the gift and how the animal was captured. Magol, it seems, was less interested in establishing a relationship with the Museum than in making a gesture that would remind the minister of the interior of Magol's existence and his desire to be named to the Legion of Honor. Lyon was a major node in the new telegraph system that enabled the government to transmit semaphore messages across France with unprecedented speed. The Paris-Lyon connection was completed in 1815, and Magol, as director of the telegraph in Lyon, felt the accomplishment was worthy of the Legion of Honor.

To secure a chamois, Magol commissioned a *montagnard* living near the village of Suze, high in the Alps on the French/Italian border. The man captured a baby chamois by surprising and killing its mother just as the infant was being born. For six months, aided by a nurse goat, the man reared the baby chamois, and the young creature became so attached to him that it followed him everywhere. Magol decided the man should escort the chamois to Paris and stay a few days in the capital while the chamois became accustomed to a new guardian.[11]

The montagnard, chamois, and nurse goat left Suze for Lyon in late November. A blizzard all but blocked the Mont-Cénis pass, and the man had to carry the young chamois through the snow. Once they reached Lyon and were resting there, Magol wrote to assure the minister that everything had been done to guarantee the chamois's safe and economical arrival in Paris; indeed, that he'd been even more frugal on the government's account than when handling his personal affairs. He hoped the minister would reward him "with an act of your high justice and your benevolence relative to the request I have made for the decoration of the Legion of Honor." The Museum, looking out for its own interests, seized the occasion to convey to Magol via the minister that the Alpine animal the zoologists most wanted was the *bouquetin*. Magol told the minister he would do his very best to get one, though it wouldn't be easy, because the *bouquetin* had grown rare in the area.[12]

In the case of the cheetah that arrived at the menagerie from Senegal in 1822, the donors were spurred by a desire for recognition of their gift. Returning from Senegal where he served as French governor from 1820 to 1822, Baron Louis-Jean-Baptiste Le Coupé and his wife brought not only the cheetah but also a genet, two rams, a jackal, and a grey-crowned crane. When these valuable animals, designated as gifts for the king, arrived at Rochefort in May 1822, accompanied by a half dozen other animals sent by the naturalist Charles Sauvigny, the Museum dispatched the keeper Magloire Rousseau to fetch them. Frédéric Cuvier advised the administration to thank Le Coupé, and the Museum sent a cordial letter expressing "sincere thanks" for a gift that would benefit science and please the public. It promised to recognize the donors with a plaque and other gestures of gratitude.[13]

The cheetah was of genuine interest to the zoologists because it differed significantly from other big cats in the Museum's collection. It was also eminently likable. Frédéric Cuvier described it as gentle as "the best dogs." It recognized its caretaker especially, but it welcomed everyone who was kind to it, and it purred like a cat when it was petted. It had been given complete liberty on the voyage from Senegal to France, but that was not possible at the menagerie because of "the fear that the curious would only abuse it." For the summer it was assigned a pen "where it could give itself some movement and frolic as much as it wanted."[14] It quickly became one of Cuvier's favorite animals, but in the meantime, its donors were forgotten. That changed abruptly when the Baroness Le Coupé visited the Museum in March 1824. She found the cheetah was there, but nowhere on its enclosure was there a plaque stating that the Baron and Baroness Le Coupé were its donors. The baroness was furious.

Writing to the Museum, the baroness presented a "long enumeration of my grievances."[15] She and her husband had gone to considerable expense to acquire and conserve the cheetah before offering it and other animals to the king. She had personally helped care for them on the voyage from Senegal and had been promised that her gifts would be recognized through inscriptions on the animals' cages and a special card granting her privileged admission to the Museum. She reminded the Museum that it had pledged to provide illustrations of the animals she had donated. It pained her that none of these promises had been kept, and that she had to ask for recognition from an institution that had a reputation for *politesse*.

Lamely claiming that an inscription identifying the cheetah as the Le Coupés's gift had "always existed" but had "momentarily gone astray as the result of changes in the domicile of this animal," the administration quickly produced a special admission pass for the baroness, good for the rest of the year. It provided colored lithographs of some of the species she had donated and had a special copy made of the painting of the cheetah that Léon de Wailly had executed for the Museum's vellum collection.[16] These gestures may have mollified the baroness somewhat, but not enough to keep her from reclaiming her grey crowned crane.[17]

The donation of two lions by Rear Admiral Henri de Rigny in 1827 was a different story. De Rigny did not need more attention or distinction; he needed the lions taken off his hands. He was already a famous and well-decorated figure, *contre amiral* in command of the French fleet in the Levant since 1825. His assignment was to rout piracy and police the maritime hostilities between the Greeks and the Turks while remaining neutral with respect to the Greeks' struggle for independence. In 1826 he received a pair of lion cubs from another French commander who had received them from the bey of Tunis. De Rigny took them aboard his flagship, the *Syrène*, and tasked a sailor named Olivier-Jean Guillet with their care. Guillet trained the cubs so well that they could be left to roam the ship freely. Only at night were they confined to the wooden cage where they ate and slept. "Sultane" (the female) and "Coco" (the male) grew gentle enough to be petted like dogs, displaying affection for not only Guillet but also his shipmates.[18]

In the summer of 1827, de Rigny decided to send the lion pair to the menagerie. Auguste de Rigny, the rear admiral's brother, announced this in a letter to Georges Cuvier, who had previously enlisted Henri de Rigny in collecting African fish for him. The motivation for the new gift went unspecified, but the lions had grown larger, stronger, and hungrier. Though still gentle, they had become "youths with a very great and very expensive appetite."[19] An even greater incentive to be rid of them, however, was that there was no room for lions on a warship in battle, and the likelihood of a confrontation with the Turco-Egyptian navy had increased.

By the Treaty of London (6 July 1827) Britain, France, and Russia agreed to intervene on behalf of the Greeks if the Ottoman Empire did not grant an armistice giving Greece a modicum of independence. The Allies realized that the Ottomans, believing in the superiority of their naval forces, were likely to reject the treaty. The battle came on 20 October 1827, at the harbor of Navarino, where the French, British, and Russian squadrons decimated the Turco-Egyptian fleet. Hailed as a great hero, De Rigny stayed in the Levant and assisted with the evacuation of Morea, in the Peloponnesian peninsula.[20] Later he was involved in retaking that territory for the Greeks.

Meanwhile, the lions reached Toulon aboard the corvette *Bayadère* on 26 August 1827. Forty-one days and 328 kilograms of meat later, their keeper, Guillet, set off with them for Paris by way of Marseille. The bill for the lions' care, lodging, and transportation to Paris from Toulon amounted to 1,130 francs, not counting the gratuity paid Guillet.[21] Sultane lived at the menagerie for approximately sixteen months, dying in April 1829. Coco died two months later.[22]

If De Rigny got rid of his lions to clear the decks, other donors to the Museum gave away exotic pets that had lost their charm. This was surely the case for É. Doumet, a high-ranking military officer and chevalier of the Legion of Honor who in 1835 sent the Museum a crocodile and a female baboon. In a letter preceding the animals' arrival in Paris, he warned that their crates should be opened with care. He specifically warned that the baboon's crate not be opened until it was in the quarters where the animal was destined to live. She was very strong, he explained, and her long trip to Paris was likely to leave her peevish and inclined to bite. That she had become a troublesome pet was evidenced by his comment, "She responds to the name of *Lolotte* and is very gentle when her jealousy is not aroused; but it does not take much for that to happen."[23]

∴

In the 1830s the menagerie's record keeping with respect to gains and losses became more systematic, offering a good sense of how animals were acquired throughout the decade. In addition to the fifteen lions confiscated in the invasion of Algiers, some six hundred new mammals arrived. Of the latter, over 40 percent came as gifts; slightly fewer than 32 percent came from births; purchases followed at 25.5 percent; and exchanges made up the remaining 2.5 percent.[24] The numbers changed appreciably over the decade. When the new monkey house opened in 1837, purchases began to outstrip donations.

How the animals were acquired also differed significantly by animal type. Of ninety-three new deer, fifty-six were born at the menagerie, thirty-three were donated, three arrived via exchanges, and just one was purchased. With monkeys, the relative number of births and purchases was the reverse. Out of 169 new monkeys, eighty-four were purchases, seventy-two were gifts, only seven were born at the menagerie, and six came via exchanges. In contrast to deer and monkeys, big cats came more from donations than any other source, except for the fifteen new specimens confiscated in Algeria in 1830. By the end of the decade, twenty-eight big cats had been donated versus ten purchased.[25]

Purchases and trades as well as donations enhanced the menagerie's holdings in the 1820s and 1830s. Sometimes a seller or trader is named in the Museum's records. Other times the identity is vague, such as "a foreigner" or "a German." Most of the names are associated with a single transaction; usually minor but occasionally more consequential. Sea captains were responsible for many of the latter, including a panther from Bengal in 1831 (Captain Luco, 400 francs); a tigress in 1835 (Captain Frion, 1,500 francs); a collection of ten more animals the same year (Captain James Burton, 3,500 francs); the menagerie's first orangutan, in 1836 (Captain Van Iseghem, 3,500 francs); and the menagerie's first chimpanzee, in 1837 (Captain Boullemer, 3,000 francs).

DEALERS AND SHOWMEN

These purchases from sea captains were all one-off affairs. The Museum was a hard bargainer, and none of the captains named seems to have made a profit for all his troubles.[26] In addition to these one-time sellers, the Museum acquired animals from animal dealers who made their livings exhibiting and selling animals. At the turn of the century, the illustrious Gilbert Pidcock in London was the dealer who sold the Museum a prize collection of carnivores in 1800 and whose efforts to palm off an ailing elephant on the Museum two years later came to naught when the elephant died. When Pidcock would not sell his prize kangaroos to the Museum's representative, Dufresne, the Frenchman turned to a rival dealer for a kangaroo pair.

The first example of the Museum's acquiring live animals from a French exhibitor/dealer seems not to have occurred until early in the Restoration. The entrepreneur in question was a small-time operator named Réaux who exhibited exotic animals and other curiosities to members of the public willing to pay the price of admission. He traded animals with the menagerie, sold animals to it, had the menagerie keep some deer for him, and sought to rent one the menagerie's lions to put on display. His animals, however, were not what interested the Museum most about him. He owned the contract of a Khosian woman, Saartjie (Sarah) Baartman, commonly known as "the Hottentot Venus." In 1815 she was Réaux's star attraction.[27] Réaux's Hottentot Venus show represented the kind of lowbrow entertainment of fairs and such that the Museum was keen to distance itself from, but it turned to Réaux for access to his star exhibit. The Museum's zoologists and anatomists wanted a closer look at her.

Readers will have heard of the exhibiting of exotic peoples in European zoos, but that was a practice of a later era. Sarah Baartman was not put on display at the menagerie. The significance of her story here is rather in the oblique ways it touches on the menagerie's history, first through Réaux, the animal dealer, then in how it gives us a glimpse of the animal shows of Paris (a source for scientific information about exotic animals, even if the Museum sought to keep its distance), then in another "distance" question of the day—the distance between animals and humans, and how questions of race were implicated in it. That Georges Cuvier's brutal "Hottentot Venus" lecture to the Academy of Sciences was republished in Frédéric Cuvier's *Natural History of the Mammals*, then left out of the second edition of the same work, adds another opportunity to consider the complex interplay of ideas and agendas in which the menagerie was implicated.

When the "Hottentot Venus" arrived in Paris (early September 1814), she was not advertised as a sight for the masses only. The press heralded her as having "long been the delight of London," where "the Prince Regent and the princes of his numerous family often honored her with their visit."[28] More information accompanied the "Venus Hottentot" show that soon opened on rue Neuve-des-Petits-Champs. According to *Le Quotidienne*, the woman's name was "Sartjée" and she came from the Houzouana tribe that lived sixty miles northeast of the Cape of Good Hope. A young Dutchman brought her to the Cape and subsequently to England, where for three years she had been shown in "the principal cities." The report continued: "She speaks Dutch, English, and her native language, in which she sings some songs. She has a very gay and pleasant air with the people who go to see her."[29] Trading on this latest sensation, by mid-October the Vaudeville Theatre mounted a play titled "The Venus Hottentot," with the usual mix of rhyming couplets and lame plot lines that delighted audiences.[30]

Sarah's first manager, Hendrik Cesars, invited the Museum professors to a special showing in September 1814, but the administration learned of the offer too late.[31] Réaux bought Sarah's contract from Cesars in January 1815 and moved the show to la rue Montesquieu, where she performed from 11:00 a.m. to 11:00 p.m. daily, with the seats at two francs or three, per the view.[32] In February 1815, the professors requested that Réaux bring Baartman to the Museum so that artists could record "the distinctive characters of this curious race."[33] The zoologists were interested in the woman's anatomical features, especially her protruding buttocks and the fabled "Hottentot apron" that according to travelers' accounts served to hide the genitals of women of her race.

In March, Réaux conducted Sarah Baartman to the Museum, where Georges Cuvier, Étienne Geoffroy Saint-Hilaire, and Henri de Blainville studied her anatomy and the artists Léon de Wailly, Nicolas Huet, and Jean-Baptiste Berré made portraits of her. Against her wishes she posed nude, but she forbade the professors to observe her as closely as they wished.[34]

Réaux promptly sought to profit from his association with the Museum through a cheerful press release about his star stating that "this beauty has been visited and sketched for three whole days at the Muséum d'histoire naturelle by the top physicians and the most celebrated painters of Paris, and they have declared unanimously that they have never seen a better made woman. The Venus Hottentot has been so flattered by this declaration, that in order to respond to the desires of people curious to see her, she has reduced the price of the seats, which are no more than 2 fr., 1 fr. 50 cent. and 1 fr. It is impossible for a Venus to be more accommodating."[35]

This was only the flimflam of a sideshow barker. Réaux had requested and received a report from the Museum, but what Geoffroy Saint-Hilaire provided was not an ode to beauty. The most distinctive feature of her head, said Geoffroy, was "the beginning of a muzzle even more considerable than that of the red orang-outang which inhabits the larger islands of the Indian Ocean." He likened her famous buttocks to those of female pig-tailed monkeys and mandrills at the onset of menstruation.[36]

Réaux's bogus explanation for the cheaper ticket price likely disguised his awareness that crowds were being drawn to a new curiosity in the neighborhood, a "superb rhinoceros" recently installed just off the rue de Rivoli, near Réaux's "Hottentot Venus" show on rue Montesquieu. The *Gazette de France* gushed about the "very beautiful live rhinoceros, recently arrived in Europe" and explained that it had been captured on the Malabar Coast at the age of six months. Now five years old and weighing 4,800 pounds, it was in good health and remarkably gentle in demeanor, and there was "nothing wild" in its behavior. "It allows itself to be approached and touched by strangers and lavishes on its master the most affectionate caresses." Noting that this rhino was much larger than the stuffed one at the Museum, the *Gazette* concluded with the hope "that this animal, precious by its rarity, will be purchased by the government and will complete the beautiful collection of the menagerie of the Jardin des Plantes."[37]

The Museum did not acquire the rhino, nor is there evidence that it even tried, but it did the next best thing. It commissioned Nicolas Huet to paint its portrait, paying him one hundred francs for the painting and fifty more for a copy for the animal's owner. Huet later complained that the Museum's standard remuneration for an animal painting was too little for work executed outside the Museum in places subject to "the public's curiosity; because then we are subject to losing much time due to the difficulty of having the model at an hour or in a suitable orientation, and we only obtain the compliance of the keepers or proprietors through other negotiations that are always onerous for us."[38] Huet's rhino painting and his nude profile of Sarah Baartman both became part of the Museum's collection of scientific illustrations on vellum.

Writing about this rhino five years later, Frédéric Cuvier noted that "all of Paris" saw it in 1815. He too visited it, and he came away with a strong impression of how the severely constrained conditions of its existence had stunted it both physically and mentally. He reported that while the animal was normally gentle and affectionate toward its master, it was subject to moments of fury, which had no apparent cause but perhaps arose from "a blind sense making it desire a liberty it had never known." The rhino's keeper knew how to control it through a regimen of treats, using food and especially fruits and sweets. In its cramped cage, the rhino could do little more than the small things its master demanded of it for a reward: open its mouth, move its head left or right, or lift a leg. Cuvier saw scant opportunity for the animal to develop its intelligence under such conditions of confinement, but he thought he detected some intelligence in the rhino's ability to distinguish people likely to indulge its desire for food. "Under more favorable circumstances," he suggested, the rhino's intelligence might have developed much further. Had an elephant been so confined, Cuvier believed, it would have been no more astute. In addition to his observations on the rhino's behavior, he described its physical features in detail, including that "it was knock-kneed, which was undoubtedly due to the constraint it was under and to the little strength that its joints could have acquired owing to the inactivity in which it was kept."[39]

The Museum viewed the "Hottentot Venus" and the rhino alike as scientific specimens. To the public they were spectacle and entertainment to be enjoyed at something of the same level. The *Gazette de France* considered them together in a review titled, "On the tamed rhinoceros and the civilized Venus Hottentot." The writer joking elided humans and animals with the observation that "Sarah . . . dances nearly as well as the bear Martin," but he then went on to signal her knowledge of different languages and other aspects of her humanity. She was gentle, he wrote, and her whims were no more than those "suitable for a pretty woman." In addition, she seemed "very attached to the man to whom she has hired herself," and "knows very well the value of money, guards carefully that which is given to her, and proposes to return to her own country after levying from European curiosity a sufficient tribute." Displaying a more generous and subtle perspective than the Museum professors,' the author imagined that Sarah would undoubtedly "recount to her compatriots our customs, inventing what she would not have been able to understand, following the privilege of all voyagers."[40]

Réaux knew the professors wanted to examine Sarah further. When she died during the night of 29–30 December 1815, after an illness of just a few days, Réaux was quick to report it to Geoffroy Saint-Hilaire before informing the municipal authorities.[41] Geoffroy immediately wrote to both the mayor of Paris and the prefect of the police asking permission for Baartman's body to be transported to the Museum's comparative anatomy laboratory.[42] It was a special opportunity, Geoffroy told the prefect, "to acquire new information on this singular race of the human species."[43] The request was granted. On 2 January 1816 the *Moniteur* was able to announce, "at this moment, in a room of the Museum of Natural History," a cast was being made of the Venus Hottentot's body. The *Moniteur*'s note explained the scientific value of the anatomization in the same way that Geoffroy had done: "The dissection of this woman will furnish an extremely curious chapter in the history of the varieties of the human race."

The story of Sarah Baartman's dissection, the treatment of her remains, and the assessment of her zoological place form an instructive chapter in the history of scientific racism. Georges Cuvier acknowledged Sarah Baartman's gaiety and good memory, but he emphasized her simian features, proclaiming "I have never seen a human head more similar to monkeys and apes than hers."[44] The comments in the press were equally dehumanizing, while seeking to be witty. Referring to Réaux as Sarah's *cornac*, or mahout, the *Journal général de France* remarked that it would be "a long time before he will have as extraordinary an animal to show the public."[45] The *Journal de Paris* reported that "at the same site, Cour des Fontaines, where the Venus Hottentot was seen, one now sees the sea lion: this locale seems destined for strange animals."[46] Two months later the Museum paid Réaux 200 francs, evidently for his help securing Sarah Baartman's body.[47]

It is likely that Réaux first became known to the Museum as the manager of Sarah Baartman, but by the fall of 1815 he and the Museum had an additional relationship through the menagerie. Reports by Frédéric Cuvier in November of that year indicate that the menagerie had been lodging three deer belonging to Réaux for long enough that he owed the Museum 80 francs for what they had eaten. In compensation, he presented the menagerie a white fallow doe and a fawn, an arrangement that pleased Cuvier because it enabled the menagerie to display simultaneously the fauve (fawn), black, and white variants of the species.[48] Another exchange with Réaux was set up in April 1816. In this case he was to receive a water buffalo and a young sheep, worth 210 francs, "to be acquitted," as Frédéric Cuvier put it, "in animals of our choice of a similar value."[49] At the end of the year Réaux was again of interest to the Museum, this time because he had a live crocodile in his collection. The artist Huet was sent to make a painting of the animal for the vellum collection. As with the rhino, executing the painting at a site outside the Museum posed special difficulties, leading Huet to ask the administration to pay him double the normal rate.[50]

In March 1817, Réaux retrieved the three deer the menagerie had been keeping for him.[51] Then in April Frédéric Cuvier entertained a proposal from Réaux that both Frédéric and his boss, Geoffroy, deemed "in the interests of the menagerie." Réaux offered to pay the Museum 600 francs to lend him a male lion to exhibit for six months. Geoffroy saw no problem with the arrangement, and the menagerie superintendent observed that beyond the financial gain for the Museum, it would allow him to move one of the North American black bears from the bear pits to the lion's vacant cage. That way, the next time the female gave birth, the male would not devour the cubs. The administration considered the proposal and the zoologists' endorsement of it, acknowledged the advantages, but rejected the plan as incompatible with "the dignity of the establishment."[52]

This was not the first time the Museum refused a proposal on the grounds that it would be detrimental to its image. We have seen, for example, how in 1800 it dismissed as unworthy of the institution Lucien Bonaparte's idea of charging visitors to see the elephant. On the other hand, the Museum had no qualms about turning the body of Sarah Baartman into multiple objects for scientific analysis and exhibition. The professors assured authorities that the body's transport to the comparative anatomy laboratory and its dissection would be done with total respect to "decency." This did not preclude, however, making a full plaster cast of her body and putting it on display. In his October 1816 address to the Academy of Sciences on the "Hottentot Venus," Georges Cuvier triumphantly made Saartjie Baartman's private parts public, holding them up in a jar as he explained the anatomy of the "Hottentot apron."

Cuvier's essay on Baartman appeared in the *Mémoires* of the Museum (1817) under the title "Extract of observations made on the body of a woman known in Paris and London under the name Venus Hottentot." It was republished in 1820 as "Woman of Boschisman Race," paired with Léon de Wailly's full frontal nude and profile paintings of Sarah Baartman, as one of the offerings in the twentieth installment of Frédéric Cuvier and Geoffroy Saint-Hilaire's *Natural History of the Mammals*. The other essays in the installment, all by Frédéric Cuvier, were on the Hanuman langur, leopard, brown rat, wapiti, zibeth, and common fox.

No rationale was given for including Georges Cuvier's "Hottentot Venus" essay in this work. The reason was presumably self-evident: humans are mammals. When the publisher, in 1824, produced a table indicating the order in which the first 120 essays should be bound, "Woman of Boschisman race" came first, the only representative offered of "the human species." Next came the monkeys (quadrumanes), then the opossum (under the heading "insectivores"), carnivores, herbivorous didelphs (the kangaroo and the wombat), rodents, edentates, pachyderms, ruminants, and finally amphibians.

Neither was any rationale given when Georges Cuvier's "Venus Hottentot" essay and Léon de Wailly's accompanying illustrations were left out of a smaller-format, two-volume edition of *Natural History of the Mammals* that Frédéric Cuvier brought out in 1833. Heading the mammals this time was the orangutan. "The human species" was omitted. Did this reflect a consideration of what was involved in "the dignity of the establishment"? It may be that Frédéric felt so, but the fact remains that the Museum continued to display Sarah Baartman's remains and the plaster cast of her body. Or it may be that Frédéric decided it was more important at this point to maintain a strong gap between humans and other animals than to give a nod to the idea that the distance between humans and apes was less for the "lower" human races than for the "higher" ones. Whatever the reason, the new edition of the work was downsized in number of species as well as format, and the human species was not included.[53]

The next individual to feature in the menagerie's ongoing experience with animal dealers was of a wholly different order from Réaux. Edward Cross, the world's most distinguished animal dealer in the 1820s and 1830s, was the purveyor of animals to King George IV of England.[54] Frédéric Cuvier referred to him in 1826 as "director of the Strand menagerie in London, [a man] who exercises his honorable profession with a rare liberality, and to whom the natural history of the mammals is already indebted for the knowledge of several entirely new species."[55]

What Cuvier called the Strand menagerie was the same enterprise as Gilbert Pidcock's Exeter 'Change. Upon Pidcock's death in 1810, it was purchased by another animal entrepreneur, Stephen Polito. Cross became its owner in 1814, acquiring the male elephant called Chunee. Among the creatures Cross subsequently added to his collection were a female orangutan in 1817 and a young chimpanzee in 1819, nearly two decades before the Paris menagerie first had these species.[56]

Cross boasted of his attempts to own "every living production" on earth, but he was more merchant than exhibitor and preferred selling animals to having them. He sold parrots and monkeys as pets. He sold pheasants, foxes, and deer to aristocrats and landed gentry as game to hunt on their estates. From 1825 until King George IV's death in 1830, Cross purchased animals for him and managed the royal menagerie at Windsor.[57]

Cross came to Paris in 1821 to trade for birds in the menagerie.[58] Another bird exchange was arranged three years later.[59] Cross's most consequential transaction with the Museum, however, was a drawn-out affair involving a pair of zebralike creatures he hoped to sell or swap. He arrived with them in late May 1826, calling them quaggas. Frédéric Cuvier identified them as dauws (now known as Burchell's zebras). The Museum had not asked for them, but Cross knew it did not have any and likely would be interested in them. He hoped to sell or exchange them for birds or other animals. He dared not mention to Cuvier his heart's desire—the menagerie's African elephant—but he told Royer, the Museum's accountant, that he wanted to replace his own elephant, which he had "unfortunately lost."[60]

Royer would no doubt have heard about that "loss" already. The gruesome demise of the elephant Chunee had been widely reported in the British and French press. Chunee had become increasing intractable, owing to a combination of ill treatment and onsets of musth. After Chunee in a rage killed a man, Cross concluded the elephant was too dangerous and had to be put down. It took a fourteen-man firing squad over an hour and a half to fell the elephant: 152 bullets were delivered to Chunee's body. A saber attached to a rifle barrel finally finished him off. The agonizing execution ranks as one of the most infamous incidents in menagerie history, but Cross did not miss the chance to profit from the elephant's death. Attracted by the rifle shots, a crowd had gathered outside the building, and it clamored to see Chunee's remains. For two shillings a head, gawkers were allowed inside to view the bloody cadaver.[61]

Cross told Royer he was willing to trade his two quaggas, a pair of leopards, and a pair of hyenas for the Museum's elephant, but it was no deal. The Museum would not part with the elephant, and of the animals it was willing to exchange with Cross, none struck his fancy. He nonetheless left his two quaggas at the menagerie, where they were installed in the rotunda.[62]

Months later, Cross reminded the Museum of his quaggas and mentioned additional specimens the Museum might want, but it was not until May 1827 that the Museum's interest in the quaggas quickened when Frédéric Cuvier discovered that the male quagga, contrary to first appearances, had not been gelded. With the expectation that the quaggas would reproduce, the administration authorized the menagerie superintendent to negotiate for their permanent acquisition. Before the year's end, the deal was concluded, with Cross receiving his full asking price of 4,000 francs.[63] The following year, a new guide to the Museum mentioned not only that the quaggas were "very gentle" and "easy to feed" but also that the female seemed to be pregnant.[64]

When the Zoological Society of London was founded in 1826, with the explicit intention of establishing a menagerie to rival the one in Paris, Cross offered to sell his whole collection to the ZSL and come along as manager of the new menagerie. He was turned down on the grounds that his asking price of three thousand pounds was too high. He could not remain at Exeter 'Change because it was scheduled for demolition as part of an urban renewal plan for the Strand. In 1829 he found temporary space for his collection at the King's Mews in Charing Cross.[65] Later that year he sent the Museum in Paris a long list of "those of his animals [that] he would be able to sell without diminishing his collection" and that he judged would be "useful for the Ménagerie du Roi." His offerings were impressive, as were his prices and his knowledge of what the menagerie in Paris needed. A pair of Persian lions headed the list at 400 pounds (the equivalent of 10,000 francs or two professors' salaries). They were followed by a young male lion and two leopard pairs of different species—all at a time when the Museum's carnivore collection was woefully depleted.

Additionally on the list were a sloth bear, diverse monkeys, a black swan, an emu, and more. All the animals, Cross indicated, were in "perfect health." The prices did not include delivery from London to Paris. The administration invited Geoffroy to reply, but the Museum's records say no more on the subject, and nothing came of Cross's proposal.[66]

In 1832, Cross established the Surrey Zoological Gardens, which for a while rivaled the Zoological Society's establishment at Regent's Park. Two years later he was back in Paris hoping for new exchanges with the Museum.[67] None occurred at the time, but Cross later provided the menagerie with a kangaroo (1842) and an Indian elephant (1843).[68]

A different sort of animal entrepreneur whose path crossed the Museum's was the daring young lion tamer and menagerie owner Henri Martin. The showman and his menagerie arrived in Paris in 1830. He set up his show near Porte Saint-Denis on the Boulevard Bonne Nouvelle, taking a six-month contract on a suitable space.[69] At a time when the Museum had no big cats in its menagerie, Martin had one and sometimes two lions; for a while he also had a tiger. He was better able to replace animals than the Museum was because his gate receipts were strong and his connections were substantial.[70] Unlike the Museum, where admission was free, Martin charged his audiences from fifty centimes to two francs per head to see his astonishing performances (his "exercises") with his animals, along with their feeding. The shows took place daily at 7:30 p.m., with an extra show at 4:00 p.m. Thursdays and Sundays.[71] The public was enthralled to see a lion, tiger, and hyena obey Martin's every command. He made his lion (Coburg) jump through a hoop; he played with his tiger (Atir); and he put his arm inside the mouth of his hyena. The duchesse de Berry was among those enthralled by his courage and talents as an animal trainer.

Martin possessed a variety of animals and natural history curiosities. In April 1830, keen to have the Museum's white alpaca for his show, Martin obtained it in exchange for a panther, a striped mongoose, a black coati, a chamois, the lower jaw of a sperm whale, and the skull of a dolphin.[72] Late in the year, when the second six-month lease for his exhibition site was about to end, he conceived a plan to install himself in Paris permanently instead of going on tour again. Not announced officially until February 1831, the plan was to forge an intimate relation with the Museum. He proposed ceding his menagerie to the government if it would appoint him director of the Museum's menagerie with an annual salary of 6,000 francs.[73]

The Museum professors were indignant about the idea. When the minister of the interior asked their opinion, they replied that they already had a menagerie with a director, and that the salary Martin was requesting was more than their own. But that wasn't the half of it. His proposal was inconsistent with the way the Museum was administered, and his plans for the menagerie were "foreign to science," which was "the unique purpose of the menagerie's establishment."[74] Martin intended to attract visitors by continuing to offer the "spectacle" of his skills as an animal trainer. The Museum adamantly insisted that Martin's plan was out of the question.

Though he had failed to find a home at the Museum, Martin soon became a hit at another of the city's attractions, the Franconi brothers' Cirque olympique, famous for its equestrian performances. Beginning in April and lasting three months, Martin and his animals drew enthusiastic crowds for their starring roles in a production titled "The Lions of Mysore." The Museum heard of the show because the prefect of police asked if someone from the menagerie would review its security measures. To that end, Frédéric Cuvier attended the last rehearsal.[75] He, the prefect of police, and architects from the prefecture attested that all precautions had been taken for the audience's safety before the show opened.[76]

Martin, his animal protégés, and the elaborately staged "Lions of Mysore" drew ecstatic reviews. *La France Nouvelle* called the show "the most miraculous spectacle one could imagine!"[77] *Le Messager des chambres* observed that the lion (Coburg) and the lioness (Fanny) displayed a greater comedic sense than many biped stage actors. "All of Paris," it noted, "will want to see to what point civilization has brought the ferocious beasts of the nineteenth century."[78]

The ameliorative effect of civilization on savage beasts was a theme evoked by the Paris menagerie's early exponents, and Frédéric Cuvier's subsequently planned to teach a course on animal "education"—that is, their training. Nonetheless, in 1831 the Museum professors were certain that what Martin was doing was "foreign to science" and inconsistent with its image of its own menagerie. Elsewhere the view would be different. In 1840 Martin became the founding director of the Rotterdam Tiergarten (zoo).

∴

Martin's show and the Museum's menagerie represented different opportunities for the public to encounter exotic animals. Another way people could engage with an exotic animal was to buy one from an animal dealer as a pet. The menagerie inevitably interacted with the animal dealers of Paris as well. A bird dealer named Perrot appears repeatedly in the menagerie's registers from 1826 to 1836 thanks to to small-scale transactions that benefitted Museum and dealer alike.[79] Perrot's contributions to the menagerie paled in comparison, however, with those of Parisian animal dealer François Vaillant, who in the late 1830s and the 1840s came to be the menagerie's most prolific supplier. Vaillant's enterprise was huge. He had more than twice as many animals as the menagerie did, though most likely not so many species. At the beginning of 1838, Frédéric Cuvier tallied slightly fewer than four hundred animals under his charge. In September of the same year, Vaillant estimated he had upward of a thousand. He provided the Museum with birds and with mammals, mostly monkeys.

Vaillant's relations with the menagerie date back at least to March 1830, when he gave the menagerie a pair of silver pheasants in exchange for some guinea fowl. The silver pheasants reproduced so well that it was at least a decade and a half before the menagerie needed to acquire new ones. Beginning in the summer of 1837, Vaillant's contributions became more regular. After filling a request for golden pheasants, he sold the menagerie a female baboon, a female Chinese bonnet monkey, and a pair of jackals.[80] In the spring of 1838, at the recommendation of Frédéric Cuvier (recently promoted to the rank of professor), Vaillant was provisionally hired to replace the bird keeper Rémond, who had become too ill to oversee the pheasant house.[81] Vaillant told Cuvier he could not leave his establishment immediately because of the size of his inventory; his loss would be enormous if he had to sell his animals all at once. It was agreed he could spend four or five hours daily with the menagerie's birds, leaving time for him to attend to his business.[82] Later in October 1838, when the Museum gave Vaillant a full appointment, it still allowed him to maintain his operation as an animal dealer.[83] Vaillant had much to offer the menagerie in terms of skills and specimens alike. He had kept serpents and crocodiles as well as mammals and birds, and he knew how to read, write, and maintain records.[84] Beyond that, his resources as an animal dealer were extraordinary.

The menagerie's bird registry for the 1830s and 1840s shows Vaillant selling, giving, exchanging, or purchasing more than fifty species, including raptors, owls, macaws, grouse, multiple kinds of pheasants, guinea fowl, quails, pigeons, moorhens, coots, water rails, white storks, swans, geese, various species of ducks, a cassowary, a marabou stork from Senegal, and Venezuelan troupials.[85] Vaillant also sold the menagerie at least one hundred live mammals representing more than twenty species. They included specimens from Africa, Southeast Asia, and South America. Most were monkeys, but there were also hedgehogs, coatis, a badger, a mongoose, a palm civet (*Paradoxurus*), a dasyure (a small marsupial), two species of lemurs, a flying squirrel, and other small mammals that the menagerie wanted.[86]

In maintaining his animal dealership while serving as a keeper, Vaillant was unique. The menagerie's earliest keepers in effect gave up being animal proprietors when their animals were confiscated and they came to work for the Museum. Later, when Henri Martin offered to give the Museum his collection in exchange for running the menagerie, he was turned down. A similar case was a man named Callégari, proprietor of an itinerant menagerie of thirty-three animals, who offered himself and his animals to the Museum in 1835. The Museum was not interested in Callégari's proposal, but it did want his polar bear, for which it gave him a lioness and 500 francs.[87]

Yet it was not out of the question for the Museum to bring in new animals and their proprietor alike if the specimens were sufficiently attractive, the price was right, the prospective keeper was a man of talent and good character, and a position was open or could be created for him. In October 1838, when the professors appointed Vaillant as keeper of birds, they also hired a thirty-one-year-old animal dealer/showman named Jean Vallée. What made Vallée special were the "magnificent reptiles" that he owned and the Museum wanted: two "enormous" pythons and three caimans. The larger python was an estimated fifteen to eighteen feet long.[88] The Museum agreed to pay Vallée the substantial sum of 1,500 francs for his reptiles, his tools for handling them, and five mammals and four birds he had also been exhibiting.[89]

The Museum appointed Vallée to the new position of reptile keeper at a starting salary of 800 francs per year. The reptiles, installed in the old monkey house, became the Museum's newest attraction. Vallée went on to have a long and much appreciated career as their keeper, earning special praise for his skill at raising, nurturing, and experimenting with his charges.[90]

Vaillant and Vallée represent one end of the spectrum of the individuals who gave, sold, or traded animals to the Museum in terms of their sustained engagement with the menagerie (Vallée as a keeper and Vaillant as both a dealer and an animal keeper at the same time). They provided the menagerie with a host of important animals, and they came on staff with a close knowledge of the animals' habits and requirements. The other donors and dealers/exhibitors encountered in this section had briefer interactions with the menagerie and were very diverse in motivation, but in their respective ways, smaller or larger, they helped feed the Museum's hunger for new and interesting specimens. Although the Museum's acceptance of a gift sometimes depended on the status of the giver, more often it was the animal in question that mattered most.

[3.7]

Tapirs and Travelers

When Buffon wrote about the tapir he disparaged it, citing it as a prime example of his thesis that New World (North and South American) animals were small and ill-formed compared with their Old World counterparts across the Atlantic. The tapir, by his account, was the biggest animal of the Americas but puny compared with the "colossal masses" of the Old World's elephant, rhinoceros, and hippopotamus. He represented the poor creature as gloomy and fearful. To add insult to injury, he claimed its meat was "tasteless and coarse." When he went on to say, "however, the Indians eat it," it was still not a compliment.[1]

Both Buffon and Linnaeus believed that the tapir of South America was the only such animal in the world. Georges Cuvier agreed, describing it in his *Règne animal* (1817) and placing *Tapir americanus* in a separate genus, in the second family of the order of pachyderms. Its closest cousins were the hippopotamuses, pigs, rhinoceroses, damans (hyraxes), and two "lost" (extinct) genera, the *Anoplotheria* and *Palaeotheria*. Its most distinguishing characteristics were its teeth, its trunklike nose, and the numbers of digits on its front feet (four) and hind feet (three). The tapir lived in the warmer countries of South America, he explained, in humid areas and along riverbanks; natives ate tapir meat, and the animal's young were spotted like fawns.[2]

When one of his protégés reported from Calcutta in 1819 that the Old World had a tapir of its own, Cuvier was taken by surprise. Buffon's theory about the Old World's superiority to the New World in its productions of nature had already been contested, but to find a large Old World mammal unknown to Western scientists was shocking. It opened the exciting possibility that more new species might yet be found in India and Asia.

Cuvier's disciple in Calcutta was Pierre-Médard Diard (1794–1863), a native of Tours who had survived his army service in 1813 and 1814, then studied medicine in Paris under the wing of the brilliant plant physiologist René Dutrochet. When Dutrochet sent Diard to the Museum to demonstrate

Figure 3.7.1. "The Indian tapir or maïba." Based on a drawing sent to Paris. C. de Lasteyrie after Jean-Charles Werner. É. Geoffroy Saint-Hilaire and F. Cuvier, *Histoire naturelle des mammifères*. General Research Division, the New York Public Library

some recent experimental findings, Cuvier was sufficiently impressed to invite him to study zoology and comparative anatomy while completing his medical degree. Early in 1817, Diard announced he planned to travel to India to pursue a commercial opportunity in Chandernagor, France's tiny foothold in Bengal. Figuring Diard could collect specimens on the side, Cuvier had the Museum recommend to the minister of the navy and colonies that Diard be given free passage on a government ship.[3] The minister obliged, and Diard's transportation was assured, though he delayed the trip to help dissect the menagerie's lately deceased female elephant.[4]

Diard sailed in August 1817 bearing a letter of introduction and other materials from Cuvier to present to the Asiatic Society of Calcutta (a British colonial organization with no Indian members). Eighteen months passed before the Museum heard from Diard because of how slowly ships, people, and mail traveled between India and France. His letter to Cuvier from Calcutta (dated 6 January 1818) did not reach Paris until the end of November or beginning of December.[5] Diard began by telling Cuvier how well he had been received at the Asiatic Society, because "all the members hold your name in veneration." The society promptly named Cuvier an honorary member and treated Diard as his protégé. Diard declared himself unworthy of the designation, but he pledged to Cuvier that he intended to earn it.[6] He reported how his prospects as a naturalist had improved through meeting the marquess of Hastings, governor general of India.[7] Hastings had created a small menagerie at Barrackpore, and his nephew, whom Diard had befriended, was helping to manage it. It was there that Diard first saw the Malayan tapir, a species European scientists knew nothing about. He referred to it as the Sumatran tapir because that was where it was captured.

The British governor of Malacca, Major William Farquhar, had recently sent a tapir skull and detailed notes on the species to the Asiatic Society of Calcutta, but his description had not yet been published. The society's secretary shared the notes with Diard, who sent some to Cuvier. Farquhar introduced the animal by citing the features of most interest to naturalists:

> The tapir is called *Tinou* by the Malays. It is as common in the forests of the peninsula as the rhinoceros and the elephant and very well known by the inhabitants. The musulmans do not eat its meat because they consider it a kind of pig. Its trunk in the adult male is seven to eight inches long. In form and size it resembles the [tapir] of America. It differs from it 1° in its color, because it is black everywhere, except for the eyes, which are bordered with white, and the underside of its body, which is a pale gray. 2° by its teeth [here Farquhar detailed the numbers of molars, canines, and incisors of the upper and lower jaws].[8]

Diard, demonstrating how well Cuvier had taught him the importance of comparative anatomy, said Farquhar had mistaken the large first upper incisor for a canine. Thus, the Sumatran tapir's dentition was in fact identical to its American counterpart's. As for its skull, it was just like the American tapir skull in Cuvier's collection. Diard sent a picture of the Barrackpore tapir, drawn by "a native." The colors were correct, he said, but not the proportions: the tapir was heavier, its body shorter, and its neck and legs both shorter and thicker. It was "a very gentle animal that loves very much to be caressed." Malays had captured it as a baby in the mountains near the Sumatran coast, and an English ship's captain brought it to Hastings's menagerie, where it had lived for two years.

Diard did not say which commercial interests lured him to India, only that they would favor his specimen collecting. He was well positioned geographically, he told Cuvier, and he had many contacts. With sufficient funds, "nothing would be easier than to bring here all that one could desire from China, Australia, the Malay Archipelago, and the Moluccas."[9] The promise of new specimens likely was music to Cuvier's ears, and even more so when his stepson, Alfred Duvaucel, decided to go and assist Diard in India.

In 1820 the Asiatic Society's journal, *Asiatick Researches*, published Farquhar's account of the tapir followed by observations by Diard, "a distinguished pupil of our illustrious colleague M. Cuvier." Diard diplomatically corrected Farquhar's mistake about the tapir's teeth, noting that many distinguished naturalists had similarly erred. He gallantly concluded, "to Major Farquhar alone belongs the honor of having first given, with the trifling exception I have noticed, a correct description of the interesting animal that forms the subject of these observations."[10]

But the Asiatic Society's papers of 1820 were not the first publications on the animal. On receiving Diard's letter, Georges Cuvier reported the tapir's discovery to the Academy of Sciences (7 December 1818), a résumé of which was published in the December 1818 issue of the *Journal de Physique*. Then, based on Diard's letter and picture, Frédéric Cuvier wrote an essay on the animal, calling it the "Maiba," which was published in a March 1819 installment of the *Histoire naturelle des mammifères*. In announcing that a new kind of tapir was now known to Western science, the younger Cuvier, like Diard, credited Farquhar with the species' first description. He expressed the Museum's optimism about Diard, "a student of the Jardin du Roi led to India by his love for natural history, who, in combining with much intelligence and activity an accurate and penetrating mind, ought to enrich zoology with important discoveries."[11]

In May 1818 Alfred Duvaucel (1793–1824), Georges Cuvier's twenty-five-year-old stepson, joined Diard in Calcutta. To his mother, Duvaucel wrote of the esteem in which his step-father was held:

> The good name of M. Cuvier has opened all doors and hearts. The admiration one has for him is beyond words: one loves all that comes from him, one welcomes all those who know him, there is no more glorious recommendation than his. No prince or sovereign receives such sincere or disinterested praise. M. Cuvier in Bengal has one hundred times more weight than the King of France. With him one can do everything, ask everything, obtain everything.[12]

Cuvier's renown placed Diard and Duvaucel in good standing with India's British governor general and his wife, who both looked with favor on the young French naturalists. Whenever a rare animal arrived at Barrackpore they were informed, and when they traveled, they carried Hastings's invaluable letters of introduction.[13]

Deciding that Calcutta was too expensive and too busy for quiet research, Diard and Duvaucel moved a few kilometers away to the French station of Chandernagor. In a small house, they kept one room for sleeping and turned the rest into galleries for displaying specimens and housing live animals. Rajahs sent specimens, hunters sent captured animals, and the two Frenchmen personally collected animals and plants and cultivated a garden with a pond to attract wading birds. Like a microcosm of the Jardin des Plantes, their menagerie drew people from Calcutta and nearby.[14] They mostly sent dead specimens to Paris, but they also sent seeds and several live animals, including a male cashmere goat, a gift from Hastings's menagerie.[15]

In 1818 the two naturalists were planning an expedition in Bengal when a more enticing possibility emerged. Sir Stamford Raffles, recently named lieutenant governor of Bencoolen on the southwest coast of Sumatra, came to Calcutta to arrange a mission in the Malay Archipelago with the East India Company. Diard and Duvaucel met Raffles several times, probably through Hastings, who was one of the five members of the East India Company's Supreme Council. It was decided that the French naturalists would travel with Raffles in Malaysia, collecting specimens while Raffles went about his affairs. Diard and Duvaucel were certain the region held untold marvels, and they thought their zoological studies would be freed from the financial constraints they had experienced in India, where they had to rely on their own funds.[16] With the riches of the East India Company backing them in Malaysia, their dreams could be achieved. They would travel under the company's protection, and Raffles was offering to create in Bencoolen, at the Company's expense, a zoological facility like the one Hastings had established at Barrackpore.

Raffles's political triumph on this mission was acquiring Singapore and establishing the colony that became the eponymous city, enabling the English to challenge Dutch mercantile supremacy in the region. Over eight months of travel the Frenchmen had meanwhile collected half a boatload of specimens on their stops at various isles, buying specimens when possible and hunting and collecting as necessary. Hunters, artists, and taxidermists in their entourage assisted them. Malacca and Singapore were excellent collection sites. Duvaucel later recalled the bewilderment of the locals in Malacca, who were familiar with commerce related to opium or pepper but "could not figure out what we wanted to do with the monkeys and birds that we bought. In two hours, we were able to procure a bear, an argus [pheasant], and other birds."[17] In Singapore their most precious acquisition was the dugong, a marine mammal that lives on algae and plants, browsing on these (as Raffles put it) "much in the same manner as a cow in a meadow." Cuvier had urged them "to neglect no occasion to extend our knowledge of such a remarkable animal."[18]

Diard and Duvaucel sent the Museum a description of the dugong that Frédéric Cuvier used as the basis for an article in the *Histoire naturelle des mammifères*, but this time the British beat the French by publishing first. Using Diard and Duvaucel's notes, Raffles sent a memoir on the dugong to Everard Home, president of the Royal Society of London, who read it to the

Society in May 1820 and swiftly published it in the Royal Society's *Transactions*.[19] As Raffles's letter to Home explained, "At Singapore, in June last, I had the good fortune to meet with one of these animals [the dugong], and Messrs. Diard and Duvaucel, two French naturalists employed under my authority, undertook the dissection of it; and have sent a dissertation upon it to Sir Joseph Banks. This does not interfere with my sending to you, as I promised, an account of it."[20]

Early relations between the French naturalists and Raffles were highly amicable. Raffles appointed Diard as his physician and treated treated Diard and Duvaucel like family members. He provided their salaries and saw to it that their travel, lodgings, draftsmen, and other forms of assistance were handled by the East India Company. Relations frayed early in 1820, however, when the Company decided to stop supporting the Frenchmen's efforts. Diard and Duvaucel argued that their original contract with Raffles and the Company was broken, and that they were entitled to at least half the specimens they collected. Raffles refused. According to Georges Cuvier's later account, based on what he had heard from Diard, the two naturalists embarked with Raffles in December 1818, "under the condition that the result of their researches would be shared equally between them and the governor; the latter would see to reimbursing the expenses by the East India Company, and these men promised their work, their time and their cooperation to the scientific description that Raffles wanted to publish of the countries that he was to visit and the one he was called to govern."[21]

Unfortunately for the Frenchmen, Cuvier continued, "this collection, made with so much care, pains, and dangers, instead of being divided equitably, which seems to have been the promise, was sent almost entirely to England, with a copy of the illustrations, descriptions, and notes assembled by our young naturalists."[22]

Cuvier was right about the fate of the collections and accompanying notes, but what "[seemed] to have been the promise" was viewed differently by the two sides. The documentary evidence Raffles produced favored Raffles, as did his status and that of the East India Company. Raffles could show a letter he had written early on to the Frenchmen stating that their researches were to be "made for and on account of the Honourable the East India Company, and that your collections, &c. are considered as their property."[23]

The Frenchmen felt they had an oral agreement with Raffles about sharing the specimens and had been duped by his blithe assurances that the document in question was a mere formality. In the end the Frenchmen were obliged to surrender their specimens or have them taken by force. The professions of goodwill and respect in the formal letters exchanged at the time between the naturalists and Raffles did not reflect underlying sentiments. Diard and Duvaucel came to believe that Raffles's display of friendship toward them had been calculated and duplicitous all along and that he had deployed "perfidious means . . . during the whole course of our discussion to make us fall into the trap he had laid for us."[24]

The affair boiled down to power politics, with natural history specimens as the prize. Raffles took nearly the whole collection, leaving only some duplicates (or triplicates) to the French naturalists. He hurried to be the first to publish descriptions of the collected animals. By 1 June 1820 he had completed a "descriptive catalogue" for the *Transactions of the Linnean Society of London*, so as not "to allow the result of all my endeavors and exertions to be carried to a foreign country."

Raffles downplayed the importance of the notes and papers he had forcibly extracted from the Frenchmen. These, he alleged, "were for the most part so speculative and deficient in the kind of information required, that I could make no use of them myself, nor give them to the world under the sanction of my authority." He claimed to have returned them to Diard and Duvaucel "and left these gentlemen at liberty to publish or amend them as they think proper." With a final arrogant flourish, Raffles observed, "They are young men not deficient in zeal, and though misled for the moment by private and national views, will, I doubt not, profit by the means I have afforded them, and eventually contribute to our further knowledge of the zoology of these islands."[25] Frédéric Cuvier's essays on some of the new species Diard and Duvaucel collected testify to the lingering French resentment regarding the affair. All the animals in Raffles's descriptive catalog, Cuvier stressed, had been procured by the two Frenchmen.[26]

Raffles sailed from Bencoolen to London in February 1824, carrying innumerable natural history specimens, drawings, notes, and "a living tapir, a new species of tiger, splendid pheasants, &c., domesticated for the voyage." The ship caught fire fifty miles after departure; the human passengers survived, but all else was lost. Back in London, Raffles drew up a prospectus for a grand zoological collection with live animals that would go "far beyond the Jardin des Plantes at Paris." National pride and international competition were very much a part of the Zoological Society of London's founding in 1826, with Raffles as its president. It was the beginning of the London Zoo.[27]

When Diard and Duvaucel left Bencoolen on 1 April 1820, Diard headed south to Batavia and Duvaucel went north to Padang to maximize the diversity of their collections. However painful their experience with Raffles, it had taken them to lands they might otherwise not have reached.[28] The quantities, diversity, and novelty of their newly collected specimens were impressive. In a letter from Batavia (October 1820), Diard listed all he was about to embark to Georges Cuvier, estimating that he and Duvaucel had collected some four hundred different species of birds and nearly as many types of mammals.[29]

Diard was also sending two live animals: a slow loris (a distant relative of the lemur) and a mongoose he had kept in Batavia for "a very long time." As he told Cuvier, he had found a cooperative sea captain who was prepared to take preserved and live specimens on the long trip from Java to Antwerp, then to Paris, and he hoped to find other obliging sea captains in the future. What he and Duvaucel needed, he said, were a depot at Chandernagor where they could assemble live specimens and the authorization "to embark them and have them properly cared for aboard ships of state that come each year to Pondicherry." This way "the tapirs, rhinoceroses, elephants, and ruminants of all kinds would come in quantity to Paris and the curious would no longer need to travel four to five thousand leagues to admire the marvels of this beautiful country."[30]

Diard's shipment from Java arrived early in May 1821. Days later, reporting to the Institut on Diard and Duvaucel's travels, Cuvier numbered twenty-four stuffed mammals; seven mammals preserved in alcohol; nine mammal skeletons; 126 species of birds; and nearly 100 species of snakes. Species new to science, or at least to the Museum, included the "so remarkable discovery" of "the tapir of the Indies," plus "a new species of one-horned rhinoceros from Java."[31] Cuvier made no mention of the slow loris and mongoose Diard embarked, nor do they appear in Frédéric Cuvier's menagerie reports. Like many other animals from overseas, they evidently failed to survive the journey.[32]

Diard's shipment arrived at the Museum about the same time as news of the deaths of naturalists Havet and Godefroy. The professors asked the minister of the interior to reallocate the deceased voyagers' funding to Diard and Duvaucel. The minister awarded them 5,000 francs each for 1822: Duvaucel for research in Bengal and Tibet, Diard for research in Cochinchina (Vietnam).[33]

After his successful expedition to Padang, Duvaucel returned to Chandernagor (West Bengal) to organize his notes and prepare a trip to Sylhet in East Bengal (now Bangladesh). He set forth equipped with letters from the marquess of Hastings; four native assistants from Sumatra: a Malabar skilled in hunting and taxidermy; a young Malay he called "Friday" (Jumahat), à la Robinson Crusoe; a skilled mulatto painter; and a cook whom Diard felt was better at dissecting animals than cooking them.[34] Though he did not report his entourage's views on his collecting, he noted the local sensibilities he excited when seeking to obtain specimens of the Houlmann, or entelle monkey, otherwise known as the Hanuman monkey, the gray langur, or the Bengal sacred langur.

In Chandernagor, seven or eight of these monkeys frequented the neighborhood for food. The Bengalis urged Duvaucel not to shoot them, telling him they were gods and anyone harming them would die before the year was out. Not convinced they had persuaded him, Duvaucel recounted, "My garden was surrounded by a guard of pious Brahmins who beat the tam-tam to chase away the god when he came to eat my fruits."[35] En route to Sylhet, traveling up the Hooghly River, Duvaucel visited the holy site of Guptipara, where the trees were filled with langurs, ripe for killing and collecting, but again locals argued against it, though eventually to no avail:

> The Hindus, upon seeing my rifle, guessed, as well as the monkeys, the motive of my visit, and ten or twelve of them came to tell me of the danger of firing upon animals that were nothing less than metamorphosed princes, or practically first cousins of one of their most revered gods. I didn't want to listen to these generous advocates, but half convinced, I started to go on, when I encountered, on my route, a princess so seductive that I could not resist the desire to consider her more closely. I let go a shot at her and was then witness to [a truly touching deed]. The poor princess, who carried a little prince on her back, was struck near the heart; she felt herself mortally wounded, and gathering all her forces, seized her baby, clung to a branch, and fell dead at my feet. So maternal a deed made more impression on me than all the lectures of the Brahmins, and the pleasure of having a beautiful animal was not able this time to outweigh the regret at having killed a being that seemed to hold on to life because of what makes it most respectable.[36]

In Bengal on 24 January 1823, Duvaucel was writing to his mother about how difficult it was to accomplish anything in India owing to people's superstitions and habits, to insufficient funds, and to a frightful climate, when he learned that a rhinoceros was in the area. He put down his pen and set off to shoot it, but the adventure did not go as planned. He attracted the fury of a second rhinoceros that gored him in the thigh. He lost a great deal of blood but suffered no broken bones or torn muscles. Continuing his letter days later, he made light of the episode. "If [the rhinoceros] had put less brutality and less anger in his action, I would have believed he only wanted to put me off my profession. However, dear mother, M. Cuvier [Georges] must not think me capable of renouncing it for so small a thing." Vowing to continue collecting as soon as he could walk, Duvaucel wrote, "I have scarcely begun my voyage and have already assembled several precious objects. I have done three times more than the English with five times fewer expenses. What ought I not expect when I find myself in countries that up to now were explored only through war and commerce?"[37]

In April 1824, in a public lecture at the Institut de France, Georges Cuvier noted the sacrifices behind natural history's advancements since the restoration of maritime peace: the deaths of Havet in Madagascar and Godefroy in the Philippines, and Duvaucel "dangerously wounded by ferocious beasts on the banks of the Ganges" and "for a long time on a bed of pain." He told how European zoologists had assumed the world's large animal species were all known, but thanks to the efforts of Diard and Duvaucel the continent had yielded many new to Western science, including "four or five deer, as many bears, two rhinoceroses, and a tapir, a genus not believed to exist outside of America."[38]

That same year, Frédéric Cuvier reviewed in the *Journal Asiatique* the stunning array of mammals Duvaucel and Diard had sent to Paris: two thousand specimens representing eighty-four species including new species of gibbons and other monkeys, roussettes, toupayes, the telagon, the binturong; three new species of bear; two new species of rhinoceros; the dugong, and more. They sent preserved skins, skeletons, animals in alcohol, paintings, and descriptions of behavior. No voyager naturalist working alone in as little time, even with a hundred times as many resources, had assembled as remarkable a collection as Duvaucel, F. Cuvier affirmed, allowing one to appreciate "the richness and power of nature." He concluded with a paean to voyager naturalists, stressing "all that must be suffered by men far from their country, under the equatorial sky, hunting rhinoceros, elephants, water buffalo, tigers, in a word, animals that seek to escape them by ruse or by violence." Continued Cuvier:

> When through care and precautions they discover the [animal's] hiding place, and kill it, they have fulfilled the easiest part of their task: it remains for them to make a study of it, to describe it with detail, to draw it, to take its skin, to separate the soft parts, to detach the flesh from the bones, to collect all that must be saved and preserve it from the thousand and one causes of destruction gathered around and commanded by nature to act incessantly. . . . perhaps no other situation demands so great a combination of physical, moral, and intellectual efforts as that of the voyager naturalist.

Frédéric Cuvier presented Diard and Duvaucel as models of what the profession demanded.[39]

The voyager naturalists would have appreciated the praise, but their letters testify that what they mostly craved from the Museum was news, instructions, and funds. Duvaucel complained (1 June 1824) that while he had done the groundwork to supply the Museum with all the plants and animals of the area, he could afford to acquire only half. He owed his collections from Tibet, Nepal, and Bhutan, he said, largely to the support he had received from the British East India Company.[40]

Duvaucel's June 1824 letter took six months to reach Paris. It was read at the professors' administrative meeting of 7 December 1824. Seeds Duvaucel sent from Hindustan arrived at the end of January 1825.[41] On 22 March 1825, the professors learned that the first seven of fourteen new crates of specimens from Duvaucel had just reached the Museum. Georges Cuvier and Geoffroy promised to prepare a catalog of the specimens for the next meeting.[42] What none of the professors knew then, but were soon to learn, was that Alfred Duvaucel had died in Madras nearly seven months earlier.[43] Georges Cuvier announced Duvaucel's death at the next assembly. The minutes do not record the sentiments and condolences that must have been expressed, only the professors' recognition of all that Duvaucel had done for the Museum and their decision to ask the minister of the interior to pay 4,000 francs from the voyager naturalists' fund for 1825 for expenses accrued by Duvaucel in his collecting.[44]

Frédéric Cuvier expressed his grief over Duvaucel's loss in an essay on an animal Duvaucel had collected, the "Croo" (crab-eating macaque). He turned the piece into a lament on his step-nephew's death. "Ah! What voyager naturalist deserves to be more regretted than he!" None, said Cuvier, had been more selfless, more oblivious to danger, and more intelligent in his efforts, and "never has the most important branch of zoology, the mammals, been increased by works more fruitful than his." Cuvier allowed that if his own latest work on mammals "has merited some interest because of the new species it contains, I have to recognize that it is especially to Duvaucel that it is indebted." Frédéric bemoaned the untimely loss: "He was ready to return to his country, and friendship awaited him, when in his stead arrived the frightful news of his death."[45] Describing this essay as "the first testimony of my sorrow," Cuvier dated it March 1825.

∴

The first live tapir displayed at the menagerie came from South America. It arrived in Paris in May 1825, courtesy of the Museum's old friend Baron Pierre Bernard Milius, then governor of French Guiana (1823-25).[46] Returning to France, Milius brought a few live animals and numerous live plants despite the Museum's warning that they often suffered greatly in transit and that it lacked the funds to cover their transport.[47]

Frédéric Cuvier was coincidentally in Le Havre when Milius's ship arrived from Cayenne, there to coordinate with the Museum's agents, Eyriès frères, the disembarking and transport to Paris of a small elephant sent by the pasha of Egypt. The steamboat that was to take the elephant was waiting for the arrival of huge crates containing a whale skeleton. Meanwhile, Eyriès frères received two cases of specimens sent from Senegal by the voyager naturalist Samuel Perrottet, who worked for the Ministry of the Navy and Colonies.[48] Cuvier was writing to the Museum about the elephant when Milius's ship arrived; he set aside his pen to see what it had brought, and hours later reported: "I return from the corvette commanded by Mr. Millius [*sic*]. It has on board a zebu bull and cow, a Tapir [the double underline emphasis was Cuvier's], a large species of owl, a royal bird, some aras [macaws] and other parakeets, some hoccos [curassows], a marail guan, some monkeys, &c. &c. I saw these animals poorly deep in the hold, but the tapir is certainly there."[49]

Cuvier had not seen Milius, but he had heard that not all the mammals were destined for the menagerie and that the birds were to remain at Milius's disposition. Still, as he told the Museum, "provided we get the tapir, I think the rest will be of little interest to you."[50]

Milius waited until he reached Paris to write to the Museum about the collection awaiting steamboat transport in Le Havre: thirty crates containing two hundred to three hundred live plants; ten crates of seeds; two casks containing five two-toed sloths preserved in alcohol; and six live animals, including the tapir and an eagle. Live sloths he'd embarked had perished from the cold at sea. The tapir, a very young specimen, had suffered greatly from the rough crossing, but, said Milius confidently, "it has recovered from its fatigues, and I am certain it will arrive at its destination without accident."[51]

The tapir arrived in good health. Initially kept inside owing to the cold, it was put outside in a pen each morning once the weather improved.[52] After that, the tapir was no longer a novelty or cause for concern and absent from Cuvier's weekly reports, though he did write about it in *Histoire naturelle des mammifères*. The young tapir was gentle and trusting, more disposed to renounce its own wishes than any domesticated animal. It was "greedy" with its food but nonetheless shared with some menagerie dogs and goats instead of fighting over it. The tapir played with young dogs, and it enjoyed running about its pen after being shut up inside. It knew where it slept, and when it felt cold, it demanded to be let inside.[53]

Cuvier offered no opinion on the taste of tapir meat, but he was certain the species "would be very easy to render domestic."[54] In an essay on domestication (1825), he suggested that just as the wild boar had given rise to different races of domestic pig, so the tapir could produce domestic descendants. The issue was whether this could be accomplished swiftly enough. Humans were hunting tapirs for meat, and with the human population growing, it seemed the American tapir would "disappear from the face of the earth."[55] It was an uncommon observation for the time. Naturalists in the early nineteenth century were only beginning to register fears about contemporary species becoming extinct.

∴

Diard's life, like Duvaucel's, had been eventful. In April 1820, after bidding his colleague farewell in Bencoolen, Diard went to Batavia, where he continued collecting and was soon employed by the Dutch government studying how to make indigo production economically viable. An attractive project that helped fund his other work, it advanced the cause of developing colonial plantations, the commercial interest Diard had in mind when he headed to India to seek his fortune. News that the French government would support his and Duvaucel's research gave him the confidence to plan a voyage from Batavia to Cochinchina with the newly appointed (and first) French consul to Hué, Jean-Baptiste Chaigneau, whose mission was to foster French trade relations in the region.

Diard lived three years in Cochinchina, completely out of contact with Paris; the Museum professors feared he was dead. His collecting efforts were relatively unproductive, in part because the authorities limited his travels. He was meanwhile assessing the possibilities of establishing a prosperous French colonial presence through the introduction and development of coffee, spice trees, cinnamon, cotton, indigo, sugarcane, silk, and such.[56] The coronation of a new ruler (Minh Mạng) who was suspicious of French motives signaled the end of such endeavors. Diard's movements were restricted, and he was able to leave only by building a boat. When the British thwarted his plan to visit Pégu (today part of Myanmar), Diard returned to Singapore, paid a visit to Batavia, and by spring 1824 he settled in Malacca.[57]

Diard expected Malacca to be a much better place for collecting than Cochinchina. He soon acquired two tapirs, but these proved "unable to survive their loss of liberty." Hopeful of capturing a healthier pair, he had tapir traps set out, but a friend informed him of a possible appointment as director of agriculture in Java, and Diard departed for Batavia, arriving at the beginning of December 1824. He left behind his collections, his boat, and the domestic servants caring for his traps.[58] The Dutch governor general, Baron Van der Capellen, appointed Diard to the post of inspector of agriculture the next month. However effective he was as a naturalist and colonial administrator serving the Dutch government, Dutch naturalists in Java mistrusted him and suspected he was collecting specimens primarily for the French.[59]

In March 1826 Diard returned to Batavia and, still exhausted from a long journey to Bangka, began writing a detailed letter to the Museum.[60] Having been out of touch for three years, he had much to say, starting with his miseries as a voyager naturalist, "absolutely deprived of all correspondence and any kind of instructions on your part." He was depressed, he said, by the slim results of his years collecting in Cochinchina and by the news of Duvaucel's death, which did not reach him until nearly a year after the event. He was worn out from several days of arranging to send his collections, including a live female tapir, and on top of that was suffering from fever. He told of his decision to cast his lot with the Dutch as a colonial administrator.

The tapir featured prominently in his account. He had returned to Malacca to find his domestic helpers caring for "a well-tamed young female tapir," and days later, he captured a male tapir himself. He saw to the repair of his boat and sent the two tapirs and a young Malaccan elephant to Batavia, where they arrived in December 1825. He then went off on an agricultural mission and returned to Batavia in January 1826. "I had the greatest hope of being able to send you two tapirs still in good health," he wrote, but the male had died twelve days before. "His prepared skin and skeleton," promised Diard, "will reach you soon." He continued, "I dare to hope that the female—younger, tamer, and to which Mr. Wattier the captain of *La Glaneuse* from Nantes is preparing to give the greatest care—will arrive in good health at the Museum. That is the greatest compensation for which I can hope for all the difficulties, inquietudes, fatigues, and for more than two thousand piasters that this animal has cost me."[61]

Diard elaborated on his poor health, the debts he had accrued while collecting, the difficulties of getting news or funds from the Museum, and his appointment as director of agriculture for Java. He had wanted to serve France exclusively, he said, but having no communication with the Museum rendered it impossible. He hoped the Museum would not sever ties and asked to be named an official correspondent. He promised to send two dugong skins, a dugong skeleton, and more than two hundred species of birds "in a very good state of preservation." He mentioned notes and a memoir that he would send later and wondered if the Museum might graciously give him some sort of indemnity to cover the tapirs' transportation costs (more than four hundred rupees per month).

Diard was twelve pages into his letter when calamity intervened. When he resumed writing two days later, he recounted "the most devastating and heartbreaking event that could be reserved for me." He explained:

> My tapir that I had with great difficulty in the evening embarked on a large dinghy and taken to the mouth of the river to be put on board ship the following morning, abandoned during the night by all the guards that I had put around it (prevented by fever from guarding it myself), broke its cage, threw itself into the water, and disappeared in the surrounding swamp, where for two days, although it has been seen and is neither vicious [*méchant*] nor wild, has been absolutely impossible for me to recapture despite some efforts and expenses I have devoted to that.

Diard felt "exhausted, defeated, and totally discouraged." Not only was the tapir gone, but also, amid all the chaos, Diard had mislaid the list of objects he was sending, critical for identifying the contents of the crates. He apologized for the disorder. The privations he had suffered and the expenses he had accumulated in his years of working for the Museum, the relative paucity of his collections after a brilliant start, the death of his dear friend Duvaucel, his decision to cast his lot with the Dutch, and the poor state of his health combined to leave him in despair.

Diard's March letter did not reach Paris until late in June 1826. The month before, the Museum had decided to suspend his stipend, having had no news from him for years.[62] When it learned he was working for the Dutch, it removed him from the payroll but granted him the honorary title of correspondent.[63] In addition to the crates of specimens Diard promised in his letter of March 1826, he sent another shipment with a live tapir and the skin of a dugong.[64] This time he successfully embarked the animal on a Dutch ship departing from Batavia, but it made it only as far as Île Bourbon, and only its remains reached the Museum.[65] A subsequent shipment of specimens from Batavia was lost in a shipwreck in spring 1827.[66] Perhaps the Museum recalled Diard's woes with his tapirs when in its latest (1829) *Instruction for Voyagers* it noted that transporting live animals was much more difficult than transporting the remains of dead ones, and the latter were of more scientific consequence anyway.[67]

∴

A live Malayan tapir finally reached France, making port in Nantes on *Le Mélayo* in April 1834. The ship's commander, a Nantes merchant named Salaün, had purchased the tapir in Malacca and kept it healthy during seven months at sea. The Nantes newspaper *La Loire* covered the tapir's arrival, describing its appearance and diet in an article republished days later in Paris in the first issue of a scientific journal for the educated public, *L'Echo du monde savant* (Echo of the Scientific World).[68] The agricultural botanist Samuel Perrottet, who was in Nantes preparing to sail to India, was also interested in the tapir. Like Diard, Perrottet had close connections with the Museum and a career of colonial service (for France). He'd worked at the Museum as a gardener, supplied it with specimens from his travels to Cayenne and the Philippines, and further impressed the administration with his botanical work in West Africa. He'd lately been corresponding with the Museum about his upcoming travels to Pondicherry, which like his earlier trips were in the service of the Ministry of the Navy and Colonies.[69] He was sure the Museum would want Salaün's "curious pachyderm," as the menagerie had never had a tapir from Sumatra.

Perrottet boarded *Le Mélayo* in Nantes harbor on the morning of 3 April 1834, saw the "superb, live tapir" in Salaün's possession, and hastened to write to Geoffroy about it. Perrottet related his impression of Salaün's animal:

> I found it standing and eating cooked beets and potatoes. Up to now it was fed yams and manioc. It is strong and very healthy. Its fur is of a grayish white from the beginning of its very short tail up to nearly its shoulders. The rest of its body and head are blackish. Its size is approximately that of the individuals of the species from Cayenne that you have stuffed in your galleries. The digits of its feet are long and very distinct; there are four on each forefoot and three on those behind. Its trunk is scarcely apparent in the state of repose, but when the animal moves, it is several inches in length.[70]

Eager to demonstrate his commitment to the Museum's interests, Perrottet reported that some Englishmen had offered to buy the tapir. Salaün preferred selling to the Museum but wanted to know how much it was prepared to pay. Perrottet warned Geoffroy that the merchant's expectations might be "a little elevated." In a follow-up letter Perrottet observed that the tapir would be an especially "precious" acquisition because it was unlikely so healthy a specimen would be seen again in France or even Europe. He told Salaün to consider the Museum's limited budget in his negotiations. Perrottet put his advice in a letter (29 April 1834) that he gave Salaün to take to the Museum by way of introduction. The next day he sailed for Réunion, carrying collecting instructions and letters of introduction from the Museum.[71]

Salaün went to Paris to bargain, without the tapir. He told Geoffroy it had cost him 6,000 francs to bring the animal to France and asked the Museum for that amount plus 1,000 francs for all his trouble and the risks of keeping the tapir alive for seven months at sea.[72] The Museum wanted the animal but declared 5,000 francs was the most it could pay. Salaün thought this too low, but wishing to conclude the transaction, he agreed. On 15 May 1834 a contract was drawn up with Adrien de Jussieu, Museum director, specifying that for 5,000 francs Salaün would sell the Museum "a live tapir, originally from Sumatra, presently at Nantes." Salaün attested to the animal's "very good state of health," but the contract stipulated that a Museum delegate would go to Nantes to confirm the tapir's condition. If the delegate "left any doubt about this good state of health," the administration would be free to annul the contract.[73]

On the Museum's behalf, Henri de Blainville contacted a colleague and former student in Nantes, Pierre Martin Marion de Procé, asking him to visit the tapir and provide a report. No stranger to foreign travel and its hazards, Marion de Procé had been in Manila in October 1820 and barely escaped the massacre that took the life of the Museum's voyager Godefroy.[74] Fourteen years later, established in his native Nantes as a physician and naturalist of distinction, he was an ideal tapir inspector. When he arrived at Salaün's house, however, Salaün's wife informed him that the tapir had just died and been turned over to a taxidermist.[75]

Finding his and the Museum's interests thwarted by the tapir's death, Salaün offered to sell the animal's skin and skeleton. The Museum responded that the live animal would have been "an acquisition of the greatest interest," but it already had a tapir skin and skeleton. It advised him to offer the remains to the natural history museums of Lyon, Rouen, or Geneva, none of which had such a specimen. Thanking the Museum for its suggestion, Salaün acknowledged that the tapir's skin had already begun to deteriorate. He sought to enlist Florent Prévost, the head of the Museum's zoology laboratory, to serve as an intermediary with the Lyon, Rouen, and Geneva museums, judging they would pay more attention to Prévost than to him.[76]

The menagerie would receive more tapirs from Brazil (one in 1838 and more in the 1840s), but the tapir of Southeast Asia continued to elude its grasp. There was more to the theme of controlling specimens, however, than a Malayan tapir's escaping from Diard by jumping ship. Diard and Duvaucel had to surrender to Raffles and the British East Company the specimens they collected in Malaysia while traveling in the Company's employ. Duvaucel found natives protecting the sacred langurs he wanted to shoot for his collection. The question of control extended beyond the possession of specimens to the matter of who first described them in print. In exchange for the benefits of traveling at the East India Company's expense and under its protection, Diard and Duvaucel promised not to write to France, even to their relatives, to discuss their findings before Raffles and the Company made them public. This was not standard protocol, where collectors had the right to be the first to publish the results of their research. Diard and Duvaucel relinquished that right in order to obtain greater support for their collecting than France was able to provide.

Yet while Diard and Duvaucel's collaboration with Raffles ended badly from the Frenchmen's perspective, the support they received from him, the British East India Company, India's governor general, and other British officials enabled their achievements in regions under British control where they could not have been so successful otherwise. A burgeoning British imperial infrastructure served them where French support could not reach. In the end, the Museum could not determine its naturalists' fate any more than it could ensure the safe collection and transfer to Paris of the specimens it wanted. It lost Duvaucel to dysentery in Madras and Diard to the Dutch colonial administration in Java, and never in the golden age of French zoology did a live, Sumatran tapir grace the menagerie.

[3.8]

The Demands of Diplomacy

Had Diard managed to get a Malaysian tapir for the menagerie it would have been a great triumph. As it was, his and Duvaucel's collecting in India and Southeast Asia benefited the Museum's galleries greatly but did little for the menagerie. The menagerie's most impressive gains in this period came instead courtesy of the voyager naturalists Milbert and Dussumier and the rulers of North Africa, though motives of the latter were very different from those of the former. The naturalists were keenly attuned to the Museum's interests, but the rulers of North Africa were not. Their animal gifts came in the service of international diplomacy. Bernardin de Saint-Pierre had foreseen this when he argued that having a menagerie would allow France to receive with dignity the animals foreign rulers sent as diplomatic gifts. What he had not anticipated was what a mixed blessing such gifts would be. The French government repeatedly made the Museum pay the high intranational transportation costs of animals that it neither needed nor wanted, wreaking havoc with the institution's limited budget.

Instances of animals sent to France as diplomatic gifts went back for well over a century. The king of Portugal sent Louis XIV an elephant in 1668, and the king of Siam sent three crocodiles in 1687.[1] The bey of Tunis sent Louis XV a tiger and a lion (in 1772) and sent Louis XVI a pair of lions (in 1777), along with other treasures. The gift list of 1772 included "six horses, an embroidered saddle, six golden sword belts, ten plush coverlets, two dagger scabbards, two scimitars, six pairs of embroidered slippers, six embroidered portfolios," and finally, "three tiger skins, two lion skins, a tiger, and a lion." The list of 1777, equally elaborate, began with "three Corsican slaves" and "a French deserter," followed by six horses, a saddle adorned with gold and silver, an array of coverlets, sabers, and more, and, finally, the two lions."[2]

The items topping the lists testified to the bey's political and military might, his wealth, and his refinement. The lions and tiger signified power over wild nature and the ability to provide a European ruler with something

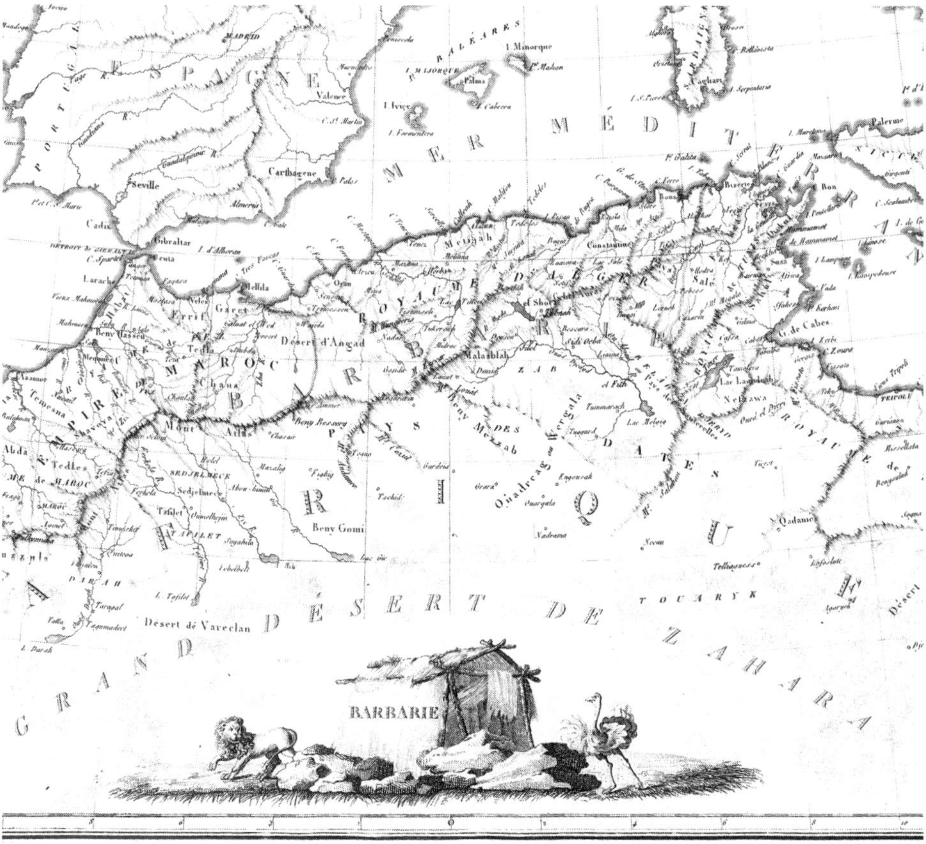

Figure 3.8.1. Detail from a map of North Africa adorned with icons of the European view of "Barbarie"—a lion, a primitive hut, and an ostrich. From Conrad Malte-Brun, *Géographie universelle* (Paris: François Buisson, 1812). R. Burkhardt collection.

his realm lacked. The gifts were meant to glorify the donor as much as to honor the recipient.[3] At the same time, the North African rulers expected reciprocity, and the French government, at least before Napoleon's ascendancy, cooperated to maintain peaceful relations.[4] Matters of international diplomacy were none of the Museum's business, but it was left to reap the benefits and suffer the consequences of animals caught up in diplomatic gift giving.

The menagerie's first animals from a North African ruler came in 1803 from Hammuda ibn Ali, bey of Tunis since 1782. The minister of the interior informed the Museum that the French Republic was about to receive "three lions, some ostriches, some gazelles, and other animals," obliging the professors to discuss how to house them as promptly and economically as possible.[5] The animal convoy reached Lyon without difficulty, but it could go no further than Châlons because a lack of water made the Canal du Centre unnavigable.[6] Judging that travel by land would be too hard on the animals, the conductors decided to wait until October before trying again.

Passing through Châlons in September, the botanist Jussieu stopped to find out for the Museum how the animals were doing. He was especially impressed by the young male lion, said to be about eight months old. He wrote back to the Museum, "it plays like a young cat with its keeper." It was content to be kept on a long chain attached to its cage, which was left open, allowing it ample air and exercise. Jussieu attributed its good health to this arrangement. He saw "no reason to try to accelerate its entry into the menagerie" by making it suffer the rigors of travel by land. Boat travel would be better. Once the waterways were navigable, then the lion "ought to be put on a boat, as he is in his yard, with the freedom to come out of his cage."[7]

The animals arrived at the Museum on 4 December 1803.[8] The deer and gazelle were housed with the herbivores; Josephine Bonaparte claimed the ostriches for Malmaison,[9] and the lions were installed in wooden cages at one end of the menagerie used for *animaux féroces*.[10] Later, by Frédéric Cuvier's account, when the Spaniard who escorted them to Paris came back to visit them, the two male lions displayed affection and "joy," but they responded with rage when their new keepers approached. The young lioness, in contrast, was "perhaps the gentlest" of all the animals in the menagerie.[11]

When the professors first heard from Chaptal about this gift, Antoine-Louis Brongniart, serving as Museum secretary, recorded incorrectly that the animals came from the dey of Algeria, not the bey of Tunis. It was a harmless slip that epitomized how distant the Museum's concerns were from those of French consuls.[12] The professors knew of France's difficulties with the Barbary Coast states (the sultanate of Morocco and the regencies of Algeria, Tunisia, and Tripoli) owing to the notorious depredations of Barbary pirates on the high seas and their enslavement of Christian captives. But whether it was the dey of Algeria, the bey of Tunis, or some other ruler sending Napoleon a trio of lions, what mattered most for the Museum was the port where the gift was arriving and how to ensure the animals' safe and thrifty transport to Paris.

To a French consul, in contrast, the details of diplomatic gift giving were of the utmost importance. France wanted to develop commercial interests in Africa, persuade North African rulers to subdue the Barbary pirates, and secure the release of sailors and others the pirates held prisoner. As of 1803, the consul in Tunis, Jacques-Philippe Devoize, was doing at least marginally better than his counterpart in Algiers, Charles-François Dubois-Thainville, who had to deal with the fits of rage of the volatile Mustafa Pasha, dey of

Algeria. Neither France's pressure on him to stop Algerian privateers nor France's substantial financial debt to Algeria (dating to the previous decade, when Algeria had supplied Napoleon's armies with wheat) was angering Mustafa Pasha as much as a question of diplomatic gifts. He had not received from Napoleon Bonaparte the gifts he believed were his due as a matter of custom.[13] He was ready to give gifts to Napoleon, but Napoleon had to give gifts first. Thainville told the Algerian ruler that Napoleon sent gifts only as signs of friendship and that he refused to be forced to give them. Mustafa Pasha threatened war if France did not promise to honor him with presents. Bonaparte answered that France wanted to live in peace with Algeria but was not afraid of war.[14]

Naphtali Busnach, an immensely wealthy Jewish trader and the dey's chief adviser, was skilled at calming him down. Not a disinterested figure, Busnach co-owned a trading firm (Bacri and Busnach) that dominated commerce in North Africa and the Mediterranean in the early 1800s.[15] It had a monopoly on the lucrative North African grain trade and was responsible for the transactions through which Algeria supplied France with grain in the 1790s. France was consequently indebted to both the Regency and the firm. Busnach, who was integral in negotiating the armistice of 1800 and the 1802 peace treaty, did not want war to upset the complicated ecology of his far-reaching business concerns. To this end, he continually assuaged the dey's ire, assuring him he was "the greatest prince on earth."[16]

Sending animals as diplomatic gifts was problematic in multiple respects, especially in the Napoleonic era, owing to the difficulties of crossing the Mediterranean without interference. The lions offered by the emperor of Morocco (1807) and the dey of Algeria (1808) illustrate the point. In the first case, the Museum learned from the minister of the interior in September 1807 that the French consul general in Tangier was planning to embark two lions from the emperor of Morocco. They were to leave on the next boat to Marseille, the agent of foreign relations would receive them there, and it was up to the Museum to arrange with him the lions' transportation to Paris.[17]

The lions never showed up, and it was five months before the Museum had an inkling of what had happened to them. The Greek vessel they embarked on in Tangier was captured by a Spanish corsair. The lions were subsequently sold in a Spanish port to a Mlle Padouane, who owned a traveling menagerie. Whether she acquired them legitimately was uncertain, as was what the French government should do next.[18] When the prefect of the department of Pyrénées-Orientales pursued the situation at the minister of the interior's behest, he found that the animal exhibitor, identified now as "Dame Victoire Padouane," with two colleagues, had purchased the lions legitimately from a Majorcan animal dealer. They were prepared to sell the two lions to the French government along with the rest of their animal troupe: a small elephant, six monkeys of three species; and three kinds of macaw. The Museum threw up its hands and recommended the minister drop the whole matter.[19]

The attempt to send gift lions from Algeria to France in 1808 exemplifies the hazards of wartime travel and political turmoil alike. France's relations with Algeria were unsettled, as were Algeria's internal affairs. In June 1805 Naphtali Busnach had been assassinated. The massacre of a significant portion of the Jewish community soon followed.[20] In August the militia assassinated the dey, Mustafa Pasha, and installed Ahmad Pasha, whom French consul Thainville found no more amenable to French demands than his predecessor.

After two and a half years of threats from both sides, in April 1808 the *Moniteur* optimistically reported that "all the differences that existed between France and the Regency of Algeria since the elevation of the new dey" had been resolved. Thainville secured Ahmad's agreement to release more than a hundred Italian seamen enslaved by Algerian privateers, an agreement that came only after Napoleon's ultimatum that he would declare war if Algeria did not free all prisoners from territories under French control.[21] The following August, Ahmad destined a pair of lions and several monkeys for Napoleon. Whatever the rationale, circumstances intervened, and after considerable delay the only remaining animal, a lion, arrived in Marseille. In the meantime, Ahmad Pasha fell from favor in Algiers. Soldiers killed him, cut off his head, and paraded it around the city.[22]

What we know of the lions' and monkeys' voyage comes not from the papers of the Museum but, curiously enough, from the autobiography of the French mathematician and astronomer François Arago (1786–1853).[23] In May 1808 the eighteen-year-old Arago was on the island of Majorca measuring a meridian line as part of a project to determine the precise length of a meter (defined as one ten-millionth of the distance from the North Pole to the equator through Paris). Arago's measuring techniques involved taking sightings of fires lit on mountaintops.

Unfortunately for the young scientist, Napoleon's invasion of Spain raised Majorcans' fears that the French would invade their island. The locals interpreted Arago's bizarre mountaintop activities as the work of a spy signaling the enemy. Arago managed to dodge angry vigilantes coming up the mountain to kill him, but the safest refuge he could find was a prison in Palma. He remained there for two months before being released, whereupon he slipped away to Algiers on a small fishing boat. In Algiers, Dubois-Thainville arranged his passage to Marseille. The ship left port on 13 August

1808 carrying a rich cargo, including the animals the dey was giving to Napoleon, but three days out, nearing its destination, a Spanish corsair captured it.

Algeria and Spain were not at war, but wanting an excuse to confiscate the cargo, the Spanish claimed the Algerian ship was attempting to violate their blockade of the French coast. Escorted to a small Spanish port near the French border, Arago and his fellow travelers were quarantined, then held as prisoners. The situation was resolved only after Arago managed to get a letter to Algiers informing the dey (most likely via Dubois-Thainville) that the Spanish had intercepted his ship and one of the lions had died. According to Arago, the lion's death infuriated the dey, who threatened Spain with war if the ship was not released.

Spain relented, and in November 1808 the ship again sailed for Marseille. This time, just as the cliffs of Marseille came in sight, furious mistral winds drove the vessel south all the way to the Algerian coast. Fearing their perilously exposed position, Arago opted to try reaching Algiers by land. He bid good-bye to the remaining lion, with which he had "lived in very good harmony," and to the monkeys, companions in misfortune that had provided him and his fellow prisoners with the welcome service of picking lice out of their hair.

Arago reached Algiers in late December, where Dubois-Thainville informed him of the old dey's decapitation. Still there in February 1809 when the next dey was executed, Arago at last left for France in late June, this time aboard a ship carrying Dubois-Thainville and his family (but not the lion, which had reached France some months earlier). The vessel successfully evaded British blockades off the French coast, and on 2 July 1809 Arago arrived in Marseille. After quarantine, he hastened to his family in Perpignan. By early September he was in the Jardin des Plantes visiting the lion with whom he'd shared his travels and travails, only to find that his sense of camaraderie was not shared. Notwithstanding other accounts of lions welcoming the return of former companions, this one, by Arago's account, "greeted me with a very unfriendly gnashing of teeth."[24] This was not how the lion behaved generally. Upon its arrival, it displayed an "extreme gentleness" toward everyone, and one could enter its cage without danger. It loved to be petted. Frédéric Cuvier believed that with good treatment it would maintain this gentleness, not only to the benefit of its keepers but also for its own health. Gentleness, he believed, was good for a menagerie animal's longevity.[25]

The Muséum ended up reimbursing transportation costs for the lion amounting to 992 francs 70 centimes. This did not count the twelve-franc gratuity it gave to François Parat, the wagon driver who contracted with the prefect of Bouches-du-Rhône to bring the lion from Marseille (730 francs). Included in the total were the 290 francs for the freight and feeding of the two lions during the months they were aboard ship plus the remaining lion's care and feeding in Marseille (172.80).[26] If the Museum expressed reservations about reimbursing these expenses, they were not recorded in the professors' meeting minutes.[27] The Museum was happy to have the lion. The menagerie had no other male lions and only two females: the long-lived Constantine, from Cassal's mission to North Africa over ten years before, and the female sent from the bey of Tunis in 1803.[28] The new male, as Frédéric Cuvier's hoped, proved notable for his longevity, living until 1825.[29]

A pride's-worth of lions followed the Algerian lion of March 1809. In September 1809 the menagerie acquired a lion, a lioness, and a hyena sent by the emperor of Morocco.[30] These were followed in months by three more lions, though not from foreign rulers. A pair of lion cubs were brought to Paris in November 1809 as a gift to Napoleon from the former French commander in chief in Senegal. While awaiting the emperor's official acceptance, the minister of the interior invited the Museum to take them.[31] Four months later, the minister informed the museum he'd had word from the minister of foreign affairs that yet another gift of a lion had arrived in Marseille and that it was the Museum's job to get it to Paris.[32] This one came from a Tunisian ship captain.[33]

The flow of gifts to the menagerie halted during the Napoleonic Wars, when battles were waged on the open sea and maritime traffic was tightly controlled. From the Museum's standpoint that was just as well, since it had more lions and lionesses to house and feed than it needed or wanted. Geoffroy traded one lioness away in November 1810, reported the death of another in February 1811, and traded yet another two weeks later.[34] Menagerie censuses taken by Frédéric Cuvier in 1813 and 1815 showed that the Museum still had three lions and three lionesses.[35]

The cost of transporting gift animals was a burden. The Museum paid approximately 1,400 francs for the lion from Algeria; 1,500 francs for the two lions and hyena from Morocco; and 800 francs for the lion from Tunis. The professors were consequently shocked in the summer of 1810 when the minister of the interior, Montalivet, asked the Museum to reimburse 6,372.79 francs that the consul in Morocco claimed as expenditures for the two lions and hyena. Montalivet said he would have liked to pay, but his 1809 budget would not allow it. He hoped that through economies in its own budget the Museum would be able to cover the bill.

The professors objected. They reminded Montalivet that he had already reduced their budget by 25,000 francs and that they had additionally been obliged to commit 6,000 francs to publishing their *Annales*. Unable to meet their most urgent needs, they felt it fairer "to have this new unanticipated expense borne by the minister of foreign relations."[36] Drafting the Museum's response, Georges Cuvier observed that "His Excellency knows perfectly the point to which the administration is constrained by the retrenchments it has experienced this year, that it has not at all requested the animals, that it has already been obliged to pay the carriage from Marseille to here." In this case and in all cases involving "a present from sovereign to sovereign," he continued, the minister of foreign Relations should pay.[37]

Ultimately the Museum did not shoulder the costs claimed by the consul in Morocco, and it was agreed that the minister of foreign relations should be responsible for expenses associated with gift animals before their arrival in France. But the Museum was still to be saddled with costs of transporting gift animals to Paris once they reached French shores.

When the wars ended and freedom of the seas was restored, diplomatic gift giving recommenced. Occasionally this brought the Museum an animal of great interest; but, wanted or not, the diplomatic gift could not be refused. The emperor of Morocco was the first of the North African rulers in the postwar period to show his beneficence, sending Louis XVIII three lions in 1816.[38] They were of no consequence from a zoological standpoint, but they set a precedent for how gift-related expenses should be apportioned. Lainé, the minister of the interior, instructed the Museum to work with César Famin, head of the office of foreign affairs in Marseille, to arrange safe transport for the animals, a male and female lion pair, aged six, and a three-year-old lioness.[39] The professors urged Famin to keep costs down, but 6,000 francs was already promised to the Swedish captain who brought the Moroccan lions to France. Considering the costs of caring for the animals through twenty-eight days of quarantine and transport to Paris, the Museum estimated the total bill at about 9,500 francs.[40]

That figure was more than half again as much as the combined annual salaries of the menagerie's six animal keepers. It was likewise more than a quarter of the menagerie's 36,000 franc budget yearly budget, which had to cover staff salaries, provisions for one hundred mammals and two hundred birds of diverse species, buildings and grounds upkeep, purchases of new specimens, and more. But Lainé knew all about the Museum's budget limitations. He had cut it as part of an across-the-board reduction in the funding of institutions under his direction, making permanent the 25,000-franc cut that had been instituted as a temporary measure earlier in the decade.[41] The Museum argued that the lions' projected costs far exceeded their worth and that it needed to devote its limited funds to more urgent issues. But the Museum's objection that the lions were "useless to us" did nothing to derail events already under way.[42]

The lions reached Marseille on 9 August and were quarantined for twenty-four days. Famin fed them beef (which they preferred, he said, to the mutton they had been eating) and advised the Museum on how transport costs to Paris could be kept down. The minister of the interior conveyed the Museum's thoughts on travel expenses to the minister of foreign affairs, who agreed to pay for the lions' transport from Morocco to Marseille, but no more. His ministry, he said, was responsible only for expenses incurred before the animals arrived in France. Five days after leaving quarantine, the lions were loaded into cages on a wagon. Famin prepaid the wagon driver and the animals' keeper for the journey to Paris. The expenses at Famin's end were more than 4,000 francs, but the minister of foreign affairs covered the quarantine costs. The final bill the Museum received from Famin was for 1,179 francs, 50 centimes, plus several hundred francs for the keeper and wagon driver's return to Marseille.[43]

The next diplomatic gifts, a pair of mountain sheep and an ostrich from the pasha of Tripoli, seem not to have caused consternation, but then the Museum learned (October 17, 1817) that the Moroccan emperor had sent another pair of lions for the French king, and they had already reached France. While apologizing for foisting them on the Museum, the minister of the interior made it clear that the Museum was responsible for all their expenses since arriving in France. "The consul knows very well that these animals will be superfluous in the king's menagerie" wrote Lainé, but he would have been "unable to refuse the gift without upsetting accepted conventions."[44] The lioness died in transport but the Museum paid nearly 2,000 francs for the male, which it neither needed nor wanted.[45]

In a matter of months, yet another Moroccan lioness was heading to the king of France, this time from Michel Busnach, brother of the assassinated Naphtali Busnach. The gift was intended as a reminder of the debt from grain deals in the late 1790s that France still owed the firm of Bacri and Busnach and Algeria, estimated at seven million francs. Negotiations ensued, with the French government agreeing to pay its debt to Bacri-Busnach, but this left Algeria's current ruler, Hussein Dey, out in the cold.[46] By the time the Museum heard of the animal, it was in transit from Marseille to Paris, sent by the very capable prefect of Bouches-du-Rhône, the comte de Villeneuve Bargemon.[47] The expenses on the prefect's end included disembarking, lodging, and feeding the lioness and her keeper (identified by Villeneuve as "the African"), pieces of equipment, food for the lion en route, the expenses of "the African," and half the sum contracted with the driver for his round trip. Villeneuve's total of expenses needing reimbursing was 864.55 francs. The Museum additionally paid the driver 200 francs in Paris for his return.[48]

The Museum did not contest Villeneuve's meticulous accounting, but it wanted to ask the minister of the interior if it really had to pay such a sum for what would be the fifth lioness in the menagerie, "an animal that it does not need and that offers little interest." The professors hoped he could cover the costs through special funds for unexpected expenses, but the minister was obdurate, saying the bill could "only be paid on the funds of the establishment."[49]

Meanwhile, the Museum was receiving live animals that were not diplomatic gifts and that it wanted very much, notably Milbert's mammals from North America, a chamois from the French Alps, and a young elephant sent by Leschenault de la Tour from India. The Museum paid for their transportation, though it asked the minister of the interior for help with the elephant, whose journey from Pondicherry to Bordeaux and on to Paris cost 5,000 francs. The minister said he had nothing to give and the Museum must find other ways to economize.[50]

In the summer of 1822, the Museum happily received a male ostrich, sent from Egypt by Bernardino Drovetti, the French consul, and it did not protest the charge of 693 francs.[51] Likewise, it paid without complaint for the transport of a pair of gazelles from Algeria via Toulon the same summer.[52] The Museum balked the following year, however, when it was asked to pay for the care of a panther and a lion sent by the French consul at Annaba (formerly Bône, Algeria) to Louis XVIII that died during quarantine in Marseille.[53] The professors asked the minister of the interior if the Museum was seriously expected to pay for animals "that do not even arrive at the Jardin du Roi" and "from which it has not in the least profited."[54] The minister (Jacques-Joseph Corbière) responded that the Museum had to pay for the lion and panther from the time they reached France and should expect to pay all such expenses in the future. It was not only fair, he argued, but in the Museum's interests to receive such gifts with all their "risks and perils," while "to adopt other measures would be to stray in the future from occasions to enrich the king's menagerie."[55]

The next lion to arrive at the menagerie was again from North Africa, yet another gift from the emperor of Morocco. This lion traveled through Cadiz, Tangier, and Le Havre, arriving in Paris by steamboat on 23 August 1824 accompanied by a conductor named Giraud assigned to it by the French consul in Tangier.[56] Frédéric Cuvier described it as "a young animal of two years, of the greatest size, in the most beautiful state of health, with its teeth entirely formed."[57] The Museum assured Minister Corbière that the lion was "indeed of the greatest beauty, and it is in good health. It is a very precious acquisition for the Jardin du Roi."[58] However precious, the Museum did not want to be stuck with the bill for its transport and appealed for help to the king's minister, le duc de Doudeauville. He replied graciously without explicitly promising anything, but this time it seems the minister of foreign affairs paid up.[59]

Diplomatic gifts continued, including a deer and a pair of "capricornes" (Alpine Ibex) sent by the French consul in Alexandria for "His Royal Majesty Monseigneur le duc d'Angoulême."[60] The female ibex died in Lyon, before reaching Toulon, and only the male ibex made it to Paris. Then the French consul general in Tunis, de Guys, wanted to do his part and sent two young lions and a deer as a present of his own for the menagerie. He was willing to pay the animals' passage across the Mediterranean.[61] Before embarkation, however, the small young deer from the African interior expired, frightened to death, according to de Guys, when the young lions tried to play with it.[62]

In January 1825, when de Guys's lions reached Toulon, the professors expressed concern to the port intendant about the health of the lions, "of which the Museum has at this moment a great need, because there is only a single lion in the king's menagerie."[63] This was all but true, for aside from the new lion from Morocco the menagerie's only other lion (the one that came from Algeria in 1809) was very old and on its last legs. The intendant de la marine in Toulon followed the Museum's instructions, keeping the animals there until warmer weather came in April. The professors suggested the lions travel in separate cages, but both the ship captain who delivered them from Tangier and the intendant felt they should be sent to Paris in the same cage, because they were accustomed to living that way and the captain worried that separation might harm their health.[64] In addition to their mutual affection, the lions were attached to the French sailor (Juquet) who cared for them aboard ship and accompanied them by wagon to Paris. When he left them at the menagerie, the press reported they seemed "overcome by sadness" and scarcely ate. When he came back to see them, the lions "cried out with the liveliest joy" but then "let out the most frightful roars" when he left again.[65]

The Museum reported to de Guys that the lions had been placed in the same cage, and the crowds that came to see them were pleased to find two such animals living together on "the most perfect terms." The Museum also told him that the lions' cage bore a sign indicating their provenance and his name as their donor.[66] The Museum then sent a notice to the *Journal des débats*, which published the information that the animals were the personal gift of de Guys and fifteen to sixteen months old, in good health despite travel fatigue and teething problems, and "already nearly accustomed to their lodgings and their keepers." The notice remarked further that the lions were crowd pleasers because "they live on very friendly terms in the same cage, and their habits are exactly those of young cats."[67]

Two months after the lions' installation at the menagerie, the Museum wrote to thank de Guys for sending a genet but also to regretfully report that the young lions had died from teething problems. The professors implored him to send no more, since lions rarely survived their teething travails around age two and because the costs of transporting large animals was prohibitive. Small animals were fine, not only because they could be sent cheaply by mail coach, but also because foreign animals were typically little known to naturalists and thus even more valuable to the Museum's collection.[68]

In the meantime, Charles X's ascension to the throne (1824) inspired new diplomatic gifts. In March 1825 Corbière wrote to the Museum to signal the arrival at Marseille of an elephant and a mouflon from the pasha of Egypt.[69] Two weeks later, the minister announced a large shipment of animals from the bey of Tunis: two lions, four ostriches, four gazelles, two sheep from Guinée, six young camels, and a male vulture (a female having perished in transit).[70]

The minister's letters never mentioned if these were part of larger gift assemblages, but the press filled out the picture. The pasha of Egypt's gift also included a dozen horses of different breeds and a fourteen-ton ancient Egyptian monolith made of red granite.[71] The bey of Tunis's offering included six prize horses, two lions, sheep, six camels, an ornate saddle, coverlets of fine wool, lion and tiger skins, and also "a complete and extremely rich Moorish outfit for Madame la dauphine." Among the animal gifts, the male lion was exceptionally handsome, the sheep were noteworthy for their size and special characteristics of their wool, and the six camels (two black, two white, and two red) were "a very beautiful breed."[72] The gifts from Tunis were valued at 56,000 francs, perhaps more, depending on how much gold adorned the saddle. The ship carrying the ostriches and the camels sank in a storm, but the animals were rescued.[73]

The Museum encouraged its contacts to do everything possible to minimize the cost of transporting the animals to Paris. When the intendant de la marine in Toulon learned that *La Cybèle* had been instructed to carry the Egyptian monolith on to Le Havre, he judged it best to send the elephant that way too, reasoning that its transport to Paris by boat would be easier than overland.[74] The Museum's agents in Le Havre, Eyriès frères, then had to decide how to ship the elephant. The eight-month-old female elephant was small but strong. Loading her aboard a steamboat for travel all the way to Paris would be tricky; a boat to Rouen would require her to be loaded and reloaded; traveling by wagon would require a specially built cage; and traveling on foot might pose problems for the young animal.[75]

Dispatched to Le Havre to decide, Frédéric Cuvier chose the steamboat to Paris, while warning that costs were mounting. The elephant's growing strength required a new cage costing over 100 francs, in addition to a freight charge of at least 250 francs. Food for the elephant and her sailor attendant was also required. The ship would also be carrying a whale skeleton for the Museum in four huge crates, costing another 300 francs.[76] The elephant arrived at the menagerie in May 1825 in excellent health. The Museum's financial responsibility for her trip was 1,397.59 francs. The sailor who accompanied her from Egypt asked for work as an animal keeper or anything else at the Museum. Forty-five years old, he had spent more than half his life at sea and hoped to settle down with his wife. But the menagerie had no job for him.[77]

The care and transport of the lion and lioness, eight sheep, and two gazelles from Tunis likewise caused difficulties. Toward the end of their quarantine in Marseille, the lions were so agitated that the staff feared they might break out of their cage. Villeneuve decided to load all the animal cages, with the animals inside, onto three wagons to take them to Paris. He provided two keepers with sufficient funds for as far as Lyon and sent additional funds to the prefect of the Rhône to give to them in Lyon. He alerted the prefects of Vaucluse, Drôme, and Isère that the convoy would cross their departments and invited them to help with its progress and with ensuring the public's safety.[78] The convoy arrived in Paris on 20 May 1826 with all the animals in perfect shape.[79] When the Tunisian ostriches and camels cleared quarantine in Toulon, local gendarmes accompanied them throughout the trip, one well in front of the camels and one behind, to forestall encounters with horse teams or carriages that might make them bolt.[80] They all made it to Paris on July 17, though some camels showed signs of a skin disease, and an assistant keeper was injured halfway from Toulon and obliged to go home.[81]

The professors calculated the domestic transport of the gifts from the pasha of Egypt and bey of Tunis would add up to 8,000 or 9,0000 francs.[82] They decided to pay the reimbursement requests coming from Marseille, Toulon, and elsewhere but to insist that the minister of foreign affairs acquit all the expenses of gift animals up to the time they reached the Jardin du Roi.[83] Once all the expenses were tallied (9,596.03 francs), the professors lodged their protest with the minister of the interior, arguing they had not asked for these diplomatic gifts, had no way to either anticipate or refuse them, and had no place for them in the budget. They were a product of the "circumstances and motives" relating to diplomatic relations between

France and foreign countries. That they happened to be animals was no reason to make the Museum responsible for their transport. If the minister of foreign affairs could not handle their expenses, the minister of the interior ought to do so from one his accounts, or else he should furnish the Museum with a supplement. The Museum spelled out how thin its annual funds were already stretched.[84]

The minister of foreign affairs, Baron Damas, adamantly rejected the museum's reasoning, citing the precedent established with the gifts sent by the emperor of Morocco in 1816: "Foreign affairs had to provide only for the expenses of shipping and quarantine; all the others, from the instant of disembarkation, pass to the charge of the minister of the interior."[85] The Museum politely replied that it knew the precedent, but that "it is precisely because this precedent has the greatest disadvantages for the administration of the Jardin du Roi in imposing absolutely unforeseen expenses entirely foreign to the credits that are open to it and often, as in the present case, very considerable, that it is important and consequently the duty of the administration to protest against the introduction of such a precedent." Expressing hope "in [your] interest in the sciences and in an establishment that is consecrated to them," the Museum urged Damas to assume an expense that was minor for his ministry (and naturally belonged to it), but "so weighty that [the Museum] would be unable to satisfy it without letting go other services the costs of which are calculated on the strictest scale."[86]

While drafting their letter to Damas, the professors read "with inquietude" a notice in the *Moniteur* describing a ceremony in Morocco where the French consul delivered presents from Charles X to the emperor, including a diamond diadem. The emperor announced he would reciprocate by sending "a lion, an ostrich, and four horses." Once again, it seemed, the Museum's budget was about to be upended by animals it did not need. Sovereigns' gifts to one another, the professors wished to tell Damas, were the Crown's business, and budgets that allowed for lavish gifts from France should also bear the expenses of gifts to France.[87]

In November 1825, when Corbière as minister of the interior told the Museum to pay all the transportation demanded of it, the Museum replied that it was "materially impossible" and that diverting resources from planned and useful purposes would hurt the institution.[88] Corbière renewed his demand in December and again in February, by the end of which the Museum had paid the entire amount charged but reserved the right to reimbursement and pleaded that it no longer be forced to bear such large and disruptive expenditures.[89]

Yet the same scenario occurred in June 1826 when the Moroccan emperor offered a new lion to Charles X. Corbière told the Museum that the minister of foreign affairs had paid for the animal's transportation as far as Marseille but "positively refuses to pay those that have taken place in France and thereby conforms to the practice observed for all time." Corbière told the Museum to pay 525 francs for the transport to Paris out of the 2,000 francs it had budgeted for animal transportation.[90]

The Museum reluctantly obeyed, insisting it was unjustly caught up in a matter of international diplomacy. "We must call your attention once more to the injustice," the professors wrote, "of (1) the administration having to pay for objects that it does not request and (2) the disturbances that the slightest fantasies of the powers of Barbary can bring to all the expectations in our budget." It was even worse, it said, when the Museum did not learn of an animal until it arrived in Paris, at which point it was too late to reduce the costs of the animal's transport or food en route. (The lion had eaten forty pounds of meat per day, when ten pounds were sufficient.) The situation was exacerbated when the animals in question were superfluous and increased the menagerie's consumption of meat and fodder. This required cutbacks in other parts of the institution that were "already too limited in their resources," thereby destroying "any kind of equity" among them.[91] The professors urged the minister to consider their concerns and take action to protect the establishment in his mandate. Diplomatic presents belong to the office of the minister of foreign aAffairs, they insisted, no matter what they consisted of. "If [they were] furniture, carpets, cloth, or even furs, he would certainly pay the transportation. How does the allocation change because it involves living animals?"[92] The Museum wrote a similar letter to Damas, omitting the phrase "the slightest fantasies of the powers of Barbary."[93]

Remarkably, Corbière came around to the Museum's position, reporting (30 June 1826) that he had just written to the minister of foreign Affairs saying, "I very much desire that this minister appreciates the justice of your complaint, and that this establishment may in the future profit completely from the presents destined to it by the munificence of foreign powers."[94]

By the fall of 1826 the Museum's situation looked more hopeful than it had for a decade. The minister of the interior seemed to have agreed that expenses related to foreign diplomacy should be paid by the Ministry of Foreign Affairs. But it was uncertain that the issue had been definitively resolved. When the Museum learned of a young female giraffe arriving in Marseille as a gift to King Charles X from the pasha of Egypt, it experienced joy and anxiety. The prospect of gaining a marvelous new animal for the menagerie was cause for rejoicing, but having to pay the assuredly great cost of transporting it to Paris from Marseille would be more than the Museum's budget could possibly bear.

[3.9]

The Giraffe from the Pasha

The most famous animal in the history of the Paris menagerie is the giraffe that walked from Marseille to Paris in 1827. A gift to the king of France from the pasha of Egypt and the first live giraffe to set foot in France, she elicited wonder at first sight. "An air of joy, of satisfaction combined with astonishment, rings out from everywhere," wrote Étienne Geoffroy Saint-Hilaire, describing the reactions of citizens of Aix-en-Provence when she came into view.[1] Of the impression she made when she reached Paris, *Le Figaro* recalled six years later,

> The capital was taken by a sudden vertigo. All Parisians were Cuviers, Buffons, they trembled with joy like a savant upon sight of a wondrous beast. The reception for the giraffe was magnificent; one would have said it was for a foreign potentate coming to visit one of his cousins, or an Oriental queen arriving amazed amid an enraptured people. It was a festival of a thousand and one nights, a mixed cortège of Negroes and whites, of Africans and Europeans, in truth, a superb entrance.

The giraffe, by this account, dominated the press for two weeks, making her "the most powerful auxiliary in Parisian polemics." People rushed to see her at the Jardin des Plantes, and she was the subject of everyone's conversation. "There is no other example," claimed *Le Figaro*, "of such a vogue.[2]

Le Figaro's point in 1833 was that the giraffe's moment of glory had long since passed. Already in 1830, Honoré de Balzac had cited the giraffe as the epitome of fleeting fame: "Literature, the theater, lithography, fashion exploited her celebrity. What name, finally, had more fame? What animal had ever had more popularity? Today she is scorned, forgotten, and visited only by backwards provincials, unoccupied nannies, and the slow-witted and naïve."[3]

The giraffe's story has often been told.[4] Captured south of Sennar, a province of Sudan, the young female was sold with another young female giraffe

Figure 3.9.1. "Female giraffe, two-and-a-half years old," pictured with her two African handlers. 1827. Henry Gauguin after A. Prevost. © Albert Harlingue/Roger-Viollet.

to Mouker Bey, governor of Sennar, who subsequently presented both as gifts to Mohammed Ali (1769–1849), Ottoman viceroy and de facto ruler of Egypt. Mohammed Ali gave the first giraffe to the king of France and the second to the king of England. The "French" giraffe's odyssey has been amply documented, beginning with her journey from Sudan to Alexandria, including the fact that the Sardinian brig that carried her to Marseilles in 1826 had a cushioned opening cut into its deck so she could stick her head out from the hold. Her winter in Marseille and every city, village, or stopping point on her forty-one-day trek to Paris can be reconstructed.[5]

The historian Eric Baratay has broken new ground in animal studies by pursuing history from "the animal's point of view," focusing on how the giraffe might have perceived the changing and stressful circumstances she experienced.[6] At the level of diplomatic history, there is a different perspective to consider. Mohammed Ali sent the giraffes to France and England with the thought that these marvelous diplomatic gifts might dispose the French and British governments not to intervene in his efforts to help the Ottoman Empire quell the Greek independence movement.

Other perspectives also merit attention. Whereas the citizens of Aix-en-Provence and Paris greeted the giraffe ecstatically, the Museum zoologists found their excitement mixed with anxiety. Aside from how to house the giraffe, the professors worried that the minister of the interior would make them pay for its transport. They doubted the reliability of the giraffe's primary caretaker, a Sudanese, and worried that the giraffe might die before reaching Paris, or shortly thereafter. News of the giraffe came in a letter from Corbière, the minister of the interior (16 November 1826) reporting that she had recently arrived in Marseille as a gift from the viceroy of Egypt and was under the care of two domestic servants who lacked funds and instructions about what to do next. Corbière attached copies of the relevant correspondence and told the Museum to immediately contact the minister of the Maison du Roi to make the necessary arrangements.[7] The papers were turned over to menagerie superintendent Frédéric Cuvier, who enthusiastically predicted that the giraffe's arrival in France would create as great a sensation as the first elephants and tigers displayed in ancient Rome. It was a moment that naturalists had long awaited. Voyagers and historians had previously conveyed little more than "astonishment and admiration" regarding the giraffe, and the remains in natural history collections afforded a "very imperfect idea" of the living creature. Being able to study the giraffe's nature, organization, and place in the zoological order "would have advantages without number for science."[8]

Cuvier was right about the public's response, but his prediction about the "advantages without number" that the animal would bring to science proved an exaggeration, given the relatively little the Museum's zoologists made of it, and he was quite wrong when he went on to suggest "several centuries will perhaps elapse before such a fortunate occasion presents itself again." Within a dozen years, giraffes had been put on display at two sites in England and in New York, and Paris had been offered opportunities for further specimens.

The French consul in Alexandria, Bernardino Drovetti, who had suggested that Mohammed Ali present the giraffe to France's king, advised keeping it in Marseille for the winter and sending it to Paris in the spring. Frédéric Cuvier agreed and specified that the giraffe would need a warm stable with southern exposure at the menagerie, a park sheltered from the north, and scrupulous attention to her food and cleanliness.

Regarding the animal's care in transit, Cuvier expressed the same anxieties the Museum and its agents had expressed eight years earlier about the African American bringing the menagerie its first bison. In Alexandria, Drovetti had entrusted the giraffe to two of his domestic servants: an Egyptian named Hassan and Hassan's Sudanese subordinate, Atir, who was immediately responsible for the giraffe's care. Cuvier told the administration, "I agree again with Mr. Drovetti that it will be necessary to keep the Negro who has the habit of looking after [the giraffe]; but who will watch over the Negro?" Noting that Hassan would have to return to Drovetti in Egypt, Cuvier observed of Atir, "It would be imprudent perhaps to rely entirely on the intelligence and good faith of this African for the conservation of the animal that Mr. Drovetti has not judged appropriate to confine to him entirely." Cuvier proposed sending a trusted man from the menagerie staff to Marseille to supervise "the Negro" caring for the giraffe.[9]

The Museum fretted about transportation costs, which Cuvier estimated at 8,600 francs: 4,600 francs from Alexandria to Marseilles (per the contract made by Drovetti); 1,000 francs for food and lodging in Marseille; another 1,000 francs for provisions and pay of two keepers; and 2,000 francs for the trip from Marseille to Paris. Georges Cuvier, as Museum director, wrote to Corbière to forestall attempts to hand the Museum the bill. He recited the usual objections about the unfairness of having to finance unexpected and unrequested animals sent as diplomatic gifts. The Museum had been unable to set aside funds to handle such a large expenditure, especially so close to the end of the fiscal year. Housing and feeding the animal and keepers during their months-long stay in Marseille was an additional burden. The Museum would, however, be very happy to have the animal if some ministry could pay for it, because it was "one of the most interesting" animals and "the government doubtless would not want the pasha of Egypt to think that his present had been received with contempt."[10]

The prefect of the Department of Bouches-du-Rhône, le comte de Villeneuve-Bargemon, the skillful administrator who had previously sent animals to the Museum, shared these concerns. Writing to Corbière to announce the giraffe's arrival and entry into quarantine, he noted that the related expenses would likely be "of some importance." The giraffe's entourage included Drovetti's two men, three milk cows for nourishing the giraffe, and a pair of antelopes. Villeneuve was arranging for the giraffe's accommodations and feeding, seeking the proper balance between minimizing costs and ensuring her health.[11] In a follow-up letter, after the giraffe had exited quarantine and been installed in a stable at the prefecture, Villeneuve evidenced how smitten he was by "this beautiful and precious animal." She stood eleven-and-one-quarter feet tall, artists were already depicting her, and the Marseille Academy had established "a commission to keep a journal of all the facts" related to her habits.[12]

The Museum sent its instructions, encouragement, and thanks to Villeneuve emphasizing that it was of "the highest importance" to keep the giraffe alive for "several years" to study it as thoroughly as possible. Signed by the Museum's "bureau" (the director, secretary, and treasurer), but drafted by Frédéric Cuvier, the letter repeated the latter's cost estimate and concern that "in addition to the Negro" there should be a second person, "first to oversee [the Negro], then to replace him in case of sickness, for we presume that the animal would let itself be better cared for by familiar people than by those to whom it is unaccustomed."[13] The Museum stressed the importance of keeping the giraffe's surroundings clean and ensuring her keeper's vigilance.[14]

The professors (and the menagerie superintendent) still knew nothing about the animal, whether "young or adult, large or small, weak or healthy, wild or tame, etc. etc." Lest she die before reaching Paris, the Museum asked that colored drawings be made to show "exactly its proportions and the forms of its head viewed frontally and in profile." These, and the giraffe's skeleton and skin, would be "a small compensation for us if we had the misfortune to lose it before its arrival."[15]

Though the giraffe lived much longer than anticipated (sixteen and a half years), her closest study was performed not in Paris, but by a man named Salze, a member of the Academy of Marseille, professor of *physique* and botany at the Collège Royal and the secondary medical school.[16] His report, written at Villenenuve's request, included information from Atir and Hassan about how giraffes were hunted, the hands through which this specimen had passed, and observations on features of the giraffe's anatomy and behavior.[17]

Salze paid close attention to what the giraffe ate (primarily cows' milk, but she also loved mimosa tree leaves); her excrement; her remarkably long and dexterous black tongue; how she behaved toward keepers and strangers; and how she refused to be touched around the head (she did not allow him to examine her teeth). Her quarters had two windows and a door. She shared the space with her milk cows, the two antelopes, and a pair of horses. Thanks to the animals' body heat and good insulation, a stove or other warming device was unnecessary.

Salze saw "nothing elegant or gracious" in the giraffe's form, finding instead something "shocking" in the contrast of her short body, her long, close-together legs, her excessively long neck, and her poorly rounded rump. People were nonetheless "seized by astonishment with her appearance, and one finds her beautiful without being able to say why." Even after one "considered her attentively," he noted, it was difficult to recall more than an uncertain memory of her form and bearing. He liked to see her often, because "each time she gives rise to some new observations."[18]

Villeneuve sent Salze's report and a sketch of the giraffe to the Museum in January 1827 along with an account of its expenses in Marseille. The professors' reply brimmed with cheer because the minister of the interior had just agreed to take on all the costs of the giraffe's care and transport. They thanked Villeneuve warmly for his efforts and published Salze's report in the Museum's *Mémoires*, omitting the long last paragraph in which he reflected on the giraffe's classification and the lack of a satisfactory comparison.[19] As he saw it, it was nothing like the horse, resembled the camel only in height, and lacked the nimbleness of deer and gazelles (and likewise the heaviness of the ox). The most comparable animal, Salze declared, was the ostrich:

> Take away [the giraffe's] back half, its ears, and horns and you have the type of the ostrich. Give to the latter the back legs, the ears, and the horns and you have the type of the giraffe. These two animals have a small head in proportion to their bodies, the large and black eye, little expression in their physiognomy; they both have a gentle, timid nature; they never attack other animals; they defend themselves with forward blows of their feet; they are herbivores and granivores; they drink little, have no voice, and are the most rapid animals of the desert.[20]

For the Museum's zoologists, who looked to comparative anatomy, not behavior or ecology, to decide which animals were most alike, Salze's giraffe-ostrich comparison was too fanciful for publication.

Villeneuve kept the Museum apprised of giraffe-related observations and costs throughout winter, and in March it was time to move.[21] Drovetti (in Alexandria) deemed land travel dangerous for the giraffe and navigation on the Rhône problematic, so he recommended shipping her to Paris from Le Havre.[22] The Museum thought that travel by land or river was preferable and asked Villeneuve for his advice.[23] He initially preferred river travel but worried the giraffe might sustain fractures while being loaded on or off boats. She was full of vigor, he said, and fine weather had made it possible to walk her outside the city. The exercise was doing her good, and she walked calmly with her cows as companions and six men holding her by long leads. "Neither the noise, the harnesses, nor the crowd of sightseers pressing around her bother her in the least," he wrote, convincing the prefect that the best way of getting the giraffe to Paris would be to walk her there in short stages. Whatever method the Museum chose, Villeneuve urged that it send someone to Marseille to learn the giraffe's habits before escorting her to Paris.[24]

When the professors invited Geoffroy Saint-Hilaire to report on how best to transport the animal, he focused on who should oversee the mission, namely an administrator of the menagerie, "either Mr. Frédéric Cuvier or me." Geoffroy decided to "charge myself with this task," explaining that "Mr. Frédéric, who would accept this task, has motives for remaining here and judges his presence necessary in Paris whether for stocking the new pheasant house or having lodgings prepared for the giraffe."[25] Geoffroy left Paris several days later, stopping along the way to meet other scientists, acquire specimens, and give lectures, leaving Villeneuve to worry that it would soon be too hot for the giraffe to be out walking in the South of France.[26]

Once in Marseille Geoffroy hurried to see the giraffe. Then he wrote a long letter to his colleagues reporting that the animal was "truly superb." She was "in perfect health," younger than anticipated (twenty-two months, by his estimate), and "imposing." Geoffroy was familiar with the giraffe skeleton and skin at the Museum, but he was "involuntarily astonished" by the live animal, as if seeing "something entirely new." Villeneuve told Geoffroy the giraffe had become "the spoiled child" and a "subject of entertainment" for the prefecture. Geoffroy wanted to see one of her promenades before deciding if Villeneuve's suggestion that she walk to Paris might be viable. He worried about the animal's reported capriciousness and the milk cows' ability to match her pace. He told his colleagues that the quandary of the giraffe's transport made him "truly afraid" but that they could count on him to "do my best."[27]

Accompanying the giraffe on a few perambulations with Hassan and Atir quickly persuaded Geoffroy that walking was the best way to get her to Paris. Geoffroy described Hassan as "the Egyptian, who is reflective, patient, attached to his animal." As for Atir, Geoffroy characterized him in racial terms as "the Negro, of a profound black, of a form different from the Negroes of the Congo."[28] Flanking the giraffe, the two men led her by ropes attached to rings on her halter. Behind walked two more handlers, with ropes attached to a neck collar.

Geoffroy would have left for Paris sooner had Drovetti not sent a pair of antelopes (of which only the male survived) of a species he considered new to science. With its spiral horns and characteristics of the gnu, wild sheep, and antelope, Geoffroy thought it might even represent a new subgenus. Had it not been so precious from a zoological standpoint, he would have gladly transported it separately from the giraffe, but he wanted to keep it close by. Departure was delayed because the antelope was too strong and dangerous to be walked. It needed a special cage to be built, a wagon, horses, and a driver. A pair of wild sheep also were given cages and loaded on the wagon as part of the convoy that set off on 20 May.[29]

Accompanying the animals, along with Geoffroy and the wagon driver, were Hassan and Atir, who spoke only Arabic; Barthélemy, a stable hand Villeneuve hired to oversee them who spoke only French; and Youssef, an eighteen-year-old born in Marseille of Egyptian parents, who spoke French and Arabic and could translate. Geoffroy considered cutting costs by excluding Youssef, but Hassan and Atir were so adamant about needing him they offered to pay him from their own scanty wages.[30]

A torrential downpour fell on the convoy a half-hour after departure, but the handlers, not the giraffe, were the problem. "I had a very difficult world to lead. Each one wanted to act as the head," Geoffroy reported, "there was a storm of wills, of languages, of temperament and colors of all kinds." He was nonetheless confident things would eventually work out.[31] On Villeneuve's advice, he had purchased a light cabriolet for the trip, but he assured the Museum that far from luxurious, its poor suspension made it a "a real seesaw." He trusted his fellow professors would not reproach their fifty-five-year-old colleague for using the carriage to keep up with the convoy and oversee the giraffe's safe arrival in Paris.

At Aix-en-Provence, the convoy's first stop, a surging crowd broke through a cordon of eight guards in its eagerness to see the giraffe, but no one was seriously injured. After two nights in Aix, the convoy spent one night in Lambesc and another in Orgon. Before leaving for Avignon (May 24), Geoffroy reported to the Museum that the voyage was going smoothly, the animals were in "perfect health," and the giraffe was doing "wonderfully well on the route." People "from one village to the other" accompanied the convoy, in "a kind of triumphal march." Geoffroy's sole complaint (other than not having heard from the Museum) was that they were being overcharged for their nighttime lodgings.[32]

When the convoy stopped for a two-hour break at the departmental border, Geoffroy wrote elatedly to Villeneuve that "it is difficult to proceed any better. Everyone knows what he must do and each is at his job. I say that of all the animals and people." The giraffe had been calm in the morning at Orgon, he said, wearing her grand saddle blanket decorated with the King's coat of arms. When the cows started to move so did she, even before being prompted by her chief handler, Hassan, who was "proud as a peacock" to hold the lead rope, for he had feared that Barthélemy would claim that role and wished to be seen as his superior. "The Negro stable hand" (Atir) held the right-hand restraining rope, while Barthélemy was given the left. "The Negro boy Youssef" led the cows. "The order is followed very exactly," Geoffroy beamed. "Each is comfortable at his post, each one, animals and people, having taken well to the sense of his condition as a voyager."[33]

Before leaving Marseille, Geoffroy had mapped out the convoy's stops, specifying where it would stay for an additional day or more of rest. He figured the trip would take fifty-two days, twenty-three from Marseille to Lyon and twenty-nine from Lyon to Paris.[34] After twelve days on the road, he was able to write from Valence that all was well. The handlers were regulating the giraffe's pace, slowing her down in the morning and speeding her up in the afternoon, by the mere display or the light tap of a baton. Striking her tail made her behave when she indulged in "some caprices." Her flagging energy toward day's end confirmed the need for the planned rest in Lyon. Geoffroy suggested she would need to be cared for in Paris somewhat like a camel, though her quarters would obviously have to be of greater height. In addition, the Museum needed to think about crowd control.[35]

The Museum approved the plan for a halt in Lyon and instructed Geoffroy to arrive in Paris by night to minimize the crowd problem. Geoffroy typically rode several hours ahead of the convoy to arrange for accommodations at rest stops, but he arrived in Lyon several days in advance and spent a night there before rejoining his companions on the road. He anticipated that three days' rest in Lyon would be good for the giraffe and would please the local citizens. The *Gazette de Lyon* later reported that large crowds gathered around the Hôtel de Provence, where the giraffe stayed. Her every appearance elicited excited shouts and exclamations. According to the *Gazette*:

> The giraffe showed herself several times in public. She was preceded by a detachment of cavalry and surrounded by a large guard that held the

> crowd in check. Four Negroes, as Negro as one could imagine, walked at her sides. They gave the appearance more of forming her court than holding her captive and contributed not just a little to the picturesqueness of the whole. Finally, so that nothing was lacking in the glory of the beautiful African, a member of the Institute [Geoffroy] followed her closely, charged with overseeing such precious days.[36]

For most of her stay the giraffe remained tranquil amid the noisy multitudes. Sometimes she let people approach, so long as they did not touch her. She occasionally favored them by licking their faces or hands, something she liked to do with her caretakers.[37]

There were two mishaps, however, on her last full day in Lyon. In a large crowd at the Place Bellecour, a cavalier's horse took off at gallop, frightening the giraffe. The reins to her halter broke, but those attached to her harness held, and her handlers were dragged along behind her. When the cavaliers rode after her, Geoffroy was knocked down by a charging horse and slightly injured.[38] That evening, when the giraffe was returning to her stable and Geoffroy was elsewhere having dinner, she frightened a horse that bolted, injuring an octogenarian woman and a young man.[39]

Geoffroy had considered transporting the giraffe from Lyon to Châlons by boat, so as not to further tire her, but the weather turned unseasonably cold, and he worried that a river trip would hurt her lungs.[40] Feeling she had rested sufficiently, he plotted the convoy's itinerary for the next twenty-four days.[41] He planned a July 2 arrival, but the convoy moved more swiftly, reaching Paris late in the afternoon of 30 June. Geoffroy by then was already in Paris, having hastened there owing to an acute health issue. His son replaced him, joining the convoy at Auxerre.[42]

The daytime entry to the city was decided at the last minute, when Atir and Hassan insisted the giraffe would resist walking at night. Word was transmitted to Geoffroy, who accepted their assessment, and a letter was quickly drafted instructing the prefect of police to arrange for fifteen to twenty mounted policemen to escort the animal to the Jardin des Plantes and protect her from any unruliness by the public.[43] The convoy entered Paris at 5:00 p.m. *Le Constitutionnel* reported that the giraffe was led by "a Negro from Darfour named Atir and a Moor from Sennaar named Hassan," both coiffed in turbans, followed by two other Africans:

> It was truly an extraordinary spectacle to see her approach her new dwelling with her escort of Egyptians, gendarmes, and curiosity seekers. Her elegant head was raised to the height of the leaves of the chestnut trees. Her long neck balanced itself gracefully above the crowd; her large black

> eye . . . was full of gentleness and gaiety. Hassan and Atir, proud of the beauty of their compatriot, have decorated her neck with a vast bouquet of flowers, and Arab amulets hung beside the bouquet. There was something in this spectacle reminiscent of the *Thousand and One Nights*.

The giraffe was installed in the *orangerie* with the other animals sent from Egypt. The newspaper noted that on Monday, 2 July, when the giraffe was first displayed, 10,000 people came to see her.[44] The Museum had previously announced the arrangement for public viewing: the giraffe would be walked daily from ten until noon, weather permitting, inside the large, fenced-in area of the Museum's School of Botany so the public could see her "unobstructed and without danger."[45]

A royal visit was planned for Thursday July 5, but instead of the king, his daughter-in-law, Madame la Dauphine, came to the menagerie and told Geoffroy it would be more dignified for the giraffe to be brought to the king than for the king to visit the giraffe. Geoffroy communicated this to his colleagues and the next day Georges Cuvier informed the minister of the new plan for the giraffe, her handlers, Frédéric Cuvier, and the professors to go to the royal château at St. Cloud.[46]

The press reported on the giraffe's July 9th audience with the king, and Geoffroy wrote a longer, more personal account to Villeneuve, noting proudly that the king addressed questions to him alone and expressed complete satisfaction with his answers. Whereas the press said the interview lasted half an hour, Geoffroy wrote it was "an entire hour," during which he imparted information about the giraffe but also some of his ideas on the taxonomy of living beings.[47] He gave the king a short paper on the giraffe that he had just presented to the Institute and published as a brochure.[48] The king asked about the giraffe's handlers, and on Geoffroy's recommendation instructed the minister to reward Hassan with 2,000 francs and Atir with 1,000. The giraffe was presented to the court at noon and later put through her paces, much to the royal family's delight. At 3:00 p.m. the giraffe and her escorts left St. Cloud.

On July 3, at the same session of the Institut where Geoffroy presented his paper on the giraffe, Antoine Mongez, another member, presented a long, scholarly account of literary references to the giraffe going back to the Old Testament.[49] The publisher Moreau meanwhile produced a less weighty brochure titled "Notice on the giraffe, with curious observations on the character, habits, and instinct of this mammal" that sold for sixty centimes. Another author's "New Notice" soon followed for seventy-five.[50]

The Vaudeville Theatre likewise moved quickly to profit from the giraffe's popularity, presenting on Saturday July 7 its first performance of "The

Giraffe." Music, rhyming couplets, and the semblance of a plot often kept audiences happy, and the reviewer for the *Journal du Commerce* predicted the piece would "live as long as the eagerness of Parisians to go cram themselves into the Jardin des Plantes." What made the piece especially apropos was not only the giraffe's arrival in Paris but also the political context of the pasha's gift. The audience responded with enthusiasm to the authors' championing "the cause of the Greeks against the giraffe's donors," and as the reviewer put it, "as a good deed is never lost, from this point on the weakness of the piece was forgotten and its success was assured."[51]

The Museum did its best to accommodate the giraffe's flocks of avid visitors. In addition to her morning promenades, she could be seen from 1:00 to 5:00 p.m. in the orangery, though this required a (free) ticket. When it became apparent that so much attention tired the animal, the daily walks at the School of Botany were reduced to Sundays, Tuesdays, and Fridays, and the afternoon visits at the orangery were limited to Mondays, Fridays, and Saturdays.[52]

The Museum instructed the artist Nicolas Huet to paint portraits of the giraffe's head and body, both in profile, and along with the government, the administration started addressing giraffe-related bills.[53] Wishing to assess the animal's value in terms of taxes, the government invited the Museum to report on revenues from the levy paid by people crossing the Austerlitz Bridge, where traffic had increased owing to an uptick in menagerie visitors.[54] Concerned with its own budget, the Museum administration learned from Frédéric Cuvier that once the giraffe was moved to the rotunda, she would need a full-time keeper assisted by a rotunda employee,[55]

Historians have detailed the myriad ways the giraffe engaged the popular imagination, inspiring descriptive brochures, poems, fables, lithographs, musical compositions, and vaudeville performances. Deemed fashionable, she fueled a burgeoning consumer culture. Women's hats and clothing, porcelain plates, bonbons, and all manner of knickknacks bore giraffe motifs (often with her African handlers) reflecting and amplifying enthusiasm for the wondrous creature.[56]

Domestically, the legislative elections of spring 1827 had dealt the king's government a serious blow. Liberals favoring a constitutional monarchy gained more than 150 new deputies while ultraroyalists lost over 200. The king responded by attacking the press, but while newspapers were severely censored, pamphleteers had more leeway. Among them was politician and writer Narcisse Achille Salvandy, responsible for most of the editorial work at the *Journal des débats*.[57] His pamphlet, *Letter from the Giraffe to the Pacha of Egypt* (July 12), adopted the perspective of a far-seeing foreign visitor (the giraffe's height naturally made it far-seeing), to contrast the integrity and forward-looking aims of the French people and the perfidy, tyranny, and backwardness of the ultraroyalist prime minister, Jean-Baptiste, comte de Villèle.[58] Three weeks later, Salvandy wrote another equally scathing giraffe-themed pamphlet.[59] Other writers and artists (including François-René Chateaubriand) likewise saw in the giraffe an opportunity to castigate the government.

The giraffe's reign over Parisians' attention was cut short by the arrival in mid-August of six Osages, a North American First Nations people. Writers delighted in pointing out similarities between the experiences of the giraffe and those of the "Indians," concocting imaginary dialogues between them to criticize the French government and society.[60] Unlike the giraffe, the Native Americans came to France voluntarily. They were surely misled in their expectations, however, by their "manager," David Delaunay, a French expatriate who sought to profit from them. According to tribal lore, the chief of the Osages had visited the king of France many years earlier and had been very well received. The Osages considered France an old ally, and they venerated the French king as "grandfather of the Osage nation."[61] Intent on having an audience with the current monarch, the group arrived in Le Havre in July 1827.[62] Pamphlets and newspapers noted their names and ages.

The French consul in New Orleans, François Guillemin, had given Delaunay a letter of recommendation to take to Eyriès frères in Le Havre. In addition to introducing Delaunay, Guillemin proposed that the Museum host the Osages. He invited the Museum's agents to help Delaunay arrange their visit to Paris. He thought, as Eyriès frères reported to the Museum, "that the presence of these Osages could be extremely interesting for science etc." Eyriès frères sought the Museum's instructions before making any commitments to Delaunay. They presumed the Museum would want to have the police escort the Osages as they entered the capital, just as they had done for the giraffe.

Delaunay planned to conduct the Osages "directly to the Jardin des Plantes, where they could tranquilly await their audience with the king and royal family and meanwhile satisfy the curiosity of the learned professors, etc." Eyriès frères supposed that "a place in one of the orangeries, with some cots, will suffice for the chief of the Osage family, his two aides de camp, and the two women." Then there was a question of finances. Delaunay's calculation of the trip's cost had been "a little short," and he asked the Museum's agents for an advance of "twelve to fifteen hundred francs." Eyriès frères indicated they were willing to give him the money. Thanks to Guillemin's testimony and their own observations they were confident the Osages were authentic, but first they needed the Museum's permission. Eventually, they planned to send the Osages to Paris by steamboat.[63]

The Museum's response to all this was an immediate and resounding no. "We hurry to inform you that we consider this affair entirely foreign to the Jardin du Roi and that we can consequently give you neither instructions nor authorization nor ourselves take any part in this measure." The Museum instructed its agents to tell Delaunay that it was up to the minister of foreign affairs to issue any orders or authorizations regarding Delaunay's plans.[64]

In the later 1800s in Europe's capitals, the display of people from foreign lands at zoos and other venues became immensely popular. The Berlin Zoo, founded in 1844, received a huge boost late in the century when throngs of paying visitors came there to see "exotic" humans.[65] But the Museum of Natural History in Paris did not charge admission, and it wanted nothing that smacked of the sensationalism of public fairs. Having the Osages camp in one of the Museum's great greenhouses would detract from the institution's dignity.

Before the Osages even got there, Parisians were talking about whether they would replace the giraffe as the city's most exciting new attraction.[66] En route from Le Havre, the Osages stopped in Rouen, where they drew large crowds at a theater. Reaching Paris on 13 August 1827, they felt it improper to tour the city before meeting the king, but they appeared frequently on their hotel balcony, delighting the multitudes waiting for a glimpse of them. Their first Paris outing (August 18) was a forty-guest luncheon, hosted by the minister of foreign affairs, Baron Damas, who accompanied them three days later to Saint-Cloud, where they had a half-hour audience with the king.[67] On August 27 they were escorted to the opera, where at intermission they drank the health of the audience, which responded with applause and cheers. The judgment in *Le Constitutionnel* was clear: "The savages have decidedly dethroned that poor giraffe, with which almost no one occupies himself anymore."[68]

The Osages supported themselves largely from their appearances at musical and theatrical performances or other venues, aided by press announcements of dates and times. Proprietors of dance halls likewise advertised when the Osages would visit their establishments as a ploy to attract more audience.

According to *Le Figaro*, on August 29, at 7:00 a.m. (before the public was admitted), the Osages visited the menagerie, where they were "very pleased to see the giraffe and birds, but the ferocious animals especially elicited their joy." The visitors then amused themselves watching the monkeys before returning to their hotel for lunch.[69] Whether *Figaro*'s piece had any truth to it or not, the idea of an encounter between Native Americans and the giraffe provided writers with a framework for lampooning European civilization and the French government. Two anonymous brochures (one seven pages, the other twelve) quickly appeared, each embellished with a lithograph of the giraffe and the six Osages. In the first, the giraffe explains to the Osages that they are being exploited and warns them of the perils of civilization. In the second, "The Giraffe's discourse to the chief of the six Osages (or Indians)," the giraffe questions them about government in America in such a way as to imply the failings of the Villèle ministry, press censorship, and so forth. The brochure is attributed to Honoré de Balzac.[70]

Critics of the government were not the only ones to co-opt the giraffe and the Osages. As Salvandy observed, government supporters were skilled at diverting the public's attention from serious matters to frivolities: "When it is a question of elections, they weigh in on the giraffe; they sacrifice the lists of jurors for the visits of the Osages."[71]

Like the giraffe, the Osages served as an excuse for a vaudeville production—in fact two—but by the time these debuted some six to seven weeks after the Osages arrived in Paris, the reviewers were ready to express boredom with the Osages as well as dissatisfaction with the performances.[72] Both shows folded within days, more because of the lameness of the productions, it seems, than their loose connections with the theme of the Osages. Still, as *La Quotidienne* remarked (1 October 1827), there were events and people that the public no longer cared to hear about. "Today it would be more than bad taste to pronounce the name of the giraffe; the Osages are abandoned to the obscure crowd of curious subordinates."[73]

But the Osages were not finished with Paris. People were invited to join them on Tuesday October 2 on a riverboat trip on the Seine to St. Cloud and back.[74] A few days later, the Osages dined at the restaurant/dance hall "La Grande Chaumière" in Montparnasse and were greeted enthusiastically by the crowd, whose members each paid a one-franc entrance fee to hear them sing their native songs. At the Tivoli that month, the management staged more than one "very great festival" where for three francs one could dance, see fireworks, and watch the Osages go up in a tethered balloon with a famed balloonist (for diplomatic reasons, the government prevented the Osage chief from riding untethered).[75]

As for the giraffe, the press's enthusiasm for her had ebbed. It mentioned her only incidentally early in October in connection with Hassan's departure for Marseille, where he hoped to board a ship to Alexandria.[76] It mentioned her again to announce that as of Sunday October 14, she would occupy her new quarters in the rotunda and could be viewed daily.[77] A week later, Geoffroy boasted in a letter to Villeneuve, "We were in danger of being overcome by the Osages, but we have put up a good fight." The giraffe, he said, was still being courted.[78] He did not acknowledge that she and the Osages had both slipped from the public's favor.

The Museum installed the giraffe and her handler, Atir, in one of the five tall hexagons forming the rotunda's corners, surrounded by a large outdoor space with a shelter. Her quarters could be closed and heated to protect her from the winter cold. Whether inside or outside, the public could see her without endangering her or the other rare animals in the rotunda. The total construction bill for the minister of the interior was 6,569 francs.[79]

The Museum was disinclined to tell the minister that some features of the giraffe's new accommodations were perhaps de trop, but Geoffroy proudly described them for Villeneuve. With parquet floors and walls upholstered with straw mats in an elegant arrangement, Geoffroy called the giraffe's lodgings "truly the boudoir of a little lady." To add to the "picturesque effect," Atir reached his bed by way of two ladders and a landing between them. He and the giraffe could view each other face to face high in the enclosure. There were stoves in the room's antechambers, but the warmth generated by the giraffe's bovine companions raised the room's temperature six degrees above normal. For cold weather, Geoffroy said, "we will double and triple the number of animals."[80]

On October 27 the Osages were presented to the duchesse de Berry; the next day their manager, Delaunay, was arrested for a substantial debt dating to 1799. During the two weeks he was jailed, the Osages traveled to the outskirts of Paris in search of audiences until November, when they left for Brussels.[81] Their subsequent press mentions were increasingly dismal. They wanted to go home, and in January 1829 they were in Munich, abandoned by Delaunay (but not their loyal interpreter), hungry, miserable, and having to rely on charity. Delaunay's schemes had generated more debts than credits; he had absconded with the receipts and gone bankrupt. By the summer of 1829 there were only five Osages: three of the original men, one of the original women, and an infant born in Liège. The American consul in France took up a subscription to send them home. Two men and the woman reached the United States in spring 1830. The infant and one man died from smallpox during the crossing.[82]

The press scarcely commented on the giraffe in the 1830s, and when it did, it was disparaging. The *Mercure de France* (1836) noted that the formerly exciting giraffe now attracted no more attention than the zebra or the llamas (the orangutan was the new wonder). Author S. Henry Berthoud, on a visit to the menagerie in December 1838 wrote: "The giraffe profits from an hour of sunshine to air her stupid head and slender legs."[83] Subsequent animal stars in Paris and elsewhere would follow the same trajectory. Beginning as sensations, they ended up as matters of public indifference or even derision.[84]

Frédéric Cuvier's weekly menagerie reports offered little giraffe news after her 1827 heyday. In April 1828, while jumping around she lightly skinned her thighs against a trellis.[85] One morning in December 1828 she suffered a "slight indisposition . . . she trembled, her hair bristled, and she held her head horizontally," but by evening, she was fine.[86] In July 1829 Cuvier diagnosed "a slight inflammation of the pituitary membrane" from which the giraffe had "completely recovered" and which had never "in no way diminished her good health."[87] In February 1830 he explained that she had not been taken outside during a warm spell because the paths were slippery.[88] Two months later he remarked that she and various other animals were fine.[89] He did not mention her again in a menagerie report until 9 February 1836 when he indicated she could no longer be walked in the mornings because she had grown too strong for her keepers to hold her. Her park should be extended, he said, to allow her to exercise properly.

Though the giraffe had lost her cachet by late 1827, the press remembered her when it came to international rivalries. It appreciated that the giraffe France received in 1826 was much healthier than the specimen sent from Egypt to the king of England. It remarked on a giraffe sent from Egypt's ruler to the emperor of Austria in 1828 that died at Schönbrunn in June 1829, and it noted the demise of the British giraffe the following October.[90] The latter, sickly from the start, lasted as long as it did only because of the interest George IV took in its care. When it died, France could boast of having the only live giraffe in Europe.[91]

The giraffe competition between France and England reemerged in February 1835 when the Museum learned of the possibility of acquiring a young male specimen that also interested the British. The news came from the minister of public instruction (François Guizot), who received it from the minister of the navy, who heard it from Jean-François Mimaut, Drovetti's successor as French consul general in Alexandria. The giraffe was in Ethiopia, owned by a French officer named Vaissière who had served with distinction in Mohammad Ali's campaign against the Wahhabis and was involved in commerce in the African interior.

By Vaissière's account, his Abyssinian giraffe was a nine-month-old male of an exceptionally beautiful variety never seen in Europe. The English wanted it, he said, but he did not wish to deprive France of the animal. He noted further that the English were simultaneously trying to acquire six giraffes from Sennaar. Vaissière wished only to be compensated for his pains and reimbursed for transporting the giraffe from Ethiopia to Cairo. Guizot asked the Museum whether it would accept the giraffe as a gift and pay for its travel.[92] Finding itself in a familiar quandary, the Museum told Guizot that a male giraffe would be "of very great interest" and that it would spare no effort "to profit from the occasion of obtaining it." It noted, however, that the previous giraffe had cost the government 20,000 francs, a sum too great for the Museum's limited budget to handle.[93]

Months later, Consul Mimaut was still urging the giraffe's acquisition and worrying about British attempts to purchase giraffes, Vaissière's included.[94] The Museum remained interested but unable to pay.[95] After that, Vaissière's trail all but vanishes from the Museum's archives, though his name reappears in the French press. In April 1836 newspapers relayed the news from Cairo that Vaissière, identified as a chevalier of the Legion of Honor and an "ex-aide-de-camp" of Mohammed Ali, had "five superb giraffes" he'd collected in Ethiopia and was sending to New York. He was also said to be planning a new expedition.[96] An October 1836 notice in the *Evening Post* (New York) indicates that Vaissière's "five very beautiful giraffes" "from Calabat, in Abysinnia" were indeed sent to New York, but it does not explicitly state that they arrived, and it appears they didn't. New Yorkers did not see their first live giraffes until 1838.[97]

Meanwhile, a French explorer and hunter, Georges Thibaut, provided the British with the specimens that gave them giraffe bragging rights. Four giraffes, three males and a female, made port in east London on Tuesday, May 24, 1836, and were delivered to the Zoological Gardens the next morning. Leading the procession were officers of the Zoological Society and "M. Thibaut, attended by three Nubians and one Maltese in their native costumes." *The Mirror* observed, "The appearance of M. Thibaut and the Nubian attendants is very picturesque. The former, although a Frenchman, has the dress of an Arab. The three Nubians wear their native dresses, and, although all are of the class of peasants, two are evidently of superior caste, from linear gashes on the face, a description of tattooing."[98]

The comparative anatomist Richard Owen described the giraffes' arrival at Regent's Park: "When they entered the Park and first caught sight of the green trees, they became excited, and hauled upon the reins, waving the head and neck from side to side, with an occasional caracole and kick-out with the hind-legs. M. Thibaud [*sic*], their captor and chief conductor, contrived, however, to coax them along with pieces of sugar, of which they are very fond."[99]

Two months later Thibaut was in Paris visiting family and attempting to persuade the Museum to commission him to capture live animals for the menagerie. He wrote that he had supplied the four giraffes to London and for years had pursued commercial ventures and "zoological research" in Nubia, Sennaar, and Kordofan (central Sudan). In 1830 he had several animals, including two giraffes he had wanted to offer to France, but Egyptian agents confiscated them and the French official to whom he complained (probably Mimaut) maintained that the viceroy should not be bothered with a matter involving mere beasts. The British general consul in Egypt, in contrast, knew that the London Zoological Society was interested in giraffes, and in 1834 he commissioned Thibaut to collect some. Along with the four giraffes he delivered to London in May 1836 were a lynx, four gazelles, and a lion.[100]

Fresh from his success with the British, Thibaut offered his services to the French, saying he hoped to provide the Museum "a companion for our beautiful giraffe." He said he could collect various species of antelopes and planned to bring the hippopotamus to Europe. He admitted he had discussed the hippo with London, but he wanted to know the Museum's intentions. News of Thibaut's visit appeared in *L'Echo du monde savant*, mentioning that a French traveler who had supplied the London Zoological Society with four giraffes would soon return to Africa, where he hoped to obtain a hippopotamus. Not ceding giraffe superiority to the British, *L'Echo* noted, "The giraffes of the menagerie of London are younger and consequently less large and less beautiful than that of the Museum in Paris."[101]

Through zoologist Henri de Blainville, the Museum informed Thibaut that certain animals were of interest if the price was right. A young hippopotamus accustomed to living out of the water topped the list. An Abyssinian rhinoceros was also highly desirable (if not a live one, then a skin or skull). Also on the Museum's wish list were a colobus monkey and a canid distinguished by its very large ears (evidently the African wild dog). As for a male giraffe, the Museum thought it would be easier to arrange to get one of those already in London than to commission one from Africa.[102] Despite these discussions, the Museum's records mention no live animals received from Thibaut. He continued to earn renown as a traveler and participated in Mohammed Ali's scientific exploration of the White Nile. He later became French consul in Khartoum.

Interactions with Vaissière and Thibaut introduced the Museum to a new breed of entrepreneurs, hunters capturing big game to sell to menageries and other collectors. Giraffes were conspicuous among the early, difficult to obtain, prizes. Atir and Hassan described the dangers of capturing them. Only those of nursing age could be taken alive, they said, and even then, attempts at capture them were "nearly always" fatal to the animal, because in struggling to free themselves they broke limbs, dislocated their necks, or simply refused to eat. To these fatalities must be added the mothers killed in capturing their offspring.[103]

Reporting to the London Zoological Society, Thibaut stated that he'd killed the mother to capture his first young giraffe but did not mention adult deaths related to the next four young specimens he acquired, all of which died en route to Dongola in the December 1834 cold. It took three months to capture three new giraffes to bring his total to the four he finally delivered to London.[104] One can conclude that in Thibaut's bringing four giraffes to London, twelve other giraffes died—the eight nursing mothers killed for their young and the four young giraffes that died in transit.

Significantly, the last leg of Thibaut's journey from Malta to London was expedited by steamship after the London Zoological Society learned that a competitor, the Surrey Zoological Gardens, was importing giraffes from Africa. The Surrey enterprise, founded by Edward Cross in 1831, was about to receive a set of giraffes and a host of other African animals collected on commission in Egypt by the naturalist John Edington Warwick. These animals arrived at the Surrey Zoological Gardens six weeks after Thibaut's animals arrived at Regent's Park. Londoners suddenly had seven giraffes to admire.[105]

A press account of the French giraffe's arrival in Paris back in 1827 had mused on the difficulty of imagining how an animal so lacking in defenses, so awkward, and without any reason to be domesticated could still "perpetuate itself in the middle of deserts in the dreadful neighborhood of tigers and lions."[106] In 1836 a different French piece on the arrival of the giraffes in London suggested that their greatest threat might be humans. The author did not implicate European hunters in the giraffe's diminishing numbers. He identified the problem as an increase in the native population: "Who will protect the giraffes from African hunters, since, unfortunately for these animals, their meat is a sought-after dish?"

In the wake of this lament, the author offered an entirely novel suggestion: that menageries could serve as refuges for giraffes destined to become extinct in their native lands. He saw the small troupe of giraffes at Regent's Park as an avatar of the future. He imagined that one day in Europe giraffes might decorate large parks in the same way that swans adorned large ponds.[107]

When Frédéric Cuvier contemplated the prospect of France's first live giraffe in 1826, he predicted it would bring untold insights to science. But the Museum's zoologists seem not to have derived many insights or observations from her tenure. Geoffroy published one brief article within days of her arrival.[108] Not until 1842 did Frédéric Cuvier's son write the article for his father's and Geoffroy's *Natural History of the Mammals*, by which time the latest and best studies of the species were British.[109]

Ironically, one reason the French giraffe did not advance science was her longevity. She outlived Georges Cuvier, Frédéric Cuvier, and Étienne Geoffroy Saint-Hilaire. By 1845, when she died, Richard Owen in London had already had ample opportunity to conduct anatomical studies on giraffes that had died at Regent's Park. Furthermore, that was not the only respect in which Paris was behind. When Owen's giraffe study appeared (1841), the London Zoo could boast of the birth of two baby giraffes, one in 1839 and another in 1841.[110]

By the late 1800s, giraffes were common fixtures in the zoological gardens that sprang up in Europe and elsewhere following the lead of Paris and London. However, the idea that flickered through the mind of an anonymous writer in Paris in 1836, that a zoo could serve to harbor an animal in danger of extinction in its native land, was yet to flourish with them. No one seemed to be worrying about the long-term fate of the giraffe—"this beautiful and precious animal," as Villeneuve called her. When first seen in France, she elicited exuberant joy. After a few months the initial ecstasy was gone, and she came to be taken largely for granted. Her value was reasserted in the context of competition with the British, but it was too early for her to be taken as symbolic of a growing threat to the existence of species around the globe.

[3.10]

The Lions of Algeria

The Museum professors were excited in July 1830 when news of the fall of Algiers to French troops reached Paris. Their interest in the North African incursion was fueled by their perpetual thirst for new specimens but also by an embarrassing fact: for a full year the menagerie had not a single lion to show to the public. The situation's awkwardness was compounded by the presence of lions elsewhere in Paris. For months, Henri Martin had been attracting crowds by the spectacle of his lions performing to his commands. It was reasonable for the professors to think that the invasion of Algeria might resolve the menagerie's lion problem. North Africa had been its primary source of lions in the past, largely because they were the diplomatic gift of choice for rulers there. This time, though, the Museum got more than it bargained for.

News of the surrender of Algiers took just four days to reach Paris, thanks to two modern technologies: the steamship and the telegraph. Algiers fell to the French at noon on 5 July 1830. The government steamship *Sphynx* left port that afternoon and in three and a half days was in Toulon, where the news was relayed to Paris by the new semaphore telegraph system, station-by-station, in less than six hours. The firing of the cannon at Invalides at 4:00 p.m. on July 9 signaled the victory to Parisians, and that evening the triumph was announced on placards and at theaters and celebrated by the illumination of public buildings. The following day, *Le Constitutionnel* proclaimed, "WE ARE MASTERS OF ALGIERS."[1]

The Museum's response to the news was swift. On July 12 the professors dispatched letters to the ministers of the Interior, war, and navy. To the latter two it hailed "the immense advantages that the Museum could derive from a circumstance as favorable as the sojourn on the coast of Africa of a French army and administration." Little was known about the region's natural history and especially its animals, they said, and if French officers were encouraged by their superiors to employ their "moments of leisure" in

Figure 3.10.1. "The Lion Hunt 1836." C. Horace Vernet. Exhibited at the Paris Salon in 1839, Vernet's depiction of a lion hunt in Algeria includes the capture of lion cubs as part of the scene. Wallace Collection, London. © Wallace Collection, London, UK/ Bridgeman Images.

procuring specimens, whatever they found would be of interest to science.[2] The administrators did not confess their lion problem to the ministers of war and the navy, acknowledging only to the minister of the interior that the collection of live carnivores was currently "very poor." With its limited budget, the Museum could not pay the exorbitant prices of animal dealers and was therefore unable to keep the menagerie well stocked. It had thrived in the past, they said, only when the government assisted during special circumstances.[3]

Given the number of lions the menagerie had received in the early years of the Restoration, their dearth in 1830 may seem strange. The uncomfortable truth was that the new pavilion for *animaux féroces*, completed in 1821, had not improved the lions' longevity. The carnivore population was so low that when an assistant carnivore keeper (Richard) died in 1825, Frédéric Cuvier postponed hiring a replacement.[4] When a lion died in October 1828, only one male and one female remained: Coco and Sultane, the pair donated by Admiral de Rigny the year before. When Sultane died in April 1829 and Coco followed shortly afterward, the menagerie was without a single lion for the first time in thirty-two years.[5]

The Museum's letters of July 1830 to the ministers of war and the navy were each accompanied by twenty-five copies of instructions for voyager naturalists. The minister of the navy passed his set along to Admiral Duperré, commander of the maritime side of the Algiers campaign.[6] The minister of war, Louis-Auguste-Victor, comte de Bourmont, was in Algeria himself, leading the expedition. He was sent the instructions by an aide who wondered to the Museum if instructions more specific to North Africa might be in order.[7]

It turned out that the opportunities for natural history work on the North African coast were not as plentiful as the Museum imagined. "The actual circumstances," as General Bourmont explained in a letter of 2 August 1830 to Prince de Polignac, France's prime minister, were that in the weeks since the French troops had landed in Algeria, they had not penetrated inland. None of the military posts was more than two miles from the sea. Moreover, a survey taken near Blida, at the foot of the Atlas Mountains, revealed that the plants and insects of the region differed little from those of southern Spain. "As for quadrupeds," Bourmont continued, "those that belong especially to Africa have been almost entirely destroyed over a zone of 18–21 miles from the seashore." Even if there had been an abundance of specimens, Bourmont did not have the personnel to collect them. "Only a very small number" of army officers, he wrote, could conduct the requested research, and the medical officers, though better suited, had no free time because of the number of sick troops.[8]

Yet Bourmont had some positive news for the Museum. He had embarked on a French ship of state a collection of seventeen ostriches, some of which were "of a remarkable size." In addition, he had fifteen lions and lionesses that he promised to send to France at the next opportunity. He did not specify where the animals came from, but they had been appropriated from the dey's menagerie in the Kasbah. The ostriches never arrived in France, but the lions did.[9] Fifteen lions! Over the previous three decades of French/Algerian relations, the Museum had received only three or four from Algeria, all as diplomatic gifts. These new ones, in contrast, were confiscated, testimony that diplomatic negotiations had failed.[10]

Bourmont did not mention difficulties with the animals' care or transport, but an officer friend of Geoffroy Saint-Hilaire reported from the Kasbah that feeding the lions was a great burden. "They eat cattle that we would infinitely prefer were consumed by our brave soldiers," he wrote. Furthermore, Admiral Duperré, who disliked Bourmont, was refusing Bourmont's request to transport the lions to France on his ship. Geoffroy's correspondent hoped the Museum could urge the minister of the navy to command Duperré to cooperate.[11]

Though news traveled faster than it had in previous decades, when Bourmont wrote to Polignac on the second of August, word of the latest events in the French capital had yet to cross the Mediterranean. The general had not heard that French citizens had revolted against the government and the king had abdicated. The day Bourmont composed his letter, Charles X and the royal family went into exile in England.

The invasion of Algeria and the July Revolution were interrelated. The government's official reasons for the incursion into North Africa involved French honor and the intention of bringing the Barbary States to heel. On the first score, Charles X wanted to teach an unforgettable lesson to Hussein, the dey of Algiers, for having insulted France twice: in 1827 when he slapped the French consul in Algiers with a fly whisk, and in 1829 when he fired on a departing French ship after refusing to talk to the ship's captain, who had come flying a flag of truce.

Aside from the pretext of French honor, France wanted what Polignac said the allies wanted: "the abolition of the enslavement of Christians, the destruction of piracy, and the end of humiliating tributes that the European states are having to pay to the Regency."[12] Polignac had another, tacit, reason for sending French troops to conquer Algiers: the political situation at home. He believed that a small, winnable war on the north coast of Africa would distract French citizens from their deep dissatisfaction with the government's attacks on the Charter of 1814, which established France as a constitutional monarchy. Critics recognized the invasion as the government's scheme to drum up constituents' support for the upcoming elections. *Le Temps* fiercely attacked the incursion because it would give the Polignac ministry the chance to bask in a military victory at the expense of brave French soldiers' lives, while doing nothing to improve France's North African affairs.[13]

The success of the Algerian campaign led Charles X to overplay his hand. He issued four "ordinances" that eviscerated the Charter of 1814 by dissolving the Chamber of Deputies, suspending freedom of the press, making members of the business class ineligible for election as deputies, and calling for a new election. The public's anger erupted in the "three glorious days" of the July Revolution (July 27–29), during which barricades were thrown up in the streets; revolutionaries waved the tricolor flag; soldiers battled citizens; key sites like the Palais de Justice, the Hôtel de Ville, and the Tuileries Palace were captured; Polignac's ministry fell; and 1,000 lives were lost. Charles X tried to abdicate in favor of his grandson, but the provisional government refused and installed Louis Philippe as "king of the

French" instead. When the news reached Algiers, the ultraroyalist General Bourmont refused to pledge allegiance to the new king. The new government relieved him of his command, appointing General Bertrand Clauzel in his place.

Unlike Bourmont, the Museum professors embraced the reign of Louis Philippe. Two weeks after his coronation, they visited his court to pay homage. They moved quickly to establish relations with the new minister of the interior and long-time critic of Charles X, François Guizot, to explain the Museum's nature and needs. Foremost among the latter was being exempted from paying to transport fifteen hungry lions from Marseille to Paris. Drafting the Museum's letter to Guizot, Georges Cuvier referred to the animals as "this beautiful part of the trophies of a glorious war" and noted that a previous minister had covered all the giraffe's expenses from Marseille to Paris. The Museum wanted Guizot to do the same for the lions, projecting a cost of 15,000 francs.[14]

Guizot was sympathetic. He granted the Museum's request and offered an additional 1,000 francs to purchase rare animals for sale at Rosny, the estate of the duchesse de Berry, now exiled with the rest of the royal family.[15] That purchase, vetted by Geoffroy, gave the menagerie two deer, two kangaroos, a mouse-deer (chevrotain), and a "tiger cat."[16] Guizot also informed the Museum that more animals were arriving from Algeria: two ostriches, a panther, and another "tiger cat," the collection given by the dey of Algiers to the Duchess of Angoulême before the invasion. The dey had since been dethroned, and the duchess had gone into exile, but the animals had arrived in Marseille in August, cleared quarantine, and were ready to be sent to Paris.[17]

The Museum dispatched its head carnivore keeper, Magloire Rousseau, to fetch the animals, and Frédéric Cuvier requested changes at the carnivore pavilion to accommodate them and the public.[18] Cuvier had to sell two wolves and move five bears from the pavilion to make space for the lions.[19] The nine-wagon convoy from Marseille arrived on the morning of 3 November 1830 carrying thirteen lions and lionesses, two panthers, and an ostrich.[20] Rousseau and three Arab handlers, Omar, Salem, and Harby, had performed their mission admirably. Only one cub had died en route. Rousseau preserved its skin and skeleton. One lion appeared extremely thin, but the others appeared healthy.[21]

Despite the convoy's size and provenance, the press scarcely noticed its arrival. It had paid greater attention two weeks earlier to a more diverse and exotic collection of animals from India and Southeast Asia collected by J.-J. Dussumier. Nor were the few press comments on the Algerian animals favorable. "These innocent animals are going to fatten themselves at the expense of the Museum's budget" remarked the *Le Figaro*.[22] *La France Nouvelle* recommended selling some.[23] With a tiger, two lion cubs, and a jackal from Dussumier, and the thirteen lions and two panthers from Algeria, the carnivore pavilion population swelled from twenty-four to forty-three. Cuvier called for filling the empty post of assistant carnivore keeper, and the administration promptly agreed.[24]

Though the professors frowned on Henri Martin's lion shows as unscientific, they failed to formulate any scientific program for their feline acquisitions. The primary concern was to keep them alive, but the results were mixed. Three of the males were dead by February 1832; in May, Cuvier predicted another would die soon. He was therefore quick to protest when Geoffroy ordered Rousseau to sell two young male lions lent eighteen months earlier by a sea captain who had acquired them at the Cape of Good Hope. The death of another male from Algeria, Cuvier told Geoffroy, would leave the menagerie on the verge of being once again bereft of male lions, and "to the public, a menagerie of fierce animals without [male] lions is a menagerie without interest, without animals!"[25]

To ensure the menagerie's stock, Cuvier pressed the administration to buy the captain's two lions. In their year and a half at the menagerie, he said, they had acquired a rare strength, health, and beauty, having survived the dangerous teething period when their canines came in. The 2000-franc asking price was half the animals' worth, Cuvier argued. The Museum accepted the purchase.[26] As he predicted, one of the Algerian male lions soon died, but Cuvier was wrong about the good health of the young males from the Cape of Good Hope. One died in September 1832 and the other in April 1833, leaving the menagerie with only one male, the last from the conquest of Algiers. When it died in March 1835, the menagerie endured six months without male lions. Four Algerian lionesses remained. One died in November 1835; another was exchanged in 1836 for a polar bear and 500 francs. The last two lived at the menagerie until 1843 and 1844.[27]

With the French occupation confined to the coast and the immediate surroundings of Algiers, Annaba (formerly Bône), and Oran, few new specimens from farther inland reached the Museum. After Admiral Duperré gave a pair of crowned cranes in 1831, the menagerie received no more Algerian animals for two and a half years. In spring 1834, a Polish officer serving in the French army gave the Museum a jackal. In the fall, another officer donated another jackal. In January 1835, Frédéric Cuvier paid fifty francs for a fox reportedly captured near Algiers.[28] At last, on 1 July 1837 the menagerie received two Algerian lions, both young, and about a year apart in age. Their combined stories exemplify the haphazard interconnections of human and animal actors in the menagerie's historical development.

The older lion arrived in France as the gift of Jean-Baptiste Drouet, comte d'Erlon, the first to hold the title of Governor General of the French Possessions in Algeria. He had received the lion as a diplomatic gift from the young emir and military commander Abd el-Kader, leader of the Arab resistance. A man of piety, courage, and intelligence, Abd el-Kader had gained control over much of western and inland Algeria by uniting Arab tribes in a jihad against the French and by outfoxing the enemy in a series of treaties.[29] The donor of the second lion was Yusuf, a young Arab military adventurer who fought on France's side, later called Yusuf Bey despite the failed French attempt to install him as bey of the Algerian city of Constantine. In short, the two lions that arrived at the menagerie in 1837 from Algiers were the gifts of men who would have gladly killed one another in battle. According to the French press, Yusuf was about to do just that to Abd el-Kader near Tlemcen in 1836, but his horse fell from fatigue as he was about to deal the mortal blow.[30]

D'Erlon, a distinguished general under Napoleon, went into exile when the Restoration began and was called back into military service in 1832. He helped quell the revolt in the Vendée instigated by the duchesse de Berry and commanded by the comte de Bourmont. But d'Erlon was not a good choice for governor general of the French possessions in Algeria, a post created in 1834 as the French began to formalize plans for colonization. Content to have peace in his immediate area, d'Erlon vacillated while Abd el-Kader energetically consolidated power over large swaths of the country. D'Erlon probably received the lion cub as part of an exchange of diplomatic gifts during a meeting at Abd-el-Kader's camp near Blida in May 1835. In June, Arab forces routed French troops under General Camille Alphonse Trézel, a defeat that triggered both Trézel's and d'Erlon's recall to France and left the lion cub from Abd-el-Kader in the hands of the intendant civil in Algiers.[31]

In the fall, Guizot, then minister of public education, contacted the Museum about the cub.[32] He'd heard from the intendant civil wanting to know if he should send it to France or wait until another lion could be sent with it. The Museum counseled waiting until spring (1836), when the weather would be warmer and the lion stronger, having passed through its teething stage, which would give the Museum "infinitely more chance of conserving it."[33] Two lions, not one, eventually embarked from Algiers on a steamship for Toulon. The Museum sent its assistant keeper of carnivores, Letellier, to collect them plus a lion cub that had arrived in Marseille.[34] The cub died before Letellier got there, but the lions in Toulon were fine. The larger one's donor was listed in the menagerie's animal register as le comte d'Erlon. The smaller lion was entered as a gift from "Jousouf Bey."[35]

Yusuf's life, as biographers have remarked, seems straight out of *The Thousand and One Nights*. Born on the island of Elba in 1808 or 1809, he was about six when he was sent to school in Italy aboard a ship that was captured by pirates. Offered as a slave to the bey of Tunis, he was raised in the royal court. He distinguished himself as a warrior in the bey's guard, but his romance with one of his daughters infuriated the ruler, causing Yusuf to flee for his life. Luckily a French envoy was in Tunis hoping to persuade the bey to support the French invasion of Algeria, and Yusuf was spirited away on the envoy's ship with the help of a young Ferdinand de Lesseps, son of the French consul general in Tunisia. In Algeria, Yusuf met General Bourmont and joined the French expeditionary force, rising quickly in the ranks and gaining fame for the victorious battles where he beheaded his rivals.[36]

By the time General Clauzel returned to Algeria in 1835 to replace d'Erlon, Yusuf's exploits were legendary. Clauzel wanted to take Constantine and install Yussef in place of the existing bey, but his long drawn-out campaign failed. Yusuf's reputation did not suffer appreciably, and when he visited France in the spring of 1837, he was given a hero's welcome.[37] The press enthusiastically reported his triumphs, including two audiences with the king; attendance at the wedding of the duc d'Orléans and Helene-Louise of Mecklenberg (marking an alliance between France and Prussia); and appearances at salons and the opera (in General Clauzel's box). He was the talk of the town, as encapsulated in the witticism: "Have you seen Youssouf? . . . He is very nice for someone who cuts off heads."[38] But there is no indication that he visited the menagerie, even though he was in Paris when his and Abd el-Kader's lions arrived.

In April 1837, before his visit to France, Yusuf sent two very young lion cubs (male and female) from Annaba to Marseille, this time as homage to the French queen. Only the female, named "Jaballa" after the hunter who killed her mother, survived the crossing. Jaballa was turned over to the prefect of Bouches-du-Rhône, who gave her to the director of the Marseille Museum of Natural History, a man named Barthélemy, who cared for her assiduously until her death twenty-five days later. When Barthélemy begged to keep her skin and skeleton for the Marseille Museum, the Paris professors obliged.[39]

What motivated Yusuf's lion gifts? The first, given to the intendant civil in Algiers (in 1836), was perhaps a gesture of thanks to the French government for granting him the title of bey of Constantine. Or perhaps Yusuf knew that Abd-el-Kader had given a lion to the intendant civil, prompting him to match his archrival's act. The cubs for the queen may have been meant to divert attention from the failed Constantine campaign, or simply a means of burnishing his image before his trip to France.

Abd-el-Kader meanwhile controlled the whole of western Algeria. The Treaty of Tafna of June 1837 obliged him to recognize France's sovereignty, but this was only a formality, and the concessions he gave were not as great as those he received. The French were nonetheless committed to expansion. Their taking of Constantine in fall 1837 signaled their flouting of the agreement to remain in their coastal enclaves. Abd-el-Kader kept reminding them that peace was good for all parties, but France was pursuing a more aggressive course, claiming the right to control the land between Constantine and Algiers.

Abd-el-Kader proceeded to do what other Algerian and North African rulers had done to smooth relations with France. He sent a splendid assemblage of diplomatic presents including nineteen horses, a beautiful saddle, carpets, perfumes, women's cloaks, lion and tiger skins, and live animals. Louis Philippe accepted the presents from Abd-el-Kader's envoy, Ben Arrach, in April 1838. He responded to Ben Arrach's pledge of the emir's commitment to peace with the observation that it was up to the emir to observe the provisions of the treaties made with him and that that was essential if he was going "to raise his nation to the level of civilized nations."[40] The king passed the exotic animals on to the Museum (with the instructions that he himself be identified as the donor).[41] There were no lions this time. Instead, there were the two other most frequently given North African animals: gazelles (three) and ostriches (two). The ostriches, hungry on their arrival, immediately began devouring the quail in the pen they were placed in.

The professors conveyed their thanks to the king and their excitement that one of the ostriches was a male. For years the menagerie had only females whose eggs had failed to hatch. Having a male raised hopes that that the Museum's ostriches might reproduce.[42] "Fashionable Paris" hastened to the menagerie to see the new animals.[43]

But neither Abd el-Kader's gifts, nor the Museum's and the public's appreciation of them, nor the previous peace treaties changed France's plans for territorial expansion. In May 1838 the Chamber of Deputies voted to bolster funding for France's occupation, the first "real budget" for Algeria, as Minister Molé wrote to the governor general.[44] The bid for the land between Constantine and Algiers put France on a collision course with the Arab resistance. The war that broke out in 1839 did not end until Abd-el-Kader's surrender in December 1847.

The simple story about the relation between the menagerie and the wider world after 1815 was that the Museum's quest for ever more specimens was greatly enhanced by the freedom of the seas and new or renewed commercial and colonial enterprises abroad. Geopolitics influenced where French (and rival) administrative officials and collectors went, and they were also manifested in the diplomatic gifts the Museum received as the result of what Georges Cuvier described as the "fantasies of Barbary." Cuvier was complaining of the burdens put on the Museum by having to pay for part of gift animals' transportation costs, but the menagerie did receive some prize specimens as diplomatic gifts, the elephant and the giraffe from the pasha of Egypt being the most prominent. The fantasy on the part of the North African rulers may have been that the gifts would make a difference. The giraffe notwithstanding, the French navy soon combined with allies to destroy the Egyptian-Ottoman navy and thereby aid the Greek independence movement. Likewise, the emir Abd-el-Kader's gifts in 1838 did not dissuade the French from their plans for colonial expansion in Algeria. In the public's appreciation of the animals of the menagerie, what counted most was having the animals, with political considerations to be downplayed or emphasized according to the context. The French loved their giraffe, while disapproving of the pasha's political intent in giving her; and they also loved her because having a giraffe when the English didn't (or had one that was sickly), was a matter for French national pride. What had begun to emerge at a different level with respect to natural history and colonialism, exhibited especially in the writings of British naturalists, was a discourse of geographical expansion where species were described as colonists, and where an appreciation of the struggle for existence was being sounded before Darwin's *Origin of Species* brought an even sharper focus on its biological importance.[45]

The menagerie was in its very essence a scene of geographical dislocation, even if none of its inhabitants (except the brown rats that installed themselves there) had come there on their own initiative. I have not determined whether the quail the ostriches devoured, sent by Abd-el-Kader from Algeria, were descendants of the quail Milbert took such care to import from North America.

4

Family Ties and a Science of Living Animals

Figure 4.1.1. The public visiting the elephant in its park at the rotunda. Andrew, Best, Leloir after C. Marville. From François Boitard, *Le Jardin des Plantes* (1842).

[4.1]

Family Ties

The advancement of science was one of the prime justifications for establishing a menagerie at the Jardin des Plantes, but it was an idea that needed cultivating. Whatever was achieved in this regard in the menagerie's first half century was largely thanks to its superintendent, Frédéric Cuvier, who addressed a range of questions that the animals might illuminate, while setting the broader goal of creating a science of animal behavior based on his observations. His attempts met resistance of different kinds. There were the difficulties inherent in studying animals in a public menagerie, and the problem of where the menagerie and Cuvier's work ranked in the hierarchy of the Museum's scientists' interests. If the animals were not always disposed to cooperate with his plans, the same was true of people.

In the complex and shifting web of interpersonal relationships, it was not easy to make headway. Frédéric Cuvier's career unfolded within the Museum's richly textured social ecology, where kinship, friendship, and patronage were the prevailing, determinative forces.[1] The proximity of the apartments of professors, superintendents, naturalists' assistants, and other salaried workers added to the goldfish-bowl character of Museum life. Intense competition when a position in the institution became vacant was another factor. An aspirant's chance was enhanced if he or she had relatives or a supportive friend or patron at the Museum. This applied as much to exalted positions like professor as to modest ones like *marchande de gateaux* (vendor of cakes and pastries). The administration reserved such concessions for workers' spouses or widows, recognizing that animal keepers, gardeners, porters, and other low-level staff earned barely enough to support their families.

Family ties and friendship were interwoven in Frédéric Cuvier's appointment as menagerie superintendent. By befriending Frédéric and promoting him to *garde de la menagerie*, Étienne Geoffroy Saint-Hilaire, back from several years in Egypt, sought to rejuvenate his friendship with Frédéric's brilliant brother Georges. Geoffroy later recalled that "the day after [Frédéric's] installation [as superintendent], I said with friendship to M. Frédéric that I acquired in him a collaborator and that there would be no question of boss and subordinate between us." The following year in his desire for the friendship of all his colleagues Geoffroy invited them to sign his marriage license at an administrative assembly. The minutes record that his colleagues readily provided this mark of their "esteem and affection."[2]

But maintaining friendships, mutual respect, and one's position was difficult. While Geoffroy craved friendship, he also hated feeling that his authority was being usurped. He was vexed in 1809 when it appeared that Frédéric Cuvier had bypassed him by communicating directly with the professors, the architect, and the Museum's director and by making decisions and running up bills without his approval. To relieve his pique, Geoffroy wrote a stern new set of rules for the menagerie's administration and sent them to Georges Cuvier, lately elected to a two-year term as Museum director. Georges, replying to "my dear friend," sagely met Geoffroy's impetuousness with a measured approach, offering a slight rewording of the proposed rules and assuring him that Frédéric had done nothing to subvert his authority. Georges invited Geoffroy to point out anything he might have missed in rephrasing Geoffroy's original draft. He concluded by asserting, "what I will not at all forget is your friendship, and the desire to prove mine to you on every occasion."[3]

Geoffroy again evoked the themes of kinship, friendship, and respect fifteen years later when he asked that his colleagues appoint his eighteen-year-old son, Isidore, to the post of *aide-naturaliste* in zoology for birds and mammals, vacant since the death of Pierre-Antoine Delalande. Geoffroy frankly asked his colleagues to consider his own feelings as well as his son's qualifications:

> In granting my presentation the approval I am soliciting, my colleagues will gratify me with a favor the kindness of which I will feel personally; I will consider it forever and ever as a very precious proof of their goodness toward me and as the sweetest reward for the work with which one knows I am constantly occupied.[4]

Geoffroy's fellow professors approved his proposal unanimously, and Isidore was named his father's assistant.[5]

But this "sweetest reward" came at a time when Geoffroy's relations with his colleagues, the Cuvier brothers in particular, were souring. Early in 1828, Frédéric Cuvier drew Geoffroy's ire with what Geoffroy perceived as more insubordination. At the professorial assembly of 5 February, Geoffroy complained that both a keeper and the menagerie superintendent had failed to recognize his authority, orally and in writing. Geoffroy demanded they be brought back into line. Having obtained no results, he repeated his complaint at the next meeting. How was it, he asked his colleagues, that a subordinate could try to boss his boss?

He fumed over the idea, evidently voiced at the first meeting, that Frédéric Cuvier deserved the Museum's special consideration because of family ties and his membership in the Academy of Sciences. Was one to pay more attention to Frédéric Cuvier's standing, Geoffroy asked, than to his? More than insubordination, he regarded Frédéric's actions as a betrayal, recalling the La Fontaine fable "The bitch and her friend" ("La lice et sa compagne") that warns against helping wicked people, because they will take more and more, and finally you will have to fight. Yet Geoffroy did not want to fight so much as to make Frédéric gracefully return to his proper place in the hierarchy. It was again left to the Museum director (now botanist René Desfontaines) to mollify Geoffroy by promising to enforce the rules governing relations between the menagerie superintendent and the professor of mammals and birds.[6]

Neither Geoffroy's nor Frédéric's situation was improved by the changes in the assembly following the deaths of the Museum's illustrious zoologists Jean-Baptiste Lamarck and Georges Cuvier. Lamarck's demise on December 18, 1829, came as no surprise. He died blind and impoverished at eighty-five after years of ill health. Georges Cuvier's death (13 May 1832) was wholly unexpected. He was struck down at sixty-two, while intellectually vigorous, at the height of his prestige, and thriving financially thanks to the numerous posts he occupied.

Replacing these giants of French zoology inevitably led to contests where kinship, patronage, friendships, and enmities were in play. Lamarck's replacement was drawn out over his last decade when blindness and ill health prevented him from teaching. The first to fill in for him at the lectern was Pierre-André Latreille, the eminent entomologist whom he had mentored since 1798. When Latreille fell ill in 1826, the knives came out. The comparative anatomist Henri Ducrotay de Blainville, angling to be Lamarck's successor, sought the position of substitute lecturer, believing it would boost his claim on Lamarck's chair when the time came. He recognized, though, that family and personal alliances stood in his way; his competition was Victor Audouin, protégé of Georges Cuvier and son-in-law of the professor of mineralogy, Alexandre Brongniart.

Lobbying Lamarck (by letter) for his endorsement, Blainville railed against the "despotism" and "nepotism" of Audouin's patrons, claiming that George Cuvier hated him and had done everything he could to block his career. He acknowledged that it was natural for Brongniart to support his son-in-law, but in Blainville's view (and in actuality), his credentials were superior to Audouin's. Cuvier and Brongniart feared him, Blainville told Lamarck, because like Lamarck, he was committed to truth and no sycophant.[7] Blainville did more than flatter Lamarck and appeal to his sense of justice; he offered to pay him, promising the aging scientist a pension of 1,800 francs "to whomever of your children you would like me to designate." He claimed Georges Cuvier had done the same thing to persuade Jean-Claude Mertrud to name him as substitute years earlier.

Despite Blainville's entreaties, Audouin was chosen to teach Lamarck's course until Latreille recuperated and returned, as he did a year later. When Lamarck died in 1829, the assembly unanimously voted to give Latreille the professorship. It also arranged to help support Lamarck's two daughters, who had cared for their father throughout his infirmity. But Blainville's day would soon come. The Museum had proposed to the minister of the interior to split the chair of invertebrate zoology into two: one for the *articulates* (the insects, arachnids, crustaceans, and other segmented creatures) and the other for the *inarticulates* (the mollusks, worms, jellyfish, and so on). When the government agreed, Latreille became professor of the articulates and Blainville was elected professor of "the natural history of mollusks, worms, and zoophytes." The unanimous vote suggests that whatever Georges Cuvier and Alexandre Brongniart may have felt about him, they acknowledged Blainville's qualifications for the post.[8]

Unlike Lamarck, Georges Cuvier did not die in semiobscurity. At his death on 13 May 1832, Georges Cuvier was one of two "perpetual secretaries" of the Academy of Sciences; a professor at the Collège de France and the Museum of Natural History; a holder of major posts in the French educational administration; a baron (since 1820); and a peer of France (since 1831). His authority had lately been challenged, notably (and independently) by Geoffroy and by Blainville, but neither enjoyed his level of prestige and command. He did not die from the cholera outbreak that enveloped Paris in the spring of 1832 and claimed the lives of many Museum workers, including the chemist André Laugier and Laugier's replacement, Georges Serrulas. Instead, he suffered from a progressive paralysis that began in his right arm and throat and in four agonizing days spread throughout his body, finally reaching his lungs.[9]

Cuvier's vacant professorship was swiftly filled. Applicants included Louis Duvernoy, a distant cousin who had assisted Cuvier with the comparative anatomy collection before becoming professor of zoology and rector of the faculty at the University of Strasbourg; Pierre Flourens, who taught Cuvier's comparative anatomy course in Cuvier's stead for two years; zoologist Constant Duméril, professor of reptiles and fish since 1825; and Henri de Blainville, professor of inarticulates since 1830. Blainville insisted he was happy with his mollusks but said his colleagues made him apply, judging he would provide a greater service to the Museum in the chair of comparative anatomy. On 25 May 1832, twelve days after Cuvier's death, B;ainville was elected to the post, receiving eight out of ten votes cast.[10]

Frédéric Cuvier didn't want his brother's chair, he wanted his own, a chair that in 1832 did not exist at the Museum but that he hoped to see established, devoted to "the nature of animals and their education." He addressed the Museum's professor/administrators late in July 1832, writing that he would have asked that the chair be created long ago had his brother not thought it inappropriate for them to be Museum professors at the same time. He asked the professors to request the minister of public Instruction's permission to create a chair that would make the menagerie genuinely useful to science, not just a source of specimens for the anatomy and zoology galleries or of amusement for the public.[11]

Cuvier was confident his proposal would be well received, he explained, because three months earlier Blainville, "on his own initiative, judging of what utility a course similar to the one I am proposing would be, had the project of asking you to create [the chair]." Blainville had indeed written to Cuvier to that effect on 20 May 1832, and Frédéric, replying with thanks, admitted that such a chair had been "the object of my dreams" and suggested that if Blainville mentioned the idea at the assembly, Frédéric's friends there would bring it to fruition.[12] But Frédéric was too naive to recognize that Blainville's letter to him had been written from self-interest. Blainville wrote it the same day the professors were to meet to choose Georges Cuvier's replacement. His suggestion to promote Frédéric's case seems to have been intended to defuse hostility that Georges's supporters may have felt toward Blainville's candidacy for the comparative anatomy chair. Blainville did not proceed to lobby the administration to create a chair for Frédéric Cuvier. To the contrary, when that chair was finally put up for a vote five years later, Blainville argued strongly against it.

Cuvier's proposal gained no traction in the summer of 1832. The assembly decided not to consider the matter until it replaced the professor of human anatomy, Antoine Portal, who had just died at age ninety. That appointment came and went without further mention of Frédéric Cuvier's request for a professorship devoted to the study and training of living animals.[13]

If Frédéric Cuvier felt unappreciated, so did his supervisor Geoffroy, professor of mammals and birds. The deaths of Lamarck and of Georges Cuvier made Geoffroy the Museum's senior zoologist, but his standing in his colleagues' eyes did not increase accordingly. He quarreled bitterly with Blainville. They were already engaged in a heated debate about the classification of the platypus when the surgeon/naturalist Fortuné Eydoux returned from Australia with a collection of specimens he left with Blainville for safekeeping. Geoffroy wanted to see them, but Blainville insisted that Eydoux, as their collector, had the right to study and describe them first.

In Geoffroy's eyes, Blainville was behaving as unfairly as Georges Cuvier had when he was Museum director: whenever new specimens arrived, Cuvier "*unduly* creamed off" [Geoffroy's emphasis] those he wanted for his own work. Now, in January 1833, it seemed Blainville was squirreling away specimens everyone had the right to see. Geoffroy told his colleagues that he would skip their next meeting for reasons of health. The two previous meetings, he said, had so upset him he had been unable to sleep and had suffered from "apoplectic agitations."[14]

Geoffroy's conflict with Blainville over access to specimens (and more generally over what to do when zoologists and comparative anatomists claimed the same specimens) led the Museum to establish a commission to hear both sides. Geoffroy, however, wearied of the affair, telling his colleagues to drop it and to please burn all his related missives. He greatly desired, he said, that no sharp or bitter feelings remain between him and his colleagues.[15] But any sense that collegial relations had returned to a firm footing was short-lived. Time and again in the 1830s Geoffroy became enraged when he felt ignored by his colleagues. They in turn, at the Academy of Sciences as well as the Museum, became increasingly embarrassed or irritated by Geoffroy's polemics, bizarre notions, and erratic behavior.[16]

Questions regarding replacement of professors, a chair for Frédéric Cuvier, Geoffroy's relations with his colleagues, and competition for specimens all figured directly or indirectly in the history of the menagerie. Meanwhile, competition for vacancies in the lower ranks of the Museum's workers could be as intense as at the top. Such was the case in 1834 of the widow Robin, who sold cakes at the menagerie's entrance. Within five days of her death, the administration received six applications for her spot.[17] Two came from the widows of deceased gardeners and a third from a gardener on behalf of his wife. The other three were from animal keepers, two on behalf of their wives and one, whose wife was dead, on behalf of his eldest daughter.

The applications from the animal keepers offer a glimpse of their straitened circumstances. Géan had six children. Féret, working as a handyman and a substitute for other keepers as needed, had two infirm, septuagenarian parents who depended entirely on him and his wife for their subsistence. The case of Jean Baptiste Roux was the most desperate. His wife (who had also sold cakes at the menagerie) had died ten months earlier, leaving him to care for their five children. He pled at the time for the cake seller's position to be awarded to his eldest daughter, who was only twelve. In the deliberations over that vacancy and the post that later became available with Madame Robin's death, that the Roux family had enjoyed the patronage of the Cuviers may have been a hindrance rather than a help in the Museum's political ecology. Geoffroy displayed an animus against Roux. Ultimately the keeper's daughter was awarded a post, but an even lesser one than cake seller.[18]

Family and interpersonal relationships, alliances, and rivalries extending over decades and generations were dynamic forces in the life of the Museum. For Frédéric Cuvier, who supervised the menagerie throughout virtually his whole career, the benefits, obligations, and consequences of family ties were inescapable. He owed his appointment to them, but also the awkwardness and antagonism that developed when Geoffroy, his superior, ceased to be a friend of his family. It didn't help that Blainville bore a strong animus toward Georges Cuvier and only momentarily pretended that his posture toward Frédéric was any different. Through it all, Frédéric sought to steer his own course by prioritizing the menagerie animals, placing himself at odds with his cabinet naturalist colleagues, including his brother, Blainville, and Geoffroy, all content to work with dead specimens. In 1832 the Museum passed over Frédéric's request for a chair devoted to the study of living animals, their faculties, and how they could be improved through breeding and training. Five years later, his prospects were enhanced when a friend of the family became minister of education. Schooled in the importance of family connections, Frédéric did not let the opportunity pass him by.

[4.2]

The Animal Closest to Humans

At the end of the eighteenth century, the animal positioned at the very boundary between the animal kingdom and humankind, "the animal closest to man," was the orangutan. Not that it had been privileged over the chimpanzee: the distinction between orangs and chimps was still being sorted out. Dutch naturalists had pioneered the study of the great apes, thanks to specimens available to them through the intersection of Dutch trade routes and the apes' natural habitats. French naturalists were also keen to have specimens. In the summer of 1795, hoping to reform ape taxonomy, Étienne Geoffroy Saint-Hilaire and Georges Cuvier studied ape skeletons and skulls, some from the great natural history collection confiscated from the Dutch Stadtholder. Then, thirteen years later, Frédéric Cuvier had the chance to study a young orangutan live. The three naturalists each approached the species according to the methods they preferred. Geoffroy and Georges Cuvier focused on bony structures; Frédéric Cuvier focused on behavior.

The first European to describe the orangutan, Jacobus Bontius (Jacob de Bondt), was a physician/naturalist working for the Dutch East India Company. Appointed overseer of the company's medical affairs in 1626, Bontius traveled to Java, where his knowledge and skills were taxed to the utmost. He lost family and friends to local diseases. He too succumbed in 1631, but not before writing several treatises on medical topics and a natural history of Java. The latter, published posthumously in 1658, included his account of seeing hairy, reddish apes that walked on two feet—what the Malays called *ourang-outang* ("man of the woods"). Bontius provided an illustration of a female orangutan, noting that females seemed to display a sense of modesty. According to Bontius, the natives believed orangutans could speak but that they hid it so humans would not try to enslave them.[1]

In the mid-1700s Buffon remarked that the Dutch scientist's illustration of a female orangutan looked more like a woman than an ape.[2] He questioned Bontius's suggestion about the creature's modesty and gave no credence to native beliefs that it could speak.[3] But, as Buffon acknowledged,

Figure 4.2.1. "The orangutan." S. Fokke after P. C. Haag. From A. Vosmaer, *Description de l'espèce de singe aussi singulier que très rare, nommé orang-outang, de l'isle de Borneo* (1778). Courtesy of Linda Hall Library of Science, Engineering and Technology.

making sense of the wealth of reports on monkeys and apes in the literature of natural history was far from straightforward. Much that had been written was unreliable, "obscure, faulty, exaggerated." He did his best to clarify matters but was ultimately unable to sort out the differences and similarities between the animals that had been called "jockos," "pongos," and "ourang-outangs."[4] He initially thought they belonged to one species, but twenty-five years later he declared that there were at least two species of orang-outang, a smaller (the jocko) and a larger (the pongo). Unwittingly, he had identified juvenile orangutans and juvenile chimpanzees as belonging to one species and adult orangutans and adult chimpanzees as belonging to another. These, his final thoughts on the matter, remained in manuscript form at his death in 1788. Lacépède published them in volume 7 (1789) of the supplements to Buffon's *Histoire naturelle, générale et particulière*.

In the meantime, two Dutch naturalists, Arnout Vosmaer and Petrus Camper, provided new observations on orangutan specimens from the East Indies. Vosmaer, director of the Stadtholder's natural history collection and menagerie, had gone to great lengths but failed to confirm reports of orang-outangs five feet tall or more.[5] In the summer of 1776, when he received a live young female orangutan from Borneo, her height corresponded to that of all the other specimens Europeans had encountered: two to two-and-a-half feet. Vosmaer believed she was only the second orangutan seen alive in Holland, the first being a specimen from Angola presented to the prince of Orange and Nassau (the Stadtholder) and described by Nicolaes Tulp in 1641.

Tulp's animal was most likely a chimpanzee, but Vosmaer believed it and his orangutan were "certainly of the same species." He cared for the orangutan for a month before the crush of individuals wanting to see her persuaded him to give her to the Stadtholder's menagerie (28 July 1776). Her health held up for several months, but she fell ill in November, "according to the fate," said Vosmaer, "that seems to be attached to this menagerie." She shivered and had severe dysentery. "A small dose of rhubarb" seemed to bring a recovery, but it was brief. She died on 22 January 1777 from a malady of "languor and consumption."[6]

In his published account (1778), Vosmaer credited the orangutan with a very gentle disposition, noting how she responded to caregivers and visitors, how she walked, used her hands and feet, unbuckled boots, untied knots, and so on. He paid particular attention to her food preferences and dining habits. She learned to use a knife and fork, and Vosmaer, like other ape owners, enjoyed watching her: "When she was given strawberries on a plate, it was a pleasure to see how she stabbed them one by one and carried them to her mouth with a fork, while she held with her other hand or paw the plate or saucer." The orangutan usually drank water, but she liked wine. "Given a bottle, she pulled the cork from it by hand and drank very well from it just as from a beer glass, and that done, she wiped her lips like a person, either simply with a hand, or with a linen."[7]

The orangutan's preparations for sleeping left the greatest impression on

Vosmaer. "When she wanted to go to bed, she arranged with care her customary bedding, shook it well, brought more of it to make her bolster, put herself most often on her side, and covered herself warmly with a blanket, being very sensitive to cold, although being accustomed at the Cape to a colder climate than that of the East Indies." On occasion, she folded a piece of linen to use as a pillow.[8]

Vosmaer did not confront the issue of whether the orangutan was capable of thought or mention the earlier suggestion that orangutans could talk. He may have felt no need since Buffon had already pronounced on the subject, comparing orangutans and humans:

> The tongue and all the organs of the voice [of the orangutan] are the same as in man, and yet the orangutan does not talk; the brain is absolutely of the same form and the same proportion, and it does not think. Is there more evident proof that matter by itself, although perfectly organized, is unable to produce neither thought nor the word that is a sign of it, unless it is animated by a superior principle?[9]

In the 1770s while Vosmaer was trying to find a five-foot-tall orangutan, his multitalented countryman, the physician and anatomist Petrus Camper, undertook the anatomical examination of several ape specimens. Dissecting at least five representatives of what he termed the "true" orangutan or "orang-outang of Bontius" (orangutan of Borneo), Camper concluded they were of a different species from the creature Tulp described in 1641, or from the "orang-outang" or "pygmy" Edward Tyson described in 1699. As to whether the orangutan could talk, after dissecting the vocal organs of orangs and similar species Camper declared them incapable of speech.[10]

Geoffroy Saint-Hilaire and Georges Cuvier had a copy of Camper's work when they coauthored "The Natural History of the Orangutans" (1795), but their immediate inspiration was the hoard of natural history treasures in the Stadtholder's collection just arrived in Paris.[11] These included the skeleton of the "pongo of Wurmb," or "large orangutan," and the skeleton and stuffed skin of a smaller orangutan, which Faujas noted the pongo "did not at all resemble."[12] A conspicuous conclusion that Geoffroy and Cuvier reached in their paper was that Wurmb's pongo, Camper's "large orangutan," was not an orangutan at all but was of a wholly different genus. Rather than positioning Wurmb's pongo with the animals closest to humans in structure—the orangutans, jockos, and gibbons—they placed it much further down the scale, with the baboons. Their mistake was realized only twenty years later, when Wurmb's pongo was recognized as an adult orangutan and the other specimens as juveniles.[13]

Geoffroy and Cuvier's paper omitted many topics that had vexed earlier naturalists. They scarcely mentioned the animals' habits and did not linger over the fact that some specimens naturalists had described were badly mounted, poorly illustrated, and possibly even fake. They did not grapple with whether orangutans, jockos, and pongos constituted a single species or two or more, nor did they address the orangutan's powers of speech. They did acknowledge that apes and monkeys had embarrassed philosophers by the difficulty of identifying "a definite limit between those of their actions that depend on organs resembling ours and those of our actions that derive from a superior faculty." These animals had likewise embarrassed naturalists when it came to distinguishing species and grouping them according to their "true natural relations."[14]

Another embarrassing problem, though the two naturalists did not characterize it that way, was the origin of species. Surprisingly, considering the firm stand Cuvier would take against the idea of species mutability, the young authors in 1795 referred to two models for the introduction of new organic forms. Impressed by what they took to be the great many kinds of monkeys and their apparent fecundity, they wrote:

> It is a rather general law of nature that the number of species in a genus is approximately in proportion to the fecundity of each of them, whether what we call species are only the diverse degenerations of a single type, which have had to multiply themselves in proportion to more or less frequent births; or whether many among them are born from the coupling of neighboring species, with the efficacy of these sorts of mixtures depending on the generative force of the species mixed.[15]

They did not name the advocates of each view, but naturalists of the day knew them well. The idea of the production of new species via "degeneration" from earlier, primordial forms was Buffon's. The idea of the production of new species via the hybridizing of existing forms was Linnaeus's.

Geoffroy and Cuvier's paper was not about the introduction of new species. They were offering a new arrangement of the various groups of apes and monkeys according to their "natural relations." Identifying "natural relations"—the great goal of eighteenth-century naturalists—involved classifying animals and plants so that the species most alike were grouped together while those least alike were placed far apart. This also applied to the taxonomic levels: genera, families, orders, and classes, and it was no simple task.

Georges Cuvier's early fame as a naturalist came from his success in developing and deploying the notion of the subordination of characters, the idea that some characters should weigh more than others in deciding which species, genera, and families are most alike. In addressing "the great family of monkeys," Geoffroy and Cuvier explained, it was necessary to begin by making subdivisions, or "cuts," in this general grouping, but these could not be arbitrary: "[So as not to] bring together very different species while separating them from very similar ones, it was necessary to seek the bases [for these divisions] in the most important differences of structure, the ones that have the greatest influence on the whole animal system."[16]

Buffon, in his efforts to group the various species of monkeys, had used several characters: the tail (whether it was long, short, prehensile, or absent), cheek pouches, and callosities on the buttocks. Geoffroy and Cuvier thought that Buffon had done the best anyone could have done at the time, but his choice of characters still produced groupings that were obviously not natural. The young naturalists suggested that a more reliable preliminary guide to the natural relations of the monkeys was the size and especially the shape of their skulls, particularly how elongation of the muzzle affected the relation between "the volume of the skull and that of the face." This was "of the greatest importance in the animal economy, since it seems that the magnitude and convexity of the skull indicate sensibility, as much as the elongation and thickness of the muzzle indicate brutality."[17]

Geoffroy and Cuvier confidently maintained that the significance of skull shape was displayed just as clearly in apes and monkeys as it was in diverse human races. The orangutans, with their round, vaulted heads and short, flat faces, were remarkable for "their gravity, dexterity, and intelligence." As one went down through the other genera, however, one found less intelligence and more petulance, until finally one reached the mandrills and baboons, those animals with doglike muzzles, disgusting colors, ferocious temperaments, revolting brutality, and infamous vices. "The same series of relations" was likewise to be found among "the diverse races of humans"; or, as they put it: "We do not see at least that any of the peoples with depressed foreheads and protruding jaws have produced accomplishments equal to the general level produced by Europeans by way of mental faculties. We are so accustomed to this connection between the proportions of the head and the qualities of mind that the rules of physiognomy that are founded upon it have become a matter of common opinion."[18]

Rather than resting their case on comfortable Western assumptions about the relationship between head shape and intellect, Geoffroy and Cuvier sought a more rigorous way to identify the natural relations of the members of the monkey family. To this end, they turned to the work of the anatomist Petrus Camper and the measurement he called the "facial angle," to which his artistic interests had led him.

Camper wanted to understand how to accurately portray the faces of people of different races while considering the ideal of beauty as represented in the heads of Greek statues. In pursuing these aims, he hit on the facial angle. He produced the angle by drawing two lines, one from the base of the nose to the ear canal, the other from the upper lip or base of the nose to the most forward part of the forehead. By his account, he found that the facial angle in humans ranged from 70 to 100 degrees, "from the head of the Negro to the sublime beauty of the ancient Greek. If you descend below 70 degrees you have an Orang Outang, a monkey; if you descend lower still you will have a dog, finally a bird, a woodcock where the facial line is found nearly parallel to the horizontal line." Camper presented tables showing how the facial angle became increasingly wide as one considered successively the skulls and heads of a monkey, an orangutan, a "Negro," a "Kalmouk" (Asian), and four "Europeans."[19]

Geoffroy and Cuvier put the facial angle to work to divide the order Quadrumana (apes and monkeys, animals with "four hands") into seven genera that they ranked hierarchically. The first genus (with the widest facial angle) they named "Orang or *Simia*." Here they placed the orangutan, jocko, and gibbon. The other genera, arranged in order of decreasing facial angle (considering the number of molars and presence or absence of a bony crest above the eyebrows), were the capuchin monkey or sapajou (*Callithrix*); the guenon (*Cercopithecus*); the macaque (*Pithecus*); the Barbary ape or magot (*Cynocephalus*); the baboon (*Papio*); and the howler monkey or alouatte (*Cebus*). Geoffroy and Cuvier decided that the "Pongo of Wurmb," which Camper had called "the large orangutan" and they called the "pongo of Batavia," had to be placed among the baboons because of its facial angle of thirty degrees.[20]

Cuvier and Geoffroy did not bother to remark on inconsistencies between their results ranking monkey genera by their physical characters and their results ranking them by the animals' temperaments. In considering physical characters, they put the howler monkeys in the lowest genus; in considering temperament, they portrayed the mandrills and baboons as even worse than the howlers. They may have thought these assessments were close enough, since both genera displayed the same low facial angle.

Although Camper had used facial angle to characterize the head shapes of individuals of different human races, he had not insisted on the correlation between head shape and intellect or morality, or on the permanence of racial types. He believed that all humans descended from the first couple, Adam and Eve, that racial differences arose from long-term exposure to different environmental circumstances, and that they were potentially subject to further change (or reversion to an earlier state) depending on such circumstances. In their 1795 paper on the broad family of orangutans, Geoffroy and Cuvier presented characters they evidently viewed as more fixed. Furthermore, the correlation they saw between skull shape, intelligence, and morality (or vice) in monkeys and apes seemed to be mirrored in what they observed in diverse races of humankind.

As of 1795 neither Geoffroy nor Cuvier had seen a live orangutan. It was not until thirteen years later that a living representative of the species, a young female, arrived in Paris. She was a gift to Empress Josephine from a French naval officer named Decaen, brother of the governor of Isle de France and Bourbon. The orangutan's journey lasted eight months and included a bout of frostbite in the Pyrenees. Too sick and exhausted to be taken directly to Malmaison, the young animal was entrusted to Decaen's friend, a barrister named Godard, who invited Frédéric Cuvier to study her behavior during an attempt to nurse her back to health.[21]

In his description of the orangutan (1810) Frédéric Cuvier was enthusiastic about her intelligence, notwithstanding her illness (which prevented any serious experiments) and her youth (her intellectual capacities had not yet fully developed). He did report an experiment (without saying who conducted it) "to see what impression our music would make on this animal." The unsurprising result was that the orangutan was as unmoved as might be expected, said Cuvier, since there was nothing about the organization or the lives of mammals to suggest their susceptibility to music's charms. "Even for us," Cuvier elaborated, "[music] is only an artificial need . . . never does it have any effect on savages other than that of noise."[22]

While assuring his readers that the orangutan was "surely not a human," Frédéric Cuvier concluded that its astounding faculties placed it closest to humans in intellect.[23] The young female impressed Cuvier with her ability to learn from simple trial-and-error experiences. Furthermore, she seemed able to generalize. When she climbed a tree and wanted no one to bring her down, she would shake the branches violently as if to frighten whoever tried climbing up after her. She repeated the behavior whenever the situation was replicated. "However one envisages [this action]," wrote Cuvier, "it will be scarcely possible not to see the combination of rather high-level ideas, and not to recognize in the animal capable of them the faculty of *generalizing*" (Cuvier's italics).

Cuvier supposed that the orangutan had reasoned from her own experience (including on the ship that brought her to Europe) that the fear of falling could deter others from climbing. Having observed that the violent shaking of whatever she was climbing could make her lose her balance and fall, "she judged that, in analogous circumstances, others would experience what she had experienced: that the fear of falling would prevent them from climbing higher. She transferred to other beings that were strangers to her an idea that was personal to her: from a personal experience she ultimately made for herself a general rule."[24] While Cuvier saw this as evidence that the orangutan could generalize, he was in effect crediting the animal with something more extraordinary, what primatologists today call "theory of mind," the ability to reason about the minds of others.[25]

Another example Cuvier gave of the animal's ability to generalize was the way she moved a chair so she could reach a door latch to open the door into the next room. She had never been taught to move a chair so as to open a door, nor had she ever seen anyone do so. She learned, however, that chairs were movable. She thereby displayed the ability to combine separate pieces of knowledge to arrive at a solution. Wrote Cuvier, "How could one not recognize in this action the faculty of generalizing? . . . I do not believe that any other animal has carried the force of reasoning further."[26] The way the orangutan wrapped herself in a cover for sleeping struck him as another "very nice proof of her intelligence, and to prove that she can both generalize her ideas and be aware of a future need, a sense of foresight."[27] He did not mention that Vosmaer had described the same behavior in his orangutan.

Cuvier also credited the orangutan with exhibiting intelligence in moments of pique when her desires were frustrated. Often when refused something she wanted, wrote Cuvier:

> She struck her head on the ground as if to excite interest or pity. This manner of expressing sadness or anger is not seen, I believe, in any other animal, and it is found in man. Would this orangutan have been led to act thus by motives that lead us sometimes to act the same way? That is what we are permitted to believe, since in her anger she lifted her head from time to time and suspended her cries to look at the people nearby and see if she had produced any effect on them, and if they were disposed to give in to her; when she believed she noticed nothing favorable in their looks or gestures she began crying again.[28]

Pleased with what he had learned about the orangutan, Frédéric Cuvier concluded that one would not have to observe and experiment much more "to obtain general and exact ideas on the intellectual faculties" of the whole class of mammals. Having studied the animal "that comes closest to man," he said it would be his point of comparison for "all the other species of mammals" if circumstances allowed him "to continue the research

I commenced long ago on the intellectual characters that distinguish one species from another."[29]

Once she was moved to Malmaison, Cuvier never saw the orangutan again. The memoir of Empress Josephine's personal attendant notes that she was dressed in a frock coat and received visitors "with a demeanor full of decency and modesty." She ate at table with a knife and fork, and she liked cutting turnips, one of her favorite foods. "When she had dined, one of her great joys was to cover her face with her napkin, then pull it away, making a thousand funny grimaces." When she fell ill, she was put in a bed and dressed in a long-sleeved camisole, "as a woman would be." She pulled the blankets to her chin, leaving her arms outside the covers. She shook hands with people she knew, amusing them with an air of connoisseurship as she drank her tea. When a favorite beverage did not come soon enough, she would grow agitated, "gesticulate with all the little rage of a child, throw her cover to the foot of the bed, pull her keeper's clothes," and fuss until she was given what she wanted.

Judging from these recollections, the orangutan was a source of amusement at Malmaison, a marvelous human impersonator, aided by humans' outfitting her with the trappings of human society. It seems unlikely anyone at Malmaison would have disagreed with Frédéric Cuvier's high opinion of the young female's mental capabilities.[30] But Cuvier himself soon backtracked on the subject. It's easy to imagine that someone, probably his brother, took him aside and cautioned him about attributing too much intelligence to the orangutan. In addition to believing in a qualitative difference between humans and animals and the importance of making that distinction crystal clear, Georges Cuvier was aware that his colleague, Lamarck, had recently speculated on how an orangutan or similar creature, in taking up new habits, could have developed the faculty of speech. The elder Cuvier may have impressed on Frédéric the importance for their reputation of not letting any scientifically, politically, or theologically questionable position be attached to the family name.

Lamarck's introduction of the orangutan in his own writings had been much bolder. In 1802 he capped the first major presentation of his ideas on evolution with "some considerations relative to man." He prefaced these by allowing that "man alone is endowed with reason, and that by this consideration he is rightly distinguished from the other productions of nature," but he then went on to suggest that the distinguishing characters of man's physical organization vis-à-vis other primates are characters that could have developed from an apelike ancestor if favorable circumstances led the creature to take up new habits.[31] Lamarck elaborated on these ideas in his *Philosophie zoologique* (1809):

> If man were distinguished from animals only by his organization, it would be easy to show that the characters used to form a separate family of man and his varieties are all products of ancient changes in his actions and of habits he took up that have become particular to the individuals of his species.

Lamarck speculated that if an ape like the "the orangutan of Angola" [the chimpanzee] had come down from the trees and acquired the habit of walking, it would, over time, have stood upright. If it no longer used its teeth as weapons but only for chewing, its facial angle would have become broader and its muzzle shorter. As its communities grew larger, its needs would multiply, leading members to communicate by signs and eventually articulate sounds. The creature would effectively have evolved the ability to speak.[32]

Lamarck also explained why there would be a gap between this dominant race of animals and all the rest. The gap would be produced and maintained by the highest form of life's driving away its competitors to environments unsuited to the development of the kinds of habits needed for their higher mental faculties to advance.

When Frédéric Cuvier next addressed the subject of the mental faculties of animals, in January 1812 at a session of the Société philomathique, he took care to identify "the principal differences" between humans and the highest animals "with respect to intelligence." Acknowledging that "the animals have, like us, the faculties of being attentive, feeling, judging, remembering, etc., and that all our actions in which we use only these faculties are absolutely similar to theirs," he located the difference between humans and animals in the "faculty of meditating or reflecting," which he claimed was unique to the human species.[33] This faculty, he said, was what allowed humans to have language, abstract ideas, and the ability to combine ideas and judgments "to infinity."[34]

In his later writings, Cuvier took care to stress the fundamental differences between humans and animals. In an 1822 article on instinct, for example, he wrote, "all animals, without exception, are deprived of the intimate sense of self-perception and the faculty of reflection."[35] But when he wrote about the orangutan in the mid-1820s, the question of its intellectual capacity's similarity to human's still troubled him. Likewise, the idea of species change refused to go away, and the question of where best to study animal behavior remained up for debate.

While Cuvier championed the menagerie, others promoted field study, including for orangutans. This was the case for the writer J. J. Virey, an earlier believer in the close proximity between apes and humans who observed of orangutans in 1803, "to recognize how much they are like man, and at what point they differ," one would have to observe them in the "ancient and silent forests of Asia, under tufted palm trees full of fruits," and "to follow them in their games, their loves, their natural habits."[36] Likewise, the naturalist Jean-Baptiste Bory de Saint-Vincent extolled the virtues of studying orangutans in Africa. Fame, he indicated, would forever accrue to the naturalist who dared undertake "a voyage whose purpose would be to penetrate the vast forests where the chimpanzees live in society, to observe their habits [*moeurs*], to deepen their [natural] history, to attempt the moral improvement of diverse individuals in constituting just how far it could be taken, and finally, to attempt the crossing of [the chimpanzee] with neighboring species, whether higher or lower."[37]

Crossing a chimpanzee with a higher species? Could Bory have meant anything other than humans? And anything other than what he considered the lowest race of humans (for him, the Hottentots)?[38] His startling suggestion reveals how far his racism extended. His point about the value of field studies, however, is one on which modern primatologists would agree: to understand the lives of the higher apes, they must be studied in their own communities and native habitats. For a European naturalist in the early nineteenth century, however, this could be no more than a pipe dream. The hazards of travel and tropical disease militated against it. Frédéric Cuvier's position was that animal behavior was best studied not in the field but under conditions over which the observer had control, specifically, in a menagerie. The problem with monkeys and apes as subjects for study in the Paris menagerie was how soon they succumbed to the diseases and other conditions of their existence in north temperate Europe. Their fate in effect mirrored that of a high percentage of the Museum's voyager naturalists in the tropics.

[4.3]

Toward a Science of Living Animals

By the spring of 1808 when Frédéric Cuvier studied the orangutan destined for Empress Josephine, he had supervised the menagerie for four years, and his confidence regarding what it could accomplish scientifically had grown. In his 1804 guide, he proposed that menageries could aid zoologists' efforts to distinguish between species and host domestication projects. In his first published scientific paper (1807) on animals' mating seasons, he more boldly proclaimed, "menageries can be for zoologists what the laboratory of the chemist is for those who study inorganic bodies. We see there not what happens in nature, but what could happen." One could learn the laws of animal behavior better in a menagerie than in the wild, he maintained, because one could control conditions in a way that was otherwise impossible.[1]

The leading eighteenth-century writers on animal behavior would not have agreed. Buffon, Hermann Samuel Reimarus, and Charles-Georges Leroy all doubted that a menagerie was an appropriate setting for animal behavior studies, despite their different perspectives. Buffon viewed animals as unthinking, Cartesian automatons. Reimarus (professor of Oriental languages at the Hamburg Academic Gymnasium) thought God equipped animals with the instincts necessary to satisfy their needs. Leroy (keeper of the king's game at Versailles and Marly) saw animals as intelligent creatures, capable (within the limits of their constitutions) of feeling, comparing, judging, reflecting, and making choices. All dismissed the idea of making a practice out of studying domestic, tame, or captive animals to learn about animal lives. Reimarus put it this way:

> The histories of tame or locked up animals offer only uncertainties. . . . It is not based on the actions of individuals that have lost their natural liberty that one can judge their true instincts. In leading a way of life that is not natural to them, their instincts vary, degenerate, or die out entirely. It seems to me much more certain to speak only of animals that enjoy their

Figure 4.3.1. "Genet of Senegal." C. de Lasteyrie after Jean-Charles Werner. É. Geoffroy Saint-Hilaire and F. Cuvier, *Histoire naturelle des mammifères*. General Research Division, the New York Public Library.

> natural state of liberty; discomfort and constraint have no influence on their actions.[2]

Buffon expressed a similar view in 1771:

> If some sovereign having conceived the truly grand idea of contributing to the advancement of this beautiful part of science by forming vast menageries and gathering there under the eyes of observers a great number of living species, one would still get imperfect ideas of Nature there. The majority of animals, intimidated by the presence of man, bothered by his observations, tormented moreover by the uneasiness that comes with captivity, would only display habits altered, constrained, and scarcely worthy of the attention of a philosopher.

The only Nature worth contemplating, Buffon said, was a Nature "free, independent, and, if you will, wild."[3]

As for Leroy, the keeper of the king's game, he was first and foremost a field observer who, as we have seen, characterized himself as "a dedicated hunter" who had "taken his course of philosophy in the woods."[4] Studying captive animals in a menagerie to learn about their lives would have been completely foreign to him.

If Frédéric Cuvier was optimistic about what a menagerie could become, he was realistic about its existing state. His 1807 paper acknowledged that living conditions for some animals were dreadful. The carnivores, he said, are "always subject to the variation of the seasons, always fed the same food, always constrained in the narrowest of prisons, alternatively exposed to the extremes of heat, cold, humidity and disgust, and their life is in fact only a succession of suffering."[5]

Cuvier followed his first paper with a rush of others, some dealing with zoological classification, others relating to menagerie animals, including two (1808) case studies of interspecific mating. The first ended badly. Mated with a horse, the menagerie's female zebra died in the eighth month of gestation, a major loss owing to both her beauty as a specimen and the scientific hopes regarding the crossing. She had previously mated successfully with a donkey, and Cuvier wanted to try mating the zebra-donkey hybrid with a zebra-horse hybrid to see if an additional hybrid generation could be produced.[6]

The mating of a wild whooper swan with a domestic goose produced a viable hybrid but was nothing special, since wild birds often mate with domestic ones. Cuvier nonetheless was enthusiastic that the cross was an indication that scientists could learn enough about animals' intimate lives in menageries to recognize the laws governing them, which could then be applied "according to our needs." Understanding why some varieties either shunned or mated with each other, he said, would benefit natural history and agriculture alike. It would help resolve "the very important question of the distinction of species" and provide the means of "reducing [wild animals] to the state of domesticity."[7]

In another paper (1808), Cuvier showed how a menagerie could help untangle issues of species identification and classification.[8] The gull specimens Louis Baillon sent to the Museum had matured, enabling Cuvier to illustrate how forms that naturalists had mistaken for two different species were developmental stages of the same species. He offered a similar example of such errors in an 1809 study of a sea eagle.[9]

Studying reproduction and hybridization, advancing the grand project of animal domestication, and sorting out problems of species identification were familiar activities for naturalists, but Cuvier had a more novel goal in mind. He effectively wanted to create a science of comparative animal psychology based on observations and experiments with the intellectual faculties of the menagerie's animals.

He addressed the topic in an 1808 paper on a young female dog brought from Australia by the Baudin expedition. Audacious and incorrigible, she stole everything she wanted aboard ship, and punishment did not cure her of the habit. (A young male of her type taken aboard behaved similarly and died from the beatings it received.) At the menagerie she continued to be intractable. Seemingly unable to measure her own strength against that of other animals, she fought with much larger dogs and would hurl herself against the cages of a panther, a jaguar, and a bear when these animals menaced her. Cuvier suspected this temerity was characteristic of her race, not just a lack of experience. He was furthermore impressed that, "quite unlike our domestic dogs, this one has no idea of human property."[10]

He introduced his paper on the Australian dog with "some reflections on the moral faculties of animals."[11] In the study of the mental faculties of higher animals, he maintained, it was no longer a question of whether they were intelligent. "It is generally admitted," he wrote, "that at least those that come closest to man feel, judge, and choose to act." The question was how "to establish with precision" the role instinct played in the life of each animal "and to distinguish exactly the [animal's] natural qualities from those that are only acquired."[12]

The literature on the subject, Cuvier said, generally agreed that instinct impelled rabbits to dig burrows, beavers to build lodges, hamsters to store provisions for the winter, and so forth. But these behaviors could be modified by circumstances. Rabbits chased by ferrets would cease to dig burrows and would live like hares. Likewise, rabbits domesticated for many years would lose the faculty of burrowing entirely. Beavers living in isolation would excavate simple dens in riverbanks instead of building lodges. Conversely, behavior that was originally learned could become as fixed and mechanical in nature as behavior that was instinctive. Different breeds of dog, said Cuvier, displayed this phenomenon clearly. The retriever needed no training to jump into the water, but it was only with great difficulty that one could train a wolflike dog to do so.

Cuvier believed there were enough such instances to take as "a truth" the inheritance of acquired characters, the idea that "some of the qualities regarded as belonging to instinct in mammals are under the same laws as those depending on training [*l'éducation*]," and that the latter finally become instinctive or hereditary after they have been exercised over a sufficient sequence of generations but would wear away and disappear when no longer strengthened and sustained by use.[13]

Cuvier did not mention here the name of Lamarck, the zoologist and evolutionist with whom the inheritance of acquired characters is now most associated. Instead, he credited the idea to Charles-Georges Leroy, the philosophically inclined gamekeeper, a man recognized for his "long experience and profound sagacity." He cited Leroy's observation that fox pups, emerging from their dens for the first time, are "more defiant and more circumspect in places where people make much war on them than are old foxes in those where one does not set traps for them."[14]

Despite his remarks about using experiments to study behavior, Cuvier suggested no protocols for determining which actions a species performs instinctively and which depend on experience. He turned instead to a discussion of the conditions conducive to modifying animals' intellectual faculties. Animals that lived in societies, as he saw it, had the most opportunities "to perfect themselves" (either as individuals or as species). However, lacking control over their own circumstances, they had no way to develop for themselves "an artificial education." That happened only when they became associated with humans. Even then, not all animals were "susceptible of the same education." The dog stood out as the species that was both most educable and most submissive to humans.[15]

Believing that dogs domesticated for only two or three centuries would have different characteristics than those domesticated for three or four millennia, Cuvier proposed a comparative study of the mental faculties of the different dog races of "barely civilized people," specifically the dogs of "Australia, Patagonia, New Zealand, Siberia, Lapland, Iceland, etc." Naturalists knew these races primarily by their physical forms, he said, but it was worth studying their behavior: "All these races, well studied, in offering us the successive development of the intelligence of their species, in giving us the means of constructing their comparative psychology, would lead us to precious results, even for the psychology of man."[16] Cuvier did not specify how this might inform human psychology, but he seems to have expected that the levels of intelligence displayed by different dog races would correlate with the intelligence of their owners.

In producing a new edition of his guide to the menagerie, Cuvier developed his rationale for menageries as sites for scientific research: "Menageries have been criticized on the grounds that the animals are greatly constrained compared to the liberty they enjoy in nature, and it has been concluded that [these animals] were useless for knowledge of the most important phenomena of the animal kingdom—the ways of life, the industry, the character, and, in word, the intellectual faculties in general." That criticism was groundless, he explained, because "nature reigns everywhere. The laws that direct animals in the middle of the desert or in the forests are still the same they obey when they are among us." Where Cuvier had previously drawn an analogy between zoology and chemistry, he replaced chemistry with physics. We can subject animals to all sorts of circumstances, he said, but we cannot force them to behave in ways other than those prescribed for them, any more than we can "prevent a falling stone from following the laws of gravity, although otherwise we can modify its motion infinitely." The challenge for science consisted in "appreciating these modifications and measuring them with exactitude."[17]

Warming to his physics/zoology analogy, Cuvier cited the apocryphal tale of Newton and the apple, from which, purportedly, "Newton deduced the motion of the universe," and speculated that "he would perhaps have needed only the slightest act by his dog to assess the intellectual faculties of the entire species, and perhaps even all animals." More plausibly Cuvier asked,

> Does one believe that Galileo would have recognized the law of gravitation, if he had not manipulated bodies himself, if he had not subjected them to modifications, perceived in advance, prepared by reason, and without which the phenomena, having no connection between them, would have remained perpetually for us in the most profound obscurity? Why then should the animals of menageries not also become a source of experiments?[18]

Experiments on menagerie animals would "compel them to offer us observations allowing us to understand, explain, and connect" the phenomena displayed by animals living freely in nature." Without the benefit of experimentation, ideas on the intelligence of animals had ranged from too limited to too extravagant. The solution, he urged, was to study animal intelligence under controlled conditions.[19]

Yet Cuvier's studies of animal behavior in this period were not sufficient to prove his point. Neither his observations on the empress's orangutan in 1808 nor his studies of the seals Baillon sent to the menagerie in 1809 involved the rigorous experimentation Cuvier called for in his confident rhetoric about "the menagerie as laboratory." The orangutan's illness left her too weak to be experimented with; the seals of 1809 did not last long enough to permit sustained experiments. Cuvier recognized the need for healthy specimens that could be studied over extended periods but also for special sites for studying them under controlled conditions, a requirement that rose to the fore in the 1815 contest over the use of the elephant's former park.[20]

Cuvier's bid for the space obliged him to spell out to the administration the issues he hoped to study experimentally. The novelty of his vision contrasts sharply with Geoffroy's report for the government on the menagerie's past development, current needs, and long-term plans. Geoffroy rightly enough considered the menagerie his domain both historically and administratively, but his zoological practice did not depend on it. His report surveyed the menagerie's founding and early development. He noted that whereas the bears and herbivores now had their "definitive quarters," the *animaux féroces* still did not; but on balance he felt the menagerie had enough space to fulfill its basic functions.[21]

In writing of those functions, he made no mention of the menagerie's scientific utility. The primary purposes of a menagerie in a large city, as he represented it, were "the amusement and instruction of the public." The deer in their parks were simply the counterparts of birds in cages serving for "the amusement of households." Key to the public's enjoyment was always having the animals in view. As Geoffroy put it, "l'animal doit en tout temps être offert à sa curiosité." Beyond that, the Museum tolerated the public's desire to feed the animals (and even, to some degree, to tease them). As for providing information about the menagerie's creatures, however, the Museum's help was minimal. A visitor could expect little more than an animal's name on a cage or what might be gleaned from a talkative keeper. On balance, the Museum was content to have visitors come away from the menagerie happy to have had the chance to see diverse live, exotic animals side by side.[22]

Geoffroy was not troubled by the complaint that animals' behavior in menageries was altered by confinement. He admitted that a deer in a forest would behave more freely and naturally than a deer kept to a small paddock, but he cheerfully suggested that menagerie animals enjoyed benefits of society that their wild counterparts lacked. Substantial expansion was unnecessary, he said, because a menagerie did not need to be a game preserve or even the size of the extensive former royal menagerie at Versailles.

Frédéric Cuvier's ideas for the menagerie were more far-reaching. When the elephant park was vacated, he wrote a seven-page request "to make experiments there" and "to transform it into a kind of laboratory where animals would be placed under conditions appropriate for producing the results one would seek."[23] With the menagerie's animals as his subjects, he wanted to study not only physical phenomena like bodily development and function but also mental phenomena. He was particularly interested in how far individual development could be influenced by environmental circumstances. The issues were how far changes could go and whether experimental studies in this regard could fix more precisely the meaning of the word species.

Along with the influence of external causes on bodily changes, Cuvier wanted to study changes related to age and sex. He wanted to compare the senses of different animals to learn, among other things, whether animals born with their eyes open were immediately able to see things and judge their distances. Experiments, Cuvier maintained, would "definitively resolve" this important question and hundreds more like it, including those related to "rutting, copulation, gestation, etc."

When it came to intellectual phenomena, the crucial task, Cuvier said, was to distinguish between those depending on intelligence and those arising from instinct. Previous authors had typically written to support one intellectual system or another, depending on whether they were Aristotelians or Cartesians, atheists or Christians, or inclined to attribute an animal's behavior entirely to instinct or to intelligence. Cuvier called for unbiased observations on which to build the foundations of a new science, "fertile in its applications to our industry, and even useful for the knowledge of human intelligence."[24] Intellectual phenomena, he insisted, were as appropriate for experimental study as the functioning of any organ, since they were the products of a specific organ, the brain.

The administration asked Geoffroy to review Frédéric Cuvier's proposal, but events intervened. Napoleon was defeated, foreign troops entered Paris, and the Museum had to focus on the occupation and on responding to countries demanding the return of their confiscated natural history collections. The professors' assembly's minutes mention no action regarding the experimental park, but an understanding seems to have been reached, since the minutes of 30 August 1815 note Cuvier's request that an order be given "to open the bay leading from the old stable of the water buffalo to the park for experiments."[25] However, Cuvier did not designate a place for his park on the map he prepared in October of the menagerie and the new parks it needed.[26] Nor did Gabriel Thouin's meticulous map of the Museum in 1819 include a park for experiments. Where the elephant park used to be, there is a blank space. J. P. F. Deleuze's history of the Museum (1823) includes a slightly revised version of Thouin's map that identifies this space as "Jardin pour les expériences," but this evidently represented an idea, not a reality.[27] A Museum map of 1842 shows the same area simply as green space with some trees.[28] For the rest of his career, neither Cuvier's menagerie reports nor his publications mention an experimental park.

Why did the experimental park never materialize? Lack of support from Geoffroy was perhaps one thing. The difficulty of identifying experiments to do along with the setup needed for them could well have been another. Beyond that, Cuvier had other responsibilities and projects in addition to those of menagerie superintendent to keep him busy.

From 1810 onward he was actively involved in French educational administration.[29] In 1816, after a hiatus of more than a decade, he resumed editing the *Dictionnaire des sciences naturelles*. In 1818 he began publishing with Étienne Geoffroy Saint-Hilaire their *Natural History of the Mammals*. Appearing in installments and handsomely illustrated with engravings from the Muséum's collection of vellum paintings, it was a production like the earlier work, *La Ménagerie*, produced by Lacépède, Georges Cuvier, and Geoffroy. Frédéric Cuvier wrote virtually all the articles of the new work.[30] He also continued classificatory work based on anatomical characters. His substantial volume *The Teeth of Mammals Considered as Zoological Characters* (1825) was the culmination of work begun in 1808.[31] He would also produce a volume on the natural history of cetaceans (1836).

Though Cuvier's papers from 1815 to 1822 did not feature animal behavior or intelligence, he still hoped to establish a science of living animals.[32] In a long article on instinct for the *Dictionnaire des sciences naturelles* (1822) he acknowledged that the science of animal intelligence was in its infancy and that the principles for studying it did not yet exist.[33] He was nonetheless prepared to describe the kinds of facts he deemed indispensable for such a science, insisting again on the importance of distinguishing between instinctive actions and those resulting from experience. He also stressed the importance of establishing "the presumed point of separation between the intelligence of the human species and the intelligence of animals." This, he said, "ought to be the principal goal" of the kinds of studies he was discussing.[34] He had clearly learned the lesson that the line between animals and humans must not be allowed to seem blurry.

Naturalists, said Cuvier, had no trouble defining instincts. Whether simple or complex, they were actions performed without prior experience, brought into being by "a blind and necessary force." These included "the cries of the child who suffers and needs help; the newborn's search for the nipple and the action of sucking; and the flight, determined by fear, of a young animal that has not yet learned to recognize the danger."[35] Conversely, the "elective" actions, were those modified by experience and adjusted according to varying conditions. An example was "when the dog runs before his master and covers him with caresses, if it sees him prepare to leave and wants to accompany him."[36] Instinct, on the other hand, was what makes a dog seek to bury the remains of a meal.

Instinct, Cuvier continued, made cows threatened by wolves protect their calves by forming a circle with their horns facing outward; it made beavers build dams; horses and reindeer paw away snow to reach the grass or moss beneath; birds build nests; and so forth. The essential features of these behaviors were invariable, nor were they adjusted to fit different situations: "These cows, so ingenious in defending their young when in a herd," wrote Cuvier, "change nothing in their means of defense when reduced to a smaller number and these means become insufficient."[37]

Cuvier had probably never seen cows form a circle to protect their calves from a wolf, though he likely witnessed the behavior of a dog hoping to be taken for a walk. Many of the Museum staff had pets. As for experimental studies on menagerie animals, the closest he came were his experiments with a beaver.

In his article on instinct Cuvier repeated what he had said in 1808 about the inheritance of acquired characters—how instincts could be subject to a certain amount of modification, while learned behaviors, maintained over time, could become so habitual and mechanical as to be basically instinctive. He did not despair of distinguishing clearly between what was necessary, instinctive behavior and what was contingent, learned behavior, even though he acknowledged that the "author" of nature had created "a harmonious whole where all the parts are tied" and "all the transitions are softened."[38] As a naturalist, he considered it his job to identify relations and draw distinctions. This brought him to the question of the "point of separation" between animals and humans.

Cuvier wrote that the "entire solution" of that question was not yet possible without a "methodical and complete classification of the modifications that our mind can experience." Humans could know only "our own intellectual actions," while those of animals were "eternally hidden to our perception." The animal mind could be studied only by induction, by attending to the animal's movements, never forgetting that similar movements could arise from different causes.[39]

What one could reasonably tell from studying animal behavior, Cuvier said, was that animals of the higher classes could pay attention, and they received via their senses impressions analogous to those received by humans. Such impressions left traces that animals could remember and from which they formed "numerous and varied associations." From these associations animals could make judgments and recognize "several relations." But that, he said, was all that humans could say with any certainty about animal faculties. We could not know the actual forms or ways by which an animal's perceptions were organized. It was thus impossible to know which mental attributes they might possess, and which belonged to humans alone.[40]

Yet Cuvier was certain there was a fundamental difference. He assured his readers: "What seems beyond doubt, is that all the animals without exception are deprived of the intimate sense of perception of self and of the faculty to reflect [back upon themselves]. . . . They ignore that they receive the impression of external bodies, that they think, that they act; the acts of their mind, like the movements of their body, have only external causes."

The lack of sense of a reflecting self, Cuvier confidently asserted, was the foundation of animals' inferiority to humans. "Grant to them all the other faculties that we recognize in ourselves," wrote Cuvier, "they would still be far from equaling us."[41]

In another article on instinct (1823) Cuvier said he was composing a "general work on the origin of animal actions" but did not expect to publish soon.[42] In an essay on animal domestication (1825), he again addressed the contributions that a menagerie could make to science and felt compelled to respond "to what has been said, relative to my research on living animals, that captive animals could offer no useful observation."[43] He did not deny that field studies were valuable or that the study of free-living animals was necessary to understand their role in the economy of nature. But field studies were not the most promising venue for understanding animal behavior, Cuvier insisted; rather, only with captive animals could one examine the broad causes of animals' behavior and discover the full potential of their intellectual faculties.[44] He went so far as to suggest that for any given species, menagerie studies could render field studies unnecessary. In his words, "As soon as we come to know the general faculties of [a] species, and its dispositions, we will be able to determine, even in advance, its individual actions [wherever we might find it]." As he saw it, "to determine the way of life of this or that species in any country whatever," it would no longer be necessary to locate individuals of the species and follow them through "all the details of their existence." One would simply need a precise knowledge of the conditions in which they lived, a knowledge he described as "much easier [to ascertain] and much less subject to error."[45]

It was a bold claim, but unrealistic. How could one know the exact conditions in which a species lived without being there, studying over an extended period the lives of animals in relation to other members of their species and the other complex ecological relations they were a part of? Cuvier's claim appears to have been an overreaction to his peers' continuing doubts about what one could learn in a menagerie.[46]

When Frédéric Cuvier petitioned the professors in 1832 to have the minister of public instruction create a professorship for him, he explained, "Ever since I was put in charge of the oversight of the menagerie, all my research, all my work has been directed principally toward the knowledge of living animals, toward the means of observing them, of subjecting them to methodical experiments to recognize the laws of their existence, as well as the nature of the modifications that these laws can undergo." Arguing that these concerns were integral to science, he cited the studies he had published relating to this new field and gave a detailed overview of what his lectures on "the nature of animals and their education" would comprise. He said the Museum had underestimated the menagerie as a resource, viewing it as an accessory to anatomy and zoology, and benefitting from the animals only after they died. Alive, they had been considered little more than a burden of 40,000 francs per year. Everything about their lives that might have been useful for an entire new science, "their sensibility, their drives [forces], their penchants, their actions," had been "lost without recovery." Only the public had benefitted from the menagerie, but that was simply by indulging people's "vain curiosity."[47]

Cuvier's letter was read aloud at the professor's assembly of 31 July 1832, but the assembly postponed acting on it and then seemed to forget it entirely. In November Cuvier reminded the administration of his request, but the assembly's pledge to address it at its next meeting went unfulfilled.[48] Cuvier let his feelings show in an article he completed for his *Histoire naturelle des mammifères* not long afterward (February 1833). Writing about a genet from Senegal (a picture of which introduces this section), he lamented the conditions under which it was kept. When it was growing up, it enjoyed greater freedom. It was never hostile and behaved endearingly toward those who had reared it. Its demeanor changed dramatically once it was installed at the menagerie:

> Today, locked in a cage, it threatens everyone who approaches, and refuses to become familiar even with those who care for and feed it. . . . Unfortunately, one cannot give it back its freedom and thereby restore the most favorable conditions for exercising the pleasing faculties that had been carefully developed in it, *the arrangement of the menagerie never having been conceived with the view of facilitating the study of the native disposition of animals* [italics added].[49]

It was a rare expression of bitterness on Cuvier's part. Notwithstanding his service to the institution and the importance of the subject he wanted to pursue, the administration had ignored his request for a chair devoted to the study of living animals. Lacking the facilities to investigate animal behavior in a systematic way and lacking the professorial position that would have provided the occasion and authority to develop a comprehensive program of instruction on his subject, he felt that his hopes for a science of comparative behavior were going nowhere. His comments in effect represented a triple condemnation of the current state of the menagerie: it was not good for the animals, it was not good for science, and it was not good for his career.

He did not give up. Instead, in a new venue the same year, he broadened his critique to say that natural history had come to be dominated by anatomy and was impoverished as a result of it. He did not lay the charge directly at the feet of the Museum, but insofar as the Museum embodied natural history, he did not have to. The charge came in his introduction to a new, two-volume version of his and Geoffroy's *Histoire naturelle des mammifères*. Given that there were some 1000 to 1200 species of mammals, he said, it was necessary to have a method of ordering them all, but for this their external features were as significant as their internal ones, and for their zoology as a whole knowing their internal organs was less important than learning their external features, their actions, their lives, their "*psychologie*," and their place in the economy of nature. The preoccupation with anatomy, he insisted, had stood in the way of acquiring new observations on all these other things, and as a result, zoology was only half of what it should be.[50]

He offered this critique without reference to his personal situation, but it pained him that the Museum's professors were not prepared at this point to agree that his research agenda and accomplishments were worthy of making him one of them. His special interest, animal "psychology," the study of animal mental faculties, was too elusive, too unsuited to the practices of the Museum's cabinet naturalists, for the Museum to consider it scientific. Cuvier nonetheless remained convinced that the menagerie was a place to study the topics of animal instinct and intelligence. Could instincts be modified by experience? Could long-continued habits become instinctive? The consideration of animal instincts also impinged on the question of which animals were subject to domestication and how to achieve that goal. And the question of the changes undergone by domesticated animals was linked to the question of the origin of species.

[4.4]

The Beaver

On the banks of the Gardon River in spring 1810, a fisherman from the commune of Montfrin in the Department of Gard (southern France) found a newborn beaver, still half covered with afterbirth. He cut the umbilical cord, removed the placenta, and took the beaver kit home without knowing how he'd feed it. His wife offered to nurse the newborn, which indeed suckled when she put it to her breast. She fed it this way for two weeks until its front teeth began to come in. She then thought of letting the beaver suck on a milk-soaked sponge, and later she fed it with bread soaked in milk, and after that fish and willow bark. The young beaver thrived and slept on a cloth in the room with its adoptive parents. It socialized with the family dog, exchanging caresses with it while the dog helped rid the beaver of fleas. The fisherman and his wife named the beaver Garçon (boy), and the kit came to whoever called its name. The couple bathed the beaver daily in the Gardon River, but it preferred being on land. It never sought to escape and return to freedom.[1]

When the prefect of the Department of Gard, baron d'Alphonse, heard the story, he went to verify it. The animal's tameness astonished him. Beavers were known for their social relations, but to accustom one to a new way of life seemed to him a "discovery" that "ought not to be without interest for natural history." He communicated the finding to the minister of the interior and announced that the beaver's owners were willing to give the animal to the menagerie in Paris if the Museum was interested. The professors promptly sent the prefect instructions on sending the beaver to Paris, along with a reinforced crate for it. Before the crate reached the prefect in Nîmes, however, the beaver had died.[2]

The following March, another young beaver was captured on the banks of the Gardon. Dispatched in the special crate intended for the first beaver, toward the end of the month it arrived in Paris via stagecoach, alive and well. The only information that accompanied it were the instructions to the driver and a record of the expenses incurred over the two-week, nearly

Figure 4.4.1. "The beaver." C. Baquoy after De Seve. Buffon, *Histoire naturelle, générale et particulière*, 1749. Alfredo Dagli Orti/Art Resource, NY.

seven hundred kilometer, trip. The driver had been told to feed the kit fresh young willow and poplar branches and to keep the animal cool by periodically placing it in water or at least splashing it with water. He was also warned to keep close watch that it did not chew through its crate. The beaver arrived in "very good health," and the driver received a gratuity. The total cost to the Museum for food, transportation, and tip came to 131 francs, 80 centimes.[3] The Museum sent thanks to Lord d'Alphonse, not realizing that it was his successor as prefect, Jean André Louis, baron Rolland, who had dispatched the beaver.[4]

The minutes of the administrative assembly of 27 March 1811 recorded: "M. le Prefect du Department de Gard sends a male beaver, tamed, that was taken very young in the Gardon, and nursed by a woman." The last part was surely wrong. Apparently the story of the previous beaver had left the professors with the image of a beaver at a woman's breast, but a subsequent letter from Lord Rolland states that the new beaver was put on the stagecoach two days after capture.[5] Given the fear that if might chew its way out of the crate, together with its surviving on a diet of willow and poplar branches on its trip to Paris, it must have already had its front teeth when it was captured. It would no longer have needed nursing, and no prospective nursemaid would have risked the task. Once ensconced in the professors' imagination, however, the story continued to have traction. Writing about this beaver in 1817, Frédéric Cuvier unhesitatingly asserted that it had "the remarkable distinction of having been nursed by a woman."[6] He repeated the claim in 1825.[7]

The Museum was pleased with the live beaver, and Lord Rolland, eager to remain useful, asked (via the minister of interior) about the beaver's sex so that he could send an appropriate mate. To that end, he requested advice on how to tell male and female beavers apart and asked that the transport crate be returned to him.[8] The Museum was slow to respond, but when Lord Rolland announced in August that another beaver had been taken, the Museum reported that the first one was a male (as was a second recently received from Vienna) and that the Museum would be glad to have the new beaver if it was a female. Geoffroy Saint-Hilaire prepared a note to help Lord Rolland with beaver sex identification, and the transport crate was sent back to Nîmes.[9] This beaver never reached Paris. Perhaps it was another male. Or perhaps as a full-grown beaver, which a newspaper report suggests, it did not tolerate life in captivity.[10]

Contemporaneously with the beaver from the Gardon, the menagerie received a beaver from the Danube, sent from Vienna.[11] Only one of these beavers remained when Frédéric Cuvier conducted a menagerie census in April 1813. The next beaver arrived from Canada in February 1818 and died the following October.[12] Two more Canadian beavers came in 1828, and one of them survived nearly ten years. Another was acquired in 1839.[13]

Whereas trappers and traders prized beavers for their pelts, naturalists had long believed their most interesting characteristic was their prodigious talent for building. They routinely cited beaver dams and lodges as prime examples of what animal industry could accomplish. Curiously, or so it seemed to European writers, the most impressive of these constructions were not in Europe but in the New World. According to Buffon, the North American beaver was a highly social and productive animal that lived in large colonies and accomplished major communal projects, whereas the European beaver was a fugitive, isolated creature that never teamed up with fellow beavers to achieve a common goal.

Buffon found this scenario unusual. He had written at length comparing the animals common to both the Old and New Worlds, and the New World form almost always came out badly. New World forms, he famously claimed, tended to be smaller, weaker, and fewer in number than their Old World counterparts. He went so far as to propose that there was something in the North American climate "that stunted the development of all the creatures that lived there."[14]

Such effects were visible not only among animals, he insisted, but among humans: "The [North American] savage is feeble and small in his reproductive organs. . . . He has no hair, no beard, no ardor for his female. Though nimbler than the European because he is more accustomed to running, he has much less bodily strength. His sensations are also less keen; and yet he is more fearful and cowardly. He has no vivacity, no activity of mind." Buffon ascribed "the cause of the scattered life of savages and their estrangement from society" to their lack of "the most precious spark of the fire of Nature." They loved only feebly their women, their parents, and their children.[15]

How then to explain the great industry of North American beavers? Were they an exception to Buffon's general rule about the inferiority of the New World's productions to those of the Old World? Buffon had a way to keep his theory intact. The reason the beavers of the New World were able to accomplish great engineering feats was that there were no significant human populations to interfere with them. By his account, the American Indians' failure to rise above savagery was what allowed the American beaver to develop its building talents unmolested. The European beaver, in contrast, had ceased building lodges and dams because of the pressures it experienced from the advance of European civilization that encroached on its territory.

Buffon saw here an opportunity for a grand generalization regarding the relations between humans and animals: "As much as man has risen above the state of nature, the animals have sunk below it. Submitted and reduced to slavery, or treated like rebels and dispersed by force, their societies have vanished, their industry has become sterile, their feeble arts have disappeared; each species has lost its general qualities."[16]

When the earth was young and man did not yet dominate the scene, Buffon claimed, animals had been able to develop their capacities to the fullest and displayed a "marvelous industry." In modern times, however, this was possible only in remote places, far from human disturbance. Buffon speculated that beavers might be "the only example that remains, like an ancient monument, of that kind of animal intelligence that, although infinitely inferior to man's, supposes common projects and relative views." The projects he had in mind, "a dike to construct, a village to build, or a republic to found," all supposed some manner of "understanding one another and acting in concert."[17] Buffon identified here "a glimmer of intelligence, which, although very different from that of man in principle, produces rather similar effects."[18]

Buffon emphasized that this was a matter of social action. Having kept a Canadian beaver for nearly a year, he concluded that the individual beaver "has no more intelligence than the dog, or sense than the elephant, or finesse than the fox, etc."[19] What made beavers remarkable, he insisted, was their capacity for communal action with other beavers in the construction of dams and lodges.[20]

Frédéric Cuvier identified Buffon as the most astute naturalist ever to have written on the lives of animals. Cuvier was unwilling, however, to credit the beaver with the glimmer of intelligence that Buffon did.[21] Regardless of what beavers built or where they built it, he was persuaded, their work was a product of instinct, not intelligence. He based this conclusion on observations of the menagerie beavers from the Gardon and Danube rivers. Despite being captured and tamed while very young, they displayed habits and inclinations appropriate to their species. They liked being in water, and they took their food to water to wash and eat it. Most important, they displayed a strong instinct to build.

Cuvier described these behaviors in an 1817 entry on beavers for the *Dictionnaire des sciences naturelles*. He had his first hint of their building instinct when he noticed how they piled sections of willow branches just beyond the bars of their cage, a large, square affair with bars on two sides. Beyond one set of bars was a shutter with empty space between the shutter and bars that the beavers could reach with their paws and muzzles. Given willow branches to gnaw, they ate the bark and reduced the peeled branches to smaller fragments. In the morning, the fragments were found piled between the bars of the cage and the shutter. Believing this indicated an innate disposition to build, Cuvier decided to test the matter. He provided the beavers with more material: earth, straw, and tree branches. The following morning, he found it heaped in the same space between the cage bars and the shutter.[22]

To observe the beavers in action, Cuvier put shutters with peepholes all around the cage so he could watch without their seeing him. The beavers set to work as soon as they received the materials. They threw soil, straw, and sticks behind them using all their paws. They pushed masses forward with their forepaws and chin. They carried branches in their mouths and smaller sticks in a forepaw. They pressed the assembled materials together with their muzzles. Sometimes they mixed bits of bread or roots with the soil, to retrieve later when they were hungry. Cuvier frequently saw a beaver bring another stick to the mass and work hard to drive it in, without any apparent goal other than to put the stick in place. He also observed that they "nearly always" worked alone."

Cuvier recounted much the same information in an essay on "the beaver of Europe" (1825). Here, however, he added another reason to believe that the beaver's building efforts were based on instinct, not learning: the animal had had no opportunity to watch other beavers build. Its parents, instead of constructing a dam or a lodge, had simply lived in a hole in a riverbank. Nonetheless, the beaver in the menagerie "sought to build every time it was given appropriate materials." Based on his observations of the creature over time, Cuvier concluded:

> It is thus by an entirely instinctive and mechanical movement that the beaver is led to build; no external circumstance determines it; its intelligence plays no role; it blindly satisfies a need that itself is blind. The space that it filled with earth was no better closed as the result of its work, and it could derive no well-being from all the pains it devoted to that, in all seasons as in all weather.[23]

In his essay on the domestication of animals (1825), Cuvier agreed with Buffon that beavers in Europe did not build much because there were too many people. But he rejected Buffon's claim that the dams and lodges of New World beavers were, in Buffon's words, "the result of communal projects based on reasoned arrangements, of natural talents perfected in repose." The beavers' building achievements, Cuvier insisted, were "the fruits of an entirely mechanical industry, the results of a purely instinctive need." It was in the beaver's nature, he said, to build things. Had Buffon experimented on any of the beavers he kept as pets, Cuvier suggested, he would have seen that even when solitary and in the vicinity of humans, the beaver remained industrious. It was still disposed to build.[24]

The Canadian beaver that arrived at the menagerie in 1818 was the gift of a sea captain who brought it to Brest from Newfoundland.[25] In an article written some months after its death, Cuvier described it as "a very gentle animal, accustomed to the presence of man." It "allowed itself to be touched and transported by hand from one place to another with the greatest trust" and "even ended up living very familiarly with some dogs." Yet it displayed no such gentleness when brought into contact with the menagerie's other beavers that, like itself, had not grown up in the company of members of their own species. Placed together, said Cuvier, the beavers fought furiously, and the menagerie staff had to pull them apart for fear they would be mortally wounded.[26]

This Canadian beaver not only was deficient in social instincts, it showed no inclination to build and didn't like going near water. A pool was constructed, but it never went to wash its food or to bathe. It mostly slept and, when awake, seemed intent on staying close to its caretakers. It showed some intelligence in recognizing its name (or at least its keeper's voice). Cuvier attributed its weak instincts and intellectual inactivity to its lifelong captivity. With its needs and desires fulfilled by others, he explained, it could feel completely secure, and therefore it lacked the stimulation necessary to develop its instincts or intellect. For a zoologist hoping to lay the foundations of a science of comparative behavior by distinguishing what was instinctive in an animal's actions from what was acquired, this beaver was a disappointment.

Cuvier's recognition of the constraints of the menagerie's beavers' living conditions figured into his proposal in 1820 to develop a space at the menagerie suitable for small animals like "beavers, marmots, porcupines, wombats, racoons, etc." that could survive outdoors in Paris winters. Their current narrow cages, he told the administration, prevented scientists from observing various features of their natural behavior, including mating and reproduction. The professors agreed to the proposal, but it seems nothing came of it. When two beavers from Canada arrived eight years later, Cuvier lamented that he lacked a good place for them.[27]

The themes of individual versus communal action in the lives of beavers can be remarked in the behavior of Frédéric Cuvier and Étienne Geoffroy Saint-Hilaire as well. In the 1790s, Geoffroy treasured his collaboration and friendship with Georges Cuvier. When he engineered Frédéric's appointment as superintendent late in 1803, he imagined his close relationship with Georges Cuvier would be revived and further that he (Geoffroy) and Frédéric Cuvier would also become collaborators and friends. By the 1820s that hope had evaporated. Étienne Geoffroy Saint-Hilaire and Georges Cuvier were rivals rather than friends, and Geoffroy felt that his subordinate, Frédéric, was seeking to command the menagerie independently. Perhaps this lay behind a curious one-page piece that Geoffroy published in the Museum's *Mémoires* in 1825. It regarded the behavior of the menagerie's first beaver, though Geoffroy did not specifically identify the specimen as such. He simply noted that it came from the vicinity of the Rhône, where beavers live in isolation rather than in large communes, and that it had lived at the menagerie "several years ago."

Geoffroy reported that even in captivity the beaver could regain its native tendencies if circumstances demanded. The beaver had only had the straw in its cage for warmth, but one bitterly cold night when the shutters were closing poorly and snow was accumulating in a corner of the cage, the animal set about improving its situation. Having been given fresh branches to chew along with fruits and vegetables, it interlaced the branches in the bars of its cage, "corresponding perfectly" to the way basket makers weave straw in and out around stems. The beaver filled the gaps in this interlacing with the materials at its disposal ("its carrots, its apples, its litter"), each piece cut to fill a corresponding opening. "Finally, as if the animal had understood it was necessary to cover the whole with a more compact cement, it used snow to fill the smallest gaps remaining." Everything the beaver had, including its food, was used for the construction. In the morning, when the keeper discovered the beaver's remarkable accomplishment, he brought Geoffroy to see the work before any of it was disturbed.[28]

Noteworthy about this account, beyond the beaver's achievement, is that nowhere in it did Geoffroy mention Frédéric Cuvier, who had written not only about beavers' building habits but also about the very specimen whose behavior Geoffroy was describing. (It had to be none other than the beaver from the Gardon discussed at the beginning of this section.) Nor did Geoffroy mention why he was writing about the incident, years after the experience. He may have known that Cuvier's essay on "the beaver of Europe" was about to appear as an installment of the *Natural History of Mammals*, an enterprise that bore Geoffroy's name on the title page as the lead author. Perhaps he felt that his report cast some doubt on Frédéric's claim that the beaver's building behavior was solely the result of blind instinct. In Geoffroy's account, the beaver seemed to display at least a hint of intelligent action, and with a definite purpose: keeping out the cold. I interpret Geoffroy's piece as reflecting his unhappiness with Frédéric Cuvier, who seemed to be usurping Geoffroy's intellectual and administrative authority alike. It may have given Geoffroy some satisfaction to report that when the keeper discovered what the beaver had done, years before, he had come to *him* to report it.

The keepers had their jobs to do regardless of whether an animal's behavior was instinctive or learned and whether the superintendent of the menagerie and the professor of mammals and birds were getting along with each other. Among the keepers' assignments, caring for the beaver from the Gardon was one of the easiest. The animal seemed content with life at the menagerie. It viewed those who fed it as family. The keepers might have been granted it greater freedom had they not feared the damage it would do. Its most distinguishing behavioral trait was to gnaw on everything it encountered.[29]

[4.5]

Acclimatization and Domestication

Early calls for a menagerie in Paris featured litanies of candidates for acclimatization and domestication identified from around the world. Bernardin de Saint-Pierre offered a list (1792) from Louis Jean-Marie Daubenton that began with native candidates for domestication: the red deer, the fallow deer, and the roebuck. Turning to Africa, the zebra seemed worth importing to France for riding or working in harness. The Americas promised new meats for the dining table, including the tapir, peccary, paca, agouti, acouchi, tatou, and white-tailed deer. Birds also made the list, among them the common pheasant and American turkey from North America and several large, alleged-to-be tasty, birds from the South: the *hocco* (the crested curassow), the marail guan, and the seriema. Another South American prospect was the agami (the gray-winged trumpeter), said to have "the instinct and the fidelity of a dog" and to be capable of controlling and leading a flock of fowl or a herd of sheep, despite its being "no larger than a hen."[1]

When Jussieu drafted a recommendation to the National Convention in 1794 to make the menagerie at the Museum official, it included many of Daubenton's suggestions and more: some wild animals (vicuña, Barbary sheep, beaver, and eider duck) and others under domestication elsewhere (water buffalo, llama, dromedary, and camel).[2] In 1794 André Thouin likewise produced a long list of animal and plant species and varieties that could be profitably introduced and developed in France, including diverse kinds of goats and sheep, the vicuña and the llama, camels and dromedaries, water buffalo, axis deer, nilgai, various races of pigs, the zebra, and the quagga. As of May 1794 the menagerie had a quagga, an animal Thouin characterized as very similar to the zebra and "a beast of burden for the Hottentots [that is] stronger than the donkey, [and] could be substituted advantageously for the same labor."[3] Likewise, Daubenton in his first lecture to the École Normale in 1795 identified acclimatization as one of the projects of natural history and listed many of the most promising animal candidates.[4]

Figure 4.5.1. "The hemione," a target for domestication and acclimatization, shown with its shelter and park. Andrew, Best, Leloir after Karl Girardet. Boitard, *Le Jardin des Plantes* (1842).

Lacépède's 1795 portrait of the "new menagerie" included many of the animals mentioned above.

Acclimatization continued to be promoted as a function of the menagerie for the new century. Outlining the responsibilities for the post of menagerie superintendent, Geoffroy Saint-Hilaire wrote, "He will neglect no occasion to attempt the acclimatization and naturalization of foreign species that would be of some benefit for the rural or domestic economy."[5] Three decades later, when Sir Stamford Raffles was launching the

Zoological Society of London, one goal was to establish a museum with a collection of specimens that would help "the general study of zoology." But the Society's primary mission was "To introduce, by means of a public Establishment, new varieties, breeds, or races of living Animals, such as Quadrupeds, Birds, Fishes, &c. which may be judged capable of application to purposes of utility, either in our Farm Yards [*sic*], Woods, Wastes, Ponds, or Rivers."[6]

Buffon wrote extensively about domestication.[7] He insisted that just as humans are superior to animals because only humans think, humans are the only species that domesticates others. The organization of his magisterial *Natural History* testified to the importance he assigned to domestication. Man, unsurprisingly, was the first species discussed, but the next, out of all those he might have chosen, was the horse, because it represented "the most noble conquest that man has ever made."[8] After the horse, Buffon opined on other domesticated animals (donkey, cow, sheep, goat, pig, dog, and cat) before turning to wild animals, starting with the red deer. Several volumes later, treating the camel and the dromedary, he again highlighted the triumph of domestication: "Gold and silk are not the true riches of the Orient, it is the camel that is the treasure of Asia."[9]

But if Buffon wrote of domestication as ennobling or perfecting animals, he acknowledged that this was a matter of perspective, for "slavery" came with it. Regarding the domestic pigeon, to the extent that "slave races" (*races esclaves*) of pigeons are "all the more perfected for us," they are "degenerated, more vitiated for Nature."[10] More generally he wrote of domesticated animals, "one will find on all enslaved animals the stigmata of their captivity and imprint of their irons," with their "wounds" all the greater and more incurable the longer their time in this state.[11]

To Buffon, man's mastery of nature and of himself went hand in hand. "Savage man," he said, was scarcely above the animal state because he was incapable of appreciating or improving nature.[12] In other words, the domestication of animals took place "only among peoples who were already civilized."[13] And again: "Nothing demonstrates better the immense distance between savage man and civilized man than the latter's conquests over the animals."[14]

Along these lines, if people were ranked from the most savage to the most civilized, one would expect to find a correlate in the domestication of their animals. Frédéric Cuvier made this point in his 1808 essay on the unruly dog of the Australian Aborigines, transported to France by the Baudin expedition, which stole whatever she wanted. He found it unsurprising that "the poorest and least industrious people of the earth" should "possess the dog most inclined to rapine that is known, and the most incorrigible in this regard."[15] When he again wrote about this dog in 1825, he once more suggested that the state of domestication of a people's dogs might be an index of their level of civilization.[16] He described the natives of Australia as "the most brutish and coarse" known to science, those "who have remained closest to nature" and "developed the fewest needs," and whose "intellectual and moral qualities," as a result, remained "the least developed." It made sense, he said, to consider their dog "as the one closest to the savage state, the least modified, and that presents most faithfully to us the characteristics of its species," short of ever finding the species "in the state of pure nature."[17]

These thoughts resonated with Buffon's identification of animal domestication as the triumph of civilized peoples and with Lacépède's historical schema where increasingly enlightened types of menageries correlated with higher stages of civilization. They were also reminiscent of Geoffroy Saint-Hilaire and Georges Cuvier's observation that the accomplishments of "peoples with depressed foreheads and protruding jaws" did not equal those of Europeans.'

But Frédéric Cuvier's take on domestication differed from Buffon's. Where Buffon represented the domestication of animals as conquest, Frédéric Cuvier viewed it as seduction. Instead of treating domestication as a matter of civilized man's forcing animals to obey his will, Cuvier denied the efficacy of force, eschewed self-congratulatory discourses about civilized versus savage man, and focused instead on why some animals were susceptible to domestication and others were not. Rather than thinking domestication degenerated an animal's original nature, Cuvier generally treated it as an improvement. Domestication involved training (*éducation*) that encouraged animals to feel that depending on humans was the natural thing to do.

Cuvier developed these views at length in an essay on domestication (1825).[18] By then he had spent more than twenty years overseeing the menagerie and was attuned to the differences in susceptibility to human influence and domestication among animal species. He began his analysis with the claim that domestication is not the same as human slavery, which was achieved only through force and where the enslaved recognized their lack of freedom. Animals, he maintained, are unable to reason about their state and simply respond according to their needs and natural dispositions. They cannot be forced into domestication because their natural response is to flee humans. Their innate disposition is to remain averse to them, even when captured. Lacking free will, they are unable to behave other than as their dispositions incline them. Consequently, the only way to domesticate an animal is not by force but by seduction, by either satisfying or weakening its needs.[19] Most important, one needed to treat an animal kindly to make it confident in human company.

Throughout his career, Frédéric Cuvier preached kindness as a requisite of animal care, but he did so as much for the benefit of humans as for animals. Kindness was his preferred means of eliciting the animal's cooperation with human goals. His essay prescribed a series of techniques, beginning with the need to start with a very young animal. Hand feeding would encourage its sense of familiarity, but to make it perform other services it was necessary to develop its needs or create new ones to build reliance on humans and "to make the society of man necessary."[20] The first technique was to induce hunger, because an animal would appreciate being fed by a human more if it was starving. Depriving an animal of food would weaken its body and will. Those who tamed wild horses, Cuvier remarked, knew that feeding them small amounts at long intervals helped to subjugate them and develop their affection. Inducing hunger combined with awarding treats, especially sugar, was a remarkable incentive in training herbivores. It was used, Cuvier explained, to command the kinds of animal performances seen in circuses.

Cuvier cited the caress as another means of binding animals to humans. Knowing little of animal behavior in the wild and overlooking the way the menagerie's elephants had warmly caressed one another, Cuvier conjectured that animals acquired their need for caresses in interacting with humans. The emotional effect of the caress, he said, was enhanced by speaking to the animal in a soft and gentle voice, and, at least in some cases, by touching the nipples. He acknowledged, however, that while some animals were moved by caresses to an almost pathological point, others were virtually indifferent.[21]

Only when an animal had come to require human society, Cuvier continued, should constraint and chastisement be employed to teach it to associate fear with the inclinations for which it was being punished. Blows were needed in circumstances the animal could not escape from, but only for diminishing or extinguishing the feelings at the root of the unwanted behavior. Humans should be careful not to use force beyond the bounds of the authority they had carefully achieved; otherwise the animal would rebel. Levels of resistance to authority varied among species and individuals, Cuvier said, but animals of all kinds revolted against brutal treatment and returned it with a vengeance. An example was the elephant so mistreated by its keeper that its former affection for him turned to such hatred that the elephant wanted to kill him.[22]

Domesticated elephants, Cuvier reported, could be used to capture and tame their wild counterparts; a wild elephant under their influence would work for a human master within just a few months. But there were species and individuals where good treatment, corrections, and the tutelage of conspecifics were not enough to overcome imperiously felt needs. Two practices remained. The first was sleep deprivation, "prudently associated with good treatment and chastisements," achievable via blows of a whip, disruptive noises (like those of drums and trumpets), and hunger. The second was castration for males that would otherwise have remained unruly. "The bull, the ram, etc., only truly submit to man after their mutilation," Cuvier wrote, and in those species, it was "only the female that is domestic."[23]

Cuvier supposed that other methods of domestication might yet be discovered. He also acknowledged the inconsistencies in what domestication had so far accomplished. The dog was "attached, obedient, grateful, faithful, devoted," but the water buffalo was "lacking in kindly and affectionate feelings and all docility." Situated on a continuum between these two poles were "the elephant, pig, horse, donkey, dromedary, camel, llama, reindeer, goat, ram, and bull."[24]

Cuvier thought that certain additional animals could be domesticated, but others were out of the question, including the Old World monkeys. Notwithstanding their high intelligence and the way they lived in troops, adult male monkeys never fully obeyed humans, no matter how well treated. He attributed their intractability to their constant physical activity. Because their needs were always changing, you could never be sure how well disposed they were toward you: "[Just] when they seem to give you the most affectionate signs, they can be ready to tear you to pieces." This was "not a matter of treason," only "their excessive mobility." Rhesus monkeys that did tricks for their masters and their audiences, he said, did so only under a regime of force.[25] In an earlier paper he had written that only the fear of punishment could control adult macaques. He had furthermore characterized the macaque as "disgusting in its filthiness, indecency, and effrontery." In other words, it was not just their "excessive mobility" that made the Old World monkeys unsuitable for domestication.[26]

As for carnivores, Cuvier reported, they were much more reliable than monkeys if treated well. "One hundred times the apparent gentleness of a monkey has been followed by a betrayal; almost never have the outward signs of a carnivore been misleading; if it is disposed to harm, everything about its posture and its look announce it, and it will be the same if a good sentiment animates it."[27] By satisfying their needs when they felt them most strongly and by always being kind and never threatening with voice or movements, you could get carnivores to approach you confidently and affectionately.

Cuvier contemplated the services lions or bears might perform if humans could use them like dogs, but he decided that of all the carnivores, seals might best be developed, through good treatment, to human advantage. He also had hopes for pachyderms like the tapir, solipeds like the zebra, and at least some female ruminants that lived in herds, feared pain, recognized good treatment, and could distinguish among caregivers.[28] The alpaca and the vicuña, for example, could serve both as beasts of burden and as sources of a fine wool worthy of the development of a new industry.[29]

What made it possible for alterations in individuals to lead to alterations in species, Cuvier explained, was that one did not have to start from the beginning with each new generation. Animals had been domesticated by controlling the circumstances they were subjected to, affecting the same changes in both sexes, and carefully breeding them to maintain the purity of lines or create judicious mixtures. The more generations devoted to this, the better the results. This was partly why the dog was more domesticated than the cow, Cuvier said. The elephant might become one of man's most obedient and affectionate servants, he supposed, given enough generations working to domesticate it.[30]

The transmission of individual modifications over generations via selection and the inheritance of acquired characters was still not enough, Cuvier believed, to ensure a species' domestication. It was necessary that the species have from the beginning "the instinct of sociability carried to a high degree" and the qualities needed to further that instinct. Domestication, Cuvier proposed, was only a modification of a preexisting tendency. It involved redirecting toward humans the social animal's innate tendency to be close to the members of its own species. Animals were in effect domesticated when they came to recognize humans as members of their own species and the leaders of their troops.[31]

Summing up his analysis, Cuvier maintained that the animals most susceptible to domestication were those that lived in groups and formed attachments. Domestication was achieved only by a kind of seduction that involved enhancing the animals' needs and satisfying them, aided by "one of the most general laws of life: the transmission of organic or intellectual modifications by way of generation." He portrayed "the transformation of a fortuitous modification into a durable form, of a passing need into a fundamental propensity, of an accidental habit into an instinct" as "one of the most astonishing phenomena of nature."[32] He nonetheless balked at the notion that such changes passed down from one generation to the next could ever go so far as to produce entirely new species.

Among the readers of Cuvier's essay on domestication was one of the most influential scientists of the nineteenth century, the geologist Charles

Lyell (1797–1875), promoter of the view that the earth's geological features are to be explained by the same forces we see today, operating at the same intensity over immense periods of time. He thus firmly opposed Georges Cuvier's idea of huge catastrophes' causing massive extinctions and shaping the earth's surface. Lyell's disagreement with the elder Cuvier did not prevent him, however, from describing Frédéric Cuvier's domestication essay as "admirable" and its author as "ingenious."[33] He did so in volume 2 of his *Principles of Geology*, the volume where he explicitly addressed Lamarck's theory of evolution with a directness and general fairness neither Cuvier brother ever hazarded.

It is curious in a book of such scientific consequence, and in the very passages where he credited Frédéric Cuvier's acuity, to find Lyell blithely stating, "A pig has been trained to hunt and point game with great activity and steadiness; and other learned individuals, of the same species have been taught to spell." The geologist credited a lodge keeper in Hampshire with the information on the pointing pig, but he provided no reference for the assertion of pigs' spelling. Frédéric Cuvier would not have repeated such a story in his writings on animal behavior. He made a point of reporting only things that he had determined for himself. It is also highly unlikely he would have been inclined to believe the story or think it worth testing. But Lyell's point was not about pigs anyway. As the Englishman went on to explain, such "fortuitous acquirements [as pigs learning to point or spell] never become hereditary, for they have no relation whatever to the exigencies of the animal in a wild state and cannot therefore be developments of any instinctive propensities."[34]

What Lyell liked about Cuvier's domestication essay was his claim that humans had managed to domesticate only animals that already had a propensity to be domesticated because of an instinct to sociability and a readiness to follow the chief of their herd. Lyell agreed that some acquired characters could become hereditary, but he put this in a providential framework, maintaining that the variations of which domestic animals were susceptible were *preplanned*. It seems "reasonable to conclude," he wrote, "that the power bestowed on the horse, the dog, the ox, the sheep, the cat, and many species of domestic fowls, of supporting almost every climate, was given expressly to enable them to follow man through all parts of the globe—in order that we might obtain their services, and they our protection."[35] At a time when Frédéric Cuvier was highlighting the need to create a new science of variation, Lyell was confidently asserting that variation was strictly limited, hence a continued divergence from a species' originally endowed attributes was impossible. With these observations, along with objections he had already registered, plus some sarcastic remarks on

Lamarck's speculations about an orangutan-like creature's potential for further development, Lyell felt he had successfully wrapped up his refutation of Lamarck and the "Lamarckians."[36]

∴

In the menagerie's experience with domestication, theory and practice remained far apart for diverse reasons. One was the sheer difficulty of getting various species from their native lands to Paris alive. Another was that the animals in question often proved a poor fit for the environmental conditions that the acclimatizers had in mind, rendering them susceptible to injury or illness. A third was the dangers the animals posed their keepers, either by physical injury or by transmitting disease. These problems surfaced in experiences with llamas and alpacas, with camels and dromedaries, and with the dauw (the Burchell's zebra).

When European conquerors witnessed the services that the llama rendered Peruvians, they dreamed of making it one of "the treasures of the conquest," and despite the difficulty of obtaining live specimens, the idea continued to be promoted.[37] Buffon in 1765 wrote that the llama and its relatives "would be an excellent acquisition for Europe, and would produce more real benefits than all the metal of the New World."[38] In 1802, Joséphine Bonaparte persuaded the king of Spain, Charles IV, to help her import a large herd. Thirty-six llamas were collected in the Andes but were halted in Buenos Aires for six years because of war. Nine reached Spain in 1808, the year Napoleon elicited Spanish hatred by installing his brother Joseph as king of Spain. None of these llamas made it to Paris.[39]

Josephine eventually acquired a pair of llamas for Malmaison and permitted Georges Cuvier to observe them. In an article of 1805, he described their prospects for acclimatization as excellent, reasoning that their native climate was no hotter than France's, their temperament was suited to domesticity, and they required neither special food nor special care. If the problems of transporting them to France could be solved, all else (or almost all else) should readily follow.[40] Cuvier hedged with respect to the llama's usefulness. Its most distinctive attraction was sure-footedness in the mountains, but horses or mules could carry heavier loads. As a beast of burden, it would likely benefit only some parts of the Alps and the Pyrenees. Otherwise, it could provide meat and wool.

In 1821 the Museum was given its first llama-like creature, an alpaca. Frédéric Cuvier identified it as a distinct species from the llama with domestication potential for France.[41] It was gentle with its keeper, a trait Cuvier attributed to its being well treated when it was young. He predicted that producing alpaca wool in clear, bright, and deep colors would create new needs in the market and "give us new riches." He also identified alpaca meat as good for human consumption.[42]

But a lone alpaca could not do the job. When the professors learned in January 1824 that the minister of the marine was sending ships to Chile, they wrote to Corbière, the minister of the interior, asking for support in making the expedition profitable not only for natural history but also and especially for "the public fortune." As the letter drafted by Frédéric Cuvier explained, Chile was home to "the alpaca, a ruminant species as large as the donkey and covered with a very thick and very fine fleece." The menagerie had been able to maintain an alpaca specimen in "the most perfect health" for several years. It had handled the winter's cold even better than the summer's heat. Acclimatization attempts in Spain had been successful, indicating that "this precious species could be naturalized in France without difficulty." The Museum asked Corbière to invite the minister of the marine, the duc de Clermont-Tonnerre, to have the French Chile expedition procure alpaca specimens.[43] Clermont-Tonnerre agreed and turned the task over to Rear Admiral Claude du Campe de Rosamel, who was about to take command of the French naval station for Chile and Peru.[44]

The Museum's records track this long story in detail, but in short, from an original herd of five alpaca pairs, only two males survived. The ship's captain, emphasizing the positive, said that the voyage had been a learning experience, for it revealed that alpacas are particularly susceptible to humidity. Humidity, he reported, made the alpacas' feet swell, provoking a disease that killed eight of them before the ship reached Rio de Janeiro. Reducing the remaining alpacas' exposure to humidity had kept them healthy.[45]

In 1826 Admiral Rosamel sought again to procure alpacas for the Museum. Of twelve animals purchased in Peru (ten alpacas and two llamas) only two—one male of each species—survived the tropical heat and a skin disease the animals suffered during the journey. The Museum ultimately received detailed information on the animals' epic journey from high in the Andes to the Chilean coast; their considerable care aboard ship; their suffering from the heat; the captain's recommendations for the future; and the ship's surgeon's account of the skin disease and the animals' deaths. From Peru, it took five months for the ship to reach Brest, where the two survivors were cared for at the garden of the Ministry of the Marine. They were subsequently moved to a paddock at the Navy hospital in Cherbourg. It was seven and a half months from the time they reached France before they made it to Paris.

When the alpaca and the llama arrived at the menagerie Frédéric Cuvier reported they seemed "full of strength and life." They were, however, afflicted with a skin disease resembling the mange that the menagerie's dromedaries and their keepers had already suffered. Cuvier informed the administration that the keepers were busy caring for existing animals and would "formally refuse" to care for the new ones. He proposed hiring a man who had been overseeing the recovery of a pair of sick dromedaries at the veterinary school at Alfort and had avoided contracting their disease. The administration approved, and the new keeper took charge of the llama, the alpaca, and the now-healthy dromedaries back from Alfort.[46]

The new alpaca and llama did not seem to excite Cuvier much. Having first identified the alpaca as an alpaca, he called it a vicuña the next week, and the week after that he simply wrote of "the llamas."[47] Tracking the Museum's llamas and alpacas over the years is difficult because of inconsistencies within the menagerie's animal registry and between that registry and Cuvier's menagerie reports and censuses. We at least learn from one of Cuvier's reports how keen the showman Henry Martin was to acquire a white alpaca from the menagerie. He exchanged a host of specimens for it.[48]

Cuvier's interest in llamas was rekindled in January 1836 when an opportunity arose to purchase "a beautiful llama female" for just 400 francs. Here, finally, was a chance to have the species reproducing at the menagerie. Requesting the administration's approval, he noted that in the past fifteen years "numerous attempts have been made to have females of these animals," and that "the government's very modest price was worth taking." "The menagerie possesses two males," said Cuvier, "and it is almost beyond doubt that one or the other, united with this female, will reproduce their species." The female llama joined the menagerie on 21 January 1836 and lived for twelve years, but her time with the males was short. One died in August 1836, the other in November 1838, and no llama babies were produced. Other llama-related problems included the keeper, Perrin, contracting mange from one of them, leaving him in January 1837 "in a very sad state," what with the costs of treatment and his very limited resources.[49] In February, the menagerie's two llamas were attacked and wounded by a bulldog that forced its way into their pen.[50]

In brief, by the late 1830s, four decades of thinking about acclimatizing and domesticating llamas and alpacas had come to nothing. Importing them in sufficient numbers of both sexes was a hurdle the Museum never overcame. Nor had hopes for domesticating other animals been realized, though not always for the same reasons. The American bison, for example, unlike the alpacas and the llamas, had not been difficult to transport. It was robust, and its trip to France from North America was shorter and not subject to the tropical heat of a journey from the far side of South America. But the bison turned out to be a dangerous creature, and the attempt to hybridize it with a cow ended, as we have seen, in misfortune (see section 3.5).

Camels and dromedaries were a different story but likewise not a successful one. Acclimatizing them was on the to-do lists of both Jussieu and Thouin in the 1790s. Before the decade was out, the menagerie had a pair of each species. Bernard Louzardy collected a pair of two-humped Bactrian camels in Brussels in 1795. They were put to work operating the Museum's main water pump. The public enjoyed watching them work (and enjoyed throwing stones at them when they were at rest).[51] After Cassal brought a pair of white dromedaries back from North Africa in 1798, viewers had the chance to compare the one-humped dromedary with its two-humped camel relative.[52] Over time, the dromedaries mated and successfully reproduced, causing their parks to need enlarging so that their newborn could exercise and grow strong enough to work the pump.

The government took an interest in acclimatizing dromedaries and camels when an entrepreneur named Porte addressed the minister of the interior (Decazes) with a proposal to import them. Decazes turned to the Museum for an opinion on the animals' suitability as a beast of burden and power source. Frédéric Cuvier and Geoffroy Saint-Hilaire drafted responses noting that a camel/dromedary was as strong as two horses, could be fed for half the cost of feeding a horse, and could survive throughout France. The scientists were skeptical, however, about acclimatizing these creatures. Cuvier noted that camels tended to slip and injure themselves in damp or rocky terrain. Both remarked further that French farmers were likely to resist changes in the rural economy. Geoffroy specifically reminded Decazes of the government's frustration when trying to get farmers to accept Merino sheep.[53]

The menagerie's experience with the animals was also mixed. The dromedaries and camels occasionally injured keepers. Equally vexing was when they gave their keepers mange, a highly contagious skin disease caused by a parasitic mite. This happened in 1825, courtesy of the six dromedaries sent as a diplomatic gift from Tunisia in 1825. The keepers caught the disease and transmitted it to their families. One of the keepers, Géan, was infected for nearly a year and almost died, despite receiving considerable medical attention. He was on the mend by January 1827, but by then two of the animals had died and another was close to doing so.[54]

Cuvier took the menagerie's remaining dromedaries (five adults and two young) to the Veterinary School of Alfort for treatment. Four and a half months later, they had racked up 425 francs of medical expenses and were still not cured.[55] Cuvier believed that even when they were "cured," the disease would return. Before any of the animals were brought back to the menagerie, he wanted a new stable prepared for them. He feared that if they were returned to the rotunda, they would infect the other animals there. At the end of July 1827, he reported that two of the adult dromedaries at Alfort were in "very good shape," but the others were not, and their ongoing expenses would exceed their value. Geoffroy thereupon recommended to the administration that the three sickest animals be killed and their skeletons used for exchanges. He called for getting rid of the other two.[56]

The question of what to do with the two recovered dromedaries was solved when the baron F.-E. G. Lefébure Saint-Ildephont offered to buy them. Eager to help in the colonies, Ildephont wanted to send them to Martinique and Guadeloupe, and he agreed to purchase two adults and one juvenile for 1,000 francs.[57] He retracted his offer, however, when he learned that importing camels to the Antilles had failed several times in the late 1700s. The animals all died from inflammations of their feet caused by chigoe fleas. Fortunately for the Muséum, the Théâtre du Cirque olympique was willing to pay the same amount, evidently intending to make the animals part of its equestrian show.[58]

While camels and dromedaries were proving questionable candidates for acclimatization as well as difficult creatures for the menagerie, Frédéric Cuvier welcomed the zebralike specimens that Edward Cross brought from London in 1826 as promising candidates for domestication. Cross called them quaggas, but Cuvier recognized them as the species midway between the quagga and the zebra known as dauw or Burchell's zebra. In an entry for his *Natural History of the Mammals*, he noted they displayed "a large part of the feeling of independence of their race; however, they know their keeper very well and treat him kindly as long as he limits himself to the care that their cleanliness requires." They even tolerated having their wounds treated. Wrote Cuvier, "It is quite certain that with a bit of care one could derive domestic races from this species, as from all others of the genus."[59]

Cuvier had expressed the same certainty when writing about what the dauw's close relative, the quagga, could do for the rural economy. He explained, "to judge by what voyagers report, this animal is endowed with a very great courage, and far from fleeing ferocious animals of medium size like the hyenas and the wolves, it attacks them and ordinarily succeeds in making them flee; thus one raises it with flocks, which, under its protection, can travel through pastures more securely than they could without this aid."[60] Recalling the menagerie's quagga, he observed that it "was not savage [*farouche*], but it was ill-natured [*méchant*] and obstinate; it tried to bite and kick when opposed and dropped to its knees when one wanted to make it move by force."[61]

Cuvier acknowledged a hurdle to acclimatization and domestication beyond those posed by the animals themselves. He and Geoffroy had pointed it out in 1819 to the minister of the interior with respect to importing camels to France: the agricultural class of French society was resistant to almost any innovation. French farmers in their ignorance carried natural prudence to such an extreme, said Cuvier, that it became "a simple instinctive response analogous to that of an animal to challenge and flee every unknown object."[62]

The Museum eventually purchased Cross's dauws, which proved fertile. Male foals were born in 1829 and 1833 and a female in 1835. But getting the dauw to breed in captivity was not sufficient to make the species a good choice for domestication, as the temperament of the first male demonstrated. In April 1837 it bit keeper Hippolyte Martin so severely he had to have his arm amputated. Keepers regarded the dauw as the menagerie's most dangerous animal. To avoid further "deplorable accidents," it was kept muzzled except when eating, and then it was forced to wear a training collar.[63]

The experience with the dauws in the summer of 1837 did not keep French scientists, Isidore Geoffroy Saint-Hilaire in particular, from dreaming of acclimatization and domestication projects. Isidore had already weighed in on the virtues of the hemione in 1835, after J.-J. Dussumier brought the menagerie its first specimen.[64] Two decades later, notwithstanding how dangerous the male dauw had been on Isidore's watch as menagerie superintendent, Isidore continued to identify the species as a domestication possibility. When he launched the Société Zoologique d'Acclimatation in 1854, he maintained it was primarily for lack of trying seriously that none of the animals on Daubenton's list of prospects in 1792 had been acclimatized.[65] Isidore credited Frédéric Cuvier with having brought more scientific precision to the acclimatizing of useful animals than anyone before him, but he faulted Cuvier for lacking sufficiently broad views or enough conviction and boldness to pursue acclimatizing animals with the same vigor and success that Thouin had shown with plants.[66]

In 1865 the British polymath (and cousin of Charles Darwin) Francis Galton revisited the topic of domestication. Without mentioning the French authors who had treated it, he upended Buffon's views on the correlation between domestication and civilization and poured cold water on Isidore Geoffroy Saint-Hilaire's rosy thoughts about bringing new species into the domestic fold. In the course of human history, Galton maintained, domesticating animals was the achievement of neither civilized societies nor "extraordinary geniuses" but emerged instead through "the every-day habits of rude races." He located the origins of domestication in "wild animals being nurtured in savage encampments, either as pets or as sacred animals." He considered the process only "half-conscious." An animal needed so many characteristics that he found it no wonder "only a few varieties of a few species should have become the companions of man." His conclusion, quite the contrary of the optimists before him, was that there were no more significant domestications to come, that "all domesticable animals of any note have already fallen under the yoke of man."[67]

The acclimatization and domestication of animal species had served early on as a rationale for establishing the menagerie in Paris and later the zoological garden of the Zoological Society of London, but success with animal acclimatization, to say nothing of domestication, proved elusive. Six years after Isidore Geoffroy Saint-Hilaire founded in Paris the Société zoologique d'acclimatation, it opened in the Bois de Boulogne its own zoo, the Jardin zoologique d'acclimatation. Half a century later, however, it had no remarkable successes to justify its name.[68]

Isidore Geoffroy Saint-Hilaire was no doubt right that Frédéric Cuvier's commitment to acclimatization projects was not as great as Isidore's own. Be that as it may, in 1837 Frédéric and the minister of public instruction were still identifying the study of acclimatization as one of the main functions of the new professorship Frédéric wanted (see section 4.11). At the same time, Frédéric expected that the professorship, with the menagerie as part of its purview, would address the comparative study of animal faculties, and he continued to believe that the menagerie could be a site for studying how far changes in a species could go.

[4.6]

The Species Question

In 1816, a year after Jean-Baptiste Lamarck published his last major exposition of his theory of biological evolution, Georges Cuvier stood before the Academy of Sciences and, ignoring the issue of species change, proclaimed, "There is nothing more celebrated in natural history than the apron of the Hottentots, and at the same time there is nothing that has been the subject of more numerous disputes."[1] Could this have been true? Could an anatomical anomaly have attracted more attention and debate than the issue of whether species are essentially the same as they have always been or have evolved over time?[2] Cuvier would certainly have preferred it that way. He did not want to dignify the idea of species change by making it the subject of extended scientific consideration. But the idea could not be totally repressed.

When Frédéric Cuvier in 1815 outlined for the professors of the Museum the issues he hoped to address by having an "experimental park" at the menagerie, one issue was how far species can be modified by the "circumstances" they are subjected to. He assumed change was limited, but he wished the point could be established experimentally. "We have only the smallest part of the observations science would need," he said, "to fix the precise sense of the word species." He admitted, "Our means would be insufficient to make big experiments in this regard," but he still considered it "reasonable that one would be able to obtain something precise on the qualities of fur and its colors, and in general on all the modifications of the parts most subject to the action of external causes."[3]

A few years earlier, at his brother's bidding, Frédéric Cuvier had investigated the skeletal differences of the dog varieties in the comparative anatomy collection. The resulting paper (1811) showed that despite variations across breeds, the relations of the bones remained essentially the same. Frédéric concluded with a carefully worded admonition not to be seduced by "the lure of a false analogy" and think that the variations of animals under domestication were a warrant for believing in species change.[4]

Figure 4.6.1. "Jean-Baptiste Lamarck" (1744–1829). Lithograph by Julien-Léopold Boilly. *Archives du Muséum national d'histoire naturelle*, 6th ser., 6 (1930), pl. 3.

The following year, Georges Cuvier had triumphantly cited Frédéric's paper and proclaimed, "there are thus characters in the animals that resist all influences, whether natural or human." Adding immense amounts of time to the equation, in Cuvier's view, changed nothing, notwithstanding the efforts of "some naturalists" who relied "on the thousands of centuries they gather with the stroke of a pen" to claim otherwise.[5]

Georges Cuvier was sniping here at his colleague Lamarck, who believed in species change and did indeed count on thousands of centuries for it to happen. Lamarck believed that beginning with the simplest forms of life and over eons of time, life on Earth in all its astonishing diversity and complexity had developed. Although he was the Museum's professor of invertebrate zoology, most of the examples he used to exemplify his ideas came from vertebrates: the long neck of the giraffe, the rudimentary eyes of the mole, the long necks of wading birds, and so forth. He drew supportive evidence from the variation of animals and plants under domestication, the difficulties naturalists frequently encounter in distinguishing species, and the existence of interspecific hybrids.

Frédéric Cuvier was deeply interested in all these topics, but he interpreted them differently than Lamarck did. He was confident that the variation of animals under domestication is limited, that species do not blend into each other, and that new species cannot arise through hybridization. He believed that species are fixed and unchanging. It is thus ironic that he championed what biologists today regard as the signature idea of Lamarck's evolutionary theorizing: the inheritance of acquired characters. Frédéric Cuvier was in fact the first zoologist to use the phrase, "the inheritance of acquired characters," something Lamarck never did.[6] But Frédéric never mentioned sharing this idea with Lamarck. Following the lead of his brother Georges, he mentioned Lamarck's speculations as little as possible. He owned a copy of *Philosophie zoologique* (1809), Lamarck's most famous evolutionary treatise, but he never cited it.[7]

Lamarck, on the other hand, was not shy in touting the novelty of his thinking. In *Philosophie zoologique*, he contrasted two ways of thinking about the origin of species. The first, which he identified as "the conclusion accepted up to the present," was:

> Nature (or its Author) in creating the various animals foresaw all the possible kinds of circumstances in which they would have to live and gave to each species a constant organization, a form that is precise and invariable in its parts, obliging each species to live in the places and climates where it is found and to maintain there its characteristic habits.

The second, which he identified as "my particular conclusion," was:

> Nature, in producing successively all the species of animals, beginning with the most imperfect or most simple in order to end her work with the most perfect, has gradually made their organization more complex; and with these animals spreading generally throughout all the habitable regions of the globe, each species received from the influence of the circumstances in which it is found the habits now recognized in it and the modification of its parts that observation shows us.[8]

The dichotomy could not have been clearer. The claim, however, was historically inaccurate. Lamarck's model of change was new, but others before him had allowed for species change over time, including the two leading naturalists of the eighteenth century, Linnaeus and Buffon. Their respectives models of change were undoubtedly what Étienne Geoffroy Saint-Hilaire and Georges Cuvier had in mind in their 1795 paper, "Natural History of the Orangutans," when they referred to new species arising through either the hybridization or the degeneration of primordial types.[9]

Though early in his career Linnaeus pronounced that "there are as many species as the Creator produced diverse forms in the beginning," by the end of his career he had a more complex view of life's origins. He thought the Creator began by creating only one species for each *order*. Fertile with each other, these primordial species then hybridized to produce all the different *genera*. These generic forms then hybridized again to produce all the different *species*.[10]

Buffon likewise said initially that the species were fixed but later proposed that new species had arisen over time. By his account, the two hundred species of modern mammals had descended from thirty-eight primordial families spontaneously generated early in the Earth's history from "internal molds" that were part of the fundamental structure of nature. A process of "degeneration" then took place within these original family types as the forms in question were exposed to different climates and environments. This resulted in the modern mammalian types recognized as species.[11] Linnaeus's and Buffon's respective views were well known. Lacépède, writing in 1800, was open to both the degeneration and the hybridization models of the history of species.[12]

Lamarck's model of change was appreciably different from both those models. Where they posited complex organisms from the beginning, his model began with the very simplest forms of life, which took an immensity of time to develop into forms of great complexity. However, his vision of organic change did not negate the mechanisms of change Linnaeus and

Buffon had suggested. He gave interspecific hybridization no role in his theory of evolution, but he cited it as an argument against the idea of species fixity. As for his idea of how animals changed in responding to different environments, it was very similar to Buffon's thoughts on degeneration.

Lamarck's theory of evolution mirrored his image of how best to represent the order of nature. He arranged the animal classes in a straight line according to the complexity of their organization. He insisted that this arrangement represented the basic path nature had followed in bringing into existence all the different forms of life. By 1809 his classificatory efforts had led him to distinguish ten invertebrate classes that he lined up with the four vertebrate classes in ranked order beginning with the infusorians ("infinitely small animals" with gelatinous, contractile bodies but without internal organs)[13] and proceeding through the polyps, radiarians, worms, insects, arachnids, crustaceans, annelids, cirripeds, and mollusks. Above them were the fish, reptiles, birds, and mammals.[14] But he acknowledged that this progressive chain of being existed only at the level of the animal classes. At lower taxonomic levels, especially in the case of the genera and species, the situation was more complicated. Genera and species had to be seen as lateral ramifications from the main series.

Lamarck explained nature's order as the product of two competing causes: "On the one hand the product of the increasing composition of organization, which tends to form a regular gradation, and on the other hand that of the influences of a multitude of very different circumstances that continually tend to destroy the regularity in the gradation of the increasing composition of organization."[15]

Lamarck attributed the tendency to increased complexity to what he called "the power of life," but he was not thinking of some mysterious vitalistic power. He explained it in wholly materialistic terms: as the result of fluids in motion acting mechanically on the soft parts of living bodies. The more familiar part of Lamarck's theorizing is the way he explained changes at the species level. This took place as animals responded to a virtually "inexhaustible" array of "circumstances," including changes in climate, atmospheric temperature, environment, and habitats. Such changes influenced animals' habits, ways of life, means of self-preservation, multiplication, and so forth. The adoption of new habits led to the greater or lesser use of specific organs and body parts, altering them accordingly. Such changes, Lamarck told his students as early as 1800, "conserve and propagate themselves by generation." It was his first voicing of the idea of the inheritance of acquired characters and was key to his account of the way that "little by little nature has arrived at the state where we see it at present." He did not consider the idea controversial.[16]

In 1802 Lamarck said more about the conditions conducive to the inheritance of acquired characters. Citing numerous examples of animal structures (wading birds' long legs and necks, herbivores' hooves, cats' claws, etc.) and relating these to the animals' habits, he explained:

> Each change acquired in an organ by a habit of use sufficient to have caused it is then conserved by generation, if it is common to the individuals that in fecundation cooperate in the reproduction of their species. Finally, this change propagates itself and thus passes to all the individuals that follow one another and that are submitted to the same circumstances, without their having been obliged to acquire it by the way that really created it.[17]

That he saw no need to put this to a real-world test is evidenced by the thought experiment he offered to illustrate the concept:

> If, with two newborn infants of different sexes, one masked their left eyes throughout the course of their lives; if then they [the two individuals] were united, and one did constantly the same thing with respect to their children, I do not doubt that at the end of a great number of generations, their left eyes would come to disappear naturally and to be gradually obliterated. Following an immense amount of time, the necessary circumstances remaining the same, the right eye would come little by little to shift its position.[18]

In 1809, Lamarck presented these ideas in the form of two laws:

> *First Law*: In every animal that has not reached the end of its development, the more frequent and sustained use of any organ will strengthen this organ little by little, develop it, enlarge it, and give to it a power proportionate to the duration of its use; while the constant disuse of such an organ will insensibly weaken it, deteriorate it, progressively diminish its faculties, and finally cause it to disappear.
>
> *Second Law*: All that nature has caused individuals to gain or lose by the influence of the circumstances to which their race has been exposed for a long time, and, consequently, by the influence of a predominant use or constant disuse of an organ or part, is conserved through generation in the new individuals descending from them, provided that these acquired changes are common to the two sexes or to those which have produced these new individuals.[19]

Lamarck considered the inheritance of acquired characters self-evident, writing in 1815: "The law of nature by which new individuals receive all that has been acquired in organization during the lifetime of their parents is so true, so striking, so much attested by the facts, that there is no observer who has been unable to convince himself of its reality."[20] What was new was Lamarck's claim that the cumulative effect of the use or disuse of organs could proceed indefinitely, bringing wholly new species into being. He claimed credit not for the idea of the inheritance of acquired characters but for having been the first to enlist it in understanding "the causes that have brought about the astonishing diversity of animals."[21]

Frédéric Cuvier for his part enlisted the idea of the inheritance of acquired characters in his very first paper (1807), where he reported on the rutting periods of animal in the menagerie. With respect to the altered rutting periods of domesticated animals, he suggested it was likely that "qualities that are at first only accidental, can finally become hereditary, if the causes that produce them exercise their influence over a certain number of generations."[22] A year later, presenting "some reflections on the mental faculties of animals," Cuvier judged it probable "that some qualities regarded as belonging to instinct in animals are subject to the same laws as those that depend on training." Learned qualities, he proposed, "become finally instinctive or hereditary as soon as they have been exercised over a series of sufficient generations." By the same token, "they become obliterated or wear away more or less, after their exercise ceases to fortify or sustain them." Cuvier credited the idea not to Lamarck but to Charles Leroy, the former royal gamekeeper at Versailles.[23]

Cuvier next used the word "hereditary" in a paper (1811) on physical rather than behavioral traits, referencing characters affected not by use and disuse but by animal breeders' choices in propagating their herds, flocks, and such. This was the study his brother asked him to conduct on the osteology of domestic dogs, but it was with respect to dogs' different coat colors that Frédéric pointed specifically to the effects of selection: "When one takes care to reunite the individuals of the same color, the race ordinarily perpetuates itself, and it is the same for the majority of the characters that we have already examined: new proofs that accidental modifications end up always by becoming hereditary."[24]

In pointing to how breeders' selecting of traits shaped the history of dog breeds, Cuvier took care to insist that the variation of a species, wild or domestic, was ultimately limited. The differences wrought in domesticated animals, he explained, were an indication of "all the modifications of which these animals are susceptible *for us*," but these changes could only go so far. They could not be used "in support of the systems by which one has wanted to deduce the different forms of animals from the diverse circumstances that have influenced them."[25]

In January 1812, Frédéric Cuvier gave a paper to the Société philomathique "on the intellectual faculties of brutes." A summary published in the society's journal states that he presented "the rule, which he has long believed able to establish, that acquired faculties propagate themselves by generation and become hereditary." Cuvier used the rule to show "the cause of the existence of races and what they owe to this heredity." Additionally, he suggested how knowledge of this "law" might be applied to "the management of animals in general."[26]

In a very different context, in his capacity as an inspector of the French educational system in 1815, Cuvier stressed the importance of inculcating proper habits in schoolchildren.[27] By his account, different habits arising from diverse circumstances had, over time, produced the different mental attributes of different peoples: "The dispositions most difficult to deracinate are those that have propagated themselves from generation to generation up to ours, and it is for this reason that it is impossible for savages to acquire ideas that become familiar to us from our childhood."[28]

Cuvier apparently believed that "savage" races were disadvantaged by their ancestral mental habits and that civilized races had the benefit of better ones. In any case, he considered inculcating good mental habits in schoolchildren would improve their brains in ways that would be passed on hereditarily for generations. The distinction between characters transmitted hereditarily and characters transmitted culturally would not be clarified until the early twentieth century. It was not a distinction that could be clearly made if one believed in the acquired effects of habit.[29]

By Frédéric Cuvier's later account, his brother Georges put ideas of species transformation "in the rank of those frivolous games of the imagination with which the truth has nothing in common; with which one may amuse oneself when they are skillfully and gracefully presented, but which lose all their charm when taken seriously."[30] That was a gentle reading of Georges Cuvier's attitude. Georges regarded speculative "systems" as inimical to the development of natural history, and he insisted they should not be dignified by public attention. Privately, he was happy to ridicule them.[31]

The primary occasion on which Georges Cuvier deigned to address the idea of species change in print was the "introductory discourse" of his *Researches on Fossil Bones* (1812), where he referred to Lamarck's theorizing but belittled it by linking it with that of German nature writer Johannes Christianus Roedig and French naturalist and philosopher Benoît de Maillet. The latter had infamously suggested that humans descended from mermaids and mermen and that flying fish had given rise to birds (i.e., when fishes were stranded on dry land their scales cracked and turned into feathers). Cuvier identified Lamarck as one of the writers who had "extended prodigiously the ideas of Demaillet." These writers, Cuvier elaborated, "say that all was fluid in the origin; that the fluid engendered at first some very simple animals such as the monads or other infusoria and microscopic species; that, through time and in taking up diverse habits, the races of these animals became more complex and diversified themselves to the point where we see them today." For a scientist who scorned theoretical "systems," it was a backhanded compliment for Cuvier to state, "Monsieur de Lamarck is the one who has recently developed this system most fully and with the most sustained sagacity."[32]

But Cuvier had scientific objections as well as guilt by association to offer in combating the idea of species change.[33] For one thing, the mummified forms Geoffroy brought from ancient Egypt showed that species had not changed between the pharaonic era and the present. For another, there was what Cuvier called the principle of the "conditions for existence," otherwise known as "final causes." He explained: "Since nothing can exist if it does not reunite the conditions that render its existence possible, the different parts of each being must be coordinated in such a way as to render possible the whole being, not only in itself, but also in its relations with those that surround it."[34] Integral to this principle was the concept of the correlation of parts, which Cuvier touted as the basis of his ability to identify extinct species based on no more than an isolated tooth or bone.[35] As Cuvier saw it, the correlation of parts made species change impossible. No single part of an organism could vary appreciably without destroying the coordination of the organs of the entire organism and thereby rendering the organism unviable.[36]

Cuvier also had paleontological evidence to offer against the idea of species mutability. The fossil record showed no signs of transitional forms between such extinct creatures as the *Palaeotherium* (a large herbivore) and any modern forms.[37] Similarly, Cuvier's efforts in systematics (classification) worked against any broad system of organic evolution. In 1812 he identified four distinct "provinces," "branches" (*embranchements*), or "general plans" of animal organization (the Vertebrata, Mollusca, Articulata, and Radiata) and maintained that these basic structural types were wholly distinct with no "nuances" between them.[38] Nor could these four types be arranged in a single line, and this was true for most of the classes of animals within them. Finally, with specific reference to change at the species level, Cuvier cited the experience of animal breeders that variations in domesticated species never went beyond specific limits and that hybridization of existing species never led to new forms that could maintain themselves longer than a few generations.[39]

Georges Cuvier never assembled all these arguments in one place. In *Research on Fossil Bones*, he pointed only to three: the lack of transitional forms in the fossil record; the limited nature of the variation seen in both wild and domestic species; and the identity between animals from the tombs of ancient Egypt and species of the present. In doing so, he did not call out Lamarck. It was only after Lamarck died that he directly attacked his theorizing, doing so in the "eulogy" for his deceased colleague that it was his responsibility to present at the Academy of Sciences. In writing the piece, he took the extraordinary approach of presenting Lamarck's theorizing as an object lesson in what scientists should not do. As it happened, Cuvier died before he could deliver the eulogy himself, and it was read in his place by the linguist Baron Antoine Isaac Silvestre de Sacy.[40]

Frédéric Cuvier, like his brother, only rarely mentioned Lamarck's name when taking potshots at the idea of species change or sketching a more extended refutation of it. He mentioned Buffon's name more often, treating the famous naturalist with respect but concluding that his speculations about "degeneration" were wrong. Frédéric also attacked with some frequency the idea that hybridization could produce new species. When it came to addressing the broader idea of life's gradual development from the simplest forms to the most complex, however, what he took for his target was a curious hodgepodge of ideas corresponding neither to Lamarck's views nor to those of any of his contemporaries.

The younger Cuvier's critiques of species mutability appeared primarily in essays for the *Natural History of Mammals*, the work published under his and Geoffroy's names that began appearing in installments late in 1818. Sometimes the criticism was merely implied. In an essay on the brown coati (March 1819), for example, he stressed the necessity of studying variation, which remained a "totally neglected branch of science." Not only did it need to be studied experimentally, he said, one also needed to know the range of varieties in nature, so that one could "establish the modifications that each species undergoes, the organs where these [modifications] manifest themselves, and *the limit at which they stop* [emphasis added]."[41]

In a September 1819 essay, "The Eskimo Dog of Baffin Bay," in contrast, Cuvier launched into what he took to be a rebuttal of contemporary notions of species change. Describing species mutability as an idea going back to "the deepest antiquity" and "having served as text for today's most celebrated scientists," he characterized the modern version as one where

> nature originally created only some germs that developed according to surrounding influences in the environment in which they found themselves, in such a way as to produce the species we know. The development of the germs would have been all the greater according to the degree that the influences, in the environment where they found themselves, were more favorable; and the animals whose organization is the simplest, such as the polyps, would have been much closer to the immediate work of the Creator than the less imperfect animals, man and the mammals. The latter would only be the product of secondary causes, and would have reached the state of perfection, or rather the complexity they have attained, only having passed through all the intermediaries that separate them from the simple beings.[42]

Cuvier charitably allowed that this "system" was not incompatible with how one ought to conceive of "the power that created the universe," "that is, as simple in its means as [it is] immense in its purview and infinite in its results," but it had already been "combatted with enough force to render it doubtful, and even to cause it to be rejected by those austere minds that see greatness only in the truth and never allow the imagination to take the place of experience." All that the advocates of species change offered in the way of evidence, Cuvier said, were the variations of animals under domestication, which could be attributed to "secondary and accidental causes." From there, it was simply a matter of drawing upon the influence of diverse environmental circumstances (which the Earth's history indicated to be "very probable") and conclude that it "is no more difficult to show that the orang-utan is the same species as man, than to show that the differences that characterize our diverse horses or diverse dogs are only differences of race."

But Cuvier thought he saw "a difficulty" that rendered the whole system invalid. In a system where original creation was limited to "germs," he wrote, the influence of environmental circumstances should be greatest on animals nearest the germ stage, just as environmental influences are strongest when a seed is sprouting, not when the plant is mature. One should thus find more varieties among the simplest species and fewer varieties among those with a more advanced organization; but naturalists had observed just the opposite, that mammals displayed the greatest number of varieties.[43] He wrote: "One must necessarily conclude, against the system that is the subject of these observations, that the simplest animals do not at all contain in germs the organs of those that are more complex, and that all the facts tend to confirm the common opinion, that the species have always been and will always be what we see today."[44]

Cuvier's identification of a fatal flaw in the system was itself flawed, however, because no one was advocating the system he described. Lamarck's theory was about evolution from the simplest species to the most complex, but it was wholly incompatible with the idea of primitive germs containing within them the germs of the organs of all species to come. The latter notion belonged instead to the late eighteenth-century Genevan naturalist and philosopher Charles Bonnet (1720–93). Bonnet's preformationism was the opposite of Lamarck's transformism. The sloppiness of Cuvier's rendering of contemporary ideas of species change stands in marked contrast to his painstaking analyses of how new specimens at the Museum and menagerie compared with descriptions of species in the scientific literature.[45]

For the most part, Frédéric Cuvier's arguments against species mutability related primarily to the limits of variation within species and the failure of interspecific hybrids to sustain themselves over many generations. Writing on a cross between a zebra and a donkey (March 1820), he was happy to endorse what he called "Buffon's rule": that all the animals that copulate, that reproduce by copulation, and that give birth to fecund individuals, are of the same species." He acknowledged a practical problem with the rule: it was difficult to apply to animals in the wild, where constant observation was not possible. He was nonetheless confident that a natural repugnance between wild animals of different species prevented them from mating with each other. Only under conditions of domestication or captivity did animals of different species occasionally mate and produce offspring. And when this happened, it always involved at least one domesticated animal (the dog with the wolf; the ass with the zebra), if not two (the horse and the ass; the goat and the sheep). The fecundity of the resulting hybrids, said Cuvier, depended on the species in question. In some cases they were sterile, in others they reproduced for a few generations before their ability to do so ran out. On balance, Cuvier agreed with Buffon that "the phenomenon of reproduction is the only one on which one can establish the distinction between species."[46]

Just as influential as Buffon's rule for distinguishing between species was his long essay "On the Degeneration of Animals" (1766), where he promoted the idea that species had changed over time.[47] Frédéric Cuvier sniped at this idea in various places, including a piece on the domestic pig of the Cape of Good Hope (1821). He explained how Buffon had traced all the different forms of mammals to a small number of primordial types, taking readers back "to the secrets of the Creation to show us the immensity of the All-Powerful in the simplicity of the means that He employs." Acknowledging that Buffon's "hypothetical views" and "brilliant speculations" had modern advocates, Cuvier insisted one should ground one's thinking in facts rather than being led astray by systems "that abuse us more often than they enlighten us." The proponents of species change, he said, in basing their claims on the extent to which climate influenced domestic animals, had extrapolated "well beyond the limits that experience has indicated."[48]

Later in 1821, when puppies were born from the mating of a jackal from India with a jackal from Senegal. the menagerie superintendent had to reassess his claim that there were no known cases of two nondomesticated species' mating and producing viable hybrids, but he did not take their successful reproduction as indicating that they were varieties of the same species. Instead, he promoted this as the first documented case of two wild species' mating and producing viable offspring (albeit in captivity).[49] In 1824, when a lion and a tigress at Windsor Castle produced a pair of cubs, Cuvier offered it as "a new proof" that the propagation of individuals did not establish their specific identity. At the same time, he dismissed the notion that it showed there were no distinct species in nature. This was disproved, he said, by the fact that interspecific hybrids reproduced at best only feebly in their first generation and that successive generations became progressively weaker. No one had ever seen interspecific hybrids reproduce beyond the fourth or fifth generation.[50]

Although Cuvier discussed the hybrid jackals of 1821 cheerfully and was equally positive about the lion-tiger cross of 1824, his tone was very different in 1830 when he reported a baby produced by two species of macaque: a bonnet monkey and a common macaque. This time he disparaged the significance of hybrids, describing them as foreign to "the ordinary paths of Providence" because they were not destined to contribute to the natural economy of the world, did not genuinely belong in natural history, and carried within them only "principles of death."[51]

What accounts for this change of attitude? It evidently had nothing to do with the monkeys, for Cuvier spoke favorably of their behavior. The parents had gotten along well, and when the baby was born (December 1829) the mother nurtured it carefully and the father did it no harm. The mother's solicitude was "absolutely the same" as a mother rhesus monkey had displayed with her own baby in the menagerie, and the male too was solicitous, though not to the same degree. What angered Cuvier was the recent proliferation of ideas where hybrids served as the takeoff point for erroneously promoting what he called "this idea of the origin of species," that is, the idea that early in Earth's history hybrids were produced that then reproduced themselves and gave rise to "durable species that today populate the surface of the globe." Unless this hypothesis could be supported by facts, he said, "all that one will conclude will be conjectural, imaginary, and more proper in a work of fiction about nature than its history."[52]

Frédéric Cuvier completed his article on the Chinese bonnet monkey-macaque hybrid in May 1830, shortly after the famous heated and protracted debate between his brother and Geoffroy Saint-Hilaire at the Academy of Sciences over how to interpret animal structure. The main issue was Geoffroy's "philosophical anatomy" and his idea of a unity of plan throughout the animal kingdom versus Cuvier's insistence on the relation between structure and function. The mutability of species was not central to the debate, but it lurked in the background as part of an assemblage of Geoffroy's views that the Cuvier brothers found unacceptable.[53]

Geoffroy publicly embraced species change in a long paper in the Museum's *Mémoires* (1825) suggesting that the Teleosauruses and Steneosauruses of antediluvian times were the ancestors of present-day crocodiles.[54] In doing so, he recommended Lamarck's evolutionary theorizing "to the meditation of young people" and endorsed Lamarck's "two laws" on the effects of use and disuse and the inheritance of acquired characters.[55] He had come to think of Lamarck as a kindred spirit, though their views of the mechanism of change differed. Geoffroy proposed that major changes in the physical conditions of the Earth in the past had direct pathological influences on the arterial systems of fetuses, producing animals that were monstrosities of a sort but nonetheless viable enough to start new generations. He challenged the conclusions Georges Cuvier had drawn from the mummified specimens he'd brought from Egypt. When he returned, Geoffroy explained, he had been unable to give full attention to his "very painfully acquired" riches and derive from them "the philosophical results" they had to offer. Cuvier had claimed that Geoffroy's collections, when compared with modern forms, showed "the rigorous identity of the species that lived three thousand years ago with the animals of today." But Geoffroy now presented a mummified crocodile skull that not only differed from the modern form but also varied along the same lines as the differences between the antediluvian Teleosaurus and modern crocodiles.[56]

Geoffroy returned to the issue in 1828, reviewing for the Academy of Sciences François Roulin's study of the changes undergone by domesticated animals that had been introduced into the New World, escaped their domestic state, and returned to the wild. Geoffroy took the opportunity to signal support for the transformation of species and selected features of Lamarck's thinking. He credited Georges Cuvier with demonstrating that fossils from Earth's antediluvian past represented different species than those of the present, but Geoffroy was prepared to think that past species were the ancestors of present ones. Noting that Roulin had discussed "the transmission by way of generation of certain acquired habits," Geoffroy approvingly cited Lamarck's discussion of how the environment influenced the actions and habits of animals and how new habits in turn modified animal organization and structure.[57]

Frédéric Cuvier's hostile words in 1830 about hybrids appear to have been a reaction to the reappearance of the idea of species change. In the same period, he was responding to an older voice as well—that of Buffon—as he edited a two-volume supplement to Buffon's *Natural History*. The first volume, on mammals, appeared in 1831; the second, on birds, came out in 1832. In his introduction to each volume and in commentaries on certain species, Cuvier complained not only about Buffon's thoughts on the production of new species through degeneration and hybridization, but also about how this hypothesis had been "carried to its extreme limits" by modern authors. He insisted that "all the changes experienced by the animals submitted to the action of the most numerous and most powerful causes are limited to an augmentation or diminution of size, or in the proportion of some parts, to some of the qualities of fur, etc."; in other words, changes involving the characteristics that naturalists normally employed to distinguish varieties, not species.[58]

Discussing the American bison again, the younger Cuvier declared that fecundity could no longer be considered sufficient to identify two forms as belonging to the same species. Had Buffon appreciated this, he said, he would not have treated the bison and the cow as different varieties of the same species. Cuvier claimed he had twice seen a cow that had mated with a bison die in labor.[59] The cow's inability to bring a hybrid calf into the world was evidence, he said, that "the bison and the cow were not at all destined to produce together."[60]

From a different direction, with respect to human races, Cuvier tentatively questioned Buffon's rule for determining species identity by interfertility plus his ideas about how quickly human racial change could occur and how, as Cuvier put it, "a European would be able, under some sort of circumstances, to become a Negro or an American, to pass from the race to which he belongs to another race." Cuvier suggested it may have been Buffon's sentiments or his sense of his religious duties that led him to take that position, but since that time the human species had "lost the absolute character it had previously." Cuvier observed, "several authors have been able, with some foundation, to multiply the species of man more than Buffon had [multiplied] the varieties." This was but a fleeting remark on Frédéric's part. For others who were arguing that the human races represented different species, it was evident that Buffon's definition of species had to be qualified or rejected, given the ability of different races to produce fertile offspring.[61]

Georges Cuvier's death in May 1832 rendered moot his injunction to his brother that they should not both occupy Museum professorships at the same time. In July, when Frédéric presented his case for a professorship to the Museum administration, one of his research goals was "to give to the idea of species a precision it does not have." His observations and experiments on living animals at the menagerie, he said, had been devoted to recognizing the "laws of existence" of living animals "as well as the nature and extent of the modifications these laws can experience."[62] He had articulated the same goal in 1815 when requesting that a park be established for experiments. It never came into being.

Frédéric returned to the species question in the introduction he wrote to the fourth edition of his brother's *Research on Fossil Bones*, where he defended Georges's stance on species mutability and the age of the human species. Georges maintained that there were no human remains to be found together with the remains of animals destroyed by the Earth's last great catastrophes, and he asserted that the species of the present are fixed, unchanging, and unconnected with the fossils of antediluvian times. Frédéric described the question of species mutability as "a hypothesis renewed from ancient times that consists in deriving from one or several primitive species all the animal species, those that existed formerly as much as those that exist today."[63] He allowed that he wished to keep his comments scientific, presided over by "the sentiment of affection I hold for the scientist I see myself forced to combat," an obvious reference to Geoffroy.

This time Frédéric portrayed the theory as beginning with a single species, "represented without doubt by an unspecified number of individuals, if not by the living organic molecules supposed by Buffon," which then developed under conditions that had changed thousands of times over thousands of years, ultimately producing "this world of animals that covers our globe today, from the microscopic beings, whose whole existence appears vegetative, up to man, intelligent and free."[64]

Characterizing this as a seductive system that "with each stroke of a wand" called into being creatures "endowed with life and sensibility," Cuvier insisted it was based not on facts but on false analogies. One knew nothing about the conditions under which these animals had originally lived, he said, and nothing about causes in the present that could have brought about the profound changes in question: "The truth is that science produces no fact of this nature. Never has one seen, under any limit or any form, a species change its conditions of existence to transform itself in whole or in part into another species." Those who based their arguments on the variation of animals under domestication, he said, had failed to appreciate that induction carried beyond its proper limits was "only a source of trouble and confusion"; it was the cause of "all these false, bizarre, obscure" ideas "that naturalism has put in vogue."[65]

When Frédéric Cuvier turned to the causes of variation in animals under domestication, he didn't mention the inheritance of acquired characters. He focused instead on "accidental variations," noting that "all the accidental modifications that reproduce themselves for several generations end up being transmitted from the fathers to the children and become characters of the race." He cited as the "most profound" example of such a "hereditary organic modification" the appearance of a fifth digit on the hind feet of some races of dogs. But he admitted that scientists did not know what caused such organic modifications and had no way of eliciting them. He noted that giving animals more food seemed to influence their height, but he offered no other character of domesticated animals that man could directly influence. He did not discuss, as he had earlier, how breeders could develop traits in animals by selecting for them. He simply insisted that no one had ever seen any species of domesticated animal develop the characteristics of another species of its genus, either domestic or wild. One never saw "a dog become a wolf, a jackal, or a fox" or a horse become an ass, or an ass become a zebra, or a goat become a sheep.[66]

Ratcheting up his argument, Cuvier said that if the theory of species change consisted of no more than a species becoming another species of the same genus, it would "crumble at its foundations," because the theory also required that species pass from one genus to another genus, from one order to another order, and from one class to another class.[67] As he saw it, the appearance of a fifth hind digit in certain races of dogs was of no value for the theory of species change because the difference between the genus of dogs and comparable genera remained entirely clear-cut. More broadly, he rejected the legitimacy of extrapolating "from the facts of domestication to the facts of the state of nature."[68]

Cuvier dismissed any appeals to authority, saying that the ideas of Buffon and Lamarck had been "generally regarded as games of their imagination, or as the consequences of principles admitted without proof." The system that Buffon developed "with as much lucidity as eloquence" and Lamarck's system that relied on "so many arbitrary suppositions" were both unacceptable because they were inconsistent with the facts, said Cuvier, and not based "on any precise observation, on any rigorous experiment." Furthermore, he claimed, ever since these systems were enunciated, no new facts had been brought to their aid.[69]

At this point Frédéric Cuvier paused to offer a statement that would prove prophetic, though he obviously did not imagine things developing this way. “Ah!” he wrote, “If there existed the feeblest proof, I would say not even of the transformation, but of the possibility of the transformation of one species into another species, how could an anatomist, a physiologist, or a naturalist be able to direct his attention thereafter to any other phenomenon?” Confident that proof of species change would never come, he acknowledged that if the idea were ever recognized as possible, then at that “very instant” there would be “a fundamental revolution in all the sciences which, directly or indirectly, have animals as their subject.”[70]

Frédéric Cuvier was not wrong to say that Lamarck had failed to prove the case for evolution. But neither did he advance the case for species fixity via observations or experiments. The animals of the menagerie were poor subjects for experimenting on how far variation in a species could go because they reproduced too slowly. If three millennia since the ancients mummified their sacred animals was not enough time to decide the issue, how could one expect it to be tested in a single scientist’s lifetime? And if Lamarck’s theory did not rely on the hybridizing of different species, what difference did it make if experiments at the menagerie showed that interspecific hybrids failed to reproduce themselves after a few generations? Frédéric Cuvier never managed to perpetuate dog-wolf hybrids beyond three generations, but that did not disprove the idea that wolves and dogs evolved from a common ancestor.[71] Considered from a different angle, how decisive was it that hybrids failed to reproduce for more than a few generations at the menagerie when only a very few species ever did?[72]

Lamarck fit Frédéric Cuvier’s portrayal of a thinker who having once conceived the idea of the transformation of species could no longer resist it, ignoring all that seemed miraculous about it. But Lamarck was not oblivious to how hard it was to imagine a natural explanation for some of nature’s remarkable phenomena, including the instincts of animals and especially of insects. He posed the question to his students in 1803:

> Would one not be stopped . . . by the sole consideration of the admirable diversity one notices in the *instinct* of the different animals, and by that of the marvels of all sorts that their diverse sorts of *industry* present? Would one dare carry the spirit of system-building so far as to say that it is nature that has, by herself, created this astonishing diversity of means, ruses, skills, precautions, and patience of which animal industry offers us so many examples! What we observe in this regard, in the class of insects alone, is it not a thousand times beyond what is necessary to make us feel that the limits of nature’s power in no way permit her to produce by herself so many marvels! And to force the most obstinate philosopher

> to recognize that here the author of all things has been necessary, and has alone sufficed to bring into existence so many admirable things?

While saying it was unwise "to assign limits to the power of the primary author of all things" (a creator god), Lamarck maintained that his admiration of the "primary cause of all things" would not be diminished if it turned out that, instead of attending to all the details of creating, developing, and changing everything in existence, the "primary cause" had delegated these matters to nature.[73] In this he was echoing a famous passage from Buffon's "Discourse on the Nature of Animals": "Who, in fact, has the greatest idea of the Supreme Being, he who envisions him creating the universe, setting in order the forms of life, founding nature on invariable and perpetual laws; or he who looks for him and wants to find him intent upon managing a republic of flies, and very much occupied with the way in which a beetle must fold its wing?"[74]

Instincts, no less than animal structures, Lamarck told his students, could be explained as arising from habits elicited and then maintained for long periods in different environmental circumstances. His confidence in being able to explain, at least in principle, such fine details as the instincts of insects as the consequence of natural causes led some writers to raise the cry of "atheism."[75] This happened both in France and in Britain, but it was more in Britain than France in this period that naturalists strived to make the case that the adaptation of organisms to the conditions of their existence constituted "the wisdom of God manifested in the works of creation."[76]

Frédéric Cuvier did not level against Lamarck the charge of atheism or materialism, but for that matter, aside from noting Lamarck's name in his introduction to the fourth edition of his brother's *Researches on Fossil Bones*, he seems never to have mentioned him in print at all. When it came to the special adaptations various species exhibited, such as the hedgehog's ability to roll itself into a ball with spines projecting outward to protect itself from predators, Cuvier suspected their explanation was beyond the reach of science. Why was the hedgehog provided with this defense when its behavior, appetites, and instincts were much the same as those of similar animals like moles, shrews, and other omnivorous rodents? Wrote Cuvier, "It is not given to us to resolve this question, undoubtedly because it is still not included in the circle of proximate causes that alone can be the object of our knowledge."[77]

Like his brother, Frédéric Cuvier focused his arguments against species change on the insufficiency of the supporting facts and the multiplicity of facts against it. He did not appeal directly to a wise creator to explain how well suited animals are to the conditions of their existence. He evidently felt, nonetheless, that was where the answer ultimately lay. In the 1830s, the word "Providence" occasionally crept into his writings, primarily when he referred to the close fit between animals' needs on the one hand and their structures and instincts on the other. In 1833, in introducing his new, two-volume edition of his and Geoffroy's *Natural History of the Mammals,* he described the study of natural history as "expanding our ideas on the views of Providence." Regarding animal instincts he wrote, "Instinct has been formed for what surrounds it, as what surrounds it has been formed for it." Thus, "when the beaver was created to build with wood and earth," there had to be "trees appropriate of cutting" and "soil appropriate for kneading." He summed up this train of thought saying, "We cannot conceive the existence of the universe otherwise."[78] Three years later, when lobbying the minister of education to create a professorship for him at the Museum, he signaled that he would teach students to recognize the hand of Providence in the relations between the faculties of animals and what they needed to live.[79]

He expressed the same perspective that same year (1836)in his book titled *On the Natural History of the Cetaceans.* He took as his primary task describing and comparing the physical characteristics and organs of the various species of whales and porpoises then known. He used teeth, bones, and books to relate the different species to each other. It was very much the work of a cabinet naturalist. Cuvier acknowledged, however, that the only way to appreciate the faculties that whales derived from their organs and the actions they were capable of would be to study them living and animated by intelligence and will. "Only then," he wrote, "can the animal show us what it really is and unveil the views of Providence in making for it a place in the creation."[80]

His aim at this point, however, was not so much to extol the wisdom of the Creator as to discuss the lives of whales and the difficulties of studying them. He noted whales' strong instinct for sociability and their high degree of intelligence and emphasized the importance of learning and understanding more about their geographical distribution. He then reflected on how human influence bore on the latter two subjects. The history of whaling would not normally be part of a naturalist's purview, he admitted, but given that "the history of nature consists . . . in the reciprocal influence of beings subject to its laws," he felt it appropriate to say something about the influence "of man on the cetaceans and the cetaceans on man." In doing so, he offered reflections on habits and change consistent with his general views on species variability and its limits.

The hunting of whales, Cuvier explained, had changed their geographical distribution. Whales had left the seas where they were once abundant "to take refuge in those that the ice renders nearly inaccessible to our vessels." He attributed this to the inheritance of acquired behavior. As he saw it, the intelligence, habits, and latest instincts of whales were a consequence of being hunted by humans. "[As] with all the other animals," he wrote, the intelligence of whales arose in relation to their needs: not only had the strength of their intelligence increased through exercise, so too had the exercise of that intelligence transformed what "had at first been only accidental and transitory modifications" into "durable habits" and "natural dispositions."[81] Had Lamarck discussed the intelligence of whales, he would doubtlessly have said the same.

Cuvier recognized that the financial profits of whaling had led to the reduction of sperm whale populations, "if not their complete destruction." But rather than reflecting on the species' possible extinction, he rhapsodized on how whaling had developed man as well as whales. The conditions to which whale hunters were subjected, he wrote, had an influence "100 times more powerful" on them than on the whales.[82] Not only did whalers have to deal with the whales' efforts to escape or defend themselves, they also had to journey to the most inhospitable parts of the globe to continue the chase. From the comfort of his study, Cuvier offered a vivid recital of the challenges whalers faced:

> At the least carelessness of the crew [whales could] drag to the bottom of the sea the frail small craft that attack them, or with a blow of the tail or head throw them in the air and break them to pieces; but also . . . this hunt scarcely takes place except in the least explored and the most dangerous seas, where no hospitable shore offers help to the vessel disabled by the tempest, where the mountains of floating ice, in gliding one against the other, can crush the ship pushed or dragged in the middle of them; where the shipwrecked whaler, without shelter from the snow and the winds, without fire to combat the cold or water to combat thirst, is left to fend against ferocious beasts and to try to take the beasts' lives to maintain his own.[83]

Such difficulties, as Cuvier saw it, had exercised and honed the whalers' mental faculties. In maritime nations, whaling served as "a nursery for experienced, courageous, and prudent seamen."

Cuvier did not specifically state that such qualities became hereditary. Nor did he reconcile the fact of the diminution of whale populations with "the views of Providence" in providing each species "a place in the creation."[84] His emphasis was on how man's activity improved himself and nature alike. His confidence and optimism in this regard paralleled his view that the changes in animals under domestication, even if termed *dégénerations*, often constituted improvements over the original forms.[85]

Frédéric Cuvier's case illuminates some of the primary difficulties when it came to addressing the question of the origin of species in the early nineteenth century. He was a strong advocate of the idea of the inheritance of acquired characters but a dedicated opponent to the idea of species change. He called for the study of variation and emphasized how breeders selected small differences among their animal populations for the purpose of developing new traits or accentuating existing ones, but he did not find the solution to the origin of species there either. Breeders had never transformed one species into another. Charles Darwin would conclude there was a parallel between the breeders' selection of variations in domestic animals and nature's "selection" of variations in the wild, but that was not an easy step to take. Cuvier was certainly right to urge that variation be studied scientifically. He was prophetic in his observation that if ever there was proof of even the possibility of the transformation of one species into another, it would produce a revolution in the life sciences. But he had no expectation that it would ever happen. He viewed humans (especially civilized ones) as improvers on nature, but within limits that were ultimately set by Providence. Species change, he believed, could go only so far.

[4.7]

Dogs, Wolves, and Jackals

Canines blurred the boundary between the menagerie and its geographical setting. Paris was full of dogs, wolves were native to France, and jackals, though foreign, were recognizably canine. Over time, the menagerie had them all. They were not the greatest attractions, but dogs, wolves, and jackals could still be of public and scientific interest. A dog living in the same cage as a lion was a pleasant surprise for the public, as was a tame wolf, and dogs featured prominently in Frédéric Cuvier's scientific efforts on multiple fronts. They figured in his thoughts on animal psychology and domestication and in the studies of intraspecific variation and hybridity he brought to bear on the species question. In addition, his efforts to distinguish among "the least well known" of the canines, those "united vaguely under the name of jackal," are an excellent reminder of the challenges he faced as a zoologist trying to put specimens in their proper taxonomic places.[1]

The dogs that influenced the life of the menagerie most dramatically and negatively were visitors themselves. Some were strays, but most came with their owners. They were a thorn in Mordant Delaunay's side throughout his tenure as the menagerie's first overseer and continued to be so for his successors. Delaunay complained to the administration time and again of the misbehavior of dogs and their owners alike, but to little avail. When the Museum required that dogs be leashed, the owners simply waited until they were out of the sentries' sight to unleash them.[2]

Canine depredations also inspired the move to make the menagerie's spatial definition more precise. Before 1804, the herbivore parks were not fenced off from the rest of the gardens. Geoffroy proposed that fencing in June 1804, explaining that it would keep dogs out and thereby give the herbivores more chance to rest.[3] His colleagues agreed to an enclosure with two entrances guarded by sentries, but the new arrangement did not prevent mishaps.[4] In April 1807 a large dog entered a deer park and seriously injured a white-tailed deer given by Josephine Bonaparte.[5] The Museum

Figure 4.7.1. "The jackal of Senegal." C. de Lasteyrie after Jean-Charles Werner. From É. Geoffroy Saint-Hilaire and F. Cuvier, *Histoire naturelle des mammifères*. General Research Division, the New York Public Library.

responded by reaffirming that dogs were forbidden in the menagerie and by reducing the public entrances to just one, with a man stationed there to keep dogs out and "take care of them until the return of their masters." The results were not perfect. The gatekeeper bore the brunt of angry donors, who chafed at being separated from their pets, and the din of the dogs howling when left with the gatekeeper upset the neighbors. The administration then sought to exclude the public's dogs from the establishment entirely. It appealed to the minister of the interior for staff to make the policing possible.[6]

Stray dogs were another problem for the Museum, as they were for the whole city, so much so that in 1807 the prefect of police ordered them destroyed. An ominous entry in the professors' procès-verbaux of 4 March 1807 records the administration's writing the police "to ask that the cadavers of dogs killed during the night not remain exposed in the vicinity of the Museum."[7] The police program was not definitive, however, judging from the professors' 1810 observation that "the majority of people living at the Museum" were "exhausted by the perpetual howling of stray dogs in the night." The administration tasked the nighttime security guards with "chasing and even killing those that make the most noise."[8]

In 1814 the commander of the sentries complained again of dogs entering the menagerie; once again the Museum forbade their entry.[9] Dog-related accidents continued periodically, however, and some were serious, such as in February 1837 when a bulldog broke into the llamas' park and grievously injured both specimens. A frustrated Frédéric Cuvier appealed to the administration that "an effective measure be finally taken to prevent the too frequent accidents caused by dogs."[10]

The incident of the bulldog in the llama park was followed within a day or two by one of the more sensational and embarrassing (and perhaps exaggerated) incidents in the menagerie's history. As reported in the press, a man from the neighborhood who wanted to be rid of his dog asked a policeman to do the job for him.[11] The policeman elected to toss the dog into a bear pit occupied by a mother bear and her two large cubs. The dog was about to be devoured when its pitiful trembling awakened the mother bear's defensive instincts. She placed herself between her cubs and the unfortunate canine and protected it until morning, when the keepers came to feed the bears and rescued the dog. The policeman was summarily fired for his act.[12] Poking fun at what it took to be the other papers' gullibility with respect to this story, *Le Figaro* embellished it by reporting that the prefect of police, after taking away the policeman's badge, made a medal of honor for the mother bear and personally pinned it on her.[13] It was not the kind of attention the Museum desired.

The bulldog incident led the administration to restate its rules regarding dogs and to insist once more on their enforcement, but another instance just a few months later, this time in the cassowary's park, led the administration to decree that dogs in the garden that were not leashed and muzzled would be seized and destroyed.[14] Later, when the administration relented on muzzles for the off season, it acknowledged that dogs belonging to the Museum's own employees had been part of the problem. It forbade employees housed at the Museum from letting their dogs wander freely on the grounds, under penalty of loss of lodging.[15] Four decades after Delaunay complained about them, dogs on the premises were still a problem.

The dogs that entered the Museum's grounds with their masters, or as strays, were not the only ones on site. From the Museum's earliest days, watchdogs guarded its grounds and some buildings (e.g., the pheasant house).[16] Dogs were also used to control some of the menagerie's more dangerous animals. An 1831 census of the Museum's animals describes two mastiffs as "useful when one goes down into the bear pits."[17] (We saw how dogs were used in 1820 to drive Martin the bear away from the body of the man who had jumped into Martin's pit.)[18] But these dogs could be problems themselves. In 1825, a dog belonging to Geoffroy broke its chain at the rotunda and killed two ewes. After being attached to a heavier chain, it broke that as well and killed two more ewes. Frédéric Cuvier turned the animal over to the proprietor of the animal fights.[19]

Another specialized capacity in which dogs served the menagerie was as companions for lions. Viewers treated the first such instance—pairing a hound with the menagerie's first lion, "Woira"—as a rarity replete with social and political significance, but it proved to be common enough. There were two such duos at the menagerie in 1804 and another in 1827.[20] The menagerie also kept dogs that were interesting as special types. The first was the wild dog of Australia, brought back by the Baudin expedition. Later breeds ranged from greyhounds to Newfoundlands and huskies. Frédéric Cuvier liked the latter two, especially huskies, because of their intelligence, courage, and gentleness.[21]

Wolves' fierce reputations interested city dwellers, but wolves were not as impressive as lions or elephants or as entertaining as bears. No one felt the menagerie would be incomplete without them. Thus in July 1810, when the menagerie had no wolves but was given the chance to get one from the royal menagerie of Amsterdam, the Museum refused.[22] A few months later, however, it accepted from the prefect of the department of Basses Pyrénées a pair of young black wolves because they represented a distinctive variety.[23] Then it acquired the wolf from Amsterdam after all, taking it in a trade for a porcupine and a vulture it did not need. When another offer arrived before the year was out, it rejected it on the grounds that three wolves were enough.[24]

In subsequent years the Museum rejected most offers of wolves and wolf-dog hybrids because they provided nothing new. When Frédéric Cuvier needed to make room for other species, he sometimes managed by selling or trading wolves or sending them to the "combat master."[25] In 1818, however, when the veterinarian Huzard requested wolves so he could experiment on ways to destroy them (in light of the regrowth of the country's wolf population), Frédéric Cuvier said he had none to spare, because all three of his specimens were of scientific importance.[26] The first, successfully bred with a dog to produce wolf-dog hybrids, was extremely tame and was one of the menagerie's oldest animals. The second was so gentle it could be experimented on (Cuvier was surely planning studies of the wolf's faculties and behavior, not how to destroy the species). The third wolf had already been promised in exchange for a pair of Newfoundland dogs from Dr. Leach at the British Museum.[27]

Over the next two decades, the menagerie's experience with wolves followed familiar patterns. Some offers were rejected because the Museum did not consider the specimens interesting enough.[28] Others were accepted because of the donor's status (e.g., Georges Cuvier's stepdaughter Sophie

Duvaucel).[29] Several were swapped, sold, or given to the combat master when space was needed in the pavilion of *animaux féroces*.[30] The menagerie apparently had no difficulty acquiring them. Indeed, at the end of December 1837, it found itself with seven (three males, three females, and one hybrid).

Jackals arrived at the menagerie much later than dogs and wolves. They exemplified what had been touted as a special advantage of having a menagerie: enabling zoologists to study similar, live individuals up close over time to determine whether they were the same or different species. The menagerie's first jackal was a male from Bengal. Sent by the botanist Jean-Baptiste Leschenault de la Tour, it arrived in March 1818.[31] When a female jackal from Senegal arrived in 1820, Cuvier was able to compare her closely with the male from Bengal and conclude she represented a previously unknown species. Compared with the common jackal, her proportions were more elegant, her form lighter, her cry softer, and her odor less strong.[32]

But the two animals offered Cuvier more than a study in species differences. They were kept in the same cage and mated when the female came into heat, resulting in a litter of five viable puppies. Cuvier celebrated the event as the first known case of two wild species successfully breeding. He was worried that this first-time mother would, under the stress of captivity, devour her puppies rather than care for them, but the strength of "maternal love" overcame her initial "extreme inquietude" and prompted her nursing. To help the mother feel secure, Cuvier left the animals largely unbothered for two weeks. The two puppies that lived beyond the weaning stage displayed different temperaments from the start. One was unafraid of humans, the other extremely wary. Cuvier's impression was that these natural dispositions would not change and that even the best treatment would never tame the wilder one.[33]

Between 1820 and 1825, the menagerie received four jackals, all from Senegal. In 1826 it received a pair from the Malabar Coast, courtesy of J. J. Dussumier, and a male from Senegal, purchased from Dominique Schmitz, a German trader from Trier (Trèves) who wanted to start an itinerant menagerie.[34] Frédéric Cuvier fretted for several years over the species identity of Schmitz's jackal, concluding in 1830 that it was the same species as the female jackal he had described a decade earlier. Their place of origin and similar coloration, he decided, were more important than the considerable structural differences between the male and female forms.[35]

The following year, questions of provenance and classification arose regarding yet another jackal, this one collected by the French traveler Jean-Baptiste Douville, just back from an expedition to Africa. In Le Havre, Douville gave the Museum's agent, Alexandre Eyriès, a jackal and a jaguar, which Eyriès sent to Paris by steamboat.[36] The Museum wrote to thank Douville for his jackal and female "panther from *Africa*."[37] The menagerie's animal register, however, lists them as a jackal and jaguar from *Peru*.[38] The discrepancy can be explained; Douville returned to France from Africa via South America, arriving at Le Havre on the brig *Cécile* from Buenos Aires.[39] But the question of the jackal and jaguar's provenance remained, as did that of the jackal's specific identity.

The menagerie census compiled by Isidore Geoffroy Saint-Hilaire on 25 July 1831 lists three kinds of jackals: "the black-backed jackal" (designated as "precious"), "the ordinary jackal," and "the jackal, another variety." We can identify the "ordinary" jackal as a common golden jackal of south-central Asia and the remaining specimen of the pair Dussumier brought from the Malabar Coast in 1826. Isidore's "jackal, other variety" was the male from Senegal purchased from Schmitz in 1826. Werner's portrait of the animal corresponds to the species identified today as the African golden wolf. As for the specimen Isidore called the "black-backed jackal" and "precious," it was the individual donated by Douville, but if it came from South America, it could not have been the species now known as the black-backed jackal, which is native to southern and eastern Africa. Such uncertainties were the basis for Frédéric Cuvier's observation in 1832 that jackals were "the least well known of the canines."[40]

Given that an animal's provenance was important for determining its identity as a species, the Museum needed its information to be reliable. In the case of Douville, he returned from his African expedition with considerable fanfare, largely generated by self-promotion. In 1832 the Paris Geographic Society awarded him its annual gold medal, valued at one thousand francs, for the most important geographical discoveries of 1830, based on what he claimed to have found in the Congo and the interior of equatorial Africa.[41] Soon afterward, however, reviewers of his three-volume travel account descried its flaws, inconsistencies, impossibilities, and examples of flagrant dishonesty.[42] The reviewer for *The Foreign Quarterly Review* chastised the Geographic Society for being fooled.[43] The reviewer for *Revue de deux mondes* described the work as "a tissue of incoherent fictions."[44] Both portrayed Douville as a liar and a swindler.

The scientific community remained silent. Douville had asked the Academy of Sciences to name a committee to "take cognizance of the productions [of nature] he collected on his route." The Academy did so in August 1831, choosing Georges Cuvier, François Arago, Étienne Geoffroy Saint-Hilaire, and Frédéric Cuvier as committee members. After two months without a report, Douville said he no longer wanted one and asked for his specimens back.[45] According to the exposé in the *Revue des deux mondes*, the committee readily recognized that objects Douville pretended to have collected in Africa were from South America, but he chose to remain mum on the matter. Though the reviewer said he could imagine "the sentiment of disgust" that prompted the committee to follow this course, he did not disguise his dismay at its decision.[46]

Douville's lack of integrity notwithstanding, the menagerie had two specimens he provided, a jackal and a jaguar. His name was presumably on their cages, since the Museum had promised to display it.[47] Stripped of his claims as a credible traveler in Africa, he may have found this scant consolation, even if the name on an animal's cage was the recognition many donors aspired to. The Museum promised this in 1834 to the donor of the menagerie's first jackal from Algeria, a Polish officer named Joseph Smolenski who fought with the French in Algeria.[48] Later that year, when Dussumier's last Indian jackal died, the menagerie was left with two jackals, the recent donation from Smolenski and the earlier one from Douville.[49]

This set the stage for a contest over specimens at the Museum in January 1835, when the owner of a deformed calf offered it to Geoffroy in exchange for a jackal and a big-tailed sheep. Geoffroy, devoted to studying "monstrosities" scientifically, sought the calf because it displayed a new kind of malformation, and he asked the administration to authorize the trade.[50] Because Geoffroy was unable to attend the next administrative assembly, Frédéric Cuvier, in his weekly menagerie report, passed along Geoffroy's request. In doing so, Cuvier observed that the menagerie had only two jackals, one from Barbary and one from Senegal, representing different species. They had mated the previous November, Cuvier said, and the female had just given birth to a litter of puppies. The puppies had died, but Cuvier attributed this to its being the mother's first litter and her unfamiliarity with caring for offspring, and he assured the administration she would do better next time.[51]

Cuvier did not explicitly portray this as a contest between his interests and those of his superior. He did not spell out that keeping the two jackals would benefit his own studies of interspecific hybrids, while securing the malformed calf would benefit Geoffroy's work on malformations. The administration handled the situation deftly. Observing that the two jackals were of different species and the menagerie had no duplicates of them, it invited Geoffroy to construct a new proposal.[52]

That progress in jackal taxonomy continues to this day underscores the challenges Cuvier faced in assessing the identities of the individual specimens at his disposal. His 1820 distinction between "the jackal of Senegal" and "the jackal of Bengal" has held up. Recent DNA analysis has led to the reclassification of Cuvier's "jackal of Senegal" (the African golden jackal), as the African golden wolf.[53] As for Douville's jackal, though Cuvier referred to it in 1835 as Senegalese, it appears in his menagerie census of 1 January 1837 as "jackal from Peru."[54] Later that year, assistant zoologist Louis Rousseau, son of the keeper Magloire Rosseau, listed it in a guide to the Jardin des Plantes not as a jackal but as "a fox from South America, given by M. Douville."[55] It is plausible that the specimen was from Peru and belonged to the species today called the Andean fox (*Lycalopex culpaeus*), a creature that, despite its name, is more closely related to jackals and wolves than to foxes.

Though dogs, wolves, and jackals, were not the animals at the menagerie most likely to inspire the public's curiosity, their history there resonates with many of the themes I have encountered with respect to the intersection of animals and humans in the life of the menagerie. Over time, the canines in the collection represented five continents: Australia, Asia, Africa, South America, and North America. That the Museum on occasion did not know the provenance of a specimen was part of the challenge of ordering nature. Canines figured in the competition between scientists for specimens. They also featured prominently in the Museum's efforts to civilize animals and humans. Taken as a family, dogs and their kin displayed the different degrees of control the Museum exerted over the animal world, from the intermittently controlled strays and unleashed dogs that wreaked havoc on the grounds, to the guard dogs that warded off disorder but sometimes caused it, to the two cases of jackals of different species, incarcerated in the same cage, producing puppies. Even in the latter case, which united animals from different parts of the world that would otherwise not have had the occasion to mate, the success of keeping the puppies alive depended entirely on the mother. Furthermore, of the two puppies that survived, the temperament of the wilder one was such that the menagerie superintendent concluded that even a regime of kind treatment would not transform it. But if the menagerie's control across the spectrum of canine forms was not absolute, these limitations were not sufficient to shake the naturalists' confidence that nature could be ordered and transformed for human benefit.

[4.8]

The Keepers Revisited

The job of menagerie superintendent was created to oversee the animal keepers as much as the animals. On balance, Frédéric Cuvier found the keepers more vexing. In 1814, infuriated by the latest transgression of the worst of the keepers, he railed to the administration that nearly all the men in his charge were "totally deprived of intelligence and probity," ill born in all respects, and "puppets of the crudest passions" who never thought beyond their momentary well-being. The menagerie's rules had little effect on them, he said, because they were not "reasonable and desirous of the good." Too many of them, he felt, had been badly chosen.[1] When Cuvier lobbied for keeping the menagerie open daily, it was less to accommodate the public than to thwart the keepers' practice of taking bribes from people seeking admittance when the menagerie was closed.[2]

The other side of the coin was that the keepers worked from early in the morning until after the menagerie closed, their jobs were arduous and dangerous, they had no days off, their pay was mediocre, and most of them had families to feed.[3] Being bitten, clawed, gored, or butted came with the territory, depending on the animals one worked with. In the mid-1820s, contracting the mange from the camels was a hazard not only for the herbivore keepers but also their families. Cuvier understood that whenever they did display "zeal" and "intelligence," it needed to be encouraged. He supported their appeals for pay increases, reimbursement for medical expenses, healthier living quarters, less onerous posts when they grew infirm, and, at the end of their careers, a pension, a place in a hospice, or both.

The fact was that the keepers were a heterogeneous lot. This section seeks to give voices to them—or at least to name them and point to their individuality with respect to their talents, work, living situations, and relations with their superintendent.

When Cuvier took charge in January 1804, he supervised eight men, and by year's end it was ten. Later the number would grow to about a dozen, depending on the menagerie's needs and the Museum's budget. Typically,

Figure 4.8.1. A keeper feeding the gazelles from Algeria. Andrew, Best, Leloir after Karl Girardet. From H. Boitard, *Le Jardin des Plantes* (1842).

Cuvier had one or two keepers for the carnivores, three for the herbivores, one or two for the elephants, and one or two for the birds and monkeys. Also on the staff were a provisioner, a cart driver, and a porter. The giraffe's arrival in 1827 added yet another keeper to the payroll. In the late 1820s and early 1830s an additional worker did small carpentry jobs like building equipment, repairing wheelbarrows and fixing the roofs of the animal huts. Sometimes family members assisted the men in their jobs.

The workers varied considerably in age, experience, enthusiasm, and integrity. Cuvier was fortunate at the start to have two trustworthy young men who would grow old with him in their jobs. One was assistant carnivore keeper Magloire Rousseau, hired by Geoffroy to second Bernard Louzardy, whom Geoffroy had made head carnivore keeper after the disgraced Félix Cassal. Geoffroy knew Rousseau's parents and could testify to the family's character. The zoologist recommended young Rousseau as scrupulous in following orders, "very strong, very hard-working," and intelligent.[4] The other man that Frédéric Cuvier could reliably count on was Guillaume Guichenot, the menagerie's provisioner as of 1803. At the other end of the spectrum were the two elephant keepers, Joubert and Antoine. I have already recounted how the new superintendent failed in his efforts to manage Joubert and how the saga ended with an elephant's death and the keeper's dismissal.[5] Antoine stayed on, and he remained a thorn in Cuvier's side.

In 1804, three men cared for the herbivores.[6] Geoffroy considered Ballet, the fifty-something head of the service, to be one of the better staff and the sixty-two-year-old Nivet to be one of the worst.[7] Geoffroy appreciated young Perrin's zeal, but he had been exempted from military service because he was too short and had bad eyesight.[8] Other staff in 1804 included Antoine Bibron as the new porter and Jean Louis Richard as the first bird keeper.[9] A man named Rémond was later added as a cart driver.[10]

Injured first by the stag and then by the gnu, Ballet in 1808 applied for the less strenuous and dangerous job of menagerie porter, a post he held until 1821, when he was in his seventies and had worked at the Museum for twenty-six years. By then the duties of porter had become too demanding for him and his wife, and his wound from the gnu still ailed him. He'd saved some money, he told the administration, but not enough to live on, and he asked to be retired with a small pension. His request was granted.[11]

Nivet had been a bad choice for the menagerie from the beginning. He had been put to work as a herbivore keeper in the late 1790s after he proved too feeble for other service at the Museum. Then in November 1809 a camel bit him viciously on the arm and left him even more incapacitated. To fill in for him, the Museum in January 1810 appointed Auguste Persy, former coachman for the late Antoine Fourcroy.[12] Nivet asked to retire with a pension, but the Museum wanted to reduce its obligation by having the government provide him a place in a hospice. That process dragged on for five years, during which time Nivet, now virtually useless as a keeper, took to drink, arriving home drunk late at night, heaping abuse on his neighbors.[13] The Museum finally retired him with a small pension in 1814 but did not get him out of his apartment until February 1815, when the minister of the interior arranged a place ("a bed") at the hospice of Bicêtre.[14]

With Nivet gone, Antoine took his place as the least reliable staff member. For a decade he cared for Parkie, the elephant, but in May 1814, a near fatal incident precipitated his reassignment. Antoine introduced a military officer into the elephant's enclosure, undoubtedly after the man paid for the privilege, whereupon Parkie pressed the intruder up against the bars, smashing his lower jaw and shoulder. The officer stated that the fault was his own, not Antoine's, but the administration felt otherwise, transferring Antoine to the herbivore service and deciding once again to post signs indicating that "all parts of the Museum are seen for free."[15] Cuvier's report on this incident included his broader remarks on the general unreliability of the keepers, cited above, but he soon acknowledged that Antoine was in a class by himself as a miscreant.

Antoine's new coworkers, Perrin and Auguste Persy, resented how poorly he performed his duties. In October 1814, Frédéric Cuvier explained the situation to the administration:

> When, because of his negligence, we had to take Antoine away from the elephant, he was charged with the pump, the sheep, and the stable litter of the two dromedaries, work that was without comparison much lighter than that of his comrades. Since then, he has pretended that the service of the pump was too tiring for him and he abandoned it; it thus only remained for him to prepare the litter of the two dromedaries, which does not require half an hour, and to feed the sheep, which requires about as much time. The rest of the day he employs only to conduct companies [of visitors], and it is incontestable that he only abandoned the pump because the service that it required happened during the hours when the public comes for a walk, since the service of the pump is done today by the child of Auguste who is twelve years old. . . . His comrades complain, and with reason, that he does nothing while they do all the work, and they [consequently] do it grudgingly, and when some circumstance augments it, no one believes himself obliged to do more than anyone else beyond his job. Thus, the young dromedary needing to be walked for its health, I have not been able to charge anyone with this job, and it is again to the child of Auguste that I have been obliged to have recourse.[16]

The keepers were forbidden to lead tour groups around the menagerie for tips. While complaining of how Antoine carried his cupidity "to the point of indecency," Cuvier acknowledged his entrepreneurial skills: "[Antoine] has pushed this art of conducting companies of visitors to a point of perfection that is very remarkable." However lazy regarding his duties, he displayed "an uncommon activity" when it came to touring visitors:

> He multiplies himself, so to speak, in order to respond to everyone, and since he could lose one company while attending to another, he takes

> care to leave them liberty only when he is with them. He closes them up while he leads one of his companies to the dromedaries and he takes his key for it to the pond of the water buffaloes, to the large building, or to the parks of the deer, and he only frees them successively to shut them in elsewhere.[17]

Cuvier begged the administration to give Antoine his pension and be done with him. If it could not afford to replace him with an adult, he said, Auguste's son could do the work for much less pay. "Prayers and remonstrance," Cuvier told the Museum, had "not done a thing" to improve Antoine's behavior.

Nothing came of the administration's request to Geoffroy that he make a recommendation on the matter. Antoine remained on staff. The professors next heard of him when Frédéric Cuvier reported a transgression on Antoine's part that was so repulsive Frédéric hesitated to mention it. He was doing so, he explained, because he feared he would be compromised if he said nothing and Antoine did such a thing again.

Once more the story involved a man wanting to see the elephant up close. Though Antoine was no longer in charge of her, he led the visitor into her stable. The elephant knocked the stranger over with such force, Cuvier wrote, that the poor man could do no more than "return to his family to put himself to bed and give up the ghost." Exclaimed Cuvier, "If it is absolutely necessary that Antoine's cupidity be practiced, it would at least be desirable that it was not at the expense of someone's life!"[18]

Despite this event (which seems to have escaped the press's attention), Antoine continued to work at the Museum. In February 1821 he asked to be retired with a pension, or else to have the minister of the interior provide him and his wife with a place in the "hospice for married couples." His age and his infirmities prevented him from doing "the painful work of the menagerie." Cuvier supported him, albeit without praising him, stating, "I take the liberty to recommend Antoine's request to the administration's justice and goodwill," but the retirement was not finalized for three years. Antoine continued at the Museum with his annual salary of 1,000 francs through 1823.[19]

Cuvier's practice when the administration distributed annual bonuses was to recommend that every worker receive one, but he was never satisfied with all of them. In 1820, for example, out of twelve workers, he was pleased with five: the head keeper of *animaux féroces* (Rousseau), the provisioner (Guichenot), the cart driver (Leblanc), and the head and assistant keepers of birds and monkeys (Rémond and Rémond's nephew). They were, the superintendent said, "the menagerie's only good subjects." While suggesting small bonuses for others, he succinctly assessed their failings. Antoine "has nothing to do . . . and is capable of nothing." The porter Bibron "occupies himself all day only with his own interests, and neglects those for which he is responsible, as light as they are." Cuvier called Auguste Persy "excessively negligent" and said his son "would be a very good subject if he did not have the example of his father." The keeper Jean Louis Richard "instead of doing his duties goes hunting."[20]

Among the men whom Cuvier identified as shirkers in 1820 at least two, Auguste Persy and Jean Louis Richard, had come highly recommended. Persy, Fourcroy's former coachman, was endorsed by Fourcroy's widow, and Geoffroy himself had cited Persy's "morality, intelligence, and assiduity."[21] A decade after Persy's hiring, however, Frédéric Cuvier was disgusted with Persy and his whole family, complaining to the administration that neither prayers, punishments, nor advice had improved his performance. The keeper was disobedient, drunken, and dishonest, and his wife was worse, since it was she, Cuvier believed, who encouraged his knavery. Cuvier recommended that the Museum no longer house the family.[22] The administration hoped to reform Persy,[23] but two years later Cuvier insisted he be dismissed. Auguste Persy routinely showed up late for work, only to behave irresponsibly. Cuvier told the administration he had done his best: "I have been the first to accept him, to support him in his needs, and to mitigate his faults. Unfortunately, nothing has been able to prevail against a base and vile nature, against a taste for drunkenness, and I have the too well-founded fear of not seeing the son conduct himself better than the father. He already follows his deplorable example nearly in every way."[24] Father and son were both fired.[25]

Jean Louis Richard also came to the menagerie highly recommended. When he was hired 1804, he was thirty-six years old with four children. His experience with falcons, pheasants, and wild animals qualified him to be the Museum's first keeper of birds. An advocate described him as having "the best character [*moeurs*] possible," and he impressed the menagerie superintendent as strong, industrious, agile, honest, and gentle.[26] After Bernard Louzardy's death in 1813 and Rousseau's promotion to head keeper of the *animaux féroces*, Richard became Rousseau's assistant, but he wasn't always in Cuvier's favor. The twenty-franc bonus Cuvier recommended for him in 1820 was the smallest he proposed for any staff member that year.

But Cuvier's discontent with Richard at that point was nothing compared to the fury he felt in September 1821 when Richard's behavior contributed to the death of a precious and newly arrived animal of exceptional interest, an electric eel. Cuvier fumed to the Museum's director, "I had given the most positive orders to the keepers of *les animaux féroces*, under whose guard I had confided the electric eel, that this animal be shielded from the public, and especially not be subjected to any experiment, without proper authorization." Richard's disobedience had been "so categorical" and its consequences "so grave" that Cuvier found it necessary "to denounce him" officially so the director could punish him appropriately. The man was "one of the most negligent and undisciplined" of the menagerie employees and the time had come, Cuvier said, to make an example of him for the other workers. "A large number of losses that the menagerie has suffered, "he wrote, "have no other cause than this indifference for duties that arises always in these low classes of society from a great security in means of existence." The menagerie's tasks were structured in such a way that workers who had "even the weakest attention and the most limited intelligence could perform them. The service's only need is zeal, but it is indispensable, and every man who is not capable of it cannot be charged with caring for the animals."[27]

Whether or not Richard was punished (the records are silent), he was not fired, and remained at the menagerie until his death in 1825. His widow wrote to the administration saying he had never received a raise, though he had fulfilled his duties at the Museum for twenty-one years "with zeal and intelligence." Several times, she observed, animals in his care had wounded him, and an injury from a large monkey nearly cost him his life. Left with 50 francs in the household, not enough to cover the cost of her husband's burial, she begged the "just and humane" professors for help. They granted her three months of her husband's salary and promised to do what they could.[28]

With Antoine's retirement, Richard's death, and the departure of Auguste Persy *père* and *fils*, Cuvier had considerable staff turnover in the 1820s. A new keeper, Warmet, stayed just a year (1823–24), his commitment having plummeted after being rammed by the bison. Two longer-term keepers, Perrin and Rémond, were showing signs of age. Rémond applied for the position of porter on the rue de Seine when it opened again in February 1825. Reminding the professors that he was old and had been at the menagerie for eighteen years, he pointed out that not only was his work with the monkeys and birds exhausting, "the bad odor that one breathes there alters one's health." Cuvier testified to Rémond's "perfect honesty," stating that his "long and good services" merited "all the administration's good will."[29] But he did not get the porter's position.

The following year, Cuvier remarked how much Rémond had declined; he was losing his eyesight and could scarcely work. His wife helped, but Cuvier was not confident in her.[30] Rémond's woes increased in 1828 when she had a mental breakdown and he had to put her in the hospice de Charenton at a cost of more than half his salary (650 of 1,168 francs). He urged the Museum to ask the minister of the interior to arrange for her stay to be free.[31] Four years later, with Rémond's salary no higher but the cost of care up nearly 30 percent, his wife again needed to go to Charenton, and Rémond again sought the Museum's aid.[32]

Despite his own misfortunes, Rémond was attentive to the needs of others. In 1834 he volunteered to exchange two of his rooms on the second floor of the monkey house with Daboncourt, the new head monkey keeper, whose wife was pregnant. Rémond's offer was motivated by his belief that the Daboncourts' humid quarters on the first floor were unhealthy for the future mother. Cuvier, recognizing that Rémond would lose his kitchen in the move, arranged for Daboncourt to get an additional room on the first floor, vacant at the time, that had recently housed an assortment of monkeys, crocodiles, and snakes. The administration agreed, with the proviso that these arrangements could be revisited if the service of the animals required it.[33] Rémond labored on for several years caring for the birds in the pheasant house and the parks. By April 1838 his health had reached a "grievous state," and Cuvier had the bird dealer Vaillant tend the birds provisionally in Rémond's stead.[34] By the end of the year, that change became permanent.[35]

The case of Magloire Rousseau was happier. Promoted to head keeper of the *animaux féroces* in 1813, his performance was admirable, and Frédéric Cuvier appreciated it. Though Cuvier's early menagerie reports say little about Rousseau (because there were no complaints), they provide a glimpse of his living situation. In 1814 Rousseau petitioned the Museum for additional space because his tiny apartment, little more than a main room and an entry, was not enough to accommodate a family with five children.[36]

When Cuvier needed a keeper to travel to a French port to fetch animals arriving from abroad, he typically chose Rousseau. In December 1819 he dispatched him to Le Havre to collect one of Milbert's large shipments from North America. In May 1822 Cuvier sent him to Rochefort to accompany live animals from Senegal to Paris.[37] In July 1826, when the frigate *La Thetis* arrived at Brest with a rich cargo of specimens from Australia including kangaroos, black swans, emus, and a cassowary, Cuvier told the administration that Rousseau was "the only man of the menagerie" he could count on to conduct the animals safely from Brest to Paris.[38] That same year, Rousseau's son Louis, just turning fifteen, was hired to prepare insects in the zoology laboratory,[39] and Rousseau's daughter Louise married Gabriel Bibron, son of Antoine Bibron, the menagerie's porter, who had served as a voyager naturalist for the Museum in Italy and Sicily.

Cuvier registered disappointment with Rousseau only twice; in 1821 when he failed to stop Richard from taking visitors to be shocked by the electric eel, and in 1827 when Cuvier found Rousseau and a new keeper, Letellier, secretly fattening a pig in a tunnel that held the heaters in the carnivore pavilion. Assuming the pig was eating food destined for the menagerie animals, Cuvier was furious because he felt it represented a larger pattern of misbehavior by the keepers.[40] Once again they were soliciting bribes from the public. The Museum had instituted a system of tickets to allow students to view the animals in the carnivore pavilion during certain hours on days when the weather prevented them from staying in their outer cages.[41] The keepers exploited the system, letting visitors into the pavilion almost any time for some small gratuity.

The keepers' pursuit of their illicit gains, Cuvier believed, was leading them to neglect the animals. The pig episode was the latest affront. Cuvier blamed it on Letellier, writing, "[Rousseau], who has been with the menagerie for twenty-four years, is one of the most honest workers (*garçons*) we have, and he has always shown modesty in choosing his means of profit. It is since Tellier has been associated with him that the trouble has grown, Tellier having a greed that lacks all shame."[42]

Cuvier's provisional solution was to move Letellier to the rotunda where staffing was short because Géan, a herbivore keeper, was in hospital with mange contracted from the camels. When Géan returned, Letellier would be moved to the raptors, where he would have no occasion to exercise his greed. Cuvier wanted to keep Letellier because of his "precious qualities" caring for animals, but if he proved incorrigible, Cuvier said, he would call for his discharge.

Within weeks of this outburst, Cuvier calmed down enough to back Letellier's request for a raise. A recent father, Tellier was finding it impossible to live on two francs per day, the same wages he received when he was hired six years earlier. Pleading to the administration he wrote, "I have always done the best that it was possible for me to do."[43] Though Cuvier had reason to think Tellier had not always done his best, he was ready to stick up for him and acknowledge that the wages the menagerie paid were pitiful. He agreed that the keeper and his family could not survive on two francs a day, and he told the administration that even "the most miserable day laborers" earned more than that. If this could "scarcely suffice for the existence of the people of the lowest class and whose work requires the least intelligence," Cuvier asked, how could the Museum pay so little to the men responsible for the lives of its animals?[44]

Cuvier said he did not want men who lacked a work ethic or an interest in acquiring one. Being unable to hire staff at better wages than those of the simplest untrained laborer was untenable. He described "the meager salary of the men I employ," as "an injury" with the most grievous consequences. The only way he could adequately compensate his men, he said, was to allow them certain "illicit benefits." Without spelling it out, he may have meant the *bénéfice* of permission to collect and sell the dung of the animals in their charge. Cuvier represented Letellier as a "very good stablehand who understands very well how to care for animals" but who in his role as keeper of the birds of prey had no *bénéfice* available to him.[45] Letellier's situation had clearly made a strong impression on the menagerie superintendent.

As for Rousseau, Cuvier had only good things to say late in 1829 when he backed his request for a raise, describing him as "one of the oldest servants of the menagerie, who has never ceased to fulfill his duties with zeal and intelligence."[46] In 1830, when Cuvier needed someone to go to Marseille to fetch the fifteen big cats from the conquest of Algeria, Rousseau was the obvious choice.[47]

While Rousseau was away Cuvier proposed to fill the assistant carnivore keeper's position, vacant since Richard's retirement. He chose thirty-six-year-old Hippolyte Martin, who had fifteen years' experience in the cavalry.[48] Before long, he moved Martin to the rotunda in charge of several herbivores, including "the little African elephant," as the specimen was called. Geoffroy later listed the animals under Martin's charge as "the African elephant, four dauws, the zebra mule, the hemione, the ass, and the wild boar."[49] Unfortunately, Martin ran afoul of one of the male dauws, which bit him so badly he lost an arm.[50]

The administration decided to pay Martin's salary throughout his recovery, assured him "employment at the Museum for the rest of his life," and provided 100 francs for medical expenses. Martin asked the administration for a porter's position whenever one became available.[51] Later he asked to be outfitted with a mechanical arm.[52] Eventually, the 400 to 500 francs he paid the prosthesis maker was reimbursed.[53] When he asked to be lodged in one of the Museum's buildings (October 1837), the administration agreed, if Martin returned to work in the menagerie as assistant to the keeper of the monkeys (Daboncourt), a condition with which he readily complied.[54]

In June 1837, when there were lions needed to be collected in Toulon, Rousseau was ailing, and Letellier, who was brought back to the carnivores about 1830 when Martin went to the herbivores, was picked for the job.[55] Whether Rousseau's problem was due to illness or to age was not identified. It does not appear to have been an injury. Ironically, though Rousseau spent his career with fierce carnivores, his closest brush with death on the job came when he accidentally cut himself while helping to slaughter an animal in the menagerie's butchery and the wound became gravely infected. He was laid up for three months.[56]

Rousseau was the keeper Cuvier most relied on, but the keeper most in the public's eye from spring 1827 was Atir, the Sudanese man who arrived with the giraffe. His very dark skin and exotic clothing made him, like his animal charge, of special interest. Jests about him appeared in the press.[57] Cuvier registered no complaints about him until April 1836, when he reported, "Atir, the giraffe's Negro, who has never had a very regular conduct, neglects his service, gives himself up to debauchery, runs up debts, and if he does not return to a better path, I will be led inevitably to demand his discharge."[58]

The following February, Cuvier tersely introduced a menagerie report with the observation: "Athier [*sic*], the Negro, has not reappeared for two days. The inquiries made in the neighborhood have turned up nothing."[59] The professors assigned Geoffroy (who had been grousing about Cuvier) to report on the state of the menagerie, and of Atir. The zoologist confronted Atir and relayed his excuse that he was "shut up at home suffering from the grippe." Geoffroy reported that Atir's comrades testified to the story and that Atir for the present was doing his job "very well."[60]

The scene that brought Atir's career as a keeper to an end played out on New Year's Day 1838. As Cuvier described it in a letter to the Museum's director, Atir arrived "drunk, with four or five *débauchés* like him, in the rotunda, where his acts of violence frightened the public who were there, and where, without the patience of his comrades [fellow animal keepers], some regrettable rows could have occurred."[61] At this point, "well convinced of Atir's incorrigibility" and unable to tolerate his behavior any longer, Cuvier took away his keys and forbade him entry to the menagerie. Considering Atir's "extreme negligence" as a keeper and "his dissolute life," Cuvier wanted the administration to make the decision permanent.

As Cuvier explained, Atir had been on a downward slide for some time: "His absences were frequent and unauthorized. It was not very rare for him to show up drunk in the rotunda. Rather serious complaints came to me about his conduct on the outside. Friendly counsel, severe reproofs, threats—nothing was able to stop the development of his penchant toward the degenerate life [*la vie crapuleuse*]."[62] In the fall of 1837, Atir disappeared for two days, but this time no medical excuse was forthcoming. Atir had been badly beaten in a brawl with "the women and the men of the port." Cuvier warned him that his next transgression would result in dismissal, but he couldn't carry through: "The pity he inspired in me, despite his frequent debaucheries, made me half close my eyes to his disorderliness, which renewed itself almost daily. He did not ignore, however, that I was informed of his conduct, and I did not neglect to remind him of the fatal consequences. Despite all that, his degeneration only increased. Each week he went several days without appearing at the menagerie. He lived in debauchery with the most vile and abject people of our *faubourgs*." The New Year's Day debacle was for Cuvier the last straw.[63]

Cuvier told the director that Atir did not dispute the decision to fire him: "The poor fellow feels the feebleness of his character and the impossibility of tearing himself away from his fateful habits. He himself asks to be returned to Egypt, convinced that as long as he remains in Paris, he will only plunge deeper into the corruption [*la crapule*] where his bad connections have led him."[64]

At the first assembly of the new year, the professor/administrators endorsed Cuvier's recommendation that Atir be dismissed. The meeting's minutes called for advice on "the most convenient means of effecting the Negro Atir's departure from Paris and embarkation for Egypt."[65] Ten days later, Atir left Paris by stagecoach, arriving in Marseille in time to board a steamer to Alexandria. The Museum paid for the 500-franc trip, and its agents in Marseille instructed the boat's captain to look after Atir during the crossing. His travel expenses included 120 francs for when he reached home.[66] Cuvier's reports do not state who replaced Atir as the giraffe's keeper, but the role likely fell to one of the other keepers in the rotunda: either Géan (keeper of the large elephant) or Roux (keeper of the small one).

The Museum's records that year do indicate that in the fall Daboncourt, the monkey keeper, and Hippolyte Martin, Daboncourt's assistant, asked and were granted permission for their wives to sell cakes outside the new monkey house. Martin later applied for the post of the recently deceased Pelois, porter at the Museum's north entry for thirty-three years.[67] He got the job, but his wife lost the right to sell cakes outside the monkey house, a privilege conferred on the widowed Madame Pelois, to share with Madame Daboncourt.[68] In this way everyone had a niche in the microecology of the Museum, and life went on.

There were two examples of upward mobility in the keepers' families, beginning with Magloire Rousseau's son Louis, who rose to the position of naturalist's assistant in zoology under Valenciennes, professor of mollusks, worms, and zoophytes. The other was Louis's brother-in-law, Gabriel Bibron, son of the menagerie's porter. Bibron became both naturalist's assistant for the professor of reptiles and fish, Constant Duméril, and a distinguished herpetologist in his own right.

Bibron *fils* coauthored with Duméril the period's leading treatise on that subject. In the 1830s, Bibron helped the young Charles Darwin and the

British herpetologist Thomas Bell with reptile specimens Darwin collected on his *Beagle* voyage.[69] Darwin was particularly excited to meet Bibron in London (February 1838) when Bibron was there lecturing at the Zoological Society and examining reptiles in the society's collection. During their talk, Darwin learned that Bibron thought the Galápagos Islands had two species of tortoise.[70] The information was included in Darwin's addenda to his narrative of the *Beagle* voyage, identifying Bibron as "one of Europe's best authorities on reptiles."[71]

The keepers were thus not as hapless a lot as Cuvier suggested in his moments of greatest frustration. They varied greatly in talents and industry and in the degree to which they sought to align these qualities with the Museum's interests. Cuvier's ongoing challenge was making them consider the Museum's needs their top priority.

[4.9]

The Electric Eel

In August 1821 the menagerie came into possession of an electric eel (*Gymnotus*), a fish that Étienne Geoffroy Saint-Hilaire had studied in Egypt and Alexander von Humboldt had studied in South America. The botanist-agriculturist Georges Samuel Perrottet brought it all the way from Cayenne, French Guiana, along with a collection of plants that André Thouin praised as the richest to have arrived in Europe so far that century.[1] Compared with this treasure trove of plants, the few live animals Perrottet delivered (eight birds, a coati, and the electric eel) were only an afterthought on a colonial mission with other objectives, like introducing vanilla trees in the Mascarene Islands and bringing Chinese laborers from the Philippines to work in Cayenne. Still, a live electric eel was an exciting acquisition. It was said on the best authority to be able to generate an electric shock powerful enough to stun a horse or kill a man. France had never seen a live *Gymnotus*.

During his South American travels with Aimé Bonpland (1799–1804), Humboldt was eager to examine the electric eel's physiology and understand how it produced shocks. Getting natives to collect specimens was not easy. "Money loses its value as you withdraw from the coast," he explained, "and how is the imperturbable apathy of the ignorant people to be vanquished, when they are not excited by the desire of gain?"[2] Eventually, natives captured specimens for him by the astonishing method of "fishing with horses." The procedure involved driving wild horses and mules into a marsh filled with electric eels, preventing the horses from escaping, and allowing the eels to shock the horses so many times that their ability to emit painful electric discharges was temporarily reduced and they could be caught. In the violent melee Humboldt and Bonpland witnessed, two horses drowned after being stunned, and five five-foot-long eels were captured. Of the hazards of dealing with such creatures Humboldt wrote:

> It would be temerity to expose ourselves to the first shocks of a very large and strongly irritated gymnotus. If by chance a stroke is received before

Figure 4.9.1. An artist's rendering of Humboldt's famous account of "fishing" for electric eels with horses. Lizars after Stewart. Frontispiece for *The Naturalist's Library, Ichthyology*, part 2, 1847. Courtesy of the Thomas Fisher Rare Book Library, University of Toronto (Vnh 01403 v. 2).

> the fish is wounded or wearied by long pursuit, the pain and numbness are so violent that it is impossible to describe the nature of the feeling they excite. I do not remember having ever received from the discharge of a large Leyden jar, a more dreadful shock than that which I experienced by imprudently placing both my feet on a gymnotus just taken out of the water. I was affected during the rest of the day with a violent pain in the knees, and in almost every joint.[3]

Humboldt detailed his experience and experiments with these electric fish in a long memoir delivered to the French Institute in the fall of 1806.[4] He recounted the eels' capture in his popular *work Pictures of Nature* (1808).[5] In 1819, *Annales de chimie et physique* republished Humboldt's account from his narrative of his South American travels.[6]

How Perrottet acquired his *Gymnotus* specimen is unknown, but he managed to keep it alive on his seven-week crossing from Cayenne to Le Havre, arriving 1 August 1821.[7] In Paris the eel was entrusted to the carnivore keepers at the new pavilion for *animaux féroces*, where it died before the month was out. As indicated previously, an angry Frédéric Cuvier announced the eel's demise in a letter to the Museum's director, explaining, "I had given the most explicit orders to the keepers of *les animaux féroces*, under whose guard I had confided the electric eel, that this animal be shielded from the public, and especially not to be subjected to any experiment, without proper authorization."[8] Both aspects of his orders were ignored. The creature evidently died from a wound to its neck.

Cuvier blamed the keepers, but Geoffroy was apparently not guilt free. By Cuvier's account, a physician in the company of Geoffroy subjected the eel to "a very long experiment" four days before its death. The next day the keeper, Richard, disobeyed Cuvier's instructions by leading several people to the eel so that they could "[have] themselves electrified by it." Cuvier acknowledged that the eel may have been injured on either day, but there is evidence to suggest Geoffroy's party was responsible, insofar as that scene degenerated into bedlam.

Le Drapeau blanc reported the story on 26 August 1821, based on several eyewitness accounts, including that of the physician who almost became "a victim of his learned curiosity."[9] Present with Geoffroy were Jean-Louis Alibert (professor at the Faculty of Medicine); the famous military surgeon Dominique-Jean Larrey; the young physician and zoologist Étienne Serres; and the physician Janin de Saint-Just, who was foolhardy enough to reach into the eel's tub, seize the creature with both hands, and lift it out of the water. Had he read Humboldt's *Personal Narrative* and discussion of *Gymnotus* experiments, he surely would not have done so.[10] By Saint-Just's account, however, only the day before, several scientists ("Cuvier, Ampère, Serre, and Geoffroy himself") had visited the eel, touched it, and received only "very light shocks."[11]

Given their experience, Saint-Just was unprepared for what befell him. According to *Le Drapeau blanc*, upon grasping the fish and pulling it from the water, the physician received "several shocks of an extraordinary force, such as no one ever experienced from the Voltaic pile." What made it especially frightening, even life-threatening, was that he found it impossible to release the eel. "An involuntary contraction made him actually squeeze it with a force much greater than normal for him, and the more he squeezed, the more violent the electric shocks became."

Saint-Just's cries of pain terrified his companions. "They feared for his life, and it is not doubtful that had this state been prolonged, it would have led to his death." The bystanders shouted, "Let go! Let go!" but Saint-Just could not. "Fortunately, it occurred to him to plunge the eel back into its tub. As soon as his hands were wet, the contact with the water, an excellent conductor of the electric fluid, gave him the freedom to disencumber himself of his enemy." It was only thus, the newspaper said, that the doctor "escaped a certain death."

While grasping the eel, Saint-Just experienced pains throughout his body and especially in his shoulders. For twenty-four hours afterward, the discomfort prevented him from moving his right arm, "which was painful to the touch, as if he had been attacked by a violent rheumatism." He continued to ache for several days.[12] It may be that the eel's neck wound was its greatest injury sustained during the commotion, but Janin de Saint-Just thought his squeezing it had perhaps damaged its internal organs.

As Geoffroy's direct subordinate, Frédéric Cuvier was reluctant to criticize him or the other scientists who wanted to experience the eel's electric charge. Blaming his own underlings was more politic. Frédéric was disappointed that Rousseau, head keeper of *les animaux féroces*, had done nothing to prevent Richard's imprudence, but it was toward Richard that he directed his fury. Geoffroy was absent from the administrative meeting where Frédéric Cuvier's letter announcing the eel's death was read, otherwise he might have said something. Cuvier asked that Richard be punished for his negligence. The assembly left the matter to the director to resolve.[13]

[4.10]

Jack the Orangutan

The letter to the Museum from Captain Andronic Van Iseghem (12 April 1836), recently returned to Nantes from Sumatra, announced "an individual of the masculine sex, aged approximately nine months, belonging to the species of monkey called Ourang outang (or man of the woods)."[1] The captain knew he had a treasure. "You will no doubt be aware," he wrote, "that this species of monkey is extraordinarily rare in its country, and it must be even more so in France." His confidence was justified. Isidore Geoffroy Saint-Hilaire had lately placed "the large orang-outang or black pongo of Sumatra" at the top of a wish list of mammals for a naturalist sailing to the Malay Archipelago. He was thinking of a dead specimen, "either of young age, or especially in the adult state," to add to the Museum's collection of orangutan remains.[2] He was not expecting a live one. European naturalists had managed to study only two live orangutans, one in Holland (1765) and one in France (1808). The Museum professors were immensely interested in Van Iseghem's acquisition, but they wanted to know if it was truly an orangutan, and if so, in what condition, and how much it would cost to get it.[3]

Van Iseghem's letter said he had purchased the two-month-old animal from Sumatrans, along with the skin of its mother. It was perfectly tame, healthy, and happy to eat whatever it was given. Whatever habits it had acquired living with the ship's crew for seven months, the captain judged "entirely inoffensive." With respect to price, he expressed confidence that the Museum of Natural History in Paris would recognize the importance of an appropriate financial sacrifice to ensure the advancement of science, and he invited the professors to make an offer.

To ascertain that the animal was indeed an orangutan, the professors asked their colleague Henri de Blainville to write to a former student in Nantes, the physician Marion de Procé, and ask him to examine the specimen (as he had the tapir), ascertain its condition, and determine Van Iseghem's price.[4] Meanwhile the captain wrote to the Museum to say he had

Figure 4.10.1. "The orangutan." Jean-Charles Werner, 1836. Collection des Velins. Muséum national d'histoire naturelle. Portfolio 102, folio 1.

numerous attractive offers for his orangutan and needed to wrap up the business before he sailed on the tenth of May.[5] Marion de Procé visited Van Iseghem on 26 April 1836 and wrote to Blainville the next day confirming that the animal was an orangutan and describing it at length, beginning with its head and face:

> His forehead is very high and convex in the median line, so that it is very similar to the forehead of certain men. It is entirely devoid of long hair, as is the rest of its face, except for the sides of the cheeks where long red hair looks very much like whiskers. His nose does not jut out at all; his eyes have an expression of intelligence and a remarkable softness; his eyelids are adorned with long lashes; his muzzle is not at all prominent, but his lips are very mobile and can extend for about two inches. His ears are well defined and would resemble those of man if they were provided with the lobe that characterizes human ears.[6]

The doctor described other features, including the animal's teeth, the lack of callosities on its buttocks; the smallness of its thumbs and big toes compared to its other digits; the long red hair covering most of its body; and the way the hair on its head looked just like a wig. What impressed Marion de Procé most was its manner: "I was struck by the slowness of the orangutan's movements, which contrast with the turbulence of the other monkeys. I was even more struck by its calm and reflective air, its apparent sociability, and the indefinable human character spread over its physiognomy. It is of the greatest gentleness and seeks caresses, even from strangers."[7]

Marion De Procé was likewise struck by the orangutan's obvious intelligence. He watched it descend from its chair, pick up the chair with both hands, move it to a place from which it could get food that its master was offering, and climb back onto the chair to get the food. He further observed the orangutan take its chair to a door that it wanted to open, climb onto the chair, and turn the doorknob the way it had seen someone turn it to open the door.

Von Iseghem told the doctor he had been offered 2,000 francs for his animal at Îsle Bourbon, and in Nantes buyers were offering 1,600. He wanted 3,500 francs form the Museum: 3,000 for the animal and 500 to bring it to Paris. He insisted he was "the only one capable of delivering his orangutan safely," which would require an entire coach "for him and his traveling companion." Marion de Procé judged the price reasonable and appropriate and urged the Museum professors "to conclude a transaction that will permit you to study from both the mental and physical side the animal that is closest to the lower races of the human species."[8]

When the Museum director solicited the menagerie superintendent's opinion on the value of the purchase, the advice was the same, though lacking the racist dimension of the doctor's remarks. Replied Frédéric Cuvier,

> I do not believe that a menagerie today could make a more useful acquisition than that of an orangutan, that no animal could be a subject

> of observations more extensive and more important for science. This *quadrumane* is placed on the limit between the animal races and the human race. The evaluation of its faculties can thus alone resolve one of the most contested questions in natural history and philosophy. If some observations have already had this question in view, they are inadequate to solve it completely, and they furthermore need verification and extension.

Cuvier did not name the question he had in mind, no doubt because he felt it so obvious. It was the question whether the differences between the mental faculties of animals and humans are merely of degree or are of a more fundamental, unbridgeable kind.[9]

While Cuvier was enthusiastic about the orangutan's potential for science, he was less optimistic that the Museum would be able to keep it alive. He knew too well that the mortality rate for the menagerie's apes and monkeys was pitifully high. He recommended lodging the orangutan in a garret adjoining the apartment of Daboncourt, head keeper of the monkeys, on the second floor of the monkey house. The space was well situated and illuminated, said Cuvier, and seemed to offer the conditions essential for maintaining such a delicate animal, at least "for some time."[10]

Marion de Procé's report and Frédéric Cuvier's recommendation were read at the professors' assembly of 3 May 1836. The professors decided on the spot to pay the captain's asking price of 3,500 francs, a substantial sum compared with annual Museum salaries (5,000 francs for a professor, 3,500 francs for Cuvier as menagerie superintendent, and 1,100 francs for the head monkey keeper).[11] The professors cobbled together the funds from the acquisition budgets of the menagerie (2,000 francs), zoology (1,000 francs), and mineralogy (500 francs). The professors grumbled to Van Iseghem about "the elevated price you request" but accepted his terms, recognizing the need to encourage voyagers who took the pains to procure live animals abroad and bring them to France. The captain was left to select the surest means of transporting the animal safely to Paris.[12] Three days later a letter composed by Frédéric Cuvier reached the minister of public instruction describing the "extraordinary" opportunity to purchase the orangutan and asking for financial help.[13]

That the Museum's director, Michel-Eugène Chevreul, consulted Frédéric Cuvier on the matter and not Geoffroy may be because Geoffroy was feeling deeply alienated from his colleagues and had not attended weekly assemblies for over a year, finding them too stressful. He was absent when his colleagues agreed to the orangutan's purchase and later claimed the price was too high relative to the needs of the Museum's zoologists, who only had to confirm "many good observations" of the species that had already been made.[14]

The Museum's first live orangutan arrived in style on 15 May 1836, riding in the cab of a stagecoach from Nantes with Van Iseghem and an aide from the captain's household. It was a beautiful and festive Sunday morning, with the citizens of Paris anticipating a near-total solar eclipse that afternoon. The orang was promptly installed on the second floor of the monkey house in a garret adjoining Daboncourt's apartment. From a pair of windows, the ape could look down on the crowds gathered below. A newspaper account published the day of his arrival described him at a window, dunking a biscuit in a glass of sugared water. With the biscuit in one of his forward hands and the glass in one of his rear hands, he offered a compelling image of the order of mammals he belonged to: the Quadrumana, or animals with four hands.[15]

The eclipse began precisely as astronomers predicted at 2:06 p.m. At 3:33 p.m., nothing but a thin, brilliant crescent of the sun remained uncovered, and the sky was dark enough for stars to show. Vendors sold smoked glass that the public used to view the event. According to *Le Figaro*, the orangutan, installed at his window at the Jardin des Plantes, "seemed to regard the eclipsed sun quite tranquilly," but the resemblance of its face to a human's was "frightening."[16]

Van Iseghem related the young orangutan's history, habits, likes, and dislikes to the press, along with the gruesome details of its capture as a baby with his mother. In attempting to flee the hunters, the mother lost one of her hands to the blow of an axe. Clutching her offspring in her other arm, she was unable to climb through the trees and escape. A few days after capture she died, her wound having turned gangrenous.[17] What the press did not mention, but naturalists regretted, was that her skull and skeleton had not been preserved for science.[18]

The baby orangutan was about six weeks old and totally hairless when captured. The hunters fed him the same sort of pap a human baby might eat. Van Iseghem recounted at length how the young animal loved company; how he sulked and cried when chastised for misbehavior; played and was careful with children; liked dogs (but not cats or other apes or monkeys); and much more. Parisians were presented with an animal they could readily relate to emotionally. The great question that remained was proximity. Just how close was the orangutan to humans? To the public, the press, scientists, and artists of all kinds, this was not just another monkey, but a special creature at the very boundary between animals and humans.

Frédéric Cuvier reported the orangutan's delivery in "a perfect state of health" and was confident that with proper care it would remain healthy.

It was entrusted to the Daboncourts, husband and wife, who Cuvier knew would treat the animal affectionately and intelligently. Likewise crucial, he said, was to adhere to the captain's recommendations to allow the orangutan "a great deal of rest and never to bother it in its sleep." Cuvier wanted to make the rule, at least provisionally, that no one excite the orangutan or visit it while it was sleeping. He asked the administration to confirm this order, "the execution of which seems indispensable to me in order to preserve so precious an animal."[19]

The administration agreed that special orders be given to the animal's keeper and posted inside the monkey house. No one was to be admitted to the orangutan's living space without explicit permission from the professors. Blainville was named to work with Cuvier in drafting these rules.[20] After that, Cuvier left for his annual two-month leave to perform his duties as inspector general of the French university system, leaving Isidore Geoffroy Saint-Hilaire in charge. But it was Geoffroy, not his son, who kept the orangutan on the minds of the professors.

On 10 May 1836, Geoffroy attended a professorial assembly after his long hiatus. He missed the next two meetings but attended that of 31 May, where he expressed anxieties about the inadequacy of the orangutan's quarters and the large crowds of visitors. He wanted the quarters rearranged and only ticket holders admitted during specific hours. Worried about the coming winter, he proposed that clothing be made for the animal.[21]

The administration deferred the question of rearranging the orangutan's quarters, but it decreed that (free) tickets were required to see it, strictly between nine and eleven o'clock in the morning and four and six o'clock in the afternoon. Geoffroy's proposal to clothe the menagerie's new resident was not taken seriously. The professor serving as secretary, Adolphe Brongniart, chose not to record the idea in the meeting's minutes.[22] The minutes do mention, however, a proposal that has escaped the attention of historians of science, primatology, and music. Geoffroy conveyed to the assembly a request from "Monsieur Litz [*sic*] and several other artists . . . to give a concert to the orang-outang to determine the impression of music on the animal." 'Mr. Litz' was none other than the twenty-four-year-old virtuoso pianist Franz Liszt (1811–86). The professors left it to the "bureau" (the director, secretary, and treasurer) and Geoffroy to consult with "Monsieur Litz" to determine whether the project could be accomplished without causing inconveniences.[23]

Alas, the Museum's records say nothing more of this initiative. Had the concert been held it would surely have attracted the kind of attention Liszt no doubt imagined it might. Liszt's showmanship is a matter of record, and his proposal may have been a briefly entertained idea for an attention-grabbing stunt, but insofar as the orangutan was seen as the animal closest to humans, the notion of testing its response to music was entirely reasonable. The same thinking had motivated the concert for the elephants, which the musicians of Paris had not forgotten.

One reason Liszt's concert did not occur may be that he was living in Switzerland and only passing through Paris.[24] Or it may be that Liszt's proposal coincided with a time in Geoffroy's career when he was so enraged with his colleagues, that he feared he would have a stroke. He left the professors' meeting of 31 May in a fury because his recommendation that protective clothes be made for the orangutan had been dismissed. At that moment, Geoffroy may have been of no mind to discuss any further orang-related arrangements with his colleagues.

A Monsieur Mesnager also wished to conduct special studies on the orangutan, based on his success teaching deaf and mute individuals to speak. As he explained to the professors, "the pensive eye and grave physiognomy" of the orangutan seemed more like a human than an inhabitant of the woods, leading him to suggest "that I might be of some utility from the point of view of the intellectual development of this half-man." Mesnager wanted to apply to the orangutan "a simple and easy method by means of which I have made the most completely deaf of deaf-mutes learn to understand and speak words." He did not expect to make the orangutan speak, as he imagined it did not enjoy the same intellectual faculties as humans (deaf and mute individuals included), but he felt it worth a try. If he could direct the efforts of the orangutan's keeper to this end, he expected that the orangutan—"this young savage"—would be able to understand "all that is said to him."[25]

The administration put Mesnager's request on hold saying the orangutan had too recently arrived for any experiment to be conveniently conducted but that he might approach the menagerie's superintendent later. This seems to have been the last the Museum heard from him.[26] The professors had not heard the last from Geoffroy, however. After the contentious meeting of May 31, his presence at the assembly was irregular. Rather than attending the meeting of July 5, he fired off an angry letter headed "Request from the undersigned professor of zoology, acting according to his duties, in favor of the orangutan." Clearly distressed, he wrote:

> Messieurs and Honored Colleagues: You did not intend to acquire at the price of 3500 the orangutan from Sumatra for it to become in the days of autumn a cadaver and a subject of anatomy. After Camper and Cuvier, what research is to be done! We do not have their talent.

> For I affirm that with the first cold weather the animal will perish [Geoffroy underlined this word twice] if nothing in his regime is changed.
>
> When, a month ago, I came to ask you to consider clothing the animal, someone who makes decisions among you was content to take pity on my proposition, and none of you committed himself to a serious discussion.
>
> At the risk of being greatly ridiculed twice, I renew my proposition. I know my duty: *honi soit qui mal y pense* [evil be to he who thinks it].
>
> One of the great physicians of our academy believes it urgent to consider this. On his own initiative, our colleague Dutrochet writes me of it *en professo* on July 2. I have received advice from two nearly equally notable persons tending to support the measure. The zoological garden of London only had recourse to this measure too late for its chimpanzee and was sorry because of it.[27]

Geoffroy went on to explain that it would be the greatest benefit to science if the Museum could keep the orangutan alive until adulthood, to see the animal develop, so to speak, into "a second species." If they were to observe this extraordinary transformation, the animal needed better quarters. Its present location, Geoffroy maintained, was bad for the animal and public alike. He proposed that an empty two-story house that was available be renovated for the purpose. The keepers and the animal would occupy the top floor. The first floor could be an antechamber for visitors waiting their turn to see the animal. Geoffroy suggested that separate rooms be created for "two publics" (one more refined), received differently and at different times.

Geoffroy did not expect his colleagues to agree. Unhappy with how they had treated his previous proposals, he wrote:

> I regret, my dear colleagues, to be unable to explain my ideas to my colleagues in person, but there have been many occasions where I have recognized that I too visibly ran the chance of a fatal seizure after discussion amid your assembly. I will refer simply to the last occasion, when nothing happened because of the care I took to swallow the insult without succumbing: but I withdraw from combat, this time in a definitive manner. Always the insults. I do not have the heart to endure them.[28]

Geoffroy's letter was read aloud in his absence. The professors chose to defer action until Frédéric Cuvier's return so they could hear his opinion on Geoffroy's propositions. Geoffroy attended the meeting the following week, insisting again on the need to clothe the orangutan, and complaining about the decision to wait for the menagerie superintendent. His words were not recorded, but he likely reminded his colleagues that he was the professor of mammals and birds, and Cuvier was his subordinate.[29]

Geoffroy had meanwhile been holding forth on the orangutan at the Academy of Sciences, where some Museum colleagues were also members. Beginning on 20 June 1836, for four consecutive Mondays, he presented his "memoir on the orangutan living at the menagerie."[30] Despite the title, the memoir offered few observations on the menagerie's specimen, focusing instead on philosophical considerations regarding the species. Geoffroy's Monday presentations gave him another forum for insisting on clothing and better housing to help the ape survive the winter. He also used them to signal that his views aligned with those of the public, if not the scientific establishment.

He began by hailing the orang's arrival in Paris and the public's response. All ranks of society had flocked to see it, he said, and whether they belonged to the idle rich or to business or artisan classes, they reached the same untutored yet profound conclusion: that the orangutan was neither a man nor a monkey. Geoffroy eventually came over to the public's side, he said, because the public's instinctive appreciation that the orangutan was neither man nor *singe* was ultimately sounder than naturalists' reasoning when they classified it with the monkeys/apes and established for man an order apart.[31]

The naturalists had separated the highest mammals into two orders, the Bimana (mammals with two hands) and the Quadrumana (mammals with four hands), with man alone in the first order, and all apes and monkeys in the second. Geoffroy viewed the distinction as trivial and faulty. Naturalists had made it out of deference to "certain privileged branches of society" and "to soothe the moral dignity of our species!"[32] But he assured his audience that his memoir would not compromise man's dignity. He concluded his first presentation by showing paintings of the orangutan in four poses executed by the Museum's artist, J.-C. Werner.

In his second presentation Geoffroy displayed four additional paintings by Werner and described how the chimpanzee and the orangutan had come to be distinguished from each other. In his third and final offering, instead of pursuing the theme of "neither man nor monkey," he considered the broad philosophical perspective of "organic unity." There was nothing about the organization or function of the orangutan's parts, he said, to set it apart from all other animals in its group (man included). "In a general sense," the orangutan was "synthetically and everywhere similar to man, without the omission of any part constitutive of the human being."[33]

On the other hand, the organic elements of the two forms showed different modifications. Tellingly, the orders of development in humans and orangutans were the inverse of each other. The skull of a young orangutan could almost be taken for that of a human, but as it matured the orangutan became a long-muzzled brute whose cranium, as it grew thicker and more crested, shut down the development of its brain. In brief, orangutans and humans became less like each other as they developed over time. Man became "more man," that is, "more susceptible of intelligence, more capable of yet more elevated functions, more disposed also to continual progress," while the orangutan developed in the opposite direction, gaining bodily force while losing intellect.[34] Geoffroy had taken the mistake he made early in his career, of not recognizing that the juvenile and adult forms of the orangutan represented the same species, and used it to shed light on the developmental differences that shored up the animal/human divide.

In the final installment of his memoir to the Academy, Geoffroy revisited the theme of the orangutan's mental faculties' degrading over time but expressed a glimmer of hope that the process might be tempered if the menagerie's specimen were given the proper treatment and training. The ape might then retain some intelligence and gentleness and not be overwhelmed by the brutish effects of its physical development. The fact that the elephant "allows himself to be restrained by his mahout, whom he would break to bits, if that was his wish" suggested to Geoffroy that it might be possible to train the orangutan, through punishment, to be docile as an adult. For this to happen, he said, the Museum would need to do everything in its power to ensure that the orangutan survived to adulthood and did not succumb to the lung diseases that so often proved fatal to animals brought to Europe from tropical climes. This meant providing the orangutan with better and warmer quarters—and clothing.[35] Such were Geoffroy's considerations during the first two months of the orangutan's residence at the menagerie. He promised he would follow up with a work titled "Studies on the orang-outang, observed alive in Paris in 1836," but it never appeared.[36]

Frédéric Cuvier produced no new studies based on the menagerie's orangutan, despite having urged its purchase, as we have seen, on the grounds that "no animal could be a subject of observations more extensive and more important for science."[37] Nor did he use it to pursue the question of an animal/human boundary. The time he might have spent studying the orangutan proved limited, but he also may have had misgivings about reexamining a species whose mental faculties situated it so close to humans. In his 1810 paper about the orangutan that he observed before it was sent Malmaison, Cuvier enthused about its seeming mindfulness of others and ability to reason and generalize ideas. Since then, he had quietly backed away from crediting the species with such powers.

Cuvier never acknowledged his backtracking, but it is clearly documented. Other than a new concluding paragraph, he took the text of his 1824 article "Orangutan, Female," written for his and Geoffroy's *Natural History of the Mammals*, nearly verbatim from his paper of 1810, with some important exceptions.[38] Most significantly, every instance in the earlier text where Cuvier explicitly credited the orangutan with powers of reasoning was expunged. He included his descriptions of the orangutan shaking tree branches as if to discourage climbers, moving a chair in order to reach the doorlatch, and taking care to cover itself at night, but in each case, he left out his original claim that the action demonstrated the orangutan's ability to generalize ideas.[39] In conclusion, Cuvier noted he had not reported anything that others had said about orangutans because, "when it comes to phenomena that must establish the last limit between the intelligence of the brute and the intelligence of man, one can only, one must only, give as certain what one has seen, what one has observed oneself. Errors can no longer be inconsequential when their consequences are not."[40]

What stakes did Frédéric Cuvier have in mind? One, evidently, was the idea of human evolution. In a piece on orangutans published in the *Dictionnaire des sciences naturelles* the following year, Cuvier referred dismissively, without naming names, to "those thinkers whose imagination is more active than their judgment is critical" who had wanted to portray humans as descending originally from orangutans.[41] It is likely his retreat from earlier enthusiastic claims about orangutan intelligence came after being warned that these might lead him to be associated with Lamarck. But even without the question of descent, there was the question of proximity, of how close the orangutan and man are to each other. His brother would have stressed the importance of insisting on an unbridgeable, qualitative difference between animals and humans. The issue of proximity attracted the attention of both scientists and the public.[42]

Cuvier warned observers of animal behavior not to assume that animal minds work like those of humans, but he left the orangutan in a kind of intellectual limbo between humans and the rest of the animal kingdom. Crediting the orangutan with "the highest degree attainable in the intellectual faculties allotted to animals," he called for careful study of these faculties in and of themselves and as a measure of the intellectual faculties of other beings. Then, in a puzzling formulation, he characterized the orangutan's faculties as being "mixed neither with reason, as in man, nor perhaps with instinct, as in the animals of a lower rank."[43] His meaning was unclear, as contemporaries observed.[44] In seeking to make sense of the creature at the very boundary between humans and animals, he seemed to have boxed himself into a corner.

Cuvier acknowledged that orangutans were capable of a remarkable level of learning things that were not natural to them. It was not just that they could learn to drink from a glass, but that they could, when thirsty, take the glass themselves and fill it with water. If they felt cold, they would look for something to cover themselves; they could arrange their bedding to make it more comfortable. Likewise, "if the place that contains their food or something else they need is closed, and the key normally comes from your pocket, they do not limit themselves to showing they know what is in this place, they come to ask you for the key and then go open the door."[45]

Cuvier interpreted these actions as indicative of higher-level associations that were remarkable in comparison to what other mammals could do. He nonetheless insisted, "Nothing in them yet announces on the part of orangutans the faculty of knowing and willing freely; the faculty, in a word, that gives morality to actions, and that up to the present belongs exclusively to the human species." He seemed to hedge slightly, however, on the idea of the uniquenesss of human moral actions when he suggested that it was not until later in its life that an orangutan's "moral forces" were overwhelmed by strong passions "like fear, anger, and hate."[46] The intelligence of the young orang diminished, it seemed, as the animal matured into an adult. Cuvier's 1825 "orang" essay thus previewed what would be a main theme of Geoffroy's memoir of 1836. Though Cuvier's piece suggested he had lingering uncertainties regarding orangutan intelligence, he did not attempt to resolve them when the menagerie received its first live orangutan eleven years later.

Geoffroy had turned his early mistake about juvenile and adult orangutans being different species into a powerful tool for talking about the orangutan's extraordinary transformation in its development to the adult stage. It also gave him a way to emphasize human and orangutan (adult) differences. Frédéric Cuvier's "mistake" in 1810 of attributing great intellectual faculties to the orangutan was of a more philosophical or political sort. The safest way to handle it was to ignore that it had happened. In any event, neither Geoffroy nor Frédéric Cuvier published any new observations on the mental faculties or behavioral repertoires of the menagerie's first live orangutan.

Information regarding the orangutan's behavior comes primarily from newspaper accounts and two books on the Jardin des Plantes (1837 and 1842).[47] L'*Echo du monde savant* led the way, reporting in mid-July 1836, two months after the orangutan's arrival, on its intelligence and the public's appreciation of it. On Sundays the crowds were so large that the Museum had to station sentries and erect barriers to direct the traffic flow. The article's author saw the orangutan's intelligence manifested in its gentleness in playing with the keeper's children and its solicitude for a child that fell. The writer also cited an incident where someone sought to give the keeper a cane to hand to the ape, but it refused to use it and returned it directly to its owner, "as if guided by a kind of sense of property."[48]

Just a few days later, *Le Courrier français* wrote at length about the orangutan, beginning with the announcement that his name was Jack.[49] Elsewhere he was called Jacques, Jacques du Sumatra, and Jackie, but Jack was the name that stuck.[50] The reporter noted that Jack was respectful to ladies (without having to specify that other apes and monkeys notoriously were not). The author also remarked on how well Jack interacted with visitors and children. More than other writers, the author reflected on Jack's "singular destiny," to be "taken away, very young, from his forests, his country, his family," tempest-tossed on his voyage to France, and thrust into Paris and human civilization, where his resemblance to humans was enhanced by giving him a name and turning him from a simple frugivore (as the writer supposed) to an omnivore. The reporter acknowledged the fact of Jack's "exile" and "captivity" but pictured the young orang without any great angst, dreaming of his homeland while enjoying the comfort of his swinging cot and his wool coverlet. The author admitted that Jack's quarters were somewhat cramped, and that the monkeys suffered from poor lodgings, but he portrayed the rest of the animals as having splendid quarters, and he observed that the monkeys would soon have a new building that, while not a palace, as some said, would be on a par with the other animals' accommodations.[51]

Further discussion of Jack's good nature and intelligence appeared in *Le Temps* in September and was reprinted in *L'Echo du monde savant*. Readers learned that Jack's normal dinnertime was 5:00 p.m. When the clock sounded, he would swing on the rope from the ceiling of his room to the dining room door and shake the lock until it opened. One evening Daboncourt tied three knots in Jack's rope, shortening it and preventing the ape from swinging to the door lock. In the presence of Geoffroy and another visitor, Jack handled the situation with ease. Discovering that he could not reach his goal with the knots in the rope, he untied one knot and found the rope still too short, then untied a second knot which still wasn't enough. The third knot was tied high on the rope. Jack climbed up to it and instead of trying to untie it from below, which his weight would have prevented, he climbed above it and untied it easily. Then he was able to swing to the door lock, open the door, and go to dinner. In the reporter's words, this represented "more than dexterity," it involved "attention and reasoning." Remarking that "the psychological state of this young ape is very worthy of the observation of philosophers," the author was concerned for the Jardin des Plantes to find it larger, warmer quarters before winter. "The orangutan needs continuous warmth; the cold and the humidity would kill him."[52]

Other sources tell how Jack ate dinner with the Daboncourt family, how he was displeased if not given the same food as the children; how he managed well with a spoon but not a fork; how much he liked playing games. He was happy to have visitors. Swinging from the rope attached to the ceiling of his room, he enjoyed touching visitors' hands, grabbing their knees, or taking someone's hat and putting it on his own head.[53] Daboncourt was present to oversee the behavior of the orangutan and his visitors.[54]

The naturalist Achille Comte attested to Jack's intelligence, describing what he did when given a salad he found unpalatable because it was dressed with vinegar. Jack retrieved a large scrap of paper from the corner of his room that biscuits had come wrapped in, spread it out like a napkin, folded it several times, put a salad leaf between the folds, pressed the paper like a sponge, then proceeded to do this with all the leaves. With the offending dressing removed, Jack ate the salad with apparent content. On another occasion, after Daboncourt relocated the lock on Jack's door to place it out of his reach, Jack found a way to open the lock by draping his rope over the latch, bracing himself against the legs of his table, and pulling on the rope until the latch slid up, as if on a pulley, and drew the bolt.[55]

Comte cited Georges Cuvier's observation that while the orangutan could imitate numerous human actions, its intelligence was less than some said and was not much greater than a dog's. But Comte felt there was more to the story. He called for studies of Jack's intellectual faculties that would go beyond his imitative abilities to "the attentive and difficult examination of the phenomena of reflection and spontaneous decision that one believes to exist in him."[56]

The Museum continued its efforts to accommodate public enthusiasm for Jack without endangering his health. Yet not everyone enjoyed observing him. Anxiety about the orangutan's status at the border between human and ape appears to have motivated a writer for the *Journal des débats* to list the features of its physiognomy he found revolting:

> And you will be struck by this appearance of repulsive bestiality that one has wanted to compare to man, and of which some ancient naturalists have not been afraid to make a degenerated child of Adam, having an origin like ours! This fat stomach, these long front paws hanging to the ground, this horrible, flattened muzzle, this enormous mobile and wrinkly mouth, this squashed nose, these eyes round and close to the nose, these long red and stiff hairs, this squatting position, these movements of revolting brutality, and a certain air of lowness spread over its monkey like physiognomy, all seem to me to put more distance between this animal and man than between my dog and me.[57]

When Frédéric Cuvier returned to the menagerie at summer's end, he promptly supported Geoffroy's proposal to clothe the orangutan to protect it from the cold, writing, "it is certain that a garment during the winter is one of the surest ways of those indicated for this purpose." Cuvier hoped this would be done by the time the weather turned, pointing out there was no reason to think the orangutan would be afraid of a garment, since it sought to wrap itself in a cover whenever the temperature slightly dropped.[58] But he did not think Geoffroy's proposal to build a new home for the orangutan was possible. There was no better lodging than the current one, he said, and he felt in no position to offer a specific opinion on Geoffroy's proposition. He asked the administration (twice) to have a committee examine Geoffroy's proposal, but once formed, the committee only dithered. As of mid-October, nothing had been decided regarding new quarters for the orangutan, but clothing had been made for him.[59]

Then what had long been anticipated happened. Frédéric Cuvier announced "the orangutan has been ill; he has shown much despondency and refused all food. This state was accompanied by fever and difficulty in breathing." Blainville bled the animal, which "at first appeared to produce very good effects." The orangutan snapped out of his torpor and began behaving almost normally again. However, the next day (the day of Cuvier's report), the animal was downcast again, if not quite as much as before. Medical protocol called for plasters, Cuvier said, but the orangutan would not tolerate them. Daboncourt was ordered not to leave him for a moment, and the public was forbidden admission.[60]

The newspapers anxiously covered Jack's condition and medical travails, reporting that Blainville was assisted by the anatomist Antoine Serres and an internist from la Pitié hospital named Forget, and that the lung specialist A. P. E. Bazin was called in to listen to the animal's breathing. Where in the same period the Museum called upon a veterinarian to treat a bison, it called upon physicians to treat Jack. They deployed the same therapies used for humans: bloodletting, poultices, and enemas. They bled Jack by the arm. They applied leeches to his chest. When his skin proved too thick for the leeches to draw blood, incisions were made to ease their work. They also gave Jack enemas.[61]

The papers generally agreed that Jack found the enemas the most objectionable aspect of his treatment. He responded stoically to the doctors' procedures at first, reported *La Presse*, but at the sight of the enema syringe he made such grimaces and contortions that the physicians were at a loss as to how to handle him. The image of Jack objecting to the enema (known since Molière's time as "the weapon of M. Fleurant") so delighted the editors of *Le Figaro* that they concocted a long, satirical piece sympathizing with Jack, lampooning the professors attending him, and calling up the image of the dance scene from Molière's comedy ballet, *Monsieur de Pourceaugnac,* where physicians armed with syringes chase the title character around a room.[62] *Le Constitutionnel* and the *Journal des débats* introduced an additional theme, alleging that Jack was terrified by the physicians' procedures and displayed confidence only in Atir, the black African in charge of the giraffe, whom Jack beseeched for "protection against those he regards as his oppressors."[63]

The press accounts of Jack's illness demonstrate the Museum's inability to control what writers said about its specimens, or its professors. The story of Jack's appealing to Atir for help was nothing more than a play on the trope of orangutans having an affinity with the "lower races" of humankind.[64] That *Le Constitutionnel* and *Journal des débats* were not the best sources on Jack is shown by the fact they both reported (as did *La Presse*) that he died on 28 October. The first two papers added the authoritative-sounding detail that death came at noon, but Jack was still very much alive.

On 25 October Frédéric Cuvier told the administration that the orangutan had recovered entirely from his illness and regained his normal "strength and gaiety." Cuvier warned, however, that despite all the Daboncourts' care, the crowds that visited Jack "each day and throughout the course of the day" were a threat to his well-being. The administration agreed to Cuvier's proposal that the public be admitted only three times a week, from 1:00 until 3:00 p.m., on those days when visitors were already required to have a ticket for the galleries.[65]

On 8 November 1836, Cuvier remained pleased with the orangutan's health: "Judging from his games and his appetite, he seems to have never been better. Up to the end of his illness he tolerated the garment that had been made for him. Afterwards he began to endure it only with distress, and for three days each time one wanted to put it on him, he tore it to shreds." Cuvier indicated they would try to make the orangutan sturdier clothing.[66]

What displeased Cuvier was the administration's failure to announce the new viewing schedule of three days a week and to print new admission tickets specifying the days and hours of admission and that a ticket was good for the entry of no more than four people.[67] When the administration failed to act, Geoffroy pressed the request at the next assembly, suggesting that admission be 11 a.m. to 3 p.m. on Mondays, Thursdays, and Saturdays, and that if too many ticket holders showed up at once, there should be a number system to handle their admittance. Groups would wait their turn in a room (then occupied by Rémond, the assistant monkey keeper) on the ground floor of the building, then ascend to the next floor and cross through the Daboncourts' apartment via a corridor created by folding screens. Visitors would enter from one side and leave by the other. Two groups could visit at a time, each for five minutes. The assembly agreed to Geoffroy's rules.[68]

La Presse had meanwhile corrected its report of the orangutan's death with news of its recovery thanks to Blainville (and leeches) and to the clothing Cuvier made it wear. It helped, the paper added, that "M. Geoffroy Saint-Hilaire came each day to the patient and brought to him musicians whose harmonies produced the happiest effect on his convalescence."[69] Other sources confirm the bloodletting and the clothing, but what about the music? Jack's recovery inspired a madcap riff in *Le Figaro* featuring Geoffroy's idea of clothing him but also serenading him. It was the story from *La Presse* spun out in gleeful buffoonery. By *Le Figaro*'s account, Geoffroy had decided that Jack's illness was primarily a matter of morale and the best way to cheer him up was to dress him fashionably. The Museum duly provided Jack with breeches, a frock coat, a felt hat, yellow gloves, and spurs. Jack liked this treatment far more than bloodletting and purges, and a rosy hue began to infuse his muzzle, but he pursed his lips from time to time to show he did not feel quite right.

At that point, *Le Figaro* continued, Geoffroy "had an idea even more triumphant than the first . . . of giving a serenade to Jack." A first effort at the Théâtre des Funambules failed because its musicians were dreadful, and Jack, who was a composer of brilliant piano fantasies for four hands on the piano, was understandably livid. Geoffroy then arranged the finest Italian and French opera singers to perform for Jack, which pleased and fully healed him. In *Le Figaro*'s telling, Jack had been last seen with his Negro companion Atir in the Bois de Boulogne, where he was writing, in gratitude to Geoffroy, a symphony to be performed under Geoffroy's balcony by an orchestra comprised of Jack and his monkey friends (baboons, mandrills, macaques, etc.).[70]

Le Figaro's mirthful flight was not wholly fanciful. Jack, Atir, and Geoffroy were all highly visible personages of the Museum. A pantomime titled "Jack, l'orang-outang" was performed at the Théâtre des Funambules.[71] Geoffroy's desire to dress Jack and Jack's illness had been well publicized. As for treating Jack to music, Franz Liszt thought about it, but *Le Figaro* most likely heard about a concert proposed by Antoine Elwart, professor of counterpoint and fugue at the Paris Conservatory of Music.

The twenty-seven-year-old Elwart had lately won the Rome Prize for the best work by a young composer. In November 1836 he was planning a concert for the orangutan, and like Liszt, first approached the Museum through Geoffroy.[72] Noting that "similar experiments" with other of the menagerie's animals had already benefited the science of music, he detailed his plans to the administration to give a concert of vocal and instrumental music to see its effects on the animal. He envisioned seven musicians performing in the presence of the Museum's professors: three "beautiful human voices" plus "our leading virtuosi on piano, oboe, French horn, and violin."[73] The experiment would be conducted "in the interest of the art that I teach at the Conservatory and not in the futile goal of a simple curiosity," with the composer and several journalists in attendance (which likely gave *Figaro* wind of the project). He had visited the orangutan's quarters, he said, and concluded they were entirely appropriate for the experiment.[74]

Like Listz's, Elwart's musical plans for the orangutan never materialized. Just as the composer was making his plans known, Jack fell ill again. On 6 December Cuvier reported of the animal, "since Sunday he has stopped eating and fallen into a deep despondency." The Museum told Elwart his experiment would have to wait until the animal recovered, and the administration asked the newspapers to announce that the orangutan was ill and could not be visited.[75]

Blainville and Duméril once again started treating the animal.[76] On 20 December Cuvier could report that some of his appetite and gaiety had returned and he was sleeping through the night. Several times a day, however, he made "a humid cough."[77] Jack's gaiety did not last. On 27 December 1836, Cuvier described the poor animal as dejected: "The orangutan continues to eat an appropriate amount; but he seems depressed; he is habitually sad and inactive. The cold and snow seem to augment this state, though his cage is always sufficiently heated. For three days he has not left his cover."[78]

The last glimpse we have of Jack alive is from the correspondence of author George Sand (Amantine Lucile Aurore Dupin de Francueil). On 29 December Sand visited the menagerie with Geoffroy. She wrote to her sister the next day: "The orang is very ill. He looked at me with a dying air, and since I wanted to help him wrap himself in his covers, he gave me a big pat on the hand."[79] Jack died three nights later. Cuvier's final report on the orangutan, dated 9 January 1837, was but a single sentence: "The orangutan died in the night of the second to third of this month."[80]

Jack had arrived at the menagerie on a beautiful, sun-bathed day in spring. He died in the harshest winter Paris had experienced for some time.[81] It was an all too familiar story. Apes and monkeys transported to cold, damp climates were extremely susceptible to lung diseases. Jack's stuffed skin ended up in the Museum's zoology gallery in the same display case as the chimpanzee that had belonged to Buffon and the orangutan that had belonged to Empress Josephine.[82] Jack expired before the scientists of the Museum did any systematic studies, before he could develop into an adult, and before hearing a concert. Seen in context, the musicians' plans for Jack were no more whimsical than Frédéric Cuvier's unrealized notion of conducting experiments on the orangutan's mental faculties or Geoffroy's hopes of seeing him survive to maturity.

Newspapers provided most of the published accounts of the orangutan's behavior, while offering a range of observations, critical or satirical, on the Museum professors' ideas and practices. *Le Figaro*'s zany scenario of Jack moving freely beyond the Museum's gates with Atir and composing a symphony for Geoffroy stands as an irreverent reminder that despite the efforts of the Cuvier brothers to render the animal/human border inviolate, skirmishes there involving scientists, nonscientists, and animals were far from over.

[4.11]

The Old Professor and the New

Frédéric Cuvier had more than a sick orangutan on his mind in December 1836. He was pondering his prospects for a professorship devoted to the study of living animals. On 6 December 1836, the day he reported Jack's relapse, he wrote to François Guizot, minister of public instruction, urging him to create the new post. His original appeal four years earlier had come to naught, but recent signs seemed auspicious. A confidant told him that the Ministry of Public Instruction was open to "new developments" in the Museum's instructional program. This time, Cuvier believed, enough professors appreciated the importance of his subject to support his bid to become one of them.[1]

Cuvier's carefully crafted letter to Guizot began by explaining that the Museum's course offerings did not encompass all that the study of animals should entail. The original teaching program, he said, was a consequence of contingent events. The courses were chosen with a view to existing collections. He was too politic to add that this scheme perpetuated the roster of naturalists who previously served as officers at the old Jardin du Roi.

The gist of Cuvier's argument was that the scientific study of animals encompassed more than their comparative anatomy and classification. It also involved their actions, the function of their organs, and the place they occupy in the economy of nature. But the study of animal behaviors and the faculties giving rise to them had never been a serious part of natural history curricula. Without knowing "the faculties with which nature has endowed them to conserve them and to provide for their needs," one could not know how these faculties functioned "in the general harmony of this world." It would be like studying humans only through their anatomy and physiology.

Cuvier proposed that just as the understanding of human anatomy had been enriched by studying animal anatomy, human psychology would be enriched by comparing it "with the numerous psychological phenomena that the long series of animals presents." "In what they have in common with us," Cuvier averred, "the animals truly offer a natural analysis of human

Figure 4.11.1. "Étienne Geoffroy Saint-Hilaire." Portrait by Antoine Chazal. Muséum national d'histoire naturelle. MS 2671–1, fol. 4.

intelligence." Furthermore, the relative simplicity of psychological phenomena in animals made it easier to study them than those of humans. The additional utility of understanding animal psychology was that it would aid in "the formation and cultivation of domestic animals," an area still dominated by "the grossest routine" and "blind empiricism."

Cuvier assured the minister that creating the chair would not require substantial new resources, for the menagerie left "little to desire" in this regard. The chair, he further suggested, would help justify the sums the government was already spending on the menagerie. "The study and teaching of the mental faculties of animals and the actions they produce," he said, represented "the only means of making the life of animals useful to science." Until now, he maintained, the menagerie's animals had been useful only after they died and were dissected or stuffed in the comparative anatomy and zoology labs. Otherwise, the "unique result" of spending 60,000 to 80,000 francs annually on the menagerie was to satisfy "the vain curiosity of the public"—and that would be better served if the menagerie acquired the scientific importance it deserved. As a final argument for establishing a chair for teaching about "the psychological nature of animals," Cuvier cited national pride. No other country offered such a course, and it would bring honor to France to be the first. Having presented these general arguments

Figure 4.11.2. "Frédéric Cuvier." Portrait by Antoine Chazal. 1826. Muséum national d'histoire naturelle. MS 2671–21, fol. 3.

for a chair of animal psychology, Cuvier reviewed his credentials and included a copy of the program of instruction he planned to deliver.

Having reviewed Cuvier's letter, Minister Guizot invited the Museum's professors to consider "a chair on the psychology of animals." He simultaneously sent documents from the physicist Antoine-César Becquerel requesting "a chair of physics applied to natural history." At the professorial assembly of 20 December 1836, the director read the cover letters for Cuvier's and Becquerel's requests along with Cuvier's letter to Guizot and his proposed program of instruction. Geoffroy proposed creating a new chair for Cuvier, a chair of "general zoology." The meeting's minutes do not specify what Geoffroy expected the chair to entail, but overseeing the menagerie was certainly not part of it.[2]

The assembly appointed two committees, one to review each of the proposed chairs. Though Guizot had referred to Cuvier's request as being for a chair of animal *psychology*, the professors called it instead "the chair of comparative physiology." They likely judged physiology to have a more secure scientific foundation than psychology, and thus a more promising future. The committee to review this chair consisted of Blainville, professor of comparative anatomy; Flourens, professor of human anatomy; and Valenciennes, professor of the natural history of the mollusks, worms, and zoophytes. Flourens and Valenciennes had been Georges Cuvier's protégés and were Frédéric's friends. Blainville, an enemy of Georges Cuvier, only pretended to be sympathetic to Frédéric's case.

After both committees reported, yet another committee was formed to identify the Museum's most pressing instructional needs. It recommended (31 January 1837) establishing the chair of physics applied to natural history but questioned whether the facts related to studying living animals had yet been satisfactorily assembled in a body of science. The question was tossed back to the collective wisdom of the assembly. After considering whether "to create the new chair defined by the program of Mr. Frédéric Cuvier and designated by him under the title 'course on the nature of animals and on their training,'" the professors rejected Cuvier's proposal by a voice vote of seven to five.[3] It was a crushing blow to the menagerie's superintendent. He had been wrong to think most professors would support him, and he knew that the administration's composition was unlikely to change soon.

Frédéric Cuvier's role as superintendent of the menagerie was not in jeopardy, but he was unhappy with his subordinate status, which had become more difficult as his relations with his superior deteriorated, as had Geoffroy's relations with nearly all his colleagues. He had not conspicuously clashed with Frédéric since 1828, but he had frequent disagreements with the others. Aside from his famous debates with Georges Cuvier at the Academy of Sciences, Geoffroy fought often with Blainville; was piqued when the administration did not approve special celebrations for the Jardin des Plantes' bicentenary; boycotted the Museum meetings for over a year; and quarreled over dressing the orangutan. The voyager naturalist Jean René Constant Quoy in 1836 wrote to a friend saying that on some matters, Geoffroy was considered almost "half insane."[4]

Frédéric Cuvier sided with Geoffroy on the matter of clothing the orangutan, but his efforts to gain a professorship put them back on a collision course because he wanted the menagerie to come with the new chair. Geoffroy saw this as an affront—more proof of Cuvier's disrespect. A week after the assembly voted down Cuvier's proposed chair, Geoffroy had more to complain about. He announced angrily to his colleagues that the menagerie's affairs were being conducted without his knowledge, against regulations.[5]

His fury arose from his sense of having been publicly embarrassed by being the last to know about a series of awkward incidents in the menagerie. It was partly his own fault; if Geoffroy had read Frédéric Cuvier's menagerie report (14 February 1837) or seen to its presentation at the assembly that day, he would have learned that the giraffe's keeper, Atir, was missing, and that a bulldog had attacked both lamas. Likewise, when a policeman threw a dog into the bear pits (16 February), Geoffroy first heard about it from friends and associates. When people asked about the story, he assumed, having had no news of it at the Museum, that it was a fabrication, and said so. When he questioned a keeper, he was dismayed to discover the story was true, and that this and other incidents had not been reported to him.

Outraged that he, the person officially in charge, had been left in the dark, Geoffroy feared people would think he had lost control of the menagerie and that it had fallen into chaos. Certain that the superintendent's insubordination was at the root of the problem, Geoffroy fulminated against Cuvier and demanded a return to the rules that had created the superintendent's post and placed it under the authority of the professor of mammals and birds. He called for a committee to collect the facts and restore the old order.[6]

Geoffroy claimed that he had done all he could to elevate Cuvier to a professor's position and that if it had failed, it was Frédéric's fault, not his. Frédéric, Geoffroy said, had refused the chair that Geoffroy had proposed: "He backed us into a corner in requiring his whole program, all or nothing." Geoffroy regarded Frédéric as "guilty of an atrocious miscalculation, guilty of ingratitude, for who more than I showers him with kindness?" However, Geoffroy said, if Frédéric apologized and the old order was reestablished, he was ready to pardon him.[7]

Geoffroy went on to remind his colleagues of other indignities he had suffered, such as that he and his son had had to cover Frédéric's three-to-four-month summer absences. Was Geoffroy to think, he asked, that Frédéric took him for a hapless ruler he could abuse, like king Soliveau (the ineffectual king in la Fontaine's fable)? Geoffroy acknowledged that he had been signing the menagerie's account books, trusting Frédéric's accounts were accurate, but he would stop signing them if nothing changed, because his trust was gone. Professing the continuing respect he had for "the great Cuvier," Geoffroy said Frédéric would do well to think of his brother and not degrade the family name he was fortunate to bear.[8]

Over the next several days Geoffroy and Frédéric achieved a truce, if not a reconciliation. Geoffroy wrote to remind Frédéric that he'd tried to secure him a professorship, proof that he was "capable of an infinite forbearance," but his "natural benevolence" could not extend to destroying his own professorship by having the menagerie removed from its purview. "We were informed," wrote Geoffroy, "that you expressly wanted your program, all or nothing, and you know the decision taken." The time had come, said Geoffroy, for them to return to the arrangements of the past.[9]

Geoffroy asked for more information about recent incidents at the menagerie and Cuvier's plans for handling them. He wanted Cuvier to provide the administration with an up-to-date list of all the menagerie animals, something called for in the menagerie's regulations but rarely realized. In addition, Geoffroy asked that Cuvier send him the weekly menagerie reports the night before the professorial assembly, not the day of, to allow him time to peruse them. Geoffroy had his letter hand delivered to Cuvier's apartment.

How Frédéric responded is not known, but he presented Geoffroy with his census of menagerie animals as of January 1, 1837. Geoffroy nonetheless toured the menagerie in late February, compiling a list of the animals present (or at least those the keepers reported). At the next professorial assembly, he announced that he'd discussed menagerie regulations with Cuvier, and he presented Cuvier's census. He recalled the need to enforce the rules about keeping dogs under control, especially as these applied to the menagerie.[10] He seemed to think the old order was returning. The remnants of his subsequent exchanges with Frédéric suggest calmer relations and a joint acknowledgment that the administration should not be troubled with menagerie affairs.[11]

In spring, Cuvier was overseeing the menagerie's daily operations and preparing brief accounts regarding its needs and news (mostly births, deaths, acquisitions, and exchanges) for Geoffroy, who passed them on to the administration. A whiff of Cuvier's simmering anger can be detected in remarks he made to Geoffroy regarding the good work of the keeper Roux, who filled in for the injured Hippolyte Martin. Cuvier resented the bias Geoffroy had long shown against Roux.[12]

In May Cuvier wrote to the Museum's director requesting a two-month leave to conduct his annual inspections for the university. He reported that his colleague, Isidore Geoffroy Saint-Hilaire, had graciously agreed to replace him, and he had "every reason to hope that Monsieur Geoffroy will consent to this arrangement." The professorial assembly approved that same day.[13] Isidore's service proved impeccable. He followed up on requests Cuvier had made to the administration and provided detailed accounts (more detailed than Cuvier's had lately been) of events including the births, deaths, offers of animals for sale, needs for repairs, and measures needed to keep the male dauw from causing more harm.

Frédéric Cuvier returned to his post at the beginning of August, writing short menagerie reports on the seventh and fourteenth of the month and a slightly longer one on the twenty-first.[14] The most interesting feature of this document is an annotation Geoffroy scrawled at the top maintaining that Cuvier refused to see him, a behavior he found incomprehensible, but there was a reason. Frédéric felt Geoffroy had maligned Georges Cuvier's name in a paper presented at the Academy of Sciences.

Geoffroy's paper was ostensibly about Earth's diverse and changing environments and how these had influenced the planet's surface and life on Earth over time. In developing his argument, he broached the subject of Georges Cuvier's promoting "the false principle of the immutability of species." Geoffroy claimed that Cuvier used all his knowledge and stunning talents as a writer and observer to obstruct the idea of species change with an eye to "political suitability." Feeling it would be disruptive to society should the revelations of paleontology clash with sacred texts, the elder Cuvier, said Geoffroy, tailored his writings to stay close to "biblical and historical truths."[15]

Frédéric rose in his brother's defense at the Academy's meeting of 21 August. His angry and sharply focused rebuttal did not address the mutability of species but concentrated instead on the paragraph where Geoffroy claimed that Georges, in his discourse on the revolutions of the globe, had "disguised the truth" and "capitulated with his conscience" for political reasons. Frédéric said he had never before felt the need to respond to Geoffroy's attacks on his brother's works or reputation, feeling totally secure in "the judgment of our peers." In this case, though, said Frédéric, it was a question of his brother's honor, and he could no longer remain silent. He challenged Geoffroy to prove his claim, to review the diverse facts in Georges's famous discourse, from the "traditions of all the principal peoples of the earth" to geography and to all the fossils by which one identified the Earth's age. He dared Geoffroy to identify any conclusions that Georges had drawn "that he [Geoffroy] would not have wanted to draw from them [himself]." If Geoffroy could not, said Frédéric, he was in the right (and generous at that) in treating Geoffroy's words as "rash, thoughtless, and totally lacking in foundation."[16] It is little wonder Frédéric Cuvier was disgruntled. The puzzle is that Geoffroy seemed baffled as to why. He rejected Frédéric's interpretation of his words, claiming his paper rendered homage to Georges Cuvier's talents and achievements, but that is hard to square with the paragraph Frédéric objected to.[17]

In other respects, the menagerie activities continued apace. The most exciting August news was the arrival at Nantes of a live chimpanzee that its owner wished to sell. September saw negotiations to purchase the animal. But Cuvier's menagerie reports were terse. When Geoffroy returned from a trip to Caen where he acquired various nonliving specimens, he shared with the administration Cuvier's reports from the previous three weeks, suggesting that their limited information constituted all he knew about menagerie affairs. He told his colleagues, "Administer if you can."[18]

In October the chimpanzee arrived in good health. In November the polar bear was placed in one of the bear pits; staff were forbidden to allow their dogs to run free on the Museum's grounds; and monkeys continued to have health problems. During this period Cuvier supplied both Geoffroy and the director with a copy of his reports, apparently to ensure that they were properly deposited with the administration.

Frédéric was meanwhile in touch with the new (since April) minister of public instruction, Narcisse Achille de Salvandy, a family friend who had delivered a eulogy for Clementine Cuvier, Georges's daughter, in 1827.[19] As inspector of the Université de France, Cuvier had ready access to Salvandy, the head (*grand maître*) of the institution. In December, Salvandy wrote to the Museum's director (Chevreul) asking the professors to advise "on the creation of a chair including the menagerie in its attributions, having as its object the general and physiological study of the instinct, educability, and acclimatization of animals." Salvandy suggested calling it the chair of "comparative physiology" or "natural history applied to the cultivation [*culture*] and acclimatization of animals."

The minister's letter stimulated "an extended discussion" when read to the assembly and a decision to consider it at the next meeting, since some members were absent, including Geoffroy, who was on a brief trip to Oxford. On December 12, with all thirteen professors present, "the proposition to create a course under the title of comparative physiology" was put to a vote and adopted with ten votes in favor, one against, and two abstentions.[20] The vote against was not from Geoffroy, but from Blainville; Geoffroy's protests came later when he realized that menagerie oversight was being taken from him.

With characteristic combativeness and misanthropy, Blainville disapproved of the new chair primarily on the grounds that it infringed on his own privileges. The functions of the new chair were already subsumed, he said, under the chair of comparative anatomy, and neither the Museum nor the minister had the right to alter Blainville's chair in his lifetime. He saw conflicts between the two chairs as inevitable. A chair of comparative physiology, he maintained, must include anatomy. Moreover, the new chair's teaching would be speculative, whereas the Museum's teaching should be "rigorously positive."[21]

Blainville's objections notwithstanding, the assembly appointed a committee to draw up a program for the course of comparative physiology and to fix the attributes of the chair. The results mirrored Cuvier's suggestions. The study of living animals would be the principal subject of the course, pursued through direct observation and experimentation. In addition to teaching about the physical action of organs (where this could easily be demonstrated), the professor would consider the faculties of intelligence and instinct. The study of "the training [*l'éducation*] of animals" and "the modifications that the attentions of man can bring about" would also be covered. The final sentence of the program announced the change that devastated Geoffroy. "The direction of the operations of the menagerie" was to be "specially confided to the professor of comparative physiology without prejudice to the scope of attributions of the other professors."[22]

The Museum sent Salvandy the course program, and on 24 December 1837 King Louis Philippe, in response to Salvandy's report, signed the royal ordinance declaring Frédéric Cuvier professor of comparative physiology at the Museum of Natural History.[23] Cuvier submitted his final report as menagerie superintendent. When the professors next met, on 9 January 1838, he was a member of the assembly. Geoffroy did not attend. Cuvier's first presentation addressed Atir's latest misconduct and called for his expulsion. He deposited one table showing the menagerie's expenses for 1837, another indicating the year's consumption of meat, and a list of the animals in the menagerie as of 31 December. He recounted the arrival of animals from the voyage of the *Bonite* and shared the happy news that the chimpanzee, which had been ill, was doing fine. He additionally called for immediate repairs to the new monkey house's heating system and related the details of the possible acquisition of a hamadryas baboon.[24]

Geoffroy was elsewhere, awash in fury and sorrow. Called to a meeting of a committee charged with assessing work needed in the zoology galleries, he told his colleagues to do without him in their deliberations "because you have destroyed me." He planned, he said, to take his case to the public.[25] Then came a long letter of complaint read at the assembly on 16 January 1828. Geoffroy began by praising the Museum's decision to honor the name of Georges Cuvier by creating the chair of comparative physiology, which was worthy of "the ideas of our glorious colleague." The problem, he said, was that a phrase representing an "injustice with regard to your old colleague, your doyen" had slipped into the course program, namely, the phrase that gave the professor of comparative physiology direction of the menagerie. This was unconstitutional and unjust, he insisted, for how could all that he had created be passed to a new naturalist? Could a lieutenant be allowed to rob his captain? "Me, your old doyen, who through hard and incessant works aspires only to public consideration. Me, bent under forty-four years of ardent research, you condemn me to such a painful disrepute." Geoffroy wanted the decision rescinded. He demanded "*justice* [Geoffroy's emphasis] for everyone."[26]

A week later Geoffroy wrote to the administrative complaining that he'd had no response to his previous letter. Justice was being denied, he said, setting a dangerous precedent. He likened his situation to that of Lamarck, who instead of being recognized as a great innovator, had been credited only for his early descriptive and nomenclature-related work:

> Lamarck appeared as a powerful and dangerous adversary in this theoretical question: *Did species come forth from the hand of God in the form in which they exist in the present world?* Yes, maintained your great head of school [Georges Cuvier]: they are *ab ovo* [from the beginning] immutable. Certainly *not, no*, contradicted Lamarck. Their philosophical character resides in the indefinite variability of their forms, by reason of the conditions of their different ambient environments. [Emphasis by Geoffroy.]

Geoffroy admitted that Lamarck's examples of organic change were "so maladroit" that "they became the subject of *vicious ridicule* [Geoffroy's emphasis]," but he nonetheless believed that Lamarck's idea was "fundamentally that of a man of genius," and the current age had come to recognize it. Lamarck's fate, however, had been to withdraw from his colleagues, suffering blindness and misery in isolation, as if in a prisoner's cell.[27]

Geoffroy commiserated with Lamarck. He too was an innovator, he said, but people complained that his discussions caused trouble, portraying him as a helot at the Academy and the Jardin des Plantes, where his contributions should have commanded respect. He urged his colleagues to confront him "conscientiously and courageously." He was about to publish a book on Buffon and the progress of the natural sciences, and he would make his case there. He would not avoid combat. It was no time, he said, to adjourn the whole matter and be suspected of covering up an injustice.[28]

Geoffroy's letter was read to the professors at their 23 January 1838 assembly. They decided that their original ruling had been made after careful consideration and that no new deliberation would be permitted. Through February and March, Frédéric Cuvier reported weekly on the menagerie while Geoffroy remained absent from the assemblies. On 3 April, Geoffroy wrote a six-page diatribe voicing grievances, offering arguments, and listing those to whom he had taken his case. He blamed his loss of the menagerie more on his colleagues than on the minister, who told Geoffroy (or so Geoffroy reported), that he had heard Geoffroy wished to be relieved of the menagerie so he could spend more time developing ideas. To underline the weakness of this excuse, said Geoffroy, his friend Jean Reynaud had just written an article on the Museum's founding and on Geoffroy's role in it, to be distributed widely. It would demonstrate, Geoffroy said, that he was perfectly capable of administering the menagerie.

Everything had been fine, Geoffroy wrote, when he did not have a lieutenant who constantly opposed him. Arguing again that it was illegal to remove the menagerie from his purview, he observed that in creating the chair of comparative physiology, the job of menagerie superintendent had been suppressed, purportedly to economize. Geoffroy believed that the post should remain independent of the new chair and that it should be awarded to his son Isidore, who was eminently well qualified. Geoffroy indicated that the injustices he'd suffered would not go unnoticed, for they were covered in his forthcoming book of biographical sketches of Buffon and others.

Reynaud's portrait of Geoffroy as the heroic founder of the menagerie was exaggerated. There is no evidence of Geoffroy's stepping boldly forward to open the Museum's gates to admit the animals confiscated from the street shows of Paris. Nor did the Convention decree that Geoffroy be menagerie director. Reynaud did not mention that when the menagerie's supervision became too burdensome for Geoffroy in 1795, it was turned over to Delaunay, who did the job for the next six critical years of the menagerie's development. Reynaud alluded to injustices' having befallen Geoffroy in his old age and concluded by saying that Geoffroy's name "will be inviolably attached to the glory of the foundation of our national menagerie."[29] The last part has proved true (aided in part by Reynaud's account), as is the case that under Geoffroy's general aegis the menagerie became a much admired Paris institution.

On 17 April 1838, Cuvier announced his imminent departure to perform his yearly duties as university inspector, noting that Achille Valenciennes, professor of mollusks, worms, and zoophytes, would oversee the menagerie in his absence. On 22 May 1838, Geoffroy also took leave of his colleagues, without saying where he was going and intimating that he might not be coming back. In his parting letter he said he had left a copy of his book on Buffon in the Museum library. The book, *Fragmens biographiques*, featured portraits of French naturalists, Geoffroy's laments on how badly his fellow professors had treated him in his old age, and Reynaud's article on the founding of the menagerie.[30] Geoffroy also left a pair of anatomical drawings of crocodiles, which he said he no longer needed.

Mournfully he wrote, "I am really and truly leaving, and it is very possible that death, which finishes many journeys, will overtake me on the way." His only request to the Museum was that it employ his faithful domestic servant as a sweeper in the new galleries. Geoffroy closed his letter, "I salute you, messieurs, with sorrowful feelings but without the least resentment."[31]

[4.12]

The Chimpanzee and the "Palace of the Monkeys"

The Museum's first live chimpanzee, like its first live orangutan, arrived at the port of Nantes, courtesy of a merchant sea captain. The Museum's go-between was again Marion de Procé, who heard of the animal's arrival before the Museum did. He went to see the young female ape and wrote to Blainville, enclosing a drawing of her face. The letter was read at the administrative assembly of 1 August 1837, and the professors agreed that the portrait of the chimp was "very pretty."[1] Captain E. Boullemer wanted 8,000 francs for the chimp, a price the Museum found exorbitant. The orangutan had cost less than half that (3,500 francs) just the year before, and that had been judged expensive. The professors decided the chimp was worth 3,000 francs and no more. Blainville proposed seeking an additional 3,000 francs from the minister of public instruction if the owner would not lower his price, but his colleagues rejected the idea.[2] Nor did they inform the minister of the matter, fearing that the higher purchase price would be extracted from their budget later, as had happened once before.[3]

Did the professors also worry that the public might hold the high price of the ape against the institution? The thought was not registered in the minutes of their meetings, but the chimp arrived in France just when the finishing touches were being put on a new monkey house at the Museum that critics had dubbed "the palace of the monkeys," judging it inexcusably expensive and luxurious at a time when the local human population was inadequately housed.

The new monkey house had been a long time coming. The need for one had been apparent ever since the first monkey quarters were created by repurposing a building on the rue de Seine just outside the menagerie's perimeter.[4] Like the first makeshift structure for *les animaux féroces*, the first monkey house was deemed provisional but took decades to replace. It was not a salubrious dwelling. An 1828 guide to the Museum (1828) described it as a place where crowds reflected on the similarities between primates and humans, where children were amused by the monkeys' capers and

Figure 4.12.1. The new monkey house, also known as "the palace of the monkeys." H. Boitard, *Le Jardin des Plantes* (1842).

grimaces, and where mortality was high because of the incurable illnesses produced in winter by the accumulation of noxious exhalations.[5] Not only did the building stink, it was also dark and humid. Frédéric Cuvier in 1826 blamed the darkness and humidity for the deaths of four peccaries he had lodged there.[6] In 1828 he attributed monkey deaths primarily to the stench.[7]

In February 1833 Cuvier outlined for his colleagues the appropriate features of an improved monkey house: fourteen to eighteen cages, four to six large enough to accommodate big monkeys like baboons; the other dozen smaller for medium-sized monkeys like macaques, guenons, and capuchins. He proposed arranging the cages in a straight line, with southern exposure and sliding glass windows; having a service corridor behind the cages; using stoves to heat the cages and the corridor, and including a large space on the north side to serve as a buffer against the cold. There would be two chambers devoted to service needs, one at each end of the building. Sliding wooden doors would permit communication between the cages when desired, but these would need to be designed to prevent "inconveniences" caused by the monkeys' mischievousness and intelligence.[8] The administration approved, and Charles Rohault de Fleury, the Museum's architect, incorporated them in architectural drawings that were likewise approved. Rohault projected a rectilinear arrangement with eleven midsized cages in the middle flanked by three large cages on each end.[9] However, the plans were set aside for other priorities.

That same year, the government decided to devote large sums to restoring the monuments of Paris and embellishing the city with new constructions. When the Museum was allotted some of these funds in 1834, it thought to use the money for new greenhouses and a grand new edifice for the geology and mineralogy collections, the botanical galleries, and the library.[10] Not thinking of a new monkey house at the time, the professors were taken by surprise in January 1835 when Rohault informed them that the Building Council and the director of public works had approved his plans for the new monkey house and it was about to be built.[11]

The administrators found themselves in the embarrassing position of having to tell the minister of the interior that they had not given the design sufficient thought. They sought to finesse the situation by annotating the drawings and calling for a major addition, a large grilled courtyard where monkeys could roam freely outdoors when the weather permitted and the public could have the pleasure of watching them. While the animals' health and public curiosity would be served, there was also a matter of national pride involved. The London Zoo already had such an arrangement that had proved to have "great advantages" and "very few inconveniences."[12]

The administration urged the minister to prevent the architect from starting construction. They wanted him to consider their proposed modifications of the original plan, with the new courtyard, or to create a new plan if needed.[13] The plan that came back looked nothing like the old one. It called for twenty cages, all the same size, arranged in a semicircle around a great circular wrought-iron cage. The individual monkey cages opened onto the caged courtyard. The monkey house itself consisted of three concentric semicircles, with the animal cages on the inside, a corridor in the middle, and storerooms on the building's north side. The corridor provided service access to the cages as well as an inner gallery for the public. The storage rooms helped protect the corridor from cold drafts from the north.[14]

Funds for the monkey house and other Museum improvements became items in the government's budget for Paris monuments, alongside allotments for refurbishing the Madeleine, the Collège de France, and the two grand public buildings of the Quai d'Orsay. Of a total budget of 4,580,000 francs, the Museum's share was 1,200,000.

When the monuments budget was argued at the Chamber of Deputies in May 1836, François Arago was its most prominent critic. The young scientist who had once shared a ship from Algeria with a lion for the menagerie had become a renowned astronomer and a distinguished figure in science and politics. His 1830 election as permanent secretary of the Academy of Sciences for the mathematical sciences burnished his reputation. His political career began the next year as a deputy in the legislature, often standing in opposition to the president of the Republic, Adolphe Thiers. In attacking Thiers's monuments budget, Arago targeted recent expenditures and design flaws at the Jardin des Plantes, pointing to what he saw as hasty and costly choices with respect to the Museum's new greenhouses, the mineralogy gallery, and the monkey house. Thiers countered Arago point for point, providing the Chamber with information on costs, construction, and the scientists who helped design the projects and whose areas of expertise made them better authorities on the matters in question than an astronomer like Arago.[15]

The deputies approved the funding, and the debate between Arago and Thiers was swiftly forgotten. What captured the public's imagination was another deputy's characterization of the new monkey house as the "palace of the monkeys." Considerable laughter attended Pierre René Auguis's phrase, but his words were not entirely in jest. Reminding his colleagues of the many homeless citizens and other unfortunate poor near the Museum, Auguis insisted it would be better "to occupy ourselves with relieving misery before working on the rearing of monkeys."[16]

It was a serious point, but people find monkeys and apes amusing, and Auguis became the butt of primate jokes. The *Revue de Paris* remarked that if the monkeys weren't given a house, he might be embarrassed to find some ill-advised ape taking his place in the chamber.[17] As it happened, Jack the orangutan arrived at the menagerie shortly after Auguis's remarks in the legislative chamber, inspiring the *Revue de Paris* to observe that although Auguis might be displeased that the Museum had purchased the animal, "many persons of high distinction" had already visited it. Mirthfully expressing the hope that Auguis would visit too, the *Revue* predicted, "M. Auguis will be well received, the orangutan being an animal entirely free of grudges."[18]

Auguis did have supporters, however. In November 1836 *Le Charivari* revived his complaint with a page titled, "Ah! If I were an orangutan!" The critique was arranged in two columns, with the left-hand column offering humorous exaggerations about the orangutan's fine treatment at the menagerie and the right-hand column detailing the desperate situation of some twenty thousand homeless people in the quarter.[19]

Whether or not critiques of the monkey house's cost influenced the decision, the Museum was unwilling in 1837 to offer more than 3,000 francs for Captain Boullemer's chimp. The captain informed the minister of public instruction "with regret" that he would begin dealing with the London Zoological Society. When the minister asked the professors why the matter had not been brought to his attention before, they invited him to contribute 1,000 to 2,000 francs to the purchase, but he declined. The Museum stood firm, and Captain Boullemer eventually accepted the offer.[20]

The chimpanzee arrived at the Museum in mid-October, just a week after monkeys were installed in the new monkey house. She did not join them. Instead, she was given Jack's former quarters, the garret attached to the apartment of the Daboncourt family on the second floor of the old monkey house.[21] Frédéric Cuvier's menagerie report (17 October) affirmed that the animal arrived in good shape and was comfortable in her new lodgings. He insisted her visiting hours be stricter than Jack's. She was to receive two meals a day, and visitors were not allowed to feed her. They would be admitted only with reserve, he said, because the chimp would be cared for by a woman (Mme Daboncourt). He did not elaborate on this explanation.[22]

The press greeted the chimpanzee's arrival with great interest. Captain Boullemer and his crew called her *la vieille* (the old lady), but within days the press named her Jacqueline. An account in the *Echo du monde savant*, reprinted in other papers, began by noting the chimp's gentleness, declaring her proportions more human than the orangutan's and predicting these features would "certainly not delay her attaining a great celebrity." That her intelligence was "highly worthy of interest" was suggested by her powers of observation and her efforts to imitate an artist who was sketching her. When an onlooker "had the idea of putting a pencil in her hand," she "perched beside the artist and set about passing the tip of her pencil on the paper like a child who wants to try to write or draw."[23]

The *Echo*'s account included a note by Blainville saying that no chimpanzee had been seen in Paris since 1740, and furthermore that when Buffon wrote about that specimen in 1766, chimpanzees and orangutans were still confused with each other. Naming chimpanzees and orangutans the animals "closest to man," Blainville explained that chimps came from the west coast of Africa, Congo, and Guinea, while orangutans came from Borneo and Sumatra. Captain Boullemer had purchased the chimpanzee "from a young Negro who had brought it in his canoe, arms and legs tied, like an object of commerce, without saying where or how he obtained it." The chimp had only four incisors above and below, which made Boullemer think she was five or six months old, and she was about a year and a half when she reached the menagerie. She was considerably smaller than the orangutan, but Blainville considered her better proportioned, meaning "much more human."[24]

The most engaging descriptions of Jacqueline and her behavior came from Samuel Henry Berthoud, a prolific writer whose articles and commentary on art, architecture, literature, natural history, fashions, customs, and goings-on in Paris and the provinces appeared in numerous newspapers and journals, including *Musée des familles: Lectures du soir*, where he was editor. His depictions of Jacqueline can be seen as a verbal counterpart to the sculptor Dantan le Jeune's bust of Jack. Indeed, he and the sculptor were close friends. They and other friends met regularly in the late afternoon at Dantan le Jeune's apartment to discuss the news of the day and play dominoes.[25] Jack and then Jacqueline, as the latest attractions at the menagerie, readily caught their eye.

Berthoud's first brief account of Jacqueline, in one of the "evening lectures" of the *Musée des familles*, included the observation that she had "the forehead, eyes and nose of an Ethiopian negro." He remarked how the different colors of the chimp and orangutan paralleled the skin colors of the people of their lands. He added that Jacqueline was smarter than Jack, though he supported the claim only with the observation that she was more patient with the "indiscretions" of the children she played with.[26]

In a second "evening lecture," Berthoud emphasized Jacqueline's intelligence and gaiety and noted that she hated to be left alone. When her cries failed to attract attention, he said, she made as much noise as she could and peeked through the keyhole to see if anyone was coming to her rescue. Berthoud described how she romped with the keeper's children and laughed, and "made sounds similar to the voice of a child who does not yet speak but knows how to make himself understood." The emotions motivating her were easily recognizable. "Joy, boredom, tenderness, or displeasure" were always accompanied by the same respective sounds. Wrote Berthoud, "it is nearly speaking."[27]

Jacqueline stayed at the old monkey house for a month before being moved (15 November 1837) to new quarters in the rotunda, which was well heated to accommodate the giraffe, elephants, dromedaries, and other animals from warm climates.[28] Berthoud described her new room as furnished with straw and with a tree in the center to accommodate her "perpetual games of gymnastics." A visit brought him pleasure but left him with a particular worry: "[There is] nothing more amusing than to see her ceaselessly in motion, grabbing hold of the ropes hanging from the ceiling, coming with an unrestricted confidence, to jump into the arms of her keeper or of the scientists whose visits she receives daily. May she not succumb to the cold from which she coughs already."[29]

With the coming of colder weather, illnesses among the monkeys had risen. Just before Jacqueline was moved to the rotunda, Cuvier reported that "a great number" of the monkeys in the new monkey house had head and chest congestion and stayed huddled around the heating drums.[30] A week later he recorded the deaths of a Barbary ape and a white-faced capuchin and acknowledged that "the majority of the monkeys are ill."[31] In the first week of December he reported that a female Barbary ape had died, "the majority of the monkeys are unwell," and "the chimpanzee has a cold in the head and shows difficulty breathing."[32] The next week Cuvier had more deaths in the monkey house to announce. The first animals to contract the malady were the ones that had died. Those that initially resisted the disease continued to do so. The losses were four baboons, a grivet (African green monkey), a Barbary ape, and a macaque. Autopsies of several indicated that their lungs had been "profoundly altered."[33]

Cuvier noted that there was "some reason to think" the cause of the lethal illness was in or around the pituitary membrane. He proposed conducting an experiment on the best means of heating the monkeys' quarters. The monkey house would be divided into two sections, half heated with the *calorifère* (the system already in place) and half heated with a stove. The architect, he said, was ready to try this immediately. Cuvier additionally reported that the chimp had recently had "fever and much difficulty breathing" but that she was "well advanced in her recovery." She had been free from fever for several days, and the congestion in her bronchial tubes had been reduced to "very little."[34]

A week later, Cuvier wrote that the chimp's improvement continued. She was playing again and had an appetite. The menagerie meanwhile had received a Barbary ape and a grivet but lost two Barbary apes and a baboon.[35] On 26 December 1837, the superintendent related more bad news: "For the past two days, around two or three in the afternoon, [the chimp] has been sad and depressed. Up to then she displayed her normal gaiety. These crises of breathing [*d'oppression*] that come over her from time to time without any appreciable cause make one fear an alteration of the lungs."[36]

With the help of Cuvier's menagerie reports, the censuses he took at the beginning and end of 1837, and the menagerie's registry of mammals, one can track closely the changes in the primate collection over the year. The numbers at the beginning and end were nearly identical (thirty-two and thirty-four), but the turnover was staggering.[37] Eighteen of the specimens in the collection on 1 January 1837 died before the year was out. Of the twenty-two new specimens acquired during the year as gifts, purchases, exchanges, or births, six had died as well.[38] As could be expected, more animals died in cold months than in warm ones. Still, it had to disappoint the administrators that fifteen of the year's twenty-four deaths occurred at the new monkey house after it opened in October.[39] Seven more monkeys died in January 1838. The Museum had hoped the new building with its outdoor park would benefit the animals' health, but the figures suggest a death trap more than a palace.[40]

Some species lived longer than others. New World capuchin monkeys did particularly well. None of the menagerie's six or seven died in 1837, and all but one survived the following year. But four Barbary apes and a rhesus monkey present at the year's start didn't make it to the end. Although five new Barbary apes were purchased from a traveling menagerie in November, their time at the monkey house was short, one lasting just sixteen days, and none surviving longer than the following June. Cuvier's first menagerie report of 1838 announced that a lion-tailed macaque and a baboon had died, but he was able to say the chimpanzee was recovering.[41]

In February Cuvier proposed that a grilled outdoor park be constructed adjoining the chimpanzee's quarters at the rotunda so that she, like the other monkeys, could exercise outside in good weather and be viewed by the public.[42] The idea was approved, and in March the architect's plans were sent to the minister of the interior with a cover letter written by Cuvier expressing his hope that with the grilled courtyard the chimpanzee could be conserved "for some time yet."[43] On 17 April 1838, at his last assembly before going off on his annual rounds as inspector general for the Université de France, Cuvier insisted on "the necessity of having a provisional cage for the chimpanzee made promptly." The administration agreed to have this done immediately.[44]

Construction of the cage began, but in a piece for *La Presse* on 15 May 1838, Berthoud was still describing the chimpanzee in her chamber at the rotunda: "Jacqueline is doing wonderfully well, has never been more gay, gentler, more spiritually joyous; she lavishes her flirtations on the small number of visitors admitted near her." He found the monkeys in their new house less obliging. Given the chance to leave their narrow cages to enjoy the greater space of a less confining prison, most quickly returned to their cages. Only the baboons spent much time outside, and they were neither rare nor valuable nor of much more interest than the monkeys displayed on the streets of Paris. Only street urchins and idlers, wrote Berthoud, came to see the monkeys. No advantage had been derived from "the rather elevated sums spent in constructing this *palace*." The animals would die there in winter, he said, and eat each other in summer.[45]

Achille Valenciennes, professor of reptiles and fish, stood in for Cuvier as menagerie superintendent after Cuvier left Paris in late April. Valenciennes's menagerie reports mentioned monkeys that were acquired and some that had died but said nothing about the chimpanzee. His terse report of 29 May 1838 thus came without warning: "The female chimpanzee died 29 May. The animal entered the menagerie on 18 October 1837. It lived there seven months and eleven days. The orangutan entered there on 15 May 1836, died there on 3 January 1837, it was alive there seven months and eighteen days."[46]

Valenciennes offered no cause for Jacqueline's death. Several newspapers alleged she died of "a crisis of dentition," but they were just echoing the standard explanation of why young lions died.[47] Berthoud's suggestion was that instead of dying of a fluxion of the chest or of consumption (as had always been feared), Jacqueline died from a case of dysentery that was inexplicable because it bore no relation to the healthy conditions provided by the Museum. Whereas Valenciennes simply reported to his fellow professors the fact of Jacqueline's death, Berthoud broadcast it publicly with all his literary skills, proclaiming: "Jacqueline is dead! Yes, the chimpanzee that ought to resolve so many important problems for the naturalists, by her unexpected death leaves yet undecided and without solutions the grave questions that her arrival in Paris raised in the scientific world." It was "a scientific catastrophe."[48]

Berthoud explained that Jacqueline had passed the winter in the warmth of the rotunda without suffering too much from the cold, and that "the spring found her healthy, joyful, alert, and having singularly grown." He described in detail how she entertained visitors, "[extending] a hand to those whose physiognomy pleased her" and "[letting] them caress her gently." If visitors bothered her, she escaped by swinging out of their reach on the rope that hung from her chamber's ceiling. And it was on this rope that she executed her remarkable gymnastic feats. Her inventive capers, wrote Berthoud, would have "dumbfounded and discouraged" the greatest English clowns and France's most famous acrobat. When she finished, she would leave her rope, descend by way of her window, and cover herself with a woolen wrap. Berthoud likened her behavior to that of a ballet dancer who after her performance went offstage where an attentive mother placed a cashmere shawl over her shoulders.

Berthoud noted that the Museum had nearly completed the outdoor cage where Jacqueline in summer would have shown the greater public the remarkable performances that only a few had been able to see in the rotunda. "Alas," he wrote, "all these cares, all these preparations will remain vain and a total loss." He was explicit about what that loss entailed. "One will not know if the chimpanzee is definitely the first being in the family of apes and monkeys or the last rank of the family of men; one will not know how her intelligence could have been developed, and what training could have produced in the progress of this gentle, sociable animal whose caressing habits did not resemble in the least the brusqueness of other monkeys." All that was left of Jacqueline was the skin that had been stripped from her, plus her cadaver, "on which the scalpels of two naturalist's assistants are pitilessly at work."[49]

Berthoud was right that the Museum scientists had not unraveled the profound questions the chimpanzee raised with respect to the similarities and differences between apes and humans. What he did not mention was that they had not even tried. Like Jack, Jacqueline's intellectual faculties were not experimentally tested. The professors worried about her health, her housing, and how to help the public enjoy watching her, but as was often the case with menagerie specimens, they did not know what to do with her until she was dead. Blainville duly requested and was authorized to have "two or three" vellums of her anatomy painted.[50]

Dismayed as he was by Jacqueline's death, by summer's end Berthoud was pleased to report that life at the monkey house had improved dramatically. Thanks to the monkey keepers, it was possible to let roam free in the same space forty monkeys of greatly different strength and character.[51] Writing again in September, he portrayed activity at the monkey house as the most uplifting feature of his latest visit to the Jardin des Plantes. Other sights were depressing: the kangaroo had lost its mate and was plunged into a profound melancholy; the Australian and Asiatic cassowaries had become enemies; the stags were fighting; and a lion given by the Prince de Joinville was suffering from its confinement and a debilitating disease. But the palace of the monkeys was a pleasure to behold. In the great monkey cage, one could see "the whole face-making and crazy family of *quadrumana*," from "the sober baboon to the frolicsome Barbary ape," leaping, climbing, tumbling, dancing, prancing, crying, and clashing with each other. Equally animated were the six hundred human spectators who stood on the sloping amphitheater outside the great cage laughing, applauding, and rooting for one or another actor as if at "a real spectacle." Far from oblivious to the public's interest, wrote Berthoud, "[the monkeys] appear very satisfied to see themselves the object of such keen curiosity, and there are no extravagances they do not invent to attract still more attention."[52]

Berthoud singled out several actors, including a recently acquired *Cynopithecus* monkey that he thought looked like a smaller chimp. Another he described as a frightful black beast with a long tail, disproportionately long arms, a bright red face, large mouth, and horrifying nose. "One would not know how to invent an uglier demon," he said, noting that it had the appropriate name of Beelzebub.[53] Berthoud's devilish monkey often chose "an unfortunate pig-tailed monkey" to victimize, but the latter had a protector, "an old baboon who is constituted commissioner of police of the republic, and who uses his strength to protect the oppressed against the oppressors." The baboon's deputy was "a handsome small dog from Iceland" that occasionally bit but never injured the most aggressive monkeys, "eliciting from them sharp, pitiful, and comical cries." Berthoud's account of the social life in the great monkey cage is the most explicit description on record. None of the Museum's naturalists ventured an account, just as they left no observations of Jacqueline's behavior.

Berthoud noted that the Jardin des Plantes was bustling with strollers "hurrying to enjoy the last beautiful days of autumn." The coming seasonal change meant fewer visitors at the menagerie and more illness and death among its occupants, especially the primates. Like the year before, deaths at the monkey house began to increase in mid-October. Four monkeys died from July 1 to October 15, but eleven perished from October 15 to December 31, Berthoud's demon among them. The menagerie's records and Berthoud's description suggest it was not *Ateles belzebuth*, the belzebuth or white-bellied spider monkey, but a recently acquired coaïta (*Ateles paniscus*), also known as the black spider monkey or red-faced spider monkey. Purchased from Vaillant for sixty francs on 20 June 1838, the monkey died on November 5.[54] A palace was no guarantee of longevity for the primates. Keeping animals alive and healthy remained the menagerie's primary challenge.

[4.13]

Quarreling over the Bones

The weather was oppressively hot in Strasbourg on Saturday, July 14, 1838, when Frédéric Cuvier arrived there as inspector general of the University of France. He left Paris equipped with a passport and visas for an academic tour through France and on to Baden, Prussia, Hesse, and Switzerland. Now he was weary, and the intense heat added to his fatigue. But tired as he was, he slept fitfully in his hotel that night. Over the next several days the heat continued, as did his sleeplessness. On the evening of Thursday July 19, having endured eight insufferably hot hours in conferences at the École Normale, he returned to his hotel exhausted. That night he began to experience weakness and muscle pain.[1]

Early the next morning Cuvier went out for a bath, but it did not ease his discomfort. Back at the hotel, he met his childhood friend and distant cousin Georges Louis Duvernoy, a physician, comparative anatomist, and chancellor of the University of Strasbourg. Duvernoy found Cuvier's condition worrisome, but aside from his muscle pain and exhaustion, plus some redness of the eyes, he saw nothing wrong. Duvernoy hoped his friend's symptoms were due to fatigue, and he recommended rest and diet. That evening Cuvier ate a light supper. Once again, he slept poorly.

Cuvier's condition worsened the next day. The fatigue and muscle pain intensified. When he tried to walk, his gait was unsteady, as if he were drunk. He complained of great pain in his back and prickling sensations in his fingers and toes. The lack of animation in his face and how cold his skin was to the touch led Duvernoy to suspect a circulatory problem. He urged Cuvier to let himself be bled, but Cuvier refused, arguing that bloodletting had not helped him before and was unlikely to work this time. He interpreted the pain in his back as a malady of the spinal column. He was content to drink lemonade and chicken broth and be given an enema.

Throughout the day, Cuvier's condition deteriorated. Duvernoy continued to recommend bloodletting, and Cuvier continued to resist. In the afternoon when he rose to relieve himself, he collapsed and had to be lifted

Figure 4.13.1. Section of a map of the Museum's grounds, showing the menagerie extended all the way to the Seine. H. Boitard, *Le Jardin des Plantes* (1842).

back into bed. That evening another physician joined Duvernoy. They bled the patient—"sixteen ounces, at least." Early the next morning (Sunday, July 22) he was given a "purgative laxative." Then small bleeding cups were arranged on each side of his dorsal spine, drawing twelve more ounces of blood from him.[2]

Later that morning, Duvernoy wrote to Cuvier's son, informing him of his father's illness. Putting the situation in the best possible light, Duvernoy expressed optimism that the aggressive medical treatment would halt the progress of the disease. The immediate results seemed positive: the patient was less restless, his pains less intense, his eyes no longer red, his face healthier in appearance, and he had a relatively calm night. Cuvier's robust constitution was an additional reason for hope. Duvernoy urged Frédéric *fils* to come immediately to boost his father's morale, knowing, however, that it would take two days for his letter to reach Paris and another two days for Frédéric *fils* to reach Strasbourg. Ideally, when the younger Frédéric arrived, he would find his father recuperating.[3]

It was not to be. Cuvier's condition worsened that afternoon. His muscular and spinal pain became severe. The next day, Monday, his agonies increased, leading him to say, "I am on the rack, it is impossible there are greater pains than these I experience." Nonetheless he remained clear-headed. He talked to those at his bedside about his condition, and he sought to analyze each of his symptoms. By evening the pains in his loins and lower abdomen were excruciating, and his upper and lower extremities were completely paralyzed except for some motion in his hands. At 10:00 p.m. his physicians applied to his torso an unguent supposed to act simultaneously as a blistering agent and a calming lotion. It contained distillations of cherry laurel and bitter lettuce, both classified by medical manuals of the time as "narcotic poisons."[4]

By Tuesday morning July 24, Cuvier's paralysis had advanced further. His breathing was feeble. He said that he expected to die from suffocation in a matter of hours. By then, his medicines included an hourly decoction of arnica flower and every two hours a pill containing one-sixteenth of a grain of strychnine, which his physicians hoped might reactivate his nervous system. In the afternoon, they gave up hope. They stopped the strychnine, carried on with the arnica, rubbed lotions over Cuvier's body, and continued with enemas.

That evening Cuvier had a horrible fever. He asked how long it would last. Later he observed, "It is really hard to die." At 9:45 p.m., when Duvernoy took his pulse, Cuvier asked, "How do you find it?" According to the official report of Cuvier's illness and death, these were his last words. He closed his eyes; his breathing and heartbeats slowed, and he expired at ten o'clock in the evening, 24 July 1838, "without struggle or death rattle, by the gradual and imperceptible extinction of the respiratory forces."

Cuvier's autopsy was performed within twenty-four hours, and his body taken to the Academy of Strasbourg. On July 26, "all the scientific and literary notables, as well as the civil and military authorities of the city" came to pay their last respects, after which the funeral cortège proceeded to the church of Saint-Nicolas for religious rites. Cuvier was buried in Strasbourg's cemetery of Sainte-Hélène. News of his death began appearing in the Paris papers on 28 July. He was said to have died from the same malady that had claimed his brother six years earlier, an inflammation of the dorsal spine.[5]

The information regarding Cuvier's illness and death presented here is only a sketch of that available in the physician's report, but enough to provide a sense of contemporary medical practices like the bloodletting and enemas prescribed for Frédéric Cuvier and Jack the orangutan alike. The physician's report also calls into question the standard account of Frédéric Cuvier's last words. A claim made at his funeral, repeated two years later by Pierre Flourens in his eulogy of Cuvier for the Institut de France, was that his final words were, "Let them put on my tomb: 'Frédéric Cuvier, brother of Georges Cuvier.'"[6] It's a good story, but it seems that Cuvier's last words were to ask about his pulse.

Cuvier's death created vacancies at the Museum that needed filling. The most obvious was his professorship of comparative physiology. Before the administration addressed the matter, Isidore Geoffroy Saint-Hilaire asked that the position of *garde de la menagerie* be resurrected and restored to the remit of the professor of birds and mammals. Having substituted as *garde* for five straight summers, he felt eminently qualified for the position and wanted the pay upgrade he expected to come with it. After fourteen years as an *aide-naturaliste*, his salary was still 2,000 francs, whereas the menagerie superintendent's pay had been 3,500. The professors tabled Isidore's request until the special meeting that would decide the fate of Cuvier's chair. That meeting (10 August 1838) did not go as Isidore had hoped. The responsibility for the menagerie was returned to the professorship of mammals and birds, but the position and title of *garde de la menagerie* were suppressed. Instead, the naturalist's assistant for mammals and birds (Isidore) was burdened with the additional charge of supervising the menagerie, with no extra remuneration.[7]

Of the several candidates for chair of comparative physiology (including Duvernoy from Strasbourg) the professors chose Pierre Flourens, provided he vacate the professorship of human anatomy. Flourens agreed, pleased with the perks the Museum offered. He was allowed to keep his naturalist's assistant, his prosector, and his laboratory, plus he was promised "some experimental parks in the menagerie," whose size and location would be determined later.[8] In leaving the chair of human anatomy, he recommended it be renamed the chair of "the anatomy and natural history of man," to reflect the growing interest in race.[9]

Frédéric Cuvier's apartment was awarded in October to the botanist Adolphe Brongniart, the Museum's newest professor. His place at the Academy of Sciences went to Henri Milne-Edwards, whose election was delayed by a literal quarrel over bones.[10] The ever pugnacious Henri de Blainville insisted that the discussion of Cuvier's successor be postponed until Blainville concluded his debate with the zoologist Achille Valenciennes (Georges Cuvier's former student and a candidate for the post) over whether a jawbone from the slate quarries of Stonesfield in Oxfordshire was from a reptile or a mammal. Valenciennes ended up on the winning (mammalian) side of the argument, but it was not enough to gain him Cuvier's vacant Academy post.[11]

The routine business of the menagerie was scarcely affected as responsibility for its operation switched from one professorship to another and back again in 1838. The collection had swelled from some 390 specimens at the end of 1837 to approximately 500 on 1 September 1838. The numbers of species (or "principal varieties") increased at the same pace, from 75 to 114 types of mammals, and from 55 to 81 types of birds.[12] Much of that increase came in summer, on the heels of a significant loss and another missed acquisition. Jacqueline the chimpanzee died at the end of May. In June the Museum could not move quickly enough to get a young female rhinoceros offered by a sea captain who had brought it to London from India. By the time the Museum learned that the duc d'Orléans was willing to pay for the rhino, the captain had sold it to the newly established Zoological Society in Manchester and was sailing back to India.[13]

The loss of the chimpanzee was not unexpected, given the difficulty of maintaining primates in an inhospitable climate. The loss of the rhino was largely balanced by the arrival in July of two important collections.[14] The first was from J.-J. Dussumier, back from the Malabar Coast with a panther, five deer of an unknown species, a lion-tailed macaque, and a pair of hemiones. The second was a gift from the nineteen-year-old naval lieutenant François d'Orléans, Prince de Joinville, home from a diplomatic mission to Brazil. History remembers him primarily for bringing Napoleon's ashes to France from Saint Helena in 1840, but in 1838 the Museum was excited about "one of the most notable additions that the menagerie has received for some time."[15]

Joinville delivered a bear from the American Cordillera (the menagerie's first); a brown coati from Rio de Janeiro; a female jaguar from Cayenne; a margay (a very small spotted cat) from Brazil; eight acouchis (rodents like the agouti and guinea pig, but rarer); and several birds and land tortoises. Along with them were a pair of baboons and a duiker antelope originally from Africa but acquired by Joinville in Brazil. The same was apparently the case of a young lion from Senegal.[16] That African animals could be acquired in Brazil shows that specimens collected in the animal trade were not destined solely for European capitals. The lion was the least rare of the animals from the menagerie's perspective, but its character was distinctive. It was "so gentle and friendly" that the menagerie staff, "to preserve its habits," could walk it about and give it some exercise during the hours closed to the public.[17]

Geoffroy Saint-Hilaire rejoined the professorial assembly for the first time in several months on August 21, the first meeting since the menagerie had been returned to his chair. His sense of order and justice restored, his primary contribution to the meeting was to deposit his son's menagerie report, a document testifying that Isidore had assumed his new duties enthusiastically, notwithstanding his disappointment about salary.[18] In the next two weeks Isidore prepared a register of the menagerie and proposed that a new census be performed every year, which had not been the general rule. His census showed the menagerie to be "very rich in monkeys, carnivorous and herbivorous mammals, birds of prey, and *échassiers brévipennes* [ostriches and cassowaries]." Its holdings were more modest in rodents, parrots, gallinaceous birds, ordinary waders, and turtles. Beyond that, it had little of interest, Isidore stated, speaking as a zoologist wanting the menagerie to hold the widest possible range of higher animals.[19]

Isidore's menagerie reports, like Frédéric Cuvier's, recorded gains and losses. His 9 October report expressed excitement about a jaguar from Bahia, gift of the prince de la Moskowa (Napoléon Joseph Ney). The menagerie, wrote Isidore, had not had a specimen of this "beautiful species" for several years. Exulting that it completed "a magnificent series of the genus *Felis*," he listed the whole array: "The lion, varieties from Barbary, Senegal, etc.—the cougar—the tiger—the jaguar, several varieties—the panther, several varieties–the ocelot, two very distinct varieties."[20] It was a blow, three weeks later, when Isidore had to report among the newly deceased menagerie animals: "the cougar, probably dead as the result of a crisis of dentition, has been found lifeless in its cage, the day after a day when its health seemed perfect."[21]

On 16 October 1838 Isidore reported the addition of twenty animals versus only two losses. The acquisitions included fourteen specimens purchased from the former animal proprietor Jean Vallée, who was joining the menagerie staff as reptile keeper, a post made possible and necessary by the most exciting of the specimens coming with him, his two enormous pythons and three small caimans. The challenge for Isidore was where to put these creatures while their assigned destination, the old monkey house, was being renovated. His choice of provisional housing, the new monkey house, could not have pleased the monkeys.[22]

Isidore addressed the menagerie's housing problems periodically in his reports, usually without eliciting attention from the professors, who had problems housing their own specimens. The professors of mammalogy and ichthyology had lately declared they needed twice as much gallery space as their collections occupied; the professor of entomology called for his display space to be tripled; and the professor of mollusks and zoophytes requested four times what he had.[23] Later in 1838, Isidore and Valenciennes submitted a plan for a new edifice with a reptile menagerie at one end, a parrot aviary at the other end, a menagerie for small rodents in the middle, two ponds for mammals (beavers in one, seals and otters in the other), and another pond for crocodiles. The administration approved the plan, but the structure was never built.[24]

Balancing the interests of the animals and of the public was an ongoing issue. Frédéric Cuvier had called for banning the practice of spectators' teasing the bears with cakes suspended on strings, but Isidore was unwilling to follow up on Cuvier's urgings because he knew an outcry would ensue, stating of the practice, "I have never seen it result in any inconvenience either for the health of the bears or anything else."[25] The public continued to tease the bears.[26]

Remarks on the animals' suffering were as likely to come from the press or visitors as from the Museum's naturalists. Joinville's lion elicited the sympathies of journalist Samuel Henry Berthoud. Usually the menagerie's advocate, he was dismayed during his mid-September 1838 visit to see the young lion had taken poorly to captivity in Paris. Aboard ship it had been allowed to roam freely, and at the park in Neuilly it was likewise left off the leash. Berthoud may not have known about the early morning walks the lion enjoyed at the menagerie. In its cage, it was miserable.

Since being locked up, the lion had been subject to "horrible convulsions." Wrote Berthoud, "One is moved by pity seeing the unhealthy thinness of the poor beast, which utters soft groans and presents its inoffensive head to the caresses of the curious." When people recoiled in fright from its advances, it retreated to the far corner of its cage, "sad and discouraged." If a keeper passed a hand through the bars to pet the lion, "a light tremor of joy" would pass through its body, and then the "melancholy thanks of a gentle and long sigh" would escape the lion's "burning and suffering chest" where, according to Berthoud, a "pulmonary phthisis [tuberculosis] is beginning to carry away the unfortunate prisoner." After such a "melancholy spectacle," said Berthoud, it behooved a visitor to hurry to the "palace of the monkeys" to be cheered up by their exuberant frolicking.[27]

Isidore had a different story to tell of the young lion, provoked by an incident where the public perception of the Museum's control of its animals was at stake. As he explained to the administration, every morning before the menagerie opened, the animal was walked in the gardens "at the end of a strong chain or a thick rope, and then attached for a little while to the trunk of a tree." The staff permitted the lion this semiliberty because of its "extreme gentleness" and how weak it was. It was suffering from "la danse de S$^{t.}$ Guy" (the neurological affliction also known as Sydenham's chorea or Saint Vitus dance).[28] Recently the lion had featured in an event that the public could have misunderstood, and Isidore wanted to tell the administration how he had handled it. Early one morning, when the lion was chained to its tree and Isidore approached, the animal rushed forward to receive his caresses, whereupon the ring attached to its collar broke. This meant the lion was totally free. Isidore called the keeper, Letellier, who quickly grabbed the lion by the collar and in minutes returned it to its cage. But a dozen or so bystanders had gathered outside the menagerie fence, and Isidore worried what they would make of what they'd seen. He wanted to prevent the spread of any unwelcome "commentaries or discourses" about an escaped lion, so he had the lion immediately equipped with a new collar. He took the animal for a conspicuous walk, and he again attached it to its tree, thus conveying to the eyewitnesses "that it was not a question of an escaped lion, but of an ill lion, walked and cared for by its keepers to afford it the beneficial influence of the sun." The gawkers left and the lion remained peacefully in place until returned to its cage, half an hour before the menagerie opened. Isidore hoped the Museum would appreciate his efforts at managing public perceptions. The incident also offered him the chance to remind the professors that *les animaux féroces* needed a courtyard.[29]

Come December 1838, Isidore reported that a worker (trellis maker Pacot) was seriously injured by one of the "peaceful" herbivores (a Kolb antelope), and that the menagerie had received an elephant shrew, another species never seen in Paris and perhaps even in Europe.[30] Life at the menagerie was continuing apace.

Late in 1839 Blainville awaited the imminent death of an Algerian lion whose bones he wanted to use for his upcoming course on myology. When the time came for his assistant to fetch its carcass, Blainville was nonplussed to learn that Étienne Serres, the new professor of the anatomy and history of man, had already claimed it for the zoology lab, along with a dead monkey. An irate Blainville recovered the lion, but in what he soon considered a weak moment, he allowed the monkey's carcass to remain at the lab, where it was wanted for research on the causes of mortality in the new monkey house.

Blainville wrote the administration a long, aggrieved letter insisting that rules for divvying up dead menagerie specimens had been determined long ago and that in the attempt to co-opt the dead lion, action had been taken by which he found himself "suspended of my rights as professor."[31] A committee assigned to resolve the problem failed to put it to rest for long. Thirteen months later, Blainville again rose in combat over the matter of cadavers coming from the menagerie.[32] The control of specimens, living and dead, remained central to the missions of the Museum and its menagerie.

∴

Marie Jean Pierre Flourens's eulogy of Frédéric Cuvier at the Academy of Sciences in 1840 described the post of menagerie superintendent as a "peaceful retreat" where Cuvier "found the two things that enable profound work: time and meditation."[33] Cuvier's equanimity apparently had misled Flourens about the rigors of the position. A professor's chair would have better favored Cuvier's meditations, giving him the opportunity and responsibility of teaching a course on his subject. His career as professor, alas, proved too brief. In December 1837, when his professorship became official, the Museum's January-May courses were about to begin. The next term (May-August) was not an option because of his duties as inspector general of the university, and he did not live to see the following term.

What Cuvier planned to teach is no mystery. In 1832, when he first requested a professorship, he submitted a prospectus for a course on "the nature of animals, and their training (*éducation*)." He presented the same prospectus in December 1836 when he again raised the issue of the chair, this time to the minister of public education, François Guizot. As Cuvier explained, the course covered "psychological phenomena in animals; the actions in which these phenomena manifest themselves in the animal series, as well as the laws governing them; and finally, the application of these laws to developing animals, rearing them, attaching them to our species, and training them according to our needs."[34]

Cuvier wanted Guizot to know that his course had no heretical implications. Distinguishing between the cognitive, instinctive, and organic faculties of animals would provide the foundation for understanding the "absolute distinction that separates man from animals." Having established general truths about animal faculties, he would offer details illustrating nature's care in establishing "the most perfect harmony" between animals' faculties and their "conditions of existence." These considerations would lead students to appreciate "the wisdom and profundity of the views of Providence."[35]

The second section of Cuvier's course was to treat how instinct and intelligence function throughout the entire animal series. The final section would address how to modify the nature of animals, whether mentally or physically. Discussing domestication, he would explore the changes already produced in domesticated animals and identify those wild creatures that might yet be made "more or less domestic." Finally, he would reveal "the means at our disposal" for creating new races and improving those in existence and the "rules" that needed to be followed in "perfecting" animals for human purposes.[36] Had Cuvier delivered his course of lectures at the Museum and published them as a book, it would likely be seen today as the founding text of comparative psychology.

Cuvier explained that he based the course on the papers on animal psychology he had begun publishing in 1808, starting with his memoir on the dog of the Australian Aborigines. His subsequent publications on the orangutan, the seal, and various other animals treated in his *Histoire naturelle des mammifères*, plus his more general memoirs on instinct, habit, animal sociability, and domestication, would feed into the course. Death intervened before these parts could be assembled as a coherent whole. As Flourens remarked, Cuvier devoted thirty years to "the most profound studies," but he died leaving "scattered observations rather than a body of doctrine, and materials rather than a work."[37]

Flourens did not try to synthesize these elements, but he did the next best thing, surveying them in a set of three lectures to the Académie des sciences in 1839, which were published in the *Journal des Savants* and the *Annales des sciences naturelles*. Two of the lectures additionally appeared as Sunday supplements in *Le Constitutionnel*. The three became the basis of a book that went through multiple editions over the next twenty-one years. Through these works, Flourens gave the breadth and depth of Cuvier's behavioral studies a visibility they would otherwise not have had.[38]

Though Flourens produced an admirable review of Cuvier's writing on animal behavior and succeeded him in the professorship of comparative physiology, his own scientific practices were a far cry from Cuvier's thoughts on how to study animal behavior. Both men sought to ground their work in a careful combination of observation and experimentation, but Cuvier was unwilling to submit animals to experiments that involved suffering. Flourens, in contrast, established his scientific reputation through vivisection experiments on brain function (in rabbits and pigeons).[39] One of his first moves after replacing Cuvier was to establish an experimental laboratory. Whereas Cuvier sought to comprehend the mental faculties of the full range of mammals in the menagerie, Flourens needed only a few common species for his work. The Museum's zoologists and comparative anatomists depended more on the menagerie's specimens than Flourens did.

∴

Frédéric Cuvier was half right when he said that if the idea of species transformation were ever recognized as possible, "all the sciences that, directly or indirectly, have animals as their subject" would be instantly transformed.[40] The recognition of the reality of biological evolution indeed produced a revolution in the zoological sciences. Cuvier was wrong, however, about the speed of change. Some naturalists in his day entertained the possibility of species change, and some like Geoffroy embraced it, but the tipping point did not come until well after the publication of Darwin's *Origin of Species* (1859). Darwin's carefully marshaled evidence and his sustained argument naming natural selection as the primary mechanism of organic change did not instantaneously transform the biological sciences. It took nearly a generation before most zoologists and botanists accepted the idea of evolution, with the French naturalists markedly slower to come around than their counterparts in Germany, Britain, and the United States.

Frédéric Cuvier firmly opposed the idea of species transformation, but he was an even-tempered man who generally expressed his objections in measured tones, even if some exasperation crept into his writings when the idea kept coming up. He was not as flippant, scornful, or arrogant on this score as his brother or as Pierre Flourens. Successor to Georges Cuvier as permanent secretary of the Academy of Sciences, and eventually a member of the Académie française, Flourens grew increasingly pompous in summing up scientists' careers and pontificating on the life sciences and on life itself.

In 1864 Flourens dashed off a cavalier and contemptuous rejection of the *Origin of Species*. Characterizing Darwin's work as a "system," the same word Georges Cuvier used to denigrate Lamarck's evolutionary theory, Flourens chastised Darwin for using figurative language and for personifying nature ("natural selection").[41] He faulted Darwin for failing to recognize that the variation of species is strictly limited. Insisting that the whole question boiled down to proof, Flourens thundered, "Does Mr. Darwin cite a single fact, I say a single one, from which one could conclude that a species has changed into another?"[42] By Flourens's account, the reason Darwin had concluded that species are mutable and that Flourens had not was that "he [Darwin] followed a system and I followed the facts."[43]

Flourens adopted the same fixed position and rhetorical style as Georges Cuvier in arguing against species change, and he utilized the same sets of facts. He trotted out Cuvier's points on the gaps in the fossil record; the lack of differences between modern forms and the mummified specimens from Egypt; and the limited variation of wild and domestic species. Flourens additionally cited his own experiments with hybridizing where he found that hybrids lost their fecundity by the third or at most fourth generation. He completely ignored new categories of evidence Darwin had brought to bear on the subject and failed to come to grips with diverse aspects of Darwin's argument such as why, under the theory of descent with modification, one should expect to find gaps between modern forms rather than to see species grading into one another. Flourens found Darwin's explanation of natural selection ridiculously unscientific, scoffing, "What pretentious and empty language. What puerile and outmoded personifications! Oh, lucidity! Oh, solidity of the French mind, what is becoming of you?"[44]

Remarkably, writing about how Georges Cuvier had used Frédéric's studies of variation in dogs to refute species change, Flourens referred to Frédéric, not Georges, as "the most rigorous naturalist I have known."[45] It is questionable whether Frédéric Cuvier's menagerie-based observations might have shed any light on the issue of species mutability. He was right to call for scientific investigation of variation, but the range of variation in a species of animals maintained through reproduction for several generations (something the menagerie almost never achieved) could not establish what might happen over an immensely greater number of generations under different conditions. Likewise, the decreased fertility of interspecies hybrids over the course of a few generations was no disproof of theories of change where interspecific mating had no role to play.[46]

In brief, while the menagerie did help clarify some problems in the identification of species, it did not illuminate the vexed question of species change. Nor did it become home base for a science of animal behavior or psychology. Frédéric Cuvier wrote about individual animals or species and elaborated the conceptual bases of the science he wanted to create, but he did not live long enough to establish a program of research that others chose to continue after him. At the Museum of Natural History, neither Isidore Geoffroy Saint-Hilaire nor Pierre Flourens, the two professors best situated to maintain and develop such a program, proceeded to do so.[47]

Neither did Britain's new zoos systematically pursue animal behavior studies, though they enabled visitors to make their own observations, Charles Darwin's at the Zoological Gardens in London being a case in point. Back from his voyage on the HMS *Beagle* and subsequently concluding that species changed over time, Darwin set about quietly gathering evidence for and against that hypothesis. Frequenting the London Zoo, he took a special interest in the young female orangutan named Jenny. Franz Liszt never played a concert for Jack the orangutan in Paris, but Darwin evidently played a harmonica ("harmonicon") for Jenny. He recorded in his "M" notebook (on "Man, Mind, and Materialism") that she "listened with great attention to Harmonicon & readily put it[,] when guided[,] to her own mouth."[48]

Darwin played his harmonica for Jenny shortly after Frédéric Cuvier's death. Late in the 1830s, he read Frédéric Cuvier's work, including his articles on instinct, domestication, the Australian dingo, and variation in dogs. Brief mentions of the French naturalist appear in Darwin's notebooks on species transmutation and later in his major publications. In addition to citing Cuvier on instinct and domestication, Darwin took special interest in Cuvier's idea of "fortuitous variations" that were not obviously the result of external influences.

Zoos in the latter half of the nineteenth century did not distinguish themselves for systematic studies of animal behavior, but they generated a wealth of information regarding animals' habits and the care needed to keep animals alive.[49] In the early 1900s, when the American zoologist C. O. Whitman contemplated what would be the best site for the biological study of animal behavior, he imagined a "biological farm," created for "the experimental investigation of heredity, variation, and evolution and for the study of life histories, habits, instincts, and intelligence."

Whitman regarded neither the laboratories of experimental biologists nor the haunts of field naturalists as adequate for the task. "The laboratory is too narrow, and the world too wide for the continuous study of living organisms, under conditions that can be definitely known and controlled." Significantly, he also rejected the zoo:

> Heterogeneous collections of animals, exhibited for the amusement of people, are wholly unsuited to the purposes of investigation in time, place and character. For the kind of work contemplated, the investigator must have forms of his own selection, collected, arranged and kept for his special purposes. He must have complete and permanent control of his quarters and the forms he is to study, and above all, *complete isolation from the public*. Only under such conditions could he have the unbroken quiet required in delicate observation, or expect natural behavior from the forms occupying his attention.[50]

Whitman was a direct precursor of the founders of modern ethology, the Austrian zoologist Konrad Lorenz and the Dutch naturalist Niko Tinbergen, but he never succeeded in getting the "biological farm" he wanted, and Lorenz and Tinbergen each had his own ideas about the research sites he preferred.

Tinbergen was a dedicated field naturalist. If he had an affinity with any of the naturalist/zoologists mentioned in this book, it was Emmanuel Baillon, the naturalist of Abbeville, who took pains to observe the behavior of animals in the wild and enjoyed it. Lorenz preferred raising birds (another dimension of Baillon's practice) and observing them under conditions of semicaptivity. Early in his career he nurtured a collection of nearly one hundred species, keeping most in an aviary and allowing others to live under barnyard conditions or in his home. In characterizing his practices versus Tinbergen's, Lorenz liked to say that he was by nature a "farmer," raising and breeding animals, caring for them when they were ill, and having them as companions, while Tinbergen was a "hunter" (albeit with binoculars and a camera rather than a gun), stalking animals in the field, seeking to match wits with them, and learning how their behavior contributed to their survival in their habitat.[51] Late in his career, Tinbergen published a two-volume collection of papers titled *The Animal in Its World* (1932).

Frédéric Cuvier's perspective was different. He sought to develop a science of comparative animal behavior based on the animals in *his* world, the Paris menagerie. He wanted the menagerie to be like a chemistry or physics laboratory, where controlled experiments enabled one to learn not what happens in nature so much as what *could* happen there. Although he extolled experimentation, much of his knowledge came from observations made while keeping and caring for animals, not from experimental protocols.

Konrad Lorenz was likewise not an experimenter. He observed live birds of different species side by side to compare their behavior patterns as a means of determining the birds' natural affinities. His home aviary provided this benefit, but it could also be found in a zoo. Not surprisingly, some of Lorenz's friends and colleagues were zoo men (in particular, Oskar Heinroth in Berlin and Otto Antonius in Vienna).

Might it be appropriate to consider Frédéric Cuvier a precursor of modern ethology? On balance, the search for precursors is a questionable enterprise, characteristic of an earlier era in the history of science. Identifying someone as a precursor too often extracts that person's life and work from the original historical contexts in which the whole of their activities made sense. That said, comparing Frédéric Cuvier's ideas and practices with those of the founders of ethology in the twentieth century is an instructive exercise. For the ethologists, like Cuvier, the comparative study of behavior and the distinction between instinct and learning in animal behavior were fundamental concerns.

Yet the differences between Cuvier and the ethologists were as significant as the similarities. One was a matter of practice: Whereas Cuvier focused primarily on mammals, Lorenz and Tinbergen focused on birds, maintaining that the role of instinct is proportionally greater in avian than in mammalian behavior and is more clear-cut. In addition, Lorenz and his mentor Heinroth gave special weight to "isolation experiments," raising young animals without contact with their parents or other members of their species to distinguish between what is instinctive and what is learned in an animal's behavioral repertoire. This was not something Cuvier could have readily attempted with baby mammals at the menagerie. His main hope when it came to menagerie births was that the mothers would care for their babies.

The rare chance to study a beaver reared apart from its parents gave him the occasion to develop conclusions about the instinctiveness of the beaver's building, but he was in no position to develop this as a general protocol.

There is a hint in the historical literature that Cuvier realized his beaver study was not enough and he should try controlled experiments with birds. Early in the twentieth century, the zoo historian Gustave Loisel reported being told by the French physicist Éleuthère Mascart that "Frédéric Cuvier was obliged to abandon the experiences he had begun on the instinct of chicks as the result of [Étienne] Geoffroy Saint-Hilaire's refusal to allow him to have some small huts [*logettes*] constructed."[52] This allows one to imagine that if Cuvier had conducted isolation experiments on the instincts of chicks, he would have arrived at results comparable to those reported in the 1870s by Douglas Spalding.[53] But Mascart, born in 1837, could only have heard the story at second hand, years after the alleged fact. Loisel offered no other sources to confirm this story, nor have I, and nothing in Cuvier's menagerie reports or his writings on instinct suggests that he considered such experiments. The story is nonetheless a reminder that, in addition to the difficulties inherent in zoo-based behavioral research, Cuvier was hampered by his subordinate status in the Museum hierarchy and his problematic relations with Geoffroy.

Another significant difference between the mind-sets of Cuvier and the modern ethologists was their attitude toward domesticated animals. Cuvier enthusiastically endorsed domestication and firmly believed that his studies of behavior could and should be enlisted in its support. He believed new species could be domesticated and existing domestic races perfected by working with and modifying animal instincts. Like Buffon, he regarded domestication simultaneously as an achievement of civilized peoples and a process reflecting humans' becoming civilized.

The early ethologists, in contrast, wanted nothing to do with domesticated animals. They shunned them in their research because they saw the instincts of domesticated animals, compared with those of wild forms, as disordered, weak, and not fully typical of their species. Lorenz's distaste went beyond their failings as experimental subjects into aesthetic and ideological realms. He regarded domestic forms as ugly compared with wild ones. Furthermore, whereas Frédéric Cuvier exalted the interconnections of civilization and domestication, Lorenz saw civilization and domestication as the primary source of racial degeneration. As he perceived it (finding common cause with Nazi ideology), civilization led to the breakdown of instinctive barriers against race mixing, just as animals under domestication lost the instinctive safeguards that would otherwise have prevented them from mating with other species.

Where Lorenz idealized the animal of the wild, Frédéric Cuvier had exactly the opposite perspective. Posing the rhetorical question of whether the ideal type of a species was "to be found in nature alone, or in nature aided by the care of man," Cuvier opted for the latter. In one of his last writings, an 1837 encyclopedia article on "degeneration," he presented domestication as a gift of Providence. By his account, Providence had "endowed living beings with the faculty of responding within certain limits to the diversity of numerous causes that act on the earth." Consequently, they changed as the causes acting on them changed.[54]

Cuvier disagreed with the idea that the "harmonious and perfect interaction of organs" existed "only in the savage state." Man, he explained, had learned to improve the conditions of animals, liberating them from some "perturbing causes that weighed on them in the state of nature." In effect, "by surrounding them more exclusively with beneficial causes," man had brought animal forms closer to their ideal type (at least from man's perspective) than anything found in nature. For example, man through all his care and attention had brought the "beautiful races" of horses "to a degree of perfection that nature alone would have never achieved."[55]

Cuvier acknowledged that not all human influences on domestic animals were so happy. When certain directions of development were selected over others, discord could result, as evidenced by the relative infertility of some dog breeds. The chief hurdle to developing new races, Cuvier maintained, was that the modifications or "degenerations" that man used to create new races of plants and animals had arisen, insofar as one could tell, essentially by chance. In his 1825 essay on domestication, he called these modifications "fortuitous." Breeders knew how to preserve and reproduce these modifications once they arose, Cuvier wrote in 1837, but not how to create them in the first place. "On this subject," he said, "we are nearly without any rules, and despite the innumerable advantages we could derive from them, no effort has been made to discover them." Proposing menageries as sites where experiments on animals could unveil the laws of degeneration, he continued, "All that belongs to the physical and moral amelioration of living beings, without excepting the human race, depends on the laws of this phenomenon." Man's exploitation of the degenerations of animals and plants had been "one of the most powerful influences on civilization." By implication, understanding the laws of degeneration would be a boon for human progress.[56]

Cuvier's paean to "degeneration" makes perfect sense when one recognizes that what he was talking about was "heritable variation." But while he promoted the idea of breeders' creating new races through "fortuitous modifications," he took care to insist that such variation was limited to the relatively superficial characters that zoologists use to distinguish species (size, color or length of fur, the proportions of certain parts, etc.) and did not involve the more fundamental characters that distinguish genera, such as those related to the more intimate organs of motion, sense, alimentation, and so on. Changes at this higher level, he observed, would produce organisms that could not survive. He reminded his readers that there was no foundation for "false systems" supposing that, "favored by circumstances and by time, all the living beings, but especially the animals, can transform themselves one into the other."[57]

In the same year that Cuvier's "degeneration" essay appeared (1837), the young Darwin, having come to believe in the idea of species change, was seeking to understand its causal mechanisms. Among the ideas he found in the scientific literature was Frédéric Cuvier's notion of "fortuitous" variations passed from one generation to the next. From the English translation of Cuvier's essay on domestication Darwin copied to his B notebook on transmutation the words: "But we could only produce domestic individuals & not races, without the occurrence of one of the most general laws of life, the transmission of a fortuitous modification, into a durable form, of a fugitive want into a fundamental propensity, of an accidental habit into an instinct." The British evolutionary embryologist Gavin De Beer later judged it "at least probable that Cuvier's remarks on 'fortuitous variations' started Darwin on this fundamentally important train of thought."[58]

Whatever importance Cuvier's observations on heredity may have had for Darwin, Cuvier remained firmly opposed to the idea of evolution. This stance leaves a fundamental gulf between him and the modern ethologists, whose work was thoroughly informed by that concept. That divide, together with his valorizing of domestication and endorsement of the inheritance of acquired characters, would have made his work feel foreign to the ethologists, notwithstanding his commitment to separating instinctive from learned behavior and creating a science of comparative animal behavior. To cast him as a precursor to the ethologists of the twentieth century requires so many qualifications that the characterization loses most of its historical utility.

There is another context, however, that legitimizes the claim that Frédéric Cuvier was the first ethologist and that his approach to behavior deserves to be called "ethological." The justice of this becomes clear when we compare his program on "the nature and training (*education*)" of animals with what Isidore Geoffroy Saint-Hilaire had in mind when he became the first zoologist to use the terms "ethological" and "ethology." (Historians recognize that John Stuart Mill had coined the word "ethology" earlier to designate "the science of character," with specific reference to humans.[59])

Isidore's conception of "ethology" closely mirrored Frédéric Cuvier's view of the study of animal behavior.[60] He introduced the word "ethological" in 1854 while surveying what he would cover in a sweeping multivolume general work on natural history. One of the sections would present "general facts, relations, and *ethological laws* pertaining to instincts, habits [*moeurs*], and more generally to the external vital manifestations of organized beings" (italics added).[61]

Isidore planned to begin with a distinction between "intelligence and instinct in the animals" and a discussion of "automatic acts." Regarding the habits of animals, he would address the preservation of the individual and of the species, the search for food, habitat, migrations, social species, maternal care, and training (*éducation*). His topics included "modification of habits, and, as a result, of the instincts of domestic animals" and "permanence of acquired instincts." Isidore did not mention Frédéric Cuvier's name in this regard, but Isidore's program was entirely of a piece with what Cuvier had projected for his own course at the Museum.

Isidore Geoffroy Saint-Hilaire never developed an analysis of instincts, intelligent acts, and the "ethological laws" pertaining to them. In the third volume of this oeuvre, published in 1862, the eighteenth chapter (of twenty) was on "ethological variations" but was only seven pages long. By "ethological variations" he meant variations "in disposition [*naturel*], in instincts, in habits." These, he said, showed up in animals under domestication just as did variations of a structural or physiological sort.[62] These few pages, however, fell far short of the broader discussion of ethological matters Isidore imagined when he began his treatise eight years earlier. His misfortune, like Frédéric Cuvier's, was to not live long enough to see his work through. His seven pages were still in manuscript form when he died on 10 November 1861, one month short of his fifty-sixth birthday, and was laid to rest with other luminaries of his day in Paris's Père Lachaise Cemetery.[63]

Frédéric Cuvier was buried in the cemetery of Sainte Hélène in Strasbourg at a site that was not maintained and is no longer identifiable.[64] His heart, however, was taken to Paris and interred at the tomb of his brother in Cimitière du Père-Lachaise.[65] In death as in life, Georges took up the greater public space. But Frédéric Cuvier's life's work at the menagerie, his caring for the animals and his curiosity toward them, the questions he posed and left unanswered, the contradictions he grappled with in seeking a deeper understanding of what separates and what unites us to other species—in short, his unique intellectual legacy—all belong to the history of zoos and animal behavior studies.

[4.14]

Conclusion

In 1800, seven years after haphazard circumstances brought the new menagerie in Paris into existence, a guidebook identified a visit there as "la promenade à la mode." It had been fashionable to stroll at the Jardin des Plantes before the Revolution, but having exotic animals there added a new dimension to the experience. The menagerie had quickly come to be the most popular part of the new Museum of Natural History, and in 1800 it seemed that new attractions had followed upon one another in rapid succession. Moreover, 1798 was a banner year with the arrival of the elephants, the bears from Bern, and the lions from North Africa. Then 1800 saw the arrival of big cats and other animals purchased in London and, simultaneously, the wondrous "*phénomène*" of lion cubs born on site. Two years later, after two more litters, the birth of cubs was no longer a phenomenon, but the menagerie was firmly established in the public consciousness as a place for a leisurely, entertaining, and perhaps even edifying stroll. When visitors came to Paris, the menagerie was a site to put on their agendas. Because the Museum did not charge admission or count the people who passed through its gates in the period of this book, I do not have figures for how many thousands flocked there on a busy Sunday or holiday, but a census taken on a Sunday in May 1860 counted more than fifty thousand.[1] Haphazard as its origins had been, a zoological garden had come to flourish in a great metropolitan center. It was a cultural innovation of profound and lasting consequence.

The animal life of the menagerie in its first decades included elephants and bears that were trophies of war; lions that were gifts from North African potentates; tigers and panthers bought from England while France and England were enemies; a giraffe whose incomprehensible beauty left all those who saw her smitten; an orangutan that was clothed for winter and sculpted for posterity; an assemblage of marvelous birds; and, among much more, the nucleus of a reptile collection (including a Leviathan-sized python). The humans who contributed to the life of the menagerie included

Figure 4.14.1. "The quagga." Nicolas Maréchal, Simon-Charles Miger, "Le couagga, Equus quagga." From Étienne de Lacépède, Georges Cuvier, and Étienne Geoffroy Saint-Hilaire, *La Ménagerie du Muséum national d'histoire naturelle, ou Description et histoire des animaux qui y vivent et qui y ont vécu*, Paris, Aug. Renouar, 1801–5. Muséum national d'histoire naturelle, 419 pl. 11.

the animal keepers, the Museum's professors and naturalists' assistants, the visiting public, sentries, food vendors, artists, actors, musicians, guidebook writers, donors, dealers, government and commercial agents at home, consular officials and other agents abroad, and royalty from home and abroad. Napoleon visited the first lion cubs. A decade later he rejected as "not magnificent enough" the new plans to house the menagerie's *animaux féroces*.

From 1793 to 1838 the menagerie's territory expanded dramatically; animal parks with picturesque shelters were created, and provisional structures for housing animals gave way to more permanent ones, notably, the bear pits, the rotunda, the pavilion for *les animaux féroces*, the pheasant house, and the edifice that critics called the "palace of the monkeys." The Paris example inspired emulators abroad. it came to be a matter of civic pride for cities in Britain, Europe, the United States, and beyond to host a collection of live animals. If it looks today as if the animals' living conditions were poor, that was not generally how it was perceived. The writer of 1836 who trusted that Jack the orangutan would soon have better quarters wrote generously and in part playfully about what the other animals already enjoyed: "The lions and tigers live in suitable quarters. The birds have a charming pavilion of marble, sand, flowers, and tent-cloth. The seal, arrived in Paris at the same time as the orangutan, has been placed in the graceful garden of the pheasants, and when the sun shines on him, he rolls about like a lazzaroni on the polished flagstones of a limpid pool." Likewise, the peaceful herbivores had been provided with elegant shelters, seeded lawns, green pastures and gentle hills—Switzerland and the Alps in miniature. The descendants of Martin the bear had courtyards to prowl in and trees to climb, "while waiting philosophically for a child or vaulter to fall into their home." The elephants had "monumental" quarters and a "courtyard of honor." The giraffe had a "beautiful garden" where she could smell acacias.[2]

Lacépède portrayed the development of menageries as proceeding in parallel with the advance of civilization, but this confident, self-congratulatory model of social progress was of no help in running the menagerie. A menagerie free and open to the public in a great metropolitan center was in effect a great experiment, but it was not an experiment that was planned. Circumstances beyond the Museum's control had brought the menagerie into existence, and a further complex of constraints and opportunities shaped the Museum's and the menagerie's subsequent development. That the Museum's naturalists, the government, the animals, the keepers, and the public all brought different interests to the evolving equation added to the challenges of acquiring, maintaining, and developing a collection of diverse and often dangerous exotic creatures in a climate inhospitable to them. The difficulties were compounded by the era's social and political turbulence. Yet had it not been for the turbulence, the Museum would not have found itself caring for a collection of exotic animals in the first place.

Fundamental to the menagerie's institutional ecology was that it was part of the Museum, which was an organ of the state. The menagerie's fortunes necessarily reflected those of France as the country lurched from one political regime to the next. Following the January 1793 execution of Louis XVI, landmark events came in rapid succession. The Museum was founded in June 1793 after the Jacobins took control of the National Convention. In November 1793 the menagerie's first animals were swept up from the streets of Paris and deposited at the Museum. The following spring, the Committee of Public Safety, inspired not only by its belief in the moral power of images of nature's order, but also by the immediate problem of accommodating the most recent additions to the menagerie, proposed a greater development of the Museum than its professors had ever dared imagine. Robespierre's execution and the end of the Terror in July 1794 brought calmer circumstances at home, but wars abroad continued, bringing new specimens (some live, many preserved) to the Museum as trophies of military victories. Other developments at the Museum peculiar to their place and time followed during the Directory (1795–99), the Consulate (1799–1804), the Empire (1804–14), the Bourbon Restoration (1814–15), the Hundred Days (1815), the Second Bourbon Restoration (1815–30), and the July Monarchy (1830–48).

Throughout it all, the Museum's overarching concern was with controlling specimens, which included all that was involved in acquiring them, caring for them, and deploying them. The Museum's disagreement with the minister of the navy and colonies over where to send voyager naturalists illustrates how the Museum's and the state's interests diverged. The government wanted knowledgeable naturalists aiding the colonies, while the Museum craved "the most interesting productions of the least-known countries." Helping the colonies was one thing, but for the Museum, securing new specimens from parts of the world little known to science was more important. From the Museum's perspective, funds for voyager naturalists were better spent sending them to Madagascar and the Philippines than to the West Indies and French Guyana.

If the Museum's and the government's interests sometimes diverged, so too did the Museum's and the menagerie's. The dominant zoological practices of the day focused on dead specimens rather than live ones, leaving the menagerie with a lower standing in the Museum's priorities and institutional ecology. This was amply illustrated when Lacépède, the most eloquent spokesman for the new menagerie, readily bargained with the government to forgo the menagerie's expansion in exchange for funds to complete the natural history galleries. Equally symbolic of priorities, though coming at a less consequential moment in the Museum's history, was the Museum's amending its instructions to travelers abroad in 1829 to state that their time and energies were better spent collecting dead specimens for the Museum's galleries than live specimens for the menagerie. The menagerie's superintendent, Frédéric Cuvier, had to continually battle this mind-set in his efforts to establish the menagerie as a site for scientific inquiry.

The interests of the public were also a factor in the history of the Museum and the menagerie. The professors and the menagerie superintendent were inclined to view the public's enthusiasm for the menagerie as a reflection of "vain curiosity," but public opinion had consequences. Any shortcomings the public perceived with respect to the quality of the menagerie's animals and their care reflected poorly on both the Museum and the state. This became acutely evident in 1798 when the arrival of the elephants from Holland and the bears from Bern coincided with a severe shortage of government funds. At a time when the Museum desperately needed money to prevent all its collections from going to ruin, it was able to emphasize to the minister of the interior how bad it would look to the French public and the world if France couldn't keep the elephants alive.

The complication of interests was likewise manifested at the level of the animals and their caretakers. Animals recurrently resisted their keepers, and most of the keepers had scars to show for it. The keepers for their part often acted more on their own behalf than on that of the menagerie or the Museum. When Frédéric Cuvier in 1816 pressed the administration to give the public daily access to the menagerie, it was not because the public was clamoring for it, nor was it for the good of the animals. In fact it was a setback for artists or anyone else (Cuvier himself included) who might want to study the animals in peace. The change was designed explicitly to thwart the keepers' practice of soliciting bribes from the public for entry at times when the menagerie was not open.

One can also detect a divergence of interests between the Museum and its most successful suppliers of animal specimens. It appears it was more a result of their own initiatives than of the Museum's desires that Jacques Milbert and J.-J. Dussumier collected live specimens for the menagerie. Likewise, it was of his own accord that Emmanuel Baillon formulated the project of keeping the menagerie's ponds stocked with shorebirds and waterbirds. They all took a special interest in the menagerie even while the Museum was pressing them for dead specimens. Milbert had the broader hope of seeing North American mammals, birds, fish, and trees become part of the French landscape. We do not have a clear sense of Dussumier's motives for devoting so much effort and cargo space to his live animals, but at least with the hemione, the vision of the animal's domestication was a guiding factor, and for the other animals, the satisfaction of demonstrating his and his crew's competence in bringing them back alive may have been a part of it.

Divergent interests were also at play in the matter of animals received through French diplomatic relations with North African states. The Museum had no say in these relations, and though the menagerie received a few prize specimens in this context—most notably the elephant and the giraffe given by the pasha of Egypt—its budget was severely strained by having to pay costs associated with what Georges Cuvier called in frustration "disturbances" produced by "the slightest fantasies of the powers of Barbary."

That different human and animal actors responded to the menagerie in different ways is illuminated by various documents, but the documentation is fragmentary, and there is more we would like to know than can be recovered. We know that the public was an integral part of the life of the menagerie. That the menagerie was free and open to all made it all the easier for citizens (red-hatted Jacobins and the bourgeoisie alike) to regard it as their own. Children were part of the scene. In the company of their parents or nannies they could delight in seeing live versions of the animals pictured in their ABC books, starting with A for *autruche* (ostrich). The menagerie could not offer B for *baleine* (whale), but for whale skulls and skeletons one could go to the comparative anatomy gallery. Moving along at the menagerie, C for *chameau* (camel) was easy to find.

The public craved not only seeing the animals but also engaging with them. Feeding the deer or elephants was a source of excitement, and the bear pits were always a big attraction. Watching the bears lumber about begging for food and climbing their trees to grab morsels dangled on strings was one of the standard pleasures of a visit to the menagerie. When the bears were taken from the pits and relocated to the new building for *les animaux féroces* after Martin dispatched the intruder in his pit, wild boars installed in the bears' place continued to attract onlookers. The public watched the boars root around, turning up paving stones and piling them on each other.[3] Boars were not rare, but their activity was engaging. Resting animals offered less appeal. Visitors threw stones or sicced their dogs on resting herbivores to make them move. Seals were also favorite targets of the stone throwers. The first lion was taunted by soldiers who rattled their swords on the bars of his cage to make him roar.

Mordant Delaunay and Frédéric Cuvier both complained of the public's cruelty to the animals, but that cruelty and the animals' other suffering under their confinement were only part of the violence animals experienced on the long passage from the wild to Paris. Wrested from their mothers and their native habitats and transported under conditions that were psychologically and physically harrowing, many animals died or suffered grievously en route. Often animals reached the menagerie in such poor condition that they soon died. Then too there were the cases where animals were given considerable freedom and excellent care aboard ship and arrived at the menagerie in good health only to deteriorate quickly in their cramped quarters in Paris. There, their suffering manifested itself in behaviors such as monotonous pacing (if they even had room for that) and injuries they inflicted on themselves.

For some animals, like elephants, human observers felt it was easy to perceive the creature's point of view. Joy, love, anger, hatred, and sorrow all seemed evident in their behavior. Resistance seemed easy enough to interpret, as when Parkie balked at the attempt to move her from the stables under the comparative anatomy gallery to new lodgings in the rotunda. She also did not countenance strangers intruding on her space, seriously injuring one interloper in 1814 and mortally wounded another in 1816. As for the menagerie's second male elephant, it was clearly infuriated when its keeper struck it a painful blow, and Frédéric Cuvier was certain the animal would have killed the man had it been given the chance.

It is tempting to think the elephant mind might be more accessible to the human mind than those of most other animals that populated the menagerie, but it is also true that making unwarranted assumptions about the elephant mind has a long history.[4] In this book I have by and large limited myself to reporting how individuals interpreted the behavior of animals at the menagerie. Such interpretations inevitably provide more of an opportunity to explore the human mind than the animal's.

As it is, we know all too little of the stories of the humans whose traces we find in the historical record, and we cannot hear their own voices. Consider the *montagnard* who killed a mother chamois to take her newborn baby, then ended up raising it, carrying it through an Alpine snowstorm to Lyon, and escorting it to the Museum in Paris. No one knows what he thought of the project. We don't even know his name. And how wonderfully instructive it would be to have had the first-person account of Peter Johnson, the free American Black man who accompanied a bison safely from somewhere in the heart of North America to Jacques Milbert in New York City, across the Atlantic to Le Havre (bleeding it when it was ill), and to the menagerie in Paris! What did Johnson think of the people he met along the way, including the Museum agents and naturalists who were suspicious of "the Negro" but depended on him for the bison's care and safe delivery? Johnson's voice is absent, as is that of Atir, the keeper of the giraffe. Did Atir refer to her by the Arabic word for the species (*zarafa*), or call her by a pet name? We don't know.[5] Likewise, about the details of Atir's trip from Alexandria to Paris, his life at the menagerie, and his relationship with his charge, we can only speculate.

For that matter, we would like to know more about the thinking of the professors who, according to Fourcroy's account to the government, paid from their own pockets for food for the elephants when funds from the government were delayed. We imagine the professors paid in part because they felt the elephants' fate would reflect on the Museum's competence, but they may also have cared for the elephants as the remarkable sentient beings they were. We know that when the male elephant, Hans, died in January 1802, the evident suffering of the female, Parkie, moved the professors to put a camel in her enclosure as company to assuage her grief.

But we also know that any appreciation of the animals' individuality tended to disappear in the professors' scientific writings. In an institution where comparative anatomy was the dominant zoological paradigm, animals were treated as representatives of their species, even when circumstances of their individual lives and personalities were known. The names Hans and Parkie do not appear in the report of "the circumstances of the death of the male elephant" that the Museum sent to the first consul, the minister of the interior, and the press, nor in the minutes of the meeting where it was proposed that "the female elephant" be given a camel as a companion. Nor are they in Georges Cuvier's published account of his anatomical investigations of Hans's remains.

The menagerie's record with respect to the multiple functions originally projected for it was uneven. Allowing France to receive and conserve with dignity animal gifts sent by foreign rulers was one of the functions Bernardin Saint-Pierre anticipated a national menagerie could serve. It did serve that function, but not without frequent disruptions to the Museum's operations. Numerous other functions imagined for the menagerie likewise proved less successful than desired. Bernardin, Lacépède, and the librarian Georges Toscan all believed that a well-run menagerie would demonstrate the benefits of treating animals kindly, which would in turn have a salutary effect on human conduct. Lacépède urged that "the images of constraint and the appearances of slavery be removed as far as possible from the eyes of a free people," and he went so far as to suggest the menagerie not host ferocious beasts like lions and tigers lest they convey the wrong image of nature. But that proved to be little more than rhetoric for the moment. As the new century began, the public delighted in the Museum's purchase of carnivores and Lacépède reveled in how many lions the menagerie had. By then the Revolution was over, Napoleon was first consul, and Revolutionary ideology about the moral power of nature's lessons was waning. As it was, the menagerie was essentially a prison for wild animals. What Bernardin and Lacépède had been promoting was not so much freedom for the animals as the illusion of it. Even the attempt at an illusion of freedom was rejected in the planning for the pavilion for *les animaux féroces*. Explaining why animals should not be given too much space, Geoffroy wrote that animals in captivity "must feel at every moment the harshness of their condition, so that they submit better to the bridle."[6] With its twenty-one double cages in a straight line, the new carnivore pavilion was unmistakably a prison. And yet, as the quotation cited earlier in this section suggests, conditions we would today consider grim were not necessarily perceived as such at the time.

While the menagerie's efforts to provide its charges with a modicum of freedom were mixed at best, neither can it be said that it distinguished itself in public instruction. The Museum's founding decree stated that public education in natural history as "applied particularly to the advancement of agriculture, commerce, and the arts" was the Museum's "principal purpose." However, the menagerie did not exist when the Museum was founded, and no thought was given to creating a course connected with it until Frédéric

Cuvier forced the issue in the 1830s. Meanwhile, the Museum took little responsibility for informing visitors about the menagerie's animals. Aside from Frédéric Cuvier's guidebook of 1804 (updated only once, in 1809), those seeking information on the animals had to rely on guidebooks of uneven quality or information from the keepers, which might or might not be reliable. Frédéric Cuvier's plans to make the live animals at the menagerie the subject of a course of lectures did not materialize because he died too soon.

Another of Bernardin's suggestions was that locating the menagerie in Paris's poorest and most politically animated neighborhood could benefit the area economically and calm its political passions. The menagerie did attract visitors and vendors to this part of Paris, but whatever revenues it generated did not transform the quarter. Legislators' objections to the construction of the "Palace of the Monkeys" in the 1830s referred to the quarter's continuing poverty, which was so pressing as to make the monkeys' costly accommodations seem an affront.

On the other hand, Bernardin's promise that a menagerie would allow artists, writers, and poets to base their studies on living creatures rather than illustrations or stuffed specimens was largely fulfilled. The Museum's artists painted scientifically accurate portraits of animals newly arrived at the menagerie; J.-B. Houël studied the elephants for months to portray them as faithfully as possible; and Eugène Delacroix painted lions both from life and from his observations of the dissection of lion cadavers. Sculptor Jean-Pierre Dantan le Jeune was inspired to create a bust of Jack the orangutan, and other animals formed the basis of J.-J. Grandville's caricatures of animals in the guise of humans and vice versa.[7] Still, the administration's willingness to accommodate the interests of artists had its limits. The 1816 decision to allow daily public entry to the Museum hurt artists by depriving them of hours to work uninterrupted. Artists were shortchanged again a few years later when the studios planned for the carnivore pavilion were dropped from the design. As for authors, the most conspicuous use they made of the menagerie's animals was not to provide accurate descriptions of them but to use them as vehicles for satirizing the state and contemporary mores.

One of the most ambitious aims for the menagerie was to make it a site for acclimatizing and domesticating specimens to introduce to the French economy. Four decades later, not a single species on the promoters' candidate lists could be listed as a success in this regard. The most successful acclimatization at the menagerie was achieved by the brown rat, whose resourcefulness surpassed the Museum's efforts to evict or exterminate it.

Other scientific work envisaged for the menagerie had mixed results.

Hybridization experiments took place (most of them intentional, some on the animals' own initiative), but no new "mules" of any value came of it. The Society of Natural History suggested that artificial insemination might help with breeding animals indisposed to mate in captivity, but Spallanzani's achievement with dogs was never tried at the menagerie. Having a menagerie did help settle some questions of species identity, as when Cuvier demonstrated that two birds Linnaeus had classified as separate species were simply juvenile and adult forms of the same creature, or when he decided a jackal from Bengal and a jackal from Senegal constituted two different species.

There were additional issues that the early proponents of having a menagerie did not mention but that Frédéric Cuvier believed the menagerie might help settle. One was the question of the fixity or mutability of species. Cuvier did well to call for the creation of a science of variation, and he offered incisive comments on how breeders used "fortuitous" variations to develop animal races, but he had no way of proving his assumption that variation in a species cannot proceed beyond fixed limits. He conducted experiments on the viability of hybrid crosses over successive generations, but he never published his results, and the inability to maintain crosses for more than a few generations would not have disproved the idea of species mutability anyway.

Cuvier told the administration that purchasing the orangutan would give scientists the opportunity to identify more precisely the differences between humans and the animal closest to them, but when it came to it, he never made observations or devised experiments for this purpose. Jack the orangutan brought public attention to the question of apes' proximity to humans. His arrival in Paris by stagecoach; his life with the family of Daboncourt the monkey keeper; his manifest intelligence in undoing knots tied to shorten his swinging rope; his only partial submission to modern medical practices; and his patting George Sand's hand as she helped wrap him in his blankets were all behaviors that underscored but did not answer the question of how similar an orangutan was to a person.

∴

As of the 1830s, another issue of huge consequence for the subsequent history of the biological sciences and the history of the modern zoo was just beginning to appear on the horizon: the extinction of modern life forms. In its early days, the menagerie had a quagga—a species now iconic in this regard—but no one at the time imagined the species would be extinct before the nineteenth century was out. The specimen arrived in Paris in spring

1794 with the last mammals from the old royal menagerie at Versailles. It died four years later at age twelve or thirteen, apparently from old age. Georges Cuvier reported that in the wild quaggas lived in herds sometimes numbering more than one hundred individuals.[8] Frédéric Cuvier two decades later identified the quagga as a prime candidate for domestication, notwithstanding his brother's observation that the animal was difficult to manage and prone to biting.[9] The Cuviers had no reason to think the quagga would become extinct. By the 1870s, however, free-living quaggas were nowhere to be found. The last one in captivity died in Amsterdam in 1883.[10]

Of the many purposes envisioned for the "new menagerie" in Paris, serving as a refuge for animals threatened by extinction was not one. Naturalists understood that human action had driven animals from areas where they had once been abundant to distant wildernesses where humans did not care to live, and they knew of the demise of the dodo, but the idea of modern forms becoming extinct with some frequency was only beginning to be contemplated. Adding a rare specimen to one's collection was the naturalist's heart's desire. For most of the first third of the nineteenth century it is difficult to find any naturalist expressing compunction about collecting specimens of species nearing extinction.[11]

In 1826, contemplating the Museum's acquisition of its first live giraffe, Frédéric Cuvier mused that centuries might pass before France had the chance to host another one. He was certainly not thinking about giraffes' disappearing from the face of the earth. I nonetheless found a journalist in the 1830s, in a piece on the new giraffes at the London Zoo, suggesting that if the giraffe were to become extinct in Africa, it might nonetheless be maintained in European parks. This was surely one of the first expressions of the idea that zoos could serve as sanctuaries for endangered species. Half a century later, Europe would abound in zoos, many of them stocked with giraffes, but without any apparent apprehension that within another century or two giraffes might go the way of the dodo or the quagga.

For the zoologists of Paris, the issue of modern-day extinctions was preceded by that of species' extinction in the past. In the late 1790s the attentions of both Jean-Baptiste Lamarck and Georges Cuvier were focused on it as they sought to account for the similarities and differences between fossil and modern forms (Lamarck for shellfish primarily, and Cuvier for large vertebrates). As a firm believer in the uniformity of nature's operations, Lamarck was not in the least attracted to Georges Cuvier's appeal to a sudden global catastrophe to explain the difference between fossil and living vertebrate forms. He grudgingly acknowledged that humans might have driven a few large animals to extinction, "where the multiplication of animals is slow and difficult," but he saw no way that human action or any more natural means could have brought about the extinction of marine invertebrates sheltered by the oceans. Rather than admit a global catastrophe and the extinction of countless marine forms, he found it more plausible to posit that forms now found only as fossils had evolved into the similar but slightly different forms of the present.[12] His earliest endorsement of the idea of species mutability was intimately related to his feeling that extinction was not a common element of nature's order.

Lamarck's unwillingness to see extinction as a major feature of the history of life on Earth did not prevent him, however, from continuing to acknowledge that humans could shape the fates of other complex organisms. He did so in 1809 when he attributed the gap between humans and apes to humans' driving their closest competitors away to areas less conducive to mental development. His most extended discussion of man's destructive proclivities, however, came in the essay on "man" he wrote for the *Nouveau dictionnaire d'histoire naturelle* in 1818. While crediting man with being "the most surprising and admirable being on the earth," he also credited him with the worst of qualities as well as the best, and he offered a shockingly grim picture of the future:

> By his egoism too short-sighted for his own good, by his tendency to revel in all that is at his disposal, in short, by his lack of concern for the future and his fellow man, man seems to work for the annihilation of his means of conservation and for the destruction of his own species. In destroying everywhere the large plants that protect the soil in order to secure things to satisfy his greediness of the moment, man rapidly brings about the sterility of the ground on which he lives, dries up the springs, and chases away the animals that once found subsistence there. He causes large parts of the globe that were once very fertile and well-populated in all respects to become dead sterile, uninhabitable, and deserted. Always neglecting the words of experience, abandoning himself to his passions, he is perpetually at war with his own kind, destroying them everywhere and under all pretexts, so that one sees formerly great populations become more and more diminished. One could say that he is destined to exterminate himself, having rendered the globe uninhabitable.[13]

Frédéric Cuvier's comments on human-induced species extinction were briefer, more limited in scope, and without any display of anguish over the dark side of the human condition or the fates of the animals involved. He suspected that the American tapir might die out as the result of natives' hunting it for food (before Europeans could domesticate it for their own dinner tables). He also intimated that the sperm whale might become extinct as the result of humans' hunting and killing it for its oil. But these were but tiny motes in Cuvier's vision of a world where the hand of Providence, the benefits of the inheritance of acquired characters, and the interrelations of humans and animals promised a happy future at least for civilized man and his domesticated creatures.

Buffon represented the domestication of animals as a triumph of civilized peoples. Frédéric Cuvier, along much the same lines, suggested that a people's level of civilization could be seen in the domestication of their dogs. Lacépède likewise proposed that menageries reflected the level of civilization of the people who maintained them. Georges Cuvier and Étienne Geoffroy Saint-Hilaire framed the matter in racist terms by maintaining that "peoples with depressed foreheads and protruding jaws" had accomplished much less than Europeans because of inferior mental faculties. Cuvier's feelings of cultural superiority also brimmed over as he noted how the ancient Egyptians' "stupid adoration" of sacred ibises and other animals had inadvertently bequeathed "nearly complete zoological collections in their sacred grottoes" for modern zoologists to understand scientifically. Lamarck was the odd man out when it came to Western scientists' overweening confidence in the forward march of Europeans across the globe.

But that march posed a conundrum for naturalists. On the one hand, it put more new specimens within their reach, but on the other it meant that the nature they encountered was rapidly ceasing to be pristine. In 1820, in his classic paper on plant geography, the Swiss botanist Augustin Pyramus de Candolle urged that the study of plant distribution be taken up "immediately, before civilization has changed the surface of the globe too much."[14] Via commerce, voyages, naturalization projects, and gardens, he explained, humans had both intentionally and inadvertently introduced plant species to areas of the world where the plants had not lived before. What was more, in some cases, "bad plants" (*mauvaises herbes*) were crowding out the indigenous forms. It was a specific case of the point he famously introduced in this same text: "All the plants of a country, all those of a given place, are in a state of war one against the other."[15] Among those who took special heed of these words were the Englishmen Charles Lyell and Charles Darwin.

Lyell, who read Candolle carefully, did not lament the "extirpation" of plant and animal species—or human races—in the face of imperial advance. "The kangaroo and the emu," he wrote, "are retreating rapidly before the progress of colonization in Australia; and it scarcely admits of doubt that the general cultivation of that country must lead to the extirpation of both."[16] Further, "The annihilation of a multitude of species has already been effected, and will continue to go on hereafter, in certain regions, in a still more rapid ratio, as the colonies of highly civilized nations spread themselves over unoccupied lands."[17] To Lyell's mind this was not a matter of upsetting the balance of nature but instead a demonstration of how the balance of nature worked: "Every species which has spread itself from a small point over a wide area, must, in like manner, have marked its progress by the diminution, or the entire extirpation, of some other, and must maintain its ground by a successful struggle against the encroachments of other plants and animals."[18] A "faint image of the certain doom" of a species less suited for such struggle was to be seen in "the extirpation of savage tribes of men of by the advancing colony of some civilized nation."[19]

Today the egoism of Lyell's triumphant colonialism makes us shudder. As Lyell saw it, though, the extinction of species at the hand of man was but a part of the broader picture where because of "incessant fluctuations in the physical geography" of the earth, one had to "concede the successive extinction of terrestrial and aquatic species to be part of the economy of our system." He went on to say, "There is no possibility of escaping from this conclusion, without resorting to some hypothesis as violent as that of Lamarck, who imagined . . . that species are each of them endowed with indefinite powers of modifying their organization, in conformity to the endless changes of circumstances to which they are exposed."[20]

How curious it seems today for Lyell to use the word "violent" to characterize Lamarck's theory in the same pages of *Principles of Geology* where, without using that word, he wrote of species succumbing to the onslaught of empire. Just what he meant in describing Lamarck's theory as violent is not entirely clear, but its implications were violent with respect to his own worldview in the sense that they stood to overturn it completely. For a man who felt comfortable in portraying the variations of domestic animals as part of God's plan, the idea of organic evolution was a fundamental threat.

This increased attention to the twin specters of evolution and human-induced species extinction came late in the heyday of the Paris menagerie: 1793–1838, the dates of this book. Human egoism, colonialism, and violence to animals and people would be significant dimensions of the continuing history of the modern zoo. Zoos were beneficiaries of processes entailing massive numbers of animal deaths, and they were sites of much animal suffering.[21] It is thus a major irony of the history of modern zoos that today they are inclined to identify the preservation of species as their primary mission.

Late in the nineteenth century, Alphonse Milne-Edwards, the Museum's professor of mammals and birds (and menagerie director), told the minister of public education that the menagerie's progress had stopped "nearly completely" in the mid-1830s, and that it had been "left well behind" by the new zoological gardens inspired by its example. Listing the most important of these new zoos (in London, Berlin, Amsterdam, Antwerp, Rotterdam,

Frankfurt, and Hamburg, plus the Jardin d'Acclimatation of the Bois du Boulogne in Paris), Milne-Edwards explained their advantages over the menagerie in Paris. Sustained by members' dues, admission fees, and in some cases animal sales, the new zoos were distinctive in the variety of animals they displayed, the splendor of their buildings, and their large and well-paid staffs. In contrast, the menagerie in Paris, depending on funding from the French government, was on the verge of "irredeemable decay."[22]

The London Zoo was Milne-Edwards's primary example of how far Paris had been outdistanced. Citing what premier animals cost (15,000 to 30,000 francs for a rhino or hippo; 10,000 to 12,000 francs for an elephant; 8,000 to 10,000 francs for a giraffe), he pointed to the menagerie's annual acquisition budget, a pitiful 4,000 francs per year, while the London Zoo's was 45,000. He noted further that there was no point in Paris's paying large sums for an anthropoid ape, because the hygienic conditions of the monkey and ape quarters made it impossible to keep a gorilla, chimpanzee, or orangutan alive long enough to study. In contrast, in the larger, better heated, plant-filled ape quarters in London, a little-known species of chimpanzee had lived for nine years. Milne-Edwards urged the minister to increase the menagerie's budget, not to bring it into line with the London Zoo, but to rescue it from oblivion.[23]

But while the London Zoo outshone the menagerie in Paris, its merits were only relative, at least to judge from one critic writing in 1901. Observing the cramped, repressive conditions of the animals' lives there, he dubbed it "the Bastille of the Beasts."[24] How ironic, since in the French Revolution the Bastille was the emblem of the ancien régime's repressiveness, and its promoters wanted the new menagerie in Paris to bear no relation to the institutions of the previous era. A century after the Paris menagerie's founding, the best of the new zoological gardens was still a prison.

Circumstances have changed, and zoos have evolved, even if the zoo animal's status as a captive has not been fundamentally altered. The first dramatic breakthrough in animal care did not come until the century's end, when the Hamburg animal exhibitor and dealer Carl Hagenbeck figured out that separating animals by moats rather than confining them to cramped cages appealed greatly to the public, was healthier for the animals, and was good for his business, because it allowed him to sell more specimens.[25] Other zoos meanwhile had been refashioning themselves in other ways to assure their financial viability and keep up with competitors. In 1827 the Museum in Paris wanted nothing to do with having Osage Indians camping on the grounds, but for zoos in the late nineteenth century whose functioning depended on attracting crowds and selling tickets, Carl Hagenbeck's exhibitions of Nubians and Inuits provided a major boost.[26] Financial viability and keeping up with competitors remain crucial considerations for zoos today.

The functions espoused by the best of today's institutions are a far cry from those prioritized at the Museum of Natural History in Paris in the first third of the nineteenth century. The mission statements of today's leading zoos feature the conservation of species and the preservation or restoration of the animals' natural habitats.[27] The Paris menagerie had no explicit mission statement, but in contrasting it with the useless, extravagant menagerie of Versailles, its promoters emphasized its utility for the new republic. One side of this was the human moral improvement that citizens were supposed to derive in gazing on scenes of a well-ordered nature. The other side was to improve animals to better suit human needs through training, acclimatizing, or domestication. Mission statements of zoos today tend to feature saving species from extinction through captive breeding programs, habitat conservation, and inculcating in the public an appreciation for nature. Programs for children are often part of it. The Museum's commitment to educating the public was manifested primarily in the public courses delivered by professors lecturing on their specialties. The Museum did not take it as its task to impress on children the wonders of nature and the importance of being kind to animals. That was left by default to enterprising authors like Louis-François Jauffret with his book *Voyage au Jardin des Plantes*.

In 1802, having studied the architect Molinos's plans for the whole of the menagerie, the naturalist Gotthelf Fischer von Waldheim joyfully imagined "the beautiful moment when the picture is fully finished," but that moment never really came. The problem was not simply that when the new building for the carnivores was finally ready for occupancy in 1821, it looked nothing like what Molinos had originally planned. It was also that no celebration was held to mark the achievement, no moment was observed for stopping to savor the completion of "the new menagerie." There was still a new pheasant house to build, and after that a new monkey house. The "palace for the monkeys" would prove to be the last major structure for decades, but it was not because the menagerie's advocates felt there was nothing left to do. In 1838, Isidore Geoffroy Saint-Hilaire and Achille Valenciennes proposed a new structure that would include a reptile menagerie, a small rodent menagerie, a parrot aviary, and aquatic accommodations for beavers, seals, otters, and crocodiles. The administration approved, but the project went no further.

What can we say about the future of the zoo of today, considering its past? Contingent circumstances brought into being and shaped the "new menagerie" of Paris at the turn of the nineteenth century, and contingent circumstances are certain to have a bearing on the zoo of tomorrow. With anthropogenic climate change and habitat destruction accelerating at an alarming pace, and with the human species seemingly ever more prone to exhibit "an egotism too short-sighted for its own good," it is hard not to imagine that the world will witness major political and ecological upheavals in the next century and that the consequences of these for life on Earth and for human social institutions will be enormous.

Though the promoters of the menagerie in Paris pictured it as a site for improving citizens' respect for nature and their behavior toward animals and each other, for much of their history since then, zoos depended on the practices of wrenching animals from their natural social groups and habitats, transporting them to parts of the world to which they were not naturally suited, and incarcerating them in quarters inhospitable to the development of their full natural faculties. Is it thus hoping against hope to think that "the new zoo" might live up to its mission statements about saving species from extinction and educating the public about the value of biodiversity, the importance of habitat conservation, and the importance of appreciating and respecting all living things?

As Lamarck saw it, change of habits was what enabled species to survive in the face of changed environmental circumstances. Might humans develop new habits suitable to meeting the challenges of life on Earth today? Might zoos, despite their history as sites for the display and exercise of human dominance over the animal world, play a role in such a conversion? Lamarck had had a stint overseeing the menagerie, but in his later words about "man," he did not think of the menagerie as a place for saving humans from destroying nature and themselves.

From the Paris menagerie's beginning, overarching visions of what the menagerie could be were continually contradicted at the level of the daily interplay of human and animal lives, where circumstances prevailed over master plans. It was at that level that diverse dramas of human and animal existence played out, with humans and animals intertwined in an age-old dance displaying affinities, differences, dualities, and continuing co-construction. That dance will go on, but in circumstances beyond what we can now imagine. What I do suspect, if there are still zoos a century from now, is that the continuing desire for a better zoo, a "new zoo," will accompany them.

Acknowledgments

This book has been a long time in the making. I could not have written it without the support and encouragement of many people. Family, friends, and fellow scholars contributed at every step of the way. Librarians and archivists and their staffs were unfailingly helpful. The suggestions of readers and editors made the manuscript much better than it would have otherwise been. Research boards and foundations provided material and other support at key times. I am deeply indebted to all, and it is a privilege to acknowledge it.

The institutions most important for my work have been the Muséum national d'histoire naturelle (MNHN) in Paris and the University of Illinois at Urbana-Champaign. The former has been a recurrent draw for me throughout my academic career, even when my focus was elsewhere. The latter has been a site where colleagues, students, and a magnificent library have been a constant source of stimulation and knowledge.

I first set foot in the MNHN and its library in January 1969. I came as a graduate student planning a dissertation on Jean-Baptiste Lamarck and evolutionary biology. I was welcomed by Yves Laissus, director of the Bibliothèque centrale. He would later (1993) publish an important study of the menagerie's history, but I doubt he was thinking about that subject at the time. I know that I wasn't.

After I finished the book on Lamarck (1977) that grew out of my dissertation, my research focus shifted from the early nineteenth century and French zoology to the mid-twentieth century and the creation of the discipline of ethology, the biological study of behavior. The latter story features the work of the Austrian zoologist Konrad Lorenz and the Dutchman Niko Tinbergen, but three conferences in France over a nine-year period kept me from neglecting the importance of earlier French work. The first of these (1988), commemorating the bicentenary of Buffon's death, led me to compare Buffon's ideas on animal behavior and domestication with those of the modern ethologists. It also brought me into contact with historians of science Giulio Barsanti, Claude Blanckaert, Pietro Corsi, Jean Gayon, Jacques Roger, and Roselyn Rey. The present book benefits from the work of each.

The 1988 Buffon conference, however, did not sow the seeds for this book. That happened four to five years later with the Museum's 1993 celebration of its bicentenary. Rather than trying to say something new about Lamarck, I decided to work up another paper on an ethological theme. I decided to study Frédéric Cuvier's early nineteenth-century efforts to create a science of animal psychology based on his observations and experiments on the animals of the menagerie.

I imagined that a month's work in Paris archives in fall 1992 would provide me sufficient material for a paper for the bicentenary in 1993, but I found more than I had bargained for: a rich vein of previously unexplored papers at the Archives nationales bearing on the menagerie's history and Frédéric Cuvier's role in it. I presented my early findings at a seminar in Paris hosted by Jean Gayon and Claude Debru late in fall 1992. I gave a more extended version of these at the Museum's bicentenary six months later. By then I had decided to make the history of the menagerie the focus of a full-length book, once I finished my book on twentieth-century ethology. That did not happen until 2005, but well before then, in 1996, François Delaporte invited me to a colloquium on the history of ethology to be held in 1997 at the Université de Picardie Jules Verne. There I gave a paper focused on Frédéric Cuvier's science.

In writing this book, it has been a pleasure to compare notes, archival sources, and ideas with other scholars attracted to the menagerie's history. I benefited from exchanges with Louise Robbins in the late 1990s while she was working on her marvelous *Elephant Slaves and Pampered Parrots* (2002); with Katie Hornstein, beginning in 2018, while she was researching her splendid *Myth and Menagerie: Seeing Lions in the Nineteenth Century* (2024); and with Violette Pouillard, commencing soon after she published her pathbreaking *Histoire des zoos par les animaux* (2019).

Mitchell Ash involved me in his collaborative project (2008) on the history of the Tiergarten Schönbrunn, the oldest zoo in existence. He later connected me with Dr. Gerhard Heindl, the Tiergarten's archivist, who provided me with invaluable information on the animal population at the "Kaiserliche Menagerie Schönbrunn" as the nineteenth century began.

At the Museum's Bibliothèque centrale, Pascale Heurtel and Alice Lemaire helped me on research visits and in responding to my requests about manuscripts and images. Likewise, in the last stage of gathering illustrations for the book, the staff of the Museum's *Service Diffusion et médiation des savoirs* replied to my requests for information and images with a kindness and efficiency for which I am deeply grateful. I also wish to acknowledge the encouragement of former menagerie directors Jacques Rigoulet and Michel Saint-Jalme, and likewise that of Yves Cauzinille, secrétaire général of the Société des Amis du Muséum, who helped me publish a brief account of Jack, the orangutan, in the society's bulletin. I am likewise pleased to thank historian of science Jean-Baptiste Grodwohl for kindly resolving a question about a manuscript and photographing a large and important file of letters from Pierre-Médard Diard to Georges Cuvier.

Among other historians of science who have helped me in this project, Camille Limoges lent me a treasure, his microfilm copy of the AN's microfilm of the procès-verbaux of the Museum's professors' meetings. I also thank Pierre-Yves Lacour for his help with research and publication questions and for the splendid model he set with his book, *La république naturaliste*. Fellow Lamarck scholar and longtime friend Pietro Corsi helped me gain access to the Lamarck and Buffon websites he created, and he fielded questions on various matters of French natural history. I likewise thank my art history friend and colleague David O'Brien for photographing in the Archives nationales an important letter I had failed to photograph clearly enough some years earlier. I am also happy to acknowledge the help of colleagues Harry Liebersohn and Dorothee Schneider in puzzling through certain German passages in Gotthelf Fischer's 1802–3 account of the Museum.

For their responses to my questions, I thank Tim Birkhead (on the elusive identity of Emmanuel Baillon's unknown "phalarope"), Sarah Allen (on the behavior of seals), Philippe Chavot (for finding that Frédéric Cuvier's gravesite is now unidentifiable), and Chris Herzfeld (who shared information with me on her knowledge of the history of primatology and the talents of orangutans). I also benefited from exchanges with Matt Chrulew, Sherrie Cohen, Jane Desmond, Mariam Motamedi Fraser, Oliver Hochadel, Jamie Jones, Paul Jobling, Paula Young Lee, John Lynn, Dorinda Outram, Benedetta Piazzesi, Jessica Riskin, Nigel Rothfels, and Pierre Serna.

The first complete manuscript I produced was much too long, laden with information and arguments that seemed appropriate at the time but less essential, I confess, upon reconsideration. Friends gave the initial corpus a try. Jo Kibbee tackled several sections, and Bruce Michelson, with his characteristic energy and enthusiasm, read and counseled me on the entire first half of the manuscript. My vague thought that the University of Chicago Press might consider a two volume work was wisely quashed at the press's end, but reducing the manuscript to the size the press required was not easy. It would not have happened without the help of the wonderful writer and editor Marla Golla, whom I previously knew only through email correspondence. Once she signed on to the project, we tackled it in tandem, section by section, she in Egypt and I in the States. Her editorial wisdom, encouragement, and friendship made the whole a happy experience. I cannot thank her enough. I also give special thanks to Violette Pouillard, who gave the press a thorough review of the manuscript. Her detailed comments and suggestions were of great help to me in making final revisions.

In researching and writing the book, I received support from the National Science Foundation (awards SBR-9601390 and SES-0441935), the University of Illinois Research Board, and the Camargo Foundation. Their help was critical, and I am indeed grateful for it.

As indicated, this book grew out of an invitation to give a conference paper. Over time, as work on the manuscript proceeded and other scholarly conferences arose, I turned to the evolving manuscript for material. As a result, several sections of the present volume involve topics I have treated in print previously, going back to "La ménagerie et la vie du Muséum," in *Le Muséum au premier siècle de son histoire* (MNHN, 1997). This paper reappeared in an English version, "Constructing the Zoo: Science, Society, and Animal Nature at the Paris Menagerie, 1794–1838," in *Animals in Human Histories: The Mirror of Nature and Culture* (University of Rochester Press, 2002). Other publications predating the book include "Frédéric Cuvier on Animal Behavior," *Bulletin d'histoire et d'épistémologie des sciences de la vie* (2001); "The Leopard in the Garden: Life in Close Quarters at the Muséum d'Histoire Naturelle," *Isis* (2007); "Civilizing Specimens and Citizens at the *Muséum d'Histoire Naturelle*, 1793–1838," in *Of Elephants and Roses: French Natural History, 1790–1830* (American Philosophical Society, 2013); "La voix du gardien du lion, ou Les significations multiples des animaux du Muséum d'histoire naturelle," *Annales historiques de la Révolution française*, (2014); and "An Orangutan in Paris: Pondering Proximity at the Muséum d'Histoire Naturelle in 1836," *History and Philosophy of the Life Sciences* 40 (2018).

Regarding the book's illustrations, I have already noted the MNHN's kind help. The many other institutions and agencies that provided images are acknowledged in their captions. I am grateful to the individuals at those institutions who helped in the process. I am also happy to acknowledge the friendship and genius of the late Mark Leff, who in 2005, tasked by the University of Illinois History Department with choosing a gift for me on my retirement, found the ideal item: a copy of Pierre Boitard's 1842 *Le Jardin des Plantes*. That copy has been a major source of illustrations for the present book.

At the University of Chicago Press, editors Karen Merikangas Darling and Fabiola Enríquez Flores did an amazing job overseeing the book's publication and ensuring it met the press's standards. I thank too my manuscript editor, Alice Bennett, for her keen eye in identifying and resolving glitches, stylistic idiosyncracies, and more. I am honored to have had her wise help. Plus, I thank Adriana Smith, production editor, for managing the copyediting and proofreading stages, and my hat is off to Daniel Lewis for creating the index.

In addition, I thank Jim Barrett and Nils Jacobsen for reading the conclusion of the manuscript's second version, making suggestions, and urging me to let it loose. Their long-term encouragement and that of fellow "Café Medico" members Don Crummey, Joe Love, and Doug Kibbee is greatly appreciated.

Last, but in no way least, I thank my wife, Jayne, who has been a loving companion throughout the book's gestation. We both wondered if it would ever be done. My previous book took just as long to finish as this one, and I dedicated it to her. This book I dedicate to our sons, Rick and Fritz, and our grandsons, Wells and Freddie.

Abbreviations

ABBREVIATIONS IN NOTES

AN Archives nationales de France, site de Pierrefitte-sur-Seine

Buffon, *HNGP* Georges-Louis Leclerc, comte de Buffon, *Histoire naturelle, générale et particulière*, 15 vols. Imprimerie Royale, 1749–67.

Buffon, *HNO* Georges-Louis Leclerc, comte de Buffon, *Histoire naturelle des oiseaux*. 10 vols. Imprimerie Royale, 1770–86.

HNM Geoffroy Saint-Hilaire, Étienne, and Frédéric Cuvier, *Histoire naturelle des mammifères, avec des figures originales enluminées, desinées d'après des animaux vivants*. 7 vols. C. de Lasteyrie, 1818–42.

Lacépède, *Ménagerie*. Lacépède, B., Georges Cuvier, and Étienne Geoffroy Saint-Hilaire. *La Ménagerie du Muséum national d'histoire naturelle, ou Description et histoire des animaux qui y vivent ou qui y ont vécu; par les citoyens Lacépède et Cuvier, avec des figures peintes d'après nature, par le citoyen Maréchal, peintre du Muséum*. Miger, 1801–5.

MNHN Bibliothèque centrale du Muséum national d'histoire naturelle.

PV Procès-verbaux des assemblées des professeurs. Note:

1. References to material from the PV are cited as PV / date / AN / volume number / page number.
2. Letters and other documents related to the individual professorial assemblies (Minutes et pièces annexes des procès-verbaux des assemblées des professeurs, 1793–1849, AN 577-AJ/15/668) are cited as Letter or document / date / date of meeting at which the item was presented (if different from the date of the document) / AN / file number.

ABBREVIATIONS IN BIBLIOGRAPHY

AN. Archives nationales de France, site de Pierrefitte-sur-Seine (Archives du Muséum national d'histoire naturelle).

AJ/15/95. Décisions réglementaires prises par l'Assémblée des professeurs, concernant l'administration du Muséum, 1790–1859.

AJ/15/96–AJ/15/139. Procès-verbaux des assemblées des professeurs, 1790–1839.

AJ/15/150–AJ/15/193. Traitements des professeurs et employés du Muséum, 1793–1838.

AJ/15/522–AJ/15/523. Bâtiments et terrains.

AJ/15/531–AJ/15//537. Affaires générales concernant le personnel. An III-1922.

AJ/15/541–AJ/15/576. Voyageurs naturalistes: Instructions et enseignement, rapports, indemnités, dossiers personnels. 1791–1922.

AJ/15/577–AJ/15/668. Minutes et pièces annexes des procès-verbaux des assemblées des professeurs, 1793–1849.

AJ/15/742–56. An V–1838. Correspondance.

AJ/15/836–AJ/15/841. Collections.

AJ/15/844. Acquisition des animaux et état de la Ménagerie au XIXe siècle, 1793–1901.

AJ/15/845. Administration et gestion de la Ménagerie. An II—1923, 1794–1923.

(Bâtiments civils)

F/13/873. Muséum d'Histoire Naturelle. An II—An VIII.

F/13/973A.III. Muséum d'Histoire Naturelle. Ménagerie Provisoire.

F/13/1218A, F/13/1218B, F/13/1219, F/13/1225. Muséum d'Histoire Naturelle et Jardin des plantes.

(Papiers des Comités d'Instruction Publique)

F/17/1229, F/17/3980, F/17/3975–77, F/17/13566.

Bibliothèque centrale du Muséum national d'histoire naturelle

MNHN Ménagerie MS 16. Observations particulières sur les animaux de la ménagerie, an XII [1804]—1822.

MNHN Ménagerie MS 25. Registre d'entrée des mammifères à la ménagerie, 1807–1849.

MNHN Ménagerie MS 33. Registre d'entrée des oiseaux à la ménagerie, 1814–1862.

MNHN MS 457. Documents du Comité de Salut Public sur le Jardin des Plantes.

MNHN MS 600–642. Papiers et manuscrits du baron G.-L.-Chr.-Fr. Cuvier (1769–1832):

MNHN MS 625. Lettres et pièces relatives au voyage de Diard et Duvaucel dans l'Inde.

MNHN MS 628. Notes et documents concernant les ossements fossiles. Cuvier, Mémoire sur les espèces d'éléphans vivantes et fossiles.

MNHN MS 2083. 2. Documents relatifs à Étienne Geoffroy Saint-Hilaire. La girafe de 1827.

MNHN MS 2289–295 Manuscrits provenant d'Étienne Geoffroy Saint-Hilaire (1772–1844).

MNHN MS 2290. 1. Collections des mammifères et oiseaux du Muséum, Ménagerie.

MNHN MS 2295. 3. Cahier de notes scientifiques. Notes et copies de lettres relatives à la ménagerie du Muséum.

MNHN MS 2528. Recueil d'autographes et de documents précieux pour l'histoire du Jardin du Roi et du Muséum d'histoire naturelle et celle de leur personnel scientifique.

MNHN MS 2711–36. Papiers d'Étienne Geoffroy Saint-Hilaire:

MNHN MS 2717 (2). Muséum d'histoire naturelle. Ménagerie.

Fonds Cuvier, Institut de France MS 3341.

Journals and Newspapers

Note: Accessed mostly through Gallica (https://gallica.bnf.fr) and its journal site, RetroNews (https://www.retronews.fr).

Le Magasin encyclopédique; *La Décade philosophique, littéraire et politique*; *Gazette nationale ou Le Moniteur universel*; *Journal des théâtres, de la littérature et des arts*; *Le Drapeau blanc*; *Journal de Commerce*; *La Quotidienne*; *La Presse*; *La Mercure*; *Le Courrier*; *Le Figaro*; *Journal des débats*; *Le Constitutionnel*; *L'Echo du monde savant*; *Magasin pittoresque*; *Le Temps*; *La France*; *Revue étrangère de la littérature, des sciences et des arts*; *Le magasin universel*; *Musée des familles*.

Notes

1.0 INTRODUCTION

1. On the history of the major zoos of the nineteenth century, see for London: Blunt, *The Ark in the Park*; Åkerberg, *Knowledge and Pleasure at Regent's Park*; Ito, *London Zoo and the Victorians*. For Amsterdam: Mehos, *Science and Culture for Members Only*. For Berlin: Bruce, *Through the Lion Gate*. For Hamburg: Rothfels, *Savages and Beasts*. For the Tiergarten Schönbrunn, founded in 1752 and the oldest zoo operating today: Ash, *Mensch, Tier und Zoo*. For American zoos: Hanson, *Animal Attractions*. For the Paris menagerie: Loisel, "Histoire de la ménagerie du Muséum"; Laissus, "Les animaux du Jardin des Plantes"; Burkhardt, "La ménagerie et la vie du Muséum" and "The Leopard in the Garden." For the Paris menagerie, London zoo, and Antwerp zoo, Pouillard, *Histoire des zoos par les animaux*. On the history of zoos generally: Loisel, *Histoire des ménageries de l'antiquité à nos jours*; Baratay and Hardouin-Fugier, *Zoos: Histoire des jardins zoologiques en occident*, and *Zoo: A History of Zoological Gardens in the West*; and Ash, "Zoological Gardens."

2. Procès-verbaux des assemblées des professeurs, 1790–1839, AN, AJ/15/96-AJ/15/139.

3. "Minutes et pièces annexes des procès-verbaux des assemblées des Professeurs, 1793–1849" AN, AJ/15/577–656. More than a century ago, both Ernest-Théodore Hamy and Gustave Loisel made excellent use of these manuscripts in their studies of the Museum and the menagerie's earliest years. See Loisel, "Histoire de la ménagerie du Muséum," and Hamy, "Les anciennes ménageries royales et la ménagerie nationale," "Les derniers jours du Jardin du roi," and *Les débuts de Lamarck*. Since then, few scholars have used this material to illuminate aspects of the menagerie's history. Modern exceptions include Saban and Lemire, "Les éléphants de la ménagerie du stathouder Guillaume V d'Orange," and Hornstein, *Myth and Menagerie*.

4. Laissus's excellent "Les animaux du Jardin des Plantes" has a broader chronological reach: 1793 to 1934. This book differs from that work in focusing on the menagerie's first four and a half decades, exploring its functioning and its relation to the broader institutional and scientific context and exploiting additional archival materials, especially the AJ/15/577–656 series at the Archives nationales.

1.1 THE LEOPARD IN THE GARDEN

1. The confiscation of Marchini's animals on 14 brumaire an 2 was reported in a document signed by the commissioner of police, Charbonnier. The Museum's copy is in

AN AJ/15/844 and further copied into PV, 15 brumaire an 2, AN AJ/15/96, 55–56, and reproduced in Hamy, "Les anciennes ménageries royales," 21. For other animal deliveries to the Museum soon thereafter, see Hamy, "Anciennes ménageries royales"; Loisel, "Histoire de la ménagerie du Muséum"; Laissus, "Les animaux du Jardin des Plantes"; Burkhardt, "Leopard in the Garden"; and Robbins, *Elephant Slaves and Pampered Parrots*. Loisel indicates that the Commune's decree called for the animals from the street shows of Paris to be killed or conveyed to the old royal menagerie at Versailles, but the police found it more expedient to take them all to the Jardin des Plantes.

2. Hamy, "Anciennes ménageries royales," 21.

3. "Second adresse des officiers du Jardin des Plantes à l'Assemblée nationale, en lui présentant un projet de règlemens pour cet établissement." In Hamy, "Derniers jours," 102–7 (quotation on 104).

4. On the founding of the Museum see especially Hamy, "Derniers jours"; Spary, *Utopia's Garden*; and Gillispie, *Science and Polity in France*, 165–83.

5. Burkhardt, "Civilizing Specimens and Citizens."

6. Desfontaines to "Citoyen president" (of the Comité d'Instruction Publique), 16 brumaire an 2 [6 November 1793], AN F17/3908; Hamy, "Anciennes ménageries royales," 22.

7. Étienne Geoffroy Saint-Hilaire, "Sur la ménagerie du Muséum d'Histoire Naturelle," 26 April 1815, AN AJ/15/610.

8. Henning Mankell, "The Art of Listening," *New York Times*, December 11, 2011, SR4.

9. That animal studies historians necessarily deal with *representations* of animals is discussed by (for example) Fudge, *Perceiving Animals* and *Animal*, and by Rothfels, *Representing Animals*. On recovering more of "the animal's perspective," see Baratay, *Le point de vue animal* and *Biographies animales*. For a discussion of Baratay and explicit attention to writing zoo animal biographies, see Chrulew, "Living, biting monitors, a morose howler and other infamous animals." Baratay's approach is developed to excellent effect in Pouillard, *Histoire des zoos par les animaux*. See too Pouillard, "Animal Biographies: Beyond Archetypal Figures."

10. États de traitements des professeurs et employés du Muséum, an 2, AN AJ/15/150.

11. On proprietors' compensation, PV 9 frimaire an 2, AN AJ/15/96, 64–68. For the *combat du taureau* in Paris, see Robbins, *Elephant Slaves*, 80–85, 207–10, which discusses Martin as well. The animal fights are also discussed by Serna, *Comme des bêtes*, 66–67.

12. PV 18 prairial an 2, AN AJ/15/96, p. 123.

13. For a recent study of Maréchal, see Jouy, "Portraits sensibles de carnassiers au service de la science."

14. PV 18 prairial an 2, AN AJ/15/96, p. 123.

15. PV 18 messidor an 2, AJ/15/96, p. 133.

16. For Maréchal's submission, PV 18 prairial an 2, AN AJ/15/96, p. 123. For his work soon after his appointment, PV 18 and 28 thermidor an 2 and 14 vendémiaire an 3, AN AJ/15/96, pp. 142, 145, 161.

17. PV 24 brumaire an 3, AN AJ/15/96, p. 170 (the panther died on 13 brumaire an 3).

18. Étienne Geoffroy Saint-Hilaire, *Catalogue des mammifères du* Muséum, 117.

19. Georges Cuvier, "La panthère," in Lacépède, *Ménagerie*. For the work's publication history, see Pieters, "'La Ménagerie du Muséum National d'Histoire Naturelle.'"

20. Cuvier, "La panthère," 4.

21. The "ounce" or snow leopard, known in Cuvier's day as *Felis uncia*, is now named *Panthera uncia.*

1.2 IMAGINING THE NEW MENAGERIE

1. Bernardin de Saint-Pierre, *Mémoire*, 58–59n.

2. Couturier letter to Bernardin, 19 September 1792, in Hamy, "Anciennes ménageries royales," 20.

3. Bernardin de Saint-Pierre, *Mémoire*, 8–10. On Bernardin's *Mémoire* see also Robbins, *Elephant Slaves*, 215–19.

4. Bernardin, *Mémoire*, 10–11.

5. Bernardin, *Mémoire*, 9.

6. Bernardin, *Mémoire*, 18.

7. Bernardin, *Mémoire*, 17.

8. See Loisel, *Histoire des ménageries*; Kisling, *Zoo and Aquarium History*.

9. Bacon, *New Atlantis*, 35.

10. Bacon, *New Atlantis*, 38–39.

11. Loisel, *Histoire des ménageries*, 2:102–57.

12. Perrault, *Mémoires pour servir à l'histoire des animaux*, "Préface," unnumbered p. 4. See also Loisel, *Histoire des ménageries*, 2:298; Robbins, *Elephant Slaves*, p. 45; and Guerrini, *Courtiers' Anatomists*, 202–4.

13. Guerrini, *Courtiers' Anatomists*, observes that the line between anatomy and natural history was not always clear-cut in practice.

14. Buffon, "L'outarde," *HNO* 2 (1771): 202. On Buffon's ideas on animal behavior, see Burkhardt, "Le comportement animal." On Buffon's attempt to bring the animals of Versailles to Paris, see Hamy, "Anciennes ménageries royales," 9.

15. Letouzey, *André Thouin*, 263–64, 271. On Thouin's botanical networks see also Spary, *Utopia's Garden*, chap. 2.

16. Bernardin de Saint-Pierre, *Mémoire*, 2:52–57n.

17. See Rey, "L'animalité dans l'œuvre de Bernardin de Saint-Pierre," and Drouin, "Bernardin de Saint-Pierre."

18. Bernardin de Saint-Pierre, *Études*, 1:34.

19. Bernardin de Saint-Pierre, *Études*, 396.

20. Bernardin de Saint-Pierre, *Mémoire*, 1–3.

21. Bernardin de Saint-Pierre, *Mémoire*, 4.

22. Loisel, *Histoire des ménageries*, 2:159, citing Mercier, says it was rumored that during the reign of Louis XVI a dromedary (or elephant) at Versailles was treated to six bottles of burgundy daily.

23. Bernardin de Saint-Pierre, *Mémoire*, 13.

24. Bernardin de Saint-Pierre, *Mémoire*, 14–15.

25. Bernardin de Saint-Pierre, *Mémoire*, 18–19.

26. Bernardin de Saint-Pierre, *Mémoire*, 19–20. I did not find this claim in either Pliny or Aristotle but may not have looked in the right places. For hybrids of more similar species, Bernardin could have cited various Enlightenment authors including Buffon, Maupertuis, and Réaumur. See Terrall, "Speculation and Experiment."

27. Bernardin de Saint-Pierre, *Mémoire*, 21–22.

28. Bernardin de Saint-Pierre, *Mémoire*, 22–23.

29. On Bernardin's amorous aspirations, see Drouin, "Bernardin de Saint-Pierre."

30. Bernardin de Saint-Pierre, *Mémoire*, 24–26.
31. Bernardin de Saint-Pierre, *Mémoire*, 53–56n.
32. Bernardin de Saint-Pierre, *Mémoire*, 30–33.
33. Bernardin de Saint-Pierre, *Mémoire*, 36–37.
34. Bernardin de Saint-Pierre, *Mémoire*, 37–40.
35. Bernardin de Saint-Pierre, *Mémoire*, 61n.
36. Bernardin de Saint-Pierre, *Mémoire*, 49.
37. Anonymous (1792), review of *Mémoire*.
38. Review of *Mémoire*, 114.
39. Review of *Mémoire*, 116.
40. Review of *Mémoire*, 115.
41. Millin, Pinel, and Brongniart, *Rapport*. I suspect Millin wrote the anonymous, negative review of Bernardin's proposal but was then persuaded by Pinel and Brongniart that the committee should emphasize the value a menagerie could have.
42. Millin, Pinel, and Brongniart, *Rapport*, 1.
43. Millin, Pinel, and Brongniart, *Rapport*, 2.
44. Millin, Pinel, and Brongniart, *Rapport*, 2.
45. Millin, Pinel, and Brongniart, *Rapport*, 3.

1.3 THE REVOLUTIONARY MUSEUM AND THE MENAGERIE (1793–1795)

1. Deleuze, *Histoire et description du Muséum*, 72–73. The lack of evidence for Deleuze's account and the fact that he published it during the Restoration are pointed out by Guillaume, *Procès-verbaux du Comité d'instruction publique*, 1:481.

2. On the relation between images of nature and Revolutionary ideas about improving humans and human society, see Spary, *Utopia's Garden*.

3. For Lakanal's report to the Convention and the decree founding the Museum, see Guillaume, *Procès-Verbaux*, 1:481–86, and Hamy, "Derniers jours," 135–38.

4. This explanation follows that of Hamy, "Derniers jours," 63–67, 132–62.

5. For a detailed examination of natural history cabinets and the seizures and confiscations that profited the Museum during the Revolution, see Pierre-Yves Lacour's masterful *République naturaliste*. Lacour does not stress, however, as I do, the close connection between the Museum's founding and the need to house the Chantilly collection.

6. Hamy, "Derniers jours," 132–33.

7. Hamy, "Derniers jours," 66.

8. In "Unpacking Baudin" (1997), I credited Dorinda Outram with calling specimens the "lifeblood" of the Museum, citing the version of her paper read at the Museum's 1993 bicentenary conference. The phrase did not survive in the published version, Outram, "Institution scientifique ou champ de Bataille," but I am still happy to credit her for it.

9. I emphasize the Museum's concern with controlling specimens in "The Leopard in the Garden" and "Civilizing Specimens" (2013). The twentieth- and twenty-first-century development of model organisms is a stronger form of shaping specimens for scientific purposes, but the notion of "civilizing" specimens goes well with earlier work in natural history, including preserving and displaying specimens and attempting to acclimatize and tame organisms and otherwise suit them to human purposes.

10. Spary, *Utopia's Garden*, stresses the professors' unity in the face of external influ-

ences. Outram, in *Georges Cuvier*, and "Institution scientifique ou champ de bataille," emphasizes the professors' interpersonal rivalries and alliances. Lacour, *République naturaliste*, 14, succinctly explains how the two perspectives are not contradictory.

11. The Museum divided the year into four sessions, three for teaching and a shorter one for vacation, as shown in the schedule for year V (fall 1796 to fall 1797). The two anatomy courses were taught from 23 October 1796 to 21 January 1797; geology, mineralogy, general chemistry, and the chemical arts from 21 January to 4 May 1797; and the six other courses (botany at the Jardin des Plantes, agriculture, mammals and birds, reptiles and fish, invertebrates, and iconography) from 4 May to 18 August. These met four times each ten-day week, according to a schedule by which no two classes were taught at the same time. The final course—botany in the countryside—was taught once a week during the second and third sessions of the year. PV 4, 14, and 24 ventôse an 3, AN AJ/15/97, pp. 122, 125–127.

12. PV 28 fructidor an 2, AN AJ/15/96, p. 154. Beginning 4 prairial an 10 (24 May 1802), the meetings were held every seven days instead of every ten.

13. For biographical sketches of Daubenton and the other professors and naturalist assistants at the Museum, see Jaussaud and Brygoo, *Du Jardin au Muséum*.

14. On Geoffroy see Appel, *The Cuvier-Geoffroy Debate*. The colorful story of Geoffroy's saving Haüy is told by Isidore Geoffroy Saint-Hilaire, *Vie, travaux et doctrine scientifique d'Étienne Geoffroy Saint*-Hilaire, 7–28, and more succinctly by Gillispie, *Science and Polity in France*, 181–82.

15. PV 3 July 1793, AN AJ/15/96, 21–22.

16. On Thouin see Letouzey, *André Thouin*; Spary, *Utopia's Garden*; and Lacour, *La République naturaliste*.

17. PV 16 ventôse an 2, AN AJ/15/96, pp. 95–98.

18. Jaussaud, "Antoine Portal," in Jaussaud and Brygoo, *Du Jardin au Muséum*, 423–24. Portal's apartment being occupied by his brother is noted on a list of persons lodged at the Museum, séance of 27 vendémiaire an 10, AN AJ/15/587.

19. PV 14 June 1793, AN AJ/15/96, pp. 12–13.

20. PV 14 June 1793, AN AJ/15/96, p. 13.

21. Hamy, "Derniers jours," 158. A menagerie was also mentioned in the 1790 reform plan. Hamy, "Derniers jours," 120.

22. Desfontaines letter to Romme and Prunelle of the Committee of Public Instruction, 7 (17?) frimaire an 2, AN F/17/3980. In Guillaume, *Procès-Verbaux*, 2:817–21. The questions are also cited by Loisel, "Histoire de la ménagerie," 264.

23. For the costs of feeding the individual animals, PV 9 frimaire an 2, AN AJ/15/96, p. 65.

24. PV 29 brumaire an 2, AN AJ/15/96, p. 60.

25. Desfontaines to Romme and Prunelle, 7 (17?) frimaire an 2, AN F17/3980; Guillaume, *Procès-Verbaux*, 2:817–21.

26. PV 13 nivôse an 2, AN AJ/15/96, p. 77.

27. PV 18 nivôse an 2, AN AJ/15/96, p. 79.

28. Jussieu, "Notes relatives à l'etablissement d'une menagerie dans le museum d'hist. naturelle," AN AJ/15/844. Loisel, "Histoire de la ménagerie," 264, says that a proposal to establish a menagerie at the Museum was presented to the Convention on 3 January 1794. This date seems unlikely, since on 4 February 1794 the administration charged Jussieu with providing notes on the importance of establishing a menagerie to Fourcroy, as a member of the CPI (PV 16 pluviôse an 2, AN AJ/15/96, p. 86), and this

presumably would not have happened had an official report already been made to the Convention.

29. PV 9 and 23 pluviôse an 2, AN AJ/15/96, pp. 83–84, 87.

30. Draft letter to Couturier in the hand of the secretary, Lamarck, 16 and 19 ventôse an 2, AN AJ/15/577.

31. PV 28 ventôse an 2, AN AJ/15/96, p. 101.

32. I take this date from Loisel, "Histoire de la menagerie," p. 264. I have found no archival evidence that confirms this date precisely, but comments in the PV of 8 and 12 floréal are consistent with the arrival having happened no later than Loisel's date of 7 floréal an 2 (26 April 1794).

33. The architect Verniquet initiated the amphitheater's design, but it was completed according to plans drawn by Molinos.

34. PV 12 floréal an 3 (1 May 1794), AN AJ/15/96, pp. 110–11.

35. "Rapport du citoyens Geoffroy et Lamarck professeurs de Zoologie aux citoyens professeurs composant le Directoire du Muséum d'histoire naturelle." 1 germinal an 2 (21 March 1794), AN AJ/15/844. See also PV 1 germinal an 2, AN AJ/15/96, p. 102.

36. The PV for 24 floréal an 2 (13 May 1794) indicates the Raincy animals' arrival in Paris. AN AJ/15/96: 113. Loisel, "Histoire de la ménagerie," p. 264, misdates the animals' arrival as two weeks later (27 May 1794).

37. PV 24 floréal an 2 (13 May 1794), AN AJ/15/96, p. 113.

38. "État des animaux, vivans, existans à la ménagerie provisoire du Museum national d'histoire naturelle, ce jourd'hui 26. Floréal an 2me de la République francaise une et indivisible." AN AJ/15/844.

39. The professors and CSP commissioners are named in the procès-verbal of the séance of 4 prairial an 2, AN AJ/15/96, p. 118. Fourcroy's letter to Thouin of 27 floréal an 2 is reproduced in Letouzey, *André Thouin*, 323, and referenced by Spary, *Utopia's Garden*, 224.

40. PV 4 prairial an 2; AN AJ/15/96, pp. 116–17.

41. On the projected increase in size see Lacour, *La République naturaliste*, 223–26, and Spary, *Utopia's Garden*, 224–25, which reproduces Molinos's map, 244–45. Lacour says the size would have quadrupled (from 15 to 59 hectares); Spary says it would have quintupled (from 50 to 270 *arpentes*).

42. Extract from CPS register of 27 floréal an 2, 28 floréal an 2; AN AJ/15/577 and PV AN AJ/15/96, pp. 114–15.

43. PV 28 floréal an 2, AN AJ/16/96, pp. 114–15.

44. "Premier Raport [*sic*] au Comité de Salut public fait par la Commission des travaux publics relativement au Jardin des Plantes." No date. MNHN MS 457. Spary, *Utopia's Garden*, dates the document 27 floréal an 2/May 16, 1794, but that was the date of the CPS's original decrees, and the report's text shows it was written after the CPW's visit to the Museum.

45. "Second Raport [*sic*] fait par la Commission des Travaux publics relativement au Jardin des Plantes." MNHN MS 457. Spary, 225–27, adroitly uses the report to underscore how the Jacobins associated the spectacle of nature with the future of France, but she does not emphasize the menagerie's centrality in the commission's vision.

46. The CPS's endorsement of the CPW's recommendations appears at the end of the "Premier Raport."

47. PV 14 prairial and 28 messidor an 2, AN AJ/15/96, pp. 122, 136.

48. The account of the commissioners' visit of 6 prairial an 2 is to be found in the procès-verbaux after the meeting of 4 prairial an 2 and before the meeting of 8 prairial an 2. PV AN AJ/15/96, p. 118.

49. PV 4 prairial an 2, AN AJ/15/96, p. 116. Daubenton told Fourcroy to ask the CPS for 50,000 livres instead of the 24,000 livres requested previously.

50. Desfontaines to Payan, Commissaire de la Commission Exécutive de l'Instruction Publique, 10 prairial an 2, AN F/17/3980.

51. Committee of Public Instruction to the Museum, 18 prairial an 2, responding to the Museum's request for authorization to buy animals for public instruction. AN F/17/3980. See also notes on the importance of establishing a menagerie in the letter to the Museum from Fourcade of the Commission de l'instruction publique, 18 prairial. AN AJ/15/577.

52. Chapter 3, article 8, in Hamy, "Les Derniers jours," 152.

53. PV 16 pluviôse an 2 (4 February 1794), AN AJ/15/96, pp. 84–85.

54. PV 21 ventôse an 2 (11 March 1794), AN AJ/15/96, p. 99. The minister's letter of 19 ventôse an 2 is cited in Bapst, "Histoire d'un cabinet," 446. Bapst incorrectly attributes the Museum's response to Daubenton. Its authors were instead Geoffroy and Lamarck. Bapst also mistakenly identifies Olivier Marat as Jean-Paul Marat's brother whereas he was his brother-in-law. Outram, *Georges Cuvier*, follows Bapst. Geoffroy and Lamarck's letter appears in Hamy, *Les Débuts de Lamarck*, 170–71.

55. PV 28 ventôse an 2 (18 March 1794). AN AJ/15/96, p. 100.

56. Hamy, *Les Débuts de Lamarck*, 170–71.

57. Hamy, *Les Débuts de Lamarck*, 171.

58. Gautier's full name appears in his letter to the professors of the Museum dated 3 brumaire an 3, séance of 4 brumaire an 3, AN AJ/15/578. Museum records contain variant spellings of it.

59. See Guillaume, *Procès-Verbaux*, 4:408–9n and 441–43 for the decrees and Guillaume's comments. For the Museum's response, see PV 18 and 28 prairial and 1 messidor an 2, AN AJ/15/96, pp. 124–28.

60. PV 24 prairial an 2, AN AJ/15/96, p. 125. Also arriving were five fawns from Raincy, two peacocks from Versailles, a baboon, a bonnet monkey, and another kind of macaque. PV 14 prairial and 4 and 8 messidor an 2, AN AJ/15/96, pp. 121, 129, 130.

61. The following year, three other committee members, Collot d'Herbois, Barère, and Billaud de Varenne, were sentenced to exile in French Guiana.

62. PV 14 thermidor an 2, AN AJ/15/96, pp. 141–43. The authorization from the CSP telling the CPI to pay the keepers' salaries and reimburse them for their animals is reported in PV 14 fructidor an 2, AN AJ/15/96, p. 151.

63. PV 24 thermidor an 12 and 4 fructidor an 2, AN AJ/15/96, pp. 143–44, 145.

64. Daubenton taught from 2 floréal to 14 fructidor an 2; Lamarck from 17 floréal an 2 to 11 vendémiaire an 3; and Geoffroy from 27 floréal to 27 fructidor an 2. PV AN AJ/15/96, pp. 142, 155, 162.

65. PV 8 prairial, 14 prairial, 4 messidor, 8 messidor and 28 messidor an 2, AN AJ/15/96, pp. 120, 122, 129, 131, 136,

66. PV 18 messidor an 2, AN AJ/15/96, p. 133.

67. PV 18 fructidor an 2, AN AJ/15/96, p. 152.

68. PV 28 fructidor an 2, AN AJ/15/96, p. 153–54. The professors retained Louzardy but reserved judgment on whether his status should be permanent.

69. Molinos's untitled plan is in AN F/13/1218/A. See Pouillard, *Histoire des zoos par les animaux*, 77.

70. PV 28 thermidor and 14 and 18 fructidor an 2, and 4 and 14 vendémiaire an 3, AN AJ/15/96, pp. 145, 151, 152, 158, 161. The property formerly belonged to the church of the *Nouveaux convertis*. After nationalizing it the government sold it to a man named Leger. For the Maison Leger's importance housing incoming specimens, etc., see Lacour, *La République naturaliste*, 224–26, and Deleuze, *Histoire et description*, 97–99. The "success of the Leger affair" was announced "finally" on 22 June 1795. PV 4 messidor an 3, AN AJ/15/97, p. 49.

71. PV 24 vendémiaire and 4 brumaire an 3, AN AJ/15/96, pp. 164, 165.

72. PV 14 and 24 brumaire an 3, AN AJ/15/96, pp. 167, 168. Olivier Marat's renewed efforts for a naturalist's post also appear here. For Lamarck and Geoffroy's report on the subject, see AN F/17/1229.

73. "Presentation à l'assemblée d'un candidat, pour remplir une place d'aide faite dans la séance du 4 nivôse an 3, par le C. Lamarck," AN AJ/15/578. Reproduced in Hamy, *Débuts*, 161–66. I mistakenly thought I was publishing Lamarck's candidate list for the first time when I reported it in "The Leopard in the Garden."

74. Jauffret, Voyage, 143-45.

75. Thibaudeau, *Rapport sur le Muséum*, 13.

76. Thibaudeau, *Rapport sur le Muséum*, 14.

77. Thibaudeau, *Rapport sur le Muséum*, 16.

78. Thibaudeau, *Rapport sur le Muséum*, 18. Spary, *Utopia's Garden*, 10.

79. Thibaudeau, *Rapport sur le Muséum*, 19–20. PV 24 nivôse an 3, AN AJ/15/96, pp. 186–88.

80. PV 24 frimaire an 3, AN AJ/15/96, p. 182.

81. PV 24 nivôse an 3 (13 January 1795), AN AJ/15/96, p. 189.

82. Dupuy, *École Normale de l'an III*, 94, 136–61.

83. PV 24 nivôse an 3, AN AJ/15/96, p. 188.

84. PV 4 pluviôse an 3, AN AJ/15/97. Daubenton quotation is on p. 2. For earlier references to the triage of Lavoisier's equipment, see PV 24 Brumaire and 14 frimaire an 3, AN AJ/15/96, pp. 168, 179.

85. Daubenton, "Sur la rédaction de l'histoire naturelle." 1: 288–302 (quotes from 290–91). See too Loveland, "Daubenton's Lions," and Dupuy, *École Normale*, 143–44. For the lion in political imagery, see Hornstein, "Jean-Baptiste Huet's Lions."

86. Daubenton, "Sur la rédaction," 291n.

87. Étienne Geoffroy Saint-Hilaire, *Fragments biographiques*, 173–74.

88. Dupuy, *École Normale*, 149.

89. For Daubenton's "Certificat de Civisme," dated 5 Frimaire an 2, see G. Cuvier, *Recueil des éloges historiques*, 1:69n.

90. Dupuy, *École Normale*, 134–35.

1.4 THE LION, THE DOG, AND THE LIBRARIAN

1. Drouin, "*La Décade philosophique*."

2. Anonymous, (G. Toscan), "Sciences. Histoire naturelle: Vues générales." 1:1–7, quotation on 6. The piece appeared slightly revised in Toscan, *L'ami de la nature*.

3. Toscan, "Histoire Naturelle: Essais de botanique morale."

4. Toscan, undated letter to the Museum, MNHN Ms 2528/139.

5. Toscan, *L'utilité de l'établissement d'une bibliothèque.*

6. Toscan, undated letter to the Museum, MNHN Ms 2528/139.

7. PV, 8 thermidor an 2, AJ/15/96, p. 139. Toscan later joined a circle that gathered at the Museum around André Thouin, including Faujas de Saint-Fond and the botanist Desfontaines. Drouin, "Histoire naturelle."

8. Toscan, "Histoire du lion." Reproduced in Toscan, *Ami de la nature.*

9. Toscan, "Histoire du lion," 130.

10. Toscan, "Histoire du lion," 133.

11. Buffon, "Lion," *HNGP*, 9 (1761).

12. Toscan, "Histoire du lion," 136.

13. Toscan,"Histoire du lion," 137,

14. Toscan, "Histoire du lion," 194–95.

15. Toscan, "Histoire du lion," 197–98.

16. Toscan, "Histoire du lion," 198.

17. Toscan, "Histoire du lion," 198–99.

18. Toscan, "Supplément à l'histoire du lion."

19. Pelletan, *Mémoire sur la colonie française du Sénégal.*

20. Toscan, "Supplément," 258.

21. Toscan, "Supplément," 258–59.

22. Toscan, "Supplément," 259–60.

23. Toscan, "Supplément," 260.

24. Toscan, "Histoire du lion," 193–94.

25. Toscan, "Histoire du lion," 195. Vivant Denon was reportedly furious when he witnessed people mistreating the lion. See Sørenson, "Representing Animals with Empathy," 135n12.

26. The phrase "Il faisoit toute sa consolation" appears under the engraving of Maréchal's illustration of the lion and dog pair between pages 14 and 15 in Toscan's *L'ami de la nature.*Although the lion killed the second dog, it later lived in peace with a third. See Anonymous (Miger), *Notice des animaux vivants*, p. 8. In 1804, two lionesses in the menagerie lived with dogs. See Anonymous (Frédéric Cuvier), *Notice des animaux vivans* (1804), 18–19.

27. PV 4 brumaire an 3, AN AJ/15/96, p. 165.

28. Maréchal's name was not ascribed to the engraving that appeared with Toscan's article in *La Décade philosophique*, but it was with the later version that appeared in Toscan's *L'Ami de la nature.* An illustration that appears to have been the basis of these engravings was sold by Christie's on 5 July 2005 for £10,800. See Sørenson, "Representing animals with empathy," 130.

1.5 THE LEMUR IN THE LABORATORY

1. On the Paris naturalists' interest in rare species, see Lacour, *La République naturaliste*, 263–64. F. Cuvier made the point about rare versus appealing forms in "Le Mococo," *HNM* (April 1819).

2. É. Geoffroy Saint-Hilaire, "Mémoire sur les rapports naturels des Makis Lemur," 36.

3. On keeping exotic pets in France see Robbins, *Elephant Slaves.* She mentions (p. 137) a dealer advertising an animal from Madagascar as a Mongous monkey, probably the mongoose lemur.

4. Buffon, "Les makis," *HNGP* 13 (1765): 173–80.

5. E. Geoffroy Saint-Hilaire, "Le Maki Mococo et le Maki Brun," in Lacépède, *Ménagerie*. Pieters dates this installment 13 February 1803.

6. "Rapport du citoyens Geoffroy et Lamarck professeurs de zoologie aux citoyens professeurs composant le Directoire du Muséum d'histoire naturelle," 1 germinal an 2, AN AJ/15/844.

7. The journalist Jean Reynaud heard the story from Geoffroy and reported it in "Fondation de la ménagerie du Muséum d'histoire naturelle," *Magasin Pittoresque* 6 (1838): 106–8. Geoffroy reprinted it in his *Fragments biographiques, précédés d'Études sur la vie, les ouvrages et les doctrines de Buffon* (F. D. Pillot, 1838).

8. Merlin letter to the Museum, 25 germinal; séance of 29 germinal an 2 (AJ/15/577). See too PV AJ/15/96, p. 108.

9. The entry appears in the second *décade* of nivôse an 6 (31 December 1797 to 9 January 1798) in an untitled and unpaginated menagerie record book that Delaunay kept. AJ/15/844.

10. Geoffroy Saint-Hilaire, "Le Maki Mococo," Lacépède, *Ménagerie*.

11. On Dufresne, see anonymous, "Notice sur M. Dufresne." On the history of taxidermy and Dufresne's role, see Farber, "The development of taxidermy" and *Emergence of Ornithology as a Scientific Discipline*, 53–54. Dufresne's skills are admirably described in Candegabe, *L'incroyable histoire de l'éléphant Hans*, 171–196. For a discussion focusing on earlier natural history cabinets and the problems of preserving specimens, see Findlen and Toledano, "The materials of natural history."

12. See Rookmaaker et al., "Ornithological Cabinet." Dufresne revealed the soap's ingredients in his guide to the practice of taxidermy (1803 and later editions), which established long-lasting standards for the field. Dufresne, "Taxidermie."

13. Geoffroy "Maki Mococo," Lacépède, *Ménagerie*.

14. Undated letter from Desmoulin and Dufresne to the Museum, séance of 24 messidor an 3 (12 July 1795), AJ/15/578. PV AN AJ/15/97, pp. 55–56.

15. Two undated letters from Desmoulins and Dufresne to the Museum, séance of 24 messidor an 6 (24 June 1798), AJ/15/581.

16. Geoffroy Saint-Hilaire, "Le Maki Mococo," Lacépède, *Ménagerie*.

17. Buffon, "Le Maki Mococo," *HNGP* (1765). In a later piece on the species, Buffon identified the lemur's habit of stretching its arms toward the sun as an attitude of "admiration or pleasure." He additionally described a brown lemur at his estate in Burgundy that always stayed near the hearth, extending its arms toward the fire. He attributed this to the animal's sensitivity to cold. Buffon, "Le Mococo," *HNGP, Supplément* 7 (1789): 119–20.

18. Mordant Delaunay's untitled record book, AJ/15/844, unnumbered page titled "La ménagerie étoit compôsé le 1er de Vendémiaire an 6eme." The entry for the maki states, "donné par Merlin de Thionville" and "empoisonné par le préservatif du laboratoire & mort le 7 floréal an 7."

19. Marechal's painting of the ring-tailed lemur was presented at the *séance* of 24 prairial an 7 (12 June 1799), PV same date, AN AJ/15/99, p. 122.

20. The specimen's existence in the gallery is registered in Geoffroy Saint-Hilaire, *Catalogue des mammifères du* Muséum, 33. The text was printed, but the work was never completed and never officially published.

21. F. Daudin, "Description" and *Traité*, vol. 2 (for Maugé's birds); Jaussaud and Brygoo, "Maugé" (biographical note).

22. Diard's case is treated more fully in chapter 3.7, "Tapirs and Travelers."

23. On Dufresne and Lapérouse, see Sweet, "Collection," pp. 34–36. On Dufresne's death, see Anonymous, "Dufresne." On Madame Dufresne teaching Prévost, see Dufresne zoology laboratory report, séance of 3 January 1810, AJ/15/603. Another woman artist/taxidermist of the time was Julie Charpentier. She prepared some specimens for the galleries, with Dufresne as an advocate, but she only eked out a modest existence. See Hamy's essay on Charpentier in É. Geoffroy Saint-Hilaire, *Lettres écrites d'Égypte*, 270–76.

1.6 BETWEEN GLORY AND RUIN

1. PV 24 nivôse an 3, AN AJ/15/96, pp. 186–89; 4 and 14 pluviôse and 24 ventôse an 3, AN AJ/15/97, pp. 1–2, 5, 18–19.

2. Thouin to Jussieu, 24 ventôse an 3, séance of 24 germinal an 3, AN AJ/15/578.

3. *Magasin encyclopédique*, 1, no. 2 (1795): 419. See too "Objets d'histoire Naturelle recueillis en Hollande," *Décade philosophique* 5 (an 3 [1795]): 534–37.

4. See Daudin, *Cuvier et* Lamarck, 1:31n1.

5. Cuvier and Geoffroy Saint-Hilaire, "Mémoire sur les rapports naturels du Tarsier."

6. Geoffroy Saint-Hilaire and Cuvier, "Histoire Naturelle des Orangs-Outangs."

7. Georges Cuvier, "Mémoire sur les espèces d'éléphans vivantes et fossiles." See too Burkhardt, *Spirit of System*, 128–29; Rudwick, *Cuvier, Fossil Bones, and Geological Catastrophes*; Cohen, *Fate of the Mammoth*.

8. Lamarck, "De la Seiche, du Calmar et du Poulpe."

9. On housing specimens, PV 14 fructidor an 3 (31 August 1795), AN AJ/15/97, pp. 70–71. On housing professors, PV 24 messidor an 3 (12 July 1795), AN AJ/15/97: 56. By 1800, the Museum was housing nine of its twelve professors and forty-five additional staff members. Most were accompanied by their immediate families or in some cases siblings, nieces, nephews, or parents. See too Deleuze, *Histoire et description du Muséum*, 85.

10. PV 4 brumaire an 4, AN AJ/15/97, p. 85. See the report of the séance of 27 vendémiaire an 10 (19 October 1801) AN AJ/15/587. Of the three professors who did not live on site (Mertrud, Portal, and Lacepède), only Lacepède was not making claims to an apartment. Mertrud was renting his apartment to Georges Cuvier, and Portal's apartment was occupied by Portal's brother.

11. PV 4 vendémiaire an 3, AN AJ/15/96, pp. 157–58.

12. PV 14 germinal an 3, AN AJ/15/97, p. 25.

13. On the menagerie's finally being ready, see the séances of quartidi complémentaire de l'an 3 and 4 vendémiaire an 4. PV AN AJ/15/97, pp. 75, 78.

14. PV 24 brumaire an 4, AN AJ/15/97, p. 92.

15. PV 4 nivôse an 4, AN AJ/15/97, p. 104.

16. PV AN AJ/15/97, pp. 35, 66, 81.

17. PV 14 thermidor an 3 (1 August 1795), AN AJ/15/97, pp. 61–62.

18. PV 4 frimaire an 4, AN AJ/15/97, p. 94.

19. Delaunay to the Museum, 4 pluviôse an 4, séance of same date, AN AJ/15/579. The PV indicates the librarians were granted the bread allowance. AN AJ/15/97, p. 112.

20. PV 24 pluviôse an 4 (13 February 1796), AJ/15/97, p. 117. Thouret's suggestion came two years before Edward Jenner's success in immunizing people from smallpox

with material from cowpox pustules. Following Jenner's lead, Thouret then promoted vaccination in France.

21. "Loi qui autorise l'acquisition de bâtimens et terreins destinés à être réunis au Museum d'histoire naturelle. Du 17 prairial an 4. de la République française une et indivisible." AN F/13/873.

22. PV 17 vendémiaire an 5, AN AJ/15/98, pp. 9–10.

23. Lacepède's presentation to the Directory was inserted in the PV of 24 vendémiaire an 5 (3 September 1796), AN AJ/15/98, pp. 11–12.

24. PV 24 vendémiaire an 5, AN AJ/15/98, p. 13. "si l'intention du gouvernement est que le Museum continue d'avoir une ménagerie."

25. Museum to the Minister of the Interior, 28 vendémiaire an 5, AN F/13/873. On the specific properties the Museum needed and the government's decision to provide them, see PV 4 frimaire and 24 ventôse an 5, AN AJ/15/98, pp. 28, 62–63. See too the Museum's letter to the Minister of the Interior of 8 frimaire an 5 and the accompanying map indicating the properties the Museum identified as necessary. AN F/13/873.

26. *Décade philosophique* 10 (10 thermidor an 4 [28 July 1796]): 235.

27. PV 14 fructidor an 4, AN AJ/15/97, pp. 183–84.

28. PV 14 frimaire an 5 (4 December 1796), AN AJ/15/98, pp. 33–34. The animals listed were "the lion, the panther, the caracal, the hyena, the tiger-cat, the gazelle, the bubal, the African buffalo, diverse species of monkeys, the ostrich, the strong breed of donkey [*le Baudet de forte race*] (appropriate for reconstituting our degenerate type of donkey, as well as the mule), the Barbary sheep (good to multiply in our southern provinces), a hen of the large species, etc."

29. Minister of the Interior to the Museum, 3 pluviôse an 5, séance of 4 pluviôse an 5, AN AJ/15/580. "Variétés," *La Décade philosophique* 12 (an 5, pluviôse an 5=29 January 1797), 237.

30. Delaunay's animal list dates her arrival as 6 ventôse an 5 (26 February 1797). AN AJ/15/844. A note under "Variétés" in *La Décade philosophique*, 12 (5): 496, dates the arrival a week later. The lioness was still living at the menagerie when Gotthelf Fischer wrote about the Museum some five years later. See Fischer, *Das Nationalmuseum*, 274.

31. On Baillon see 2.8.

32. On the rattlesnakes see PV 4 ventôse an 5, AN AJ/15/99, p. 54. See too "Du serpent à sonnettes, arrivé vivant au Muséum d'Histoire-Naturelle," *Décade philosophique* 14 (20 thermidor an 5): 257. The rattlesnakes appear to have died soon. Delaunay's menagerie list of 1 vendémiaire an 6 (AN AJ/15/844) does not mention them.

33. Delaunay to the Museum, 24 pluviôse an 5, AN AJ/15/580, séance of same date. See too PV AN AJ/15/98, p. 53.

34. PV 24 germinal, an 5, AN AJ/15/98, pp. 73–74.

35. Copy of letter to the Minister of Finances, 25 germinal an 5, with the minister's reply on 29 germinal an 5, AN F/17/1229. See also Bureau des Musées, Bibliothèques et conservatoires, "Rapport présenté au ministre de l'intérieur," 23 floréal an 5.

36. Geoffroy letter to the Minister of the Interior, 20 floréal an 5, AN F/17/1229. A month later, the Museum still wondered if selling precious gemstones was the best way of financing its building projects. PV 24 prairial an 5, AN AJ/15/98, pp. 91–93.

37. Minister of the Marine to the Museum, 13 frimaire an 6, séance of 14 frimaire an 6, AN AJ/15/581.

38. Lassaigne to the Museum, 21 frimaire an 6, séance of 24 frimaire an 6, AN AJ/15/581.

39. Jussieu letter to his colleagues, 12 nivôse an 6, séance of 14 nivôse an 6, AN AJ/15/581. See too PV AN AJ/15/98, p. 134.

40. Séance of 4 ventôse an 6, AN AJ/15/98, pp. 147–48. On trees of liberty in the Revolution and the Museum's distribution of them throughout France see Spary, *Utopia's Garden.*

41. "Fête civique célébrée dans le Museum national d'histoire naturelle le Dix Ventôse an six, pour la plantation d'un arbre de la Liberté," PV AN AJ/15/98, pp. 148–52.

42. Delaunay to the Museum, 14 ventôse an 6 (March 4, 1798), séance of same date, AN AJ/15/581. I discuss this letter further in section 1.8, "Theory and Practice."

43. PV 24 ventôse an 6 (14 March 1798), AN AJ/15/98, pp. 153–54.

44. Jussieu to the Minister of the Interior, 25 ventôse an 6, AN F/13/873.

45. Jussieu, 25 ventôse an 6.

46. Minister of the Interior to Jussieu, 1 germinal an 6 (21 March 1798), AN AJ/15/844.

47. PV 14 germinal an 6, AJ/15/98, p. 159. For mention of the crowds some two weeks later, see *Décade philosophique*, 17 (30 germinal, An 6): 180.

48. Minister of the Interior to the Museum, 13 germinal an 6, séance of 14 germinal an 6, AN AJ/15/581.

49. Thouin to the Museum, 18 ventôse an 6, séance of 14 germinal an 6, AN AJ/15/581.

50. PV 14 germinal an 6, AN AJ/15/98, p. 159.

51. PV 27 germinal and 4 floréal an 6, AN AJ/15/98, pp. 162–64.

52. Delaunay to the Museum 14 floréal an 6 (3 May 1798), séance of same date, AN AJ/15/581.

53. Jussieu to the Executive Directory, PV 4 prairial an 6 (23 May 1798), AN AJ/15/98, pp. 169–72.

54. Jussieu to the Executive Directory.

55. Minister of the Interior to the Museum, 22 floréal an 6; Jussieu to the Minister of the Interior, 27 floréal an 6, AN AJ/15/581. Museum to the Minister, 27 prairial an 6, AN F/13/873. PV, 14 and 27 prairial an 6, AN AJ/15/98, pp. 175, 177–78. The name Duhameau is clearly written in these documents, but I have been unable to identify him further.

56. "Lettre du C. Thouin, Commissaire français en Italie, sur les objets d'arts conquis dans ce pays," *Décade philosophique* 14 (an 5): 213. Also in Letouzey, *André Thouin*, 499–501.

57. See Mainardi, "Assuring the Empire of the Future."

58. "Fêtes de la liberté et entrée triomphale des objets de sciences et d'arts recueillis en Italie," in Neufchâteau, *Recueil des lettres*, 1:12–30.

59. Humboldt, *Journal parisien*. *La Décade philosophique* reported that the animals were the festival's biggest crowd-pleaser. See Mainardi, "Assuring the Empire of the Future," 162, 39.

60. *Gazette nationale*, quoted in Mainardi, "Assuring the Empire of the Future," 158.

61. For the French perception of the transfer of cultural power, see Mainardi, "Assuring the Empire of the Future."

62. "Variétés," *Décade philosophique* 18 (6): 301–5.

63. *Décade philosophique* 18 (6): 233–34.

64. "Variétés," *Décade philosophique* 16 (6): 113.

65. PV 4 nivôse an 7, AN AJ/15/99: 54–55. The artist in question, Beaumanoir, was hired with funds from the minister, and worked for the Museum for about three years.

66. Leroi to the Museum, 13 ventôse an 7, séance of 14 ventôse an 7, AN AJ/15/582. See PV AN AJ/15/99, p. 73.

67. PV 24 germinal an 7 (13 April 1799), AN AJ/15/99, pp. 84–86.

68. Degennes to the Museum, 14 vendémiaire an 8, séance of 24 vendémiaire an 8, AN AJ/15/583.

69. Fourcroy to the Minister of the Interior, 27 vendémiaire an 8, séance of 4 brumaire an 8, AN AJ/15/583.

70. Fourcroy, 27 vendémiaire an 8,

71. Fourcroy to the Minister of the Interior, 29 vendémiaire an 8, séance of 4 brumaire an 8, AN AJ/15/583.

72. PV 14 brumaire an 8, AN AJ/15/100, p. 33.

73. Minister of the Marine and Colonies to the Museum, 2 brumaire an 8, séance of 4 brumaire an 8; Director of the Museum to the Minister of the Marine and Colonies, 5 brumaire an 8. AN AJ/15/583.

74. PV 24 brumaire an 8, AN AJ/15/100, pp. 40–41.

75. PV 4 frimaire an 8, AN AJ/15/100, p. 43.

76. PV séance extraordinaire de 11 nivôse an 8, AN AJ/15/100, pp. 65–66.

77. For Fourcroy's letter to the minister, plus the minister's approval, PV AN AJ/15/100, pp. 68–69.

78. For the plans for the ceremony, followed by the official account, PV 14 nivôse an 8 (4 January 1800), AN AJ/15/100, pp. 70–79.

79. PV 14 nivôse an 8, AN AJ/15/100, pp. 76–78.

80. Jean Thouin to the Minister of the Interior, n.d., séance of 14 pluviôse an 8 (3 February 1800); AN AJ/15/583.

81. PV 4 and 23 messidor 8, AJ/15/100, pp. 159, 160. On the festivities in Paris see Dwyer, *Citizen* Emperor, 45–47.

82. PV 24 prairial an 8, AN AJ/15/100, pp. 155–56.

83. PV 4 messidor an 8, AN AJ/15/100, pp. 160, 164–65.

84. Minister of the Interior (Lucien Bonaparte) letter of 12 messidor an 8, séance of 14 messidor an 8 (filed in séance of 24 prairial an 8), AN AJ/15/584. The letter and the Museum's preparation of a response are mentioned in PV 14 messidor an 8, AN AJ/15/100, p. 161.

85. PV 23 messidor an 8 (12 June 1800), AN AJ/15/100, p. 165.

86. For the text of Lucien Bonaparte's 1 vendémiaire an 9 decree and the professors' first response, see séance of 27 vendémiaire an 9. For subsequent exchanges with Lucien Bonaparte, 7 and 17 brumaire an 9. For the dénouement upon Chaptal's appointment as minister, 27 brumaire an 9, PV AJ/15/101, pp. 31–34, 38–45, 48–56, and 61–62.

87. Dwyer, *Citizen Emperor*, 56–58.

1.7 THE BEARS FROM GREENLAND AND THE BEARS FROM BERN

1. F. Cuvier, *Notice des animaux vivans* (1804), 2. Georges Cuvier, "L'ours polaire," in Lacepède, *La Ménagerie*. Pieters dates this installment as 26 December 1801.

2. Another bear, sent by the Minister of the Interior and identified only as a female, was added to the collection in July 1798. PV 4 Thermidor an 6 (22 July 1798), AN AJ/15/99, p. 12.

3. *Décade philosophique* 17 (10 floréal an 6): 240–41.

4. This date (4 floréal an 6 [23 April 1798]) comes from Delaunay's menagerie record book, AJ/15/844.

5. Buffon, "L'ours," *HGNP*, new edition, ed. Sonnini de Manoncourt, 26:77–78 (note by Sonnini).

6. *Gazette des deux-ponts* 127 (vendredi 4 mai 1798) (unpaginated).

7. *Gazette nationale*, 4 May 1798.

8. Surrelot, *La presse féminine en* France, 74–81. Jensen, "Caroline Wuiet."

9. Buffon, "L'ours," 26:77–78n (Sonnini).

10. Heinzmann, *Voyage d'un Allemand à* Paris, 134–35.

11. Delaunay letter to the Museum, 14 floréal an 6, séance of same date. AN AJ/15/581.

12. Leroi letter to the Museum, 13 ventôse an 7, séance of 14 ventôse an 7. AN AJ/15/582.

13. Delaunay letter to the Museum, 27 messidor an 9, séance of same date. AN AJ/15/586. The PV identifies the bear as a female. AN AJ/15/102, p. 28.

14. Fischer, *Das Nationalmuseum*, 2:65–66.

15. Delaunay letter to the Museum, séance of 7 pluviôse an 10 (27 January 1802), séance of same date. AN AJ/15/587.

16. PV 27 ventôse an 10 (18 March 1802). AN AJ/15/103, p. 30.

17. Pujoulx, *Promenades au Jardin des Plantes*, 1:169, mentions two young brown bears replacing the recently deceased bears from Bern.

18. G. Cuvier, "L'Ours polaire," 2.

19. G. Cuvier, "L'Ours polaire," 2. Cuvier's essay, published in December 1801, says one of the polar bears died two years earlier, but it seems to have been at least two years earlier than that, since Delaunay's list of animals in the menagerie in September 1797 mentions only one polar bear. AJ/15/844.

20. Pujoulx, *Promenades*, 1:81.

21. For more on stereotyped behavior, self-mutilation, and other aberrant behaviors related to confinement, see Pouillard, *Histoire des zoos par les* animaux, 125–29.

22. Fischer, *Nationalmuseum*, II, 64.

23. F. Cuvier, *Notice des animaux vivans* (1804), 3. See more in section 2.10, "Caring for the animals."

24. PV 14 floréal an 11, AN AJ/15/104, 64–65.

25. PV 6 nivôse an 12 (28 December 1803), AN AJ/15/105, pp. 80–81.

26. PV 18 pluviôse an XII (8 February 1804), AN AJ/15/105, p. 111.

27. PV 3 fructidor an 13 (21 August 1805), AN AJ/15/107, p. 72.

28. Mention of the public's teasing the polar bear appears in F. Cuvier's later description of the bear's roar: "It only made its voice heard when its anger was excited by provocations, in which case it was strong, harsh, and always of the same tone." F. Cuvier, "L'ours polaire," 2, *HNM*.

1.8 THEORY AND PRACTICE (1795–1801)

1. Lacépède, "Sur le but auquel doit attendre le naturaliste"; G. Cuvier, "Éloge historique de Lacépède," in G. Cuvier, *Recueil des éloges historiques*, 3:281-335.

2. Loisel argued more than century ago that "the hazard of circumstances" and "the energy of some men" had more to do with the menagerie's creation and development

than did the writings of Buffon, Bernardin de Saint-Pierre, and Lacépède. Delaunay was prominent among the men he specifically credited. Loisel, "Histoire de la ménagerie," 262. Most subsequent writers on the menagerie, however, have attended more to its theorists than its practitioners. Pierre Serna, *Comme des bêtes*, acknowledges that the ideal visions of Bernardin, Thibaudeau, and Lacépède bore little resemblance to the menagerie's functioning, but he nonetheless says more about these visions than about what he calls the "chaos" of daily events.

3. For Lacépède's biography, see Cuvier, "Éloge historique de Lacépède," and Hahn, "Les carrières de Fourcroy et de Lacépède."

4. Cuvier, "Éloge historique de Lacépède," 306.

5. Lacépède's first professorial assembly was 14 pluviôse an 3 (2 February 1795).

6. Lacépède, "Lettre relative aux établissemens publics." Ménageries."

7. Lacépède, "Lettre relative aux établissemens publics," 453–54.

8. Jauffret, *Voyage au Jardin des* plantes, 126–27. The words put in Cassal's mouth are not an actual quotation, but they correspond with what is known of his interests and personality.

9. Lacépède, "Lettre relative aux établissemens publics," 460.

10. Lacépède, "Lettre relative aux établissemens publics," 458.

11. Lacépède, "Lettre relative aux établissemens publics," 459.

12. Lacépède, "Lettre relative aux établissemens publics," 460–61.

13. Lacépède, "Lettre relative aux établissemens publics," 462.

14. Lacépède's presentation to the Directory was inserted in the PV of 24 vendémiaire an 5 (3 September 1796), AN AJ/15/98, pp. 11–12.

15. Delaunay letter to the Museum, 23 pluviôse an 2séance of same date, and letter of 23 prairial an 2, séance of 28 prairial an 2, AN AJ/15/577. Henceforth in citing a letter Delaunay wrote on the same day it was read at the assembly, I cite only the day of composition. When the day he wrote the letter and the day it was presented at the assembly differ, I cite both dates.

16. PV 24 brumaire an 4 (15 November 1795), AN AJ/15/97, p. 92.

17. PV 4 nivôse an 4 (25 December 1795), AN AJ/15/97, p. 104.

18. Delaunay to the Museum, 4 messidor an 4 (22 June 1796) and 14 fructidor an 4 (31 August 1796). AN AJ/15/579.

19. Delaunay to the Museum, 14 vendémiaire an 5 (5 October 1796), AN AJ/15/580.

20. Delaunay to the Museum, 14 vendémiaire an 5, AN AJ/15/580.

21. Delaunay to the Minister of the Interior, 24 fructidor an 4, séance of 4 brumaire an 5, AN AJ/15/580.

22. Delaunay to the Museum, 24 pluviôse an 5 (12 February 1797). Six weeks later he reported that the food situation had not been resolved and only two weeks of fodder remained. Delaunay to the Museum, 4 germinal an 5 (24 March 1797). Both letters in AN AJ/15/580.

23. See, for example, Delaunay to the Museum, 4 prairial an 5 (23 May 1797). AN AJ/15/580.

24. Delaunay to the Museum, 14 ventôse an 6 (March 4, 1798). AN AJ/15/581.

25. Delaunay to the Museum, 14 ventôse an 6 (March 4, 1798). AN AJ/15/581.

26. Delaunay to the Museum, 14 ventôse an 6 (March 4, 1798). AN AJ/15/581.

27. Delaunay to the Museum 14 floréal an 6 (3 May 1798). AN AJ/15/581.

28. I have not found among the Museum's papers a document corresponding to Molinos's original design for the new menagerie. The description offered here is based

on later accounts of the emerging plan. See section 1.14, "The New Menagerie Takes Shape."

29. Delaunay to the Museum, 24 ventôse an 7 (14 March 1799), AN AJ/15/582.

30. Delaunay to the Museum, 4 germinal an 7 (24 March 1799). AN AJ/15/582.

31. See section 2.11 on provisions for the animals.

32. Delaunay to the Museum, 4 brumaire an 8,(date?), AN AJ/15/583).

33. Garnier letter to the Museum (n.d.) and list of animals, séance of 4 frimaire an 8 (25 November 1799), AN AJ/15/583, and PV AN AJ/15/100, pp. 43–44. Garnier's menagerie was small but had impressive holdings, which he valued at 7,680 francs. His five mammals and five birds included a pair of zebus, a "tiger panther" from Bengal, and a cassowary.

34. Delaunay to the Museum, 4 ventôse an 8 AN AJ/15/583. PV AN AJ/15/100, p. 103.

35. Delaunay to the Museum, 24 germinal an 8, AN AJ/15/583, PV AN AJ/15/100, p. 128.

36. Delaunay to the Museum, 24 germinal an 8, AN AJ/15/583, PV AN AJ/15/100, p. 128.

37. Delaunay to the Museum, 24 floréal an 8, plus his list of what he sold, séance of 4 prairial an 8, AN AJ/15/584. See also PV 11 prairial an 8, AN AJ/125/100, pp. 144–46.

38. Lacépède and Lamarck reported on the first installment at the Institut de France on 17 vendémiaire an 9 (9 October 1800). See Pieters, "Ménagerie."

39. Lacépède, "Introduction," in Lacépède, *Ménagerie*, quotations from pp. 2, 4.

40. Lacépède, "Introduction," p. 6.

41. PV 7 pluviôse an 9 (27 January 1801), AN AJ/15/101, p. 98.

42. PV 17 prairial an 9 (6 July 1801), AN AJ/15/101, p. 173.

43. PV 7 thermidor an 9 (26 July 1801), AN AJ/15/102, p. 9. Lacépède's colleagues also asked him to discuss with Molinos the designs for the huts in the herbivore parks.

44. PV pluviôse an 9 (15 February 1801), AN AJ/15/101, p. 108. The PV refers to a long memoir that Delaunay submitted, with a map attached, but the document is not in the file for the meeting, and I have not found it elsewhere.

45. PV 7 floréal an 9 (27 April 1801), AN AJ/15/586. See too AN AJ/15/101, p. 146.

46. PV 17 prairial an 9 (6 June 1801), AN AJ/15/101, p. 171.

47. Museum to the Minister of Interior, 25 prairial an 9, séance of 27 prairial an 9. AN AJ/15/586.

48. The main context in which Delaunay's name subsequently appeared in the Museum's records was in a year's-long struggle with the Museum (1808–9) after he was told to cede his apartment to the professor of geology. Exasperated by Delaunay's complaining and delaying, the administration threatened to fire him. Only then did he leave the apartment.

1.9 THE CONCERT FOR THE ELEPHANTS

1. Dwyer, *Napoleon: The Path to Power*, 319.

2. See, for example, *Journal des departemens de la Mozelle, de la Meurthe, de la Meuse, des Ardennes et des Vosges*, 10 prairial an 6 (31 May 1798), 3.

3. This overview of elephants in France up to Buffon's time is derived largely from Houël, *Histoire naturelle des deux éléphans*, and Guerrini, *Courtiers' Anatomists*, 202.

4. Buffon, "Addition à l'article de l'Éléphant," *Supplément à l'Histoire Naturelle* 3 (1776): 293–94.

5. Houël, *Histoire naturelle des deux éléphans*, 14.

6. The claim that there were only three elephants in Europe in 1795 comes from Lebreton, "Rapport au Comité d'Instruction Publique," undated, in séance of 14 vendémiaire an 4, AN AJ/15/579. Lebreton, the head of museums and lycées in France, cited two elephants in Holland and "an old male" at the Tower of London, but he was wrong about the last. There appears to have been no elephant at the Tower menagerie at the time, but there was one on exhibit in London at Gilbert Pidcock's "Grand Menagerie" at Exeter-'Change.

7. Thouin to Jussieu, 24 ventôse an 3, séance of 24 germinal an 3 (13 April 1795), AN AJ/15/578.

8. Pliny the Elder. *Natural History*, book 8, chapter 1:1.

9. G. Cuvier, "Pline l'Ancien," 520.

10. Pliny the Elder, *Natural History*, Book 8, chapter 5.

11. Buffon, "L'Éléphant," *HNGP* 11 (1764): 1–6.

12. Buffon, "L'Éléphant," 6–7.

13. Buffon, "L'Éléphant," 56–62.

14. Buffon, "L'Éléphant," 63.

15. Buffon, "L'Éléphant," 16–17.

16. Buffon, "Addition à l'article de l'Éléphant," *Supplément à l'Histoire Naturelle* 3 (1776): 295–96.

17. The best treatment of the animal pair is Candegabe, *L'incroyable histoire de l'éléphant Hans*. Well researched and wide-ranging, it surpasses what was long the standard account, Houël's *Histoire naturelle des deux éléphans*. Lassaigne told Houël that the female's name was "Marguerite," but that was a name Lassaigne chose. The Museum's naturalists typically wrote of animals as representatives of their species, not as individuals, and they almost never mentioned pet names.

18. Thouin's letter to Jussieu of 24 ventôse an 3 identifies Thompson as English. Likewise, an identity certificate prepared for Thompson on 20 prairial an 9 identifies him as Jacques Thompson, born in England and sixty-three years old. However, a letter of 28 fructidor an 10 from the Museum to the Minister of the Interior mentions James Thompson's desire to retire and go live in Germany "where he was born." AN AJ/15/743. See further details at the end of this section.

19. Houël, *Histoire naturelle des deux éléphans*, 22–24.

20. On the elephants' transport, see Saban and Lemire, "Les éléphants de la ménagerie du stathouder," a very valuable work based on a wealth of MNHN papers conserved at the Archives nationales. The authors miss some important documents, however, including a whole file devoted to the elephants' transport (in AN AJ/15/844) and the musician's report on the elephant concert (see below).

21. J. Lebreton, "Rapport au Comité d'Instruction Publique," undated copy, in séance of 14 vendémiaire an 4, AN AJ/15/579.

22. Undated letter from Delaunay to the professors of the Museum, séance of 14 fructidor an 4 AN AJ/15/579.

23. Letter from Nöel to the Museum from The Hague, 4 messidor an 5, séance of 14 messidor an 5 (2 July 1797). AN AJ/15/580.

24. Minister of the Interior to Jussieu, director of the Museum, 9 germinal an 6, séance of 14 germinal an 6, AN AJ/15/581.

25. On Lassaigne and the Chantilly specimens, see PV 23 Vendémiaire an 2, AN AJ/15/96, p. 42. Regarding his travels with the elephants and letters to Delaunay, see

the Museum file AJ/15/844. See also Houël, *Histoire naturelle des deux éléphans*, and Candegabe, *L'incroyable histoire de l'éléphant Hans*.

26. Houël, *Histoire naturelle*, 25–26.

27. Saban and Lemire, "Les éléphants," 285. Candegabe, *L'incroyable histoire*, cites Houël's account but sides with Saban and Lemire's explanation. Candegabe, 123 and 273n156.

28. Houël, *Histoire naturelle*, 28–29.

29. Dejean to the Minister of the Interior, 23 fructidor an 5, AN AJ/15/844. General Jean-Baptiste Eblé was also involved. See Nöel to the Museum, 4 messidor an 5, séance of 14 messidor an 5, AN AJ/15/580, and Saban and Lemire, "Les éléphants."

30. Lassaigne to Delaunay, 11 frimaire an 6, AN AJ/15/844.

31. Lassaigne to Delaunay, 1 nivôse an 6, AN AJ/15/844.

32. Lassaigne to Delaunay, 1 nivôse an 6, AN AJ/15/844. Lassaigne wrote to the administration as well as to Delaunay. His letters of 11 frimaire, 21 frimaire, 24 frimaire, 1 nivôse, 6 nivôse, 14 pluviôse, 18 pluviôse, and 17 ventôse are in the elephant transport file, AN AJ/15/844.

33. Lassaigne to Delaunay, 6 nivôse an 6, AN AJ/15/844.

34. On the tusks, see Houël, *Histoire naturelle*, 32. The military engineer Detroye wrote to Delaunay (15 nivôse an 6) reporting he was bringing two broken tusks from Cambrai but did not say not which of the elephants lost them. AN AJ/15/844. The elephants arrived tuskless.

35. Lassaigne to Delaunay, 14 pluviôse an 6, AN AJ/15/844.

36. Having delivered the elephants in excellent condition, Boucher asked the Museum to recommend him for a promotion. Boucher to Jussieu, 7 germinal an 6, AN AJ/15/844. Jussieu's responses to Boucher and DeTroye are noted in PV 14 germinal an 6, AN AJ/5/98, p. 158.

37. Lassaigne to Delaunay, 17 ventôse an 6, AN AJ/15/844.

38. *Journal de Paris* 22 March 1798.

39. Saban and Lemire, "Les éléphants," 290, state that the elephants disembarked at Port des Invalides on 20 March 1798 and were installed in their quarters at the Jardin des plantes the same evening. Candegabe follows their lead. However, the note in *Journal de Paris* (22 March 1798: 3) merely reports the elephants' arrival across from the Invalides on 30 ventôse (20 March), not their disembarkation. Explicit testimony indicates the animals arrived at the Museum on 24 March and were installed in their quarters on 25 March. This appeared under the heading "Variétés" in *La Décade philosophique* 17 (10 germinal an 6 [30 March 1798]: 44–46. We credit this anonymous piece to Toscan. It identifies the time of arrival of the elephants at the Museum as 8:00 p.m. on "last quartidi" (24 March 1798) and says the elephants were not placed in their quarters until the next morning. The notice was repeated verbatim in *Le Journal de Paris*, 2 April 1798, and *Le Moniteur universel*, 14 Germinal an 6 (3 April 1798). 779. Independent confirmation that the elephants' first day in their new quarters was 25 March comes from Lacoste, "Sur les éléphans." The four-day disembarkation in Paris is consistent with the ten days it took earlier in their trip, at Gand, to move them from one boat to another. *La Décade philosophique* 15 (6, no. 6, 20 November 1797): 364–65.

40. Another account of the elephants' reunion appears in Houël, who heard about it from Thompson. Houël, *Histoire naturelle*, 33.

41. "Variétés," *Décade philosophique* 17 (a6): 44–46.

42. Lacoste, "Les éléphans," 2.

43. "Variétés," *Décade philosophique*, 17 (10 germinal an 6): 44–46; *Le Moniteur universel*, 14 germinal an 6 (3 April 1798), 779.

44. Thouin to Desfontaines 18 ventôse an 6 (from Toulon), séance of 14 germinal an 6, mentions the incident. AN AJ/15/581. At the same meeting, the assembly agreed to the days and hours for viewing the elephant, PV 14 germinal an 6, AN AJ/15/98, p. 59. The hours were also given in *La Décade philosophique* 17 (6): 180.

45. Reneaudu letter to the Museum, n.d., séance of 24 germinal an 6, AN AJ/15/581.

46. Letourneux to the Museum, 26 germinal an 6, séance of 4 floréal an 6, AN AJ/15/581.

47. The first published account of the concert was a note in *La Chronique universelle*, 8 June 1798, by the Museum's professor of iconography, Van Spandoenck. The next and best-known account is Toscan, "Du pouvoir de la musique." It has been the source of most commentaries on the concert since then, including those in Saban and Lemire, "Les éléphants"; Putnam, "Captive audiences"; Candegabe, *L'incroyable histoire*; and Serna, *Comme des bêtes*. In fall 1992 I discovered a manuscript account of the concert previously unknown to historians. Written by the musicians themselves, it is titled "Concert au Jardin des plantes decadi 10 prairial an 6 [29 May 1798] exécuté par les citoïens Adrien L'ainé, Guichard, Rousseau L'ainé, Chol, Rousseau cadet, Chlart, De Vienne, Maitcoffer, Felix, Delcambre, Frederic, le Fevre jeune, Veillart, Rochefort tous attachés au théatre des arts, au théatre faideau et au conservatoire de Musi [*sic*]." The document is filed with the special séance of 27 prairial an 6, AN AJ/15/581. I first reported on it at the Museum bicentenary conference in 1993, the paper later published as Burkhardt, "La ménagerie et la vie du Muséum." On Devienne and Adrien's collaboration for the Festival of the Supreme Being see Pierre, *Hymnes et Chansons*, 61.

48. Saban et Lemire, "Les éléphants," 292.

49. Serna, *Comme des bêtes*, 117.

50. Burkhardt, "La ménagerie et la vie du Muséum," 493–95; Putnam, "Captive audiences," states (155–56), "Toscan makes it clear from the outset that the goal of the experiment was to gauge the effects of music on sensitive, living creatures."

51. "Concert au Jardin des plantes," AN AJ/15/581. See also PV AN AJ/15/98, p. 177.

52. Toscan, "Pouvoir de la musique."

53. Toscan, "Pouvoir de la musique," 258–59.

54. "Concert au Jardin des plantes," AN AJ/15/581.

55. Toscan, "Pouvoir de la musique," 259.

56. Toscan, "Pouvoir de la musique," 260.

57. Devienne, *Bataille de Gemmapp*.

58. Toscan, "Pouvoir de la musique," 262.

59. Toscan, "Pouvoir de la musique," 321–22.

60. "Concert au Jardin des plantes," AN AJ/15/581.

61. Toscan, "Pouvoir de la musique," 322.

62. *La Chronique universelle*, 20 prairial an 6 (8 June 1798), 2.

63. Séance extraordinaire du 27 prairial an 6, PV AN AJ/15/98, p. 177.

64. Toscan, "Pouvoir de la musique," 328.

65. Toscan, "Pouvoir de la musique," 328–29.

66. Fischer, *Nationalmuseum*, 2:55–59. Textual evidence indicates that Fischer wrote his account of the elephants in 1802, after Hans died and before the plan to purchase a new elephant in London failed.

67. Fischer, *Nationalmuseum*, 2:56.

68. Fischer, *Nationalmuseum*, 2:57.

69. Fischer, *Nationalmuseum*, 2:58.

70. Fischer, *Nationalmuseum*, 2:59.

71. Houël, *Histoire naturelle*, 101–2n.1. No mention of the trumpeters' concert (or of any other concert besides the first) appears in the Museum's PVs.

72. Houël, *Histoire naturelle*, 82.

73. Houël, *Histoire naturelle*, 85.

74. Houël, *Histoire naturelle*, plate XVI, 105–7.

75. Candegabe, *L'incroyable* histoire, 161–65, plausibly suggests that the two animals, growing up in the absence of other elephants on which to model their behavior, developed their own sexual games.

76. Houël, *Histoire naturelle*, 107.

77. Houël, *Histoire naturelle*, 73.

78. Houël, *Histoire naturelle*, 56.

79. Houël, *Histoire naturelle*, 90.

80. The supplement to the *Gazette Nationale* (26 floréal an 6 (15 May 1798) advertises a piece called *Relation d'une visite faite aux deux éléphans nouvellement arrivés au Jardin des plantes*. It appears to have offered stock information about the species rather than saying much about a visit to the elephants.

81. On Jauffret's *Voyage au Jardin de Plantes* see section 2.3, "Guiding the public." The book was announced in *La Chronique universelle*, 24 June 1798.

82. Anonymous, (Vignier), *Vertus morales des deux* éléphans. For more on Vignier's guidebooks see section 2.3.

83. PV 14 fructidor an 6, AN AJ/15/99, p. 19.

84. *Décade philosophique* 18 (20 fructidor an 6): 498–99.

85. *Le Publiciste*, 9 June 1799; *Gazette nationale*, 10 June 1799.

86. See, among others, *Almanach des spectacles de Paris* (Duchesne, an 8), 118; *Mercure de France* (8 July 1799), 135–38; *Journal de Paris*, 6 July 1799; and *Décade philosophique* 22 (30 messidor an 7): 175–77.

87. PV 17 brumaire, an 10, AN AJ/15/102, pp. 128–29.

88. PV 17 nivôse an 10, AN AJ/15/102, pp. 174–76; *Gazette nationale*, 8 January 1802.

89. The Museum did not suggest this behavior was related to the elephant's death, but Fischer did. Fischer, *Nationalmuseum*, 2:59.

90. PV 17 nivôse an 10, AN AJ/15/102, pp. 174–76; *Gazette nationale*, 8 January 1802.

91. Museum to the Minister of the Interior, 17 nivôse an 10, AN AJ/15/587.

92. PV 17 nivôse an 10, AN AJ/15/102, p. 176.

93. Cuvier, "Mémoire sur les espèces d'éléphans vivantes et fossiles," 12.

94. Cuvier," L'éléphant des Indes," in Lacépède, *Ménagerie*. Pieters dates this essay vendémiaire an 10.

95. Cuvier, L'éléphant des Indes," 10–11.

96. Cuvier, L'éléphant des Indes," 11.

97. Corse, "Observations."

98. Cuvier, "L'éléphant des Indes," 6. Cuvier also described how the female's vulva changed in form, appearance, and relative location when the animal was sexually aroused.

99. Grigson, *Menagerie*, 112–13.

100. Bertuch, *Bemerkungen auf einer Reise*, 2:138–39. I thank Dr. Gerhard Heindl for this reference.

101. Perrault, *Histoire naturelle des animaux*, part 3: 107. Cited in Houël, *Histoire naturelle*, 118–19; Loisel, *Histoire des* ménageries, 2:298; Guerrini, *Courtiers' Anatomists*, 202–4.

102. PV 12 fructidor an 10, AJ/15/103: 136. Rousseau's annual salary was two thousand francs.

103. Cuvier, "L'éléphant des Indes, femelle," in Lacépède, *Ménagerie*. Pieters dates the essay nivôse an 12 (December 1803—January 1804). Despite the title and the accompanying portrait of the female elephant, Cuvier devoted the essay primarily to anatomical findings from the dissection of the two males.

104. PV 16 messidor an 10 (5 July 1802), AJ/15/103: 90.

105. For a superb discussion of taxidermy and Dufresne and Lassaigne's work mounting the elephant, see Candegabe, *L'incroyable histoire*, 171–96.

106. PV 26 vendemiare an 12 (date?), AJ/15/105, p. 24.

107. PV 8 frimaire an 12 (30 November 1803), AJ/15/105, p. 50.

108. PV 15 frimaire an 12 (7 December 1803), AJ/15/105, pp. 56–57.

109. Lassaigne to the Museum, 13 nivôse an 12, AJ/15/591; PV 13 nivôse an 12, AJ/15/105, p. 82. Geoffroy and Cuvier were appointed to consider publishing Lassaigne's engraving and an account of the manikin's construction in the Museum's *Annales*. No report appears in the PV, and publication did not occur.

110. *Journal des débats*, 8 thermidor an 12 (27 July 1804), 3. The story of the specimen's eyes figures in Candegabe's conclusion that Hans is the stuffed specimen now in the Museum of Bourges. Candegabe, *L'incroyable histoire*, 38–39, 195.

111. Geoffroy called for the purchase of the replacement tusks at the meeting of 15 frimaire an 12, PV AJ/15/105, p. 56. Candegabe, *L'incroyable histoire*, 195, reports Dufresne's asking Geoffroy to do so.

112. On this elephant and the keeper who came with it, see section 2.6, "Kindness in context."

113. Cuvier, "Sur les éléphans vivans et fossiles" (1806).

114. Cuvier, "Sur les éléphans vivans et fossiles" (1806), 268.

115. Some six months after Hans's death, Thompson asked to retire, citing family affairs and his advanced age. He wanted to go with his wife and two children to live in Germany and said he was born there. The Museum had believed he was born in England. See Thompson to the Museum, 7 thermidor an 10 (26 July 1802), AJ/15/588; Museum to the Minister of the Interior, 28 fructidor an 10 (15 September 1802), AN AJ/15/743. Thompson agreed to postpone his departure until the elephants were accustomed to someone else. He stayed until July 1803, whereupon Louzardy became the new mahout and moved into Thompson's apartment. PV 24 messidor an 11, AJ/15/104, 111. The Museum's request to the police for a passport for Thompson (13 messidor an XI, AJ/15/743) identifies him as James Thompson, keeper of the menagerie's two elephants since germinal an 6, and as wishing to go with his family to live in Mulheim on the Ruhr.

116. Lassaigne to Delaunay, 6 nivôse an 6, AJ/15/844.

117. Lassaigne served in this capacity for at least another four decades. He was named assistant head of carpentry in 1825, then head (as of 1832). He was head at least until July 1838 (the last date for which I reviewed staff salaries), earning two thousand francs annually. AJ/15/193.

1.10 THE KEEPERS

1. MNHN MS 628, 14–15. PV 29 brumaire an 2, AN AJ/15/96, p. 61.

2. Cochon (with three monkeys) was awarded four *livres* per day. "Mémoire de la dépense occasionnée pour la nourriture des animaux," 9 vendémiaire an 3, AN AJ/15/577.

3. Guichenot's previous Museum job was as *garde bosquet*. His having twenty-six children was noted in "État des personnes logés dans le Muséum N[at]. d'histoire naturelle," séance of 27 vendemiaire an 10 (19 October 1801), AN AJ/15/587, and in a letter Delaunay wrote the Museum on behalf of Guichenot's widow, 28 pluviôse an 11, AN AJ/15/589.

4. Cassal's name first appears in the Museum's records 24 prairial an 2 (12 June 1794), PV AJ/15/96, p. 125.

5. PV 28 fructidor an 2, AN AJ/15/96, p. 155.

6. PV 14 prairial an 3, AN AJ/15/97, p. 41.

7. PV 24 prairial an 3, AN AJ/15/97, p. 47; "Traitemen . . . fructidor an 3," AJ/15/150. In 1802 Marchini resurfaced, seeking reappointment as a keeper of the animaux féroces, but he was turned down. PV 27 floréal an 10, AN AJ/15/103, p. 58. Later, when he planned a collecting trip to Africa, the Museum provided him a certificate attesting to his prior service. PV 29 frimaire an 11, AN AJ/15/104, p. 12.

8. Louzardy to the Museum, undated, read 4 floréal an 3, AN AJ/15/578. PV 4 floréal an 3, AN AJ/15/97, p. 33. The Museum awarded him a one-time bonus of fifty livres.

9. Louzardy to the Museum, undated, read 4 thermidor an 3. AN AJ/15/578. PV 4 thermidor an 3, AN AJ/15/97, p. 59.

10. On the reimbursement, see the Museum to the Executive Commission of the Committee of Public Instruction, 8 thermidor an 3, AN AJ/15/578.

11. Louzardy to "Citoien" (probably Delaunay), 4 fructidor [an 4] (misfiled in the séance for "jours complémentaires de l'an 3," AN AJ/15/578, and mislabeled as the jours complementaires of an 6).

12. Louzardy to the Museum, 2 October [1795], read 24 vendémiaire an 4, AN AJ/15/579. The full letter, more than twice the length of this sample, proceeds without punctuation. Saban and Lemire, 280n, have also tried to decipher the passage.

13. Louzardy to "Citoien" 4 fructidor [an 4].

14. PV 14 fructidor an 4, AN AJ/15/97, pp. 183–84.

15. Jussieu to the Directoire, 18 vendémiaire an 5, AN F/17/3980. Saban and Lemire, "Les éléphants de la ménagerie du stathouder," 285.

16. Someone else signed the pay sheets for Louzardy late in the year (AJ/15/150), but we have no direct evidence suggesting Louzardy was in Holland then.

17. That Louzardy subbed for Cassal is mentioned in Ballet to the Museum, 14 nivôse an 6, AN AJ/15/581.

18. Delaunay to Fourcroy, 1 brumaire an 9, AN AJ/15/742. See Robbins, *Elephant Slaves*, 228–29.

19. Louzardy to the Museum, 7 germinal an 9, AN AJ/15/586. PV 7 germinal an 9, AN AJ/15/101, p. 129.

20. Jauffret, *Voyage*, 76–88, 115–27 (quotations from 124–27). See also the passage quoted in section 1.8.

21. Reported by Cassal in a letter to Daubenton and an almost identical one to Delaunay, both dated an 6, from the Lazareth (quarantine station) at Toulon, AN AJ/15/844.

22. Cassal to Daubenton, 27 messidor an 5 (15 July 1797), read 14 vendémiaire an 6, AN AJ/15/581.

23. Copy of Beaussier to the Ministre des Relations Exterieurs, Tunis 17 floréal an 5 (6 May 1797), AN AJ/15/580. Read 4 Messidor an 5, PV 4 Messidor an 5, AN AJ/15/98, 93.

24. The details of Cassal's expedition come from the subfile of letters from Cassal to the Museum, titled "Transport d'animaux d'Afrique," AN AJ/15/844.

25. Cassal to Daubenton, 27 messidor an 5 (15 July 1797), read 14 vendémiaire an 6, AN AJ/15/581.

26. Cassal to Delaunay from Toulon (no day or month given) l'an 6, AN AJ/15/844.

27. Cassal to Delaunay from Toulon, 8 frimaire an 6, AN AJ/15/844.

28. Cassal to the director of the Museum from Toulon, 16 nivôse an 6, AN AJ/15/844.

29. Thouin to the Museum, 18 ventôse an 6, séance of 14 germinal an 6, AN AJ/15/581.

30. "Rapport de la compte de C. Cassal," 14 fructidor an 6, AN AJ/15/581, indicates Cassal departed Paris on 10 pluviôse an 5. It mistakenly records Cassal's return to Paris as 27 thermidor instead of 27 messidor an 6. He delivered his report to the administration 4 thermidor an 6. PV AN AJ/15/99, p. 9.

31. See Anonymous, *Fêtes de la liberté*; Mainardi, "Assuring the Empire"; and Burkhardt, "Civilizing specimens."

32. Cassal to Delaunay 29 brumaire an 6 (19 November 1797), AN AJ/15/844.

33. Cassal to "Citoyen," 22 frimaire an 6, AN AJ/15/844.

34. Thouin and Jussieu, "Rapport sur le comte du C. Cassal," 14 fructidor an 6, AN AJ/15/581.

35. Cassal to the Museum, 14 floréal an 7, séance of 24 floréal an 7, AN AJ15/582.

36. PV 14 fructidor an 4, AN AJ/5/97, p. 184.

37. Ballet to the Museum, 14 ventôse an 5 (written by Delaunay), AN AJ/15/580. Ballet renewed his request three times before the administration acted. Ballet letters of 4 germinal an 5, AN AJ/15/580; 14 nivôse and 14 thermidor an 6, AN AJ/15/581.

38. Ballet to the Museum, 14 nivôse an 6 (written by Delaunay), AN AJ/15/581.

39. Ballet to the Museum, 14 thermidor an 6 (written by Delaunay), AN AJ/15/581.

40. Delaunay annotation of Ballet to Museum, 7 frimaire an 9, AN AJ/15/585.

41. "État du traitement . . ." vendémiaire an 10, AN AJ/15/157.

42. PV 24 nivôse an 5. AN AJ/15/98, p. 45.

43. PV 14 vendémiaire an 6, AN AJ/15/98, pp. 6–7.

44. Jussieu report on the ménagerie's organization, 7 thermidor an 9, AN AJ/15/586.

1.11 LIONS, CUBS, AND THEIR KEEPER

1. Lacépède, "La lionne," in Lacépède, *Ménagerie*. Dated by Pieters vendémiaire an 10 (23 September to 22 October 1801). Also in anonymous [Miger], *Notice des animaux vivants*.

2. Lacépède, "La lionne." F. Cuvier, "Le lion de Barbarie, la lionne et ses petits," *HNM*, October 1819.

3. F. Cuvier, *Notice des animaux vivans* (1804), 9–12, and "Le lion de Barbarie." In the latter, Cuvier wrote that no animal ever displayed more affection for its keeper than Constantine did for Cassal.

4. Toscan, "Ménagerie," 131–32.

5. Denys Montfort, "Sur la lionne de la menagerie."

6. F. Cuvier, "Le lion de Barbaric."

7. Denys Montfort, "La lionne"; Anonymous, "Musée [*sic*] d'Histoire Naturelle," *Magasin encyclopédique* 6, no. 4 (1800): 385.

8. Denys Montfort, "La lionne de la ménagerie," 239.

9. Cassal's role is described by Lacépède, "La lionne."

10. The note appeared 13 November 1800 in *Le clef du cabinet du souverain* and *Journal des débats et des décrets*; 15 November 1800 in *Gazette Nationale*; and in *Magasin encyclopédique* 6, no. 4 (1800): 384–86.

11. Deleuze, "Maréchal."

12. On Huet's lions, see Hornstein, "Jean-Baptiste Huet's Lions" and Hornstein, *Myth and Menagerie*, 36–46.

13. D'Abrantès, *Mémoires*, 4:150–55.

14. D'Abrantès, *Mémoires*, 4:151–52.

15. D'Abrantès, *Mémoires*, 6:305.

16. D'Abrantès, *Mémoires*, 4:155. In revisiting her encounter with Cassal(6:305, the duchess said more about his storytelling, including that he had been bitten in the leg by a crocodile when he was in Constantinople. She went on to recount some of his tales about acquiring animals in Tunis.

17. F. Cuvier, "Le lion de barbarie," 3. The press (see references above) mentioned the lioness's picking up her cubs and carrying them around in her mouth, but without remarking that onlookers made her anxious. Cuvier remarked on the mother's anxiety in notes he made 10 messidor an 12 respecting her third successful litter. He said it calmed her to have her cells' shutters closed. With the shutters open, the public's presence bothered her. Cuvier called her "Marquette" instead of "Constantine." MNHN MS 16:40.

18. Séance of 7 thermidor an 9, AN AJ/15/586; AN AJ/15/102, 34–36.

19. Lacépède, "La lionne," p. 7; *Journal de Paris* and *La Clef du cabinet des souverains*, 7 March 1801; *Journal des débat et des décrets*, 31 July 1801, 3.

20. Lacépède, "La lionne," p. 1.

21. Lacépède, "La lionne," p. 1.

22. G. Cuvier, "Le lion," in Lacépède, *Ménagerie*. Pieters dates this essay fructidor an 10.

1.12 BUYING FROM THE ENEMY

1. French justifications for the confiscations, including a "physiocratic" one, are expertly analyzed by Lacour, *République naturaliste*, 87–103.

2. Charles-Jean-Marie Alquier, "Lettre du représentatif du peuple français aux États-Généraux," *Décade philosophique* 5 (1795): 112–14. Cited in Burkhardt, *Spirit of System*, 117.

3. Michel-Ange Lancret, "Description de l'île de Philae," in *Description de l'Égypte*, 1:9. J.-B.-J. Fourier's "Préface historique" to the same work (vi) likewise speaks of bringing to the Egyptians "all the advantages of a perfected civilization." Fourier's words are cited in Said, *Orientalism*, 85, and Dwyer, *Napoleon and the Path to Power*, 347.

4. Loisel, "Histoire de la ménagerie," 270; Laissus, "Les animaux du Jardin des plantes," 94–95, Robbins, *Elephant Slaves*, 228–29.

5. Grigson, *Menagerie*, 96-121.

6. See announcements in *The Times* (London), March 12, 1791, 1, and October 22, 1792, 1. Grigson, *Menagerie*, 104, gives 24 January 1793 as the date of Pidcock's purchase of "the principal part" of Clark's collection, "including the rhinoceros." Perhaps Pidcock toured with the rhino before he officially purchased it.

7. "Exeter-'Change, the Strand," *Times* (London), 18 October 1793, 4, col. A. Times Digital Archive, article CS67375442.

8. *Times* (London), 24 January 1794, 4.

9. Altick, *Shows of London*, 39.

10. *Times* (London), 26 March 1795, 3.

11. *Times*, 8 February 1802, 3, col. D. *Times* Digital Archive, Article CS51651144.

12. Visiting the Paris menagerie in 1802, British traveler Henry Richard Yorke observed, "Upon the whole, this collection is very insignificant, and by no means comparable, either in number, or variety, to Pidcock's Exhibition over Exeter 'Change." Yorke was not an unbiased witness judging by his comment regarding monkeys, "whose anticks correspond in every respect, with the manners of the French people." Yorke, *Letters from* France, 1:231. On the Schönbrunn collection, see Bertuch, *Bemerkungen auf einer Reise*, 2:124–42, and Dittrich, "Fürstliche Menagerien."

13. Garner, *Brief Description.*

14. Garner, *Brief Description*, 44.

15. Martin, "Tipu Sultan's Ambassadors."

16. Stronge, *Tipu's Tigers.*

17. *Times* (London), 6 October 1800, 3.

18. *Times* (London), 27 November 1799, 3.

19. Garner, *Brief Description*, 13.

20. Delaunay to professors, 24 germinal an 8 (14 April 1800), AJ/15/584.

21. PV 24 germinal an 8, AN AJ/15/100, p. 128; Minister of the Interior to the Museum, 12 messidor an 8, séance of 14 messidor an 8, AN AJ/15/584.

22. Delaunay to "Monsieur," London, 27 August 1800—8 fructidor an 8, séance of 24 fructidor an 8, AN AJ/15/584.

23. Delaunay to "Monsieur," London, 27 August 1800.

24. Delaunay to the Museum director (Fourcroy), Brussels, 1 brumaire an 9, AN AJ/15/742. See also section 1.10 above, "The Keepers."

25. *Journal de Paris*, 18 October 1800, publishing a note from Brussels dated 21 vendémiaire. The same paper later reported (14 brumaire an 9 [5 November 1800]) that the animals from Tipoo-Saib's menagerie were awaited in Paris, having left Brussels on 7 brumaire.

26. The cubs' birthday (19 brumaire an 9 [10 November 1800]) is recorded in the caption of Miger's engraving of Maréchal's picture of the lioness and the cubs. Lacépède gives the same date for the animals' arrival from London. Lacépède, "Le Tigre," 3, in Lacépède, *Ménagerie.*

27. Vignier, Description. Vignier's claim that three baboons were among the animals purchased in London was false.

28. "Variétés," *Magasin encyclopédique*, 6e année 5 (1801), 5:544; Pujoulx, Promenades, 1: 90; Anonymous, *Description de la vie et mœurs des animaux féroces* (1807).

29. Vignier, *Phénomène d'histoire naturelle.*

30. *Times* (London), 26 August 1801, 3, no. 5195, col. D. *Times Digital Archive 1785–1985*. Article CS51126554.

31. Minister of the Interior to the professors, 29 nivôse an 10, séance of 7 pluviôse an 10. AN AJ/15/587. PV 7 pluviôse an 10 AJ/15/102, p. 183.

32. Garner, *Brief Description*, 3–4.

33. Otto to the Museum administration, 26 germinal, séance of 7 floréal an 10, AN AJ/15/588.

34. Pidcock letters of 4 and 31 May 1802 to Louzardy, séance of 25 prairial an 10, AN AJ/15/588. The quotations are from the second letter.

35. PV 9, 16, and 23 messidor an 10, AN AJ/15/103, pp. 87, 89–90, and 97.

36. See Dufresne, "Circonstances de mon voyage en Angleterre," in séance of 3 brumaire an 11, AN AJ/15/589. This thirty-six-page report includes copies of the letters Dufresne wrote to Paris during the expedition. Most of the original copies can be found with the minutes of the séances and are also reported in the procès-verbaux.

37. Dufresne to the Museum, 11 thermidor an 10, séance of 21 thermidor an 10, AN AJ/15/588.

38. Dufresne, "Circonstances," AN AJ/15/589.

39. Dufresne letter of 25 thermidor an 10, PV AN AJ/15/103, pp. 125–27; and Dufresne, "Circonstances," AN AJ/15/589.

40. PV 21 thermidor an 10, AN AJ/15/103, pp. 112–13.

41. Desfontaines to Dufresne, 28 thermidor an 10, copied in Dufresne, "Circonstances," AN AJ/15/588,

42. Pidcock to the Museum, 27 August 1802, séance of 19 fructidor an 10, AN AJ/15/588. Pidcock's letter was in English. The minutes of the meeting suggest the assembly paid it little mind.

43. Dufresne to the Museum, 2 fructidor an 10 [?], "Circonstances," p. 11. AN AJ/15/589.

44. Pidcock to the Museum dated 27 August 1802. Pidcock included a placard headed "TO CHEMISTS, DRUGGISTS, &c." where he solicited information from anyone who had sold any poisons "to any person pretending to destroy rats, &c." Séance of 19 fructidor an 10, AN AJ/15/588.

45. Copy of Desfontaines to Dufresne, 28 thermidor an 10, in Dufresne, "Circonstances," AN AJ/15/589.

46. This animal was most likely a cheetah. It is unclear whether Dufresne's designation, "The hunting tiger of Tipoo Saib," was a claim about the animal's provenance or an identification of the kind of big cat Tipu used for hunting. Grigson, *Menagerie*, 179–80, says four of Tipu Sultan's hunting cheetahs were brought to London after the sack of Seringapatam, presented to King George III, and placed in the menagerie at the Tower of London (where they soon died).

47. Dufresne, "Circonstances," AN AJ/15/589.

48. Dufresne to the Museum from Amiens, 3 vendémiaire an 11 to the Museum. Séance of 5 vendémiaire an 11. PV AN AJ/15/103, pp. 162–63; Dufresne, "Circonstances," AN AJ/15/589.

49. Dufresne to the Museum from Rouen, 10 Vendémiaire an 11; séance of 12 vendémiaire an 11. PV AN AJ/15/103, pp. 169–71; Dufresne, "Circonstances," AN AJ/15/589. In the first account, Dufresne says he purchased the rattlesnakes and flying squirrels; in the second, he credits Louzardy with this.

50. Dufresne, letter of 19 vendémiaire an 11, séance of same date, PV AN AJ/15/103, p. 179.

51. PV 19 vendémiaire an 11, AN AJ/15/103, p. 178.
52. Dufresne, "Circonstances," AN AJ/15/589.

1.13 THE TIGER

1. Lacépède, "Le Tigre," in Lacépède, *Ménagerie*. Pieters dates the essay to an 12 (23 December 1803 to 21 January 1804).
2. Buffon, "Le tigre," *HNGP*, 9 (1761): 129–30.
3. Buffon, "Le tigre," 9:130–31.
4. Buffon, "Le tigre," 9:131–32.
5. Lacépède, "Le tigre," 9:2.
6. Lacépède, "Le tigre," 9:2.
7. Lacépède, "Le tigre," 9:3.
8. Lacépède, "Le tigre," 9:3–4.
9. Lacépède, "Le tigre," 9:4.
10. Lacépède, "Le tigre," 9:4.
11. Anonymous [Miger], *Notice des animaux vivants* (1802): 10. Likewise, F. Cuvier in his 1804 guide to the menagerie reported (p. 21) that the menagerie's male tiger sought the caresses of those it was familiar with.
12. Frédéric Cuvier, "Le tigre royal," *HNM*, August 1820.

1.14 THE NEW MENAGERIE TAKES SHAPE

1. "Nouvelles littéraires: Embellissemens du Jardin des plantes," *Magasin Encyclopédique* 9, no. 2 (1803): 518–20.
2. Pujoulx, *Promenades*, 1:76–77.
3. "Programme d'une Ménagerie Nationale," séance of 27 pluviôse an 10, AN AJ/15/845.
4. Pujoulx, *Promenades*, 1:77–78.
5. Fischer, *Nationalmuseum*. Fischer's work served both to describe an institution he believed other countries should emulate and to advance his own career. He dedicated his publication "to the First Consul, Bonaparte, the promoter of everything great, good, and beautiful." In 1801, before volume I appeared, he was appointed professor of natural history and librarian at the University of Mainz, in the part of Germany west of the Rhine that had been in French hands since 1797. In 1804 Fischer became professor and director of the natural history museum at Moscow University.
6. Fischer, *Nationalmuseum*, 1:294–95.
7. Fischer, *Nationalmuseum*, 1:295–96.
8. Fischer, *Nationalmuseum*, 1:295–96, 296–97.
9. Fischer, *Nationalmuseum*, 1:298.
10. PV 7 and 16 frimaire an 9, AN AJ/15/101, pp. 64, 69. The Museum referred to these workhouses as "ateliers de charité."
11. On early discussions of the plans, see PV 7 messidor, 17 messidor, 7 thermidor, and 17 fructidor an 9, AN AJ/15/102, pp. 20–21, 39, and 67. Chaptal announced his decision to terminate the project in a letter of 14 vendémiaire an 10, read 17 vendémiaire an 10 (3 October 1801). PV of the same date, AN AJ/15/102, pp. 102–3.
12. PV 27 brumaire an 10 (18 November 1801), AN AJ/15/102, p. 139.
13. PV 17 frimaire year 10 (8 December 1801), AN AJ/15/102, p. 152.

14. On the animal quarters' design, see PV 27 frimaire and 7 nivôse an 10, AN AJ/15/102, pp. 160–61, 168–69. Chaptal reported the CBC's complaints in a letter to the Museum, 8 pluviôse an 10, séance of 17 pluviôse an 10 (6 February 1802], AN AJ/15/587. See "Programme d'une Ménagerie Nationale," séance of 27 pluviôse an 10, AJ/15/845.

15. "Programme d'une Ménagerie Nationale," AJ/15/845.

16. "Programme d'une Ménagerie Nationale," AJ/15/845.

17. Molinos's design of small "circles" arranged around a larger circle predated the founding of the Legion of Honor by six months and by more than two and a half years the first investiture of members of the Legion when the medals were first awarded (15 July 1804). Loisel, "Histoire de la menagerie," 271, and *Histoire des ménageries*, 3:132, made the claim about the rotunda's design. Laissus, *Les animaux du Muséum*, 107, repeats the story, citing Loisel.

18. Museum to the Minister of Interior, 22 fructidor an 10 (9 September 1802), copied in PV 26 fructidor an 10, AN AJ/15/103, pp. 156–58.

19. Sédillez to the Tribunat, séance of 12 floréal an 11, *Gazette nationale*, 14 floréal an 11 (4 May 1803), p. 1002. Sédillez's address was read, to the professors' delight, at the 14 floréal an 11 assembly, PV AN AJ/15/104, p. 65. At the same meeting, the professors toured the menagerie grounds.

20. Museum letter to Eparvier, directeur des domains au département de la Seine, 30 fructidor an 11, AN AJ/15/743. Talboutier's house was to be converted into studios for the Museum's four painters: the Rédouté brothers, de Wailly, and Oudinot. Museum letter to the artists, 22 floréal an 12, AN AJ/15/744.

21. PV 26 vendémiaire an 12, AN AJ/15/105, pp. 21–24.

22. Report of the ceremony, 19 pluviôse an 12 (9 February 1804), filed with séance of 25 pluviôse an 12. AN AJ/15/591.

23. Geoffroy Saint-Hilaire, "Rapport sur l'emploi du 20,000 f. disponibles pour la construction d'une ménagerie," séance of 20 germinal an 13, AN AJ/15/594.

24. PV 27 germinal an 13, AN AJ/15/107, p. 13.

25. PV 3 fructidor an 13 (21 August 1805), AN AJ/15/107, p. 72.

26. PV 1 and 29 June 1808, AN AJ/15/109, pp. 28, 29, 41–42.

27. PV 20 June, 8, 15, and 22 August, 10 October 1810; and 20 February and 24 April 1811, AN AJ/15/111, pp. 34, 62–63, 67, 70, 100–101, 179, 219.

28. Thouin and Desfontaines's report to the Museum, 10 May 1809. AN AJ/15/601.

29. Thouin and Desfontaines's report to the Museum, 10 May 1809. AN AJ/15/601.

30. PV 24 and 31 May 1809, AN AJ/15/109, pp. 162, 165 (Geoffroy's program is mentioned here, but without details); PV 8, 15, and 22 November 1809, AN AJ/15/110, pp. 61, 62–63, 66.

31. Museum to the Minister of the Interior, 21 August 1809, AN AJ/15/845.

32. Conseil de Bâtiments civils, "Rapport présenté au Ministre de l'Intérieur," 7 August 1810, AN F/13/1218/A.

33. Museum to the Minister of the Interior, 9 August 1810, AN F/13/1218A.

34. Report of 13 August 1810 signed by two representatives from the Conseil des bâtiments civils, the architect Molinos, and Desfontaines, Thouin, and Georges Cuvier representing the Museum. AN F/13/1218A.

35. Molinos to the Minister of the Interior, 16 August 1810, AN F/13/1218A.

36. PV 20 May 1812, AN AJ/15/113, p. 4.

37. PV 19 May 1813, AN AJ/15/113, p. 172.

38. PV 2 March 1814, AN AJ/15/114, p. 107.

39. PV 6 and 13 July 1814, AN AJ/15/114, pp. 151–52, 154. F. Cuvier menagerie report, 6 July 1814, AJ/15/609.

40. PV 9 November 1814, AN AJ/15/115, pp. 20–21.

41. PV 31 May 1815. AN AJ/15/610. See too PV AJ/15/115, p. 100.

42. "Motifs sur lesquels sont fondés les demandes du programme ci joint relatif à la construction d'une ménagerie d'animaux féroces présentés le 22 novembre 1809 aux professeurs du Muséum d'histoire naturelle et adoptés par eux sur le rapport de M. Geoffroy Saint-Hilaire, professeur, directeur de la ménagerie." AN F/13/1225.

43. Minister of the Interior to the Museum, August 1810. AN F/13/1218A.

44. PV 30 January 1811, AN AJ/15/111, p. 165. The only further information offered was, "He would like the animals to have quarters separated or grouped in small number." The Museum asked for more details. The records say nothing further on the subject.

45. Loisel, "Histoire de la ménagerie," 271–72.

2.1 NATURALISTS' PRACTICES AND THE MENAGERIE IN THE GOLDEN AGE OF FRENCH ZOOLOGY

1. "Annales du Muséum d'Histoire Naturelle. Prospectus." *Annales de Chimie* 44 (1802): 100–112. Also in *Gazette Nationale*, 5 vendémiaire an 12 (27 September 1802).

2. Geoffroy, "Un nouveau genre de poisson du Nil"; "L'achire barbu"; "Anatomie comparée des organes électriques"; "Deux Kanguroos vivans," 178.

3. Pieters, "La ménagerie."

4. Lamarck, *Recherches sur l'organisation des corps vivans*, 63.

5. This conjunction of events is described in Burkhardt, "Lamarck, Cuvier, and Darwin on Animal Behavior," and "Lamarck, Evolution, and the Inheritance of Acquired Characters."

6. PV 27 floréal an 10, AN AJ/15/103, p. 57.

7. Burkhardt, *Spirit of System*, 43–45.

8. J.-B. Lamarck, *Réfutation de la théorie pneumatique*.

9. "Annales . . . Prospectus," 104. Cuvier said virtually the same in 1807, writing that the conduct of scientific societies with respect to theoretical systems should be "to encourage with their eulogies those who report positive facts and to retain an absolute silence over the systems that succeed one another." In Burkhardt, *Spirit of System*, 196.

10. Koerner, *Linnaeus*, describes Linnaeus's career and his commitment to bringing natural history into the service of Sweden and making the nation economically independent. She also describes his nature hikes, pp. 41–42. For Linnaeus's taxonomic work, see Larson, *Reason and* Experience; and for the vogue of Linnaean outdoor practices in Paris in the 1790s, see Goodman, "What it Meant to be Linnaean." See also Van Damme, "In the name of Linnaeus," and Hodacs, "Linnaeans outdoors."

11. White, *Selborne*, 115.

12. White, *Selborne*, 232.

13. White, *Selborne*, 144.

14. LeRoy, *Lettres philosophiques*, 4–5. Originally published 1768. LeRoy distinguished his views from those of philosophers who saw animals as automata.

15. Levaillant, *Histoire Naturelle des Oiseaux d'Afrique*, 2:222, 310.

16. Levaillant, *Histoire Naturelle des Oiseaux d'Afrique*, 2:222.

17. On the Museum's acquisition of Levaillant's collection in 1797, see PV 4 and 14 ventôse an 5, AN AJ/15/98, pp. 54–55, 58.

18. For similar tasks performed b Lamarck, see PV 1 germinal and 24 floréal an 2, AN AJ/15/98, pp. 102, 113.

19. As originally proposed by the committee (Lamarck, Lacépède, and Jussieu), the menagerie's surveillance would be assigned to a professor with an "inspector" working under him. The original report, in Jussieu's hand, is filed with the séance of 7 thermidor an 9, AN/15/586. For the final regulations, PV 7 thermidor an 9, AJ/15/102, pp. 34–36.

20. Two months after being named menagerie inspector, Lamarck was runner-up in the election for Museum director. His colleagues proceeded to elect him secretary, but he asked to be excused. Then they elected him treasurer, and he agreed to serve in that post. PV 27 fructidor an 9, AN AJ/15/102, pp. 75–76.

21. PV 17 thermidor an 9, AN AJ/15/102, pp. 43–44.

22. PV 17 fructidor an 9, AN AJ/15/586; see too AJ/15/102, pp. 69, 70.

23. PV 27 frimaire and 7 nivôse an 10, AN AJ/15/102, pp. 160–61, 168–69.

24. PV 17 brumaire an 10, AN AJ/15/102: 128–129 (for the male), and 7 germinal and 17 and 27 floréal an 10, AN AJ/15/103, pp. 33, 52, 57 (for the female).

25. PV 17 and 27 brumaire an 10, AN AJ/15/102, pp. 128–29, 137.

26. PV 7, 17, and 27 germinal an 10, AN AJ/15/103, pp. 33, 35. 41.

27. PV 5 fructidor an 10, AN AJ/15/103, p. 128.

28. Lamarck, *Corps vivans*. For his presentation of his book to his colleagues, PV 7 thermidor an 10, AN AJ/15/103, p. 106.

29. Lamarck, *Système des animaux sans vertèbres*, 15.

30. Lamarck, *Système des animaux sans vertèbres*, 14.

31. Most were ducks, but some were curlews. "Oiseaux vivants envoyès par le Cn. Baillon," séance of 14 floréal an 7, AN AJ/15/582.

32. Pluche, *Spectacle de la nature*; Burkhardt, *Spirit of System* (1995), xxiv, 172.

33. See Georges Cuvier's 1798 depiction of avian foot structures, reproduced in Burkhardt, *Spirit of System* (1995), xxix.

34. Burkhardt, *Spirit of System* (1995), xxiii-xxx.

35. Lamarck, *Corps vivans*, 208.

36. On mounting the giraffe skin see "Variétés," *Décade philosophique* 26 (1800): 437–38. For the Museum's praise of the three taxidermists, PV 24 thermidor an 8, AN AJ/15/100, p. 180. In a letter to Lacépède 28 thermidor an 8, Desmoulins described Lassaigne and Delalande's work on the giraffe and recommended they be awarded bonuses. Séance of 4 fructidor an 8, AN AJ/15/584.

37. Cuvier, "Mémoires pour server," 1:185. Cited in Outram, *Georges Cuvier*, 61; Appel, *Cuvier-Geoffroy Debate*, 72; Lacour, *République naturaliste*, 258.

38. *Journal des débats*, 14 pluviôse an 11 (3 February 1803), 3.

39. Cuvier, "La collection d'anatomie comparée."

40. "Rapport à l'assemblée administrative des professeurs du Muséum national d'histoire naturelle" was written in Cuvier's hand; signed by Fourcroy, Jussieu, Cuvier, Faujas, and Lacépède; and presented at the séance of 17 messidor an 11 (6 July 1803), AN AJ/15/590. Hereafter "Cuvier 1803 gallery report," AN AJ/15/590. The report is paginated, one number to each double-sided page. I add (b) to indicate material on the back of a page.

41. Prarond, "Baillon." Taquet, *Cuvier. Anatomie d'un naturaliste*, cites this passage (p. 118) but erroneously attributes it to Baillon, not Cuvier.

42. Cuvier, "La collection d'anatomie comparée," 411.

43. Outram, *Georges Cuvier*, 176.

44. Cuvier 1803 gallery report, AN AJ/15/590. Cuvier's fellow committee members were Fourcroy, Jussieu, Faujas de Saint-Fond, and Lacépède, but the text makes it clear that Cuvier wrote the report. It is in his handwriting and occasionally lapses into the first person.

45. Cuvier 1803 gallery report, 8–9. AN AJ/15/590.

46. Cuvier 1803 gallery report, 18(b). AN AJ/15/590.

47. Identified as *aide-anatomiste* instead of *aide-naturaliste*, Rousseau as of vendémiaire an 8 received from the Museum only half of the 3,000-franc annual salary that *aide-naturalistes* received (AJ/15/155).

48. On Cuvier's hiring hometown connections see Outram, *Georges Cuvier*, 57–58, 172–73. None of Cuvier's assistants from Montbéliard was on the Museum payroll, nor was Cuvier himself until he succeeded Mertrud as chair of comparative anatomy and began receiving a professor's salary of 5,000 francs. Cuvier's precise financial arrangements with Mertrud are unknown.

49. Cuvier 1803 gallery report, 3(b)-4. AN AJ/15/590.

50. Cuvier 1803 gallery report, 5(b)-6. AN AJ/15/590.

51. Cuvier 1803 gallery report, 10(b)-11. AN AJ/15/590.

52. Cuvier 1803 gallery report, 9-10. AN AJ/15/590. Had the galleries not expanded, Cuvier claimed, over twelve hundred students would have been deprived of instruction over the previous five or six years, and none of the works of natural history and natural history illustration published during this time in Paris (and several published abroad) would have achieved the level of perfection they did.

53. Cuvier 1803 gallery report, 11(b)-13. AN AJ/15/590 Approximately thirty quadrupeds were from the Stadtholder's collection. Others were mounted from skins from the storerooms, skins Geoffroy purchased from furriers, or animals that died at the menagerie. Aside from the giraffe, the large animals on display included an African and an Asian rhino, both recently installed. On previous plans to place these and other quadrupeds see PV 7 thermidor an 9, AJ/15/102, p. 36.

54. Cuvier 1803 gallery report, 13-13(b). AN AJ/15/590.

55. PV 17 messidor an 11 (6 July 1803), AN AJ/15/104, p. 97.

56. Letters from Humboldt to different correspondents appeared in three consecutive volumes of the Museum's *Annales*: 2 (1803): 170–80, 322–37; 3 (1804): 228–32, 396–404; and 4 (1804): 475–78.

57. Humboldt, *Tableaux de la nature*. Cuvier's unpublished review is fifteen pages (twenty-nine sides). Institut de France, Fonds Cuvier, MS 3159. Dorinda Outram discusses to excellent effect Cuvier's introduction to the review. Outram, *Georges Cuvier*, 62–63, and "New Spaces in Natural History."

58. Humboldt, *Tableaux de la nature*, 2:58–59.

59. Humboldt, "Observations sur l'anguille électrique," 56.

60. Lyell, *Life, Letters and* Journals, 1:249–50.

61. Duvernoy, *Notice historique*, 87.

62. Cuvier, diverse essays in Lacépède, *Ménagerie* (1801–1805).

63. Cuvier, "Le Blanc-nez," "L'Autruche," "Le Casoar," and "Le Couagga," in Lacépède, *Ménagerie*.

64. Cuvier, "Le Chameau," "L'Éléphant des Indes," "La Genette," and "Le Mandrill," in Lacépède, *Ménagerie*.

65. Cuvier, "La Civette," in Lacépède, *Ménagerie.*

66. Cuvier, "L'Eléphant des Indes, femelle," in Lacépède, *Ménagerie.*

67. É. Geoffroy Saint-Hilaire, *Lettres écrites d'Égypte,* 137–38.

68. É. Geoffroy Saint-Hilaire, *Lettres d'Égypte,* 129–30.

69. É. Geoffroy Saint-Hilaire, *Lettres d'Égypte,* 271 (E.-T. Hamy's observation).

70. É. Geoffroy Saint-Hilaire, *Lettres d'Égypte,* 155–56.

71. É. Geoffroy Saint-Hilaire, *Lettres d'Égypte,* 183.

72. É. Geoffroy Saint-Hilaire, *Lettres d'Égypte,* 180, 183.

73. Humboldt felt the same frustration, writing in November 1802 from Lima to Paris noting three years had passed since he had sent rare seeds to the Jardin des plantes but had heard nothing about them. "Copie d'une lettre de M. de Humboldt," *Annales du Muséum* 2 (1803): 172.

74. É. Geoffroy Saint-Hilaire, *Lettres d'Égypte,* 157–58.

75. É. Geoffroy Saint-Hilaire, *Lettres d'Égypte,* 184.

76. É. Geoffroy Saint-Hilaire, *Lettres d'Égypte,* 202.

77. É. Geoffroy Saint-Hilaire, *Lettres d'Égypte,* 203–4.

78. É. Geoffroy Saint-Hilaire, *Lettres d'Égypte,* 187, 189, 191. The PV refer to several Geoffroy letters from Egypt that no longer exist. PV 7 *nivôse* and 7 pluviôse an 9, AN AJ/15/101: 86, 98, and 27 thermidor an 9, AN AJ/15/102: 55.

79. In this paper Geoffroy hinted at the idea of the unity of composition, the concept that became the centerpiece of "philosophical anatomy," his greatest claim to scientific fame. On his work in Egypt see Appel, *Cuvier-Geoffroy Debate,* 72–81.

80. Isidore Geoffroy Saint-Hilaire, *Vie, travaux et doctrine scientifique,* 101–4; É. Geoffroy Saint-Hilaire, "L'anatomie comparée des organes électriques" and *Études progressives d'un naturaliste,* 148–51.

81. É. Geoffroy Saint-Hilaire, *Lettres d'Égypte,* 205.

82. É. Geoffroy Saint-Hilaire, *Études progressives d'un naturaliste,* 149–150n. I. Geoffroy Saint-Hilaire, *Vie, travaux et doctrine scientifique,* 104–8.

83. É. Geoffroy Saint-Hilaire, *Lettres d'Égypte,* 206–09.

84. É. Geoffroy Saint-Hilaire, *Lettres d'Égypte,* 209–12.

85. É. Geoffroy Saint-Hilaire, *Lettres d'Égypte,* 214–21 (quotation on 221).

86. PV 17 pluviôse an 10, AN AJ/15/103, pp. 3–4.

87. É. Geoffroy Saint-Hilaire, *Lettres d'Égypte,* 179.

88. PV 27 pluviôse an 10, AN AJ15/103, p. 7.

89. PV, 12 fructidor an 10, AN AJ/15/103, p. 137.

90. "Rapport des professeurs du Muséum sur les collections d'histoire naturelle rapportées d'Égypte," 234. The original manuscript, signed by Cuvier, Lamarck, and Lacépède (in that order), is filed with the séance of 19 fructidor an 10, AN AJ/15/588. The text is reproduced in the PV of the same date, AN AJ/15/103, pp. 147–54.

91. "Rapport des professeurs," 235–36.

92. "Rapport des professeurs," 236.

2.2 THE SACRED IBIS

1. Wasef et al., "Mitogenomic diversity."

2. E. Geoffroy Saint-Hilaire, *Lettres d'Égypte,* 228.

3. G. Cuvier, "Mémoire sur les espèces d'élephans vivantes et fossiles. Lu à la Séance publique de l'institut national, le 15 germinal an 4," MNHN MS 628: 14–15.

4. On Cuvier's understanding of species extinction as species *destruction*, see Rudwick, *Georges Cuvier, Fossil Bones*, 261–62.
5. Burkhardt, *Spirit of System*, 128–30.
6. Lamarck, "Mémoire sur les fossiles."
7. Lamarck, "Mémoire sur les fossiles," 303.
8. Lamarck, "Discours d'ouverture," 85–105.
9. Cuvier, "Mémoire sur l'Ibis."
10. Cuvier, "Mémoire sur l'ibis," 121.
11. Cuvier, "Mémoire sur l'ibis," 121–22.
12. Cuvier, "Mémoire sur l'ibis," 123.
13. Cuvier, *Recherches sur les ossemens fossiles* (1812), 79.

2.3 GUIDING THE PUBLIC

1. Lacour, *République naturaliste*," 297–331. On the site's healthiness, see Synowiecki, "Paris côté jardins," 130.
2. Thouin, Lamarck, and Vanspaendonck, "Rapport sur les marchands qui vendent aux portes et dans l'enceinte du Muséum," séance of 14 germinal an 8 (4 April 1800), AN AJ/15/584. The porters and café keeper were to deny alcohol to drunks, not dispense drinks beyond reasonable hours, restrain drinkers from singing, prevent the playing of certain (unspecified) musical instruments, and keep arguments from escalating.
3. "Rapport sur les marchands." According to John Pinkerton, the elderly ladies cry of "*plaisir*" became "a real and repeated pain in this charming promenade." Pinkerton, *Recollections of Paris*, 1: 87. On Pinkerton and the merchants, see Lacour, *République naturaliste*, 320. Pinkerton (*Recollections*, 1:89) jokingly offered the double entendre, "the male elephants perhaps fell victim to the intemperate use of *plaisir*, like many of their sex," simultaneously remarking that "almost very" visitor fed the elephants, and "elephants cannot have been much accustomed to eat *plaisir*."
4. Delaunay to the Museum, 4 germinal an 8 (25 March 1800), séance of 4 germinal an 8, AN AJ/15/584.
5. Delaunay to the Museum, 4 germinal an 8.
6. Lucien Bonaparte to the Museum, 12 messidor an 8, read at the séance of 14 messidor an 8, filed with the séance of 24 prairial an 8, AN AJ/15/584.
7. Jauffret, *Voyage*. Jauffret was active in the fledgling anthropological organization, La Société des Observateurs de l'Homme. See Reboul, *Louis-François Jauffret*; Lacour, *République naturaliste*, 316–17; and Chappey, *La Société des Observateurs de l'Homme*. Robbins, *Elephant Slaves*, 230, discusses Jauffret's *Voyage*.
8. Jauffret, *Voyage*, 86.
9. Jauffret, *Voyage*, 135.
10. See Toscan, *L'Ami*, vin, for his complaint about Jauffret's plagiarism. For the purloined pages, Jauffret, *Voyage*, 90–111.
11. Toscan, *L'Ami*, 263–64.
12. Toscan, *L'Ami*, 306–7.
13. Vignier recounted his misfortunes in an undated letter to the Museum, read 6 nivôse an 12 (AN AJ/15/591), saying his annual income of four thousand livres before the Revolution (not counting his earnings as a lawyer) had shrunk to six hundred livres.
14. Minister of the Interior to the Museum, 15 floréal an 7, séance of 24 floréal an

7, AN AJ/15/582. The PV do not mention action any action taken by the Museum in response to the minister's request. AN AJ/15/99, p. 93. Vignier mentioned his three-hundred-page manuscript in his *Vertus morales et description abrégée*, 11. In *Phénomène d'histoire naturelle*, p. 34, he said the work was complete but would not be published until the menagerie was finished.

15. Vignier, *Vertus morales des deux éléphans.*

16. Vignier, *Épitre aux éléphans.*

17. Vignier, *Description abrégé des animaux quadrupeds.*

18. Vignier, *Phénomène d'histoire naturelle*, 15.

19. Vignier, *Phénomène d'histoire naturelle*, 29.

20. Vignier to the Museum, 1 fructidor an 9, séance of 7 fructidor an 9, AN AJ/15/586.

21. Séance of 7 fructidor an 9, AN AJ/15/102: 63. Vignier renewed his complaints about Cassal the following week. Vignier to the Museum, undated, séance of 17 fructidor an 9, AN AJ/15/586.

22. Vignier to the Museum, undated, séance of 6 nivôse an 12. AN AJ/15/591, asked for a "supernumerary" appointment in the Museum's library as a "justice" the Museum owed him for having deprived him of income from his booklet. The Museum turned down the request. PV, séance of 6 nivôse an 12, AN AJ/15/105, p. 76.

23. Miger to "Monsieur," 5 floréal an 9, séance of 27 floréal an 9, AN AJ/15/586. See PV 27 floréal an 9, AN AJ/15/101, p. 162, and 7 messidor an 9, AN AJ/15/102, p. 11.

24. Anonymous [Miger], *Notice des animaux vivants* (1801).

25. At least this is what Vignier believed. See Vignier to "Monsieur," 1 fructidor an 9, séance of 7 fructidor an 9, AN AJ/15/586, and PV of same date AN AJ/15/102, p. 63.

26. [Miger], *Notice des animaux vivants* (1801), 10.

27. [Miger], *Notice des animaux vivants* (1802).

28. In 1804, shortly after Cassal was fired, Vignier wrote that he had produced a booklet about the menagerie's new elephant and planned to print three hundred copies if the Museum would allow the elephant's keeper to sell it. This would not harm Miger, he said, whose guide was being sold "at the end of the Jardin" by Bernard [Louzardy]. Vignier to the Museum, undated, séance of 11 pluviôse an 12; AN AJ/15/591; PV of same date AN AJ/15/105, p. 104. The proposed work never appeared. The PV do not mention a response to Vignier's request.

29. Fischer, *Nationalmuseum.*

30. Anonymous, *Le Guide au Jardin des plantes*, 2.

31. Anonymous, *Le Guide au Jardin des plantes*, 6–7.

32. Anonymous, *Le Guide au Jardin des plantes*, 7–12.

33. Anonymous, *Le Guide au Jardin des plantes*, 12–19.

34. Anonymous, *Le Guide au Jardin des plantes*, 25. See chapter 2.6 on the publisher's reusing this passage with the names of new keepers.

35. Anonymous, *Le Guide au Jardin des plantes*, 20–35; quotation on 25.

36. Anonymous, *Le Guide au Jardin des plantes*, 2, 36.

37. For a list of Pujoulx's writings, see Mahul, "Pujoulx (Jean Baptiste)."

38. Pujoulx, *Modernes enrichis*, iii.

39. Pujoulx, *Paris*, 12–13.

40. Pujoulx, *Paris*, 73–84. Quotation from p. 80. See also 81–82.

41. Pujoulx, *Le livre du second âge*, 1.

42. Pujoulx, *Promenades*, 1.1–5.

43. Pujoulx, *Promenades*, 1:xiii-xiv. Lacour describes Pujoulx's guide as sufficiently detailed to make it possible "to reconstitute precisely the spatial ordering of the collections." Lacour, *République naturaliste*, 245.

44. Pujoulx, *Promenades*, 1: 168–70.

45. Ibid., 1: 79.

46. Buffon, *HNO*, 1 (1770): 144; Pujoulx, *Promenades*, 1:81.

47. Pujoulx, *Promenades*, 1:83–85.

48. Pujoulx, *Promenades*, 1:86.

49. Pujoulx, *Promenades*, 1:95–96.

50. Pujoulx, *Promenades*, 1:97.

51. Pujoulx, *Promenades*, 1:158–63.

52. Pujoulx, *Promenades*, 1:114–15.

53. On this elephant, see section 2.6, "Kindness in Context." The purchase was ultimately made, but the animal did not live long, and in the meantime the Museum's zoologists conducted no studies of its intellectual faculties.

54. Pujoulx, *Promenades*, 1:166–67.

55. Pujoulx, *Promenades*, 1:173–74.

56. Not until the summer of 1804 were the herbivore parks fenced off from the rest of the grounds. Not until 1815 were the elephant quarters moved to the rotunda from their original location in the Regie under the comparative anatomy gallery.

2.4 FIRING FÉLIX CASSAL

1. Comte de Bontemps to Minister Chaptal, 1 floréal an 9, séance of 17 prairial an 9, AN AJ/15/586.

2. Boutin to the Museum, 12 messidor an 10, séance of 16 messidor an 10. AN AJ/15/588.

3. Boutin, 12 messidor an 10.

4. PV 16 messidor an 10, AN AJ/15/103, pp. 90–91.

5. Boutin to Minister Chaptal, 27 messidor an 10; séance of 7 thermidor an 10, AN AJ/1515/588.

6. Minister of the Interior to the Director of the Museum, 2 thermidor an 10, séance of 7 thermidor an 10, AN AJ/15/588.

7. Félix Cassal undated letter to the professor/administrators, séance of 7 thermidor an 10, AN AJ/15/588.

8. PV 7 thermidor an 10, AN AJ/15/103, p. 105.

9. Museum to Minister Chaptal, 10 thermidor an 10, AN AJ/15/743.

10. The widow Guichenot to the Museum, dated 28 pluviôse an 11 [*sic*], séance of 27 pluviôse an 11. AN AJ/1515/589. The handwriting is Delaunay's, with Mme Guichenot's signature.

11. Geoffroy Saint-Hilaire, "Rapport du Professeur dirigeant la Ménagerie sur une dénonciation du pourvoyeur Guichenot contre le Cit[n.] Felix Cassal." Séance of 8 frimaire an 12, AN AJ/1515/591.

12. Geoffroy, "Rapport."

13. PV 8 frimaire an 12, AN AJ/15/105, p. 51.

14. Geoffroy report to "Mes chers collègues," 29 frimaire an 12, calling for Cassal's dismissal, séance of same date, AN AJ/15/591. This report was also entered into the meeting's minutes, PV AN AJ/15/105, pp. 67–72.

15. Geoffroy report 29 frimaire an 12. AN AJ/1515/591.
16. Geoffroy report 29 frimaire an 12. AN AJ/1515/591.
17. Geoffroy report 29 frimaire an 12. AN AJ/1515/591.
18. Geoffroy report 29 frimaire an 12. AN AJ/1515/591.
19. PV 6 nivôse an 12 (28 December 1803), AN AJ/15/591; PV AJ/15/105, p. 80.
20. PV 13 nivôse an 12, AN AJ/15/591; AJ/15/105, p. 85.
21. PV 13 nivôse an 12, AN AJ/15/105, p. 84. Letter in Geoffroy's hand dated 22 nivôse an 12, AN AJ/15/744.
22. Thouin to Cassal, 20 nivôse an 12, séance of the same date. AN AJ/1515/591.
23. PV 13 January 1808, AN AJ/15/108, p. 144.
24. Cassal to the professors of the Museum, 29 June 1813, séance of 30 June 1813. AN AJ/1515/607; PV AN AJ/15/114, p. 9.

2.5 THE NEW SUPERINTENDENT

1. The first administrative meeting Geoffroy attended after his return to Paris was 17 pluviôse an 10 (4 February 1802); PV AN AJ/15/103, pp. 3–4.
2. PV 5 vendémiaire an 11, AN AJ/15/103, p. 165.
3. Frédéric Cuvier to Georges Cuvier, 26 brumaire an 11 (17 November 1802). Institut de France, Fonds Cuvier, MS 3342–49. Frédéric writes that Geoffroy mentioned the possibility of his taking the job provisionally, as a leg up for when the post was formally authorized.
4. On the tensions between Geoffroy and Georges Cuvier, see Outram, *Georges Cuvier*, 60–64.
5. On Frédéric Cuvier see Flourens, "Éloge historique de F. Cuvier." Flourens's estimate that Frédéric arrived in Paris "toward the end of 1797" is evidently wrong, because when Geoffroy left for Egypt in April 1798 he had not yet met Frédéric.
6. A previous son named Georges died at age four. The one now known as Georges was christened Jean-Leopold-Nicholas-Frédéric, but his parents called him Georges after their dead son. When the Cuviers' third and last son was born in 1773, they baptized him Georges-Frédéric but called him "Frédéric." Outram, *Georges Cuvier*, 17, notes the parents' focus on Georges.
7. Outram, *Georges Cuvier*, 17.
8. Frédéric Cuvier wrote one paper with Biot in 1801 in *Annales de chimie et de physique* and is mentioned in this context in "Nouvelles experiences galvanique," *Magasin encyclopédique*, 7, no. 3 (1801): 398.
9. Frédéric Cuvier to Georges Cuvier, 3 brumaire an 10. Institut de France, Fonds Cuvier, MS 3342–1. Full text in Taquet, *Cuvier*, 493–94.
10. Frédéric Cuvier to Georges Cuvier, 3 brumaire an 10.
11. Frédéric Cuvier to Georges Cuvier, 1 prairial an 10. Institut de France, Fonds Cuvier, MS 3342–3. Taquet, *Cuvier*, 540–41.
12. Frédéric Cuvier to Georges Cuvier, 1 prairial an 10.
13. Frédéric Cuvier to Georges Cuvier, 26 prairial an 10. Institut de France, Fonds Cuvier, MS 3342–5. Taquet, *Cuvier*, 543-44.
14. Frédéric Cuvier to Georges Cuvier, 17 brumaire an 11. Institut de France, Fonds Cuvier, MS 3342–8. Taquet, *Cuvier*, 572–73.
15. Frederic Cuvier to Georges Cuvier, 26 brumaire an 11. Institut de France, Fonds Cuvier, MS 3342–9. Taquet, *Cuvier*, 575–77.

16. Frédéric Cuvier to Georges Cuvier, 14 frimaire an 11 (5 December 1802). Institut de France, Fonds Cuvier, MS 3342–12.

17. The marriage documents give Frédéric's age as twenty-nine years, six months, and two days and Christine Charlotte Macler's age as twenty-two years, eight months, and eight days. Institut de France, Fonds Cuvier, MS 3341–6.

18. Geoffroy's untitled, undated report, addressed "Citoyens collegues," was presented at the séance of 4 ventôse an 11 (23 February 1803), MNHN MS 2717 (2). Geoffroy's colleagues discussed his proposal and invited him to provide more details. PV AN AJ/15/104, p. 38.

19. Frédéric Cuvier to Georges Cuvier, undated. Institut de France, Fonds Cuvier, MS 3342–20, apparently written on or about 9 germinal an 11, the day the animals' arrival was announced at the professors' weekly assembly (PV 9 germinal an 11, AN AJ/15/104, 49–50). Letters from Geoffroy to Georges Cuvier in this period do not mention conversations with Frédéric about the potential job. Institut de France. Fonds Cuvier. 3225.9–13. For Frédéric's letter in full, see Taquet, *Cuvier*, 620–21.

20. Frédéric Cuvier to Georges Cuvier, undated. Institut de France, Fonds Cuvier, MS 3342–20.

21. Frédéric Cuvier to Georges Cuvier, undated. Institut de France, Fonds Cuvier, MS 3342–20.

22. Geoffroy's proposed regulations for the post of garde de la menagerie. PV 29 frimaire an 12, AN AJ/15/105, pp. 63–66. The proposal came as no surprise. The Museum had expected Geoffroy to follow through on his plan of having a "chef" of the menagerie. Beginning in vendémiaire an 12, the position was listed in the Museum's payroll registers but without a name.

23. PV 29 frimaire an 12, AN AJ/15/105, pp. 63–66.

24. Geoffroy to the professorial assembly, undated, but written on or about 13 February 1828, the date the assembly discussed the matter. MNHN, in file MS 2717(2). Re: article 16, Geoffroy recalled: "Afin d'amener plus surement M. Frédéric a l'emploi qu'il désiroit, je lui communiquai le projet de reglement; cette clause fut inseré de son aveu.—pour éloigner des competiteurs." See PV 13 February 1828 AN AJ/15/127, p. 152 (also p. 150 for the 6 February 1828 meeting).

25. Geoffroy Saint-Hilaire to "Citoyens collegues," untitled report presented at the séance of 29 frimaire an 12, AN AJ/15/591.

26. These memoirs appear not to have survived.

27. PV 6 nivôse an 12, AN AJ/15/105, p. 80. That Cuvier began work the next day is indicated in the note,"Rappel du traîtement de Garde de la Ménagerie," in "État du traitement des professeurs et employés . . . thermidor an douze," AN AJ/15/159.

28. Among Cuvier's first reports was "État de la ménagerie depuis le 27 nivôse jusqu'au 4 pluviôse an 12," séance of 4 pluviôse an 12, AN AJ/15/591.

29. Five years later in the 1809 edition of his *Notice des animaux vivans de la Ménagerie*, xiii, Cuvier asked with respect to managing the animals, "Should one use force or gentleness?"

30. Anonymous [F. Cuvier], *Notice des animaux vivans* (1804), ix-x. Cuvier's authorship of this work is certain. He presented it to the professorial assembly on 20 thermidor an 12 (8 August 1804). PV AN AJ/15/105, p. 66. Judging from the guide's lack of any mention of the seal that arrived at the end of June, Cuvier finished writing it before then.

2.6 KINDNESS IN CONTEXT

1. Cuvier, "Le lion de barbarie," *HNM*, October 1819, p. 2. Burkhardt, "La voix du gardien," Pouillard, *Histoire des zoos par les animaux*, 116.

2. Anonymous (F. Cuvier), *Notice des animaux vivans* (1809), xiii.

3. Anonymous (F. Cuvier), *Notice des animaux vivans* (1804), 19.

4. Anonymous (F. Cuvier), *Notice* (1804), 17.

5. G. Cuvier, "L'Hyène," pp. 3–4, in Lacépède, *Ménagerie*. Cassal reported to Cuvier, however, that he had previously had a very tame hyena for a pet.

6. Anonymous (F. Cuvier), *Notice* (1804), 27.

7. Anonymous (F. Cuvier), *Notice* (1804), 27.

8. Anonymous (F. Cuvier), *Notice* (1804), 10, 14.

9. [F. Cuvier], "Registre des animaux de la Menagerie depuis le 7 Nivose an. XII," (Identified by the Museum as "Observations particulières sur les animaux de la menagerie, an XII [1804]-1822"). MNHN, Ménagerie, MS 16: 41. The observation is dated nivôse an 12.

10. F. Cuvier, "Mouvement de la menagerie du 9 pluviôse an 12," séance of 11 pluviôse an 12, AN AJ/15/591.

11. Anonymous (F. Cuvier, *Notice* (1804), 11.

12. PV 25 fructidor an 12, AJ/15/106, p. 81.

13. Dufresne zoology laboratory report, 18–25 fructidor an 12, AN AJ/15/592. See also PV 4 vendémiaire an 13, AN AJ/15/106, p. 85.

14. F. Cuvier, "Le lion de barbarie," *HNM*, 1.

15. Anonymous (F. Cuvier), *Notice* (1804), 12.

16. Cuvier, "Observations particulières sur les animaux de la menagerie, an XII [1804]-1822." MNHN Ménagerie MS 16. Earlier a writer had speculated that the elephant Hans's demise was caused by an unfulfilled sex drive.

17. F. Cuvier, "Le lion de barbarie," *HNM*, 3.

18. Anon., *Le Guide au Jardin des plantes* (1803), 25.

19. The 1812 version of the guidebook named "Bernard Lazardi" [*sic*] as keeper; the 1819 and 1823 versions named Rousseau and Richard.

20. Buffon, "Le loup," *HNGP*, vol. 7. Smith, *Monsters of the Gévaudan*.

21. Delisle de Moncel, *Méthodes*.

22. L. B., "Destruction des loups," *Décade philosophique* 17 (1798): 333–37. In establishing bounties, the French followed the example of the British. Centuries earlier, the British had put increasingly high prices on wolf heads until there were none left on the whole island. Because the French were not isolated by water, they could not hope to eliminate wolves entirely but believed a constant campaign could come close.

23. Delaunay's untitled menagerie record book, AN AJ/15/844. Jauffret, *Voyage*, 80–81.

24. Delaunay's menagerie record book indicates the birth took place on 4 June 1799, after the male covered the female on 31 March and 1 April. AN AJ/15/844,

25. Toscan, "Suite des faits nouveaux."

26. Repeating the story in 1800, Toscan mentioned only a single pair. Toscan, *L'Ami*, 289.

27. Anonymous (Miger), *Notice des animaux vivants* (1801), 26.

28. The Museum instructed Geoffroy Saint-Hilaire to identify a specimen for the

purpose, "provided the animal is a duplicate in the collection, and by itself of no great interest." Séance of 4 prairial an 10, AN AJ/15/103: 62. Eugène Delacroix's visits to the menagerie (and zoology laboratory) a generation later are well documented. See O'Brien, *Exiled*, 116–22.

29. Anonymous, *Le Guide* (1803), 32. Pujoulx, *Promenades*, 1:80–81.

30. Anonymous (F. Cuvier), *Notice des animaux vivans* (1804), 29–30.

31. Anonymous F. Cuvier), *Notice des animaux vivans* (1804), 30–31.

32. F. Cuvier, "Le Loup," 3, *HNM*, January 1821. This she-wolf was mated with a mastiff belonging to Cassal, producing a litter of eight pups, at least three of which survived. The hybrid pups had the physical appearance of their mastiff father but retained a wolf-like "defiance and timidity" not found in young dogs. Two pups were given to Cassal, as owner of the father, and one to Josephine Bonaparte. F. Cuvier, "Le loup femelle, 2e," MNHN Ménagerie, MS 16, 54. See also Cuvier's menagerie reports of 5 and 6 pluviôse an 12, AN AJ/15/591.

33. F. Cuvier, "Le loup," 2–3, *HNM*.

34. F. Cuvier, "Le loup," 1, *HNM*.

35. F. Cuvier menagerie report 27 nivôse an 12, AN AJ/15/591.

36. F. Cuvier, "L'éléphant mâle," in "Observations," MNHN Ménagerie MS 16:2. See below.

37. F. Cuvier menagerie report 7 pluviôse an 12, AN AJ/15/591.

38. Geoffroy to the Museum "mardi," 10 prairial an 11, séance of 12 prairial an 11, AN AJ/15/590; AN AJ/15/104: 77.

39. *Journal des débats*, 28 June 1803, 2. That Joubert continued to lead the elephant through its tricks once it was in the menagerie is indicated in F. Cuvier, "Essai sur la domesticité," 434.

40. *Journal des Arts*, 29 June 1803, 40. *Journal des débats*, 4 September 1803, 3, used the same expression about the female's "widowhood."

41. *La décade philosophique* 38 (1803): 573; Cuvier, *Notice* (1804), 39.

42. Geoffroy to "Citoyens collegues" (on staffing assignments), 13 nivôse an 12, AN AJ/15/591.

43. G. Cuvier, *Ossemens fossiles*, 2:79, describes the device and likens the procedure to that used by dentists straightening children's teeth.

44. F. Cuvier menagerie reports of 27 nivôse and 4 pluviôse an 12, AN AJ/15/591, and MNHN Ménagerie MS 16:2. Pouilllard uses the case of the elephant's displaced tusk to underscore the fact of the animal's incarceration. *Histoire des zoos*, 85–86, and "Visions," 22.

45. F. Cuvier menagerie report 4 pluviôse an 12 (25 January 1804), AN AJ/15/591.

46. *Journal de Paris*, 2 May 1804; *Journal des débats*, 3 May 1804.

47. Cuvier, "Essai sur la domesticité," 434.

48. Cuvier, "L'éléphant mâle," MNHN Ménagerie MS 16:2.

49. Anonymous (F. Cuvier), *Notice des animaux vivants* (1804), 38.

50. PV 8 messidor an 12, AJ/15/106: 24.

51. F. Cuvier menagerie report, 15–22 messidor an 12, séance of 22 messidor an 12, AN AJ/15/592.

52. F. Cuvier, "L'éléphant mâle," MNHN Ménagerie MS 16:2. When Cuvier (or someone else) went back to these notes, he misdated them as 4–8 messidor (instead of 26 messidor—1 thermidor [15–20 July 1804]). Serna, *Comme des bêtes*: 109, quotes Cuvier's notes (but misidentifies the elephant as the male from 1798 and repeats the

notebook's incorrect dates). Candegabe dates the death correctly as the night of July 18–19.

53. *Le Courrier des spectacles*, 21 and 24 July 1804; *Le Journal de Paris*, 22 July 1804. Again one hears of the pernicious consequences of an unfulfilled sex drive. Other press accounts reported that the female was taken ill and was not eating and that Hans's stuffed reincarnation was on display in the natural history gallery.

54. F. Cuvier menagerie report of 5–13 thermidor, séance of 16 thermidor an 12, AN AJ/15/591; PV AN AJ/15/106, p. 53.

55. Georges Cuvier blamed the elephant's death on an inflammation of its chest. PV 6 thermidor an 12, AN AJ/15/106, 51.

56. See also Pouillard "Visions," 22.

57. On early glimmers of ideas about animal rights in the French Revolution, see Serna, *Comme des bêtes*, and *L'animal en République*.

58. The poster is described by Théophile Gautier, "Variétés. Combat de l'ours et du taureau," *La Presse*, 25 June 1838, 3.

59. Blanvillain, *Le Pariséum*, 327–28. That this entertainment was frequented "only by the lowest class of people" was also said by Saint-Victor, *Tableau* historique, 2:427. For earlier critiques, see Serna, *Comme des bêtes*, 66–71.

60. In the fall of 1830 Cuvier arranged for two wolves andtwo bears to go to the combat master to make room for the lions coming from Algeria. F. Cuvier menagerie report, 12 October 1830, AN AJ/A5/640.

61. Geoffroy, "Motifs sur lesquels sont fondés les demandes du programme ci joint relatif à la construction d'une ménagerie d'animaux féroces présentés le 22 novembre 1809," AN F/13/1225.

62. Geoffroy, "Motifs." Pouillard, *Histoire des zoos*, 94.

2.7 THE EMPRESS, THE KANGAROOS, AND THE BLACK SWANS

1. For the official account, see Péron and Freycinet, *Voyage*. See also Bonnemains et al., *Baudin*. Napoleon's remark is cited by Horner, *French Reconnaissance*, 12. On handling the zoological specimens see Burkhardt, "Unpacking Baudin." For a contrast of Baudin's Australia expedition with his earlier voyage to the Antilles, see Fornasiero and West-Sooby, "Voyages." For a treatment of the Baudin voyage in relation to the early history of environmental thinking, see Benson, *Surroundings*.

2. The deaths of Riedlé and Maugé were reported in the letter from Baudin to the Museum dated 20 brumaire an 11, *Annales du Muséum* 2 (1803): 415–22.

3. Jussieu, "Notice sur l'expédition."

4. Antoine Guichenot to the Museum, 3 germinal an 12. See too Joseph Charles Bailly (the expedition's mineralogist) to the Museum 4 germinal an 12, séance of 10 germinal an 12. AN AJ/15/592. PV 10 germinal an 12, AJ/15/105: 145–46.

5. Geoffroy to the Museum, 18 germinal an 12 (séance of 28 germinal an 12) notes that the thefts from the *Naturaliste* had been "more considerable than we had first thought." AN AJ/15/592.

6. PV 10 germinal an 12, AN AJ/15/105, pp. 146–47.

7. On the symbolism of this moment and Peron's concerns, see Burkhardt, "Unpacking Baudin."

8. Burkhardt, "Unpacking Baudin."

9. Péron's two lists of live animals aboard ship, dated 4 germinal an 12, were presented at the séance of 21 germinal an 12, AN AJ/15/592.

10. Geoffroy Saint-Hilaire to the Museum, 18 germinal an 12, séance of 28 germinal an 12, AN AJ/15/592.

11. Chaptal to the Museum, 12 germinal an 12, séance of 14 germinal an 12, AN AJ/15/592.

12. Péron to Fourcroy, 15 germinal an 12, séance of 21 germinal an 12, AN AJ/15/592.

13. Knapton, *Empress Josephine*. For Josephine's relations with the Museum, Jouanin, *L'Impératrice Joséphine*.

14. PV 24 ventôse an 8, AN AJ/15/100, p. 117; Jouanin, *L'Impératrice Joséphine*, 24.

15. PV 7 messidor an 9, AN AJ/15/102, p. 15.

16. PV 5 vendémiaire an 11, AN AJ/15/103, p. 160.

17. PV 26 prairial an 11, AN AJ/15/104, p. 82.

18. PV 3 messidor an 11, AN AJ/15/104, p. 84. For the minister's letter to the Museum of 27 prairial an 11 and the competition between the Museum and Josephine over the black swans, see Jouanin, *L'Impératrice Joséphine*, 142–46.

19. PV 29 thermidor an 11, AN AJ/15/104, p. 114. Mirbel (1776–1854), a naturalist's assistant at the Museum early in his career, became professor of agriculture in 1828.

20. PV 15 and 22 frimaire an 12, AN AJ/15/105, pp. 55, 60. Vivant Denon to Fourcroy, 16 frimaire an 12, séance of 22 frimaire an 12, AN AJ/15/592.

21. At Isle de France two axis deer from Bengal were also embarked aboard the *Géographe*, a gift for General Moreau from his brother-in-law, a sea captain.

22. Geoffroy to the professors, 18 germinal an 12, séance of 28 germinal an 12, AN AJ/15/592.

23. F. Cuvier, undated letter to the Museum's director, séance of 28 germinal an 12, and the administration to Geoffroy, 30 germinal an 12 (copy), séance of 5 floréal an 12, both in AN AJ/15/592.

24. Mirbel to the Museum, 9 floréal, séance of 12 floréal an 2, AN AJ/15/592.

25. The two giant kangaroos were mounted 5–12 floréal. Dufresne zoology lab report, 12 floréal an 12, AN AJ/15/592.

26. Mirbel to the Museum, 9 floréal an 12, séance of 12 floréal an 12, AN AJ/15/592.

27. On the kangaroos sent to Malmaison see Jouanin, 135–38.

28. Fourcroy to Madame Bonaparte, 13 floréal an 12, cited by Jouanin, 136 (not seen by me). A draft written by F. Cuvier for Fourcroy to send to Mirbel, dated 13 floréal an 12, is in AN AJ/15/744.

29. G. Cuvier, "Le Zèbre," 4, in Lacépède, *Ménagerie*.

30. Mirbel to the Museum, 9 floréal an 12, séance of 12 floréal an 12, AN AJ/15/592. Mirbel was fired in 1806.

31. PV AN AJ/15/104–12.

32. PV 30 November 1808, AN AJ/15/109, p. 92.

33. PV 1 and 8 June 1814, AN AJ/15/114, pp. 140, 143. Jouanin, 144.

34. F. Cuvier, "Le Kanguroo géant," 2, *HNM*, October 1819.

35. Draft in F. Cuvier's hand, "Le conseiller d'état Fourcroy au citoyen Mirbel directeur des jardins de Malmaison," 13 floréal an 12, AN AJ/15/744.

2.8 COLLECTING FOR THE MENAGERIE BEFORE 1815

1. On the animal trade, its violence, and its casualties, see Rothfels, "Catching Animals" and *Savages and Beasts.*

2. G. Cuvier, "L'ours polaire," in Lacépède, *Ménagerie*. An Eyriès Frères letter to the Muséum, 12 September 1833, states that a ship's captain back from the Arctic was offering for sale a young polar bear whose "mother and father were killed in order to have it." Séance of 17 September 1833, AN AJ/15/646.

3. F. Cuvier menagerie report, 30 November 1808, AN AJ/15/600. The female and an adult male were embarked in spring 1808 from Corsica, but the male died en route. The female was so sick when she arrived in Marseille in May that the prefect, Thibaudeau, delayed shipping her to Paris. In July she gave birth to a small male. In November, Thibaudeau judged them ready to travel. Thibaudeau to the Museum, 4 May 1808 (séance of 11 May 1808) and 28 June 1808 (séance of 6 July 1808), AN AJ/15/599; PV 23 November 1808, AJ/15/109, p. 89.

4. F. Cuvier, "Le Mouflon," p. 1, *HNM*, dated December 1818.

5. Baudin, *Journal*, 469.

6. Baudin, *Journal*, 470.

7. Baudin, *Journal*, 471.

8. Baudin, *Journal*, 480.

9. Péron and Freycinet, Voyage, 1:16–17.

10. Péron and Freycinet, Voyage, 1:117.

11. For bird catching in France, see B*** [Buillard], *Aviceptologie*.

12. For Baillon's observations on birds kept in his garden, see Buffon, "Le chevalier varié," *HNO* 8 (1783): 327–28n; Vanneaux, *HNO* 8 (1783), 403n; "L'huîtrier," *HNO* 9 (1784), 42; "Le pierre-garin," *HNO* 9 (1784), 209n; "Les goélands," *HNO* 9 (1784): 256–57; "La grande mouette cendrée ou mouette à pieds bleus," *HNO* 9 (1784): 283–84.

13. Prarond, "Baillon," 622. Daubenton conveyed Buffon's request to Baillon in a letter dated 14 August 1778. Baillon soon contacted André Thouin and supplied him with plant material. Baillon's letters to Thouin date back to at least 1780. See MNHN, Fonds Andre Thouin. Spary, *Utopia's Garden*, 74, quotes a 1781 letter from Baillon to Thouin.

14. Buffon, "Le butor," *HNO* 8 (1783): 247–49.

15. The English call the bird the ruff after the elaborate neck plumage the male displays in breeding. The French call the bird *le combattant* because of the males' fights.

16. Buffon, "Les combattans," *HNO* 8 (1783): 333.

17. Buffon, "La Tadorne," *HNO* 10 (1786): 70–74. Raising birds from the egg stage was of great importance for the early ethologists Charles Otis Whitman, Oskar Heinroth, and Magdalena Heinroth. See Burkhardt, *Patterns of Behavior.*

18. Buffon, "Les pierre-garins," *HNO* 9 (1784). 209n.

19. Buffon, "L'avocette," *HNO* 9 (1784): 318n.

20. Buffon, "Les goélands," *HNO* 9 (1784): 261–63. Buffon thanked Baillon for the shelducks and Brant goose in an undated letter cited by Prarond, "Baillon," 624.

21. Buffon, "Le canard siffleur," *HNO* 10 (1786): 41–42. Baillon, using a different edition, proudly cited this from *HNO* 9 (1783), 171. See his letter of 14 brumaire an 3, cited below.

22. Prarond, "Baillon," 625, gives no date for the letter. Lacépède did not specify the

circumstances that led him to tell Baillon to concentrate on sending specimens, not on studying bird life.

23. PV 4 brumaire, AN AJ/15/96, p. 165. The Museum already had Baillon on its books as a correspondent and did not see the need for a new commission.

24. Undated "observations" by Baillon, séance of 14 brumaire an 3, AN AJ/15/578, PV AN AJ/15/96, p. 167.

25. Baillon to the Museum, 11 nivôse an 3 (31 December 1794), séance of 24 nivôse an 3, AN AJ/15/578.

26. Baillon to the Museum, 11 nivôse an 3 (31 December 1794), séance of 24 nivôse an 3, AN AJ/15/578.

27. Baillon to the Convention Nationale, undated. The earliest date in the annotations to this document is 7 pluviôse an 3. AN F/17/1229.

28. Baillon to Delaunay, 18 pluviôse an 4, séance of 24 pluviôse an 4, AN AJ/15/579.

29. Baillon to "Citoien," germinal an 4, séance of 4 floréal an 4, AN AJ/15/579.

30. Baillon meanwhile sent the Museum a memoir on the habits of the mountain linnet. Séance of 24 brumaire an 5, AN AJ/15/98: 15.

31. The Museum later reimbursed Baillon and awarded him a 300-franc annual stipend in appreciation of his services. PV 24 floréal an 5, AN AJ/15/98, p. 80.

32. Baillon to the Minister of the Interior, 20 germinal an 7 (9 March 1799), séance of 14 floréal an 7, AN AJ/15/582.

33. Séance of 14 floréal an 7 (23 April 1799). The list of "oiseaux vivants envoyés par le Cn. Baillon" is unsigned, but the handwriting is Delaunay's. A longer list of "oiseaux pour le cabinet," likewise unsigned, was probably Dufresne's. Baillon mostly sent birds, but the Museum once asked him for a pair of live badgers: PV 24 nivôse an 8, AN AJ/15/100: 86. G. Cuvier thanked him for fish. Prarond, "Baillon," 631.

34. Letters regarding gunpowder arrangements are in AN, F/17/1229, along with the Museum's inquiries about the government's providing a wagon to transport Baillon's specimens to Paris. Before the Revolution, Baillon had served as a forest inspector. In 1794 he published a *mémoire* on forest regeneration. See Goret, "Notice historique." The Museum's recommendations of Baillon for the forestry post are in AJ/15/743.

35. Baillon from Abbeville, 6 vendémiaire an 7, séance of 14 vendémiaire an 7, AN AJ/15/542 (Baillon personnel file), and PV AN AJ/15/99, p. 27.

36. Baillon letter 11 brumaire an 7, séance of 14 brumaire an 7, AN AJ/15/542, and PV AN AJ/15/99, pp. 37–38.

37. PV 14 and 24 brumaire an 7, AN AJ/15/99, pp. 38, 40–41.

38. PV 24 brumaire an 7, AN AJ/15/99, p. 41. The duck is identified in the PV as "Canard Nyreca." This was probably the ferruginous duck, *Aythya nyroca*, common to Eurosiberia.

39. PV 24 prairial an 7, AN AJ/15/99, p. 118.

40. Baillon proposal, 16 fructidor an 8, filed with séance of 27 vendémiaire an 9, AJ/15/585; PV 24 fructidor an 8 (11 September 1800), AN AJ/1/101, pp. 11–12.

41. After his first mention of the project, he returned to it frequently in letters to the Museum over the next few months. AN AJ15/585.

42. PV 27 vendémiaire, 7 brumaire, and 16 frimaire an 9, AN AJ/15/101, pp. 30, 36, 76. After the 16 frimaire meeting, the PV do not mention the project again.

43. Baillon to the Museum, 11 vendémiaire an 10 (3 October 1801), AN AJ/15/587. This letter may not have been read to the assembly. A Baillon letter was announced at

the meeting of 17 vendémiaire an 10 (PV AN AJ/15/102, p. 105), but the minutes mention seeds Baillon was going to send, not his phalarope.

44. Baillon to the Museum, 11 vendémiaire an 10.

45. The Abbé Pluche, with his *Spectacle de la nature*, was the most popular natural history writer in France before Buffon. As for Baillon's phalarope, we have no record of Lacépède's clarifying the bird's identity. Furthermore, no mention of Baillon's find appears in the *Nouveau dictionnaire d'histoire naturelle*'s entry on the phalarope, 25 (1818): 480–82.

46. Baillon's death was announced at the séance of 17 brumaire an 10 (8 November 1801), PV AN AJ/15/102, p. 133.

47. Prarond, "Baillon," 639–40, tells of Delaunay and Jussieu's efforts on Baillon's behalf to dissuade Louis Baillon from joining the Baudin expedition.

48. Louis Baillon to the Museum, 30 brumaire an 10, séance of 7 frimaire an 10, AN AJ/15/587.

49. Prarond, "Baillon." Jaussaut and Brygoo, "Baillon, Louis Antoine François," *Du Jardin au Muséum*, 53–53, mention L. Baillon's sending cetacean skeletons to the Museum in 1839 and 1842. Prarond and Jaussaut/Brygoo give 3 December 1851 as the date of L. Baillon's death.

50. Minister of the Interior to the Professors of the Museum, 22 December 1807, in séance of 30 December 1807, AN AJ/15/598. See too PV AN AJ/15/108, p. 133. Draft of Museum letter to the minister, 27 December 1807, AN AJ/15/745.

51. Minister of the Interior to the Museum, February 1808, in séance of 17 February 1808 AJ/15/599. See also PV AN AJ/15/108, p. 157. Museum to the prefect at Brest, 18 February 1808, AN AJ/15/745.

52. Th. Caffarelli to [G.] Cuvier, director, 27 February 1808, séance of 9 March 1808, AN AJ/15/599.

53. Péron, *Voyages*, 1:116–18.

54. Jussieu, "Notice sur l'expédition," 9–10.

55. Selous, *Bird Watching*, 335. On Selous, see Burkhardt, *Patterns*, 77–92.

2.9 THE GNU

1. Geoffroy Saint-Hilaire to the Museum, 18 germinal an 12, séance of 28 germinal an 12, AN AJ/15/592.

2. Jouin and Stein, *Histoire*, 277.

3. Anonymous [F. Cuvier], *Notice des animaux vivans* (1804), 62.

4. Anonymous [F. Cuvier], *Notice des animaux vivans* (1804), 63.

5. Ballet to the professors, undated, annotated by F. Cuvier, séance of 23 March 1808, AN AJ/15/599.

6. PV 27, December 1809 and 3 January 1810, AN AJ/15/110, pp. 86, 92.

7. E. Geoffroy Saint-Hilaire menagerie report, séance of 1 May 1811, AN AJ/15/604. Geoffroy gave 25 April 1811 as the date of the gnu's death.

8. F. Cuvier, "Le gnou femelle," 2, *HNM*, May 1820.

9. G. Cuvier, *Règne animal*, 1:264–65.

10. Rousseau and Lemonnier, *Promenades*, 193–94.

11. Minister of the navy and colonies to the Museum, 3 February 1825, and "Extrait d'un rapport de Mr. Bédier Sous Commissaire de la Marine à Pondichery," séance of 8 February 1825, AN AJ/15/629.

12. Museum to the minister of the navy and colonies, 11 February 1825, AN AJ/15/750. See also PV 8 February 1825, AN AJ/15/124, pp. 173–74.

13. Minister to the Museum, 21 February 1825, séance of 22 February 1825, AN AJ/15/629.

14. Minister to the Museum, 20 May 1826, séance of 6 June 1826, AN AJ/15/631.

15. Museum to the minister of the navy and colonies, 13 June 1826, AN AJ/15/750.

16. Rousseau and Lemonnier, *Promenades*, 192.

2.10 CARING FOR THE ANIMALS OF THE MENAGERIE

1. French veterinarians thought tainted water and food were behind most farm animals' illnesses. See Chabert et al., *Instructions*, 5–6.

2. PV 24 brumaire an 4 (15 November 1795), AN AJ/15/97, p. 92.

3. PV 9 frimaire an 2, AN AJ/15/96, p. 65.

4. Geoffroy Saint-Hilaire, "État des animaux, vivans, existans à la ménagerie provisoire du Museum national d'histoire naturelle, ce jourd'hui 26; floréal an 2 de la République francaise une et indivisible," AN AJ/15/844.

5. The menagerie's 1829 meat expenses were a quarter of its expenses for fodder and grain (5,500 versus 22,000 francs): 5,000 francs went for bread, fruits, vegetables, etc. "Exercice 1829. Repartition de credit de 335,000 fr. alloués au Muséum d'histoire naturelle . . . ," séance of 9 January 1829, AN AJ/15/637. For 1837, see F. Cuvier, "État des dépenses de la Ménagerie durant l'année 1837"; séance of 9 January 1838, AN AJ/15/655.

6. In Delaunay's menagerie record book see the page titled, "La Ménagerie étoit composé le 1er vendémiaire an 6eme savoir. . . . ," AN AJ/15/844.

7. New mammals in 1797 included a pair of raccoons from North America, a porcupine from the Stadtholder's menagerie, and the young lioness taken from a captured British ship in 1797.

8. The general arrangement between police and Museum seems to have begun in mid- to late 1796. See Division des Subsistances, Bureau Central du canton de Paris, to the Museum 14 pluviôse an 5, séance of same date, AN AJ/15/580. See too Préfecture de Police, "Arrêté concernant le commerce de la viande," Paris, du 9 germinal an VIII de la République . . ." séance of 24 germinal an 8, AN AJ/15/584. This decree required the police to contact Guichenot to collect seized meat.

9. Delaunay to the Minister of the Interior, 3 ventôse an 4, AN, F/17/1229, implies that the carnivores were fed beef in their first months at the Museum. By the time of the menagerie census of May 1794, however, they ate horsemeat. See Geoffroy, "État des animaux, vivans, existans à la ménagerie provisoire," AN AJ/15/844.

10. Delaunay to the Minister of the Interior, 3 ventôse an 4, AN, F/17/1229.

11. The lion's death was reported in *La Décade philosophique* 10 (10 thermidor an IV, 28 July 1796), 235. Delaunay specified the cause of death in his letter to the Museum, 14 ventôse an 6, AN AJ/15/581. See also Toscan, *L'Ami*, 47n. Re "farcinous," farcy is a contagious, glanderslike disease of horses.

12. Delaunay, untitled record book, AN AJ/15/844.

13. Delaunay to the Museum, 14 messidor an 7 (2 July 1799), filed with the séance of 4 brumaire an 8, AN AJ/15/583. Prefect of Police to the Museum, regarding its request for more seized meat, 12 germinal an 8, séance of 24 germinal an 8, AN AJ/15/584; PV AN AJ/15/100: 128.

14. Bureau central du Canton de Paris to the Museum, 17 brumaire an 7, AN AJ/15/845.

15. Delaunay to the Museum, 14 messidor an 7. As late as the winter of 1797–98, the menagerie received meat from glanders-infected horses. See Bureau Central du Canton de Paris to the Museum, 27 nivôse an 6, AN AJ/15/581.

16. For more on provisioning the carnivores, see Delaunay to the Museum, 4 germinal an 8, séance of 4 germinal an 8, AN AJ/15/584; Prefect of Police to the Museum, 5 fructidor an 11, séance of 6 fructidor an 11, AN AJ/15/590.

17. From Delaunay's record books we calculate 1,120 francs paid for butchers' meat for brumaire year 8 through vendémiaire an 9; and 1,910 francs for butcher's meet for brumaire year 9 through vendémiaire year 10 (figures rounded to the nearest 10 francs) The cost of seized meat went from 130 francs in the first year to 270 francs in the second.

18. Delaunay's records for year 10 of the Republic allow one to total the costs for purchased beef (1,742 francs), seized meat (323 francs), and horsemeat (182 francs), each rounded to the nearest franc; 5,769 pounds of beef were purchased. The amounts of seized meat and horsemeat were not recorded, but we can estimate the amount of meat the carnivore collection consumed based on their number and what their normal rations were in other years. These rations we have taken from F. Cuvier's lists of animals and what they ate in 1813 (séance of 12 May 1813, AN AJ/15/607) and 1837 (séance of 9 January 1838, AN AJ/15/655). From these we learn that in year 10 the two tigers needed about twelve pounds each; the five lions, ten pounds each; the leopard, panther, hyena, and two wolves, six pounds each; and the two lynx, two pounds each. This totals 100 pounds daily or 36,500 pounds annually. The 5,769 pounds of butcher's meat amounts to less than 16 percent of the total. The rest had to be made up by seized meat and horsemeat.

19. Museum to the Prefect of Police, 28 brumaire an 10, AN AJ/15/743. Delaunay records three trips to the markets in the first decade of brumaire, none in the second, and one in the third, AN AJ/15/844.

20. See Cuvier's "Consommation de la viande en 1837," 9 January 1838, AN AJ/15/655. A negligible amount of seized meat (36 francs' worth) was still listed in Cuvier's expenses for the year. The total meat cost was 10,500 francs. See Louis Guichenot's note on his trips around Paris to collect "une vache" and other food for the menagerie. "Note des courses faits pour la menagerie commencé le 18 messidor an 11" AJ/15/590. PV 13 fructidor an 11, AN AJ/15/104, pp. 120–21.

21. The quantities of hay, grain, etc., were recorded in *setiers*, each one the equivalent of twelve bushels.

22. See Delaunay's undated letter to the Museum, séance of 24 pluviôse an 5, AN AJ/15/580; PV AN AJ/15/98, p. 53; plus the letter in Delaunay's handwriting (but sent in Geoffroy Saint-Hilaire's name) to the Minister of the Interior, 5 ventôse an 5 (23 April 1797), AN AJ/15/844.

23. Delaunay to the Museum, 14 messidor an 7, séance of 4 brumaire an 8, AN AJ/15/583. Delaunay announced 4993.5 francs owed to Degennes for year 6.

24. Delaunay to the Museum, 14 messidor an 7, in séance of 4 brumaire an 8, AN AJ/15/583, AN AJ/15/583.

25. Degennes to the Museum, 14 vendémiaire an 8, séance of 24 vendémiaire an 8, AN AJ/15/583.

26. Séance of 24 ventôse an 8, AN AJ/15/100, p. 117. Minister of the Interior to the Museum, 26 ventôse an 8, séance of 24 germinal an 8, AN AJ/15/584.

27. See Delaunay's chart presented at the séance of 4 prairial an 8 (24 May 1800), AN AJ/15/584. For other animals sold, see AJ/15/102, 74.

28. Delaunay to the Museum, 24 floréal and 4 prairial an 8, séance of 4 prairial an 8, AN AJ/15/584; PV, AN AJ/15/100, pp. 144–45.

29. Delaunay's menagerie notebook, AN AJ/15/844. In 1815 Frédéric Cuvier noted that the female elephant ate thirty-six pounds of bread daily (plus hay, bran, and carrots). Cuvier, "État des animaux présens à la Ménagerie le 1er juin 1815 et de leur consummation journalière," AJ/15/610.

30. Delaunay menagerie register. Potatoes and cabbage dropped off the ledger the following spring.

31. G. Cuvier, "L'éléphant des Indes," 10, in Lacépède, *Ménagerie.*

32. For vendémaire an 10, Delaunay purchased 540 to 600 pounds of bread per decade. With 100 pounds of this going to the bears, this left 25 pounds of bread daily for each elephant. In 1804, when there was only one elephant, F. Cuvier identified her daily food consumption as forty pounds of bread and carrots. Cuvier, *Notice des animaux vivans* (1804). In an 1815 list of the animals and what they ate, he put the elephant's daily consumption of bread at thirty-six pounds (twelve loaves of three pounds each). F. Cuvier, "état des animaux présens à la Ménagerie le 1er juin 1815 et de leur consummation journalière," AJ/15/610.

33. In frimaire an 10 (22 November—21 December 1801) the menagerie's monthly consumption of bread (1,800 pounds) and the price of bread (17.5 centimes per pound) were at a five-year high. That month bread cost the Museum 315 francs, nearly double the year before.

34. Delaunay's bird list in his menagerie notebook, AN AJ/15/844; Delaunay to the Museum, 14 ventôse an 6, séance of same date, AN AJ/15/581.

35. Delaunay to the Museum, 24 germinal an 8, séance of same date, AN AJ/15/584. Delaunay underlined the phrase "n'ont pas manqué des secours de la médecine."

36. Cuvier's daily menagerie reports from 5 to 11 pluviôse an 12, AN AJ/15/591.

37. F. Cuvier, *Notice* (1804), 6–7.

38. F. Cuvier menagerie report, 7 messidor an 13 (26 June 1805), AN AJ/15/594.

39. In 1813 the menagerie's eight bears each received daily a three-pound loaf of bread and a bunch of carrots. See Cuvier's "État des animaux présens à la ménagerie au treize avril 1813 et nourriture consommée par eux chaque jour," séance of 12 May 1813, AN AJ/15/607.

40. The PV for 7 messidor an 13 simply notes the report's submission; AN AJ/15/107, p. 49.

41. Pouillard, *Histoire des zoos*; Pouillard, "The Silence and the Fury."

42. Delaunay's account book recorded payments for postal deliveries; for transporting things (e.g., new specimens, seized meat, deliveries of fodder and straw from Degennes); for repairing or purchasing horse carts and wagons and outfitting the horses with shoes and harness. The menagerie paid for the services of wheelwrights, farriers, coopers, harness-makers, knackers, and others. Items critical for the keepers' work included buckets, shovels, and pitchforks. The menagerie paid taxes on materials brought into the city, new specimens, medicines, and more. Some things like axle grease (*vieux oing)* were purchased weekly, others only rarely.

43. F. Cuvier to the Museum director, 24 September 1810, séance of same date, AN AJ/15/603.

44. Geoffroy's November 1809 proposal for the menagerie had featured double

cages (section 2.6), but judging from Cuvier's comments, double cages were yet to be incorporated in the architect's latest plan.

45. Lavoisier, "Altérations," 687. Reported in *Journal de Paris*, 12 May 1777, 1–2.

46. Jussieu to the Minister of the Interior, 25 prairial an 9, séance of 27 prairial an 9, AN AJ/15/586.

47. PV 8 January 1806, AN AJ/15/107, p. 130.

48. F. Cuvier to the Museum director, 24 September 1810, séance of 26 September 1810, AN AJ/15/603.

49. Leger to the Museum, undated, séance of 24 germinal an 6, AN AJ/15/581; PV AN AJ/15/98, 160–61.

50. Twenty years later an odor from the menagerie's meat storage site was so foul that it was said to spoil food. See Martinet (Commissaire de Police) to the Museum, 13 June 1822, séance of 18 June 1822, AN AJ/15/623. The previous year the prefect of police complained about noxious water from the Museum. Letter of 15 May 1821, séance of 22 May 1821, AN AJ/15/621.

51. Leger letter of 30 nivôse an 8, séance of 14 pluviôse an 8 (3 February 1800), AN AJ/15/583. PV AN AJ/15/100: 93–94. Many neighbors likewise attributed the most offensive odors to the Bièvre. PV 24 thermidor an 8, AN AJ/15/100, p. 179.

52. Museum to the Minister of the Interior, 7 floréal an 12 (27 April 1804), AN AJ/15/744. The physician Alibert, *Traité des fièvres*, 222–24, also called attention to the toxic stench from the Salpêtrière sewer.

53. The Museum corresponded with the Prefecture of the Seine and the police about cleaning the blocked sewer. See Museum to Villemsens of the Prefecture du Département de la Seine, 27 July 1811, AN AJ/15/745, and PV 31 July and 7 August 1811, AN AJ/15/112, pp. 57, 61–62.

54. Delaunay to the Museum, 4 April 1813, séance of 4 April 1813, AN AJ/15/607. See an earlier letter to the same effect, dated (incorrectly) 9 August 1808, read at the séance of 8 August 1808, AN AJ/15/603. PV AN AJ/15/111, p. 65.

55. Delaunay to the Museum 24 germinal an 8, séance of same date, AN AJ/15/584.

56. Delaunay to the Museum, 14 vendémiaire an 5 (5 October 1796), séance of same date, AN AJ/15/580.

57. PV 17 January 1810, AN AJ/15/110, pp. 98–100.

58. "Programme d'une Ménagerie Nationale," séance of 27 pluviôse an 10, AN AJ/15/845.

59. F. Cuvier menagerie report of 5–13 thermidor an 12, séance of 13 thermidor an 12, AN AJ/15/592; PV AN AJ/15/106, p. 53.

60. PV 19 August 1807, AN AJ/15/108, p. 92.

61. PV 7 thermidor an 9, AJ-1–5-102, p. 35.

62. Untitled note in Lamarck's hand, séance of 17 fructidor an 9, AN AJ/15/586. Lamarck's presentation of Cassal's complaints is noted in PV AN AJ/15/102, 70.

63. Geoffroy Saint-Hilaire, "Reglements de la ménagerie," séance of 20 thermidor an 12, AN AJ/15/592, PV AN AJ/15/106, pp. 63–64

64. PV 23 prairial an 13, AN AJ/15/107, p. 42.

65. Frédéric Cuvier, menagerie report of 17 April 1816, AN AJ/15/612. The assembly approved Cuvier's suggestion the same day. Cuvier made his recommendation with specific reference to the herbivore part of the menagerie. This may have been because the more trustworthy keeper Rousseau had charge of the *animaux féroces* and the abuses there were fewer.

66. In June 1815 Cuvier introduced a printed form for weekly tracking of the menagerie's stores. There he recorded foods on hand, amounts acquired, amounts consumed, etc. He also used the form to record animal acquisitions and losses and to offer "general observations." He continued to employ the form at least until March 1821. In 1837 he kept a month-by-month chart of the menagerie's expenditures for the year. For the same year he produced another chart of how much meat the carnivores ate and how much was paid for various kinds of meat. Both charts are filed with the séance of 9 January 1838, AN AJ/15/655.

2.11 MARTIN THE BEAR

1. Anonymous (F.Cuvier), *Notice des animaux* (1809), 59.

2. Anonymous, *Description* (1811), 9.

3. Anonymous, *L'Officer russe à* Paris, 214. For other songs and poems about Martin and an analysis of his story in the context of anxieties about human race relations and the boundary between animals and humans, see Lee, "Curious Affair."

4. PV 15 December 1813, AN AJ/15/114, pp. 80–81.

5. Museum letter of 15 December to the Minister of the Interior and letter of same date to the editors of the *Journal de l'Empire*, the *Journal de Paris*, and the *Gazette de France*, AN AJ/15/608. PV 15 December 1813, AN AJ/15/114: 82.

6. Séance of 15 December 1813, AN AJ/15/114: 82. Quotations from the Museum's summary of the report. The full report by Lacour, the surgeon, is in the file of the same séance, AN AJ/15/608.

7. F. Cuvier, menagerie report, 22 December 1813, AN AJ/15/608.

8. Cuvier, *Notice* (1804), 4.

9. The Museum traded a lion for the three bears and a monkey. PV 6 nivôse an 12 (28 December 1803), AN AJ/15/105, pp. 80–81.

10. Anonymous (F. Cuvier), *Notice* (1804), 4.

11. Anonymous (F. Cuvier), *Notice* (1804), 5.

12. The length and width of the pits is reported by Lee, "The Curious Affair." The depth is identified as fifteen to eighteen feet in the letter from the Museum to the Minister of the Interior, 15 September 1820, AN AJ/15/747.

13. M. E***, *Promenades instructives.*

14. Anonymous, *Description* (1820).

15. Séance of 6 December 1815, AN AJ/15/611.

16. E***, *Promenades instructives*, 40–42. The PV of 5 October 1814 state, without further explanation, "the bear from Poland called Martin killed an Angora goat." AN AJ/15/115, p. 8. That this event did not become part of the public lore about Martin's cruelty suggests that the Museum succeeded in keeping it quiet.

17. Frédéric Cuvier, "L'ours noir d'Amérique," *HNM*, October 1820.

18. *La Quotidienne* and *Journal des cébats*, 12 February 1820; reviewed in *Journal des cébats*, 14 February, 1820, 4.

19. *Journal des théâtres, de la littérature et des arts*, 6 September 1820, 4.

20. Accounts appeared on 9 September 1820 in *Journal de Paris*, 2; *Gazette nationale*, 3; *Journal des débats politiques et littéraires*, 2; *Gazette de France*, 1016; *Le Constitutionnel*, 2; and *La Quotidienne*, 4; and on 10 September 1820 in *Le Constitutionnel*, 2; *Journal des débats*, 4; *Le Courrier français*, 2; *Journal de Commerce*, 2; *Journal des Théâtres, de la litérature et des arts*, 4; and *Le Drapeau blanc*, 1.

21. *Gazette nationale*, 9 September 1820, 3.

22. *Journal des débats*, 9 September 1820, 2.

23. *Le Constitutionnel*, 10 September 1820, 2. The equally elaborate account the same day in *Le Drapeau blanc*, 1, credits a courageous dog with causing Martin to retreat to his cage.

24. *Journal des débats*, 10 September 1820, 4.

25. *Journal des débats*, 10 September 1820, citing eyewitnesses for its story, reported that the cadaver was still "trembling" when recovered.

26. *Le Courrier français*, 10 September 1820, 2; *Journal de Commerce*, 10 September 1820, 2.

27. Museum to the Minister of the Interior 19 September 1820, AN AJ/15/747.

28. *Le Quotidienne*, 9 September 1820, 4. Minister of the Interior to the Museum, 9 September 1820, séance of 12 September 1820; Prefect of the Police to the Museum, 19 September 1820; both letters in AN AJ/15/620. Having heard that the trees in the bear pits were rotting at their bases, the Prefect worried that if one toppled while a bear was in it, the tree could end up leaning against a parapet, allowing the bear to escape and wreak havoc.

29. Museum letters to the Minister of the Interior and to the Prefect of Police, both dated 19 September 1820, AN AJ/15/747.

30. Anonymous, *Détails exacts du malheureux événement*, n.d. (1820).

31. Anonymous, *Délibération des animaux*, n.d. (1820).

32. Museum letter to M. le Marquis, 29 October 1820, responding to a letter from him dated 27 October (not preserved). AN AJ/15/747.

33. *Journal de Commerce*, 27 March 1821, 2.

34. F. Cuvier's menagerie reports, 16 November and 7 December 1824, AN AJ/15/628. Shortly after the bears' return to their pits, one briefly escaped into the neighboring park, but it was quickly retrieved without mishap. Cuvier menagerie report, 14 December 1824, AN AJ/15/628.

35. F. Cuvier, menagerie report, 6 July 1824, AN AJ/15/628.

36. L. S., "Cirque Olympique. Première représentation des *Ours*, folie du carnaval, oeuvre posthume de Martin," *Journal des débats*, 1 March 1824, 4.

37. *Jocko, ou Le singe de Brésil* (1825), had a long run at the Théâtre de la Porte Saint-Martin beginning in March 1825 (see *Journal de Paris*, 17 March 1825). The satire, Anonymous, *Martin-Ours à la représentation de Jocko* (La Ménagerie royale, 1825) did not have the same visibility. We have found no advertisements or reviews, simply an announcement (the *Courrier*, 2 May 1825) of the play's publication as a brochure. Lee, "The Curious Affair," analyzes perceptively both works but does not persuade us that "Martin's true utility was not to further zoological enquiry but to perform racial colonial narratives." Lee maintains that the discourse about blacks, browns, and whites offered by the fictive Martin in *Martin Ours* reflected how the Museum displayed brown bears, black bears, and polar bears in the three bear pits. She cites Alphonse Milne-Edwards, *La menagerie: rapport au ministre de l'instruction publique* (G. Masson, 1891), 21, but that text speaks only of a polar bear and brown bears in the 1880s. The arrangement Lee describes was not in place while Martin lived in his bear pit (1808/1809—1820). The menagerie had no polar bears at that time. The last of its original polar bears lived not in the bear pits but in the old menagerie for *animaux féroces*, dying there in 1806. Not until April 1826 (MNHN, Ménagerie, MS 25) did the Museum acquire a new polar bear, which it installed not in the bear pits but in the new menagerie des *animaux*

féroces. See Anonymous, *Guide des étrangers* (1828). The first polar bear to live in the pits seems to have been the specimen installed there in November 1837. See F. Cuvier menagerie report, 14 November 1837, AN AJ/15/654.

38. Gosse, *Histoire Naturelle drolatique.*

39. F. Cuvier letter to the Museum, 11 September 1816, séance of 18 September 1816, AN AJ/15/613.

40. On "the terrible bear," see Eyries frères letters to the Museum of 27/11/1828 and 8 December 1828, AN AJ/15/636, and "Registre d'entrée des mammifères," MNHN ménagerie MS 25: 21, where it is identified as a brown female from North America. The bear's time at the menagerie (December 1828 to December 1831) coincides with the inclusion of "l'ours terrible" on the menagerie census of 25 July 1831, AN AJ/15/642.

41. F. Cuvier menagerie reports, 1 and 8 November 1831, séance of 8 November 1831, AN AJ/15/642. Cuvier said he was reacting to newspaper stories, but we have been unable to identify any such accounts from the period. The menagerie received the bear, a male from Russia, eight years earlier as a gift from "Monsieur Rothschild." MNHN, ménagerie MS 25, records its death as 31 October 1831.

42. F. Cuvier menagerie report, 22 June 1830, séance of 29 June 1830, AN AJ/15/639.

43. F. Cuvier menagerie report, 25 August 1835, AN AJ/15/651. PV 25 August 1835 AN AJ/15/132, pp. 158–59.

44. F. Cuvier menagerie report, 14 November 1837, AN AJ/15/654.

2.12 THE SEAL

1. Baillon to the Museum 31 January 1808, séance of 3 February 1808, AN AJ/15/599.

2. Geoffroy Saint-Hilaire, menagerie report, séance of 28 June 1809, AN AJ/15/601.

3. Dufresne, zoology laboratory report, séance of 25 October 1809, AN AJ/15/602. Geoffroy menagerie reports, PV 8 and 15 November, AN AJ/15/110, pp. 61, 63.

4. Lamarck, *Philosophie zoologique*, 2:451–66.

5. PV 16 August 1809, AN AJ/15/110:19.

6. They feed primarily on mollusks and other marine invertebrates.

7. Lamarck, *Philosophie zoologique* 2:454.

8. I thank Sarah Allen for this observation.

9. Lamarck, *Philosophie zoologique*, 2:456–66. Quotation on 462, table on 463.

10. Lamarck, *Philosophie zoologique*, 2:463.

11. Gould, "A tree grows in Paris."

12. Gould, "A tree grows in Paris," 136.

13. Gould, "A tree grows in Paris," 135.

14. Lamarck, *Philosophie zoologique*, 2:451.

15. Cuvier, *Notice des animaux vivans* (1809).

16. Cuvier, *Notice des animaux vivans* (1809), x, xii.

17. Cuvier, "Observations zoologiques . . . du Phoque commun," 392.

18. Cuvier,"Observations zoologiques . . . du Phoque commun," acknowledged that other authors, including the physiologist Franz Joseph Gall, had previously made this point.

19. Cuvier, "Observations zoologiques . . . du Phoque commun," 387–88.

20. Cuvier, "Observations zoologiques . . . du Phoque commun," 389.

21. Cuvier, "Observations zoologiques . . . du Phoque commun," 393.

22. On the species question, see below, 4.6.

23. Cuvier, Observations zoologiques . . . du Phoque commun," 397.

24. Cuvier, "Description zoologique d'un Phoque moine femelle."

25. Cuvier, "Description zoologique d'un Phoque moine femelle," 388.

26. Cuvier, "Description zoologique d'un Phoque moine femelle," 391.

27. F. Cuvier, "Le phoque commun," *HNM* (September 1819); "Le phoque commun," *HNM* (May 1824).

28. Cuvier, "Le phoque commun" (1824), 2–3.

29. F. Cuvier, "Essai sur de nouveaux caractères," 10 (1808): 114.

30. F. Cuvier, *Des dents des mammifères*.

31. F. Cuvier, "De quelques espèces de phoques.

32. F. Cuvier, "De quelques espèces de phoques, 211.

33. F. Cuvier, "De quelques espèces de phoques, 211.

34. F. Cuvier, "De quelques espèces de phoques, 212.

35. "Registre d'entrée des mammifères à la ménagerie, 1807–1849." MNHN, MS 25, 190. In 1828 the Museum purchased a seal for one hundred francs from a conductor of the Royal Mail in 1828, but it died before it arrived. PV 10 September 1828, AN AJ/15/127, p. 260.

36. Isidore Geoffroy Saint-Hilaire menagerie report 30 July 1835, AN AJ/15/650.

37. F. Cuvier menagerie report 11 August 1835, AN AJ/15/650.

38. "Registre d'entrée." MNHN, MS 25, 190.

39. Isidore Geoffroy Saint-Hilaire menagerie report of 28 June 1836 (misdated 29 June 1836), AN AJ/15/651. See PV 28 June 1836, AN AJ/15/133: 86. Registre d'entrée, 1807–49, MNHN, MS 25:190.

40. Rousseau and Lemonnier, *Promenades*, 51.

2.13 PRUSSIANS AT THE GATES

1. PV 4 messidor an 8, AN AJ/15/100, p. 160.

2. PV 14 and 23 messidor an 8, AN AJ/15/100, pp. 163–65.

3. PV 5 June 1811, AN AJ/15/112, pp. 26–27.

4. Museum to the Minister of the Interior, 4 November 1811 and 18 January 1812, AN AJ/15/745. Perhaps the Museum singled out the menagerie because Napoleon had allocated 400,000 francs for constructing the new quarters for the *animaux féroces*. The Museum hoped 10,000 francs could be taken from that fund to help cover the 1811 budget overrun.

5. PV 2 March 1814, AN AJ/15/114, p. 108.

6. Rousseau returned to Paris and the comparative anatomy lab early in 1815. PV 15 March 1815, AN AJ/15/115, p. 70. On his long and distinguished career at the Museum, see Jaussaud and Brygoo, *Du Jardin au Muséum*, 462.

7. For Prévost and Delalande, see the Museum's respective letters to Baron Chabrol, prefet du département de la Seine, 19 October and 12 December 1813, plus Duméril's 8 December 1813 certification of Delalande's illness, AN AJ/15/745. For Louzardy's son, see the letters filed with the seances of 14 and 21 April 1813 (AN AJ/15/607) and PV of the same dates, AN AJ/15/113, pp. 152, 155–56.

8. Museum to the Minister of the Interior, 14 October 1813. AN AJ/15/745 (draft), and final letter AN F17/3977.

9. Minister of the Interior to Museum, 21 July 1810, séance of 25 July 1810, AN AJ/15/603, and PV AN AJ/15/111, pp. 52–53. For the animals list, 17 October 1810, AN

AJ/15/603. For news of their transport, PV 17 and 31 October and 7 November 1810, AN AJ/15/111, pp. 103, 111–12, 116–17.

10. PV 6 April 1814, AN AJ/15/114, p. 116.

11. Draft letter in séance of 30 March 1814, AN AJ/15/609.

12. Draft letters from the Museum to Humboldt and to Goltz, both dated 8 April 1814, AN AJ/15/745.

13. Humboldt to Cuvier, dated "Vendredi," AJ/15/745; PV 13 April 1814, AN AJ/15/114, p. 114.

14. PV 13 April 1814, AN AJ/15/114, pp. 117–18.

15. On problems of maintaining the service to the galleries while the allied troops resided in Paris, see PV 22 June 1814, AN AJ/15/114, p. 147. At year's end, all staff received bonuses for overtime. PV 28 December 1814, AN AJ/15/115, p. 38.

16. PV 27 April 1814, AN AJ/15/114, p. 122. Other royal visitors came in the following weeks.

17. PV 13 April 1814, AN AJ/15/114: 118. Museum to "Monsieur le General," 11 April 1814, AN AJ/15/745.

18. Unsigned letter of 12 May 1814, in Georges Cuvier's handwriting, addressed to "le commissaire provisoire au departement de l'interieur." AN AJ/15/745.

19. PV 18 May 1814, AN AJ/15/114, p. 130.

20. PV 18 May 1814, AN AJ/15/114, pp. 130–31.

21. PV 18 May 1814, AN AJ/15/114, pp. 132–33.

22. The duchess's visit of 19 May 1814 and the delegation's visit to the minister are noted in PV 25 May 1814, AN AJ/15/114, pp. 135–36. Later, on 14 September 1814, the minister visited and assured the professors he would help with issues including purchasing more land. The king, he said, intended to "enlarge and embellish the establishment." Reported in PV 21 September 1814, AN AN/15/114, p. 180.

23. PV 17 August 1814, p. 166.

24. PV 24 August 1814, AN AJ/15/114, p. 171. The Museum considered sending Florent Prevost.

25. PV 21 and 28 September 1814, AN AJ/15/114, pp. 180–82 and AN AJ/15/115, pp. 1–2. Thouin sent 336 packets of seeds back to L'Herminier at the first opportunity. PV 2 November 1814, AN AJ/15/115, p. 17.

26. Milius to the Museum, 11 January 1815, séance of 18 Jan 1815, AN AJ/15/610.

27. PV 12 and 19 April 1815, AN AJ/15/115, pp. 79, 82–83.

28. "Notice sur la foundation, les accroissemens successifs et la situation actuelle et les besoins du Muséum d'histoire naturelle" (undated and unsigned, but in Jussieu's hand). Séance of 24 May 1815, AN AJ/15/610. In describing the library's needs, Jussieu contrasted it with the better-supported library of the University of Göttingen.

29. Geoffroy Saint-Hilaire, "Sur la ménagerie du Muséum d'histoire naturelle," dated 26 April 1815, séance of 24 May 1815, AN AJ/15/610.

30. PV 14 July 1814, AN AJ/15/114, p. 154.

31. PV April 1815, AN AJ/15/115, pp. 77–78.

32. PV 3 May 1815, AN AJ/15/115, p. 89.

33. PV 17 May 1815, AN AJ/15/115, p. 94.

34. PV 31 May 1815, AN AJ/15/115, p. 100. Molinos to Le Directeur Général Travaux Publics, 30 May 1815. F/13/1225.

35. PV 24 May 1815, AN AJ/15/115, p. 97.

36. F. Cuvier undated letter to "Messieurs," preserved with the minutes of the séance of 31 May 1815, AN AJ/15/610, but not mentioned in the PV.

37. F. Cuvier to "Messieurs," séance of 31 May 1815, AN AJ/15/610.

38. F. Cuvier to "Messieurs," séance of 31 May 1815, AN AJ/15/610.

39. PV 14 June 1815, AN AJ/15/115, p. 107.

40. PV 5 July 1815, AN AJ/15/115, pp. 112–13.

41. Museum to Humboldt, 9 July 1815, and Humboldt to the Museum, same date, AN AJ/15/746.

42. Published notice signed De Pfuel/von Pfuel dated 9 July 1815. Séance of 12 July 1815, AN AJ/15/611.

43. PV 12 July 1815, AN AJ/15/115, pp. 114–15. Museum letter of thanks to Humboldt, 12 July 1815, AN AJ/15/746.

44. See AN AJ/15/115, AN AJ/15/611, and AJ/15/745 for July 1815.

45. Brugmans to the Museum, 21 September 1815, AN AJ/15/611, copied in the PV of the special meeting of 22 September 1815, PV AN AJ/15/115, pp. 144–45. On the relations of the Dutch and French naturalists with respect to the French appropriation of the Stadtholder collection, see Lipkowitz, "'The Elephant in the Room.'"

46. PV 22 September 1815, AN AJ/15/115, pp. 146–48.

47. Brugmans to the Museum, 23 September 1815, AN AJ/15/611, copy in AJ/15/115, pp. 146–48.

48. Alexander von Humboldt to "Mon cher confrère," dated vendredi [24 September 1817], séance of 24 September 1817, AN AJ/15/611.

49. Alexander von Humboldt to "Messieurs très chers et très illustres confrères," n.d. (one or more pages missing), séance of 31 October 1815, AN AJ/15/611.

50. Lacour, *République naturaliste*, 527–33.

51. In an early draft of this letter, Georges Cuvier underlined this whole sentence, then added another line under the phrase "nothing unique" and many lines under the phrase "lost nothing essential." The undated draft was the basis of the 31 October 1815 final letter. AN AJ/15/746.

52. Museum to the Minister of the Interior, 31 October 1815, AN AJ/15/746.

53. Minister of the Interior to the Museum, 8 November 1815, séance of 15 November 1815, AN AJ/15/611.

54. PV 15 November 1815, AN AJ/15/116, pp. 176–77.

55. Receipt for the plants, séance of 15 November 1815, AN AJ/15/611. See, in the same file, Laurent (maritime prefect of Brest) to the Museum, 1 October 1815 and Milius to the Museum, 13 November 1815. PV AN AJ/15/115, pp. 178–79.

3.1 CAPTAIN MILIUS'S GREEN MONKEY

1. Van der Kuyl et al., "St. Kitts Green Monkeys."

2. Milius to the Museum, 11 January 1815, séance of 18 January 1815, AN AJ/15/610.

3. "Notices pour les médailles d'encouragement décernées par la Société, dans sa séance publique du 9 avril 1815," *Mémoires d'agriculture, d'économie rurale et domestique, publiés par la Société royale et centrale d'agriculture* (1815), 131–34. The information that Thouin received a breadfruit tree from Jamaica some twenty years earlier is in a letter from the Museum to the Minister of the Interior, 19 January 1815, AN AJ/15/746.

4. Laurent to the Museum, reported in the séance of 15 March 1815, AN AJ/15/115, p. 70.

5. Milius to the Museum, 13 November 1815, séance of 15 November 1815, AN AJ/15/611. The PV of the same date mention that Milius gave the Museum three birds from Guadeloupe along with the green monkey. AN AJ/15/115, p. 177.

6. Adanson, *Histoire naturelle du Sénégal*, 178.

7. PV 4 pluviôse an 3, AN AJ/15/97, p. 2.

8. G. Cuvier, "Le Callitriche," in Lacépède, *Ménagerie*. Pieters dates the essay floréal an 10.

9. F. Cuvier menagerie report of 18 floréal an 12, séance of 19 floréal an 12, AN AJ/15/592.

10. F. Cuvier, "Le Callitriche," *HNM*, March 1819.

11. F. Cuvier menagerie report, 1 June 1824, AN AJ/15/627.

12. F. Cuvier, "Le Callitriche."

3.2 COLONIAL AMBITIONS AND THE SCHOOL FOR VOYAGER NATURALISTS

1. Copy of Leclerc to the Minister of the Interior, 28 messidor an 10, read 19 fructidor an 10, PV AN AJ/15/103, pp. 139–41. Leclerc's letter inspired one professor to suggest constructing greenhouses for plants and animals from Saint-Domingue, Louisiana, and other French colonies.

2. Rochambeau to "Citoyen Geoffroy," 20 ventôse an 11, séance of 3 messidor an 11; Minister of the Interior to the Museum, 3 messidor an 11, séance of 10 messidor an 11, AN AJ/15/590.

3. Anonymous (F. Cuvier), *Notice des animaux vivans* (1804), 25–26.

4. The rich literature on eighteenth-century colonial natural history, particularly as it relates to France, includes McClellan, *Colonialism*; McClellan and Regourd, "Colonial Machine"; Schiebinger and Swan, *Colonial Botany*, in which see esp. Spary, "Of Nutmegs and Botanists," and Bourguet, "Measurable Difference." For voyager naturalists' activities more generally, see Bourguet, "La collecte du monde"; Bourguet and Lacour, "Les mondes naturalistes."

5. Milbert, *Itinéraire pittoresque du fleuve Hudson*, 1:ii.

6. Auguste de Saint-Hilaire, "Aperçu."

7. Leschenault de la Tour, "Relation abrégé." On Leschenault, see Deschamps, "Leschenault."

8. Savigny and Corréard, *Naufrage*, 10.

9. See Burkhardt, "Naturalists' practices."

10. Minister of the Marine to the Museum, 11 December 1817, séance of 17 December 1817, AN AJ/15/614. Many instructions for voyager naturalists were composed before this, most often for specific voyages. See Bourguet, "La collecte du monde."

11. Museum to the Minister of the Marine, 26 December 1817, AN AJ/15/747.

12. Muséum royal d'histoire naturelle, *Instruction pour les voyageurs* (1818). Before this, the Museum made a detailed set of instructions for the French consul in New York, Hyde de Neuville, "Instructions générales sur les moyens de preparer et de conserver les objets d'histoire naturelle à envoyer au Cabinet du Roi à Paris." These and the draft of the Museum's letter to Richelieu, 9 March 1818, asking him to send these instructions to Hyde de Neuville, are in AN AJ/15/747. See also PV 11 March 1818, AN AJ/15/117, p. 173. The instructions begin with a paragraph extolling the value of live plants and animals but acknowledge that most specimens sent to the Museum will be dead.

13. The *Instruction pour les voyageurs* of 1818 and its later editions are discussed by Kury, *Histoire naturelle*, 94–98.

14. Minister of the Marine and Colonies to the Museum, 30 July 1818, séance of 5 August 1818, AN AJ/15/616. PV of same date, AN AJ/15/118, p. 49.

15. Muséum royal d'histoire naturelle, "Instruction sur les recherches," 207.

16. Muséum royal d'histoire naturelle, "Instruction sur les recherches," 218.

17. Muséum royal d'histoire naturelle, "Instruction sur les recherches," 195.

18. Muséum royal d'histoire naturelle, "Instruction sur les recherches," 194–95.

19. Muséum royal d'histoire naturelle, "Instruction sur les recherches," 195.

20. Muséum royal d'histoire naturelle, "Instruction sur les recherches," 195–96.

21. Muséum royal d'histoire naturelle, "Instruction sur les recherches," 196.

22. Muséum royal d'histoire naturelle, "Instruction sur les recherches," 196.

23. PV 28 October and 4 November 1818, AN AJ/15/118, pp. 83–84. Lainé letter of 2 November, séance of 4 November, AN AJ/15/616.

24. Desfontaines and Thouin, "Projet de rapport à l'assemblée administrative sur deux propositions de S. E. le Ministre de l'Intérieur, l'une relative à des collections d'histoire naturelle et l'autre à l'introduction de l'arbre du quinquina dans les colonies." Séance of 18 November 1818, AN AJ/15/616. The project is identified as a response to the minister's letter of 2 November 1818, read at the séance of 4 November 1818.

25. On these points see I. Geoffroy Saint-Hilaire and Blainville, "Rapport sur les résultats scientifiques," 84–86.

26. Desfontaines and Thouin, "Projet de rapport."

27. Minister of the Interior to the Museum, 29 November 1818, séance of 2 December 1818, AN AJ/15/565. See too PV AJ/15/118, pp. 97–98.

28. PV 20 January 1819, AN AJ/15/118, p. 117.

29. PV 3 February 1819, AN AJ/15/118, p. 123.

30. Museum to the Minister of Interior, 10 February 1819, AN AJ/15/747. The cover letter, in G. Cuvier's hand, stresses the importance of government funding of voyager naturalists.

31. "Rapport aux professeurs administrateurs du Muséum d'histoire naturelle," undated report written by Geoffroy Saint-Hilaire on behalf of a committee whose other members were Haüy, Thouin, and Jussieu, AN AJ/15/565. PV 10 February 1819 AN AJ/15/118, p. 126.

32. "Rapport aux professeurs administrateurs," section titled "reglement." AN AJ/15/565. This differs from the final set of regulations. See "Règlement. Copie," sent to Decazes 10 February 1810 and signed by him 20 February 1819), AN AJ/15/565. Decazes approved the regulations (Decazes to the Museum, 20 February 1819), but not until just before the assembly of 23 March 1819 did the Museum have the approval in hand. PV, AN AJ/15/118, pp. 118, 144.

33. The announcement of the school and the call for applicants appeared in *Gazette Nationale*, 25 March 1819.

34. The candidates' applications, Cuvier's notes on the interviews, and other information on the school are in the file, "École de naturalistes voyageurs," AN AJ/15/565. For more on the interviews, see Kury, *Histoire* naturelle, 169–78.

35. Unsigned letter (in G. Cuvier's hand) to "Monseigneur," 11 May 1819, AN AJ/15/565.

36. Portal to the Museum, 2 June 1819, séance of 8 June 1819, AN AJ/15/617. PV AN AJ/15/119, p. 15.

37. Havet to the Museum, 26 May 1820, séance of 19 December 1820, AN AJ/15/620.

38. Marquis, "Havet." Marquis based his account on notes provided by Havet's brother, Nicole.

39. Museum to Havet, 28 August 1820, AN AJ/15/747.

40. Milius to the Minister of the Navy, 27 August 1820. F/17/3977. *Le Constitutionnel*, 18 December 1820. Museum to the Minister of the Interior, 26 December 1820, F/17/3977. *Le Moniteur universel*, 2 January 1821. Marquis, "Havet."

41. Jussieu, "Notice sur l'expédition," 3.

42. Godefroy's younger brother received a nearly fatal blow to the head. Undaunted, he asked to become his brother's replacement, but the Museum declined. Godefroy frère to the Museum, 8 November 1820, séance of 22 May 1821, AN AJ/15/621. The file contains two nearly identical versions of this letter. See too the letter from the Godefroy brothers' father of 2 May 1821 requesting that Victor François Godefroy be allowed to continue his brother's mission. See also PV AN AJ/15/121, pp. 16–17. For more on the older son, see "Godefroy (Félix-François)," in *Biographie bretonne* 1 (1852): 811–12.

43. *Gazette de France*, 10 April 1821; *Journal de Commerce*, 6 May 1821. For more on the massacre see Thesée, *Auguste Plée*, 211–15.

44. Thesée, *Auguste Plée*.

45. Victor Fontanier completed his training and departed as a voyager naturalist, but the Museum viewed his years as a naturalist in the Levant as unproductive.

46. Minister of the Navy to the Minister of the Interior, 19 April 1820; Minister of the Interior to the Museum, 27 April 1820; Museum to the Minister of the Interior 4 May 1820; all in F/17/3977. For the quotation, PV 2 May 1820, AN AJ/15/120, p. 2. The Museum's letter of 4 May 1820 notes that Decazes had earlier ordered 6,000 francs taken from the 20,000 voyagers' fund to support De Lalande at the Cape of Good Hope. The Museum insisted to the new minister (Siméon) that this money be paid back.

47. Minister of the Navy to the Minister of the Interior, 8 June 1820, F/17/3977.

48. Sauvigny to "Monsieur" (Pierre Latreille), 30 August 1820, séance of 7 November 1820, AN AJ/15/620; PV AN AJ/15/120, p. 93. For earlier mentions of plans for Sauvigny see PV AN AJ/15/120, pp. 2, 15, 38, 43, 59, and 90. Further correspondence is in F/17/3977.

49. Museum to the Minister of the Navy and Colonies, 12 October 1821, AN AJ/15/748.

50. Sauvigny to the Director of the Museum (G. Cuvier), 11 March 1822, séance of 30 April 1822; Intendant de la Marine to Rochefort to the Museum, 16 May 1822, séance of 21 May 1822; and F. Cuvier menagerie report of 4 June 1822, AN AJ/15/623. The convoy from Rochefort included animals from le Baron Coupé.

51. Bourdon, *Illustres médecins*, 94.

52. PV 13 December 1825, AN AJ/15/126, p. 15. "État de situation au 31 Décembre 1826 du fonds de 20,000. fr. alloué pour l'entretien des naturalistes voyageurs du Muséum pendant l'année 1826." AJ/15/567. On Alcide d'Orbigny (and his father) see Chambard, "Alcide Dessalines d'Orbigny."

53. PV 6 November 1827, AN AJ/15/127, p. 109.

54. PV 6 April 1830, AN AJ/15/128, pp. 247–48.

55. Ad. Brongniart, "Rapport," 104. Brongniart was right about the perils voyager naturalists faced, but his list of eight did not include all the naturalists who received support from the Museum's voyager funds. Others of great importance were Jacques Milbert and Charles Lesueur (in North America) and J. J. Dussumier and Pierre-

Médard Diard (in Southeast Asia). Milbert and Dussumier, however, turned the analysis of their specimens over to the Museum's naturalists rather than reserving the right for themselves, which was what generally happened with the specimens Lesueur and Diard sent to Paris.

56. Jacquemont to the Museum, 17 June 1831, séance of 3 January 1832, AN AJ/15/643; PV of same date AN AJ/15/130, p. 128.

57. Diard to the Museum, 6 March 1826, séance of 27 June 1826, AN AJ/15/631.

58. See H. Diard, "Pierre-Médard Diard," 199–210, for information on Diard's administrative and agricultural services for the Dutch.

59. H. Diard, "Pierre-Médard Diard."

3.3 THE EGYPTIAN MONGOOSE AND THE BROWN RAT

1. É. Geoffroy Saint-Hilaire, *Lettres écrites d'Ègypte*, 226.

2. É. Geoffroy Saint-Hilaire, "L'ichneumon," in Lacépède, *Ménagerie*. Pieters dates the publication fructidor an 10.

3. PV 19 vendémiaire an 11, AN AJ/15/103, p. 176.

4. E. Geoffroy Saint-Hilaire, "Sur trois bouquetins," 247.

5. Anonymous (F. Cuvier), *Notice des animaux* (1804), 8–9.

6. Anonymous (F. Cuvier), *Notice des animaux* (1804), 9. Geoffroy does not list the female mongoose in his *Catalogue des* mammifères. The Museum's Bibliothèque centrale dates this catalog 1803. That cannot be correct because the catalog includes the Baudin expedition mongooses, which did not arrive until 1804.

7. Molé to the Museum, 16 December 1818, séance of 23 December 1818, AN AJ/15/616. The accompanying "Extrait d'un rapport statistique sur la Guadeloupe et ses dépendances" reports that the end of the slave trade, fewer workers, and soil degradation contributed to the rat plague.

8. Portal to the Museum, 9 September 1819, séance of 14 September 1819, AN AJ/15/618.

9. Museum to the Minister of the Marine, 22 September 1819, AN AJ/15/618. The same views had been registered by the professors, PV 14 September 1819, AN AJ/15/119, pp. 59–60.

10. Portal to the Museum, 8 December 1819, séance of 14 December 1819, AN AJ/15/618; PV 14 December 1819 AN AJ/15/119, p. 99.

11. Geoffroy report to the assembly 21 December 1819, AN AJ/15/618. PV AN AJ/15/119, p. 103.

12. PV 2 January 1821, AN AJ/15/120, p. 120.

13. F. Cuvier, "La Mangouste," *HNM*, April 1819. Cuvier subsequently wrote "Mangouste de Java" (February 1821), "Mangouste à bandes, femelle (November 1830)," and "Mangouste d'Alger" (March 1834), all in *HNM*.

14. The vansire came from Polito's menagerie, in exchange for Drovetti's mongoose on 15 February 1826. It died 18 September 1837. MNHN MS 25. See too F. Cuvier, "Vansire," *HNM*, June 1826.

15. Draft letter from Museum (F. Cuvier) to Aubry Bailleul (ship's captain), 20 September 1837 (AJ/15/755). Bailleul to the Museum 3 October 1837, séance of same date, AN AJ/15/654. On 12 October 1837 F. Cuvier responded to the Museum director's request (AJ/15/654) and offered the same advice in a letter to the Commissaire de la Marine in Toulon (AJ/15/755). Quotation from the latter.

16. Espeut, “Acclimatization.”

17. The mongoose was introduced to control Martinique’s deadly snake, the *fer-de-lance*. See Hinton and Dunn, *Mongooses*.

18. Barun et al., “Review.”

19. F. Cuvier, “Le surmulot,” *HNM*, September 1820. On the Baudin expedition ship, *Le Naturaliste*, traps were set day and night to prevent rats from damaging specimens. The rats dug into planters and gnawed at plants’ roots. Jean Bertrand Renaud to the Museum, 8 thermidor an 11, séance of same date, AN AJ/15/590.

20. Delaunay to Citoyen directeur, 22 frimaire an 7, séance of 24 frimaire an 7, AN AJ/15/582.

21. PV 27 frimaire an 10, AN AJ/15/102, pp. 160–61.

22. F. Cuvier, “Mouvement de la ménagerie du 9 pluviôse an 12, AN AJ/15/591. See PV 17 pluviôse an 13 AN AJ/15/106, p. 163, authorizing Cuvier to employ Samoël Hirch as an exterminator,

23. F. Cuvier to “Monsieur” (Geoffroy Saint-Hilaire), 18 January 1809, séance of same date, AN AJ/15/601, and PV AN AJ/15/109, p. 109.

24. PV AN AJ/15/109, p. 109.

25. I. Geoffroy Saint-Hilaire menagerie report 25 July 1837, AN/AJ/15/654.

3.4 THE NEW COLLECTORS: JACQUES MILBERT AND J.-J. DUSSUMIER

1. Milbert to the Minister of the Interior, 20 April 1824, F/17/3976.

2. J. J., “Nécrologie” (Jacques Milbert), *Journal des débats*, 13 August 1840.

3. Milbert, *Voyage pittoresque à l’Ile-de-France*, 1:vii-viii, 123–24, 145–46. Louis Lebrun and Michel Garnier also quit the expedition at Mauritius. See Bonnemains, *Baudin*, 15–17.

4. Milbert, *Voyage pittoresque*; and Cheke and Hume, *Lost Land of the Dodo*.

5. Milbert, *Voyage pittoresque*, 2:360–62. The mongoose’s main flaw, Milbert said, was its inclination to steal food from the dining table, which he admitted was his fault for allowing it on the table at mealtimes. Milbert gave the mongoose’s remains to the Museum. Dufresne laboratory report for January and February 1816, séance of 28 February 1816, AN AJ/15/612.

6. Milbert, *Voyage pittoresque*.

7. Milbert, *Itinéraire pittoresque du fleuve* Hudson. 1:ii. For an insightful assessment of Milbert’s account, see Liebersohn, *Aristocratic Encounters*, 83–86.

8. Evidence of Hyde’s interactions with the Museum first appears in the PV of 11, 18, and 25 March 1818, AN AJ/15/117, pp. 173, 175, 177, 182. Hyde asked for specimens to exchange for American ones; the Museum sent specimens and instructions on how to prepare them for shipment to France. See the Museum’s letter about Hyde to Richelieu, the minister of the interior, 11 March 1818, and the Museum’s instructions, AN AJ/15/747. Lesueur, like Milbert, had come to the United States without Museum funding. Lesueur arrived in New York in May 1816 with the Scottish geologist William Maclure, with whom he’d agreed to work for two years as a naturalist/artist. On Lesueur, see Hamy, *Les voyages du naturaliste Ch.-Alex. Lesueur dans l’Amérique du Nord*.

9. Museum to the Minister of the Interior, 2 April 1818, AN AJ/15/747. For a list of the crates’ contents, séance of 5 August 1818, AN AJ/15/616.

10. Museum to Milbert, 23 March 1818, AN AJ/15/747. Draft letter in Cuvier's hand.

11. Museum (Cuvier) to Milbert, 23 March 1818, AN AJ/15/747.

12. Hyde de Neuville to the Museum, 15 June 1818, séance of 5 August 1818, AN AJ/15/616. In the same file is a detailed list of items sent by Milbert.

13. Milbert to the Museum, 25 October 1818, AN AJ/15/616.

14. Milbert to the Museum, 8 November 1818, AN AJ/15/616. No mention of this letter's being read at as assembly appears in the PVs.

15. Museum to Richelieu (Minister of the Interior), 24 December 1818, AN AJ/15/747.

16. Museum to Milbert, 24 December 1818, AN AJ/15/747.

17. Museum to Milbert, 25 December 1818, AN AJ/15/747. Draft in Cuvier's hand.

18. Milbert to "Monsieur," 18 March 1819, séance of 27 April 1819, AN AJ/15/617. Milbert's plan to send the buffalo was first announced at the assembly of 13 April 1819, PV AN AJ/15/118, p. 154. The Museum wrote to Eyries frères, 15 April 1819, AN AJ/15/747, announcing the bison's imminent arrival in Le Havre.

19. Milbert to the Museum, 18 April 1819, séance of 22 June 1819, AN AJ/15/617.

20. Milbert letter of 5 January 1820, reported in PV 15 February 1820, AN AJ/15/119, p. 128. For the Museum's requests to the minister to fund Milbert's trip, see its letters of 4 April 1820 and 20 October 1822, AN F/17/3976.

21. Eyries frères to the Museum, 24 May 1819 (séance of 25 May 1819), 26 May 1819 (séance of 1 June 1819), and 2 June (séance of 8 June 1819), AN AJ/15/617.

22. Milbert, *Itinéraire pittoresque*, 2:192.

23. Museum to Milbert, 16 August 1819, AN AJ/15/747. Draft in Cuvier's hand.

24. Museum to Milbert, 16 August 1819, AN AJ/15/747. Draft in Cuvier's hand.

25. Milbert described his process of acclimatizing the fish to fresh water in "Au rédacteur," *Journal des débats*, 17 July 1836.

26. Milbert to the Museum, 18 November 1819, séance of 28 December 1819, AN AJ/15/575.

27. Eyries frères to the Museum, 24 December 1819, séance of 28 December 1819, AN AJ/15/575.

28. PV 28 December 1819 and 11 January 1820, AN AJ/15/119, pp. 106 and 112–13.

29. Frédéric Cuvier to Milbert, 11 January 1820, AN AJ/15/747. Cuvier announced the animals' arrival in his menagerie report of the same day, AN AJ/15/619.

30. Milbert to the Museum, 26 July 1819, séance of 7 September 1819, AN AJ/15/575.

31. F. Cuvier to Milbert, 11 January 1820.

32. Milbert, *Itinéraire pittoresque*, 1:xi.

33. Milbert, *Itinéraire pittoresque*, 2:32–33.

34. Museum to Chabannon, 4 April 1820; to Milbert 4 April 1820; to Eyries frères, 13 May 1820; and to Milbert 27 May 1820, AN AJ/15/747.

35. Milbert to the Museum, 25 August 1820, séance of 14 November 1820, AN AJ/15/620.

36. Séance of 12 December 1820, PV AN AJ/15/120, p. 109.

37. Milbert, *Itinéraire pittoresque*, 2:33–34.

38. On the animals' transport, see Eyries frères to the Museum, 8 May 1821, séance of 15 May 1821; 13 May 1821, séance of 15 May 1821; and 18 May 1821, séance of 22 May 1821, AN AJ/15/621.

39. Milbert, *Itinéraire pittoresque*, 2:35.

40. Milbert, *Itinéraire pittoresque*, 2:78–79.

41. Eyries frères to the Museum, 15 September 1821, séance of 18 September 1821, AN AJ/15/622 and PV AN AJ/15/121, p. 71.

42. Milbert to Geoffroy, dated 20 March 1820; 28 May 1822; and 8 January 1823. MNHN, MS 3383/70.

43. Milbert to Geoffroy, 8 January 1823, MNHN, MS 3383/70.

44. Twenty-four years later Warren became the first surgeon to perform an operation using ether for anesthesia.

45. Milbert to the Museum 9 January 1823, séance of 25 February 1823, AN AJ/15/625; Milbert, *Itinéraire pittoresque*, 1:xiii-xiv.

46. Museum to Milbert, 6 March 1823, AN AJ/15/749. G. Cuvier was especially interested in fish that Lesueur had described in *Journal of the Academy of Natural Sciences of Philadelphia*. Leueur's articles appeared in vols. 1 and 2 (1818–22)

47. Milbert to Museum, 6 May 1823, séance of 8 July 1823, AN AJ/15/626; PV AN AJ/15/123, p. 59.

48. Milbert to Museum, 22 February 1823, séance of 25 March 1823, AN AJ/15/625.

49. PV 29 April 1823, AN AJ/15/123, p. 27.

50. Milbert to the Museum 18 April 1823, séance of 20 May 1823, AN AJ/15/625.

51. Milbert to the Museum, 11 May 1823, séance of 8 July 1823, AJ/15/626.

52. On the shipwreck, see Milbert, *Itinéraire pittoresque*, 2:253–57.

53. "Extrait d'un rapport fait à l'Administration du Muséum par une commission composée de plusieurs de MM. Les Professeurs, sur les résultats de la mission que M. Milbert a remplie aux États-Unis d'Amérique pendant sept ans (de 1817 à la fin de 1823)," *Mémoires du Muséum d'Histoire Naturelle* 12 (1825): 18–29.

54. "Extrait d'un rapport fait à l'Administration du Museum," 19.

55. Milbert to the Minister of the Interior (copy), 20 April 1824, séance of 27 April 1824, AN AJ/15/627.

56. PV 25 May 1824, AN AJ/15/124, pp. 48–50.

57. Milbert to the Minister of the Interior, undated, but probably about the time of a similar letter, 10 January 1831, F/17/3976.

58. *Moniteur universel*, 19 March and 30 May 1831.

59. Mirbel report of 20 May 1834, séance of same date, AN AJ/15/647.

60. *Moniteur universel*, 25 July 1840.

61. An invaluable overview of Dussumier's life and career is Laissus, "Note sur Dussumier." For contemporary evaluations, see G. Cuvier's 13 November 1827 report on the results of Dussumier's first five voyages, F/17/3975; G. Cuvier, "Rapport fait à l'Académie"; and Achille Valenciennes, "Rapport de Mr. Valenciennes administrateur du Museum sur les collections faites dans l'Inde de 1830 à 1838 par Mons.r J. J. Dussumier," AN F/17/3975.

62. PV 5 October 1818, AN AJ/15/118: 77. Milbert had sent twenty birds two months before. See Hyde de Neuville to the Museum, 15 June 1818, séance of 5 August 1818, AN AJ/15/616.

63. Dussumier to G. Cuvier, 13 December 1820, in Laissus, "Note sur Dussumier," pp. 388–89.

64. According to G. Cuvier, PV 17 October 1820, AN AJ/15/120, p. 84.

65. F. Cuvier, "Le zibeth," *HNM*, October 1820.

66. Gray, "Viverridae," 63.

67. Dussumier to G. Cuvier, 13 December 1820, in Laissus, "Note sur Dussumier," 388–89. PV 20 February 1821, AN AJ/15/120, p. 146.

68. For his status as *consul honoraire* for China, see *État générale de la marine et des colonies pour l'année* 1825 (L'Imprimerie Royale, 1825), 203.

69. Cuvier, 1827 report on Dussumier's collections, AN F/17/3975.

70. Cuvier, 1827 report on Dussumier's collections, AN F/17/3975.

71. Dussumier to the Museum, 14 October 1827, *Mémoires du Museum d'Histoire Naturelle* 15 (1827): 377–384; Cuvier, "Rapport fait à l'Académie," 460. For Dussumier's dolphins, see Laissus, "Note sur Dussumier," 393–96.

72. Museum to Dussumier, 18 October 1827, AN AJ/15/751.

73. Granted at G. Cuvier's suggestion. PV 23 October 1827, AN AJ/15/127, p. 102. Dussumier to the Museum 13 November 1827, AN AJ/15/634.

74. F. Cuvier's menagerie reports, 19 and 26 October 1838, AN AJ/15/640, and G. Cuvier's "Rapport fait à l'Académie." The PV for 26 October 1830 (AN AJ/15/129, p. 87) indicate the professors agreed to award the animals' caretaker three hundred francs and that the tiger and two lion cubs belonged to a Bordeaux merchant who offered to sell them to the Museum. The menagerie's animal register evidences the purchase of the tiger but not the cubs. MNHN MS 25.

75. Museum to the Minister of Interior, 22 October 1830, AN AJ/15/752. Citation from *L'Ami de la Religion* 66 (1831): 42.

76. Dussumier's requests were explained in a letter from the Museum to the Minister of the Interior, 10 January 1833, AN AJ/15/572, reproduced in Laissus, "Note sur les voyages de Dussumier," 399–401. Additional copy in AN F/17/3975.

77. Minister of Public Instruction, 25 January 1833, "Note à l'occasion de la demande formée par MM. les Professeurs administrateurs du Muséum d'histoire naturelle, au sujet de M. Dussumier, de Bordeaux." Note signed C. B. AN F/17/3975.

78. *L'Europe littéraire*, no. 28 (3 May 1833): 116. *Journal des débats*, 13 May 1833. *La Quotidienne*, 24 April 1833, reported that Jacquemont "is at this moment dangerously ill in Bombay." Jacquemont's death was announced at the Museum 14 May 1833, PV, AN AJ/15/131, pp. 5–6.

79. See the letters in Laissus, "Note sur Dussumier."

80. Séance of 3 September 1833, PV AN AJ/15/131, p. 62.

81. For the animals, see Valenciennes, "Rapport de Mr. Valenciennes" (AN F/17/3975) and the menagerie registry, MNHN MS 25. For the fish and reptile specimens, see Laissus, "Note sur Dussumier."

82. Dussumier to the Museum, 16 May 1835, séance of 26 May 1835, AN AJ/15/649. Dussumier anticipated that the male gazelle's skin he'd carefully preserved would help place it with the gazelles of "division D of the Règne animal."

83. I. Geoffroy Saint-Hilaire, menagerie report, 9 June 1835, AN AJ/15/649.

84. On the hemione see *Magasin pittoresque* 3 (1835): 235.

85. The Museum's draft letter to the directors of the *Journal des débats* and *Le Temps*, 12 June 1835, is in AN AJ/15/754. The report appeared in the *Journal des débats* (16 June), *Le Courrier français* (17 June), *La France*, and *Le Temps* (18 June), and the *Gazette de France* (19 June). It totaled Dussumier's preserved specimens as twenty mammals; three hundred birds; fifty species of reptiles; five hundred fish; 180 species of mollusks and zoophytes; sixty species of crustaceans; and "about six hundred insects and crabs of different families." These figures are not identical to those in Dussumier's letter

of 16 May 1835 and can be a source of confusion because some refer to new species and others to total number of specimens.

86. F. Cuvier, menagerie report, 27 September 1836, AN AJ/15/652. The zoologist Valenciennes, commenting in 1838 on this collection ("Rapport de Mr. Valenciennes . . .") noted that the lioness was a particularly beautiful specimen from the east bank of the Indus River, the first of its kind the Museum had, and suspected to be a different race or speciesthan the lions of Africa. He noted that Dussumier had provided both sexes of diverse species, for eventual reproduction, and that his notes and specimens allowed Frédéric Cuvier to reduce to a single species of deer what had initially been described as three different ones. Valenciennes also applauded the new preserved specimens Dussumier provided, notably the fish. AN F/17/3975.

87. Dussumier to the Museum, 17 September 1836, séance of 27 September 1836, AN AJ/15/652.

88. Laissus, "Note sur les voyages de Dussumier," 403–4.

89. A notice in the *Bulletin Colonial*, 6 August 1837, indicates that the *London*, a Dussumier and Company ship, was soon to leave Bordeaux for Maurice (Mauritius).

90. *La Presse*, 4 July 1838. For more precise details, see "Rapport de Mr. Valenciennes."

91. Museum to Dussumier, 23 August 1839. F/17/3975. The animals are listed in the menagerie register, MNHN MS 25.

92. Dussumier file, AN AJ/15/572.

93. Laissus, "Note sur Dussumier," cites Dussumier, "Note sur l'Hémione (séance du 13 avril 1855)," *Bulletin de la Société impériale zoologique d'acclimatation* 2 (1855): 260–62.

94. Muséum royal d'histoire naturelle, *Instruction pour les voyageurs*, 3rd ed. (1829), 4.

95. G. Cuvier, "Rapport sur la partie zoologique de l'Expédition Duperrey," 20.

96. G. Cuvier, "Rapport sur Dussumier," 458.

97. Writing of seeking a female bison for the male, Milbert stated, "my most ardent desire was to see this precious species multiply in France." *Itinéraire pittoresque*, 2:193.

98. On the hemione see F. Cuvier, "Le dzigtai," *HNM*, April 1823; I. Geoffroy Saint-Hilaire, "Sur le genre *Cheval*"; Frédéric Cuvier fils, "Hémione ou Dziggetai, Femelle," *HNM*, August 1842.

99. Or so it was when I was a boy out and about with a butterfly net.

100. G. Cuvier and Valenciennes, *Histoire naturelle des poisons*, 10:148.

3.5 THE BISON

1. G. Cuvier, *Règne animal*, 1:270. Milbert, *Itinéraire pittoresque*, 1:ix.

2. Museum to le comte Daru, 5 August 1809 (draft in Cuvier's hand). AN AJ/15/745. Cuvier represented this not as a confiscation but as an exchange, with Vienna receiving "more than the equivalent" of what Paris would acquire.

3. Daru to the Museum, 24 August 1809, séance of 6 September 1809, and 31 August 1809, séance of 13 September 1809, AN AJ/15/602.

4. Buffon, *HNGP* 11 (1754): 307–8; Cuvier, *Règne animal*, 1:269–70.

5. Hyde de Neuville to the Museum, 15 June 1818, séance of 5 August 1818, AN AJ/15/616. I use the words "buffalo" and "bison" in this section according to the letter writer's choice.

6. Jappie account of expenses, 11 August 1819, séance of 24 August 1819, AN AJ/15/618.

7. For a detailed analysis of Enlightenment views on this subject see Curran, *Anatomy of Blackness*.

8. Cuvier, *Règne animal*, 1:94.

9. Cuvier, *Règne animal*, 1:94–95.

10. Cuvier, *Règne animal*, 1:95.

11. Cuvier, "Extrait d'observations."

12. D'Espinville to the Commissaire Général de la Marine at Le Havre, 8 July 1819, séance of 24 August 1819. AN AJ/15/618. By "the Minister," d'Espinville may have meant Richelieu.

13. Eyriès frères to the Museum, 11 August 1819, séance of 17 August 1819, AN AJ/15/618, PV AN AJ/15/119, p. 49.

14. Eyriès frères to the Museum, 15 August 1819 (séance of 17 August 1819) and 17 August 1819 (séance of 24 August 1819). AN AJ/15/618,

15. Museum to Eyriès frères, 14 August 1819, AN AJ/15/747.

16. Eyriès frères letter of 11 August 1819.

17. Eyriès frères to the Museum, 15 August 1819 (séance of 17 August 1819), AN AJ/15/618. Museum to Eyriès frères, 14 August 1819. AN AJ/15/747.

18. Eyriès frères to the Museum, 17 August 1819, séance of 24 August 1819. AN AJ/15/618.

19. "Accord fait entre Le Sieur Noziere et Messieurs Eyriès frères," dated 16 August 1819, séance of 24 August 1819, AN AJ/15/618.

20. Frédéric Cuvier to the director of customs on the rue des Francs Bourgeois, 24 August 1819, alerting him to the bison's arrival. AN AJ/15/747.

21. Museum to Eyriès frères, 26 August 1819, AN AJ/15/747. This file contains two letters, one detailing what remained for the agents in Le Havre to pay, the other thanking them for their help in acquiring the bison.

22. For Jacques Thouin's memorandum see 26 August 1819, AN AJ/15/747.

23. Eyriès frères to the Museum, 1 September 1819, séance of 7 September 1819. AN AJ/15/618.

24. Eyriès frères to the Museum, 1 September 1819, AN AJ/15/618.

25. Museum to d'Espinville, n.d. (but ca. 24 August 1809). AN AJ/15/618.

26. Frédéric Cuvier, "Le bison d'Amérique," *HNM*, November 1819.

27. See Cuvier's menagerie report of 13 January 1824 and the physician's bill from 14 June 1823, both in the séance of 20 January 1824, and Cuvier's menagerie report of 8 June 1824. AN AJ/15/627.

28. Milbert to the Museum, 26 July 1819, séance of 7 September 1819, and Milbert to the Museum, undated, read at the séance of 11 January 1820, both in AJ/15/575.

29. Eyriès frères to the Museum, 15 September 1821, séance of 18 September 1821, and 10 October, 1821, séance of 16 October 1821; AJ/15/622; Museum to Eyriès frères, 20 September and 23 October 1821. AN AJ/15/748. See also PV 30 October 1821, AN AJ/15/121, p. 94. All we know of this conductor is his name, Grangé.

30. F. Cuvier, "Bison, âgé de 30 jours," *HNM*. May 1825.

31. F. Cuvier report, 9 March 1824, AN AJ/15/627.

32. F. Cuvier report, 30 March 1824, AN AJ/15/627.

33. F. Cuvier, "Bison, agé de 30 jours," *HNM*.

34. F. Cuvier report, 15 March 1825, AN AJ/15/629. Cuvier reported the birth date as 12 March 1825.

35. F. Cuvier report, 4 October 1825, AN AJ/15/63. Hydatid cysts are caused by tapeworms.

36. F. Cuvier report, 2–16 August 1825, séance of 16 August 1825. AN AJ/15/630.

37. F. Cuvier report, 23 August 1825, AN AJ/15/630.

38. F. Cuvier report, 10 January 1826, AN AJ/15/631.

39. F. Cuvier report, 27 June 1826. AN AJ/15/631. Five years later Cuvier wrote that cows mated with bison frequently died in labor and he had witnessed this twice; but he noted no details in his reports. F. Cuvier, *Supplement à l'histoire naturelle*, 1:107.

40. F. Cuvier, report, 17 October 1826, AN AJ/15/632.

41. F. Cuvier, report, 27 December 1826, AN AJ/15/632.

42. Frédéric Cuvier report, 22 June 1830, séance of 29 June 1830. AN AJ/15/639.

43. F. Cuvier report, 25 October 1836, AN AJ/15/652. The menagerie Registre d'entrée for 1807–49 dates the bison's death as 31 November 1836. MNHN Menagerie MS 25.

44. MNHN, MS 25, 134.

45. G. Cuvier, *Règne animal*, new ed. (Deterville, 1829), 1:280.

46. F. Cuvier, *Supplement*, 1:107.

3.6 DONORS, DEALERS, SHOWMEN, AND THE GROWTH OF THE MENAGERIE

1. See 1.4 and 1.6 for the donations of the camel and the lemur. For the other donations, see PV 4 germinal an 5, AN AJ/15/98, p. 67, and 4 and 24 messidor an 7, AN AJ/15/99, pp. 126, 133.

2. Debrun to the Museum, 3 frimaire an 10, séance of 7 frimaire an 10, AN AJ/15/587; PV AN AJ/15/102, p. 144.

3. A man named Beauchêne or Bocchini conducted the convoy from Marseille. For the animal list, see 9 germinal an 11, AN AJ/15/590. Adanson didn't mind his genets' ending up at the Museum; he just wished he'd been able to offer them himself. Adanson to the Museum, séance of 27 vendémiaire an 10, AN AJ/15/587.

4. Though the donor's identity is clear in retrospect, when the minister of the interior in 1803 alerted the Museum to the incoming gift, the Museum misrecorded the news and credited the gift to the dey of Algeria, not the bey of Tunis. See 3.8, "The Demands of Diplomacy."

5. On Decaen see 2.7. For the others, see PV 29 frimaire an 12 (re Vial and Fouché) AN AJ/15/105, p. 63, and 29 germinal an 12 (re Menou and Duroc), AN AJ/15/105, p. 194.

6. Museum to General Menou, 1 germinal an 12 (in Frédéric Cuvier's handwriting, but signed by Fourcroy), AN AJ/15/744; Menou to Fourcroy, 8 germinal an 12, séance of 21 germinal an 12, AN AJ/15/592. The difficulty of distinguishing among the region's mountain goats can be seen in Frédéric Cuvier's observation that the three goats from Montblanc that Geoffroy had called *bouquetins* were instead a different species, the *égagre*. Compare Geoffroy, *Annales du Muséum* 2 (1803): 244–48, with F. Cuvier, *Notice des animaux vivans* (1804), 49–51. See too F. Cuvier, "L'Égagre," *HNM*, July 1821.

7. The differential treatment of donors is apparent in the minutes of the professors' meetings from 1800 to 1820. Nonetheless, when General Louis Joseph Auguste Gabriel Saint-Laurent offered a live fox in 1805, the Museum declined because it was "too common to merit a place in the menagerie." PV 4 floréal an 13, AN AJ/15/107, p. 15.

8. F. Cuvier menagerie report, 18 March 1818, AN AJ/15/615.

9. Prigny (captain of *la Bellone*) to the Museum director, 2 July 1818, séance of 15 July 1818, AN AJ/15/616; F. Cuvier menagerie report (Newfoundland dog), 16 December 1818, séance of 23 December 1818, AN AJ/15/616; PV AN AJ/15/118, p. 106 (Captain Kergariou's two sheep from la Cybèle).

10. These cases are extracted from the minutes and PV of the professorial assemblies from 1819 to 1830.

11. Mirbel to the Museum, 25 September 1819, séance of 27 September 1819; Mirbel to the Museum, 30 October 1819, séance of 16 November 1819; Magol to the Minister, 15 November 1819, filed with Mirbel to the Museum, 25 November 1819, séance of 7 December 1819; Magol to the Museum, 13 December 1819, séance of 28 December 1819; all in AN AJ/15/618. For the corresponding PV, AN AJ/15/119, pp. 65, 88, 96, 106.

12. Magol to the Minister, 11 December 1819, and Minister to the Museum, 20 December 1819, séance of 4 January 1820, AN AJ/15/619; PV AN AJ/15/119, p. 109. Magol's gift was announced in *Gazette de France*, 30 October 1819.

13. Séance of 4 June 1822, AN AJ/15/122: 41. Comparing F. Cuvier's 1822 report on the animals arriving from Senegal (Cuvier to the Museum, 3 June 1822, AN AJ/15/623) and Madame Le Coupé's 20 March 1824 letter shows a discrepancy regarding which animals from came Sauvigny and which came from the Le Coupés.

14. F. Cuvier, "Le guépard," 3–4, *HNM*, March 1823. This specimen later earned a mention in Georges Cuvier's *Recherches sur les Ossemens fossils*, 4th ed. (Edmond D'Ocagne, 1835), 7:410.

15. La Baronne le Coupé to the Museum, 20 March 1824, séance of 30 March 1824, AN AJ/15/627.

16. Museum to Madame la Baronne Coupé, 2 and 15 April 1824, AN AJ/15/749.

17. F. Cuvier menagerie report, 15 June 1824, and Madame le Coupé to the Museum, 20 March 1824, AN AJ/15/627.

18. Jal, *Scènes*, 1:69–74. I thank Katy Hornstein for this reference. See her *Myth and Menagerie*, 93–97, for the lions' story and their place in the art of Barye and Delacroix.

19. Auguste de Rigny to Georges Cuvier, August 11, 1827, AN AJ/15/751.

20. An account of the battle appears in *Moniteur universel*, 5 December 1827.

21. F. Cuvier menagerie report, 20 November 1827, AN AJ/15/634.

22. Menagerie MS 25 registers the lioness's death as 27 April 1829 and the lion's as 19 June 1829, but Cuvier's menagerie report of 7 April 1829 dates the lioness' death as 5 April 1829. AN AJ/637.

23. E. Doumet to the Museum, 26 October 1835, séance of 3 November 1835, AN AJ/15/650. The menagerie register records the animal as a *papion* (baboon) female, arriving 3 November 1835. MNHN MS 25.

24. Registre d'entrée, 1807–49, MNHN, Menagerie, MS 25. Counting all the entries in this register from 1830 through 1839, there were 244 donations, 191 births, 153 purchases, and 15 exchanges, for a total of 603 specimens. Not all entries were clearly categorized. Animals listed as "deposited" we counted as donations if they remained at the menagerie without being purchased. I did not count as births animals that died as soon as they were born. I have not included the relatively small number of animals identified as "put back" (*remis*) by various professors (not being certain what this means).

25. As for other animals, most of the sheep came from births or donations (other than the purchase of a few *mouflons*). In contrast, all five seals acquired in the 1830s were purchased, as were the nine marsupials. For all the other categories where there

were at least five specimens, donations were the primary source. The registry was divided into sections, beginning with monkeys and apes and concluding with seals.

26. Captain Luco, for example, claimed it cost him 850 francs to buy his leopard and transport it from Calcutta to Le Havre. Though he hoped the Museum would display some sympathy for the financial straits his venture had put him in, it did not, and he was left to accept the Museum's offer of 400 francs. Luco's story can be tracked in the Museum's papers from 1 June to 13 September 1831. AN AJ/15/641-42, 753; PV AN AJ/15/129: 197, 214, 219, 230, 236.

27. Popular accounts of Sarah Baartman include Badou, *L'Énigme*, and Holmes, *African* Queen. The scholarly literature is huge. Scholars, writers, and playwrights have explored Baartman's story from the perspectives of race, gender, colonialism, abolitionism, and scientific practices. Fine-grained analyses have sought to consider her experiences as a person, the ambiguities of her situation, and the changing meanings of the display of her remains. See Qureshi, "Displaying Sara Baartman"; Blanckaert, *La Vénus hottentote*; and Patin, "Vies *post-mortem*."

28. *Journal Général de France*, 4 September 1814. Mentioning the famous *tablier* (apron) of the Hottentot race, a prurient dimension of the story from the start, the paper noted, "It is not said whether the Venus of the Cape [of Good Hope] is provided with this ornament."

29. *La Quotidienne*, 23 September 1814.

30. See *Journal de Paris* and *La Quotidienne* for 20 November 1814. The production is discussed by Badou, *L'Énigme*.

31. PV 14 September 1814, AN AJ/15/114: 178. "Henry Taylor" to André Thouin, 10 September 1814, in Badou, *L'Énigme*, chapter 3.

32. *Journal de Paris*, 22 January 1815 (for the show's new location).

33. Séance of 15 February 1815, AN AJ/15/115: 59. Draft letter of 16 February 1815, in Geoffroy Saint-Hilaire's handwriting, AN AJ/15/746.

34. Blainville, "Sur une femme"; G. Cuvier, "Extrait d'observations"; Kirby, "'Hottentot Venus.'" Kirby includes a translation of Geoffroy's 1 April 1815 report for Réaux. It states that "Sara" was "delineated by four artists from four points of view."

35. *La Quotidienne*, 8 May 1815; *Journal de Paris*, 25 May 1815.

36. Kirby, "Hottentot Venus," 320.

37. *Gazette de France*, 12 April 1815.

38. Huet's painting of the rhino was presented to the assembly, PV 17 May 1815, AN AJ/15/115: 93. For his complaint, see Huet to the Museum, 23 August 1815, séance of same date, AN AJ/15/611.

39. F. Cuvier, "Le rhinocéros unicorne male," *HNM*. January 1820.

40. "Revue de quelques curiosités: Du Rhinocéros apprivoisé et de la Vénus hottentote civilisée." *Gazette de France*, 30 April 1815.

41. Baartman's death was announced in *Journal général de France*, 31 December 1815.

42. See the superb analysis of the affair by Patin, "Vies *post-mortem*."

43. Geoffroy to "monsieur le prefet," 30 December 1815, AN AJ/15/746.

44. G. Cuvier, "Extrait d'observations," 271.

45. *Journal général de France*, 8 January 1816. References to Sarah's manager as her *cornac* began earlier in a clumsy satire in *Journal de Paris*, 7 November 1814.

46. *Journal de Paris*, 8 January 1816.

47. Patin, "Vies post-mortem," 83–87.

48. F. Cuvier menagerie reports, 9 and 22 November 1815, filed with the séance of 22 November 1815, AN AJ/15/611.

49. F. Cuvier, undated letter to "Monsieur," séance of 3 April 1816, AN AJ/15/612; PV AN AJ/15/116, pp. 67–68.

50. See Patin, "Vies post-mortem," 86, and PV 18 December 1816, AN AJ/15/116, pp. 173–174.

51. PV 26 March 1817, AN AJ/15/117: 26. Cuvier's printed menagerie form for the week of 5–12 March, séance of 7 January 1818, indicates the date Réaux picked up the animals. AN AJ/15/615.

52. F. Cuvier to the Museum, 23 April 1817, AN AJ/15/614. PV AN AJ/15/117, p. 44.

53. É. Geoffroy Saint-Hilaire and Frédéric Cuvier, *Histoire naturelle des mammifères* (1833).

54. On Cross, see the rich account in Grigson, *Menagerie*, 195–214, 231–33, 254–58.

55. F. Cuvier, "Le Daman d'Éthiopie," *HNM*. November 1826.

56. The orangutan lived in Cross's menagerie for about two years; the chimpanzee expired soon after arrival. See Grigson, *Menagerie*, 197–98.

57. Grigson, *Menagerie*, 209.

58. F. Cuvier note dated 18 October 1821, AN AJ/15/622.

59. F. Cuvier menagerie report, 26 July 1824, séance of 27 July 1824, AN AJ/15/628.

60. Cross to "My Dear Friend" [Royer] dated "Saturday morning" (3 June 1826 or possibly the week before). The letter is filed with a session the following year, 22 May 1827, AN AJ/15/633.

61. *Moniteur universel*, 7 March 1826, 273. See Grigson, 210–12.

62. PV 6 June 1826, AN AJ/15/126, p. 110. F. Cuvier's menagerie report for this date is filed with the session of 22 May 1827, AN AJ/15/633.

63. F. Cuvier report, 22 May 1827, AN AJ/15/633. PV 29 May 1827, AN AJ/15/127, pp. 44–45. Museum to the Minister of the Interior, 13 November 1827, AN F/17/3980 (draft in AN AJ/15/751).

64. Anonymous, *Guide des Étrangers* (1828), 99.

65. Grigson, *Menagerie*, 252–55.

66. "Liste des animaux vivants appartenant à M. E. Cross de Londres & dont il pourrait se défaire," n.d., AJ/15/638. Presented by Geoffroy at the assembly of 13 October 1829. PV AN AJ/15/128, pp. 180–81.

67. Isidore Geoffroy Saint-Hilaire, menagerie report, 30 July 1835, AN AJ/15/754.

68. "Registre d'entrée des mammifères, 1807–1849." MNHN MS 25.

69. On Martin see Pichot, *Leven en Lotgevalen van Henri Martin.*

70. *Journal des débats*, 25 February 1830.

71. *Figaro*, 1 November 1830.

72. F. Cuvier menagerie report, 27 April 1830, AN AJ/15/639. PV AN AJ/15/128, p. 256.

73. Secretary General of the Minister of the Interior to the Museum, 11 February 1831, séance of 22 February 1831, AN AJ/15/641. Martin's letter to the Minister is not in the file of the meeting, but the considerations he offered were evidently comparable to those in the anonymous 1831 pamphlet, *Notice sur M. Henry Martin.*

74. PV 22 February 1831, AN AJ/15/129, p. 141.

75. Museum to the prefect of police, 7 April 1831, AN AJ/15/753.

76. *Le Figaro* and *La Quotidienne*, 21 April 1831.

77. *La Nouvelle France*, 22 April 1831.

78. *Messager des Chambres*, 25 April 1831.

79. See Isidore Geoffroy Saint-Hilaire menagerie report, 2 June 1835, séance of same date, AN AJ/15/649. See also "Registre d'entrée des oiseaux à la ménagerie, 1822–1848," MNHN MS ménagerie 32.

80. Isidore Geoffroy Saint-Hilaire menagerie report, 5 and 20 June 1837, AJ/15/653; AN AJ/15/134, pp. 89, 99.

81. PV 3 April 1838, AN AJ/15/135, p. 74.

82. Vaillant's arrangement with Frédéric Cuvier is stated in a letter from Vaillant to "Monsieur" [Isidore Geoffroy Saint-Hilaire], undated but presented to Isidore 4 September 1838, as indicated in his menagerie report of the same date, AN AJ/15/656.

83. PV 9 October 1838, AN AJ/15/135: 180. Vaillant's salary was set at 950 francs per year.

84. Vaillant to "Monsieur," [4 September 1838], AN AJ/15/656.

85. The total of birds Vaillant provided the menagerie was not as large as the number sent by François Baillon in Abbeville over the course of half a century, but Baillon's greater numbers of specimens did not match Vaillant's contributions in numbers of different species.

86. "Registre d'entrée, 1807–1849," MNHN, Ménagerie MS 25.

87. Callégari to the Museum, 23 October 1835, séance of 27 October 1835, AN AJ/15/650, and PV 27 October and 24 November 1835 and 12 January 1836, AN AJ/15/132, pp. 187, 213, 237.

88. The phrase "magnificent reptiles" comes from Isidore Geoffroy Saint-Hilaire's menagerie report, 4 September 1838, AN AJ/15/656. On the pythons, see S. Henry Berthoud, "Deux ménageries," *La Presse*, 18 December 1838.

89. Vallée was appointed to the menagerie at the recommendation of Constant Duméril, professor of fish and reptiles, séance of 9 October 1838 (AJ/15/135: 180). Isidore Geoffroy Saint-Hilaire's menagerie report, 4 September 1838, suggests Vallée acquired some of his reptiles from Vaillant, AN AJ/15/656.

90. Isidore Geoffroy Saint-Hilaire's menagerie report, 10–16 October, séance of 16 October, AN AJ/15/656, and Duméril, "Notice historique." See also Berthoud, "Deux menageries," *La Presse*, 18 December 1838, 1. The Museum had received rattlesnakes from the United States in the 1790s, and later it received snakes, turtles, tortoises, and crocodiles, but it was not recognized as having a reptile collection per se before 1838.

3.7 TAPIRS AND TRAVELERS

1. Buffon, "Animaux du nouveau monde," *HNGP* (1761), 9:84–96; "Animaux communs aux deux continents," *HNGP* (1761), 9:97–142; "Le tapir ou l'anta," *HNGP* (1764), 11:444–50.

2. Georges Cuvier, *Le Règne animal*, (1817), 1:242.

3. Museum to the Minister of the Marine, 14 and 26 February 1817, AN AJ/15/747.

4. Champoiseau, "Voyages." Much of Champoiseau's information on Diard can be confirmed elsewhere, but the statement that "it was he who dissected the Vénus Hottentot" seems unlikely. More plausible is Champoiseau's claim that Diard did not embark on the first government ship available to him because he wanted to help dissect the elephant.

5. Cuvier announced Diard's letter and tapir discovery on 7 December 1818 at the Academy of Sciences. Cuvier, "Sur un Tapir."

6. Diard to Cuvier, 6 January 1818, séance of 20 January 1819, AN AJ/15/617. For a later naturalist's engagement with the Asiatic Society, see Brandon-Jones, "Edward Blyth."

7. Hastings is known in American history for having led the final British assault on Bunker Hill.

8. Diard to Cuvier, 6 January 1818.

9. Diard to Cuvier, 6 January 1818.

10. Farquhar, "Account of a New Species of Tapir"; Diard, "Observations."

11. F. Cuvier, "Le Maïba," *HNM*, March 1819.

12. Coulmann, *Réminiscences*, 1:147.

13. Coulmann, *Réminiscences*, 1:142–43.

14. G. Cuvier, "Notice sur les voyages de MM. Diard et Duvaucel."

15. G. Cuvier, "Notice sur les voyages de MM. Diard et Duvaucel." For the live goat from Hastings, see 139–40. In 1821 six live monkeys sent by Diard and Duvaucel from Calcutta reached Le Havre and were sent along to Paris. PV 18 September 1821, AN AJ/15, 121, pp. 70–71.

16. In April 1819 the minister of the interior gave Diard and Duvaucel an annual stipend of 3,000 francs each, PV 27 April 1819, AN AJ/15/118, p. 158.

17. G. Cuvier, "Notice sur les voyages de MM. Diard et Duvaucel," 479.

18. G. Cuvier, *Recherches sur les ossemens fossiles*, 3rd ed. (1825), 5:261.

19. F. Cuvier, "Le dugong," *HNM*, April 1821; Sir Thomas Stamford Raffles, "Account of the Dugong." Before getting a full specimen to dissect, the French naturalists had the chance to examine the remains of a dugong preserved by Farquhar. Diard sent an account of their observations to Georges Cuvier from Singapore on 18 June 1819. MNHN MS 625 (Georges Cuvier letters: "Lettres et pièces relative au voyage de Diard et Duvaucel dans l'Inde"). I thank Jean-Baptiste Grodwohl for sending me digital copies of the letters in this file.

20. Raffles, "Dugong," 174.

21. G. Cuvier, "Notice sur les voyages de MM. Diard et Duvaucel," 476–77; "Voyage de Duvaucel," 41. See also Champoiseau, 226–31, and H. Diard, "Pierre-Médard Diard," 199.

22. Georges Cuvier, "Notice sur les voyages de MM. Diard et Duvaucel," 480.

23. Lady Sophia Raffles, *Memoir*, quotation on 703. MNHN MS 625 (Papiers de Georges Cuvier) includes an undated letter from Cuvier to "Monsieur le Duc" [Decazes], former French minister of the interior, asking him to persuade the British to treat the French naturalists more justly.

24. Diard to Madame Cuvier, no date or place given [but from Batavia in 1820], MNHN MS 625 (12). The quotation comes from page 22 of a twenty-three-page letter recounting the affair with Raffles.

25. Raffles, "Descriptive catalogue." Dated Fort Marlborough, June 1, 1820, this catalog was communicated to the Linnean Society by Everard Home on 5 December 1820. Quotations on p. 240. Regarding the catalog, see Raffles's comments in Lady Raffles, *Memoir*, 439–40.

26. See F. Cuvier, "Le Cimepaye," *HNM*, July 1821.

27. Lady Raffles, *Memoir*, 568, 592–93, 698–701. Blunt, *Ark in the Park*, 24–26.

28. Diard to Madame Cuvier, 20 October 1820, MNHN MS 625 (6), 22.

29. Diard to G. Cuvier, 20 October 1820, MNHN MS 625.

30. Diard to Cuvier, 21 October 1820, MNHN MS 625 (10).

31. G. Cuvier, "Notice sur les voyages de MM. Diard et Duvaucel," 481.

32. In April 1821 Diard embarked with a ship's captain from d'Anvers a pair of small deer from Java and two macaques from Japan, but they never arrived. See Diard to Cuvier, undated, MNHN MS 625 (15), probably written the same time as Diard's letter of 28 April 1821 to Duvaucel. MNHN NS 625 (14).

33. Museum to the Minister of the Interior of 9 October 1821 and 28 February 1822, AN AJ/15/748. The government agreed to fund Diard and Duvaucel but ultimately refused Duvaucel's proposal to move the naturalization garden from Pondicherry to Goretty, despite the Minister of the Navy and Colonies' expressed interest in the project. Minister of the Navy to the Museum, of 20 December 1822, séance of 24 December 1822, AN AJ/15/624.

34. Anonymous (F. Cuvier), "Voyage de M. Alfred Duvaucel." Voyager naturalists' accounts characteristically said little about how local people aided them, but native expertise was essential to their work. For an earlier contribution to this now burgeoning literature, see Camerini, "Wallace in the Field," and more recently, Wyhe, "Wallace's Help."

35. Quoted by F. Cuvier, "L'entelle vieux," 3, *HNM*, February 1825.

36. Duvaucel, "Lettres familières," 622–23. F. Cuvier edited this version lightly in quoting it.

37. Duvaucel, "Lettres familières." For Duvaucel's treatment by a Scottish doctor and an Indian doctor, see Coulmann, *Réminiscences*, 1:152–53.

38. Georges Cuvier, "Extrait d'un rapport sur l'état de l'histoire naturelle," 43, 48.

39. F. Cuvier, "Troisième notice."

40. Duvaucel to the Museum, 1 June 1824, séance of 7 December 1824, AN AJ/15/628.

41. PV 1 February 1825, AN AJ/15/124, p. 171.

42. PV 22 March 1825, AN AJ/15/124, p. 196.

43. *Journal des débats*, 24 March 1825, 1. The notice stated Duvaucel had died in Madras toward the end of August 1824, a victim "to the influence of the climate." At the séance of 18 October 1825, the Museum learned Duvaucel's official death date from the Minister of the Marine. PV 7 September 1824, AN AJ/15/125, p. 138.

44. PV 29 March 1825, AN AJ/15/124, p. 201.

45. F. Cuvier, "Le Croo," *HNM*, March 1825.

46. *Le Constitutionnel*, 28 May 1825, 3.

47. Museum to Milius, 22 February 1824, AN AJ/15/749.

48. Eyriès frères to the Museum, 5 May 1825, séance of 10 May 1825, AN AJ/15/629.

49. F. Cuvier to the Museum, 7 May 1825, séance of 10 May 1825, AN AJ/15/629.

50. F. Cuvier to the Museum, 7 May 1825.

51. Milius to the Museum, Paris le 12 mai 1825, séance of 17 May 1825, AN AJ/15/629.

52. F. Cuvier menagerie report, 31 May and 7 and 14 June 1825, séances of the same dates.

53. F. Cuvier, "Tapir d'Amérique," 2, *HNM*, November 1825.

54. F. Cuvier, "Tapir d'Amérique," *HNM*.

55. F. Cuvier, "Essai sur la domesticité."

56. For Diard's memoir to the Minister of the Navy and the Colonies, see Champoiseau, "Voyages," 234–35.

57. H. Diard, "Pierre-Médard Diard."

58. Diard to the Museum, 6 March 1826, séance of 27 June 1826, AN AJ/15/631. A more legible copy is in F/17/3975, though it misdates the letter as 6 May 1826.

59. Weber, "Collecting Colonial Nature."

60. Diard to the Museum, 6 March 1826, séance of 27 June 1826, AN AJ/15/631.

61. Diard to the Museum, 6 March 1826, séance of 27 June 1826, AN AJ/15/631.

62. PV 30 May 1826, AN AJ/15/126: 99. A draft letter to this effect dated 19 May 1826 is in the Diard file, AN AJ/15/572.

63. PV 11 July 1826, AN AJ/15/126, pp. 126–27.

64. PV 16 January 1827, AN AJ/15/126, p. 196.

65. PV 6 March 1827, and AJ/15/127, p. 2.

66. PV 12 June 1827, AN AJ/15/127, p. 48.

67. See above, 3.4. The number of animals that died in transport is untold, but the archives reveal many instances of its happening. Examples include the naturalist Ramón de la Sagra's attempts in 1829 and 1837 to send the menagerie live crocodiles from Havana. The 1829 specimen expired just three days before reaching *l'Orient*. When the two specimens of 1837 died at sea, the captain pitched the putrefying carcasses overboard. See letters in the séances of 15 September 1829, AJ/15/638, and 8 August 1837, AJ/15/654. PV AN AJ/15/128, p. 168, and AJ/15/134, p. 131.

68. *L'Écho du monde savant* 1 (10 April 1834): 6.

69. PV 25 February 1834, AN AJ/15/131, pp. 150–51. The Museum agreed to name Perrottet an official correspondent and provide him with letters of recommendation, three hundred francs for purchasing specimens, and instructions on collecting. PV 11 March 1834, AN AJ/15/131, p. 161.

70. Perrottet to Geoffroy, 3 April 1834, séance of 8 April 1834, AN AJ/15/647. PV 8 April 1834, AN AJ/15/131, p. 175.

71. Perrottet to the Museum, 23, April 1834, séance of 29 April 1834, AN AJ/15/647. In 1837 Perrottet would be sending the Museum animal, plant, and mineral specimens from Pondicherry. PV 14 November 1837, AN AJ/15/134, p. 190.

72. Salaün to the Museum 4 May 1834, séance of 6 May 1 834, AN AJ/15/647. PV AN AJ/15/131, p. 190.

73. Salaün to Geoffroy Saint-Hilaire, 10 May 1834; contract between Adrien de Jussieu and Salaün, 25 May 1834, AN AJ/15/647.

74. Marion de Procé, "Notice," 235.

75. Marion de Procé to Blainville, 17 May 1834, séance of 20 May 1834, AN AJ/15/647. On Marion de Procé, see Laënnec [Obituary].

76. Museum to Salaün, 21 May 1834, AN AJ/15/754. Salaün to the Museum, 25 May 1834, séance of 3 June 1834, AN AJ/15/647. A note in *L'Écho du monde savant* (18 July 1834): 62, reported the tapir's death and mistakenly claimed, "This tapir, carefully mounted, will soon decorate the galleries of the Museum."

3.8 THE DEMANDS OF DIPLOMACY

1. Loisel, *Histoire des ménageries*, 2:112.

2. Plantet, Correspondence, 3:16n, 56n

3. On animals as diplomatic gifts, see Leira and Neumann, "Beastly Diplomacy." For gifts in French diplomatic relations with Tunis, see Windler, *La diplomatie*.

4. On gifts as a source of friction between Napoleon and Algeria, see Dzanic, "Civilizing Sea."

5. Minister of the Interior (Chaptal) to the Museum, 15 nivôse an 11, séance of 20 nivôse an 11 (10 January 1803), AN AJ/15/589. PV 20 nivôse an 11, AN AJ/15/104, p. 22.

6. *Journal des débats*, 4 pluviôse an 11 (24 January 1803), 2.

7. Jussieu to Desfontaines, 4 vendémiaire an 11, séance of 12 vendémiaire an 11, AN AJ/15/589; PV AN AJ/15/103, p. 169.

8. Geoffroy Saint-Hilaire, report of 14 frimaire an 12, séance of 15 frimaire an 12, AN AJ/15/591.

9. See Vivant-Denon to Fourcroy, 16 frimaire an 12, séance of 22 frimaire an 12, AN AJ/15/591.

10. PV 15 frimaire an 12, AN AJ/15/105, p. 55.

11. F. Cuvier, *Notice des animaux vivants* (1804), 13, 18. Cuvier reported the Spanish keeper accompanied the lions from North Africa; Jussieu said he joined them in Marseille.

12. PV 20 nivôse an 11, AN AJ/15/104, p. 22.

13. Dubois-Thainville's predecessor, Dominique-Marie-Moltedo, committed a faux pas by failing to arrive with the proper presents for the dey. See Devoulx, *Archives*, 129. Thainville as consul reported several of the dey's fits over presents. Dzanic, "Civilizing Sea," 159.

14. *Journal des débats*, 22 March 1803, 1.

15. Rosenstock, "House"; Kalman, *Kings of Algiers*.

16. Dubois-Thainville to Talleyrand, 23 January 1803. Archives du Ministère des Affaires Étrangères, Algiers/36. I accessed this on 7 November 2018 on the website www.le-prince-de-talleyrand.fr. This website no longer exists.

17. Minister of the Interior to the Museum, 5 September 1807, séance of 9 September 1807, AN AJ/15/598.

18. Geoffroy's report, séance of 24 February 1808 and the Museum's letter to the minister the next day are in AJ/15/745.

19. Séance of 1 June 1808, AN AJ/15/109, pp. 27–29.

20. Kalman, *Kings of Algiers*, 88–96.

21. *Moniteur universel*, 1 April 1808, 362. Dzanic, "Civilizing Sea," 166.

22. An account of Ahmed's death, dated Algiers, 17 November (1808), appeared in *Le Moniteur universel*, 19 January 1809.

23. Arago, *Histoire*. For Arago's adventures between Majorca and Marseille see J.-B. Biot, "Notice sur les operations faites en Espagne pour prolonger la méridienne de France jusqu'aux îles Baléares, lue à la dernière séance publique de la classe des sciences physiques et mathématiques," *Le Moniteur universel*, 7 January 1810, 25–26.

24. Arago, *Histoire*, 118.

25. See Cuvier's notes under "Jeune lion," 24 March 1809, MNHN MS 16, 41, and his *Notice des animaux vivans* (1809), 42–43. On the lion's transportation expenses, see Prefect of Bouches-du-Rhône to the Museum, 3 March 1809, séance of 29 March 1809, AN AJ/15/601; PV AN AJ/15/109, pp. 135–36.

26. The bill from the Préfecture des Bouches-du-Rhône, Marseille, 2 May 1809, was presented at the séance of 10 May 1809. I found it filed in AN AJ/15/603 with items from 1810.

27. PV 10 May 1809, AN AJ/156/109, p. 157.

28. Museum to Thibaudeau, 15 February 1809, AN AJ/15/745.

29. On the lion's death, see F. Cuvier menagerie report, 26 July 1825, AN AJ/15/630.

30. Minister of the Interior to the Museum, séance of 9 August 1809, AN AJ/15/602; Museum letters to Famin and Thibaudeau, both dated 7 August 1809, AN AJ/15/745; PV 20 September 1809, AN AJ/15/110, p. 39.

31. Minister of the Interior to the Museum, 25 October 1809, séance of 8 November 1809, AN AJ/15/612. The donor, Pierre Charles Le Vasseur, later died in battle in the Russian campaign. See PV 26 June 1816, AN AJ/15/116, p. 104, for the minister's decision to pay Le Vasseur's family for the one animal of the pair (the lioness) that remained.

32. Minister of the Interior to the Museum, 17 April 1810, no séance specified, AN AJ/15/603.

33. Thibaudeau to the Museum, 6 June 1810, séance of 13 June 1810, AN AJ/15/603. Thibaudeau identified the captain as Makeusos Rais, commanding the Tunisian xebec, *La Gambo*. PV AN AJ/15/111, pp. 27–28.

34. PV 21 and 28 November 1810, 13, and 27 February and 20 March 1811, AN AJ/15/111, pp. 125–26, 129, 176, 186, and 198. See Geoffroy's manuscript notes in the séances of 6 and 20 March 1811, AN AJ/16/604. Keeping track of which lion was which is difficult. The tallies would almost balance if the 1810 lion from Tunis happened to be female. The tallies would fit perfectly if the lioness sold in February 1810 was the lioness from the Sahara that Cassal acquired in 1798. However, that animal had likely died by then. It is not included in F. Cuvier's menagerie guide of 1809.

35. [F. Cuvier], "État des animaux présens à la ménagerie au treize Avril 1813 et nourriture consommé par eux chaque jour." Séance du 12 mai 1813, AN AJ/15/607. "État des animaux présens à la ménagerie le 1er juin 1815 et de leur consommation journalière." AJ/15/610. Filed with the séance of 28 June 1815.

36. PV 4 July 1810, AN AJ/15/111, pp. 40–41.

37. Minister of the Interior (Montalivet) to the Museum, 30 June 1810, séance of 4 July 1810, AN AJ/15/603.

38. Minister of the Interior (Lainé) to the Museum, 14 June 1816, séance of 19 June 1816. AN AJ/15/612. The Museum instructed Geoffroy to contact Famin. The file includes a note in Geoffroy's hand dated 1808 regarding previous costs of transport and animal acquisitions.

39. Minister of the Interior to the Museum, 16 June 1816, and Sourdeau, Commissioner General and Chargé d'Affaires in Morocco to the Minister of the Interior, 21 May 1816, séance of 19 June 1816, AN AJ/15/612.

40. Famin to the Museum, 12 July, 10 August, 16 September, and 22 October 1816, in AJ/15/613; Museum to Famin, 5 July, 29 July, 23 August, and 14 October 1816, AN AJ/15/746. Minister of the Interior to the Museum, 14 June 1816, AN AJ/15/612, and 19 August, 26 August, and 30 October 1816, AN AJ/15/613; Museum to the Minister of the Interior, 20 July and 17 August 1816, AN AJ/15/746. For the corresponding procès-verbaux: AN AJ/15/116, pp. 101–2, 114–15, 126, 128, 129–30. The sum of 9,500 francs appears in the 20 July 1816 letter of the Museum to the Minister.

41. PV 3 and 10 January 1816, AN AJ/15/116, pp. 14, 17–23.

42. Museum to the Minister of Interior, 20 July 1816, AN AJ/15/746.

43. Minister of the Interior to the Museum, 30 October 1816, séance of 6 November 1816, AN AJ/15/613.

44. Minister of the Interior to the Museum, 18 October 1817, séance of 22 October 1817, AN AJ/15/614.

45. On transportation expenses see the procès-verbaux for April and May 1818, AN AJ/15/118, pp. 5, 9, 11, 15, 18, 23.

46. Hornstein, "From Museum to Menagerie."

47. Minister of the Interior to the Museum, 27 and 30 March 1819, séance of 6 April 1819, AN AJ/15/617; PV AN AJ/15/118, p. 149.

48. Villeneuve, "État général des frais faits pour le débarquement, nourriture et envoi à Paris, d'une lionne, donnée au roi par Monsieur Busnack [*sic*], 17 April 1819, AN AJ/15/844.

49. Museum to the Minister of the Interior, 21 May 1819 F/17/3980; PV 18 May 1819, AN AJ/15/119, p. 6. Minister of the Interior to the Museum, 27 May 1819, séance of 15 June 1819, AN AJ/15/617.

50. Museum to the Minister of the Interior, 30 September 1820, AN AJ/15/747.

51. This was the second of two ostriches Drovetti sent to France in 1822. The first, a female, arrived on 23 April 1822 and died the next day. The male arrived on 7 May 1822 and completed thirty days of quarantine before being shipped to Paris. Famin to the Museum, 11 May 1822, séance of 21 May 1822; Famin to the Museum 27 June 1822, séance of 9 July 1822, both in AN AJ/15/623. See also PV 9 July 1822, AN AJ/15/122, p. 57. Museum to Famin, 12 July 1822, AN AJ/15/748.

52. Museum to the Prefect of Chalons sur Saone, 11 June 1822; Museum to M. Bardousse Vassel [?] at Chalons, 10 August 1822; and Museum to the Minister of the Interior, 2 September 1822. AN AJ/15/748. Prefect of Arles to the Minister of the Interior, 18 June 1822, AN AJ/15/623.

53. Villeneuve to the Museum, 10 June 1823, AN AJ/15/625; Villeneuve to the Minister of the Interior, 18 June 1823, F/17/3980.

54. PV 15 July 1823, AN AJ/15/123: 61. Museum to the Minister of the Interior, 25 July 1823, AN AJ/15/749.

55. Minister of the Interior to the Museum, 20 September 1823, séance of 23 September 1823, AN AJ/15/626.

56. Minister of the Marine to the Museum, 14 July 1824, séance of 20 July 1824, AN AJ/15/628. Minister of the Marine to the Minister of the Interior, 14 July 1824. AN F/17/3980.

57. F. Cuvier menagerie report, séance of 24 August 1824, AN AJ/15/628.

58. Museum to the Minister of the Interior, 2 September 1824, AN AJ/15/749.

59. The expenses totaled 727.80 francs, not including the return trip for Sr. Giraud, the animal's keeper. PV 24 August 1824, AN AJ/15/124, p. 92. See Giraud to the Museum, 26 September 1824, séance of 28 September 1824; Duc de Doudeauville to the Museum, 8 September 1824, séance of 14 September 1824, AN AJ/15/628. On the Minister of Foreign Affairs' paying for the lion, see the Museum's letter to him, 20 June 1826 (misdated 2 June 1826), AN AJ/15/750.

60. Minister of the Marine to the Museum, 3 July 1824, séance of 6 July 1824. Samson (interim commissaire general of the Marine in Toulon) in letters of 14 and 17 July 1824 asked for instructions on transporting the animals; séance of 27 July 1824. All in AN AJ/15/628.

61. On de Guys's gift, see seances of 31 August and 14 September 1824; 1 February 1825, AN AJ/15/124, pp. 95, 101, 170.

62. De Guys to the Museum, 29 December 1824, séance of 1 February 1825, AN AJ/15/629.

63. Museum to the Intendant of the Marine in Toulon, 5 February 1825, AN AJ/15/750, in response to his letter to the Museum, 22 January 1825, séance of 1 February 1825, AN AJ/15/629, PV AN AJ/15/124, p. 170.

64. Intendant to the Museum, 16 April 1825, séance of 26 April 1825, AN AJ/15/629; Museum to the Intendant, 22 April 1825, AN AJ/15/750; Intendant to the Museum, 28 April 1825, séance of 10 May 1825, AN AJ/15/629.

65. *Journal des débats*, 6 May 1825.

66. Museum to de Guys, 5 May 1825, AN AJ/15/750.

67. *Journal des débats*, 8 May 1825. The article noted that "for a long time" it was known that lion cubs played like kittens.

68. Museum to de Guys, 5 July 1825, AN AJ/15/750.

69. Minister of the Interior to the Museum, 26 March 1825, séance of 29 March 1825; PV AN AJ/15/629, AN AJ/15/124, p. 199.

70. Minister of the Interior to the Museum, 9 April 1825, séance of 12 April 1825, AN AJ/15/629.

71. *Journal des débats*, 3 March 1825.

72. *Journal des débats*, 2 April 1825.

73. *Journal des débats*, 23 April 1825.

74. Intendent de la Marine to the Museum, 16 April 1825, séance of 26 April 1825, AN AJ/15/629.

75. Eyriès frères to the Museum, 28 April 1825, séance of 3 May 1825. On Valotte's attention to the elephant see additionally Eyriès frères to the Museum, 16 May 1825, séance of 17 May 1825, AN AJ/15/629.

76. F. Cuvier to the Museum, 7 May 1825, séance of 10 May 1825, AN AJ/15/629.

77. Valotte (letter written on his behalf) to the Museum, n.d., filed with séance of 14 June 1825, AN AJ/15/629; PV 31 May 1825, AN AJ/15/125, p. 41.

78. Villeneuve to Minister of the Interior, 25 April 1825, AN F/17/3980.

79. Museum to the Minister of the Interior, 21 May 1825, AN F/17/3980.

80. Acting prefect of Bouches du Rhône to the Museum, 23 June 1825, séance of 28 June 1825, AJ/15/629. The acting prefect's instructions for Joseph Issaurat, André Malhomé, and Hyacinte Imbert, 16 June 1825, and his commission for them, 17 June 1825, are filed with related documents for the séance of 23 August 1823, AN AJ/15/630.

81. See the above, plus the acting prefect's note of 21 June 1825, granting an indemnification of 50 francs to Hyacinte Imbert, injured by the camels, AN AJ/15/629. On the camels' skin disease, see the prefect to the Museum, séance of 26 July 1825, AN AJ/15/630.

82. Museum to Minister of the Interior, 21 May 1825, AN F/17/3980.

83. PV 5 and 12 July 1825, AN AJ/15/125, pp. 61–62, 66–67.

84. Museum to the Minister of the Interior, 21 July 1825, AN F/17/3980.

85. Minister of Foreign Affairs (Le Baron de Damas) to the Museum, 1 September 1825, séance of 6 September 1825, AN AJ/15/630.

86. Museum to the Minister of Foreign Affairs, 10 September 1825, AN AJ/15/750. PV 6 September 1825, AN AJ/15/125, pp. 105–6.

87. *Le Moniteur universel*, 9 September 1825; Museum to the Minister of Foreign Affairs, 10 September 1825, AN AJ/15/750.

88. Minister of the Interior to the Museum, 12 November 1825, séance of 15 November 1825, AN AJ/15/630. Museum to the Minister, 7 December 1825, AN F/17/3980.

89. PV 13 Dec. 1825, AN AJ/15/126: 16. Minister of the Interior to the Museum, 1 February 1826, séance of 14 February 1826, AN AJ/15/631. Museum to the Minister, 24 February 1826, AN F/17/3980.

90. Minister of Foreign Affairs to the Museum, 8 June 1826; séance of 13 June 1826, AN AJ/15/631.

91. Museum to the Minister of the Interior, 20 June 1826, F17 3980.

92. Museum to the Minister of the Interior, 20 June 1826, F17 3980.

93. Museum to the Minister of Foreign Affairs, 20 June 1826 (misdated 2 June 1826), AN AJ/15/750.

94. Minister of the Interior to the Museum, 30 June 1826, séance of 4 July 1826, AJ/15/632.

3.9 THE GIRAFFE FROM THE PASHA

1. Geoffroy to the Museum, Aix, 21 May 1827, AN AJ/15/844 (giraffe file). Henceforth cited simply as AN AJ/15/844.

2. "Le Bubale," *Le Figaro*, 14 August 1833, 2.

3. Anonymous [Honoré de Balzac], "Études de philosopie [*sic*]," *La Silhouette*, 17 June 1830, 101–2.

4. The best account remains Dardaud, "L'extraordinaire aventure." See also the eminently readable and insightful Allin, *Zarafa*; Lebleu, *Les avatars de zarafa*; and Lagueux, "Geoffroy's giraffe."

5. Accounts to date have drawn on manuscripts in Marseille in the archives of Department of Bouches-du-Rhône (with copies in the Archives nationales in Paris), scientific papers, newspapers (both provincial and Parisian), pamphlets, theatrical performances, and giraffe-inspired artifacts. This account uses additional MNHN papers located both at the MNHN and at the AN.

6. Baratay, "La girafe de Charles X, une extraterrestre en France (1825–1845)," in Baratay, *Biographies animales*.

7. Minister of the Interior to the Museum, 16 November 1826, séance of 21 November 1826, AN AJ/15/632; PV AN AJ/15/126, p. 174. The Museum may have heard earlier about the giraffe's arrival in Marseille. On 1 November 1826, *Journal des débats* named its donor and indicated it was destined for the "ménagerie royale."

8. F. Cuvier, undated report to "Messieurs," séance of 21 November 1826, AN AJ/15/632. The procès-verbaux do not mention the report, but the assembly addressed some of the points in it. Previous accounts of the giraffe do not mention this report or Cuvier's other efforts regarding its care.

9. F. Cuvier to "Messieurs," séance of 21 November 1826, AN AJ/15/632.

10. Draft in Georges Cuvier's handwriting of a letter to the Minister of Interior, 21 November 1826, AN AJ/15/844.

11. Villeneuve to the Minister of the Interior, 11 November 1826, AN AJ/15/632, with copies in AJ/15/844 and extracts in Dardaud, "L'extraordinaire aventure," 56–57.

12. Villeneuve to the Minister of Interior, n.d., accompanying his letter of 11 November, séance of 5 December 1836, AN AJ/15/632.

13. Museum to Villeneuve, 28 November 1826, in Frédéric Cuvier's' handwriting, AN AJ/15/844.

14. Frédéric Cuvier, "Instructions sur les soins à donner à la girafe," n.d., AN AJ/15/844.

15. Museum to Villeneuve, 28 November 1826, AN AJ/15/844.

16. This is all we know of Salze. It is what Villeneuve reported in a letter of 5 January 1827 to the Museum, AN AJ/15/844. Dardaud, "L'extraordinaire aventure," 58.

17. [Salze], "Observations faites sur la giraffe [*sic*] arrivé à Marseille, et sortie du Lazaret le 14 Novembre 1826," AN F/17/3980. Villeneuve signed this document; there is a copy in AJ/15/844.

18. [Salze], "Observations," AN F/17/3980.

19. Villeneuve to the Museum, 5 January 1827, and Museum to Villeneuve, 19 January 1827, AN AJ/15/844. Dardaud, "L'extraordinaire aventure," 58–59, reproduces Villeneuve's letter but only summarizes the Museum's response and does not include the Museum's observations on Salze's report. On the minister's decision (29 December 1826) regarding the giraffe's expenses, see Minister to the Museum, 12 January 1827, AN AJ/15/844, and Museum to the Minister, 13 January 1827, AN AJ/15/751.

20. [Salze], "Observations," AN F/17/3980, and "Observations faites sur la girafe envoyée au roi par le pacha d'Égypte," *Mémoires du Muséum d'histoire naturelle* 14 (1827): 68–84.

21. Villeneuve to the Museum, 5 and 12 February 1827, AN AJ/15/844.

22. Drovetti to Villeneuve, 9 March 1827, in Dardaud, "L'extraordinaire aventure," 59.

23. Museum to Villeneuve, 15 March 1827, and Villeneuve to the Museum, 26 March 1827, AN AJ/15/844. The Museum asked Villeneuve about a proposition from [John] Polito, proprietor of a traveling menagerie, who saw the giraffe in Marseille and offered to transport it to Paris. Polito wanted to charge people to see it en route, a scheme the Museum found inappropriate for an animal "belonging so directly to the king." See séance of 27 March 1827, AN AJ/15/127, p. 15. On Polito, see Grigson, *Menagerie*, 214–17.

24. Villeneuve to the Museum, 19 March 1827, séance of 27 March 1827, AN AJ/15/844.

25. Geoffroy report to the Museum, 3 April 1827, AN AJ/15/844.

26. Villeneuve to the Museum, 28 April 1827, séance of 8 May 1827, AN AJ/15/844.

27. Geoffroy to the Museum, 4 May 1827, séance of 15 May 1827, AN AJ/15/844.

28. Geoffroy to the Museum, 6 May 1827, séance of 15 May 1827, AN AJ/15/844.

29. Geoffroy to Villeneuve, 19 May 1827, AN AJ/15/844. While the cages were built, Geoffroy spent three days in Lyon, where he lectured, met with scientists and students, and collected specimens. In Marseille, he performed the same scientific activities as he did in Montpellier, Nimes, Aix, and Lyon.

30. Geoffroy to the Museum, 14 May 1827, séance of 22 May 1827, AN AJ/15/844.

31. Geoffroy to the Museum from Aix, 21 May 1827, AN AJ/15/844.

32. Geoffroy to the Museum, 24 May 1827, AN AJ/15/844; Dardaud, "L'extraordinaire aventure," 63–64.

33. Geoffroy to Villeneuve, 24 May 1827, AN AJ/15/844; Dardaud, "L'extraordinaire aventure," 63–64. See also Burkhardt, "Civilizing Specimens and Citizens at the Muséum d'histoire naturelle."

34. Geoffroy to the Museum from Marseille, 14 May 1827, séance of 22 May 1827, AN AJ/15/844; Geoffroy, "Aperçu sur le transport de la giraffe," AJ/15/844; and Dardaud, "L'extraordinaire aventure," 60–61.

35. Geoffroy to the Museum from Valence, 31 May 1827, AN AJ/15/844.

36. *Gazette de Lyon*, 9 June 1827, reprinted in *La France chrétienne*, 13 June 1827.

37. *Gazette de Lyon*, reproduced in *Gazette nationale*, 30 June 1827.

38. Geoffroy to the Museum from Lyon, 9 June 1827, AN AJ/15/844. Geoffroy, "Compte que je rends, de ma mission dans le midi de la France, à l'administration du Muséum d'histoire naturelle," 24 July 1827, séance of same date, AN AJ/15/634.

39. Geoffroy to the Museum, 9 June 1827, AN AJ/15/844; *Le Constitutionnel*, 13 June 1827.

40. In a letter to the Museum from Lyon, 2 June 1827, Geoffroy described his interest both in river travel and in giving the giraffe a three-day rest at Fontainebleau before the last leg to Paris (AN AJ/15/844). He wrote a similar letter to the minister of the interior the same day (Dardaud, "L'extraordinaire aventure," 70–71). See Museum to Geoffroy, letters of 4 and 6 June 1827, AN AJ/15/751.

41. Geoffroy to the Museum, 9 June 1827, AN AJ/15/844, with itinerary. Geoffroy, "Compte . . . de ma mission," AJ/15/634.

42. Geoffroy's oral report at the assembly of 26 June 1826 is referenced in the PV but has not been preserved. AN AJ/15/127, p. 55. His illness (urine retention) and replacement are reported in *Le Constitutionnel*, 3 July 1827, and in a letter from Geoffroy to Descharmes, 27 June 1830, MNHN MS 2083, II.

43. F. Cuvier to Geoffroy, 30 June 1827, AN AJ/15/634. Museum to "Monsieur le Prefet," 30 June 1827 (draft in Geoffroy's hand, edited by Georges Cuvier), AN AJ/15/751.

44. *Le Constitutionnel*, 3 July 1827.

45. Museum to the editors of *Le Moniteur*, *Journal de Paris*, *Journal de débats*, and *Le Constitutionnel*, 30 June 1828, AN AJ/15/751. Notices appeared in the *Gazette nationale* (*Le Moniteur*) 1 July 1827 and *Journal des débats* 2 July 1872. Frédéric Cuvier worried that crowds at the rotunda might alarm the elephants. F. Cuvier menagerie report, 26 June 1827, AN AJ/15/633.

46. Geoffroy wrote to Villeneuve on 12 July 1827 reporting that the decision to go to St. Cloud was the dauphine's. See Dardaud, "L'extraordinaire aventure," 67; AJ/15/844. See too the séance of 3 June 1827, AN AJ/15/127, p. 58. For the rest, see Museum to the Minister, 4 July 1827, AN AJ/15/751; Minister to the Museum, 4 July 1827, séance of 5 July 1827, AN AJ/15/634; and the Museum secretary (Bosc) to his fellow professors, 6 July 1827, AJ15/751.

47. *Le Moniteur universel*, 11 July 1827, and Geoffroy to Villeneuve, 12 July 1827, in Dardaud, "L'extraordinaire aventure," 67, and in AJ/15/844.

48. É. Saint-Hilaire, *Sur la girafe* and "Quelques considerations."

49. Mongez, "Mémoire."

50. Saint-Ange, *Notice* (announced in *Journal des débats*, 2 July 1827, but not seen by me). Ferlus's *Nouvelle notice* advertised itself as "augmented by precious documents furnished by the Ethiopians who conducted the giraffe to Paris."

51. *Journal du Commerce*, 9 July 1827, 2.

52. Undated, unsigned note in Georges Cuvier's hand, AN AJ/15/751; *Le Figaro*, 20 July 1827.

53. PV 17 and 24 July 1827, AN AJ/15/127, pp. 63, 67, 68.

54. Dardaud, "L'extraordinaire aventure," 21, gives the increase in revenues at the bridge in 1827 over the comparable six-week period in 1826 as 13,650 francs.

55. F. Cuvier to "Monsieur," n.d., AN AJ/15/634, and PV 21 August 1827, AN AJ/15/127, pp. 80–81.

56. Dardaud"L'extraordinaire aventure." On the representations of the African handlers, see Harkett, "The giraffe's keepers."

57. On Salvandry's role in the struggle for press freedom, see Kent, *Election of 1827 in France*, 82–84.

58. Anonymous, [Salvandy], *Lettre.*

59. Anonymous, [Salvandy], *Seconde lettre.*

60. Dardaud, "L'extraordinaire aventure," 26–27; Lescart, "An Egyptian giraffe."

61. Eyriès frères to the Museum, 29 July 1827, citing a letter from Guillemin, the French consul in New Orleans dated 29 May 1827. Séance of 31 July 1827, AN AJ/15/634. Delaunay evidently carried this letter to Le Havre.

62. *Le Constitutionnel*, 23 July 1827.

63. Eyriès frères to the Museum, Le Havre, 29 July 1827, séance of 31 July 1827, AN AJ/15/634.

64. Museum to Eyriès frères. 31 July 1827, AN AJ/!5/751.

65. Bruce. *Lion Gate.*

66. *Journal des débats*, 10 August 1827, 2–3.

67. *Le Moniteur*, 15 and 20 August 1828; *Journal des débats*, 23 August 1827.

68. *Le Constitutionnel*, 29 August 1827. *Le Figaro* said the same the same day. The coverage in *Gazette de France* and *Journal des débats* did not make the comparison with the giraffe but noted the Osages' minimal interest in the opera. *La Quotidienne* credited the Osages with patience.

69. *Figaro*, 30 August 1827.

70. Anonymous, *La Girafe et les Osages*, 1827. Alibassan [Balzac], *Discours.* On Balzac's authorship see Lescart, "An Egyptian giraffe."

71. Salvandy, *Insolences*, 62.

72. For reviews of "Les Osages," which opened at the *Théâtre des Variétés* on September 27, see *Le Constitutionnel, Journal des débats*, and *Le Figaro* for 29 September 1827 and *Le Journal de Commerce* for 1 October 1827. The one-act folly was about a person who disguised himself as an Osage for amusement, but the audience was not amused and whistled its odisapproval. "Les Osages" failed almost immediately. The last notice for it appeared in *Le Constitutionnel* 7 October 1827. No more successful was "Naufrage des Artistes, ou Les Osages," which debuted at the Théâtre de Vaudeville on October 1. See *Le Figaro*, 2 October, *La France nouvelle*, 5 October, *Journal des débats*, 6 October, and *La Quotidienne*, 9 October, 1827.

73. *La Quotidienne*, 1 October 1827.

74. *Journal des débats*, 2 October 1827; *Le Constitutionnel*, 3 October 1827.

75. *Le Constitutionnel*, 6, 14. and 21 October 1827. See balloonist Dupuis Delcourt's account, *Journal du Cher*, 31 October 1827. The Osages had been performing at the Tivoli since early September.

76. *La France Nouvelle* and *Nouveau journal de Paris et des départemens*, 4 October 1827; *Le Figaro*, 5 October 1827, 635. The press said that Hassan's fear of the Parisian winter prompted his return to Alexandria, but the Museum had already decided in July that it would not need him once the giraffe was moved to the rotunda, where a regular keeper could assist Atir with her care. See F. Cuvier to "Monsieur" [Geoffroy Saint Hilaire], undated, presented at the séance of 21 August 1827, AN AJ/15/634. Illness de-

layed Hassan's departure until 20 October, causing the Museum to fear he would not reach Toulon in time to board the government ship taking Drovetti to Egypt. Museum to the Minister of the Marine and the Prefect of Police respectively, 20 October 1827, to Villeneuve, 23 October 1827, and to the Minister of the Interior, 24 December 1827, all in AJ/15/751; Geoffroy to Villeneuve, 22 October 1827; Dardaud, "L'extraordinaire aventure," 69.

77. *Journal des débats*, 13 October 1827. Cuvier recounted the giraffe's installation at the rotunda in menagerie reports 16 and 23 October 1827, AN AJ/15/634.

78. Geoffroy to Villeneuve, 22 October 1827; Dardaud, "L'extraordinaire aventure," 69.

79. Museum to the Minister of the Interior, 7 February 1828, AN F/17/3980.

80. Geoffroy to Villeneuve, 22 October 1827; Dardaud, "L'extraordinaire aventure," 69.

81. *Journal de Commerce*, 29 October 1827; *Le Constitutionnel*, 2 November 1827; *Journal de Commerce*, 17 and 29 November 1827. *Journal de commerce*, 14 November 1827, has a note about Osage impersonators trying to perform for money in Melun.

82. This account is constructed from the *Moniteur*, 26 February and 24 August 1828; *Journal des débats*, 19 July 1829; *Moniteur*, 2 June 1830.

83. S. Henry Berthoud, "Deux ménageries," *La Presse*, 18 December 1838, 1.

84. For an astute analysis of one such example, see Simon, *Obaysch*.

85. F. Cuvier menagerie report, 30 April 1828, AN AJ/15/635.

86. F. Cuvier, menagerie report, 19 December 1828, AN AJ/15/636.

87. F. Cuvier, menagerie report, 7 July 1829, AN AJ/15/638.

88. F. Cuvier, menagerie report, 9 February 1830, AN AJ/15/639.

89. F. Cuvier, menagerie report, 6 April 1830, AN AJ/15/639.

90. *La France nouvelle*, 9 May and 18 August 1828.

91. *La Quotidienne*, 4 July 1829; *La France nouvelle*, 27 July 1829. *Journal des débats*, *Le Figaro*, and *La France Nouvelle* on 20 October 1829 announced the British giraffe's death. *Le Figaro* jokingly alleged that Geoffroy would travel to England to oversee its stuffing.

92. Minister of Public Instruction (Guizot) to the Museum, 5 February 1835, and copy of Mimaut's letter to the Minister of the Marine, 15 December 1834, séance of 10 February 1835, AN AJ/15/649.

93. Museum to the Minister of Public Instruction, 14 February 1835, AN AJ/15/754. Séance of 10 February 1835, PV, AN AJ/15/132, p. 68.

94. Mimaut to the President of the Council of Ministers, filed with the séance of 13 October 1835, AN AJ/15/650. Discussion in AN AJ/15/132, p. 177.

95. Museum to the Minister of Public Instruction, 16 October 1835, AN AJ/15/754. See also PV 13 October 1835, AN AJ/15/132, p. 177.

96. *Le Moniteur*, 2 April 1836. See also *L'Echo du monde savant*, 7 April 1836; *Journal des débats*, 28 April 1836. Vaissiere's "new voyage" was described as "as interesting as it is dangerous," but without further details.

97. *Evening Post* (New York) 4 October 1836: 2. The same paper 27 June 1838 noted "Those wonderful animals, the Giraffes, at 509 Broadway . . . are attracting and gratifying constant crowds of people." John Clayton was credited with capturing them in South Africa. An advertisement in the *Evening Post*, August 16, 1838, claims they were the first giraffes brought to America. Clayton's giraffes are mentioned in Hallock, "The American Circus," 573.

98. "The Giraffes at the Zoological Gardens, Regent's Park," *Mirror of Literature, Amusement, and Instruction* 27 (June 11, 1836): 385–89. Quotation from 387n.

99. Owen "Note," quotation on 244.

100. G. Thibaut to the Museum, 17 July 1836, AN AJ/15/652.

101. *L'Echo du monde savant*, 24 July 1836. This article was probably written by Henri de Blainville, who met with Vaissière to discuss procuring animals for the menagerie.

102. Blainville to Thibaut, 17 August 1836, AN AJ/15/755.

103. Salze, "Observations faîtes sur la giraffe," 71.

104. Thibaut letter dated 8 January 1836 from Malta, read February 9, 1836, *Proceedings of the Zoological Society of London*, 4 (1836): 9–12.

105. Grigson, *Menagerie*, 257. By year's end, two of Surrey Gardens' giraffes had died.

106. *Journal des débats*, 5 June 1827.

107. "Girafes arrivées en 1836 dans la ménagerie de la Société zoologique de London," *Magasin pittoresque* (1836): 375–76.

108. Geoffroy in 1832 was writing a piece about changes in the giraffe's anatomy over time, but all that remains of the venture is a partial unpublished manuscript, "Quelques nouvelles considérations sur la giraffe," in MNHN MS 2717 (2).

109. Frédéric Cuvier fils, "La giraffe," *IINM*, September 1842.

110. Owen "Note."

3.10 THE LIONS OF ALGERIA

1. *Le Constitutionnel*, 10 July 1830: 1.

2. Museum letter to the Minister of War and the Minister of the Navy, 12 July 1830, AN AJ/15/752.

3. Museum to the Minister of the Interior, 12 July 1830, AN AJ/15/752.

4. The anonymous *Guide des étrangers* (1828), observed (p. 97) that the number of *animaux féroces* had been "diminished for some time."

5. The previous gap was between the death of the lion from Versailles in July 1796 and the arrival on 26 February 1797 of the young lioness from a captured Spanish ship.

6. Minister of the Navy to the Museum 19 July 1830, séance of 20 July 1830, AN AJ/15/640. See too PV AN AJ/15/129, p. 40.

7. Champagny (aide-de-camp de Mr le Daupin) to the Museum, 17 July 1830, urged the Museum to prepare supplementary notes tailored for the region. AN AJ/15/640. See also the Museum's letter to Mr le Maréchal de France, Gouverneur d'Alger, séance of 20 July 1830, AN AJ/15/640.

8. The Comte de Bourmont's letter, addressed simply "Prince," dated Cassauba (Casbah), 2 August 1830, was passed to the Museum by the minister of the interior and read at the séance of 14 September 1830, AN AJ/15/640.

9. The nonarrival of the ostriches is stated but not explained in *Gazette littéraire*, 11 November 1830, 783.

10. Hornstein, *Myth and Menagerie*, begins with the story of the confiscated lions and the mixed messages they evoked.

11. Sergent [name illegible], to "Monsieur et ami" [Geoffroy Saint-Hilaire] dated "Alger, du palais de la Cassauba, 20 aout 1830." Séance of 14 September 1830, AN AJ/15/640.

12. Bertier de Sauvigny, *Bourbon Restoration*, 436.

13. *Le Temps*, 20 June 1830. See also *Le Constitutionnel*, 15 June 1830, 2–3.

14. Séance of 17 August 1830, AN AJ/15/129, p. 48. The July Revolution left little mark in the Museum's records. No meeting was held the first day, July 27, because a quorum was not present. At their next meeting, the professors decided that summer courses could be declared finished, since they nearly were and students had largely dispersed. Séance of 3 August 1830, AN AJ/15/129, p. 44.

15. Museum to Minister of the Interior, 24 September 1830, AN AJ/15/752.

16. The purchase was concluded for 1,200 francs, with three mouflons and a white guinea hen added for free. PV 21 and 28 September 1830, AN AJ/15/129, pp. 68, 70. See too F. Cuvier menagerie report, 4 October 1830 (misdated 4 September 1830), séance of 5 October 1830, AN AJ/15/640.

17. Minister of the Interior to the Museum, 11 September 1830, séance of 21 September 1830, AN AJ/15/640.

18. PV 28 September 1830, AN AJ/15/129, p. 71. The Museum could cover 6,000 to 7,000 francs of improvements to the carnivore pavilion, thanks to 50,000 francs from the government for construction projects.

19. F. Cuvier, menagerie reports 28 September 1830 and 26 October, séances of same dates, AN AJ/15/640.

20. *Journal des débats*, 26 October 1830, citing the *Journal de la Drôme*, 20 October 1830.

21. F. Cuvier, menagerie report 9 November 1830, AN AJ/15/640.

22. *Le Figaro*, 22 November 1830.

23. *La France Nouvelle*, 20 November 1830.

24. F. Cuvier, menagerie report, 9 November 1830, AN AJ/15/640. The PV of 9 November 1830 incorrectly report the carnivore pavilion population as "42 species." AN AJ/15/129, 95. Hippolyte Martin's appointment was reported in PV 16 November 1830, AN AJ/15/129, p. 97.

25. Cuvier to Geoffroy Saint-Hilaire, 8 May 1832, AN AJ/15/643.

26. Cuvier to Geoffroy, 8 May 1832, AN AJ/15/643.

27. MNHN Ménagerie MS 25. The register incorrectly lists the lions from captain Geoffroy as lionesses.

28. The jackals from Joseph Smolenski and Lieutenant Court and the purchase of the fox are mentioned respectively in PV 8 and 23 September 1834 and 27 January 1835, AN AJ/15/131, pp. 174, 254, and AJ/15/132, p. 59. In 1835, anticipating improved opportunities for acquiring specimens from Algeria, it was decided that the professors should prepare a wish list to send to the governor-general. PV 2 February 1836, AN AJ/15/132, p. 249.

29. For a succinct discussion of events and consequences of the invasion of Algeria see Arthur Girault, "Algeria. History," *Encyclopaedia Britannica*, 11th ed. (1911), 1:649–53. On Abd El-Kader see Danziger, *Abd Al-Qadir, and* Brower, "The Amîr."

30. *Le Temps*, 28 April 1836.

31. On the military developments, Danziger, *Abd Al-Qadir*, 116–18.

32. PV 15 September 1835, AN AJ/15/132, p. 166. Museum to the Minister of Public Instruction 21 September 1835 (apparently responding to a letter from the minister dated 8 September), AN AJ/15/754.

33. Museum to the Minister of Public Instruction, 21 September 1835, AN AJ/15/754. Minister of Public Instruction to the Museum, letters of 25 January 1836 (séance of 26 January 1836, AN AJ/15/651; PV AN AJ/15/132, p. 243) and 9 March 1836 (séance of 15

March 1836, AN AJ/15/651; PV AJ/15/133, p. 13). Museum to the Minister of Public Instruction, 28 March 1836, AN AJ/15/755.

34. F. Cuvier draft letter to Jean René Constant Quoy in Toulon, 29 October 1836, AN AJ/15/755, requesting information on the two lions.

35. For correspondence regarding the lions see the PV for 8 November 1836, AN AJ/15/133, p. 152; 21 February 1837, AN AJ/15/134, p. 28; and 9 May—20 June 1837. AN AJ/15/134, pp. 71, 78, 83, 87, 94, 97. See too the Museum to Quoy, 18 May and 2 June 1837, and to the Maritime Prefect at Toulon, 15 May and 2 June 1837, AN AJ/15/755; plus Letellier to Isidore Geoffroy Saint-Hilaire, 9 June 1837, séance of 13 June 1837; and Quoy to the Museum, 14 June 1837, séance of 20 June 1837, both in AJ/15/653.

36. My account of Yussef's early life draws on items in the press and especially Trumelet, *Général Yusuf.* See also Rousset, *Commencements d'une conquête.* On Yusuf and decapitation see, for example, *Journal de Commerce*, 27 April 1836.

37. *La Presse*, 21 April 1837.

38. *La France*, 7 July 1837.

39. For Jaballa, see *Journal des débats*, 1 June 1837. For Barthélemy's care of Jaballa, see his letter to Geoffroy, 15 June 1837, séance of 20 June 1837, AN AJ/15/653. For more on Barthélemy's work at the Museum of Natural History in Marseille, see Barthélemy to the Museum, 24 January 1838, séance of 30 January 1838, AN AJ/15/655

40. *Le Constitutionnel*, 24 April 1838, 1–2, describes the ceremony and the presents.

41. Minister of War to the Minister of Public Instruction, 22 April 1838, AN F/17/3980.

42. PV 24 April 1838, AN AJ/15/135, pp. 85–86. Museum to the Minister of Public Instruction, 26 April 1838, AN F/17/3980 and AN AJ/15/756.

43. Le Temps, 28 April 1838, 1.

44. Danziger, *Abd Al*-Qadir, 157.

45. Janet Browne, "A Science of Empire: British Biogeography before Darwin," *Revue d'histoire des sciences* 45, no. 4 (1992): 453–75.

4.1 FAMILY TIES

1. Much information on family ties can be gleaned from Jaussaud and Brygoo, *Du Jardin au Muséum.* Outram, *Georges Cuvier*, chapter 8 and pp. 246–47n9, is highly instructive on kinship and patronage at the Museum. Among the family ties at the professorial rank, the Museum's first chemists, Antoine-François Fourcroy and Antoine-Louis Brongniart, were first cousins; Fourcroy's successor was another cousin, André Laugier. Antoine Louis Brongniart's nephew, Alexandre Brongniart, became professor of mineralogy, and Alexandre in turn had both a son (Adolphe Brongniart) and a son-in-law (Victor Audouin) who became Museum professors. André Thouin, the professor of agriculture and horticulture, had three brothers—Jacques, Jean, and Jean-André—serving in nonprofessorial capacities at the Museum.

2. PV 22 ventôse an 13, AN AJ/15/106, p. 175.

3. Georges Cuvier to Geoffroy, n.d., filed with the séance of 19 April 1809, AN AJ/15/601; PV AN AJ/15/109, pp. 146–47. The rules as adopted affirmed that the professor of zoology had charge of approving major aspects of the menagerie's business, including construction projects, paying provisioners, removing animals, and designating which superfluous ones could be killed or sold.

4. É. Geoffroy Saint-Hilaire, "Présentation à l'Assemblée administrative des professeurs du Muséum d'histoire naturelle," 9 November 1824, AN AJ/15/628.

5. Séance of 9 November 1824, AN AJ/15/124, p. 125.

6. PV 5, 12 and 19 February 1828, AN AJ/15/127, pp. 150, 152, 155. Geoffroy's undated letter to "l'assemblée des professeurs," MNHN 2717(2) was apparently written in reaction to the 12 February meeting. He cited this fable again when complaining about Frédéric Cuvier's rise in the Museum's ranks. Geoffroy Saint-Hilaire, *Fragments biographiques*, 139–40.

7. Blainville's letter is cited (undated) in Nicard, *Études*, 105–7, and further excerpted in Landrieu, *Lamarck*, 98–99.

8. For the division of Lamarck's chair, selection of Latreille and Blainville as professors, and arrangements for Lamarck's daughters, see PV from 22 December 1829 to 16 February 1830, AN AJ/15/128, pp. 208–30. For further dimensions of the intrigues in the division of Lamarck's chair, see Nicard, *Études*, 108–11.

9. Duvernoy, *Notice* historique, 99.

10. PV 20 and 25 May 1832, AN AJ/15/130, pp. 91–94. Of the two votes not cast for Blainville, one went to Flourens, the other to Duméril.

11. F. Cuvier, letter to "Messieurs," 28 July 1832, séance of 31 July 1832, AN AJ/16/644.

12. Nicard, *Études*, 248.

13. Frédéric Cuvier to "Messieurs," 28 July 1832. Pierre Flourens was elected to fill Portal's chair on 17 August 1832. The Museum had yet to replace the professor of mollusks when Blainville became professor of comparative anatomy.

14. Geoffroy to "Messieurs et très honorés collègues," 18 January 1833, séance of 29 January 1833, AN AJ/15/645. For Geoffroy's dispute with Blainville in 1833, see Burkhardt, "Naturalists' Practices." Geoffroy requested that his letters be burned, but they were not. For documents relating to this dispute, see the assemblies from 29 January to 19 February 1833, AN AJ/15/645.

15. Geoffroy letter to his colleagues, 23 April 1833, séance of same date, AN AJ/15/645.

16. For Geoffroy's increasing alienation from the scientific community see Appel, *Cuvier-Geoffroy Debate*, 178–88.

17. Séance of 15 July 1834, AN AJ/15/648. The applications came from the wives of the former gardeners Dumas and Beuchers, the gardener Champy, and the keepers Géan, Férêt, and Roux.

18. The details of the competition for the cake seller vacancies can be tracked in AN AJ/15/646–648. For F. Cuvier's view that Roux was being maligned, see his menagerie report of 8 October 1833, AN AJ/15/646.

4.2 THE ANIMAL CLOSEST TO HUMANS

1. On Bontius see Harold J. Cook, "Global Economies." Cook focuses on Bontius's botanical work. The present account of Bontius's observations on orangutans derives from Buffon and Camper (below).

2. Buffon, "Les orang-outangs, ou Le pongo et le jocko," *HNGP* (1756), 14:43n. Schiebinger, *Nature's* Body, chapter 3, analyzes the gendering of early modern accounts of apes.

3. Buffon, "Les orang-outangs," 46–47. Buffon later noted that the Dutch naturalist

Jean-Nicolas-Sebastian Allamand had reported hearing "several people who had been to Batavia" confirm stories of the female's modesty. Buffon, "Addition à l'article sur les Orang-outangs," *HNGP, supplément* (1789), 7:8.

4. For Buffon on apes, Martinez-Contreras, "Des moeurs des singes."

5. Vosmaer, *Description*. On this orangutan, see Ingensiep, "Der Orang-outang des Herrn Vosmaer." Ingensiep discusses the dispute between Vosmaer and Camper over the rights to the orangutan's remains. The same "orangutan war" is treated by Cribb et al., *Wild Man*, 42–43.

6. Vosmaer, *Description*, 9.

7. Vosmaer, *Description*, 15–16.

8. Vosmaer, *Description*, 16.

9. Buffon, "Les orang-outangs," *HNGP* (1756), 14:61.

10. Petrus Camper, "De l'orang-outang."

11. É. Geoffroy Saint-Hilaire and G. Cuvier, "Histoire naturelle des orangs-outangs."

12. Unsigned letter to "cher collègue et ami," La Haye, 24 ventôse an 3, AN AJ/515/836. I assume Faujas wrote this letter. Its handwriting and spelling differ from Thouin's letter to Jussieu, La Haye, le 24 ventôse an 3, séance of 24 germinal an 3, AN AJ/515/578.

13. On the pongo of Wurmb, see Huxley, *Evidence, and* Barsanti, "L'orang-outan déclassé."

14. Geoffroy and Cuvier, "Histoire naturelle des orangs-outangs," 451–52.

15. Geoffroy and Cuvier, "Histoire naturelle des orangs-outangs," 452.

16. Geoffroy and Cuvier, "Histoire naturelle des orangs-outangs," 454.

17. Geoffroy and Cuvier, "Histoire naturelle des orangs-outangs," 457.

18. Geoffroy and Cuvier, "Histoire naturelle des orangs-outangs," 457.

19. P. Camper, *Dissertation physique*, quotation on p. 42.

20. Geoffroy and Cuvier, "Histoire naturelle des orangs-outangs," 463. On Geoffroy and Cuvier's assessment of how to classify the "large orangutan" from the Stadtholder's collection, see Barsanti, "L'orang-outan déclassé."

21. F. Cuvier, "Description d'un orang-outang."

22. F. Cuvier, "Description d'un orang-outang," 50.

23. F. Cuvier, "Description d'un orang-outang," 54.

24. F. Cuvier, "Description d'un orang-outang," 58–59.

25. Branch shaking is a common feature of orangutan agonistic display behavior and needs no theory of mind to explain it. For the kind of evidence that would support an animal's possessing theory of mind, see Premack and Woodruff, "Does the Chimpanzee Have a Theory of Mind?"

26. F. Cuvier, "Description d'un orang-outang," 62.

27. F. Cuvier, "Description d'un orang-outang," 64.

28. F. Cuvier, "Description d'un orang-outang," 61.

29. F. Cuvier, "Description d'un orang-outang," 65.

30. Anonymous, *Mémoires de Mademoiselle Avrillion*, 2:320–22.

31. Lamarck, *Recherches sur l'organisation* (1802), 134. On ideas about humans' developing from apelike ancestors, see Stoczkowski, "Portrait." For Lamarck's sources on apes, Barsanti, "Les singes de Lamarck."

32. Lamarck, *Philosophie zoologique*, 1:349–57.

33. F. Cuvier, "Essais sur les facultés intellectuelles."

34. F. Cuvier, "Essais sur les facultés intellectuelles," 217–18.

35. F. Cuvier, "Instinct," 537.

36. Virey, "Orang-outang," 306.

37. Bory de Saint-Vincent, "Orang," 272.

38. On his view that Hottentots are the humans closest in form and intellect to the higher apes, see Bory, "Homme de Saint-Vincent," 327.

4.3 TOWARD A SCIENCE OF LIVING ANIMALS

1. F. Cuvier, "Du rut," 119. Burkhardt, "Frédéric Cuvier on animal behavior."

2. Reimar, *Observations physiques*, 1:xvii-xviii.

3. Buffon, "L'outarde," *HNO* (1771) 2:3–4.

4. See above, 2.1 "Naturalists' practices."

5. F. Cuvier, "Du rut," 121.

6. F. Cuvier, "Note sur l'accouplement d'un zèbre."

7. F. Cuvier, "Observations sur l'accouplement d'un cigne."

8. F. Cuvier, "Observations sur quelques espèces de Goélands."

9. F. Cuvier, "Du pygargue."

10. F. Cuvier, "Du pygargue," 474.

11. F. Cuvier, "Observations sur le chien des habitans de la Nouvelle-Hollande."

12. Cuvier, "Observations sur le chien des habitans de la Nouvelle-Hollande," 459–60.

13. Cuvier, "Observations sur le chien des habitans de la Nouvelle-Hollande," 462.

14. Cuvier, "Observations sur le chien des habitans de la Nouvelle-Hollande," 462–63.

15. Cuvier, "Observations sur le chien des habitans de la Nouvelle-Hollande," 464–66.

16. Cuvier, "Observations sur le chien des habitans de la Nouvelle-Hollande," 469–70.

17. F. Cuvier, *Notice des animaux vivans* (1809), xi.

18. F. Cuvier, *Notice des animaux vivans* (1809), xii.

19. F. Cuvier, *Notice des animaux vivans* (1809), xii.

20. See 2.13, "Prussians at the Gates."

21. É. Geoffroy Saint-Hilaire, "Sur la ménagerie du Muséum d'histoire Naturelle," 26 April 1815, filed with the séance of 24 May 1815, AN AJ/15/610. The manuscript of Geoffroy's report is incomplete, with the missing page(s) apparently addressing the needed carnivore building.

22. Geoffroy, "Sur la ménagerie du Muséum d'histoire Naturelle"

23. F. Cuvier to the Museum, 21 June 1815, séance of same date, AN AJ/15/610. Noted in PV AN AJ/15/115, p. 108.

24. Cuvier to the Museum, 21 June 1815, AN AJ/15/610.

25. PV 30 August 1815, AN AJ/15/115, p. 133. The assembly later agreed to pay "to open a door of communication between the apartment of Mr. Frédéric Cuvier and the old park of the elephant," but nothing was recorded about this space becoming a "park for experiments." PV 13 September 1815 AN AJ/15/115, p. 139.

26. Cuvier to "Monsieur," 25 October 1815, AN AJ/15/611. It is unclear whether "Monsieur" refers to the director or Geoffroy. This map's purpose was to address the crowding of the herbivores in the rutting season.

27. Deleuze, *Histoire*, 720.

28. Boitard, *Le Jardin des Plantes*.

29. Legée, "Georges et Frédéric Cuvier." F. Cuvier's educational writings include *Projet d'organisation* and "De l'Education des sourds-muets." In the latter he discusses how zoology has expanded to include the study of animals' mental faculties, and he sug-

gests further that these considerations can illuminate the study of deaf mutes and their education. He also promoted the teaching of natural history in schools. F. Cuvier, *Observations* (1838).

30. É. Geoffroy Saint-Hilaire and Frédéric Cuvier, *HNM*.

31. F. Cuvier, *Des Dents des mammifères*.

32. Although Cuvier's articles in *HNM* and the *Dictionnaire des sciences naturelles* included behavioral information on species and individual specimens, behavioral questions did not take precedence over other zoological questions.

33. F. Cuvier, "Instinct," 536.

34. F. Cuvier, "Instinct," 536.

35. F. Cuvier, "Instinct," 530.

36. F. Cuvier, "Instinct," 531.

37. F. Cuvier, "Instinct," 532.

38. F. Cuvier, "Instinct," 536.

39. F. Cuvier, "Instinct."

40. F. Cuvier, "Instinct."

41. F. Cuvier, "Instinct," 537.

42. F. Cuvier, "Quelques observations de Dugald-Stewart," 241n.

43. The statement appears in a letter (1826) to the physiologist Flourens, cited in Legée, *Pierre Flourens*, 1:138.

44. F. Cuvier, "Essai sur la domesticité des mammifères," 420–21.

45. F. Cuvier, "Essai sur la domesticité des mammifères," 422.

46. Cuvier to P. Flourens, 29 or 30 October 1826, notes that his essay on domestication had two parts, the first devoted to responding "to what has been said with respect to my research on living animals, that captive animals can offer no useful observation." The letter is reproduced and dated by Legée, *Pierre* Flourens, 1:136–38.

47. F. Cuvier to the Museum, 28 July 1832, and "Programme d'un cours sur la nature des animaux et leur éducation," both presented at the séance of 31 July 1832, AN AJ/15/644. The course program as well as the letter are published in Legée, *Pierre Flourens*, 1:138–43, but the program is incorrectly dated 21 instead of 28 July 1832.

48. PV 31 July 1832, AN AJ/15/130, pp. 127–28. Cuvier to the Museum 12 November 1832, séance of 13 November 1832, AN AJ/15/644; PV AN AJ/15/130, p. 173.

49. F. Cuvier, "La genette panthérine," *HNM*, February 1833, 3.

50. Frédéric Cuvier, "Introduction."

4.4 THE BEAVER

1. Baron d'Alphonse, prefect du Garde, to Montalivet, Minister of the Interior, 28 June 1810, séance of 22 August 1810, AN AJ/15/603; PV AN AJ/15/111, p. 70.

2. D'Alphonse to the Muséum, 21 September 1810, séance of 3 October 1810, AN AJ/15/603. Without elaborating, d'Alphonse said it was suspected that the beaver was poisoned. See too PV AN AJ/15/111, p. 97.

3. "Instruction pour la conduit au Museum d'histoire naturelle de Paris d'un Castor pris dans les eaux du Gardon," 13 March 1811, séance of 27 March 1811, AN AJ/15/604. PV AN AJ/15/111, p. 203.

4. Museum to d'Alphonse, 27 March 1811, AN AJ/15/745.

5. Rolland to Museum, 7 June 1811, séance of 19 June 1811, AN AJ/15/604. Rolland's letter specifies that the beaver was captured 11 March 1811 and sent 13 March 1811.

6. F. Cuvier, "Castor," 250.

7. F. Cuvier, "Castor d'Europe," *HNM*, October 1825. Cuvier incorrectly identifies the "breastfed" beaver as having come from the banks of the Rhône and being sent by the prefect of Isère. It came instead from the Gardon River (just before it empties into the Rhône) and was sent by Rolland, prefect of the department of Gard.

8. Rolland to Museum, 25 June 1811, séance of 3 July 1811, AN AJ/15/604; PV AN AJ/15/112, pp. 40–41.

9. PV 21 and 28 August 1811, AN AJ/15/112, pp. 73, 75. Draft of Museum's letter to Rolland, 21 August 1821, AN AJ/15/745.

10. *Gazette de France*, 12 August 1811, 2, reported the beaver's weight as twenty-five kilograms. See too PV 28, August 1811, AN AJ/15/112, p. 75.

11. F. Cuvier, "Castor," 249.

12. Cuvier menagerie reports of 25 February 1818 (AN AJ/15/615) and 21 October 1818 (AN AJ/15/616).

13. "Registre d'entrée des mammifères à la ménagerie, 1807–1849." MNHN, Ménagerie. MS 25.

14. Buffon, "Animaux communs aux deux continents," *HNGP* 9 (1769), 103.

15. Buffon, "Animaux communs aux deux continents," *HNGP* 9 (1769): 104–5.

16. Buffon, "Le Castor," *HGNP* 8 (1760): 282.

17. Buffon, "Le Castor," *HGNP* 8 (1760): 283.

18. Buffon, "Le Castor," *HGNP* 8 (1760): 285.

19. Buffon, "Le Castor," *HGNP* 8 (1760): 288.

20. Buffon, "Le Castor," *HGNP* 8 (1760): 289–94.

21. F. Cuvier, "Sur la domesticité des mammifères," 419.

22. I take this account primarily from Cuvier's 1817 "Castor" essay. There are discrepancies between it and his "Castor d'Europe" (1825), where he writes as if he observed only one European beaver.

23. F. Cuvier, "Castor d'Europe" (1825), 2.

24. F. Cuvier, "Sur la domesticité des mammifères," 418–19.

25. Séances of 31 December 1817 and 28 January 1818, AN AJ/15/117, pp. 144, 155. F. Cuvier, "Le castor du Canada," *HNM*, May 1819.

26. F. Cuvier, "Le castor du Canada," *HNM*.

27. F. Cuvier to "Messieurs," 25 January 1820, AN AJ/15/619. PV 25 January 1820, AN AJ/15/119: 117–18. PV 8 January 1828, AN AJ/15/127, 137.

28. E. Geoffroy Saint-Hilaire, "Habitudes des castors," 232.

29. Cuvier, "Castor," 250.

4.5 ACCLIMATIZATION AND DOMESTICATION

1. Bernardin de Saint-Pierre, *Mémoire*, 1792, 55. For the agami's reputation, see Buffon, "L'Agami, *HNO*, 4 (1778): 498.

2. "Notes relatives à l'établissement d'une ménagerie dans le Museum d'hist. naturelle." This unsigned and undated manuscript is in Jussieu's hand, AN AJ/15/844. See 1.4.

3. A. Thouin, "Notes sur les moyens d'employer utilement les domains nationaux réservés par le décret du . . . [*sic*] floréal," 16 fructidor an 2, AN F/17/1229.

4. Daubenton, "Définition,"

5. PV frimaire an 12, AN AJ/15/105, p. 64.

6. Blunt, *Ark in the Park*, 25.

7. Burkhardt, "Le comportement animal et l'idéologie de domestication."

8. Buffon, "Le cheval," *HNGP* (1754), 4:174.

9. Buffon, "Le chameau et le dromédaire," *HNGP* (1764), 11:239–40.

10. Buffon, "Le pigeon," *HNO* (1771), 2:496.

11. Buffon, "De la dégénération des animaux," *HNGP* (1766), 14:317.

12. Buffon, "Animaux communs aux deux continens," *HNGP* (1760), 8:285.

13. Buffon, "Animaux du nouveau monde," *HNGP* (1761), 9:85.

14. Buffon, "L'Agami," *HNO* (1778), 4:501.

15. Cuvier, "Observations sur le chien des habitans de la Nouvelle-Hollande," 474. The theme of dogs and civilization can also be found in Bewick, *A General History of Quadrupeds*. See Ritvo, *Animal Estate*, 20–21.

16. F. Cuvier, "Chien de la Nouvelle-Hollande," 4, *HNM*, May 1825.

17. F. Cuvier, "Chien de la Nouvelle-Hollande," 4, *HNM*, May 1825.

18. F. Cuvier, "Sur la domesticité des mammifères."

19. F. Cuvier, "Sur la domesticité des mammifères," 426.

20. F. Cuvier, "Sur la domesticité des mammifères," 427.

21. Cuvier reported that ruminants show little response to caresses, but horses and elephants do respond, and that cats seek caresses, but not as much as dogs. "Essai sur la domesticité," 429–31.

22. Cuvier, "Essai sur la domesticité," 432–34.

23. Cuvier, "Essai sur la domesticité," 438–40.

24. Cuvier, "Essai sur la domesticité," 441.

25. Cuvier, "Essai sur la domesticité," 441–42.

26. F. Cuvier, "Macaque de Buffon," 120.

27. Cuvier, "Essai sur la domesticité," 441–44. Cuvier admitted that more needed to be known about Old World orangutans, chimpanzees, gibbons, and langurs. Some New World monkeys might be candidates for domestication because they were intelligent, social, gentle, and craved caresses.

28. Cuvier, "Essai sur la domesticité," 444–45.

29. Cuvier, "Essai sur la domesticité," 452–53.

30. Cuvier, "Essai sur la domesticité," 445–48.

31. Cuvier, "Essai sur la domesticité," 449–51. In "Sociabilité des animaux" (1825), Cuvier argued that animal sociability was essentially instinctive and not a function of intelligence or morality.

32. Cuvier, "Essai sur la domesticité," 454.

33. Lyell, *Principles of Geology*, 1:41, 44.

34. Lyell, *Principles of Geology*, 1:42.

35. Lyell, *Principles of Geology*, 1:44.

36. Lyell, *Principles of Geology*, 1:46–48.

37. Lavison, "Différentes tentatives."

38. Buffon, "Le lama et le paco," *HNGP* (1765), 13:31–32. Cited and embellished in I. Geoffroy Saint-Hilaire, *Acclimatation*, 33–34.

39. I. Geoffroy Saint-Hilaire, *Acclimatation*, 35.

40. G. Cuvier, "Le Lama," in Lacépède, *Ménagerie*. Dated 1805 by Pieters.

41. MNHN Menagerie MS 25 lists an alpaca given on 31 October 1821 by M. Poydeban, deceased August 14, 1836.

42. F. Cuvier, "L'alpaca," *HNM*, October 1821.

43. Museum to the Minister of the Interior, 19 January 1824, AN AJ/15/844.

44. Minister of the Interior to the Museum, 14 February 1824, séance of 17 February 1824, AN AJ/15/844.

45. Copy of Gicquel des Touches to the Minister of the Interior, 19 May 1825, séance of 7 June 1825, AN AJ/15/629. The Museum to the captain, 16 June 1825, AN AJ/15/750.

46. F. Cuvier menagerie reports, 20 and 27 November and 4 December 1827, AN AJ/15/634. Quotation from 20 November.

47. F. Cuvier menagerie report, 4 December 1827, AN AJ/15/634.

48. F. Cuvier menagerie report, 27 April 1830, AN AJ/15/639. See 3.6 above.

49. Perrin's request to the Museum, with a supporting note from F. Cuvier, 31 January 1837; AN AJ/15/653. PV, AN AJ/15/134, p. 12.

50. F. Cuvier menagerie report, 14 February 1837, séance of 21 February 1837, AN AJ/15/653; PV, AN AJ/15/134, p. 30.

51. Delaunay to the Museum, n.d., séance of 14 prairial an 6, AN AJ/15/581.

52. G. Cuvier, "Le Chameau" and "Le Dromadaire," Lacépède, *Ménagerie*.

53. Minister of the Interior to the Museum, 27 April 1819; F. Cuvier (draft) to the Minister of the Interior, n.d.; and Geoffroy (draft) to the Minister of the Interior, n.d., all filed with the séance of 27 April 1819. Geoffroy report to the Museum 3 May 1819. AN AJ/15/617.

54. F. Cuvier menagerie report, 9 January 1827, AN AJ/15/633.

55. PV 12 June 1827, AN AJ/15/627, p. 47.

56. F. Cuvier menagerie report 31 July 1827, AN AJ/15/634. The long saga of the animals' treatment at Alfort can be tracked through Cuvier's menagerie reports of 9 January, 26 January, 2 June, 19 June 1827; 15 February 1827 report from Alfort, séance of 6 March 1827, all in AJ/15/633; plus his reports of 31 July and 7 August 1827 and the Alfort report of 3 August 1827, séance of 7 August 1827, AN AJ/15/634. See too PV 31 July 1827 AN AJ/15/127, 72.

57. Ildephont to "Monsieur," 5 October 1827, AN AJ/15/634.

58. Ildephont to "Monsieur," 12 October 1837, and F. Cuvier menagerie report 16 October 1827, AN AJ/15/634; PV 16 and 23 October 1827, AN AJ/15/127, pp. 101, 104.

59. F. Cuvier, "Le Dauw," 2, *HNM*, September 1826.

60. F. Cuvier, "Le Couagga," 1–2, *HNM*, July 1821.

61. F. Cuvier, "Le Couagga," 3, *HNM*.

62. F. Cuvier, "Le Couagga," 1, *HNM*.

63. I. Geoffroy Saint-Hilaire, menagerie report of 30 May 1837. His 20 June 1837 menagerie report included more on the dauw. AN AJ/15/653.

64. I. Geoffroy Saint-Hilaire, "Sur le genre cheval." For more on Isidore's ideas on acclimatization and domestication see Blanckaert, "Animaux 'utiles,'" and Osborne, *Nature*.

65. I. Geoffroy "Allocution."

66. I. Geoffroy Saint-Hilaire, "Fragments," 396–98. See also his "Acclimatation," 499–501.

67. Galton, "First steps."

68. Osborne, *Nature*.

4.6 THE SPECIES QUESTION

1. G. Cuvier, "Extrait d'observations," 1817, 159.

2. Péron's paper at the Academy in 1805 (coauthored with C. A. Lesueur and titled

"Observations sur les femmes hottentotes") was not published at the time, but it was the subject of public debate. See "Réponse de M. Péron, naturaliste de l'expédition de découvertes aux Terres Australes, aux observations critiques de M. Dumont, sur la tablier des femmes hottentotes," *Journal de Physique* 61 (1805): 210–17. The original paper was not published until 1883. See *Bulletin de la Société zoologique de France* 8 (1883): 15–33.

3. F. Cuvier to the Museum, 21 June 1815, AN AJ/15/610.

4. F. Cuvier, "Caractères ostéologiques" (1811), 353.

5. G. Cuvier, *Ossemens fossiles*, 1:79.

6. On Lamarck's evolutionary theory see Burkhardt, *Spirit of System*. On Lamarck and the French authors who discussed organic mutability, see Corsi, *The Age of Lamarck* and Corsi's subsequent publications on the same theme.

7. Anonymous, *Catalogue des livres*, 42.

8. Lamarck, *Philosophie zoologique*, 1:265–66.

9. See above, 4.2.

10. For Linnaeus's views on the formation of species see Ramsbottom, "Linnaeus"; Larson, *Reason*; and Eriksson, "Linnaeus."

11. Buffon, "De la dégénération des animaux," *HNGP* 14 (1766): 311–74. For an analysis of Buffon's ideas on species and organic change over time, see especially Farber, "Buffon"; and Roger, *Buffon*, chap. 19.

12. Burkhardt, *Spirit of System*, 202–6.

13. Lamarck, *Philosophie zoologique*, 1:209–12.

14. Lamarck first listed these in decreasing order but acknowledged it was "contrary to that of nature." Later in the same volume he offered a table in reverse, "according to the order most consistent with nature." *Philosophie zoologique*, 1:128, 277–82.

15. Lamarck, *Philosophie zoologique*, 1:221.

16. Lamarck, Système, 13–15. See Burkhardt, "Lamarck, Cuvier, and Darwin," and "Lamarck, Evolution, and the Inheritance of Acquired Characters." See also Gayon, "Hérédité." Gayon's chapter is thorough and incisive on Lamarck but passes quickly over Frédéric Cuvier, missing Cuvier's use of the word "hérédité" and misidentifying his first use of the word "héréditaire."

17. Lamarck, *Recherches sur organisation des corps vivans*, 61.

18. Lamarck, *Recherches sur organisation des corps vivans*, 53–54.

19. Lamarck, *Philosophie zoologique*, 1:235.

20. Lamarck, *Histoire naturelle des animaux sans vertèbres*, 1:200.

21. Lamarck, *Histoire naturelle des animaux sans vertèbres*, 1:191.

22. F. Cuvier, "Du rut," 129–30. This appears to be the first time the word "hereditary" (*héréditaire*) was used in its now familiar biological sense rather than in a legal or a medical one. On the legal and medical uses of the words "hereditary" and "heredity" see Gayon, "Hérédité."

23. F. Cuvier, "Le chien des habitants de la Nouvelle-Hollande," 462.

24. F. Cuvier, "Caractères ostéologiques," 349.

25. F. Cuvier, "Caractères ostéologiques," 353.

26. F. Cuvier, "Essais sur les facultés intellectuelles," 217–18. Like his earlier use of the word "hereditary," this was apparently the first time the word "heredity" (*l'hérédité*) was used in a biological sense in relation to the inheritance of acquired characters. The phrenologists Gall and Spurzheim had used it the year before with respect to the transmission of head shapes. F. J. Gall and G. F. Spurzheim, *Dispositions* innées, 38–39.

27. F. Cuvier, *Projet d'organisation.*

28. F. Cuvier, *Projet d'organisation*, xiii.

29. On this point see Stocking, *Race*, 265–67.

30. F. Cuvier, "Observations préliminaires" (1834), 1:ix.

31. Burkhardt, *Spirit of System*, 195–99.

32. G. Cuvier, *Ossemens fossiles* (1812), 1:28.

33. Coleman, *Georges Cuvier.*

34. G. Cuvier, *Le Règne animal*, 1:6. For an earlier statement of the idea of the conditions of existence, see G. Cuvier, *Léçons d'anatomie comparées*, 1:47.

35. Coleman, *Georges Cuvier*, 119–20.

36. Rudwick, *Georges Cuvier*, 260–67.

37. G. Cuvier, *Ossemens fossiles* (1812), 1:74.

38. G. Cuvier, "Sur un nouveau rapprochement."

39. G. Cuvier, *Ossemens fossiles* (1812), 1:76–79.

40. G. Cuvier, "Éloge de Lamarck."

41. F. Cuvier, "Le coati-brun," *HNM*, March 1819. Because of the importance of variation, Cuvier indicated, he planned to include different varieties of coatimundi in *HNM.*

42. F. Cuvier, "Chien des Eskimaux, de la baie de Baffin," *HNM*, September 1819).

43. Cuvier cited no authorities or studies to support his claim, nor did he suggest any biological reason it might be true. The category "variety" was arbitrary. His claim that mammals have the most varieties may have been an artifact of mammals' being his specialty (combined perhaps with an appreciation of the number of dog varieties produced by breeders). His argument would not have sustained a comparison between the different classes and the number of species in them.

44. F. Cuvier, "Chien des Eskimaux," *HNM.*

45. F. Cuvier, "Du macaque de Buffon"; comparable papers in Museum's *Mémoires* for the next ten years.

46. F. Cuvier, "Métis femelle d'âne et de zèbre," *HNM*, March 1820.

47. Buffon, "De la dégénération des animaux," *HNGP* (1766), 14:311–74.

48. F. Cuvier, "Cochon domestique, variété du Cap de Bonne-Espérance," *HNM*, February 1821.

49. F. Cuvier, "Mulets du chacal de l'Inde et du chacal du Sénégal," *HNM*, December 1821.

50. F. Cuvier, "Jeunes métis de lion et de tigresse," *HNM*, February 1826. Cuvier noted that lion-tiger hybrids had been born in London previously, but no illustrations of them had been provided.

51. F. Cuvier, "Mulet d'un bonnet-chinois et d'une femelle de macaque," *HNM*, May 1830.

52. F. Cuvier, "Mulet d'un bonnet-chinois et d'une femelle de macaque," *HNM*, May 1830.

53. Appel, *Cuvier-Geoffroy Debate.*

54. É. Geoffroy Saint-Hilaire, "Gavials," 144–55.

55. É. Geoffroy Saint-Hilaire, "Gavials," 150–51 and p. 151n.

56. É. Geoffroy Saint-Hilaire, "Gavials," 154–55.

57. E. Geoffroy Saint-Hilaire and Serres, "Rapport."

58. F. Cuvier, *Supplément à Buffon*, 1:25.

59. Nowhere else in his writings does he claim he saw this twice. His essay on the bison mentions only the June 1826 case.

60. F. Cuvier, *Supplément à Buffon*, 1:105–10.

61. Blanckaert, "Voyages naturalistes," 217, notes that Georges Cuvier was suspected of holding the polygenist view that the human races were separately created. On the subject generally, see Cohen, *French Encounter*.

62. F. Cuvier to the Museum, 28 July 1832, séance of 31 July 1832, AN AJ/15/644.

63. F. Cuvier, "Observations préliminaires," 1:viii.

64. F. Cuvier, "Observations préliminaires," 1:xi. Lamarck never suggested that all animals descended from a single, microscopic species. In his system, the simplest forms of life continue to be spontaneously generated at the bottom of the scale. His theory does not portray descent from a common ancestor so much as the path nature followed in producing all the different forms of life.

65. F. Cuvier, "Observations préliminaires," 1:xii-xv.

66. F. Cuvier, "Observations préliminaires," 1:xvii.

67. F. Cuvier, "Observations préliminaires," 1:xviii-xix.

68. F. Cuvier, "Observations préliminaires," 1:xix.

69. F. Cuvier, "Observations préliminaires," 1:xx-xxi.

70. F. Cuvier, "Observations préliminaires," 1:xxi-xxii.

71. The information on the number of hybrid generations is from Flourens, *Examen*, 107.

72. A point made by I. Geoffroy Saint Hilaire, *Histoire naturelle* générale, 3:229.

73. Lamarck, "Discours d'ouverture," 100–101.

74. Buffon was ridiculing his rival René Antoine Ferchault de Réaumur (1683–1757), who felt his detailed studies of insect life evidenced the work of the Creator. Buffon, "Discours sur la nature," *HNGP* (1753), 4:95.

75. Virey, "Nature," 411–12. On Virey, see Corsi, "Virey," and *Age of Lamarck*, esp. chap. 5.

76. The English entomologist and clergyman William Kirby explicitly called out Lamarck for his materialism. See Kirby, *Power*, esp. xxviii-xxix.

77. F. Cuvier, "Le hérisson d'Europe," *HNM*. March 1821.

78. F. Cuvier, "Introduction," vii-x.

79. See 4.13 below.

80. F. Cuvier, *L'histoire naturelle des cétacés*, xxx.

81. F. Cuvier, *L'histoire naturelle des cétacés*, xl-xli.

82. F. Cuvier, *L'histoire naturelle des cétacés*, xlv.

83. F. Cuvier, *L'histoire naturelle des cétacés*, xlii.

84. F. Cuvier, *L'histoire naturelle des cétacés*, xl-xliii.

85. For more on Cuvier's seeing "degenerations" in domestic animals as *improvements*, see 4.13.

4.7 DOGS, WOLVES, AND JACKALS

1. F. Cuvier, menagerie report 4 December 1832, AN AJ/15/644.

2. Delaunay to the administration, 4 germinal an 7, AN AJ/15/582.

3. The building for the *animaux féroces* and its courtyard were fenced in, but the herbivore parks were not.

4. PV 8 messidor and 20 thermidor an 12, AN AJ/15/106, pp. 21, 64.

5. PV 8 April 1807, AN AJ/15/108, 57. G. Cuvier, *Ossemens fossiles* (1812), 4:33–35, identifies the deer as *Cervus virginianus.*

6. PV 3 June, 19 August, and 2 September 1807, AN AJ/15/108, pp. 72, 92, 97.

7. PV 4 March 1807, AN AJ/15/108, 43. An official announcement of the police decision did not appear in the government newspapers until June. See *Journal de l'Empire,* 18 June 1807, 2, and *Gazette nationale,* 18 June 1807, 664.

8. PV 31 January 1810, AN AJ/15/110, pp. 111–12.

9. PV 5 December 1810, AN AJ/15/111, p. 132; PV 14 September 1814, AN AJ/15/114, p. 177.

10. PV 5 June 1832, AN AJ/15/130, pp. 102–3. Cuvier menagerie report, 14 February 1837, AN AJ/15/653.

11. *Journal des débats, La France,* and *La Presse,* all 16 February 1837.

12. *Le Constitutionnel,* 17 February 1837, embellished the story by noting that the event produced "une vive sensation" among the strollers in the garden.

13. *Figaro,* 19 February 1837, 193.

14. PV 4 July 1837, AN AJ/15/134, p. 108.

15. PV 21 November 1837, AN AJ/15/134: 196.

16. In February 1794 Lamarck offered the Museum his dog as a watchdog. PV 5 ventôse an 2, AN AJ/15/96, p. 92. For earlier notes on acquiring watchdogs, see the séances of 14 brumaire and 2 pluviôse an 2. Cuvier menagerie report, 27 November 1837, AN AJ/15/654.

17. I. Geoffroy Saint-Hilaire, "État général des animaux composant la ménagerie au 25 juillet 1831," séance of 26 July 1831, AN AJ/15/642.

18. See section 2.13.

19. I. Geoffroy Saint-Hilaire, "État général au 25 juillet 1831," séance of 26 July 1831, AN AJ/15/642.

20. They were (1) a lioness purchased by Felix Cassal in North Africa that lived with a water spaniel (*chien barbet*); (2) a lioness and hound (*chien braque*) that arrived in December 1803; and (3) a lion identified by F. Cuvier in a menagerie report of 24 April 1827 (AN AJ/15/633) as "le lion qui est avec le chien."

21. F. Cuvier, "Chien des eskimaux" (November 1819), "Le chien levrier (May 1820), "Le chien de terre neuve"(December 1820), *HNM.*

22. Minister of the Interior to the Museum, 21 July 1810, séance of 25 July 1810, AN AJ/15/603. PV AN AJ/15/111, pp. 52–53.

23. PV 12 September 1810, AN AJ/15/111, p. 84. Museum to de Vaussay, 13 September 1810, AN AJ/15/745.

24. PV 19 December 1810, AN AJ/15/111, pp. 140, 141.

25. See Le Chevalier de Schwarz to "Monsieur," 23 July 1814, séance of 27 July 1814, AN AJ/15/609 and the Museum's response to him 29 July 1814 regarding a pair of young black wolves sent to the king of Wurttemberg, AN AJ/15/745.

26. PV 11 February 1818, AN AJ/15/117, p. 161.

27. Museum to Huzard, 14 February 1818, AN AJ/15/747.

28. This was the case with a wolf cub offered by Le Chevalier de Lespinasse, lieutenant of the wolf hunt (Louveterie) for the department of Deux-Sèvres. See Lespinasse to Lacépède, 17 November 1824, séance of 23 November 1824, AN AJ/15/628, and Museum to Lespinasse, 4 December 1824, AN AJ/15/749.

29. The menagerie's animal register, MNHN Menagerie MS 25, notes the wolf's ar-

rival, given by "Mlle Duvaucel," as 24 March 1830. It died 23 November 1840 and was sent to the zoology lab. On Sophie Duvaucel, see Orr, "Keeping It in the Family."

30. Cuvier's menagerie report, 12 October 1830, AN AJ/15/640, and the animal register, MNHN Menagerie MS 25, note four wolves sent to the combat master on 6 October 1830 to make room for the lions from Algeria.

31. F. Cuvier, "Le chacal," *HNM*, January 1819.

32. F. Cuvier, "Le chacal du Sénégal, Femelle." *HNM*, June 1820. Cuvier chose the name *Canis anthus* for the new species, based on Pliny's account of an Arcadian belief in a family named Anthus whose members became werewolves.

33. F. Cuvier, "Mulets du chacal de l'Inde et du chacal du Sénégal," *HNM*, December 1821.

34. The Museum's papers sometimes refer to "Schmitz" as "Schmidt," but he signed his letters Schmitz. The Museum bought four animals from him in 1826. PV 27 December 1825, and 24 January, 21 February, and 4 July 1826. AN AJ/15/126, pp. 29, 42, 57, 123. MNHN MS 25 gives 10 July 1826 as the entry date at the menagerie for both Schmitz's jackal and Dussumier's, but Schmitz's jackal arrived in February. Schmitz to the Museum, 15 February 1826, AN AJ/15/631, and the Museum to the commandant of Senegal, 1 March 1826, AN AJ/15/750.

35. F. Cuvier, "Chacal du Sénégal. Male," *HNM*, May 1830.

36. Alex. Eyriès to the Museum, 20 June 1831, AN AJ/15/641.

37. Museum to Douville, 27 August 1831, AN AJ/15/753.

38. MNHN Menagerie MS 25 dates the arrival of the "jaguar du Pérou" and the "chacal du Pérou" as 12 July 1831.

39. *Nouvelle annales des voyages* 50 (June 1831): 392–94.

40. F. Cuvier menagerie report, 4 December 1832, AN AJ/15/644.

41. "Rapport de la commission de la Société de Géographie, qui a décerné une médaille d'or de mille francs à M. Douville, pour son voyage de découvertes au Congo et dans l'intérieur de l'Afrique équinoxiale," *Bulletin de la Société de Géographie de Paris* 17 (1832): 157–75.

42. Douville, *Voyage*.

43. *The Foreign Quarterly Review* 10 (1832): 163–206. When Douville published a "defense," the reviewer renewed his attack, *Foreign Quarterly Review* 10 (1832): 541–46. When Douville offered yet another "justification" in reply to the "infamies" launched against him, the reviewer again lambasted him. *Foreign Quarterly Review* 11 (1833): 519–27.

44. Lacordaire, "Revue," 250n.

45. Académie des sciences, *Procès-verbaux des séances de l'académie* 9 (1828–31): 682 (séance of 16 August 1831) and 707 (séance of 24 October 1831).

46. Lacordaire, "Revue," 251. The reviewer for the *Foreign Quarterly Review* 11 (1833): 527 simply commended the Institut's committee for its "wise reserve."

47. Museum to Douville, 27 August 1831, AN AJ/15/753.

48. Museum to Smolenski, 30 May 1834, AN AJ/15/754.

49. F. Cuvier menagerie report 2 December 1834, AN AJ/15/648.

50. On his father's contributions to teratology, see I. Geoffroy Saint-Hilaire's, *Vie, travaux et doctrines*, 270–304.

51. F. Cuvier menagerie report, 27 January 1835, AN AJ/15/649.

52. PV 27 January 1835, AN AJ/15/132, p. 59.

53. Koepfli et al., "Genome-wide Evidence." See too "Golden Jackal: A New Wolf

Species Hiding in Plain Sight," *Guardian*, 30 July 2015. https://www.theguardian.com/science/grrlscientist/2015/jul/30/golden-jackal-a-new-wolf-species-hiding-in-plain-sight.

54. "État des animaux vivans à la menagerie au 1er janvier 1837," séance of 28 February 1837, AN AJ/15/653.

55. Rousseau and Lemonnier, Promenades, 16.

4.8 THE KEEPERS REVISITED

1. F. Cuvier to "Monsieur" (either Geoffroy or the Museum's director), 25 May 1814, AJ/15/609.

2. F. Cuvier menagerie report, 17 April 1816, AJ/15/612.

3. The keepers' lot did not improve over the century. See Milne-Edwards, *Ménagerie*, 15–17, commenting on the hard work and poor pay of the keepers and thus the difficulty of recruiting young, intelligent, and able-bodied individuals for the job.

4. Geoffroy Saint-Hilaire report on appointments and wages for menagerie workers, séance of 13 nivôse an 12, AJ/15/591.

5. See above, section 2.6.

6. Ballet was on the staff by 1797, Nivet by 1798, and Perrin by 1799.

7. Museum to the Minister of the Interior, 18 June 1814, gives Nivet's year of birth as 1742, AN AJ/15/745.

8. Museum to the Minister of the Interior, 6 frimaire an 12, AN AJ/15/744.

9. PV 25 fructidor an 12 and 25 vendémiaire an 13, AN AJ/15/106, pp. 80, 101.

10. F. Cuvier to the Museum, 20 April 1808, AN AJ/15/599. Cuvier writes of "Rémon" and his exemplary service as *chartier* for the past two years.

11. Ballet *fils* writing for his father to the Museum, 5 February 1821, séance of 6 February 1821, AN AJ/15/621. On Ballet's retirement and pension, PV AN AJ/15/120, pp. 138, 142–43, 147.

12. Étienne Geoffroy Saint-Hilaire, "Rapport du professeur chargé de la direction de la ménagerie au sujet de l'employé Nivet," séances of 17 and 24 January 1810, AN AJ/15/603; PV 31 January 1810, AN AJ/15/110, p. 111.

13. PV 21 December 1811, AN AJ/15/112, p. 143. Nivet to the Museum, 2 January 1812, AN AJ/15/605.

14. PV 6 July 1814, AN AJ/15/114, p. 150; PV 22 February and 5 April 1815, AN AJ/15/115, pp. 61, 78. In 1813 Nivet was moved from the herbivores to the birds and monkeys. PV 30 June 1813, AN AJ/15/114, p. 8.

15. PV 18 May 1814, AN AJ/15/114, p. 132.

16. F. Cuvier to the Museum, 7 December 1814, AN AJ/15/612; PV AN AJ/15/115, p. 29.

17. Cuvier letter 7 December 1814.

18. F. Cuvier to the Museum, 6 March 1816, AN AJ/15/612; PV AN AJ/15/116, p. 46.

19. Antoine to the Museum, with Frédéric Cuvier's annotation, 13 February 1821, AN AJ/15/621. For Cuvier's summary of menagerie workers' salaries and wages from 1822 to 1831, séance of 20 September 1831, AN AJ/15/642.

20. Notes by F. Cuvier on bonuses for the menagerie workers, 1 February 1820, AN AJ/15/619. A line was drawn through the recommendation that Antoine receive twenty francs.

21. PV 31 January 1810, AN AJ/15/110, p. 111.

22. F. Cuvier to the Museum, 27 June 1820, AN AJ/15/619.

23. PV 27 June 1820, AN AJ/15/120, pp. 34–35.

24. F. Cuvier to the Museum, 21 May 1822, AJ/15/623.

25. PV 21 May 1822, AN AJ/15/122, pp. 33–34.

26. F. Cuvier, undated letter to the Museum, accompanied by undated recommendation of Richard by Gillet Laumont, PV 11 fructidor an 12, AN AJ/15/592.

27. F. Cuvier to the Museum's director, 30 August 1821, AN AJ/15/622.

28. Letter written on behalf of Richard's widow, 20 September 1825, séance of same date. AN AJ/15/630.

29. Rémond letter of 7 February 1825 annotated by Cuvier, séance of 8 February 1825, AN AJ/15/629.

30. F. Cuvier menagerie report, 1 August 1826, AN AJ/15/632.

31. The administration agreed to write on his behalf. See Rémond to the Museum, 23 April 1828, AN AJ/15/635, and PV AN AJ/15/127, pp. 189–90.

32. Rémond to the Museum (n.d.), séance of 18 September 1832, AN AJ/15/644. Rémond indicated that the cost of Charenton was now 70 francs per month. Frédéric Cuvier supported Rémond's request for assistance. The administration granted him 50 francs "because of his good services." PV AN AJ/15/130, p. 151.

33. F. Cuvier to the Museum, 7 October 1834, AN AJ/15/648. For the assembly's approval, PV AN AJ/15/131, p. 264.

34. PV 3 April 1838, AN AJ/15/135, p. 74.

35. Unsigned "Note pour Monsieur le Directeur," 24 December 1838, AN AJ/15/656.

36. Rousseau to Museum, n.d., séance of 24 August 1814, AN AJ/15/609; PV AN AJ/15/114, p. 171.

37. Jaussaud, "Bibron, Marie Gabriel," in Jaussaud and Brygoo, *Du Jardin au Muséum*, 78. In 1826 Rousseau received a raise of 160 francs per year. PV 28 March 1826 AN AJ/15/126, p. 75.

38. F. Cuvier, "Notes sur les moyens de faire parvenir à Paris les animaux arrivés à Brest par le frigate la Thetis, le 23 juin dernier." Séance of 11 July 1826. The animals are identified in a letter from Redon, Intendent de la Marine at Brest, to the Museum, 8 July 1826, séance of 8 July 1826. Per Cuvier's menagerie report, 22 August 1826, all the animals survived the trip to Paris, but one black swan died two days later. AN AJ/15/632.

39. PV 21 February 1826, AN AJ/15/126, p. 58.

40. F. Cuvier menagerie report, 1 May 1827, AN AJ/15/633.

41. PV 14 June 1825, AN AJ/15/125, pp. 51–52.

42. F. Cuvier menagerie report, 1 May 1827, AN AJ/15/633. Though Cuvier wrote Letellier's name here as "Tellier," I use "Letellier" because that is how he signed his name.

43. Letellier to the Museum, 29 May 1827, AN AJ/15/635.

44. F. Cuvier annotation of Letellier letter of 29 May 1827, AN AJ/15/635.

45. Letellier's annual salary was raised to 750 francs in 1828 (from 732 francs for the six previous years). By 1832 it had reached 900.

46. F. Cuvier annotation of a letter from Rousseau to the Museum, no date, séance of 29 December 1829, filed with séance of 26 January 1830, AN AJ/15/639.

47. F. Cuvier menagerie report, 19 October 1830, AN AJ/15/640.

48. F. Cuvier menagerie report, 16 November 1830, AN AJ/15/640. PV AN AJ/15/129, p. 97.

49. É. Geoffroy Saint-Hilaire, "Registre d'ordre relative à la ménagerie et copie de lettres." MNHN MS 2295.

50. PV 18 April 1837, AN AJ/15/134, p. 60.

51. PV 16 May 1837, AN AJ/15/134, p. 76.

52. PV 29 August 1837, AN AJ/15/134, p. 148. See too Geoffroy's support of Martin's request, AN AJ/15/653.

53. PV 31 October 1837, AN AJ/15/134, p. 185.

54. PV 3, 10, and 17 October 1837, AN AJ/15/134, pp. 167, 174, 177.

55. I. Geoffroy Saint-Hilaire responding to Museum director (n.d., but evidently by early June 1837), since Letellier wrote to Isidore from Toulon on 9 June 1837, AN AJ/15/653. Isidore's letter, filed in AN AJ/15/654, says Lettelier has been with the carnivores for the past seven or eight years.

56. Rousseau to the Museum, 14 December 1824, AN AJ/15/628. The Museum reimbursed Rousseau for his medical bills, which amounted to 390 francs.

57. Lebleu, *Les Avatars de Zarafa*, 151.

58. F. Cuvier menagerie report, 26 April 1836 AN AJ/15/651.

59. F. Cuvier menagerie report, 14 February 1837, séance of 21 February 1837, AN AJ/15/653.

60. Geoffroy report on his tour of the menagerie, 26 February 1837, MNHN MS 2295/3.

61. Geoffroy 26 February 1837 tour.

62. F. Cuvier to the Director, 3 January 1838, séance of 9 January 1838, AN AJ/15/655.

63. F. Cuvier to the Director, 3 January 1838.

64. F. Cuvier to the Director, 3 January 1838.

65. PV 9 January 1838 AN AJ/15/135, p. 18.

66. See Pastré frères to the Museum, 30 January 1838, and F. Cuvier's list of Atir's travel expenses, both in séance of 6 February 1838, AN AJ/15/655.

67. PV 22 January 1839, AJ/15/135, p. 268.

68. PV 29 January 1839, AJ/15/135, p. 276.

69. Bell, who was helping Darwin classify his reptile specimens from the *Beagle* voyage, sent some to Paris for Bibron's judgment.

70. Sulloway, "Darwin's conversion."

71. Darwin, *Narrative*, 628.

4.9 THE ELECTRIC EEL

1. Perrottet, *Catalogue*, 1–5.

2. Humboldt, *Personal Narrative*, 2:112–31.

3. Humboldt, *Personal Narrative*, 2:118.

4. Humboldt, "Observations sur l'anguille électrique."

5. Humboldt, *Tableaux de la nature*, 1:53–57.

6. Humboldt, "Gymnotes."

7. Perrottet, *Catalogue*, 5.

8. F. Cuvier letter to the Director, 30 August, séance of 4 September 1821, AN AJ/15/622. PV AN AJ/15/121, p. 66.

9. *Le Drapeau blanc*, 26 August 1821, 2. Reprinted in *Gazette nationale*, 27 August 1821, 1239. *Le Drapeau* states that Janin de Saint-Just suffered for "several days" after handling the eel. Cuvier says the eel's death came four days after the event. Given that the account published on 26 August does not mention the eel's death, Saint-Just's encounter with the eel probably occurred on or about 23 August 1821.

10. Humboldt, *Personal Narrative*, 2: 120n, quotes the gymnotus experience of the Swedish physician Samuel Fahlberg (1758–1834): "the shocks it gave were so violent, especially in the open air, that I found scarcely any means of protecting myself by non-conductors, in transporting the fish from one place to another."

11. Janin de Saint-Just, "Gymnotus electricus." The report does not specify whether Georges or Frédéric Cuvier was present, though it was probably Georges, since Frédéric did not describe the scene.

12. *Le Drapeau blanc*, 26 August 1821, 2. See also Janin de Saint-Just, "Gymnotus electricus."

13. PV 4 September 1821, AN AJ/15/121, p. 66.

4.10 JACK THE ORANGUTAN

An earlier version of this material appeared in Richard Burkhardt, "An Orangutan in Paris: Pondering Proximity at the Muséum d'Histoire Naturelle in 1836," *History and Philosophy of the Life Sciences* 40 (2018);https://doi.org/10.1007/s40656-018-0186-1.

1. Van Iseghem to the Museum, 12 April 1836, read 19 April 1836, AN AJ/15/651. For an engaging account of the animal, see Cozic, *L'orang*-outan. This book uses additional information from the Museum's correspondence, administrative assemblies, and menagerie reports.

2. "Mammifères et oiseaux. Note pour Mr Martin (Malaca, Sumatra, Ile Maurice)." In I. Geoffroy Saint-Hilaire's handwriting. Undated, séance of 15 March 1836. AJ/15/651. The Museum's orangutan specimens at this point consisted of the stuffed skin and skeleton of the empress's specimen of 1808, two or three more skeletons (including the famous "Pongo of Wurmb" from the Stadtholder), and perhaps another mounted skin.

3. Van Iseghem to the Museum, 12 April 1836.

4. PV19 April 1836, AN AJ/15/133, p. 44.

5. Van Iseghem to the Museum, 22 April 1836, read 26 April 1836 (but filed with the séance of 3 May 1836), AN AJ/15/651.

6. Marion de Procé to Blainville, 27 April 1836, séance of 3 May 1836, AN AJ/15/651. Blainville communicated a long extract of the letter to the Académie des sciences on 2 May 1836, *Compte rendu des séances de l'Académie des sciences* 2 (1836): 425–27. The extract also appeared in *Annales des sciences naturelles*, ser. 2, 5 (1836): 313–14.

7. Marion de Procé to Blainville, 27 April 1836.

8. Marion de Procé to Blainville, 27 April 1836.

9. F. Cuvier menagerie report, 3 May 1836, AN AJ/15/651.

10. Baratay and Hardouin-Fugier, *Zoos*, 144–49.

11. PV 3 May 1836, AN AJ/15/133, p. 53.

12. Museum to Van Iseghem, 4 May 1836, AN AJ/15/755.

13. Draft letter in Frédéric Cuvier's handwriting to "Monsieur le Ministre," 7 May 1836, AN AJ/15/755.

14. É. Geoffroy Saint-Hilaire, "Extrait d'un mémoire sur l'orang-outang," 3, 1–2.

15. *L'Écho du monde savant*, 15 May 1836, 85–86.

16. Eclipse reported in the 16 May 1836 issues of *Le Figaro*, *La Quotidienne*, and *Journal des débats*.

17. "L'arrivée de l'orang-outang au Jardin-des-Plantes," *Echo du monde savant*, 15 May 1836. Reprinted in *Mercure de France*, 2 (1836), 76, and *Revue étrangère de la littérature, des sciences et des arts* 18 (1836): 673–74.

18. In his letter to the Museum, Marion de Procé called it "vexing" that the mother's other parts, especially the head, were not preserved, since naturalists could have used them to judge whether the orangutan and the pongo were the same species.

19. "Registre d'entrée," MNHN, MS 25. F. Cuvier menagerie report of 17 May 1836, AN AJ/15/651.

20. PV 17 May 1836, AN AJ/15/133, pp. 60–61.

21. PV 31 May 1836, AN AJ/15/133, p. 70, and Geoffroy to the Museum, dated [incorrectly] 6 July 1836, séance of 5 July 1836, AN AJ/15/652.

22. PV 31 May 1836, AN AJ/15/133, p. 70.

23. 31 May 1836, AN AJ/15/651 and PV AN AJ/15/133, pp. 70–71.

24. Walker, *Franz Liszt*.

25. Mesnager to the Museum 10 June 1836, séance of 14 June 1836, AN AJ/15/651.

26. PV 7 June 1836, AN AJ/15/133, p. 79. Museum to Mesnager, 22 June 1836, AN AJ/15/755.

27. Geoffroy to the Museum, 6 [5] July 1836.

28. Geoffroy to the Museum, 6 [5] July 1836.

29. PV 5 and 12 July 1836, AN AJ/15/133, pp. 89, 94.

30. E. Geoffroy Saint-Hilaire, "Extrait d'un mémoire."

31. Geoffroy, "Extrait," 2 (1836): 582–83. Geoffroy's story that the public persuaded him that the orangutan was "ni l'un ni l'autre" ("neither the one nor the other") must be taken with a grain of salt. He already used the phrase "ni l'un ni l'autre" in the same context in 1828 lecturing at the Museum (lesson 7:9 in Geoffroy, *Cours de l'histoire* naturelle. There, however, he still used the orders Bimana and Quadrumana.

32. Geoffroy Saint-Hilaire, "Extrait," 2 (1836): 583, 584.

33. Geoffroy Saint-Hilaire, "Éxtrait," 3 (1836): 5.

34. Geoffroy Saint-Hilaire, "Éxtrait," 3 (1836): 7. Blanckaert identifies Xavier Bichat's "fundamental law of the distribution of forces" as the primary source of Geoffroy's thoughts on the inverse relation between muscular power and intellectual function. See Blanckaert, "'Produire l'être singe.'"

35. Geoffroy Saint-Hilaire, "Extrait," 3 (1836): 30.

36. Geoffroy Saint-Hilaire, "Extrait," 3 (1836): 31.

37. F. Cuvier menagerie report, 3 May 1836, AN AJ/15/651.

38. F. Cuvier, "Orang-outang, femelle," *HNM*, June 1824.

39. Also omitted in Cuvier's 1824 orangutan paper were his previous comments on the orangutan's medical treatment; how intelligence was related more to the brain than to the senses; and his plans to do a comparative study of animal intellectual faculties.

40. F. Cuvier, "Orang-outang, femelle," *HNM*, June 1824, 7.

41. F. Cuvier, "Orang," *Dictionnaire des sciences naturelles* (1825), 276.

42. In an excellent article on the influence of Darwin's *Origin of Species* on visitors to German and Austrian zoos, Oliver Hochadel writes that after the publication of the *Origin*, "apes and monkeys were suddenly no longer just apes and monkeys." Hochadel, "Darwin in the Monkey Cage." As we see here, however, already in the case of the Paris menagerie's first orangutan, the issue of proximity caused the orangutan to be seen as more than just an ape or a monkey.

43. F. Cuvier, "Orang," 276.

44. Bory de Saint-Vincent criticized Cuvier, saying it was impossible to conceive of

an animal with neither reason nor instinct. Bory willingly credited the orangutan with reason. Bory de Saint-Vincent "Orang," 265.

45. F. Cuvier, "Orang," 283.

46. F. Cuvier, "Orang," 284.

47. Rousseau and Lemonnier, *Promenades*, 144–45; Bernard et al., *Le Jardin des Plantes*, 84–85. For broader overviews in the press, see "L'orang-outang du Muséum national d'histoire naturelle de Paris," *Magasin pittoresque* (1836), 224, and Comte, "Orang-outang," *Le magasin universel* 4 (1836): 92–94.

48. *L'Echo du monde savant*, 10 July 1836.

49. *Le Courrier français*, 14 July 1836, 1.

50. A notice in *Journal des débats*, 15 July 1836, refers to the orangutan as "Jacques." "Jacques du Sumatra" appears in *L'Echo du monde savant*, 11 September 1836. "Jack" became common during the ape's illness in the second half of October.

51. *Le Courrier français*, 14 July 1836, 1.

52. "Jacques du Sumatra," *L'Echo du monde savant*, 11 September 1836. For an examination of how journalists and others besides scientists offered observations of a later ape in a later zoo (the gorilla Mafuka in Dresden, beginning in 1873), see Hochadel, "Watching."

53. Rousseau and Lemonnier, *Promenades*, 144–45.

54. F. Cuvier menagerie report, 17 October 1837, AN AJ/15/654.

55. Comte, "Orang-outang," *Le magasin universel*, 93–94.

56. Comte, "Orang-outang," *Le magasin universel*, 94.

57. *Journal des débats*, 22 June 1836.

58. F. Cuvier menagerie report, 30 August 1836, AN AJ/15/652.

59. F. Cuvier menagerie reports, 30 August and 13 September 1836, AN AJ/156/652; séances of 27 September and 11 October 1836, AN AJ/15/133, pp. 136, 137. No decision to clothe Jack was announced in the Museum's Procès-verbaux or Cuvier's weekly menagerie reports, but Cuvier's report of 8 November (see below) refers to Jack's having a garment at the time his illness began (mid-October).

60. F. Cuvier menagerie report, 18 October 1836, AN AJ/15/652.

61. The most extensive account of Jack's treatment, written after his recovery, appeared in *L'Echo du monde savant*, 2 November 1836.

62. *La Presse*, 19 October 1836; *Figaro*, 21 October 1836; *L'Echo du monde savant*, 22 October 1836; *Le Journal des débats*, 24 October 1836.

63. *Le Journal des debats* and *Le Constitutionnel*, 24 October 1836. The same piece appears as "Jack," in *Courrier des Théâtres*, 24 October 1836, 2–3.

64. Had Atir helped with Jack's care, Frederic Cuvier would have mentioned it in his weekly reports, since he was looking for reasons to keep Atir, despite the keeper's drinking and carousing.

65. F. Cuvier menagerie report, 25 October 1836, AN AJ/15/652. See too AN AJ/15/133, pp. 148–49.

66. F. Cuvier menagerie report, 8 November 1836, AN AJ/15/652.

67. Menagerie report, 8 November 1836.

68. É. Geoffroy St. Hilaire, "Reglement pour les visites à l'orang outang," séance of 15 November 1836, AN AJ/15/652. See too AN AJ/15/133, 157–58.

69. *La Presse*, 4 November 1838.

70. *Figaro*, 19 November 1836.

71. Cozic, *L'orang-outan.*

72. The earliest archival evidence of this affair appears in a letter dated 24 November 1836 where Elwart asks Geoffroy to respond to a proposal Elwart already made about a concert for the orangutan. Séance of 29 November 1836, AN AJ/15/652.

73. Presumably Elwart did not intend to have such a heterogeneous group of musicians performing together at the same time. Had he done so, the concert would have been experimental in more ways than one.

74. Elwart to the Museum, 5 December 1836. Séance of 6 December 1836, AN AJ/15/652.

75. PV 6 December 1836, PV, AN, AN AJ/15/133: 168. Cuvier menagerie report of the same date, AN AJ/15/652. Draft of the Museum's response to Elwart, 10 December 1836, AN AJ/15/755.

76. F. Cuvier menagerie report, 13 December 1836, AN AJ/15/562.

77. F. Cuvier menagerie report, 20 December 1836, AN AJ/15/562.

78. F. Cuvier menagerie report, 27 December 1836, AN AJ/15/652.

79. Corsi, "Revolutions," mentions Sand's visit. See Sand, *Correspondence* 3:621–23 (letter to Eliza Tourangin, 30 December 1836).

80. F. Cuvier menagerie report, 10 January 1837, AN AJ/15/653.

81. *Journal des débats*, 6 January 1837, reports that Paris had not had so much snow and such intense cold since 1830.

82. Rousseau and Lemonnier, *Promenades*, 144.

4.11 THE OLD PROFESSOR AND THE NEW

1. Frédéric Cuvier to the Minister of Public Instruction, 6 December 1836. AN F17/13566.

2. PV 20 December, AJ/15/133, pp. 179–80.

3. PV 14 February 1837, AN AJ/15/134, p. 25.

4. E. Hamy, "Notes intimes," 480.

5. PV 21 February 1837, AN AJ/15/134, p. 30.

6. Geoffroy, "Rapport sur la ménagerie," MNHN MS 2290.

7. Geoffroy, "Rapport sur la ménagerie."

8. Geoffroy, "Rapport sur la ménagerie."

9. Geoffroy, "Registre d'ordre relatif à la ménagerie et copie de lettres." Copy of Geoffroy letter to Frédéric Cuvier, 22 February 1836. MNHN MS 2290 (1).

10. PV 28 February 1837, AN AJ/15/134, p. 35.

11. Geoffroy, "Rapport sur la ménagerie."

12. F. Cuvier to Geoffroy Saint-Hilaire, 18 April [1837], AN AJ/15/653.

13. F. Cuvier to the Museum director, 9 May 1837, AN AJ/15/653. PV AJ/15/134, pp. 70–71.

14. F. Cuvier, menagerie reports, 7, 14, and 21 August 1837, séance of 22 August 1837, AN AJ/15/654.

15. Geoffroy Saint-Hilaire, "Changements à la surface de la terre" (1837).

16. Frédéric Cuvier, "Observations" (1837), 305–6.

17. Geoffroy, "Note remise" (1837), 306.

18. Geoffroy to his fellow professors, 26 September 1837, AN AJ/15/654. See F. Cuvier menagerie reports of 18 and 26 September 1837, AN AJ/15/654. PV 26 September

1837, AN AJ/15/134, p. 165, mentions that there was also a report for 11 September, but none remains in AN AJ/15/654.

19. Théodoridès, "Quelques documents inédits," 191.

20. PV 12 December 1837, AN AJ/15/134, p. 210.

21. Blainville, note protesting the new chair, read at the séance of 5 December 1837, AN AJ/15/654.

22. PV 19 December 1837, AN AJ/15/135, pp. 4–5.

23. Museum to the Minister of Public Instruction, 20 December 1837, and Ordonnance du Roi, 24 December 1837. AN F/17/13566.

24. F. Cuvier to the Museum director, 3 January 1838; menagerie report of 9 January 1838. Séance of 9 January 1838, AN AJ/15/655.

25. Geoffroy to "Messieurs," 15 January 1838. Filed with the séance of 23 January 1838. Notwithstanding his refusal to attend the committee meeting, Geoffroy later communicated his assessment that the zoology gallery needed twice its current space. PV 13 February 1838, AN AJ/15/135, pp. 44–45.

26. Geoffroy to the Museum, 16 January 1838, séance of the same day, AN AJ/15/655.

27. Geoffroy to "Messieurs et honorés collègues," 23 January 1838, AN AJ/15/655.

28. Geoffroy to the Museum, 23 January 1838.

29. Reynaud, "Fondation."

30. Geoffroy, *Fragments biographiques*.

31. Geoffroy Saint-Hilaire to "Messieurs," dated "ce mardi" (22 May 1838), séance of same date. AN AJ/15/655. See also PV AN AJ/15/135, p. 104.

4.12 THE CHIMPANZEE AND THE "PALACE OF THE MONKEYS"

1. PV 1 August 1837, AN AJ/15/134, p. 128.

2. PV 8 August 1837, AN AJ/15/134, p. 133.

3. Museum to the Minister of Public Instruction, 19 September 1837, AN AJ/15/755.

4. Rousseau and Lemonnier, *Promenades*, xxiii, date the original monkey house to 1802. Deleuze, *Histoire*, 110, says the conversion of the building to accommodate the monkeys was finished in 1805.

5. Anonymous, *Guide des étrangers* (1828), 102.

6. F. Cuvier, menagerie report, 10 January 1826, AN AJ/15/631.

7. F. Cuvier, menagerie report 5 November 1828, AN AJ/15/636.

8. "Texte du programme donné en février 1833 pour la construction d'une singerie." Filed with séance of 10 March 1835, AN AJ/15/649. The document is in F. Cuvier's hand.

9. Belin, "Architectures zoologiques." I thank Jacques Rigoulet for providing me a copy of this work.

10. At the séance of 13 May 1834, the professors learned that the Museum's budget for the year was being augmented by 20,000 francs. PV AN AJ/15/131, p. 194

11. The beginning of construction was announced 20 January 1835, PV AN AJ/15/132, p. 57. That the Museum forgot about its early plans for the monkey house is stated in Museum to the Minister of the Interior, 20 February 1835, séance of 10 March 1835, AN AJ/15/649.

12. The grilled courtyard of the London Zoo's monkey house (opened in 1833) was greeted with approval, but the monkey house itself, according to Wilfrid Blunt, was "a failure from the start." Blunt, *Ark in the Park*, 40.

13. Museum to the Minister of the Interior, 20 February 1835, AN AJ/15/649. PV 10 and 17 February 1835, AN AJ/15/132, pp. 72–74.

14. PV 14 and 21 April 1835, AN AJ/15/132, pp. 105, 106. Belin, "Architectures zoologiques."

15. *Le Moniteur universel*, 17 May 1836, 1110–13.

16. *Le Moniteur universel*, 13 May 1836, 1083.

17. *Revue de Paris*, n.s. 29 (1836): 200.

18. *Revue de Paris*, n.s. 29 (1836): 278.

19. *Le Charivari*, 2–3 November 1836, reproduced in Jobling, "Daumier's 'Orang-outanania.'" *Le Charivari* noted that Jacke [*sic*], was soon to have "a magnificent palace, constructed on the plans of an architect and under the direct supervision of M. Thiers."

20. Minister of Public Instruction to the Museum, 9 September 1837, séance of 12 September 1837, AN AJ/15/654. Museum to the minister, 19 September 1837, AN AJ/15/755. PV 9 September and 10 and 17 October 1837, AN AJ/15/134, pp. 159, 171, 178.

21. F. Cuvier menagerie reports, 9 October 1837 (séance of 10 October 1837) and 17 October 1837, AN AJ/15/654.

22. F. Cuvier menagerie report, 17 October 1837, AN AJ/15/654. The administration appointed a committee composed of Geoffroy Saint-Hilaire, Blainville, and Valenciennes to set the rules regarding the chimp. PV 17 October 1837, AN AJ/15/134, p. 178. In an annotation of Cuvier's menagerie report of the day before, Geoffroy fussed about the monkeys' staying in their cages instead of their courtyard, preventing visitors from seeing them. He suggested that people be allowed into the grilled courtyard to approach the monkeys in their cages for a closer look. The professors agreed to try Geoffroy's suggestion, but only after a barrier was built to keep the public from getting too close to the animals.

23. *Journal des débats*, 23 October 1837, citing from *l'Echo du monde savant*, 21 October 1837.

24. *Journal des* débats, 23 October 1837. Jacqueline was treated in a jesting piece by Vicomte Charles de Launay, "Courrier de Paris," *La Presse*, 28 October 1837, reprinted in Girardin, *Vicomte de Launay*. Additional stories appear in Boitard, *Jardin des Plantes*, 4–5.

25. For Berthoud's role in the domino club, see Jousserandot, *Le domino*. On the history of the club and Berthoud and Dantan le Jeune's literary and artistic caricatures of the players, see "Chronique du remplaçant," *Le Figaro*, 13 September 1869.

26. Berthoud, *Musée des familles: Lectures du soir* 5 (1837): 64.

27. Berthoud, "Histoire naturelle," *Musée des familles* 5 (1837): 95.

28. The date is from F. Cuvier, menagerie report of 14 November 1837 (Cuvier's copy for Geoffroy), AN AJ/15/654.

29. Berthoud, *Musée des* familles, 5 (1837): 95.

30. F. Cuvier, menagerie report of 14 November 1837 (copy for Geoffroy).

31. F. Cuvier, menagerie report, 21 November 1837, AN AJ/15/654. Cuvier blamed the illnesses on the heating system. He proposed abolishing the heating drums in the cages and widening the hot air vents in the corridor.

32. F. Cuvier, menagerie report, 4 December 1837, AN AJ/15/654.

33. F. Cuvier, menagerie report, 12 December 1837, AN AJ/15/654.

34. F. Cuvier, menagerie report, 12 December 1837.

35. F. Cuvier, menagerie report, 18 December 1837, AN AJ/15/654.

36. F. Cuvier, menagerie report, 26 December 1837, AN AJ/15/654.

37. The lists were not entirely accurate. The first failed to include the orangutan (still alive but at death's door), and, according to the mammal register, a Barbary ape and a malbrouck. The second included a chacma baboon that the register says died on 29 December.

38. See F. Cuvier, "État des animaux vivans à la Ménagerie au 1er janvier 1837," séance of 28 February 1837, AN AJ/15/653; and "Animaux vivans à la Ménagerie le 31 X[bre] 1837," séance of 9 January 1838, AN AJ/15/655. See also "Registre d'entrée," MNHN Ménagerie MS 25.

39. Three deaths occurred in September 1837; three-quarters of the menagerie's primate deaths in 1837 (18 out of 24) occurred in the year's last quarter.

40. For a table of the average longevity of monkeys from 1830 to 1959 see Baratay and Hardouin-Fugier, *Zoo*, 145. The authors give the overall average as 18.606 months. One table shows that monkeys at the menagerie from 1830 to 1834 lived approximately thirty-three months, markedly better than those from 1835 to 1839 (fourteen months).

41. Undated note by Cuvier, filed with the minutes of the séance of 9 January 1838, AN AJ/15/655.

42. PV 20 February 1838, AN AJ/15/135, p. 49.

43. PV 13 March 1838, AN AJ/15/135, p. 59. F. Cuvier, undated text with annotation, "Note pour la lettre à Mr. le Ministre." Séance of 13 March 1838. AJ/15/655.

44. PV 17 April 1838, AN AJ/15/135, p. 84.

45. S. Henry Berthoud, "Devant le palais des singes," *La Presse*, 15 May 1838.

46. Valenciennes menagerie report, 29 May 1838, AN AJ/145/655.

47. *Journal des débats*, 3 June 1838.

48. S. Henry Berthoud, "Faits," *Musée des familles*, 5 (1837–38): 288.

49. S. Henry Berthoud, "Faits," *Musée des familles*, 5 (1837–38): 288.

50. PV 29 May 1838, AN AJ/15/135, p. 109.

51. The keepers at the time were Daboncourt and Martin.

52. Berthoud, "Feuilleton de la Presse," *La Presse*, 18 September 1838.

53. The menagerie's records and Berthoud's description show that he misidentified the species.

54. The menagerie register, MNHN MS 25, dates the coaïta's entry as 20 June 1838 and its death as 5 November 1838. The date the assembly authorized Valenciennes's purchase of the animal was 26 June 1838. PV AN AJ/15/135, p. 126. I have compiled the number of deaths from the register. I did not include in the total a baboon that died three days after its birth on 30 July 1838.

4.13 QUARRELING OVER THE BONES

1. This description of Cuvier's illness and death is based on two documents in the Fonds Cuvier, Institut de France: (1) a fifteen-page manuscript titled, "Maladie et mort de M. Frédéric Cuvier," MS 3341.9; and (2) a letter from Duvernoy to Frédéric Cuvier's son dated "Dimanche à dix heures" [22 July 1838], MS 3341.9e. Cuvier's passport for his summer 1838 duties as inspector général des études is in the same file, item MS 3341.8.

2. Duvernoy to Frédéric Cuvier fils, 22 July 1838.

3. Duvernoy to Frédéric Cuvier fils, 22 July 1838.

4. Martinet, *Manuel de clinique*, 527.

5. See *Le Temps*, 28 July 1838, and *Le Commerce*, *Le Constitutionnel*, and others 29 July 1838. Le *Moniteur universel*, 30 July 1838, republished an article from *Le Cour-*

rier du Bas-Rhîn, one of the lengthier pieces. Quotation from *L'Hermine*, 30 July 1838. Physicians have recently suggested that Georges Cuvier's symptoms are consistent with those of Guillain-Barré syndrome. See Mathis and Vallat, "Mysterious death." Frédéric Cuvier's symptoms, though not identical with his brother's, also seem consistent with GBS.

6. Flourens, "Éloge historique de F. Cuvier," xviii. Flourens's source regarding Cuvier's last words was Gabriel Fort Dutrey's oration at Cuvier's funeral ("Discours sur la tombe de M. F. Cuvier"). As Cuvier's friend and fellow inspector general of the university, Dutrey was with Cuvier on the same administrative mission, stayed at the same hotel in Strasbourg, and was present when he died. Cuvier may well have said the line Dutrey reported, but without its being his last utterance. Flourens repeated the line in the later editions of his book on instinct and intelligence, including the final one, *De l'instinct*, 311.

7. PV 10 August 1838, AN AJ/15/135, pp. 152–54. There was no money left in the budget for menagerie supervision. In 1838, the 3,500 francs for Frédéric Cuvier's salary as superintendent became his salary as professor. It was to be raised to the standard professorial salary of 5,000 francs in 1839. For the 1838 salaries, see AN AJ/15/193.

8. PV 10 August 1838, AN AJ/15/135, p. 154.

9. Étienne Serres was appointed to the renamed chair. Séances of 13 and 20 November 1838, AN AJ/15/135 p. 206, 213–15.

10. PV 16 October 1838, AN AJ/15/135, p. 186. *Journal des débats*, 6 November 1838.

11. The debate's roots went back to 1818 when William Buckland showed Georges Cuvier a fossil jawbone in Oxford that Georges concluded was from a mammal. Blainville argued it was from a reptile. In 1838, Valenciennes and the Museum's professor of reptiles and fish, Constant Duméril, sided with G. Cuvier. The distinguished British comparative anatomist Sir Richard Owen identified the fossil as mammalian and named the genus *Phascolotherium*, ending the debate. On the debate at the Academy, see Isid. B[ourdon], "Revue scientifique. Académie des sciences. Séance du 15 Octobre 1838," *Le Constitutionnel*, 18 October 1838.

12. F. Cuvier, "Animaux vivans à la ménagerie le 31 Xbre 1837," AN AJ/15/655. Isidore's census has not survived. I take the numbers of animals from his letter to the Museum, 4 September 1938, AN AJ/15/656.

13. PV 12 and 26 June and 10 July 1838, AN AJ/15/135, pp. 117, 124, 134–35. See the letter from the Sécretaire des Commandements de S. A. R. Mgr. le duc d'Orléans, séance of 26 June 1838, AN AJ/15/655, and the letters from Richard Owen to (a) Frédéric Cuvier, 23 May 1838, and (b) Achille Valenciennes, 8 July 1838, both in séance of 10 July 1838, AN AJ/15/656.

14. PV 17 and 31 July 1838, AN AJ/15/135, pp. 139, 144–45.

15. Museum letter to the Minister of Public Instruction (written by Valenciennes), 4 August 1838, AN AJ/15/756.

16. Valenciennes menagerie report, 17 July 1838, AN AJ/15/656, and Museum letter to the Minister of the Interior, 4 August 1838, written by Valenciennes. AN AJ/15/756. Valenciennes's letter to the minister appeared nearly verbatim in *Le Moniteur*, 2 September 1838.

17. From Valenciennes's notes on the animals in the collection (Museum letter to the Minister of the Interior, 4 August 1838).

18. I. Geoffroy Saint-Hilaire, menagerie report, 21 August 1838, AN AJ/15/656. PV AN AJ/15/135, pp. 156–58.

19. I. Geoffroy Saint-Hilaire, menagerie report, 4 September 1838, séance of same date. AN AJ/15/656. See PV AN AJ/15/135: 163. Étienne Geoffroy Saint-Hilaire presented Isidore's animal register, but the assembly's minutes make no mention of the administration's encouraging an annual census, nor have I found a copy of this one.

20. I. Geoffroy Saint-Hilaire menagerie report, 9 October 1838, AN AJ/15/656.

21. I. Geoffroy Saint-Hilaire menagerie report, 30 October 1838, AN AJ/15/656.

22. On accommodating new animals, see Isidore's menagerie reports, 9 and 16 October 1838; AJ/15/656.

23. PV 13 February 1838, AN AJ/15/135, pp. 44–45.

24. Isidore Geoffroy Saint-Hilaire and Achille Valenciennes, "Programmes relatifs à diverse ameliorations pour la ménagerie," 14 December 1838, read at the special séance of 16 December 1838. AJ/15/656. The animals intended for the small rodent house included pacas, agoutis, squirrels, porcupines, etc.—but also edentates (e.g., anteaters, armadillos), frugivorous marsupials, and small pachyderms like damans. PV 16 December 1838, AN AJ/15/135, 241–42.

25. I. Geoffroy Saint-Hilaire menagerie report, 6 November 1838. AJ/15/656.

26. "Muséum d'histoire naturelle: La Fosse aux ours," *Magasin pittoresque* 11 (1843): 306–9. See also 2.11 on Martin the Bear.

27. Anonymous, "Feuilleton de la presse," *La Presse*, 18 September 1838, 2. I attribute this piece to Berthoud because of its strong similarity to ones he signed in *La Presse* on 15 May and 18 December 1838.

28. I. Geoffroy Saint-Hilaire menagerie report, 23 October 1838, AN AJ/15/656. On the understanding of the disease at the time, see Desains, *La chorée*.

29. Isidore Geoffroy Saint-Hilaire menagerie report, 23 October 1838, AN AJ/15/656. With respect to managing public perceptions about zoo animals, see Simons, *Obaysch*, on how the London Zoo sought to hide the hippo Obaysch's aggressiveness from the public.

30. Pacot had broken ribs and a severe contusion to his right buttock from being kicked by the Kob antelope. Isidore Geoffroy Saint-Hilaire's menagerie reports, 27 November and 4 December 1838, AN AJ/15/656.

31. Blainville to the Muséum, 3 December 1839, AN AJ/15/845.

32. Blainville to the Muséum 26 January 1841, AN AJ/15/845.

33. Flourens, *Éloge historique de F. Cuvier.*

34. F. Cuvier to the Minister of Public Instruction, 6 December 1836, AN F/17/13566.

35. Cuvier to the Minister of Public Instruction, 6 December 1836.

36. There are several copies of Cuvier's course program, one he sent to his colleagues on 31 August 1832, AN AJ/15/644; one sent to Salvandy on 6 December 1836, AN F/17/13566; one published posthumously from a copy among his papers, "Considération sur l'étude des actions des animaux, suivres du programme d'un cours sur la nature des animaux et sur leur éducation," *Annales des sciences naturelles* 12 (1839): 140–48, and one in the papers of Adolphe Brongniart, MNHN, MS 660, fols. 173–75 cited by Piazzesi, *Domesticare gli istinti*, 236–37n714.

37. Flourens, *Éloge historique de Frédéric Cuvier*, i-ii.

38. Flourens's "Resumé analytique," was published in 1839 in multiple forms: by Imprimerie Royale as an independent text; in *Annales des sciences naturelles; in Journal des Savants (June, August, and September 1839)*; and in *Le Constitutionnel*, 13 and 20 October 1839. Other versions followed. Cuvier's name was dropped from the title of the

fourth edition, *De l'instinct et de l'intelligence des animaux* (Garnier frères, 1861). The fifth and last edition (1870) had the same title and publisher as the fourth.

39. Though animal suffering was an inevitable consequence of Flourens's practices, it bears note that he is credited with discovering the value of chloroform as an anesthetic agent.

40. F. Cuvier, "Observations préliminaires" (1834), 1:xxii.

41. Flourens, *Examen.*

42. Flourens, *Examen,* 31.

43. Flourens, *Examen,* 64.

44. Flourens, *Examen,* 65.

45. Flourens, *Examen,* 20.

46. For I. Geoffroy Saint-Hilaire's rejection of Frédéric Cuvier and Flourens's views on hybrids and intraspectific variation and for Isidore's concept of "the variability of type," see his *Histoire naturelle générale,* 3:135–530.

47. Isidore Geoffroy Saint-Hilaire promoted acclimatization to benefit France and her colonies, but the menagerie was not his base of operations. He founded a new scientific society, the Imperial Society of Zoological Acclimatization (La Société impériale zoologique d'acclimatation), in a new setting. His zoological acclimatization garden (Le Jardin zoologique d'acclimatation) opened in 1860 at the north end of the Bois du Boulogne. See Osborne, *Nature.*

48. Quotations from pages 139e and 153e respectively of Darwin's 1838 "Notebook M." See in Barrett et al., *Darwin's Notebooks.* Darwin observed Jenny further, and later at the London Zoo George John Romanes observed a young orangutan and baboons. See Romanes, *Animal Intelligence,* 474, 476.

49. On the contribution of German zoos to understanding living animals, see Nyhart, *Modern Nature.* More generally on science in the zoo, see Hochadel, "Science at the Zoo," which introduces a collection of articles on zoo science.

50. Whitman, "Biological Farm," 221, 222.

51. For an elaboration of these points and those that follow on Lorenz, Tinbergen, and twentieth-century ethology, see Burkhardt, *Patterns of Behavior.*

52. Loisel, *Histoire des ménageries,* 3:152.

53. Spalding, "Instinct"; Morgan, *Habit and Instinct*; Richards, *Darwin,* 387–88.

54. Frédéric Cuvier, "Dégeneration," *Encyclopédie des gens du monde* (1837), 7:673.

55. Cuvier, "Dégeneration," 673–74.

56. Cuvier, "Dégeneration," 675.

57. Cuvier, "Dégeneration," 675.

58. De Beer, "Darwin's notebooks," 34. De Beer observes that Darwin miscopied the original quotation in two respects, transcribing "concurrence" as "occurrence" and leaving out a line.

59. Jaynes, "Historical origins," credits John Stuart Mill with first employing the term "ethology" and Pierre Flourens with having founded comparative psychology, at least in name, with the publication of his *Psychologie comparée* (Paris: Garnier frères, 1864). Jaynes is incorrect to say, however, that the latter represented "the first attempt to found comparative psychology as a new science" (602). Frédéric Cuvier made the attempt, though without using the phrase "comparative psychology" for the enterprise. He did use the phrase with more limited scope, however, in writing of the "comparative psychology" of different races of dogs. On the word "ethology," see also Durant, "In-

nate character." On "ethologie" as developed by Alfred Giard and his students in the late nineteenth and early twentieth century, see De Bont, "Organisms."

60. The connection between Frédéric Cuvier and Isidore Geoffroy Saint-Hilaire is explored by Piazzesi, *Domesticare gli istinti.*

61. Isidore Geoffroy Saint-Hilaire, *Histoire naturelle*, 1:xxii. Explaining in volume 2 what ethology involved, he wrote, "to ethology . . . belongs the study of the relations of organized beings in the *family* and the *society*, in the *aggregate* and the *community*." 2:285.

62. Isidore Geoffroy Saint-Hilaire, *Histoire naturelle Générale*, 3:480–86.

63. On the state of Geoffroy's *Histoire naturelle générale* at the time of his death, see the editorial note, 3:531.

64. I thank Philippe Chavot for inquiring about Frédéric Cuvier's grave site on my behalf.

65. Jules Moiroux, *Le cimetière du Père Lachaise* (A. Maréchal, 1922), 57.

4.14 CONCLUSION

1. Milne-Edwards, *La menagerie.*

2. E., "Jack," *Le Courier français*," 14 July 1836.

3. I. Geoffroy Saint-Hilaire menagerie report, 5 July 1836, AN AJ/15/652.

4. For an elegant treatment of how human assumptions about elephants' cognitive worlds have had a closer relation to the lives of humans than to the lives of elephants, see Rothfels, *Elephant Trails.*

5. In the literature on her, the practice of calling her "Zarafa" began only in 1998 with Allin, *Zarafa*. See p. 5.

6. É Geoffroy Saint-Hilaire, "Motifs" (1809). See 2.6.

7. On Grandville see Kaenel, "Le Buffon de l'humanité"; and Grandville, Scènes. Grandville's illustrations of scenes at the Museum appear in Boitard, *Le Jardin des Plantes.*

8. G. Cuvier, "Le couagga," in Lacépède, *Ménagerie*. Dated by Pieters Ventôse an 10 (20 February to 21 March 1802).

9. F. Cuvier, "Le couagga," *HNM*, July 1821.

10. Ritvo, "Q is for Quagga."

11. The case of Baillon's "phalarope" evidences the naturalist's thrill in collecting a rare specimen. Nonetheless, objectors to collecting specimens of nearly extinct species did exist in the first half of the nineteenth century. See Fennell, "Extermination," 192–93. On emerging thoughts on the reality of human-induced species extinction and how to respond to its specter, see Barrow, *Nature's Ghosts.*

12. Lamarck, *Philosophie zoologique*, 1:77–78. Further, Burkhardt, *Spirit of System*, 128–36.

13. Lamarck, "L'Homme," 270. I have cited this passage before, including in *Spirit of System*, 214.

14. Candolle, "Geéographie botanique." Quotation on 421. See also 410 and 419.

15. Candolle, "Géographie botanique," 384.

16. Lyell, *Principles of Geology*, 2:150.

17. Lyell, *Principles of Geology*, 2:156.

18. Lyell, *Principles of Geology*, 2:156.

19. Lyell, *Principles of Geology*, 2:175.

20. Lyell, *Principles of Geology*, 2:168, 169.

21. See Pouillard, *Histoire des zoos par les animaux*.

22. Milne-Edwards, *La Ménagerie*, 9. For a closer look at the menagerie's condition from 1838 to 1890, see Laissus, *Les Animaux du Muséum*, 138–61.

23. Milne-Edwards, *La Ménagerie*, 13–15, 22–23, 28–30.

24. Anonymous, "The Old Zoo and the New." I believe the author was the British field naturalist Edmund Selous, on whom see Burkhardt, *Patterns of* Behavior, 77–92.

25. Rothfels, *Savages and Beasts*.

26. Osborne, *Nature*, describes how *the* Jardin zoologique d'acclimatization in Paris reconceived its functions from 1860 to 1920, noting Hagenbeck's human exhibits there in 1877 as an important innovation. Bruce, *Lion Gate*, reports on how the Hagenbeck and other ethnographic exhibits at the Berlin Zoo produced a critical upswing in that zoo's fortunes.

27. The San Diego Zoo has recently rebranded itself and its complement, the San Diego Safari Park, as the San Diego Zoo Wildlife Alliance. Its mission statement reads: "San Diego Zoo Wildlife Alliance is committed to saving species worldwide by uniting our expertise in animal care and conservation science with our dedication to inspiring passion for nature." https://sandiegozoowildlifealliance.org/about-us/about-san-diego-zoo-wildlife-alliance. Accessed 16 November 2023.

Bibliography

PRIMARY SOURCES

Note: Not listed here, but cited in the notes, are numerous articles by Buffon in his *Histoire naturelle, générale et particulière*; by Georges Cuvier, Geoffroy Saint-Hilaire, and Bernard de Lacépède in Bernard de Lacépède et al., *La Ménagerie du Muséum*; and by Frédéric Cuvier in É. Geoffroy Saint-Hilaire and F. Cuvier, *Histoire naturelle des mammifères.*

Adanson, Michel. *Histoire naturelle du Sénégal.* C.-J.-B. Bauche, 1757.

Alibassan [Honoré de Balzac], *Discours de la girafe au chef des six Osages (ou Indiens), prononcé le jour de leur visite au Jardin du Roi. Traduit de l'Arabe par Alibassan, interprète de la girafe.* Martinet, 1827.

Alibert, J. L. *Traité des fièvres pernicieuses intermittentes.* 3rd ed. Crapelet, an 12 [1804].

Anonymous. Review of Bernardin de Saint-Pierre, "Mémoire sur la nécessité de joindre une ménagerie au Jardin national des plantes de Paris." *Magasin encyclopédique* 1 (1792): 113–16.

Anonymous [Georges Toscan]. "Sciences. Histoire naturelle. Vues générales." *La Décade philosophique, littéraire et politique* 1 (10 floréal, an 2 [1794]): 1–7.

Anonymous. *Fêtes de la liberté, et entrée triomphale des objets de sciences et d'arts recueillis en Italie. Programme.* Imprimerie de la République, thermidor an 6 [1798]. https://gallica.bnf.fr/ark:/12148/bpt6k487328.

Anonymous [Miger]. *Notice des animaux vivants, actuellement à la ménagerie du Muséum national d'histoire naturelle, et une description historique sur la manière de vivre, et les habitudes de chacun d'eux: Avec le portrait de la lionne allaitant ses lionceaux.* Patrice et Gilbert, 1801.

Anonymous [Miger]. *Notice des animaux vivants, actuellement à la ménagerie du Muséum national d'histoire naturelle, et description historique sur la manière de vivre, et les habitudes de chacun d'eux, à laquelle on a joint quelques circonstances de la mort de l'éléphant mâle.* Patris, 1802.

Anonymous. *Le guide au Jardin des Plantes, ou Description de tout ce que les galléries, serres, etc. du Muséum d'histoire naturelle de Paris, renferment de plus curieux.* Glisau, an 11 [1803].

Anonymous [F. Cuvier]. *Notice des animaux vivans de la Ménagerie, leur origine et leur histoire dans cet établissement.* Levrault, Schoell, an 12 [1804].

Anonymous. *Description de la vie et mœurs des animaux féroces qui sont renfermés tant à la Ménagerie que dans la Vallée Suisse; suivie des curiosités qui se trouvent au Cabinet d'histoire naturelle.* Aubry, 1807.

Anonymous [F. Cuvier]. *Notice des Animaux vivans de la Ménagerie du Muséum d'histoire naturelle*. Dufour, 1809.

Anonymous. *Description de ce qu'il y a de remarquable à la ménagerie, et au cabinet d'histoire naturelle; contenant la vie et les mœurs des animaux féroces qui sont renfermés tant à la Ménagerie que dans la Vallée suisse: suivie des curiosités qui se trouvent au Cabinet d'histoire naturelle*. L. P. Sétier, 1811.

Anonymous [Révéroni Saint-Cyr, Jacques-Antoine de]. *L'Officier russe à Paris, ou Aventures et réflexions critiques du Comte de ****, vol. 1. Barba, 1814. http://catalogue.bnf.fr/ark:/12148/cb312027.

Anonymous. *Délibération des animaux. Jugement contre l'Ours Martin, rendu par les animaux de la Ménagerie du Jardin des Plantes, sous la présidence du Lion. Son interrogatoire et ses réponses*. Imprimerie de Brasseur aîné, n.d.

Anonymous. *Détails exacts du malheureux événement arrivé au Jardin des plantes, dans la fosse de l'ours*. Imprimerie de Stahl, n.d.

Anonymous. *Description de ce qu'il y a remarquable à la ménagerie et au cabinet d'histoire naturelle*. . . . Chez les Marchands de Nouveautés, 1820.

Anonymous. *La girafe et les Osages*. Ledoux, 1827.

Anonymous. *Guide des étrangers au Museum d'histoire naturelle et au jardin du roi*. F. G. Levrault, 1828.

Anonymous [Achille-Narcisse de Salvandy]. *Lettre de la girafe au pacha d'Egypte, pour lui rendre compte de son visite à Saint-Cloud, et envoyer les rognures de la censure de France au journal qui s'établit à Alexandre en Afrique*. A. Sautelet, 1827.

Anonymous [Achille-Narcisse de Salvandy]. *Seconde lettre de la girafe au pacha d'Égypte, en lui envoyant son album enrichi des dernières noirceurs de la censure*. A. Sautelet, 8 August 1827.

Anonymous [Honoré de Balzac]. "Études de philosopie [*sic*] morale sur les habitans du Jardin des Plantes." *La Silhouette* (17 June 1830): 101–2.

Anonymous *Notice sur M. Henry Martin et sa menagerie*. J.-S. Cordier fils, 1831.

Anonymous [Charles Maxime Catherinet de Villamarest]. *Mémoires de Mademoiselle Avrillion, première femme de chambre de l'impératrice, sur la vie privée de Joséphine, sa famille et sa cour*. 2 vols. Ladvocat, 1833.

Anonymous. "Muséum d'histoire naturelle: La fosse aux ours." *Magasin pittoresque* 11 (1843): 306–9.

Anonymous. *Catalogue des livres d'histoire naturelle, (particulièrement de zoologie), de philosophie, d'histoire, etc., composant la bibliothèque de feu M. Frédéric Cuvier*. J.-F. Delion, 1846.

Anonymous. "The Old Zoo and the New. I. The Bastille of the Beasts." *Saturday Review*, 16 March 1901, 330–32.

Arago, François. *Histoire de ma jeunesse*. Kiessling, Schnée, 1854.

Bacon, Francis. *New Atlantis*. Edited by Alfred B. Gough. Clarendon Press, 1915.

Baudin, Nicolas. *The Journal of Post Captain Nicolas Baudin, Commander-in-Chief of the Corvettes "Géographe" and "Naturaliste."* Translated by Christine Cornelle. Libraries Board of South Australia, 1974.

Bernard, Pierre, L. Couailhac, Paul Gervais, and Emm. Lemaout. *Le Jardin des Plantes: Description complète, historique et pittoresque du Muséum d'histoire naturelle, de la ménagerie*. . . . Curmer, 1842. Bernardin de Saint-Pierre, Jacques-Henri. *Études de la nature*. 2d ed. Didot le jeune, 1786.

Bernardin de Saint-Pierre, Jacques-Henri. *Mémoire sur la nécessité de joindre une ménagerie au Jardin national des plantes de Paris*. Didot le jeune, 1792.

Bertuch, Carl. *Bemerkungen auf einer Reise aus Thüringen nach Wien im Winter 1805–1806*. 2 vols. 1808–10.

Bewick, Thomas. *A General History of Quadrupeds*. 4th ed. Edited by S. Hodgson, R. Beilby, and T. Bewick, 1800.

Biot, J.-B. "Notice sur les opérations faites en Espagne pour prolonger la méridienne de France jusqu'aux iles Baléares, lue à la dernière séance publique de la classe des sciences physiques et mathématiques." *Moniteur universel*, January 7, 1810.

Blainville, Henri de. "Sur une femme de la race hottentote." *Bulletin des sciences par la Société philomathique de Paris*, 1816, 183–90.

Blanvillain, J. F. C. *Le Pariséum, ou Tableau de Paris*. Frères Piranesi, 1807.

Bory de Saint-Vincent, J. B. G. M. "Homme." *Dictionnaire classique d'histoire naturelle* 8 (1825): 269–346.

Bory de Saint-Vincent, J. B. G. M. "Orang." *Dictionnaire classique d'histoire naturelle* 12 (1827): 262–85.

Brongniart, Ad. "Rapport sur la partie botanique de la voyage de Mr. Dorbigny." *Nouvelles annales du Muséum d'histoire naturelle* 3 (1834): 99–103.

Brougham, Henry. *Dissertations on Subjects of Science Connected with Natural Theology*. 2 vols. C. Knight, 1839.

Buffon, Georges-Louis Leclerc, Comte de. *Histoire naturelle des oiseaux*. 9 vols. Imprimerie Royale, 1770–86.

Buffon, Georges-Louis Leclerc, Comte de. Histoire naturelle, générale et particulière. 15 vols. Imprimerie Royale, 1749–67.

Buffon, Georges-Louis Leclerc, Comte de. *Histoire naturelle, générale et particulière*. Edited by C. N. S. Sonnini de Manoncourt. F. Dufart, 1798–1801.

Buffon, Georges-Louis Leclerc, Comte de. *Supplément à l'Histoire naturelle*. 7 vols. Imprimerie Royale, 1774–89.

B*** [Buillard, Pierre]. *Aviceptologie française, ou Traité générale de toutes les ruses dont on peut servir pour prendre les oiseaux qui sont en France*. J. Cussac, an 3 [1795].

Camper, Pierre. "De l'orang-outang, et de quelques autres espèces de singes." In *Œuvres de Pierre Camper, qui ont pour objet l'histoire naturelle, la physiologie et l'anatomie comparée*, vol. 1. H. L. Jansen, 1803.

Camper, Pierre. *Dissertation physique de Mr. Pierre Camper, sur les différences réelles que présentent les traits du visage chez les hommes de différents pays et de différents âges; sur le beau qui caractérise les statues antiques et les pierres gravées*. B. Wild et J. Altheer, 1791.

Candolle, Augustin Pyramus de. "Géographie botanique." *Dictionnaire des sciences naturelles* 18 (1820): 359–422.

Cattet, Jean-Joachim-François, and Jean-Baptiste Joseph Gardent. *Essai sur la contagion. . . .* Testus, an 10 [1802].

Chabert, P., P. Flandrin, and J.-B. Huzard, eds. *Instructions et observations sur les maladies des animaux domestiques*. La Veuve Vallat-La-Chapelle, 1791.

Champoiseau, Noel. "Voyages aux Indes orientales. Notice sur les voyages de M. Diard, naturaliste français." *Journal des voyages, découvertes et navigations modernes* 43 (1829): 222–39.

Comte, Achille. "Orang-outang." *Le magasin universel* 4 (1836): 92–94.

Corse, John. "Observations on the Manners, Habits, and Natural History, of the Elephant." *Philosophical Transactions of the Royal Society of London* 89 (1799):31-55.

Cuvier, Frédéric. "Castor." *Dictionnaire des sciences naturelle* 7 (1817): 242–52.

Cuvier, Frédéric. "Dégénération." *Encyclopédie des gens du monde* 7 (1837): 672–76.

Cuvier, Frédéric. "De la sociabilité des animaux." *Mémoires du Muséum d'histoire naturelle* 13 (1825): 1–27.

Cuvier, Frédéric. "De l'éducation des sourds-muets de naissance, par M. de Gérando." *Revue encyclopédique* 36 (1827): 614–29.

Cuvier, Frédéric. *De l'histoire naturelle des cétacés, ou Recueil et examen des faits dont se compose l'histoire naturelle de ces animaux.* Librairie encyclopédique de Roret, 1836.

Cuvier, Frédéric. "De quelques espèces de phoques et des groupes génériques entre lesquels elles se partagent." *Mémoires du Muséum d'Histoire naturelle* 11 (1824): 174–214.

Cuvier, Frédéric. "Description d'un orang-outang, et observations sur ses facultés intellectuelles." *Annales du Muséum d'histoire naturelle* 16 (1810): 46–65.

Cuvier, Frédéric. "Description zoologique d'un phoque moine femelle" (*Phoca monachus*). *Annales du Muséum d'histoire naturelle* 20 (1813): 387–92.

Cuvier, Frédéric. *Des dents des mammifères considérées comme caractères zoologiques.* F. C. Levrault, 1825.

Cuvier, Frédéric. "Du genre paradoxure et de deux espèces nouvelles qui s'y rapportent." *Mémoires du Muséum d'histoire naturelle* 9 (1822): 41–48.

Cuvier, Frédéric. "Du macaque de Buffon." *Mémoires du Muséum* 4 (1818): 109–20.

Cuvier, Frédéric. "Du rut." *Annales du Muséum d'histoire naturelle* 9 (1807): 118–30.

Cuvier, Frédéric. "Essai sur de nouveaux caractères pour les genres des mammifères." *Annales du Muséum d'histoire naturelle* 10 (1807): 105–29; 12 (1808): 27–51; 19 (1812): 268–95.

Cuvier, Frédéric. "Essai sur la domesticité des mammifères." *Mémoires du Muséum d'histoire naturelle* 13 (1825): 406–55.

Cuvier, Frédéric. "Essais sur les facultés intellectuelles des brutes." *Nouveau bulletin des sciences par la Société philomathique*, February 1813, 217–18.

Cuvier, Frédéric. "Examen de quelques observations de M. Dugald-Stewart, qui tendent à détruire l'analogie des phénomènes de l'instinct avec ceux de l'habitude." *Mémoires du Muséum d'histoire naturelle* 10 (1823): 241–60.

Cuvier, Frédéric. "Instinct." *Dictionnaire des sciences naturelles* 23 (1822): 528–44.

Cuvier, Frédéric. "Introduction." In *Histoire naturelle des mammifères, avec des figures originales coloriées, dessinées d'après des animaux vivans*, by Geoffroy Saint-Hilaire, Étienne Cuvier, and Frédéric Cuvier. Vol. 1. A. Belin, 1833.

Cuvier, Frédéric. "Note sur l'accouplement d'un zèbre et d'un cheval." *Annales du Muséum d'histoire naturelle* 11 (1808): 237–40.

Cuvier, Frédéric. "Observations de M. Frédéric Cuvier, sur un paragraphe du Mémoire de M. Geoffroy Saint-Hilaire, inséré dans le Compte rendu de la séance du 14 août 1837." *Comptes rendus hebdomadaires des séances de l'Académie des sciences* 5 (1837): 305–6.

Cuvier, Frédéric. "Observations préliminaires." In *Recherches sur les ossemens fossiles*, by Georges Cuvier. 4th ed. Edmond D'Ocagne, 1834.

Cuvier, Frédéric. "Observations sur l'accouplement d'un cigne chanteur mâle et d'une oie domestique femelle. Description du mulet qui en est provenu." *Annales du Muséum d'histoire naturelle* 12 (1808): 119–25.

Cuvier, Frédéric. "Observations sur le chien des habitants de la Nouvelle-Hollande, précédés de quelques réflexions sur les facultés morales des animaux." *Annales du Muséum d'histoire naturelle* 11 (1808): 458–76.

Cuvier, Frédéric. *Observations sur l'enseignement de l'histoire naturelle dans les collèges.* Imprimerie Royale, 1838.

Cuvier, Frédéric. "Observations sur quelques espèces de Goélands." *Annales du Muséum d'histoire naturelle* 11 (1808): 283–92.

Cuvier, Frédéric. "Observations zoologiques sur les facultés physiques et intellectuelles du phoque commun (*Phoca vitulina* L.)." *Annales du Muséum d'histoire naturelle* 17 (1811): 377–97.

Cuvier, Frédéric. "Orang." *Dictionnaire des sciences naturelles* 36 (1825): 275–89.

Cuvier, Frédéric. *Projet d'organisation pour les écoles primaires.* Delaunay, 1815.

Cuvier, Frédéric. "Rapport de M. F. Cuvier . . . à M. le ministre de l'instruction publique." In V. Rendu, *La Providence révélée par ses moindres ouvrages, ou tableau des mœurs des* insectes. J. Angé, 1838.

Cuvier, Frédéric. "Recherches sur les caractères ostéologiques qui distinguent les principales races du chien domestique." *Annales du Muséum d'histoire naturelle* 18 (1811): 333–53.

Cuvier, Frédéric. *Supplément à l'Histoire naturelle générale et particulière* de Buffon. *Offrant la description des mammifères et des oiseaux les plus remarquable découverts jusqu'à ce jour, et accompagné de gravures.* 2 vols. F. D. Pillot, 1829–31.

Cuvier, Frédéric. "Troisième notice sur les voyages de M. Alfred Duvaucel, dans l'Inde, ayant pour objet plus particulier, l'histoire naturelle." *Journal asiatique* 4 (1824): 277–85.

Cuvier, Georges. "Éloge de M. Lamarck," *Mémoires de l'Académie royale des sciences de l'Institut de France* 13 (1835): i-xxxi.

Cuvier, Georges. "Extrait d'observations faites sur le cadavre d'une femme connue à Paris et à Londres sous le nom de Vénus Hottentote." *Mémoires du Muséum d'histoire naturelle* 3 (1817): 259–74.

Cuvier, Georges. "Extrait d'un rapport sur l'état de l'histoire naturelle et sur ses accroissements depuis le retour de la paix maritime." *Recueil des discours prononcés dans la séance publique de l'Institut royal de France, le samedi 24 avril 1824.* Didot, 1824.

Cuvier, Georges. *Lécons d'anatomie comparées*, 5 vols. Baudouin, an 8 [1800].

Cuvier, Georges. "Mémoire sur les espèces d'éléphans vivantes et fossiles." *Mémoires de l'Institut de France* (Class math. Phys.), 2 (1799): 1–22.

Cuvier, Georges. "Mémoire sur l'ibis des anciens Égyptiens." *Annales du Muséum d'histoire naturelle* 4 (1804): 116–35.

Cuvier, Georges. "Mémoires pour server à celui qui fera mon éloge." In Pierre Flourens, "Éloge historique de G. Cuvier." *Recueil des éloges historiques lus dans les séances publiques de l'Académie des sciences.* Garnier frères, 1856.

Cuvier, Georges. "Notice sur les voyages de MM. Diard et Duvaucel, naturalistes français, dans les indes orientales et dans les îles de la Sonde; extraite de leur correspondance, et lue à l'Académie des sciences, le 14 mai dernier." *Revue encyclopédique* 10 (1821): 473–82.

Cuvier, Georges. "Notice sur le voyage de M. Alfred Duvaucel dans l'Inde." *Journal asiatique* 4 (1824): 137–45, 200–213.

Cuvier, Georges. "Pline l'Ancien." 520. *Biographie universelle ancienne et moderne*, new ed., 33 (n.d.), 518–24.

Cuvier, Georges. "Rapport fait à l'Académie royale des sciences, sur les collections rapportées récemment de la mer des Indes par M. Dussumier, de Bordeaux." *Annales des sciences naturelles* 21 (1830): 458–68.

Cuvier, Georges. "Rapport sur la partie zoologique de l'Expédition Duperrey." *Annales des sciences naturelles* 6 (1825): 5–20.

Cuvier, Georges. *Recherches sur les ossemens fossiles de quadrupèdes, où L'on rétablit les caractères de plusieurs espèces d'animaux que les révolutions du globe paroissent avoir détruites.* 4 vols. Deterville, 1812.

Cuvier, Georges. *Recherches sur les ossemens fossiles.* où l'on rétablit les caractères de plusieurs animaux dont les révolutions du globe ont détruit les espèces. Nouv. ed. 5 vols. Dufour et D'Ocagne, 1821-24.

Cuvier, Georges. *Recherches sur les ossemens fossiles* où l'on rétablit les caractères de plusieurs animaux dont les révolutions du globe ont détruit les espèces. 4th ed. 10 vols. Edmond D'Ocagne, 1834-36.

Cuvier, Georges. *Recueil des éloges historiques lus dans les séances publiques de l'Institut royal de* France, 3 vols. F. G. Levrault, 1819–27.

Cuvier, Georges. *Le règne animal distribué d'après son organisation, pour servir de base à l'histoire naturelle des animaux et d'introduction à l'anatomie comparée.* 4 vols. Deterville, 1817.

Cuvier, Georges. "Sur les éléphans vivans et fossiles." *Annales du Muséum d'histoire naturelle* 8 (1806), 1–58, 93–155, 249–69.

Cuvier, Georges. "Sur l'établissement de la collection d'anatomie comparé du museum." *Annales du Muséum d'histoire naturelle* 2 (1803): 409–14.

Cuvier, Georges. "Sur un nouveau rapprochement à établir entre les classes qui composent le règne animal." *Annales du Muséum d'histoire naturelle* 19 (1812): 73–84.

Cuvier, Georges. "Sur un tapir découvert en Asie." *Journal de Physique* 87 (1818): 473–74.

Cuvier, Georges. *Tableau élémentaire d'histoire naturelle des animaux.* Baudouin, an 6 [1798].

Cuvier, Georges, and Étienne Geoffroy Saint-Hilaire. "Mémoire sur les rapports naturels du Tarsier." *Magasin encyclopédique* 1, 3 (1795): 147–54.

Cuvier, Georges, and Achille Valenciennes. *Histoire naturelle des poisons.* Vol. 10. F. G. Levrault, 1835.

D'Abrantès. *Mémoires de Madame la Duchesse d'Abrantès, ou Souvenirs historiques sur Napoléon, la Révolution, le Directoire, le Consulat, l'Empire et la Restauration.* Ladvocat, 1831–32.

Darwin, Charles. *Narrative of the surveying voyages of His Majesty's Ships "Adventure" and "Beagle" between the years 1826 and 1836.* Henry Colburn, 1839.

Daubenton, Louis Jean Marie. "Sur la définition et les limites de l'histoire naturelle." *Séances des Écoles Normales, recueillies des sténographes, et revues par les professeurs.* New edition. Vol. 1. Imprimerie du Cercle-Social, 1800.

Daubenton, Louis Jean Marie. "Sur la rédaction de l'histoire naturelle." *Séances des Écoles Normales, recueillies des sténographes, et revues par les professeurs.* New ed. Vol. 1. Imprimerie du Cercle-Social, 1800.

Deleuze, J.-P.-F. *Histoire et description du Muséum royal d'histoire naturelle*. Royer, 1823.

Deleuze, J.-P.-F. "Notice sur le Citoyen Maréchal." *Annales du Museum* 2 (1803): 65–74.

Delisle de Moncel, Nicolas. *Méthodes et projets pour parvenir à la destruction des loups dans le royaume*. Imprimerie Royale, 1768.

Denys Montfort, Pierre. "Sur la lionne de la menagerie." *Magasin encyclopédique* 6, no. 2 (1800): 238–39.

Desains, Ant. *Dissertation sur la chorée, danse de Saint-Guy, ou Myotyrbie*. Didot le jeune, 1811.

Description de l'Égypte, ou Recueil des observations et des recherches qui ont été faites en Égypte pendant l'expédition de l'armée française. Vol. 1. Imprimerie Impériale, 1809.

Devienne, François. *La Bataille de Gemmapp: Sinfonie à grand orchestre. Executée au concert du Théatre de la rue Feydeau*. Imbault, n.d. [1794].https://gallica.bnf.fr/ark:/12148/btv1b9058040p/f13.

Diard, P.-M. "Observations by M. Diard on the Tapir of Malacca." *Asiatick Researches, or Transactions of the Society, Instituted in Bengal, for Enquiring into the History and Antiquities, the Arts, Sciences, and Literature of Asia* 13 (1820): 422–24.

Douville, J.-B. *Voyage au Congo et dans l'intérieur de l'Afrique equinoxiale, fait dans les années 1828, 1829 et 1830*. 3 vols. Jules Renouard, 1832.

Dufresne, Louis. "Taxidermie." *Nouveau dictionnaire d'histoire naturelle*, appliquée *aux arts, principalement à l'agriculture et à l'économie rurale et domestique* 21 (1803): 507–63. Republished in *Nouveau dictionnaire d'histoire naturelle*, 2nd ed., vol. 32 (1819), and as *Taxidermie, ou L'art de préparer et de conserve la dépouille de tous les animaux*. Deterville, 1820.

Dureau de la Malle, Adolphe. "Considérations générales sur la domestication des animaux." *Annales des sciences naturelles* 27 (1832): 5–33.

Duvaucel, Alfred. "Lettres familières sur l'Inde." *Revue des deux mondes* (1833): 601–31.

Duvernoy, Georges Louis. *Notice historique sur les ouvrages et la vie de M. le B.*[on] *Cuvier*. Levrault, 1833.

E***. *Promenades instructives et amusantes d'un père avec ses enfans dans Paris et ses environs*. Guillaume, 1817. http://gallica.bnf.fr/ark:/12148/bpt6k6397260j.

Farquhar. "Account of a New Species of Tapir Found in the Peninsula of Malacca." *Asiatick Researches, or Transactions of the Society, Instituted in Bengal, for Enquiring into the History and Antiquities, the Arts, Sciences, and Literature of Asia* 13 (1820): 417–21.

Fennell, James. "On the Total and Partial Extermination of Animals." *Field Naturalist* 2 (1834): 182–97.

Ferlus, L. D. *Nouvelle notice sur la girafe, envoyée au roi de France par le pacha d'Ègypte, et arrivé à Paris le 30 juin 1827. Observations curieuses sur le caractère, les habitudes et l'instinct de ce quadrupède*. Moreau, 1827.

Fischer, Gotthelf. *Das Nationalmuseum der Naturgeschichte zu Paris*. 2 vols. Esslinger, 1802–3.

Flourens, Pierre. *De l'instinct et de l'intelligence des animaux*. Garnier frères, 1870.

Flourens, Pierre. *Éloge historique de F. Cuvier . . . Lu à la séance publique du 13 juillet 1840*. Didot frères, 1840.

Flourens, Pierre. *Examen du livre de M. Darwin sur l'origine des espèces*. Garnier frères, 1864.

Flourens, Pierre. "Histoire naturelle des mammifères." *Journal des savants* (1839): 321–33, 461–79, 513–27.

Flourens, Pierre. *Psychologie comparée*. Garnier frères, 1864.

Flourens, Pierre. "Résumé analytique des observations de Frédéric Cuvier sur l'instinct et l'intelligence des animaux." *Annales des sciences naturelles. Zoologie et biologie animale* 12 (1839): 235–86.

Flourens, Pierre. *Résumé analytique des observations de Frédéric Cuvier sur l'instinct et l'intelligence des animaux*. Imprimerie Royale, 1839.

Gall, Franz. *Discours d'ouverture lu par M. le docteur Gall, à la première séance de son cours public, sur la physiologie du cerveau, le 15 janvier 1808*. Didot, Lefort, F. Schoell, 1808.

Gall, F. J., and G. F. Spurzheim. *Anatomie et physiologie du système nerveux en général, et du cerveau en particulaire, avec des observations sur la possibilité de reconnoître plusieurs dispositions intellectuelles et morales de l'homme et des animaux, par la configuration de leurs têtes*. Vol. 3. Librairie Grecque-Latine-Allemande, 1818.

Gall, F. J., and G. F. Spurzheim. *Des dispositions innées de l'âme et de l'esprit, du matérialisme, du fatalisme et de la liberté morale, avec des réflexions sur l'éducation et sur la législation criminelle*. F. Schoell, 1811.

Galton, Francis. "The First Steps toward the Domestication of Animals." *Transactions of the Ethnological Society* 3 (1865): 122–38.

Garner, T. *A Brief Description of the Principal Foreign Animals and Birds, Now Exhibiting at the Grand Menagerie, over Exeter-'Change, the Property of Mr. Gilbert Pidcock, Chiefly Extracted from the Works of Buffon and* Goldsmith. T. Burton (printed for G. Pidcock), 1800.

Geoffroy Saint-Hilaire, Étienne. *Catalogue des mammifères du Muséum national d'histoire naturelle*. n.p., n.d.

Geoffroy Saint-Hilaire, Étienne. "De l'achire barbu, espèce de Pleuronecte indiquée par Gronou." *Annales du Muséum* 1 (1802): 152–55.

Geoffroy Saint-Hilaire, Étienne. "Histoire naturelle et description anatomique d'un nouveau genre de poisson du Nil, nommé Polyptère." *Annales du Muséum* 1 (1802): 57–68.

Geoffroy Saint-Hilaire, Étienne. "Mémoire sur l'anatomie comparée des organes électriques de la raie torpille, du gymnote engourdissant, et du silure trembleur." *Annales du Muséum d'histoire naturelle* 1 (1802): 392–407.

Geoffroy Saint-Hilaire, Étienne. "Mémoire sur les rapports naturels des Makis Lemur, L. et description d'une espèce nouvelle de Mammifère." *Magasin encyclopédique* 1 (1796): 20–50.

Geoffroy Saint-Hilaire, Étienne. "Note sur deux kanguroos vivans, acquis en Angleterre." *Annales du Muséum* 1 (1802): 178–80.

Geoffroy Saint-Hilaire, Étienne. "Quelques considérations sur la girafe." *Annales des sciences naturelles* 11 (1827): 210–23.

Geoffroy Saint-Hilaire, Étienne. "Recherches sur l'organisation des gavials, sur leurs affinités naturelles . . . et sur cette question, si les gavials (Gavialis), aujourd'hui répandus dans les parties orientales de l'Asie, descendent, par voie non interrompue de génération, des gavials antédiluviens. . . ." *Mémoires du Muséum d'histoire naturelle* 12 (1825): 97–155.

Geoffroy Saint-Hilaire, Étienne. *Sur la girafe*. C. Thuau, 1827.

Geoffroy Saint-Hilaire, Étienne. "Sur les habitudes des castors." *Mémoires du Muséum d'histoire naturelle* 12 (1825): 232.

Geoffroy Saint-Hilaire, Étienne. "Sur trois bouquetins et un ichneumon nouvellement acquis pour la ménagerie nationale." *Annales du Museum d'histoire naturelle* 2 (1803): 244–48.

Geoffroy Saint-Hilaire, Étienne. *Cours de l'histoire naturelle des mammifères*. Pichon et Didier, 1829.

Geoffroy Saint-Hilaire, Étienne. "Des changements à la surface de la terre qui paraissent dépendre originaire et nécessairement de la variation préexistante, incessante, lente et successive, des milieux ambiants divers et consécutifs, du globe terrestre." *Comptes rendus hebdomadaires des séances de l'Académie des sciences* 5 (1837): 183–94.

Geoffroy Saint-Hilaire, Étienne. *Études progressives d'un naturaliste pendant les années 1834 et 1835*. Roret, 1835.

Geoffroy Saint-Hilaire, Étienne. "Extrait d'un mémoire sur l'orang-outang vivant à la ménagerie." *Comptes rendus des séances de l'Académie des sciences* 2 (1836): 581-85, 601–3; 3 (1836): 1–8, 27–31.

Geoffroy Saint-Hilaire, Étienne. *Fragments biographiques, précédés d'études sur la vie, les ouvrages et les doctrines de Buffon*. F. D. Pillot, 1838.

Geoffroy Saint-Hilaire, Étienne. *Lettres écrites d'Égypte à Cuvier, Jussieu, Lacépède, Monge, Desgenettes, Redouté Jeune, Norry, etc., aux professeurs du Muséum et à sa famille*. Edited by E.-T. Hamy. Hachette, 1901.

Geoffroy Saint-Hilaire, Étienne. "Note remise par M. Geoffroy Saint-Hilaire," *Comptes rendus hebdomadaires des séances de l'Académie des sciences* 5 (1837): 306.

Geoffroy Saint-Hilaire, Étienne. *Notions synthétiques, historiques et physiologiques de philosophie naturelle*. Dénain, 1838.

Geoffroy Saint-Hilaire, Étienne, and Frédéric Cuvier. *Histoire naturelle des mammifères, avec des figures originales enluminées, dessinées d'après des animaux vivants*. 7 vols. C. de Lasteyrie, 1818–42.

Geoffroy Saint-Hilaire, Étienne, and Frédéric Cuvier. *Histoire naturelle des mammifères, avec des figures originales coloriées, dessinées d'après des animaux vivans*. 2 vols. A. Belin, 1833.

Geoffroy Saint-Hilaire, Étienne, and Georges Cuvier, "Histoire naturelle des Orangs-Outangs." *Magasin encyclopédique*, 1, vol. 3 (1795): 451–63.

Geoffroy Saint-Hilaire, Étienne, and Étienne Serres, "Rapport fait à l'Académie des sciences sur un Mémoire de M. Roulin, ayant pour titre: 'Sur quelques changemens observés dans les animaux domestiques transportés de l'ancien monde dans le nouveau continent.'" *Annales des sciences naturelles* 16 (1829): 34–44.

Geoffroy Saint-Hilaire, Isidore. *Acclimatation et domestication des animaux utiles*. 4th ed. Maison Rustique, 1861.

Geoffroy Saint-Hilaire, Isidore. "Allocution." *Bulletin de la Société impériale zoologique d'acclimatation* 1 (1854): vii-xiv.

Geoffroy Saint-Hilaire, Isidore. "Fragments historiques sur la domestication et la culture des animaux." *Bulletin de la Société impériale zoologique d'acclimatation* 1 (1854): 155-61, 241-44, 283-96, 389-98.

Geoffroy Saint-Hilaire, Isidore. *Histoire naturelle générale des règnes organiques, principalement étudiée chez l'homme et les animaux*. 3 vols. Masson et fils, 1854–62.

Geoffroy Saint-Hilaire, Isidore. "Sur le genre *Cheval* et spécialement sur *l'Hémione*." *Nouvelles annales du Muséum d'histoire naturelle* 4 (1835): 97–120.

Geoffroy Saint-Hilaire, Isidore. *Vie, travaux et doctrine scientifique d'Étienne Geoffroy Saint-Hilaire*. P. Bertrand, 1847.

Geoffroy Saint-Hilaire, Isidore, and H. M. D. de Blainville, "Rapport sur les résultats scientifiques du voyage de M. Alcide d'Orbigny dans l'Amérique du Sud, pendant les années 1826, 1827, 1828, 1829, 1830, 1831, 1832 et 133. Partie zoologique." *Nouvelles annales du Muséum d'histoire naturelle* 3 (1834): 84–99.

Gosse, Isid. S. [Isidore Salles] de. *Histoire naturelle drolatique et philosophique des professeurs du Jardin des Plantes, des aide-naturalistes, préparateurs, etc., attachés à cet établissement, accompagnée d'épisodes scientifiques et pittoresques.* Gustave Sandré, 1847.

Grandville, J.-J. *Scènes de la vie privée et publique des animaux. Vignettes par Grandville. Études de moeurs contemporaines.* J. Hetzel and Paulin, 1841.

Gray, John Edward. "On the family of Viverridae and its generic subdivision; with an enumeration of the species of Paradoxurus and characters of several new ones," *Proceedings of the Zoological Society of London* 2 (1832): 63–68. [I take the title of Gray's comments from the Royal Society's *Catalogue of Scientific Papers.*]

Heinzmann, Johann Georg. *Voyage d'un Allemand à Paris, et retour par la Suisse.* Hignou, 1800.

Houël, J. P. L. L. *Histoire naturelle des deux éléphans, mâle et femelle, du Muséum de Paris, venus de Hollande en France en l'an VI. . . .* Treuttel et Wurtz, an 12 (1803).

Humboldt, Alexander von. "Observations sur l'anguille électrique (*Gymnotus electricus,* Linn.) du nouveau continent." In *Voyage de Humboldt et Bonpland. Deuxième partie. Observations de zoologie et d'anatomie comparé.* Vol. 1. Schoell; Dufour, 1811.

Humboldt, Alexander von. *Personal Narrative of Travels to the Equinoctial Regions of America, during the Years 1799–1804. By Alexander von Humboldt and Aimé Bonpland.* Edited and translated by Thomasina Ross. 2 vols. Bell and Daldy, 1869.

Humboldt, Alexander von. "Sur les Gymnotes et autres poissons électrique." *Annales de chimie et physique* 11 (1819): 408–37.

Humboldt, Alexander Von. *Tableaux de la nature, ou Considérations sur les déserts, sur la physionomie des végétaux et sur les cataractes de l'Orénoque.* Translated by J. B. B. Eyriès. 2 vols. Schoell, 1808.

Huxley, Thomas H. *Evidence as to Man's Place in Nature.* D. Appleton, 1872.

Jal, A. *Scènes de la vie maritime,* vol. 1. Charles Gosselin, 1832.

Janin de Saint-Just. "Über den Gymnotus electricus." *Notizen aus dem Gebiete der Natur- und Heilkunde* 1 (1822), cols. 209–12.

Jauffret, L. F. *Voyage au Jardin des Plantes.* Ch. Houel, an 6 [1798].

Jousserandot, Louis. *Le Domino, épître à Dantan jeune et S.-H. Berthoud.* N.p., 1848.

Jussieu A.-L. "Notice sur l'expédition à la Nouvelle-Hollande, entreprise pour des recherches de géographie et d'histoire naturelle." *Annales du Muséum d'histoire naturelle* 5 (1804): 1–11.

Kirby, William. *On the Power, Wisdom and Goodness of God as Manifested in the Creation of Animals and in Their History, Habits and Instincts.* William Pickering, 1835.

Lacépède, Bernard de. "Lettre relative aux établissemens publics destinés à renfermer des animaux vivans, et connus sous le nom de ménageries." *La Décade philosophique, littéraire et politique* 7 (1795): 449–62. Republished in *Séances des Écoles Normales,* new ed. 8 (1800): 303–18.

Lacépède, Bernard de. "Sur le but auquel doit attendre le naturaliste, et particulièrement sur les rapports de l'étude de sciences naturelles avec le bonheur de ceux qui les cultivent." *Séances des Écoles Normales.* New ed. 8 (1800): 319–46.

Lacépède, Bernard de, Georges Cuvier, and Étienne Geoffroy Saint-Hilaire. *La Ménagerie du Muséum national d'histoire naturelle, ou Description et histoire des an-*

imaux qui y vivent ou qui y ont vécu; par les citoyens Lacépède et Cuvier, avec des figures peintes d'après nature, par le citoyen Maréchal, peintre du Muséum. Miger, 1801–5.

Lacépède, Bernard de, J.-B. Lamarck, and Georges Cuvier. "Rapport des professeurs du Muséum sur les collections d'histoire naturelle rapportées d'Égypte, par E. Geoffroy." *Annales du Museum d'histoire naturelle* 1 (1802): 234–41.

Lacordaire, Théodore. "Revue de voyages. 2. Voyage au Congo, par M. Douville." *Revue de deux mondes* 8 (1832): 250–62.

Lacoste, "Sur les éléphans." *La Chronique universelle*, 10 germinal an 6 [30 March 1798], 2.

Lamarck, Jean-Baptiste. "Discours d'ouverture des cours de zoologie donnés dans le Muséum d'histoire naturelle (An VIII, An X, An XI et 1806) par J.-B. Lamarck." Edited by A. Giard. Bulletin scientifique de la France et de la Belgique, 1907.

Lamarck, Jean-Baptiste. "Espèce." *Encyclopédie méthodique, botanique*. Vol. 2. Pancoucke, 1786.

Lamarck, Jean-Baptiste. *Flore françoise*. 3 vols. Imprimerie Royale, 1779.

Lamarck, Jean-Baptiste. *Histoire naturelle des animaux sans vertèbres . . . précédée d'une introduction offrant la détermination des caractères essentiels de l'animal, sa distinction du végétal et des autres corps naturels, enfin, l'exposition des principes fondamentaux de la zoologie*. Verdière, 1815–22.

Lamarck, Jean-Baptiste. "L'Homme." *Nouveau dictionnaire d'histoire naturelle* 15 (1817): 270–76.

Lamarck, Jean-Baptiste. "Mémoires sur les fossiles des environs de Paris, comprenant la détermination des espèces qui apartiennent aux animaux marins sans vertèbres, et dont la plupart sont figurés dans la collection des vélins du Muséum." *Annales du Muséum d'histoire naturelle* 1 (1802): 299-312.

Lamarck, Jean-Baptiste. *Philosophie zoologique, ou Exposition des considérations relatives à l'histoire naturelle des animaux; à la diversité de leur organisation et des facultés qu'ils en obtiennent; aux causes physiques qui maintiennent en eux la vie et donnent lieu aux mouvemens qu'ils exécutent; enfin, à celles qui produisent, les unes le sentiment, et les autres l'intelligence de ceux qui en sont doués*. 2 vols. Dentu, 1809.

Lamarck, Jean-Baptiste. *Recherches sur l'organisation des corps vivans et particulièrement sur son origine, sur la cause de ses développement et des progrès de sa composition, et sur celle qui, tendant continuellement à la détruire dans chaque individu, amène nécessairement sa mort; précédé du discours d'ouverture du cours de zoologie, donné dans le Muséum d'histoire naturelle*. Maillard, 1802.

Lamarck, Jean-Baptiste. *Réfutation de la théorie pneumatique ou de la nouvelle doctrine des chimistes modernes . . . dans une suite de réponses aux principes rassemblés et publiés par le citoyen Fourcroy dans sa Philosophie chimique. . . .* Agasse, 1796.

Lamarck, Jean-Baptiste. "Sur la détermination des espèces parmi les animaux sans vertèbres et particulièrement parmi les mollusques testacés." *Annales du Muséum d'histoire naturelle* 15 (1810): 20–26.

Lamarck, Jean-Baptiste. "Sur les caractères distinctifs de la Seiche, du Calmar et du Poulpe." *Magasin encyclopédique*, 4th year, vol. 2 (1798), 389.

Lamarck, Jean-Baptiste. *Système des animaux sans vertèbres*. Deterville, 1801.

Lavison, Rufz de. "Note sur les différentes tentatives d'introduction et d'acclimatation des lamas et alpacas qui ont eu lieu en Europe." *Bulletin de la Société impériale zoologique d'acclimatation* 1, 2d ser. (1861): 327–37.

Lavoisier, Antoine. "Altérations qu'éprouve l'air respiré." In *Œuvres de Lavoisier*, 2:676–87. Imprimerie Impériale, 1862.

Leroy, Charles-Georges. *Lettres philosophiques sur l'intelligence et la perfectibilité des animaux, avec quelques lettres sur l'homme*. New ed. Valade, an X [1802].

Leschenault de la Tour, Jean-Baptiste, "Relation abrégé d'un voyage aux Indes orientale." *Mémoires du Muséum d'histoire naturelle* 9 (1822): 245–74.

Levaillant, François. *Histoire naturelle des oiseaux d'Afrique*. 6 vols. J. J. Fuchs, 1789–1808.

Lyell, Charles. *Life, Letters and Journals of Sir Charles Lyell, Bart*. Edited by Mrs. Lyell. 2 vols. John Murray, 1881.

Lyell, Charles. *Principles of Geology, Being an Attempt to Explain the Former Changes of the Earth's Surface, by References to Causes Now in Operation*. 3 vols. John Murray, 1830-33.

Marion de Procé, Pierre-Martin. "Notice sur plusieurs espèces nouvelles de poissons." *Journal de physique* 95 (1822): 235–40.

Martinet, L. *Manuel de clinique médicale*. 2nd ed. Gabon, 1826.

Milbert, Jacques. *Itinéraire pittoresque du fleuve Hudson, et des parties latérales d'Amérique du Nord*. 2 vols. H. Gaugain, 1828.

Milbert, Jacques. *Voyage pittoresque à l'Ile-de-France, au Cap de Bonne-Espérance et a l'Ile de Ténériffe*. 2 vols. A. Nepveu, 1812.

Millin, A. L., P. Pinel, and Alex. Brongniart. *Rapport fait à la Société d'histoire naturelle de Paris, sur la nécessité d'établir une ménagerie*. Boileau, 1792. Also in *Le Magasin encyclopédique* 1 (1792): 131–34.

Milne-Edwards, Alphonse. *La Ménagerie (Rapport au Ministre de l'instruction publique)*. Masson, 1891.

Mongez, Antoine. "Mémoire sur la girafe." *Annales des sciences naturelles* 11 (1827): 225–35.

Morgan, C. Lloyd. *Habit and Instinct*. Edwin Arnold, 1896.

Muséum royal d'histoire naturelle. *Instruction pour les voyageurs et pour les employés dans les colonies, sur la manière de recueillir, de conserver et d'envoyer les objets d'histoire naturelle*. Belin, 1818.

Muséum royal d'histoire naturelle. *Instruction pour les voyageurs et pour les employés dans les colonies, sur la manière de recueillir, de conserver et d'envoyer les objets d'histoire naturelle*. 3rd ed. Belin, 1829.

Muséum royal d'histoire naturelle. "Instruction sur les recherches qui pourraient être faites dans les colonies, sur les objets qu'il serait possible d'y recueillir et sur la manière de les conserver et de les transporter." *Mémoires du Muséum d'histoire naturelle* 4 (1818): 193–239. Neufchâteau, François de. *Recueil des lettres circulaires, instructions, programmes, discours, et autres actes publics, émanés du C.[en] François (de Neufchâteau), pendant ses deux exercices du Ministère de l'intérieur*. Vol. 1. Imprimerie de la République, an 7 [1799].

Owen, Richard. "Note on the Anatomy of the Nubian Giraffe." *Transactions of the Zoological Society of London* 2 (1841): 217–48.

Pelletan, Jean-Gabrielle. *Mémoire sur la colonie française du Sénégal, avec quelques considérations historiques et politiques sur la traite des Nègres, sur leur caractère, et les moyens de faire servir la suppression de cette traite, à l'accroissement et la prospérité de cette colonie*. Pancoucke, an 9 (1800).

Péron, F., and L. Freycinet, *Voyage de découvertes aux terres australes*. 2 vols. Imprimerie Impériale, 1807–16.

Perrault, Claude. *Mémoires pour servir à l'histoire des animaux*. Imprimerie Royale, 1671.

Perrottet, Samuel. *Catalogue raisonné des plantes introduites dans les colonies françaises de Bourbon et de Cayenne, et de celles rapportées vivantes des mers d'Asie et de la Guyane, au Jardin du roi à Paris*. L'imprimerie de Lebel, Imprimeur du Roi, 1824.

Pinkerton, J. *Recollections of Paris, in the years 1802–3-4–5*. 2 vols. Longman et al., 1806.

Pliny the Elder, *Histoire naturelle de Pline*. Vol. 6. Translated by Ajasson de Grandsagne. Panckoucke, 1829.

Pliny the Elder. *The Natural History*. Edited by John Bostock. Taylor and Francis, 1855.

Pluche, L'Abbé. *Le Spectacle de la nature, ou Entretiens sur les particularités de l'histoire naturelle qui ont paru les plus propres à rendre les jeunes gens curieux & à leur former l'esprit*. 8 vols. Estienne, 1764–70.

Pujoulx, J. B. *Le livre du second âge, ou Instructions amusantes sur l'histoire naturelle des animaux. Ouvrage orné de soixante-douze figures, représentant quadrupèdes, oiseaux, et insects*. Gillé fils pour Debray, an 9 [1801].

Pujoulx., J.-B. *Les modernes enrichis, comédie en trois acres, en vers libres*. Migneret, an 6 [1798].

Pujoulx, J. B. *Le livre du second âge, ou Instructions amusantes sur l'histoire naturelle des animaux. Ouvrage orné de soixante-douze figures, représentant quadrupèdes, oiseaux, et insects*. Gillé fils pour Debray, an 9 [1801].

Pujoulx, J. B. *Le livre du second âge, ou Instructions amusantes sur l'histoire naturelle des animaux. Ouvrage orné de soixante-douze figures, représentant quadrupèdes, oiseaux, et insects*. Gillé fils pour Debray, an 9 [1801].

Pujoulx, J.-B. *Paris à la fin du xviiie siècle, ou Essai historique et morale des monumens et des ruines de cet capitale, de l'état des sciences, des arts et de l'industrie à cette époque, ainsi que des mœurs et des ridicules de ses habitans*. Mathé, an 9[1801].

Pujoulx, J. B. *Promenades au Jardin des Plantes, à la ménagerie et dans les galeries du Muséum d'histoire naturelle*. 2 vols. Guilleminet, an 13 [1803].

Raffles, Thomas Stamford. "Descriptive Catalogue of a Zoological Collection, Made on Account of the Honourable East India Company, in the Island of Sumatra and Its Vicinity, under the Direction of Sir Thomas Stamford Raffles. . . ." *Transactions of the Linnean Society of London* 13 (1822): 239–76 (part 1), 277–340 (part 2).

Raffles, Thomas Stamford. "Some Account of the Dugong." *Philosophical Transactions of the Royal Society of London* 110 (1820): 174–82.

Reimar, Hermann Samuel. *Observations physiques et morales sur l'instinct des animaux, leur industrie et leurs mœurs*. D. J. Changuion, 1770.

Reynaud, Jean. "Fondation de la ménagerie du Muséum d'Histoire naturelle." *Magasin pittoresque* 6 (1838): 106–8. Reprinted in Étienne Geoffroy Saint-Hilaire, *Fragments biographiques, précédés d'études sur la vie, les ouvrages et les doctrines de Buffon*. F. D. Pillot, 1838.

Romanes, George John. *Animal Intelligence*. Appleton, 1892.

Rousseau, Louis, and Céran Lemonnier. *Promenades au Jardin des Plantes*. J.-B. Baillière, 1837.

Saint-Ange, Louis de. *Notice sur la girafe, avec des observations curieuses sur le caractère, les habitudes et l'instinct de ce quadrupède*. Moreau, 1827.

Saint Hilaire, Auguste de, "Aperçu d'un voyage dans l'intérieur du Brésil, la province

Cisalpine et les missions dites du Paraguay." *Mémoires du Muséum d'Histoire naturelle* 9 (1822): 307–80.

Saint-Victor, J.-M. B. B. de. *Tableau historique et pittoresque de Paris*. H. Nicolle, 1808–9.

Salvandy, N.-A. de. *Insolences de la censure, et considérations sur la politique en générale du Ministère*. A. Sautelet, 1827.

Salze, "Observations faites sur la girafe envoyée au roi par le pacha d'Égypte." *Mémoires du Muséum d'histoire naturelle* 14 (1827): 68–84.

Sand, George. *Correspondance de George Sand*. 26 vols. Edited by G. Lubin. Garnier frères, 1964–95.

Savigny, J. B. Henri, and Alexandre Corréard. *Naufrage de la frégate La Méduse, faisant partie de l'expédition du Sénégal, en 1816*. Hocquet, 1817.

Selous, Edmund. *Bird Watching*. J. M. Dent, 1901.

Société de Géographie. "Rapport de la commission de la Société de Géographie, qui a décerné une médaille d'or de mille francs à M. Douville, pour son voyage de découvertes au Congo et dans l'intérieur de l'Afrique équinoxiale." *Bulletin de la Société de Géographie de Paris* 17 (1832): 157–75.

Spalding, Douglas. "Instinct, with original observations on young animals." *Macmillan's Magazine* 27 (1873): 288–93.

Thibaudeau, Antoine Claire. *Rapport fait au nom du Comité d'instruction publique et des finances, sur le Muséum national d'histoire naturelle*. Convention nationale, frimaire an 3 [1794].

Toscan, Georges. "Du pouvoir de la musique sur les animaux, et du concert donné aux éléphans." *La Décade philosophique* 18 (an 6) [1798]: 257–64, 321–29.

Toscan, Georges. "Histoire du lion et du chien." *La Décade philosophique* 3 (an 3) [1794]: 129–37, 193–99.

Toscan, Georges. "Histoire naturelle. Essais de botanique morale." *La Décade philosophique* 1 (1794): 387–95, 451–59.

Toscan, Georges. *L'ami de la nature, ou Choix d'observations sur divers objets de la nature et de l'art; suivi d'un catalogue de tous les animaux qui se trouvent actuellement dans la ménagerie*. Crapelet, an 8 [1800].

Toscan, Georges. *Mémoire sur l'utilité de l'établissement d'une bibliothèque au Jardin des Plantes*. Imprimerie du Cercle-Social, 1793.

Toscan, Georges. "Ménagerie," *La Décade philosophique* 22 (an 7)[1799]: 131–32.

Toscan, Georges. "Sciences. Histoire naturelle. Vues générale." *La Décade philosophique* 1 (1794): 1–7.

Toscan, Georges. "Supplément à l'histoire du lion de la ménagerie, et de son chien." *La Décade philosophique* 5 (an 3) [1795]: 257–61.

Toscan, Georges. "Suite des faits nouveaux observés dans le Muséum national d'histoire naturelle," *La Décade philosophique* 22 (an 7) [1799]: 323.

Vignier, E. J. B. *Description abrégé des animaux quadrupèdes de la ménagerie de Tipoo-Saiab, nouvellement achetés à Londres pour enrichir la ménagerie du Muséum d'histoire naturelle; suivie du récit de la progéniture de la lionne que l'on peut regarder comme un Phénomène*. Quillau, an 9 [1801].

Vignier, E. J. B. Epitre aux éléphans de la ménagerie nationale. Chez les Marchandes de nouveautés. An 7 [1799].

Vignier, E. J. B. *Phénomène d'histoire naturelle. Récit de la deuxième gestation de Constantine, une des lionnes de la Ménagerie du Jardin des Plantes; précédé de la des-*

cription de tous les quadrupèdes sous la garde du citoyen Félix Cassal. Quillau, an 9 [1801].

Vignier, E. J. B. *Vertus morales des deux éléphans mâle et femelle, nouvellement arrivés à la Ménagerie nationale du Jardin des Plantes, précédées d'un traité sur le genre de ces animaux précieux et singulier en tout point, tiré du célèbre Buffon, et des meilleurs auteurs qui ont écrit sur l'histoire naturelle*. Gueffier, an 6 [1798].

Vignier, E. J. B. *Vertus morales et description abrégées des éléphans de la ménagerie: Suivie de la liste des animaux vivans du Jardin des Plantes*. Quillau, an 8 [1800].

Virey, J. J. "Habitation et station. Article deuxième. Examen de la question si l'habitation des animaux et des plantes, les soumettant toujours aux mêmes circonstances, détermine leur mode d'organisation." *Nouveau Dictionnaire d'Histoire naturelle* 14 (1817): 123–28.

Virey, J. J. *Histoire des mœurs et de l'instinct des animaux, avec les distributions méthodiques et naturelles de toutes leurs classes*. 3 vols. Deterville, 1822.

Virey, J. J. "Nature." *Nouveau dictionnaire d'histoire naturelle* 15 (1804): 358–414.

Virey, J. J. "Orang-outang." *Nouveau dictionnaire d'histoire naturelle* 16 (1803): 305–12.

Vosmaer, A. *Description de l'espèce de singe aussi singulier que très rare, nommé orang-outang, de l'isle de Borneo*. Pierre Miejer, 1778.

White, Gilbert. *The Natural History and Antiquities of Selborne*. Bensley, 1789.

Whitman, C. O. "A biological farm for the experimental investigation of heredity, variation and evolution and for the study of life-historics, habits, instincts and intelligence." *Biological Bulletin* 3 (1902): 214–24.

Yorke, Henry Richard. *Letters from France, in 1802*. 2 vols. H. D. Symonds, 1804.

SECONDARY SOURCES

Åkerberg, Sofia. *Knowledge and Pleasure at Regent's Park: The Gardens of the Zoological Society of London during the Nineteenth Century*. Umeä University, 2001.

Alberti, J. M. M. *The Afterlives of Animals*. University of Virginia Press, 2011.

Allin, Michael. *Zarafa: A Giraffe's True Story, from Deep in Africa to the Heart of Paris*. Walker, 1998.

Altick, Richard D. *The Shows of London*. Belknap Press, 1978.

Anonymous. "Notice sur M. Dufresne, aide-naturaliste au Muséum." *Nouvelles annales du Muséum d'histoire naturelle* 2 (1833): 357–59.

Appel, Toby. *The Cuvier-Geoffroy Debate: French Biology in the Decades before Darwin*. Oxford University Press, 1987.

Ash, Mitchell G., ed. *Mensch, Tier und Zoo: Der Tiergarten Schönbrunn im internationalen Vergleich vom 18. Jahhundert bes heute*. Bohlau, 2008.

Ash, Mitchell G. "Zoological Gardens." In *Worlds of Natural History*, edited by H. A. Curry, N. Jardine, J. A. Secord, and E. C. Spary. Cambridge University Press, 2018.

Badou, Gérard. *L'Énigme de la Vénus hottentote*. J. C. Lattès, 2000.

Bapst, Germain. "Histoire d'un cabinet minéralogique." *Revue des deux mondes* 2 (1892): 437–49.

Baratay, E. *Biographies animales: Des vies retrouvées*. Éditions du Seuil, 2017.

Baratay, E. *Le point de vue animal*. Éditions du Seuil, 2012.

Baratay, Eric, and Elisabeth Hardouin-Fugier. *Zoo: A history of zoological gardens in the West*. Reaktion, 2002.

Baratay, E., and E. Hardouin-Fugier. *Zoos: Histoire des jardins zoologiques en occident (xvie-xxe siècles)*. Éditions la Découverte, 1998.

Barrett, P. H., P. J. Gautrey, S. Herbert, D. Kohn, and S. Smith, eds. *Charles Darwin's Notebooks, 1836–1844: Geology, Transmutation of Species, Metaphysical Enquiries*. British Museum (Natural History) and Cornell University Press, 1987.

Barrow, Mark V., Jr. *Nature's Ghosts: Confronting Extinction from the Age of Jefferson to the Age of Ecology*. University of Chicago Press, 2009.

Barsanti, Giulio. "Linné et Buffon: deux visions différentes de la nature et de l'histoire naturelle." *Revue de synthèse*, 3rd ser., nos. 113–14 (1984): 83–111.

Barsanti, Giulio. "L'orang-outan déclassé (Pongo wurmbii Tied.). Histoire du premier singe à hauteur d'homme (1780–1801) et ébauche d'une théorie de la circularité des sources." *Bulletins et Mémoires de la Société d'Anthropologie de Paris*, n.s., 1 (nos. 3–4) (1989): 67–104.

Barsanti, Giulio. "Les singes de Lamarck," In Corbey and Theunissen, *Ape, Man, Apeman*, 101–15.

Barun, A., C. C. Hanson, K. J. Campbell, and D. Simberloff, "A review of small Indian mongoose management and eradications on islands." In *Island Invasives: Eradication and Management*, edited by C. R. Veitch, M. N. Clout, and D. R. Towns. IUCN, 2011.

Beer, Gavin de, ed. "Darwin's Notebooks on Transmutation of Species. Part I. First Notebook (July 1837—February 1838)," *Bulletin of the British Museum (Natural History). Historical Series*, 2 (no. 2) (1960).

Belin, Philippe. "Architectures zoologiques: 1662–1932." Travail personnel de troisième cycle en architecture. École d'Architecture de Clermont-Ferrand, December 1985.

Benson, Etienne S. *Surroundings: A History of Environments and Environmentalisms*. University of Chicago Press, 2020.

Berger, John. "Why Look at Animals?" In John Berger, *About Looking*. Vintage International, 1991.

Bertier de Sauvigny, Guillaume de. *The Bourbon Restoration*. University of Pennsylvania Press, 1966.

Blanckaert, Claude. "Les animaux 'utiles' chez Isidore Geoffroy Saint-Hilaire: La mission sociale de la zootechnie." *Revue de Synthèse*, no. 3–4 (1992): 347–82.

Blanckaert, Claude, "Produire l'être singe." Langage du corps et harmonies spirituelles." *Annales historiques de la Révolution française*, no. 377 (2014): 9–35.

Blanckaert, Claude, coordinator. *La Vénus hottentote: Entre Barnum et Muséum*. Muséum national d'Histoire naturelle, 2013.

Blanckaert, Claude. "Voyages naturalistes en Hottentotie." In *La Vénus hottentote: Entre Barnum et Muséum*, coordinator, Claude Blanckaert, Muséum national d'histoire naturelle, 2013.

Blanckaert, Claude, Claudine Cohen, Pietro Corsi, and Jean-Louis Fischer, eds. *Le Muséum au premier siècle de son histoire*. Muséum national d'Histoire naturelle, 1997.

Blunt, Wilfred. *The Ark in the Park: The Zoo in the Nineteenth Century*. Hamish Hamilton, 1976.

Boitard, Pierre. *Le Jardin des Plantes. Description et mœurs des mammifères de la ménagerie et du Muséum d'histoire naturelle . . . précédé d'une introduction historique, descriptive et pittoresque par M. J. Janin*. Gustave Barba, 1851.

Bonnemains, J., E. Forsyth, and B. Smith, eds., *Baudin in Australian Waters: The Art-*

work of the French Voyage of Discovery to the Southern Lands, 1800–1804. Oxford University Press/Australian Academy of the Humanities, 1988.

Bourdon, Isidore. *Illustres médecins et naturalistes des temps modernes.* Comptoir des Imprimeurs-Unis, 1844.

Bourguet, Marie-Noëlle. "La collecte du monde: Voyage et histoire naturelle (fin xvie siècle—début xixe siècle)." In *Le Muséum au premier siècle de son histoire*, edited by Claude Blanckaert et al. Muséum national d'Histoire naturelle, 1997.

Bourguet, Marie-Noëlle. "Measurable difference: Botany, climate, and the gardener's thermometer in eighteenth-century France." In *Colonial Botany: Science, Commerce, and Politics in the Early Modern World*, edited By Londa Schiebinger and Claudia Swan. University of Pennsylvania Press, 2005.

Bourguet, Marie-Noëlle, and Pierre-Yves Lacour, "Les Mondes naturalistes: Europe (1530–1802)." In *Histoire des sciences et des savoirs. Vol. 1. De la Renaissance aux lumières*, sous la dir. Stéphane Van Damme. Éditions du Seuil, 2015.

Boyer, Ferdinand. "Le Muséum d'Histoire naturelle de Paris et l'Europe des sciences sous la Convention." *Revue d'Histoire des sciences et de leurs applications* 26 (1973), 251–57.

Boyer, Ferdinand. "Le transfert à Paris des collections du Stathouder (1795)," *Annales historiques de la Révolution française* 43 (1971): 389–404.

Brandon-Jones, Christine. "Edward Blyth, Charles Darwin, and the Animal Trade in Nineteenth-Century Britain." *Journal of the History of Biology* 30, no. 2 (1977): 145–58.

Brower, Benjamin Claude. "The Amîr 'Abd Al-Qâdir and the 'Good War'; in Algeria, 1832–1847." *Studia Islamica* 106 (2011): 169–95.

Browne, Janet. "A Science of Empire: British Biogeography before Darwin." *Revue d'histoire des sciences* 45, no. 4 (1992): 453–75.

Bruce, Gary. *Through the Lion Gate: A History of the Berlin Zoo.* Oxford University Press, 2017.

Burkhardt, Richard W., Jr. "Civilizing Specimens and Citizens at the *Muséum d'Histoire Naturelle*, 1793–1838." In *Of Elephants and Roses: French Natural History, 1790–1830*, edited by Sue Ann Prince. *Memoirs of the American Philosophical Society*, 267. 2013.

Burkhardt, Richard W., Jr. "Le comportement animal et l'idéologie de domestication chez Buffon et chez les éthologues modernes." In *Buffon 88: Actes du Colloque international Paris—Montbard—Dijon*, edited by Jean Gayon et al. Vrin, 1992.

Burkhardt, Richard W. Jr. "Constructing the Zoo: Science, Society, and Animal Nature at the Paris Menagerie, 1794–1838." In *Animals in Human Histories: The Mirror of Nature and Culture*, edited by Mary Henninger-Voss. University of Rochester Press, 2002.

Burkhardt, Richard W., Jr. "Frédéric Cuvier on Animal Behavior." *Bulletin d'histoire et d'épistémologie des sciences de la vie* 8 (2001): 75–98.

Burkhardt, Richard W., Jr. "Lamarck, Cuvier, and Darwin on Animal Behavior." In *Transformations of Lamarckism: From Subtle Fluids to Molecular Biology*, edited by S. B. Gissis and E. Jablonka. MIT Press, 2011.

Burkhardt, Richard W., Jr. "Lamarck, Evolution, and the Inheritance of Acquired haracters." *Genetics* 194 (2013): 793–805.

Burkhardt, Richard W. Jr. "La ménagerie et la vie du Muséum." In *Le Muséum au premier siècle de son histoire*, edited by Claude Blanckaert, Claudine Cohen, Pietro Corsi, and Jean-Louis Fischer. Muséum national d'histoire naturelle, 1997.

Burkhardt, Richard W., Jr. "The Leopard in the Garden: Life in Close Quarters at the Muséum d'Histoire Naturelle." *Isis* 98 (2007): 675–94.

Burkhardt, Richard W., Jr. "Naturalists' Practices and Nature's Empire: Paris and the Platypus, 1815–1833." *Pacific Science* 55 (2001): 327–41.

Burkhardt, Richard W., Jr. "An Orangutan in Paris: Pondering Proximity at the Muséum d'Histoire Naturelle in 1836." *History and Philosophy of the Life Sciences* 40 (2018). https://doi.org/10.1007/s40656-018-0186-1.

Burkhardt, Richard W., Jr. *Patterns of Behavior: Konrad Lorenz, Niko Tinbergen, and the Founding of Ethology*. University of Chicago Press, 2005.

Burkhardt, Richard W., Jr. *The Spirit of System: Lamarck and Evolutionary Biology*. Harvard University Press, 1977, 1995.

Burkhardt, Richard W., Jr. "Unpacking Baudin: Models of Scientific Practice in the Age of Lamarck." In *Lamarck*, 119th Congr. Nat. Soc. Hist. Scient., edited by Goulven Laurent. Amiens, 1994.

Burkhardt, Richard W., Jr. "La voix du gardien du lion, ou Les significations multiples des animaux du Muséum d'histoire naturelle." *Annales historiques de la Révolution française*, no. 3 (2014): 145–73.

Camerini, Jane R. "Wallace in the Field." *Osiris* 11 (1996):44-65.

Candegabe, Philippe. *L'incroyable histoire de l'éléphant Hans: Des forêts du Sri Lanka au Muséum d'histoire naturelle*. Vendémiaire, 2016.

Chambard, Antoine. "Alcide Dessalines d'Orbigny. De l'école de la nature à la politique des voyages sous la Restauration." In *Terre à découvrir Terres à parcourir. Exploration et connoissance du monde XIIe-XIXe siècles*, edited by Danielle Lecoq and Antoine Chambard. L'Harmattan, 1998.

Chappey, Jean-Luc. "Révolution, régénération, civilisation: Enjeux culturels des dynamiques politiques." In *Pour quoi faire la Révolution*, collection *Passé et Présent*. Agone, 2012.

Chappey, Jean-Luc. *La Société des observateurs de l'homme. 1799–1804: Des anthropologues au temps de Bonaparte*. Société des études robespierristes, 2002.

Chappey, Jean-Luc, and Maria Pia Donato, "Introduction: Voyages et mutations des savoirs entre dynamiques scientifiques et transformations politiques: Fin XVIIIe—début XIXe siècle." *Annales historiques de la Révolution française*, no. 385 (Voyages, Voyageurs et Mutations des Savoirs entre Révolution et Empire) (July-September 2016): 3–21.

Cheke, Anthony, and Julian Hume. *Lost Land of the Dodo: An Ecological History of Mauritius, Réuion et Rodrigues*. T & AD Poyser, 2008.

Chrulew, Matthew. "Living, Biting Monitors, a Morose Howler and Other Infamous Animals: Animal Biographies in Ethology and Zoo Biology." In *Animal Biography: Re-framing Animal Lives*, edited by André Krebber and Mieke Roscher. Palgrave Macmillan, 2018.

Cohen, Claudine. *The Fate of the Mammoth: Fossils, Myth, and History*. University of Chicago Press, 2002.

Cohen, William B. *The French Encounter with Africans: White Response to Blacks, 1530–1880*. Indiana University Press, 1980.

Coleman, William. *Georges Cuvier: Zoologist. A Study in the History of Evolution Theory*. Harvard University Press, 1964.

Cook, Harold J. "Global Economies and Local Knowledge in the East Indies: Jacobus Bontius Learns the Facts of Nature." In *Colonial Botany: Science, Commerce, and*

Politics in the Early Modern World, edited by Londa Schiebinger and Claudia Swan. University of Pennsylvania Press, 2005.

Corbey, R., and Theunissen, B., eds. *Ape, Man, Apeman: Changing Views since 1600*. Department of Prehistory, Leiden University, 1995.

Corbey, Raymond. *The Metaphysics of Apes: Negotiating the Animal-Human Boundary*. Cambridge University Press, 2005.

Corsi, Pietro. *The Age of Lamarck: Evolutionary Theories in France, 1790–1830*. University of California Press, 1988.

Corsi, Pietro. "Julien Joseph Virey, le premier critique de Lamarck." In *Histoire du concept d'espèce dans les sciences de la vie*, edited by S. Atran et al. Fondation Singer-Polignac, 1985.

Corsi, Pietro. "The Revolutions of Evolution: Geoffroy and Lamarck, 1825–1840." *Bulletin du Musée d'Anthropologie préhistorique de Monaco* 51 (2012): 97–122.

Coulmann, J. J. *Réminiscences*, Vol. 1. Michel Levy frères, 1862.

Cozic, J.-C. *L'orang-outan du capitaine Van Iseghem*. Éditions Joca Seria, 2014.

Cribb, Robert, Helen Gilbert, and Helen Tiffin. *Wild Man from Borneo: A Cultural History of the Orangutan*. University of Hawai'i Press, 2014.

Curran, Andrew S. *The Anatomy of Blackness: Science and Slavery in an Age of Enlightenment*. Johns Hopkins University Press, 2011.

Danziger, Raphael. *Abd Al-Qadir and the Algerians: Resistance to the French and Internal Consolidation*. Holmes and Meier, 1977.

Dardaud, Gabriel. "L'extraordinaire aventure de la girafe du Pacha d'Égypte." *Revue des Conférences françaises en orient* 15 (1951): 1–72.

Daudin, Henri. *Cuvier et Lamarck. Les classes zoologiques et l'idée de série animale (1790–1830)*. 2 vols. Félix Alcan, 1926–27.

Daugeron, Bertrand. *Collections naturalistes entre science et empires (1763–1804)*. Muséum national d'histoire naturelle, 2009.

De Bont, Raf. "Organisms in Their Milieu: Alfred Giard, his pupils, and early ethology, 1870–1930." *Isis* 101 (2010): 1–29.

Deschamps, J. Eugène. "Leschenault de la Tour (Jean-Baptiste-Louis-Claude-Théodore)." *Nouvelle biographie générale* 30 (1859): 923–27.

Devoulx, Albert. *Les Archives du Consulat Général de France à Alger*. Bastide, 1865. https://gallica.bnf.fr/ark:/12148/bpt6k3149573/f5.

Diard, H. "Pierre-Médard Diard. Naturaliste." *Annales de la Société d'agriculture, sciences, arts et belles-lettres d'Indre-et-Loire* 56 (1877): 199–210.

Dittrich, Lothar. "Fürstliche Menagerien im deutschsprachigen Raum ab den 1760er Jahren bis zu Gründung der zoologischen Gärten Mitte des 19. Jahrhunderts." In *Die Kulturgeschichte des Zoos*, edited by Lothar Dittrich, Dietrich von Engelhardt, and Annelore Rieke-Müller. Verlag für Wissenschaft und Bildung, 2001.

Drouin, "Bernardin de Saint-Pierre." In *Du jardin au Muséum en 516 biographies*, edited by Philippe Jaussaud and Édouard-Raoul Brygoo. Muséum national d'histoire naturelle, 2004.

Drouin, Jean-Marc, "L'histoire naturelle à travers un périodique de l'époque révolutionnaire: *La Décade philosophique*." In *La Nature en revolution: 1750–1800*, edited by Andrée Corvol. Harmattan, 1993.

Duméril, Auguste. "Notice historique sur la ménagerie des reptiles du Muséum d'histoire naturelle." *Archives du Muséum national d'Histoire naturelle* 7 (1854–55): 193–319.

Dupuy, Paul. *L'École Normale de l'an III*. Hachette, 1895.

Durant, John R. "Innate character in animals and man: A perspective on the origins of 'ethology.'" In *Biology, Medicine, and Society, 1840–1940*, edited by Charles Webster. Cambridge University Press, 1981.

Dwyer, Philip. *Citizen Emperor: Napoleon in Power*. Yale University Press, 2013.

Dwyer, Philip. *Napoleon: The Path to Power*. Yale University Press, 2007.

Dzanic, Dzavid. "The Civilizing Sea: The Ideological Origins of the French Mediterranean Empire, 1789–1870." PhD diss., Harvard University, 2016. http://nrs.harvard.edu/urn-3:HUL.InstRepos:33840734.

Eriksson, Gunnar. "Linnaeus the botanist." In *Linnaeus: The Man and His Work*, edited by Tore Frängsmyr. University of California Press, 1983.

Espeut, W. Bancroft. "On the acclimatization of the Indian mungoos in Jamaica." *Proceedings of the Zoological Society of London* (1882): 712–14.

Farber, Paul Lawrence. "Buffon and the Concept of Species." *Journal of the History of Biology* 5 (1972): 259–84.

Farber, Paul Lawrence. "The Development of Taxidermy and the History of Ornithology." *Isis* 68 (1977): 550–66.

Farber, Paul Lawrence. *The Emergence of Ornithology as a Scientific Discipline*. D. Reidel, 1982.

Findlen, Paula, and Anna Toledano. "The Materials of Natural History." In *Worlds of Natural History*, edited by H. A. Curry, N. Jardine, J. A. Secord, and E. C. Spary. Cambridge University Press, 2018.

Fletcher, Alice A. "The Osage Indians in France." *American Anthropologist* 2 (1900): 395–405.

Fornasiero, Jean, and John West-Sooby, "Voyages et déplacements des savoirs: Les expéditions de Nicolas Baudin entre révolution et empire." *Annales historiques de la Révolution française*, No. 385 (Voyages, Voyageurs et Mutations des Savoirs entre Révolution et Empire) (2016), 23–45.

Fudge, Erica. *Animal*. Reaktion Books, 2002.

Fudge, Erica. *Perceiving Animals: Humans and Beasts in Early Modern English Culture*. University of Illinois Press, 2002.

Gayon, Jean. "Hérédité des caractères acquis." In *Lamarck, philosophe de la nature*, edited by P. Corsi, J. Gayon, G. Gohau, and S. Tirard. Presses universitaires de France, 2006.

Giard, A., ed. "Discours d'ouverture des cours de zoologie donnés dans le Muséum d'histoire naturelle (an 8, an 10, an 11, et 1806) par J.-B. Lamarck." *Bulletin scientifique de la France et de la Belgique* 40 (1906): 443–599.

Gillispie, Charles Coulston. *Science and Polity in France: The Revolutionary and Napoleonic Years*. Princeton University Press, 2004.

Girardin, Mme. Émile de. *Le Vicomte de Launay. Lettres parisiennes*. Vol. 1. Michel Levy frères, 1857.

Girault, Arthur. "Algeria. History." *Encyclopaedia Britannica*, 11th ed., 1 (1911): 649–53.

Goodman, Dean. "What It Meant to be Linnaean in Revolutionary France." *Isis* 111 (2020): 67–85.

Goret. "Notice historique sur Emmanuel Baillon." Société d'émulation d'Abbeville. Classe des sciences et arts. An 10 [1802]: 28-31.

Gould, Stephen Jay. "A Tree Grows in Paris: Lamarck's division of Worms and Revision of Nature." In Gould, *The Lying Stones of Marrakech*. Three Rivers Press, 2020.

Grigson, Caroline. *Menagerie: The History of Exotic Animals in London.* Oxford University Press, 2016.

Guerrini, Anita. *The Courtiers' Anatomists: Animals and Humans in Louis XIV's Paris.* University of Chicago Press, 2015.

Guillaume, M. J., ed. *Procès-Verbaux du Comité d'instruction publique de la convention nationale.* 5 vols. Imprimerie Nationale, 1891-1904.

Hahn, Roger. *The Anatomy of a Scientific Institution: The Paris Academy of Sciences, 1666–1803.* University of California Press, 1971.

Hahn, Roger. "Du Jardin du roi au Muséum: Les carrières de Fourcroy et de Lacépède." In *Le Muséum au premier siècles de son histoire*, edited by Blanckaert et al. Muséum national d'Histoire naturelle, 1997.

Hallock, E. S. "The American Circus." *Century Magazine* 70 (1905): 57.

Hamy, Ernest-Théodore, "Les anciennes ménageries royales et la ménagerie nationale fondée le 14 brumaire an II (4 novembre 1793)." *Nouvelles archives du Muséum d'histoire naturelle*, ser. 3, no. 5 (1893): 1–22.

Hamy, Ernest-Théodore. *Les Débuts de Lamarck, suivis de recherches sur Adanson, Jussieu, Pallas, Geoffroy Saint-Hilaire, Georges Cuvier, etc.* Guilmoto, 1909.

Hamy, Ernest-Théodore. "Les derniers jours du Jardin du roi et la fondation du Muséum d'Histoire naturelle." In *Centenaire de la fondation du Muséum d'Histoire naturelle, 10 juin 1793—10 juin 1893, Volume commémoratif publié par les professeurs du Muséum.* Imprimerie Nationale, 1893.

Hamy, Ernest-Théodore. "Notes intimes sur Georges Cuvier rédigées en 1836 par le Dr. Quoy pour son ami J. Desjardins, de Maurice." *Bulletin biologique de la France et de la Belgique* 41 (1907): 459–84.

Hamy, Ernest-Théodore. *Les voyages du naturaliste Ch.-Alex. Lesueur dans l'Amérique du Nord (1815-1837), d'après les manuscrits et les œuvres d'art conservés au Muséum d'histoire naturelle de Paris et au Muséum d'histoire naturelle du Havre.* Société des Americanistes, 1904.

Hanson, Elizabeth. *Animal Attractions: Nature on Display in American Zoos.* Princeton University Press, 2002.

Harkett, Daniel. "The Giraffe's Keepers and the (Dis)play of Difference." In *Of Elephants and Roses: French Natural History 1790–1830*, edited by Sue Ann Prince. American Philosophical Society, 2013.

Heninger-Voss, Mary, ed., *Animals in Human Histories: The Mirror of Nature and Culture.* University of Rochester Press, 2002.

Hinton H. E., and A. M. S. Dunn. *Mongooses: Their Natural History and Behaviour.* University of California Press, 1967.

Hochadel, Oliver. "Darwin in the Monkey Cage: The Zoological Garden as a Medium of Evolutionary Theory." In *Beastly Natures: Animals, Humans, and the Study of History*, edited by D. Brantz. University of Virginia Press, 2010.

Hochadel, Oliver. "Science at the Zoo." *Centaurus* 64, no. 3 (2022): 561–90.

Hochadel, Oliver. "Watching Exotic Animals Next Door: Scientific Observations at the Zoo (ca. 1870–1910)." *Science in Context* 24, no. 2 (2011): 183–214.

Hodacs, Hanna. "Linnaeans Outdoors: The Transformative Role of Studying Nature 'on the Move' and Outside." *British Journal for the History of Science* 44 (2011): 183–209.

Holmes, Rachel. *African Queen: The Real Life of the Hottentot Venus.* Random House, 2007.

Horner, Frank. *The French Reconnaissance: Baudin in Austalia, 1801–1803*. Melbourne University Press, 1987.

Hornstein, Katie. "From Museum to Menagerie: Théodore Géricault and the Leonine Subject." *Art Bulletin* 101, no. 1 (2019): 26–47.

Hornstein, Katie. "Jean-Baptiste Huet's Lions and the Look of the Captive in Post-Revolutionary France." In *Time, Media, and Visuality in Post-Revolutionary France*, edited by Richard Taws and Iris Moon. Bloomsbury, 2021.

Hornstein, Katie. *Myth and Menagerie: Seeing Lions in the Nineteenth Century*. Yale University Press, 2024.

Ingensiep, H. W. "Der Orang-outang des Herrn Vosmaer: Ein *aufgeklärter* Menschenaffe." In *Ich das Tier: Tiere als Persönlichkeiten in der Kulturgeschichte*, edited by J. Ullrich, F. Weltzien, and H. Fuhlbrügge. Dietrich Riemer Verlag, 2008.

Ito, Takashi. *London Zoo and the Victorians, 1828–1859*. Royal Historical Studies in History. New series. Boydell Press, 2014.

Jaussaud, Philippe, and Édouard-Raoul Brygoo, *Du Jardin au Muséum en 516 biographies*. Muséum national d'histoire naturelle, 2004.

Jaynes, Julian. "The Historical Origins of 'Ethology' and 'Comparative Psychology." *Animal Behaviour* 17 (1969): 601–6.

Jensen, Heather Belnap. "Caroline Wuiet, la Baronne Auffdiener (1766–1835)," in *Women Art Critics in Nineteenth-Century France: Vanishing Acts*, edited by Wendelin Guentner. University of Delaware Press, 2013.

Jobling, P. "Daumier's ''Orang-outanania.'" *Print Quarterly* 10 (1993): 231–46.

Johnson, James H. *Listening in Paris: A Cultural History*. University of California Press, 1995.

Jouanin, Christian. *L'Impératrice Joséphine et les sciences naturelles*. Éditions de la Réunion des musées nationaux, 1997.

Jouin, Henry, and Henri Stein. *Histoire et description du Jardin des Plantes*. Plon, 1887.

Jouy, Claire. "Portraits sensibles de carnassiers au service de la science et sa diffusion (années 1790-1800). Les vélins du peintre Nicolas Maréchal." *Le Temps des médias*, 40 (1), 15–33. https://doi.org/10.3917/tdm.040.0015.

Kaenel, Philippe. "Le Buffon de l'humanité: La zoologique politique de J.-J. Grandville (1803–1847)." *Revue de l'art* (1986): 20–28.

Kalman, Julie. *The Kings of Algiers: How Two Jewish Families Shaped the Mediterranean World during the Napoleonic Wars and Beyond*. Princeton University Press, 2023.

Kent, Sherman. *The Election of 1827 in France*. Harvard University Press, 1975.

Kirby, Percival R. "The 'Hottentot Venus' of the Musée de l'Homme, Paris," *South African Journal of Science* 50 (1954): 319–22.

Kisling, J. Vernon, Jr. *Zoo and Aquarium History: Ancient Animal Collections to Zoological Gardens*. CRC Press, 2001.

Knapton, Ernest John. *Empress Josephine*. Harvard University Press, 1963.

Koepfli, Klaus-Peter, John Pollinger, Raquel Godinho, et al., "Genome-wide Evidence Reveals that African and Eurasian Golden Jackals Are Distinct Species." *Current Biology* 25 (2015): 2158–65. DOI: 10.1016/j.cub.2015.06.060.

Koerner, Lisbet. *Linnaeus: Nature and Nation*. Harvard University Press, 1999.

Kury, Lorelai, *Histoire naturelle et voyages scientifiques (1780–1830)*. L'Harmattan, 2002.

Lacour, Pierre-Yves. *La République naturaliste: Collections d'histoire naturelle. Révolution française (1798–1804)*. Muséum national d'histoire naturelle, 2014.

Laënnec, Mériadiac. [Obituary of Marion de Procé]. *Journal de médecine et de chirurgie pratique* 25 (1854): 138–40.

Lagueux, Olivier. "Geoffroy's Giraffe: The Hagiography of a Charismatic Mammal." *Journal of the History of Biology* 36 (2003): 225–47.

Laissus, Yves, "Les animaux du Jardin des plantes," in Yves Laissus and Jean-Jacques Petter, *Les animaux du Muséum*, 73–204. Muséum d'histoire naturelle, 1993.

Laissus, Yves. "Note sur les voyages de Jean-Jacques Dussumier (1792–1883)." *Annales de la Société des sciences naturelles de la Charente-Maritime* 5 (1973): 387–406.

Laissus, Yves. "Les voyageurs naturalistes du Jardin du roi et du Muséum d'histoire naturelle: Essai de Portrait-Robot." *Revue d'histoire des sciences* 34 (1981): 259–317.

Landrieu, Marcel. *Lamarck, le fondateur du transformisme: Sa vie, son oeuvre*. Societé Zoologique de France, 1909.

Larson, James L. *Reason and Experience: The Representation of Natural Order in the Work of Carl von Linné*. University of California Press, 1971.

Laurent, Goulven, ed. *Jean-Baptiste Lamarck, 1744–1829*. Éditions du CTHS, 1997.

Lebleu, Olivier. *Les avatars de Zarafa: Première Girafe de France. Chronique d'une girafomania:1826–1845*. Arléa, 2006.

Lecoq, Daniel, and Antoine Chambard, *Terre à découvrir Terres à parcourir. Exploration et connoissance du monde XIIe-XIXe siècles*. L'Harmattan, 1998.

Lee, Paula Young. "The Curious Affair of Monsieur Martin the Bear." *Journal of Eighteenth Century Studies* 33, no. 4 (2010): 615–29.

Legée, Georgette. "La participation de Georges et Frédéric Cuvier à l'organisation de l'instruction publique (1802–1838)." *Histoire et Nature* (1974): 47–72.

Legée, Georgette. *Pierre Flourens, 1794–1867, Physiologiste et historien des sciences*. Vol. 1. F. Paillart, 1992.

Leira, Halvard, and Iver B. Neumann. "Beastly Diplomacy." *The Hague Journal of Diplomacy* 12 (2016): 1–23.

Lemire, Michel. "La France et les collections du stathouder Guillaume V d'Orange." In *Een vorstelijke dierentuin: De menagerie van Willem V (Le zoo du prince: la ménagerie du stathouder Guillaume V)*, edited by B. C. Sliggers Walburg Instituut, 1994.

Lescart, Alain. "An Egyptian Giraffe and Six Osage Indians: An Exotic Plea against the Censorship of 1827." In *Of Elephants and Roses: French Natural History, 1790–1830*, edited by Sue Ann Prince. American Philosophical Society, 2013.

Letouzey, Yvonne. *Le Jardin des Plantes à la croisée des chemins avec André Thouin, 1747–1824*. Éditions du Muséum, 1989.

Liebersohn, Harry. *Aristocratic Encounters: European Travelers and North American Indians*. Cambridge University Press, 1998.

Limoges, Camille. "The development of the Muséum d'Histoire Naturelle of Paris, c. 1800–1914." In *The Organization of Science and Technology in France (1808–1914)*, edited by Robert Fox and George Weisz. Cambridge University Press, 1980.

Lipkowitz, Elise. "The 'Elephant in the room:' The Impact of the French Seizure of the Dutch Stadholder's Collection and Relations between Dutch and French Naturalists." In *Of Elephants and Roses: French Natural History, 1790–1830*, edited by Sue Ann Prince. American Philosophical Society, 2013.

Loisel, Gustave. "Histoire de la ménagerie du Muséum." *Revue scientifique* 88 (1911): 262–77, 301–4.

Loisel, Gustave. *Histoire des ménageries de l'antiquité à nos jours*. 3 vols. Octave Doin et Fils / Henry Laurens, 1912.

Loveland, Jeff. "Daubenton's Lions: From Buffon's Shadow to the French Revolution." *New Perspectives on the Eighteenth Century* 1 (2004): 29–47.

Mahul, A. "Pujoulx (Jean-Baptiste)." *Annuaire nécrologique* 1821: 265-68.

Mainardi, Patricia. "Assuring the Empire of the Future: The 1798 Fête de la Liberté." *Art Journal* 48, no. 2 (Summer 1989): 155–63.

Marquis. "Havet (Armand Étienne Maurice)." In A. J. Mahul, *Annuaire nécrologique* (1823), 177. Reproduced in *Archives biographiques françaises* (microfiche). K. G. Saur, 1988.

Marsh, Kate. *India in the French Imagination: Peripheral Voices, 1754–1815*. Pickering and Chatto, 2009.

Martin, Meredith."Tipu Sultan's Ambassadors at Saint-Cloud: Indomania and Anglophobia in Pre-Revolutionary Paris." *West 86th: A Journal of Decorative Arts, Design History, and Material Culture* 21 (2014): 37–68.

Martinez-Contreras, Jorge. "Des mœurs des singes: Buffon et ses contemporains." In *Buffon 88*, edited by Jean-Claude Beaune et al. Vrin, 1992.

Mathis, S., and J.-M. Vallat, "The Mysterious Death of Georges Cuvier (1832): An Early Case of Guillain-Barré Syndrome." *Neuromuscular Disorders* 30 (2022): 250–53.

McClellan, James E., III. *Colonialism and Science: Saint Domingue in the Old Regime*. Johns Hopkins University Press, 1992; University of Chicago Press, 2010.

McClellan, James E., III, and François Regourd, "The Colonial Machine: French Science and Colonization in the Ancien Régime." Osiris 15 (2000): 31–50.

Mehos, Donna. *Science and Culture for Members Only: The Amsterdam Zoo* Artis *in the Nineteenth Century*. Amsterdam University Press, 2006.

Milne-Edwards, Alphonse. "Les relations entre le Jardin des Plantes et les colonies françaises." *Revue des cultures coloniales* 4 (1899): 2–11.

Moiroux, Jules. *Le cimetière du Père Lachaise*. A. Maréchal, 1922.

Nicard, Pol. *Études sur la vie et les travaux de Ducrotay de Blainville*. Ballière, 1890.

Nyhart, Lynn. *Modern Nature: The Rise of the Biological Perspective in Germany*. University of Chicago Press, 2009.

O'Brien, David. *Exiled in Modernity: Delacroix, Civilization, and Barbarism*. Pennsylvania State University Press, 2018.

Orr, M. "Keeping It in the Family: The Extraordinary Case of Cuvier's Daughters." In *The Role of Women in the History of Geology*, edited by C. V. Burek and B. Higgs. Geological Society, 2007.

Osborne, Michael. *Nature, the Exotic, and the Science of French Colonialism*. Indiana University Press, 1994.

Outram, Dorinda. *Georges Cuvier: Vocation, Science and Authority in Post-Revolutionary France*. Manchester University Press, 1984.

Outram, Dorinda. "Institution scientifique ou champ de bataille pour les familles et les groupes d'influence?" In *Le Muséum au premier siècles de son histoire*, edited by Blanckaert et al. Éditions du MNHN, 1997.

Outram, Dorinda. "New Spaces in Natural History." In *Cultures of Natural History*, edited by N. Jardine, J. A. Secord, and E. C. Spary. Cambridge University Press, 1996.

Patin, Christelle. "Les vies *post-mortem* de Saartjie Baartman: Muséology et économies morales." In *Vénus hottentote: Entre Barnum et Muséum*, coordinator Claude Blanckaert. Muséum national d'Histoire naturelle, 1997.

Peabody, Sue. "La question raciale et le 'sol libre de France': L'affaire Furcy." *Annales* 64 (2009): 1305–34.

Piazzesi, Benedetta. "Domesticare gli istinti: Una Genealogia del discorso etologica." PhD thesis, Sculoa normale superiore, anno 2019–2020.

Pichot, Pierre-Amédee. *Leven en Lotgevalen van Henri Martin, Ex-Directeur van de Rotterdamsche Diergaarde.* H. J. Ter Gunne, 1877.

Pierre, Constant. *Les Hymnes et chansons de la Révolution.* Imprimerie Nationale, 1904.

Pieters, F. F. J. M. "La Ménagerie du Muséum National d'histoire naturelle par Lacépède, Cuvier et Geoffroy." *Bijdragen tot de Dierkunde* 51, no. 2 (1981): 219–49.

Plantet, Eugène. *Correspondance des beys de Tunis et des consuls de France avec la cour, 1577–1830*, vol. 3 *(1770–1830)*. Félix Alcan, 1899.

Pouillard, Violette. "Animal Biographies: Beyond Archetypal Figures." Review of *Obaysch. A Hippopotamus in Victorian London*, by John Simons, and *Biographies animales*, by Éric Baratay. *Journal of Animal Ethics* 12, no. 2 (2022). https://doi-org./10.5406/21601267.12.2.07.

Pouillard, Violette. *Histoire des zoos par les animaux: Impérialisme, contrôle, conservation.* Champ Vallon, 2019.

Pouillard, Violette. "The Silence and the Fury: Addressing Animal Resistance and Agency through the History of Human-Animal Relationships." In *Animals Matter: Resistance and Transformation in Animal Commodification*, edited by Julien Dugnoille and Elizabeth Vander Meer. *Animal Studies* 26 (2023): 32-55. https://doi.org/10.1163/9789004528444 004.

Pouillard, Violette. "Visions of Concord: Wild Animals and the Garden of the Revolution (Jardin des plantes Menagerie, 1793-c. 1820)." *Journal of the History of Environment and Society* 4 (2019): 11–40. https://doi.org/10.1484/J.JHES.5.120674.

Prarond, Ernest. "Baillon (Louis-Antoine-François)." *Mémoires de la société impériale d'émulation d'Abbeville* 8 (1852): 620-51.

Premack, D., and Woodruff, G. "Does the Chimpanzee Have a Theory of Mind?" *Behavioral and Brain Sciences* 1 (1978): 515–26.

Prince, Sue Ann, ed. *Of Elephants and Roses. French Natural History, 1790–1830.* American Philosophical Society, 2013.

Putnam, Walter. "Captive Audiences: A Concert for the Elephants in the Jardin des plantes." *Drama Review* 51 (2007): 154–60.

Qureshi, Sadiah. "Displaying Sara Baartman, the 'Hottentot Venus,'" *History of Science* 43 (2004): 233–57.

Raffles, Lady Sophia. *Memoir of the Life and Public Services of Sir Stamford Raffles, . . . Particularly in the Government of Java, 1811–1816, and of Bencoolen and Its Dependencies, 1817–1824.* John Murray, 1830.

Ramsbottom, J. "Linnaeus and the Species Concept." *Proceedings of the Linnean Society of London* 150 (1938): 192–209.

Reboul, Robert Marie. *Louis-François Jauffret: Sa vie et ses oeuvres.* J. Baur et Détaille, 1869.

Rey, Roselyne. "L'Animalité dans l'œuvre de Bernardin de Saint-Pierre: Convenance, consonance, et contraste." *Revue de synthèse* 113 (1992): 311–31.

Reynaud, Jean. *Vie et correspondance de Merlin de Thionville.* Furne, 1860.

Richards, Robert J. *Darwin and the Emergence of Evolutionary Theories of Mind and Behavior.* University of Chicago Press, 1997.

Rieke-Müller, Annelore. "Menagerien zwischen Privatheit und Wissenschaft vom Menschen. Die Haltung exotischer Wildtiere in der zwetien Hälfte des 18. Jahrhunderts." In *Mensch, Tier und Zoo: Der Tiergarten Schönbrunn im internationalen Vergleich vom 18. Jahhunder bis heute,* edited by Mitchell G. Ash. Bohlau, 2008.

Ritvo, Harriet. *The Animal Estate: The English and Other Creatures in the Victorian Age.* Cambridge, MA, Harvard University Press, 1987.

Ritvo, Harriet. "Q is for quagga." In *Animalia: An Anti-Imperial Bestiary for Our Times,* edited by Antoinette Burton and Renisa Mawani. Duke University Press, 2020.

Robbins, Louise E. *Elephant Slaves and Pampered Parrots: Exotic Animals in Eighteenth-Century Paris.* Johns Hopkins University Press, 2002.

Roger, Jacques. *Buffon: A Life in Natural History.* Cornell University Press, 1997.

Rookmaaker, L. C., P. A. Morris, I. E. Glenn, and P. J. Mundy. "The ornithological cabinet of Jean-Baptiste Bécoeur and the secret of the arsenical soap." *Archives of Natural History* 33 (2006): 146–58.

Rosenstock, Morton. "The house of Bacri and Busnach: A chapter from Algeria's commercial history." *Jewish Social Studies* 14 (1952): 343–64.

Rothfels, Nigel. "Catching Animals." In *Animals in Human Histories: The Mirror of Nature and Culture,* edited by Mary Henninger-Voss. University of Rochester Press, 2002.

Rothfels, Nigel. *Elephant Trails: A History of Animals and Cultures.* Johns Hopkins University Press, 2021.

Rothfels, Nigel, ed. *Representing Animals.* Indiana University Press, 2002.

Rothfels, Nigel. *Savages and Beasts: The Birth of the Modern Zoo.* Johns Hopkins University Press, 2002.

Rousset, Camille. *Les Commencements d'une conquête: l'Algérie de 1830 à 1840,* 2d ed., 2 vols. Plon-Nourrit, 1900.

Rudwick, Martin J. S. *Georges Cuvier, Fossil Bones, and Geological Catastrophes.* University of Chicago Press, 1997.

Saban, Roger, and Michel Lemire, "Les éléphants de la ménagerie du stathouder Guillaume V d'Orange au Muséum d'Histoire naturelle sous la Convention nationale et le Directoire." *114e Congr. nat. Soc. sav., Paris, 1989, Scientifiques et Sociétés.* Éditions du CTHS, 1990.

Said, Edward. *Orientalism.* Penguin Books, 1995.

Schiebinger, Londa. *Nature's Body: Gender in the Making of Modern Science.* Beacon Press, 1993.

Schiebinger, Londa, and Claudia Swan, eds. *Colonial Botany: Science, Commerce, and Politics in the Early Modern* World. University of Pennsylvania Press, 2005.

Schorger, A. W., *The Passenger Pigeon: Its Natural History and Extinction.* University of Wisconsin Press, 1955.

Schroeder, Paul W. *The Transformation of European* Politics. Clarendon Press, 1994.

Serna, Pierre. *L'animal en République. 1789–1802, genèse du droit des bêtes.* Anarchis, 2016.

Serna, Pierre. *Comme des bêtes: Histoire politique de l'animal en Révolution (1750–1840).* Fayard, 2017.

Sliggers, B. C., ed., *Een vorstelijke dierentuin: de menagerie van Willem V (Le zoo du prince: La ménagerie du stathouder Guillaume V).* Walburg Instituut, 1994.

Smith, Jay. *The Monsters of the Gévaudan: The Making of a Beast.* Harvard University Press, 2011.

Sørenson, Madeleine Pinault. “Representing Animals with Empathy, 1793-1810.” In *Of Elephants and Roses: French Natural History, 1790–1830*, edited by Sue Ann Prince. *Memoirs of the American Philosophical Society* 267. 2013.

Spary, E. C. “Of Nutmegs and Botanists: The Colonial Cultivation of Botanical Identity.” In *Colonial Botany: Science, Commerce, and Politics in the Early Modern World*, edited by Londa Schiebinger and Claudia Swan. University of Pennsylvania Press, 2005.

Spary, E. C. *Utopia's Garden: French Natural History from Old Regime to Revolution*. University of Chicago Press, 2000.

Stanton, William. *The Leopard's Spots: Scientific Attitudes toward Race in America, 1815–59*. University of Chicago Press, 1960.

Stocking, George W., Jr. *Race, Culture, and Evolution: Essays in the History of Anthropology*. Free Press, 1968.

Stoczkowski, Wiktor. “Portrait de l'ancêtre en singe: l'hominisation sans evolutionnisme dans la pensée naturaliste du xviiie siècle.” In *Ape, Man, Apeman: Changing Views since 1600*, edited by R. Corbey and B. Theunissen. Department of Prehistory, Leiden University, 1995.

Stronge, Susan. *Tipu's Tigers*. V and A, 2009.

Sulloway, Frank. “Darwin's conversion: The Beagle Voyage and its Aftermath.” *Journal of the History of Biology* 15 (1982): 325–96.

Surrelot, Evelyne. *Histoire de la presse féminine en France, des origines à 1848*. Armand Colin, 1966.

Synowiecki, Jan. “Paris côté jardins: Nature urbaine, publics et sociabilité au XIIIe siècle.” In *Paris et ses peuples au XVIIIe siècle*, edited by Pascal Bastien and Simon Macdonald. Éditions de la Sorbonne, 2020.

Taquet, Phillipe. *Georges Cuvier: Anatomie d'un naturaliste*. Odile Jacob, 2019.

Terrall, Mary. “Speculation and Experiment in Enlightenment Life Sciences.” In *Heredity Produced: At the Crossroads of Biology, Politics, and Culture, 1500–1870*, edited by Staffan Müller-Wille and Hans-Jörg Rheinberger. Cambridge University Press, 2007.

Théodoridès, Jean. “Quelques documents inédits ou peu connus relatifs à Georges Cuvier, à sa famille et à son salon.” *Stendhal Club*, nos. 33–34 (1967): 55–64, 179–88.

Thesée, Françoise. *Auguste Plée—1786/1825. Un voyageur naturaliste*. Éditions Caribéennes, 1989.

Trumelet, Corneille. *Le Général Yusuf*. 2 vols. Paul Ollendorff, 1890.

Vachon, Max, Georges Rousseau, and Yves Laissus, *Inédits de Lamarck*. Masson, 1972.

Van Damme, Stephan. “In the Name of Linnaeus: Paris as a Disputed Capital of Natural Knowledge (1730–1789).” In *Linnaeus, Natural History, and the Circulation of Knowledge*, edited by Hanna Hodacs, Kenneth Nyberg, and Stephane Van Damme. Voltaire Foundation, 2018.

Van der Kuyl, Antoinette C., John T. Dekker, and Jaap Goudsmit. “St. Kitts Green Monkeys Originate from West Africa: Genetic Evidence from Feces.” *American Journal of Primatology* 40 (1996): 361–64.

Walker, A. *Franz Liszt: The Virtuoso Years, 1811–1847*. Knopf, 1983.

Wasef S., S. Subramanian, R. O'Rorke R., et al. “Mitogenomic Diversity in Sacred Ibis Mummies Sheds Light on Early Egyptian Practices.” *PLOS ONE* 14, no. 11 (2019). e0223964. https://doi.org/10.1371/journal.pone.0223964.

Weber, Andreas. “Collecting Colonial Nature. European Naturalists and the Netherlands Indies in the Early Nineteenth Century." *BMGN—Low Countries Historical Review* 134, no. 3 (2019): 72–95.

Windler, Christian. *La diplomatie comme expérience de l'autre: Consuls français au Maghreb (1700–1840)*. Librarie Droz S.A., 2002.

Wyhe, John van. "Wallace's Help: The Many People Who Aided A. R. Wallace in the Malay Archipelago." *Journal of the Malaysian Branch of the Royal Asiatic Society* 91 (2018): 41–68.

Index

Page numbers in italics refer to illustrations.

Individual animals occurring in the book are primarily noted by taxon (birds, camels, etc.), but in several cases, they are noted at the species level. Some of the common names are archaic constructions as reflected in the archival sources.

A number of people's names as they appeared in primary sources did not include full first names, and in a few cases, any first name. Because many of these names are vanishingly obscure, where possible, their primary roles are noted parenthetically.